Control Systems Engineering

CONTROL SYSTEMS ENGINEERING

(Second Edition)

I.J. NAGRATH
Birla Institute of Technology & Science
Pilani

M. GOPAL
Indian Institute of Technology
Bombay

JOHN WILEY & SONS
Singapore New York Chichester
Brisbane Toronto

CONTROL SYSTEMS ENGINEERING, 2ND EDITION
WILEY STUDENT EDITION

Copyright © 1986
Exclusive rights by John Wiley & Sons (SEA) Pte. Ltd. – Singapore for manufacture and export. This book cannot be re-exported from the country to which it is consigned by John Wiley & Sons.

Copyright © 1982 WILEY EASTERN LIMITED
Second Edition December, 1981
First Reprint April, 1982
Sixth Reprint September, 1985

Authorised reprint of this edition published by WILEY EASTERN LIMITED, New Delhi, Copyright © 1986, Wiley Eastern Limited. All rights reserved. No parts of this book maybe reproduced without the written permission of Wiley Eastern Limited.

ISBN 9971-51-056-1

10 9 8 7 6 5 4 3

Printed in the Republic of Singapore.

Preface

This book was first written in 1975 with the object of providing an integrated treatment of continuous-time linear and nonlinear control systems. This edition apart from being thoroughly revised, includes in its ambit sampled-data control systems, optimal control and generalized criteria of stability. Throughout the book the stress is laid on the interdisciplinary nature of the subject. Effort has been made to draw examples from various engineering disciplines for illustrating basic control concepts with a view to encourage the users from all relevant disciplines. Modern and classical approaches follow the natural sequence with the classical preceding the modern. The integration of these two approaches is attempted through strong emphasis on their interrelationship and elaboration of their respective merits.

The underlying concepts in control are brought home to the reader through a variety of practical systems involving hardware. Systems modelling and control strategies are kept in sharp focus throughout. To reinforce the practical aspects of control hardware a good coverage is given to electromechanical, hydraulic and pneumatic control components exemplifying control systems. In fact, this has been expanded in the present edition by inclusion of stepper motors and gyros. With the amount of material and the level of treatment of control components, the teachers engaged in teaching control systems would not be required to refer to specialized books on this topic. Control implementation using microprocessors is, of course, outside the scope of this book and necessarily requires a separate study.

While examples on feedback control systems are included immediately in Chapter 2 on mathematical modelling, the link between feedback concepts and sensitivity is established in Chapter 3, so as to answer the question 'why feedback?' at the introductory stage. These concepts have been further viewed through root sensitivity and frequency domain approaches in later chapters.

The material on stability has been progressively built in the book at three places. Chapter 6 introduces the two basic concepts of stability—BIBO (bounded-input, bounded-output) stability and asymptotic stability, as applied to linear systems and advances algebraic criteria of stability determination. Nyquist's frequency domain stability criterion of linear systems is taken up at its appropriate location in Chapter 9. The more general concepts of stability are advanced in Chapter 14. The present edition includes the two generalized stability criteria—the Liapunov's time domain criterion and the Popov's frequency domain criterion. It is also established that the Routh criterion follows as a special case from the Liapunov criterion.

Strong emphasis on design aspects of control systems is maintained all through the book. Both classical and modern design concepts have been introduced early enough in the chapters on time and frequency domain analysis. These ideas have been fully illustrated and the interplay of time and frequency domain design is brought out through exhaustive treatment of second-order systems because of their frequent occurrence, and because of the fact that many of the higher-order systems in practice can be approximated as second-order systems. Chapter 5 has now been enlarged to include analytical design of control systems of higher-order from the point of view of minimization of ISE and ITAE. The classical design of more complex systems then follows as a logical step in Chapter 10 on compensation techniques. Both cascade and feedback methods of compensation are presented and illustrated through several examples. This chapter now includes network compensation of ac systems.

Keeping in view the increasing importance of computer control systems, a new chapter (Chapter 11) has been introduced in this edition on sampled-data control systems. Assuming no knowledge of the z-transform on part of the reader, the chapter gives all the relevant details. Analysis, stability and design of sampled-data control systems is covered exhaustively with many illustrative examples.

Chapter 12 on state variable analysis and design is considerably expanded to include latest material compared to the corresponding chapter in the first edition. This chapter now deals with both continuous-time and discrete-time systems apart from exposing the problems of formulation, diagonalization, solution, controllability and observability. It also includes design of observer and pole placement design by state feedback.

While the first edition gave merely an introductory article on optimal control as part of the state variable chapter, the present edition has devoted a full chapter (Chapter 13) to optimal control. It deals with optimal control systems of both continuous-time and discrete-time type. It begins with the transfer function approach in s- and z-domains and its limitations, and then goes on to the state variable approach to optimal control. By adopting the dynamic programming approach, the solution to the optimal linear regulator problem in continuous and discrete form is quickly built up. The calculus of variation approach is particularly not adopted as it would have been highly space-consuming for a book of this level and nature.

Chapter 14 is devoted to in-depth treatment of nonlinear systems. It focusses the reader's attention to the limitations of linear modelling and treats at length the two powerful and commonly used methods of nonlinear analysis, viz., phase-plane and describing function. Jump resonance and the two generalized stability critica—Liapunov and **Popov, have** now been added in this chapter.

Illustrative examples are interspersed throughout the book at their natural locations. These have been selected so that apart from illustrating the concepts involved, they help to take the reader to a higher level of application. The same approach is followed in unsolved problems with hints given wherever necessary. Some of these problems also serve the purpose of extending the text material. Answers to problems have been given to inspire confidence in the reader.

The text presumes on the part of the reader some knowledge of the basic tools like complex variables, the Fourier and Laplace transforms and elementary matrix algebra. However, appendices on the Fourier and Laplace transforms including partial fraction expansion, determination of roots of algebraic equations and introduction to matrix algebra have been included for ready reference.

The book is intended as a text for two one-semester control systems courses at undergraduate level or one one-semester course at undergraduate and one one-semester course at postgraduate level. The material presented has been class-tested in the present form. The sequencing and the internal organization of various chapters is such that it permits flexibility of adaptation to variations in students' prior training and curricula. The book is written in such style and form that makes it easily comprehensible for all engineering disciplines that have control courses as part of their curricula.

The first author (IJN) takes this opportunity of expressing his gratitude to Prof. T.J. Higgins for introducing him to the discipline of control systems when he was a student at Wisconsin University.

We are grateful to the authorities of Birla Institute of Technology and Science, Pilani, and Indian Institute of Technology, Powai, Bombay, for providing the facilities necessary for revising this book.

I J Nagrath
M Gopal

Contents

Preface v

1. INTRODUCTION 1

1.1 The Control System *1*
1.2 Servomechanisms *4*
1.3 Historical Development of Automatic Control *5*
1.4 Sampled-data and Digital Control Systems *9*
1.5 Multivariable Control Systems *10*
1.6 Application of Control Theory in Non-engineering Fields *11*

2. MATHEMATICAL MODELS OF PHYSICAL SYSTEMS 13

2.1 Introduction *13*
2.2 Differential Equations of Physical Systems *15*
2.3 Transfer Functions *30*
2.4 Block Diagram Algebra *38*
2.5 Signal Flow Graphs *45*
2.6 Illustrative Examples *51*
     Problems *54*

3. FEEDBACK CHARACTERISTICS OF CONTROL SYSTEMS 61

3.1 Feedback and Non-feedback Systems *61*
3.2 Reduction of Parameter Variations by Use of Feedback *62*
3.3 Control over System Dynamics by Use of Feedback *65*
3.4 Control of the Effects of Disturbance Signals by Use of Feedback *66*
3.5 Regenerative Feedback *69*
3.6 Illustrative Examples *70*
     Problems *74*

4. CONTROL SYSTEMS AND COMPONENTS 82

4.1 Introduction *82*
4.2 Linear Approximation of Nonlinear Systems *83*
4.3 Electrical Systems *84*
4.4 Stepper Motor *99*

4.5 Hydraulic Systems *104*
4.6 Pneumatic Systems *116*
4.7 Gyroscopes—Inertial Navigation *121*
     Problems *126*

5. TIME RESPONSE ANALYSIS, DESIGN SPECIFICATIONS AND PERFORMANCE INDICES   132

5.1   Introduction *132*
5.2   Standard Test Signals *133*
5.3   Time Response of First-order Systems *135*
5.4   Time Response of Second-order Systems *137*
5.5   Steady-state Errors and Error Constants *147*
5.6   Effect of Adding a Zero to a System *151*
5.7   Design Specifications of Second-order Systems *152*
5.8   Design Considerations for Higher-order Systems *157*
5.9   Performance Indices *159*
5.10 Illustrative Examples *164*
      Problems *175*

6. CONCEPTS OF STABILITY AND ALGEBRAIC CRITERIA   181

6.1 The Concept of Stability *181*
6.2 Necessary Conditions for Stability *186*
6.3 Hurwitz Stability Criterion *188*
6.4 Routh Stability Criterion *190*
6.5 Relative Stability Analysis *196*
6.6 More on the Routh Stability Criterion *197*
     Problems *198*

7. THE ROOT LOCUS TECHNIQUE   201

7.1 Introduction *201*
7.2 The Root Locus Concept *202*
7.3 Construction of Root Loci *205*
7.4 Root Contours *226*
7.5 Systems with Transportation Lag *231*
7.6 Sensitivity of the Roots of the Characteristic Equation *233*
     Problems *239*

8. FREQUENCY RESPONSE ANALYSIS   243

8.1 Introduction *243*
8.2 Correlation between Time and Frequency Response *244*
8.3 Polar Plots *250*
8.4 Bode Plots *252*
8.5 All-pass and Minimum-phase Systems *264*
8.6 Experimental Determination of Transfer Functions *266*
8.7 Log-magnitude versus Phase Plots *270*
     Problems *271*

9. STABILITY IN FREQUENCY DOMAIN 274

9.1 Introduction *274*
9.2 Mathematical Preliminaries *274*
9.3 Nyquist Stability Criterion *277*
9.4 Assessment of Relative Stability Using Nyquist Criterion *286*
9.5 Closed-loop Frequency Response *297*
9.6 Sensitivity Analysis in Frequency Domain *310*
    Problems *313*

10. INTRODUCTION TO DESIGN 319

10.1 The Design Problem *319*
10.2 Preliminary Considerations of Classical Design *321*
10.3 Realization of Basic Compensators *328*
10.4 Cascade Compensation in Time Domain *335*
10.5 Cascade Compensation in Frequency Domain *351*
10.6 Feedback Compensation *366*
10.7 Network Compensation of A.C. Systems *374*
    Problems *380*

11. SAMPLED-DATA CONTROL SYSTEMS 385

11.1 Introduction *385*
11.2 Spectrum Analysis of Sampling Process *387*
11.3 Signal Reconstruction *390*
11.4 Difference Equations *391*
11.5 The z-transform *393*
11.6 The z-transfer Function (Pulse Transfer Function) *402*
11.7 The Inverse z-transform and Response of Linear Discrete Systems *407*
11.8 The z-transform Analysis of Sampled-data Control Systems *410*
11.9 Response between Sampling Instants *421*
11.10 The z- and s-domain Relationship *423*
11.11 Stability Analysis *424*
11.12 Compensation Techniques *435*
    Problems *443*

12. STATE VARIABLE ANALYSIS AND DESIGN 448

12.1 Introduction *448*
12.2 Concepts of State, State Variables and State Model *449*
12.3 State Models for Linear Continuous-time Systems *456*
12.4 Diagonalization *472*
12.5 Solution of State Equations *481*
12.6 Concepts of Controllability and Observability *493*
12.7 Pole Placement by State Feedback *504*
12.8 State Variables and Linear Discrete-time Systems *511*
    Problems *523*

13. OPTIMAL CONTROL SYSTEMS — 530

- 13.1 Introduction 530
- 13.2 Parameter Optimization: Servomechanisms 531
- 13.3 Optimal Control Problems: Transfer Function Approach 539
- 13.4 Optimal Control Problems: State Variable Approach 551
- 13.5 The State Regulator Problem 556
- 13.6 The Infinite-time Regulator Problem 566
- 13.7 The Output Regulator and the Tracking Problems 571
- 13.8 Parameter Optimization: Regulators 573
  Problems 577

14. NONLINEAR SYSTEMS — 583

- 14.1 Introduction 583
- 14.2 Common Physical Nonlinearities 588
- 14.3 The Phase-plane Method: Basic Concepts 593
- 14.4 Singular Points 595
- 14.5 Stability of Nonlinear Systems 604
- 14.6 Construction of Phase-trajectories 608
- 14.7 System Analysis by Phase-plane Method 620
- 14.8 The Describing Function Method: Basic Concepts 625
- 14.9 Derivation of Describing Functions 627
- 14.10 Stability Analysis by Describing Function Method 634
- 14.11 Jump Resonance 642
- 14.12 Liapunov's Stability Criterion 646
- 14.13 Popov's Stability Criterion 663
  Problems 667

Appendix I FOURIER AND LAPLACE TRANSFORMS 675

Appendix II ELEMENTS OF MATRIX ANALYSIS 686

Appendix III DETERMINATION OF ROOTS OF ALGEBRAIC EQUATIONS 696

Appendix IV ANSWERS TO PROBLEMS 700

Index 715

1. Introduction

1.1 The Control System

The *control system* is that means by which any quantity of interest in a machine, mechanism or other equipment is maintained or altered in accordance with a desired manner. Consider, for example, the driving system of an automobile. Speed of the automobile is a function of the position of its accelerator. The desired speed can be maintained (or a desired change in speed can be achieved) by controlling pressure on the accelerator pedal. This automobile driving system (accelerator, carburettor and engine-vehicle) constitutes a control system. Figure 1.1 shows the general diagrammatic representation of a typical control system. For the automobile driving system the input (command signal) is the force on the accelerator pedal and the automobile speed is the output (controlled) variable.

Input (command) signal / Force → Accelerator pedal, linkages and carburetter → Engine-vehicle → Output (controlled) variable / Speed

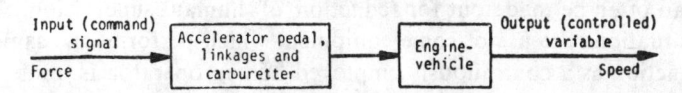

Figure 1.1 The basic control system.

CLOSED-LOOP CONTROL

Let us consider the automobile driving system. The route, speed and acceleration of the automobile are determined and controlled by the driver by observing traffic and road conditions and by properly manipulating the accelerator, clutch, gear-lever, brakes and steering wheel, etc. Suppose the driver wants to maintain a speed of 50 km per hour (desired output). He accelerates the automobile to this speed with the help of the accelerator and then maintains it by holding the accelerator steady. No error in the speed of the automobile occurs so long as there are no gradients or other disturbances along the road. The actual speed of the automobile is measured by the speedometer and indicated on its dial. The driver reads the speed dial visually and compares the actual speed with the desired one mentally. If there is a deviation of speed from the desired speed, accordingly he takes the decision to increase or decrease the speed. The decision is executed by change in pressure of his foot (through muscular power) on the accelerator pedal.

These operations can be represented in a diagrammatic form as shown in Fig. 1.2. In contrast to the sequence of events in Fig. 1.1., the events in the control sequence of Fig. 1.2 follow a closed-loop, i.e., the information about the instantaneous state of the output is fedback to the input and is used to modify it in such a manner as to achieve the desired output. It is on account of this basic difference that the system of Fig. 1.1 is called an *open-loop system*, while the system of Fig. 1.2 is called a *closed-loop system*.

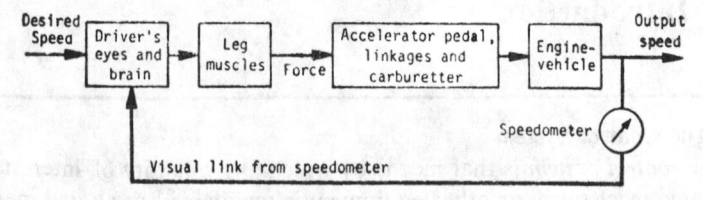

Figure 1.2 Schematic diagram of a manually controlled closed-loop system.

Systems of the type represented in Fig. 1.2 involve continuous manual control by a human operator. These are classified as *manually controlled systems*. In many complex and fast-acting systems, the presence of human element in the control loop is undesirable because the system response may be too rapid for an operator to follow or the demand on operator's skill may be unreasonably high. Furthermore, some of the systems, e.g., missiles, are selfdestructive and in such systems human element must be excluded. Even in situations where manual control could be possible, an economic case can often be made out for reduction of human supervision. Thus in most situations the use of some equipment which performs the same intended function as a continuously employed human operator is preferred. A system incorporating such an equipment is known as *automatic control system*. In fact in most situations an automatic control system could be made to perform intended functions better than a human operator, and could further be made to perform such functions as would be impossible for a human operator.

The general block diagram of an automatic control system is shown in Fig. 1.3. An error detector compares a signal obtained through feedback elements, which is a function of the output response, with the reference input.

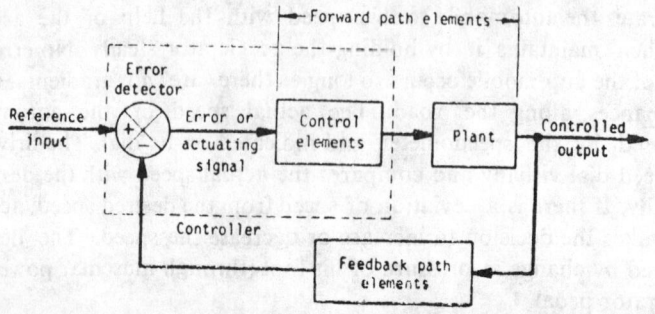

Figure 1.3 General block diagram of an automatic control system.

Any difference between these two signals constitutes an error or actuating signal, which actuates the control elements. The control elements in turn alter the conditions in the plant (controlled member) in such a manner as to reduce the original error.

In order to gain a better understanding of the interactions of the constituents of a control system, let us discuss a simple tank level control system shown in Fig. 1.4. This control system can maintain the liquid level h (controlled output) of the tank within accurate tolerance of the desired liquid level even though the output flow rate through the valve V_1 is varied. The liquid level is sensed by a float (feedback path element), which positions the slider arm B on a potentiometer. The slider arm A of another potentiometer is positioned corresponding to the desired liquid level H (the reference input). When the liquid level rises or falls, the potentiometers (error detector)

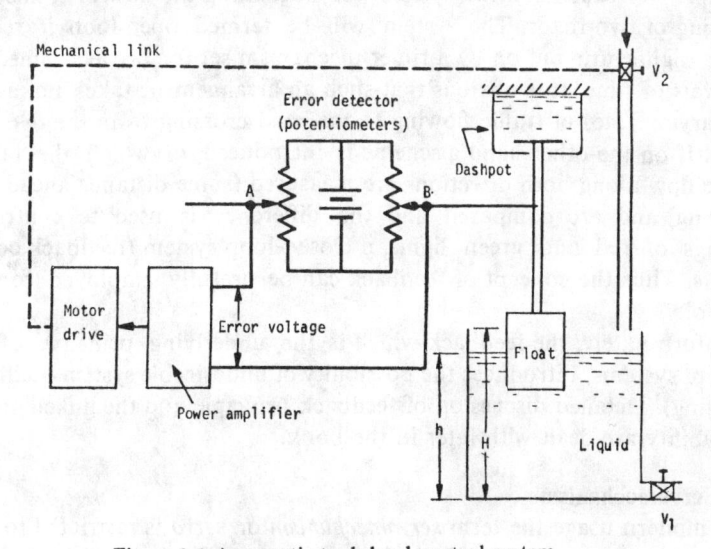

Figure 1.4 Automatic tank-level control system.

give an error voltage (error or actuating signal) proportional to the change in liquid level. The error voltage actuates the motor through a power amplifier (control elements) which in turn conditions the plant (i.e., decreases or increases the opening of the valve V_2) in order to restore the desired liquid level. Thus the control system automatically attempts to correct any deviation between the actual and desired liquid levels in the tank.

OPEN-LOOP CONTROL

As stated already, any physical system which does not automatically correct for variation in its output, is called an open-loop system. Such a system may be represented by the block diagram of Fig. 1.5. In these systems the output remains constant for a constant input signal provided the external conditions remain unaltered. The output may be changed to any desired value by appropriately changing the input signal but variations in

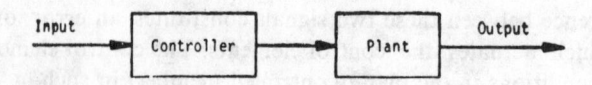

Figure 1.5 General block diagram of open-loop system.

external conditions or internal parameters of the system may cause the output to vary from the desired value in an uncontrolled fashion. The open-loop control is therefore satisfactory only if such fluctuations can be tolerated or system components are designed and constructed so as to limit parameter variations and environmental conditions are well-controlled.

It is important to note that the fundamental difference between an open- and closed-loop control system is that of feedback action. Consider, for example, a traffic control system for regulating the flow of traffic at the crossing of two roads. The system will be termed open-loop if red and green lights are put on by a timer mechanism set for predetermined fixed intervals of time. It is obvious that such an arrangement takes no account of varying rates of traffic flowing to the road crossing from the two directions. If on the other hand a scheme is introduced in which the rates of traffic flow along both directions are measured (some distance ahead of the crossing) and are compared and the difference is used to control the timings of red and green lights, a closed-loop system (feedback control) results. Thus the concept of feedback can be usefully employed to traffic control.

Unfortunately, the feedback which is the underlying principle of most control systems, introduces the possibility of undesirable system oscillations (hunting). Detailed discussion of feedback principle and the linked problem of stability are dealt with later in the book.

1.2 Servomechanisms

In modern usage the term *servomechanism* or *servo* is restricted to feedback control systems in which the controlled variable is mechanical position or time derivatives of position, e.g., velocity and acceleration.

A servo system used to position a load shaft is shown in Fig. 1.6, in which the driving motor is geared to the load to be moved. The output (controlled) and desired (reference) positions θ_C and θ_R respectively are measured and compared by a potentiometer pair whose output voltage v_E is proportional to the error in angular position $\theta_E = \theta_R - \theta_C$. The voltage $v_E = K_P \theta_E$ is amplified and is used to control the field current (excitation) of a d.c. generator which supplies the armature voltage to the driving motor.

To understand the operation of the system assume $K_P = 100$ volts/rad and let the output shaft position be 0.5 rad. Corresponding to this condition, the slider arm B has a voltage of $+50$ volts. Let the slider arm A be also set at $+50$ volts. This gives zero actuating signal ($v_E = 0$). Thus the motor has zero output torque so that the load stays stationary at 0.5 rad.

Assume now that the new desired load position is 0.6 rad. To achieve this, the arm A is placed at +60 volts position, while the arm B remains instantaneously at +50 volts position. This creates an actuating signal of +10 volts, which is a measure of lack of correspondence between the actual load position and the commanded position. The actuating signal is amplified and fed to the servo motor which in turn generates an output torque which repositions the load. The system comes to a standstill only when the actuating signal becomes zero, i.e., the arm B and the load reach the position corresponding to 0.6 rad (+60 volts position).

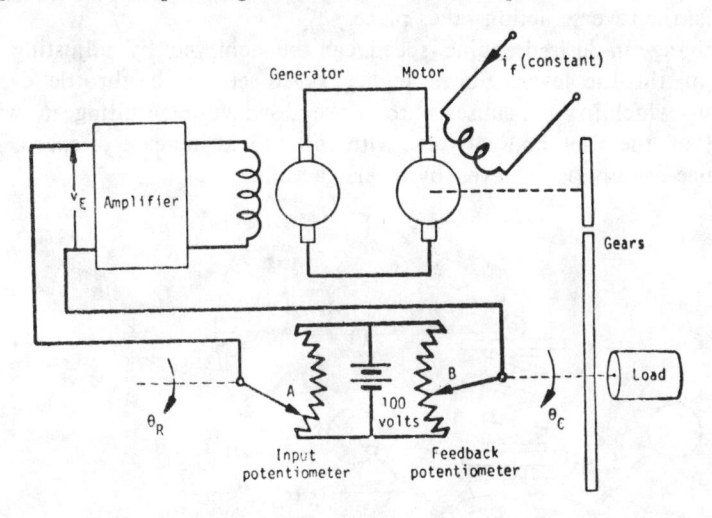

Figure 1.6 A position control system.

The position control systems have innumerable applications, namely, machine tool position control, constant-tension control of sheet rolls in paper mills, control of sheet metal thickness in hot rolling mills, radar tracking systems, missile guidance systems, inertial guidance, roll stabilization of ships, etc. Some of these applications will be discussed in this book.

1.3 Historical Development of Automatic Control

It is instructive to trace brief historical development of automatic control. Automatic control systems did not appear until the middle of eighteenth century. The first automatic control system, the fly-ball governor, to control the speed of steam engines, was invented by James Watt in 1770. This device was usually prone to hunting. It was about hundred years later that Maxwell analyzed the dynamics of the fly-ball governor.

The schematic diagram of a speed control system using a fly-ball governor is shown in Fig. 1.7. The governor is directly geared to the output shaft so that the speed of the fly-balls is proportional to the output speed of the engine. The position of the throttle lever sets the desired speed. The lever pivoted as shown in Fig. 1.7, transmits the centrifugal force from the fly-balls to the bottom of the lower seat of the spring. Under steady conditions, the centrifugal force of the fly-balls balances the spring force* and the opening of the flow control valve is just sufficient to maintain the engine speed at the desired value.

If the engine speed drops below the desired value, the centrifugal force of the fly-balls decreases, thus decreasing the force exerted on the bottom of the spring, causing x to move downward. By lever action, this results in wider opening of the control valve and hence more fuel supply which increases the speed of the engine until equilibrium is restored. If the speed increases, the reverse action takes place.

The change in desired engine speed can be achieved by adjusting the setting of throttle lever. For a higher speed setting, the throttle lever is moved up which in turn causes x to move downward resulting in wider opening of the fuel control valve with consequent increase of speed. The lower speed setting is achieved by reverse action.

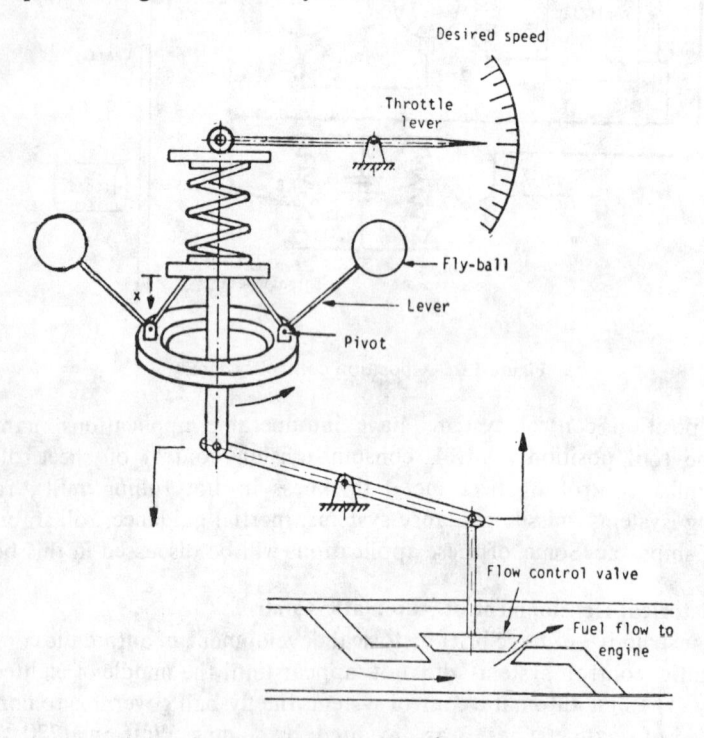

Figure 1.7 Speed control system.

The importance of positioning heavy masses like ships and guns quickly and precisely was realized during the World War I. In early 1920, Minorsky performed the classic work on the automatic steering of ships and positioning of guns on the shipboards.

*The gravitational forces are normally negligible compared to the centrifugal force.

Introduction

A date of significance in automatic control systems is that of Hazen's work in 1934. His work may possibly be considered as a first struggling attempt to develop some general theory for servomechanisms. The word 'servo' has originated with him.

Prior to 1940 automatic control theory was not much developed and for most cases the design of control systems was indeed an art. During the decade of 1940's, mathematical and analytical methods were developed and practised and control engineering was established as an engineering discipline in its own right. During the World War II it became necessary to design and construct automatic airplane pilots, gun positioning systems, radar tracking systems and other military equipments based on feedback control principle. This gave a great impetus to the automatic control theory.

The missile launching and guidance system of Fig. 1.8 is a sophisticated example of military applications of feedback control. The target plane is sited by a rotating radar antenna which then locks in and continuously tracks the target. Depending upon the position and velocity of the plane as given by the radar output data, the launch computer calculates the firing angle in terms of a launch command signal, which when amplified through a power amplifier drives the launcher (drive motor). The launcher angular position is fedback to the launch computer and the missile is triggered as soon as the error between the launch command signal and the missile firing angle becomes zero. After being fired the missile enters the radar beam which is tracking the target. The control system contained within the missile now receives a guidance signal from the beam which automatically adjusts the control surfaces of the missile such that the missile rides along the beam, finally homing on to the target.

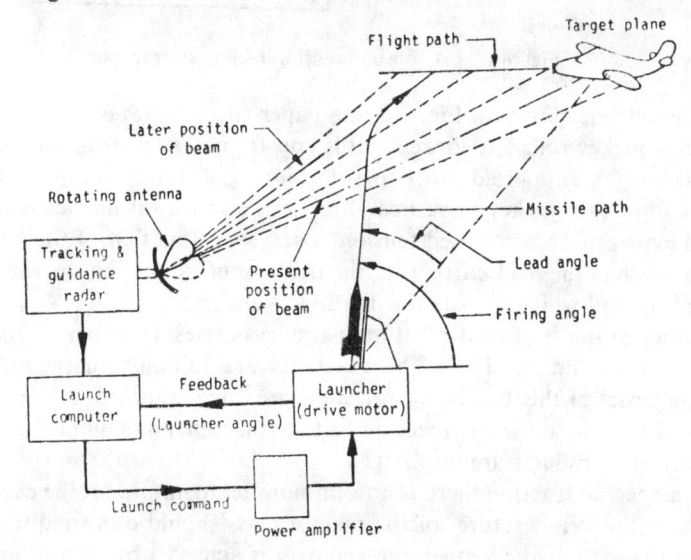

Figure 1.8 Missile launching and guidance system.

It is important to note that the actual missile launching and guidance system is far more complex requiring control of gun's bearing as well as elevation. The simplified case discussed above illustrates the principle of feedback control.

The industrial use of automatic control has tremendously increased since the World War II. Modern industrial processes such as manufacture and treatment of chemicals and metals are now automatically controlled.

A simple example of an automatically controlled industrial process is shown in Fig. 1.9. This is a scheme employed in paper mills for reeling paper sheets. For best results the paper sheet must be pulled on to the wind-up roll at nearly constant tension. A reduction in tension will produce a loose roll, while an increase in tension may result in tearing of the paper sheet. If reel speed is constant, the linear velocity of paper and hence its tension increases, as the wind-up roll diameter increases. Tension control may be achieved by suitably varying the reel speed.

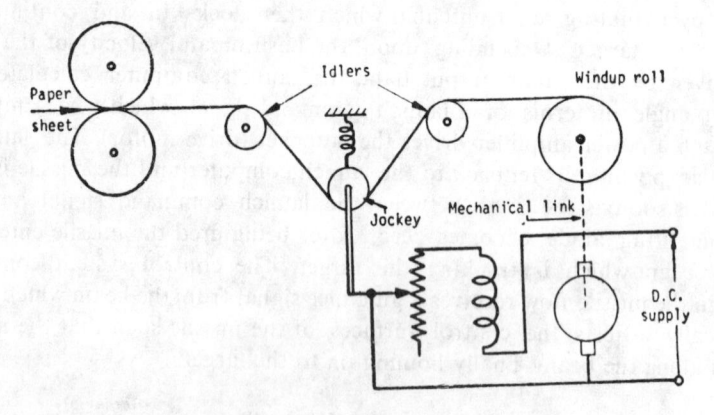

Figure 1.9 A constant tension reeling system.

In the scheme shown in Fig. 1.9, the paper sheet passes over two idling and one jockey roll. The jockey roll is constrained to vertical motion only with its weight supported by paper tension and spring. Any change in tension moves the jockey in vertical direction, upward for increased tension and downward for decreased tension. The vertical motion of the jockey is used to change the field current of the drive motor and hence the speed of wind-up roll which adjusts the tension.

Another example of controlled industrial processes is a batch chemical reactor shown in Fig. 1.10. The reactants are initially charged into the reaction vessel of the batch reactor and are then agitated for a certain period of time to allow the reaction to take place. Upon completion of the reaction, the products are discharged.

For a specific reaction there is an optimum **temperature profile** according to which the temperature of the reactor mass should be varied to obtain best results. Automatic temperature control is achieved by providing both

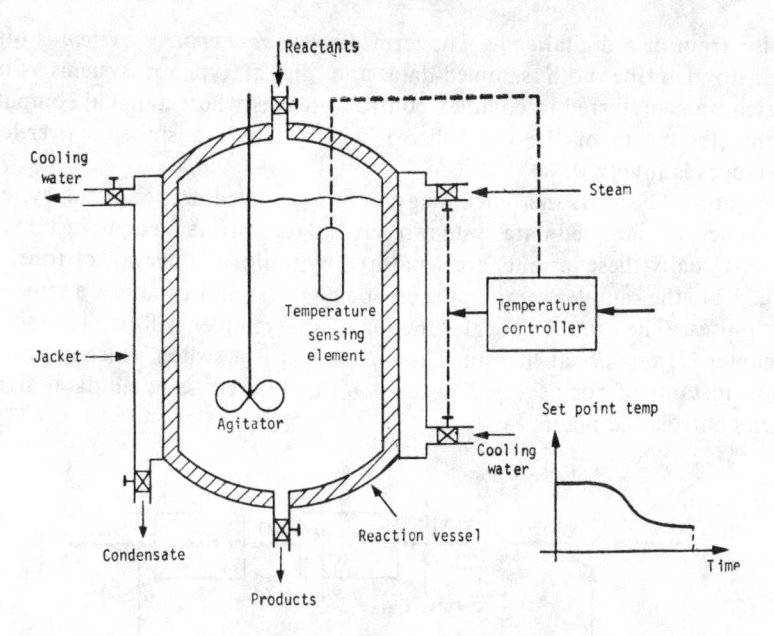

Figure 1.10 A batch chemical process.

steam and cooling water jackets for heating or cooling the reactor mass (cooling is required to remove exothermic heat of reaction during the period the reaction proceeds vigorously). During the heating phase, the controller closes the water inlet valve and opens and controls the steam inlet valve while the condensate valve is kept open. Reverse action takes place during the cooling phase.

Control engineering has enjoyed tremendous growth during the years since 1955. Particularly with the advent of analog and digital computers and with the perfection achieved in computer field, highly sophisticated control schemes have been devised and implemented. Furthermore, computers have opened up vast vistas for applying control concepts to non-engineering fields like business and management. On the technological front fully automated computer control schemes have been introduced for electric utilities and many complex industrial processes with several interacting variables particularly in the chemical and metallurgical processes.

A glorious future lies ahead for automation wherein computer control can run our industries and produce our consumer goods provided we can tackle with equal vigour and success the socio-economic and resource depletion problems associated with such sophisticated degree of automation.

1.4 Sampled-data and Digital Control Systems

In the control schemes discussed so far, the signals associated with various parts of a system are all continuous functions of the time variable. Sampled-data and digital systems differ from these continuous-data systems in that the signals at one or more points of the system are either in form of a pulse train or a digital code. The term *discrete-data control system* is often used to describe both sampled-data and digital type of systems. Such systems are preferred in complex control schemes where a digital computer forms the heart of the controller or where signal transmission over long distances is involved.

Figure 1.11 shows the block diagram of a sampled-data control system. In general a sampled-data system receives data only at discrete instants of time. Usually these instants are separated by uniform intervals of time. In Fig. 1.11, the sampler samples the error signal $e(t)$ and produces a sequence of pulses. The mathematical function of the sampler will be discussed in Chapter 11; we may at this stage assume that it is a switch which closes for short instants of time every T seconds. The filtered sampled-data signal then controls the plant.

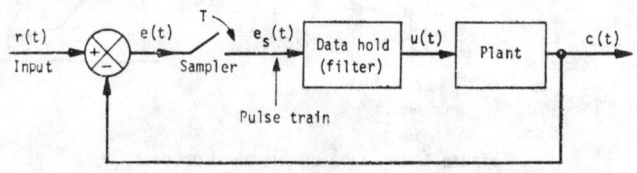

Figure 1.11 Block diagram of a sampled-data control system.

1.5 Multivariable Control Systems

In relatively simple control systems some of which have been discussed in the previous sections, we are interested in only one output quantity which may be position, speed, voltage, current, frequency, etc. Such systems are called *single variable* systems (Fig. 1.3). In many practical applications, more than one variable of interest are encountered, e.g., elevation as well as bearing for a guided antenna, speed as well as direction of an automobile driving system. In chemical plants, a number of variables like temperatures, pressures and fluid flows are required to be controlled. Such systems are referred to as *multivariable* systems. A block diagram depicting a multivariable control system is shown in Fig. 1.12.

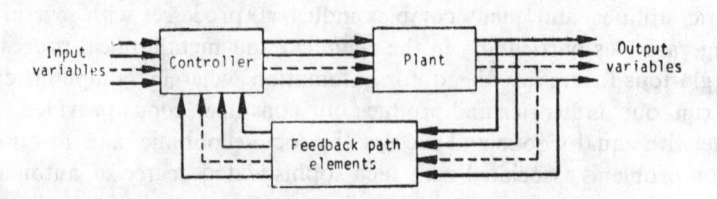

Figure 1.12 General block diagram of multivariable control system.

As an illustration, let us discuss an autopilot system which steers a rocket vehicle in response to radioed command. Fig. 1.13 shows a simplified block diagram representation of the system.

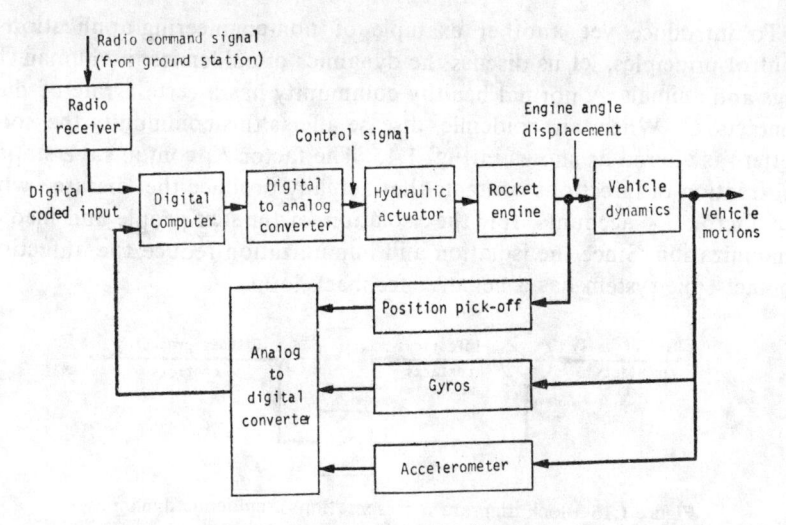

Figure 1.13 A typical autopilot system.

The state of motion of the vehicle (velocity, acceleration) is fed to the control computer by means of motion sensors (gyros, accelerometers). A position pick-off feeds the computer with the information about engine angle displacement and hence the direction in which the vehicle is heading. In response to heading-commands from the ground, the computer generates a signal which controls the hydraulic actuator and in turn moves the engine.

1.6 Application of Control Theory in Non-engineering Fields [1]

We have considered in previous sections a number of applications which highlight the potentialities of automatic control to handle various engineering problems. Although control theory originally evolved as an engineering discipline, due to universality of the principles involved it is no longer restricted to engineering confines in the present state of art. In the following paragraphs we shall discuss some simple examples of control theory as applied to fields like economics, sociology and biology.

Consider an economic inflation problem which is evidenced by continually rising prices. A model of the vicious price-wage inflationary cycle, assuming simple relationships between wages, product costs and cost of living is shown in Fig. 1.14. The economic system depicted in this figure is found to be a positive feedback system.

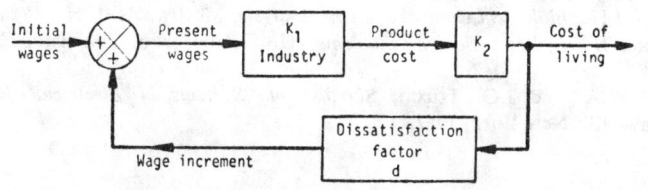

Figure 1.14 Economic inflation dynamics.

To introduce yet another example of non-engineering application of control principles, let us discuss the dynamics of epidemics in human beings and animals. A normal healthy community has a certain rate of daily contacts C. When an epidemic disease affects this community the social pattern is altered as shown in Fig. 1.15. The factor K_1 contains the statistical fraction of infectious contacts that actually produce the disease, while the factor K_2 accounts for the isolation of the sick people and medical immunization. Since the isolation and immunization reduce the infectious contacts, the system has a negative feedback loop.

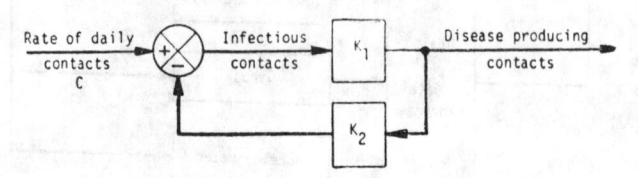

Figure 1.15 Block diagram representation of epidemic dynamics.

In medical field, control theory has wide applications, such as temperature regulation, neurological, respiratory and cardiovascular controls. A simple example is the automatic anaesthetic control. The degree of anaesthesia of a patient undergoing operation can be measured from encephalograms. Using control principles anaesthetic control can be made completely automatic, thereby freeing the anaesthetist from observing constantly the general condition of the patient and making manual adjustments.

The examples cited above are somewhat over-simplified and are introduced merely to illustrate the universality of control principles. More complex and complete feedback models in various non-engineering fields are now available. This area of control is under rapid development and has a promising future.

References and Further Reading

1. West, J.C., "Control's Limitations", *Proc. IEE*, 1971, **118**, 1, 225-238.
2. Haberman, R., *Mathematical Models: Mechanical Vibrations, Population Dynamics and Traffic Flow*, Prentice-Hall, Englewood Cliffs, N.J., 1977.
3. Sage, A.P., *Methodology for Large-Scale Systems*, McGraw-Hill, New York, 1977.
4. Gould, L.A., *Chemical Process Control: Theory and Applications*, Addison-Wesley, Reading, Mass., 1969.
5. Cannon, R.H., *Dynamics of Physical Systems*, McGraw-Hill, New York, 1967.
6. Milsum, J.H., *Biological Control Systems Analysis*, McGraw-Hill, New York, 1966.
7. Bukstein, E.D., *Basic Servomechanisms*, Holt Rinehert and Winston, New York, 1963.
8. Lynch, W.A. and J.G. Truxal, *Signals and Systems in Electrical Engineering*, McGraw-Hill, New York, 1962.

2. Mathematical Models of Physical Systems

2.1 Introduction

A *physical system* is a collection of physical objects connected together to serve an objective. Examples of a physical system may be cited from laboratory, industrial plant or utility services—an electronic amplifier composed of many components, the governing mechanism of a steam turbine or a communications satellite orbiting the earth are all examples of physical systems. A more general term *system* is used to describe a combination of components which may not all be physical, e.g., biological, economic, socio-economic or management systems. Study in this book will be mainly restricted to physical systems though a few examples of general type systems will also be introduced.

No physical system can be represented in its full physical intricacies and therefore idealizing assumptions are always made for the purpose of analysis and synthesis of systems. An idealized physical system is called a *physical model*. A physical system can be modelled in a number of ways depending upon the specific problem to be dealt with and the desired accuracy. For example, an electronic amplifier may be modelled as an interconnection of linear lumped elements, or some of these may be pictured as nonlinear elements in case the stress is on distortion analysis. A communication satellite may be modelled as a point, a rigid body or a flexible body depending upon the type of study to be carried out. As idealizing assumptions are gradually removed for obtaining a more accurate model, a point of diminishing return is reached, i.e., the gain in accuracy of representation is not commensurate with the increased complexity of the computation required. In fact beyond a certain point there may indeed be an undetermined loss in accuracy of representation due to flow of errors in the complex computations.

Once a physical model of a physical system is obtained, the next step is to obtain a *mathematical model* which is the mathematical representation of the physical model through use of appropriate physical laws. Depending upon the choice of variables and the coordinate system, a given physical model may lead to different mathematical models. A network, for example, may be modelled as a set of nodal equations using Kirchhoff's current law or a set of mesh equations using Kirchhoff's voltage law. A control system may be modelled as a scalar differential equation describing the system or state variable vector-matrix differential equation. The particular mathematical model which gives a greater insight into the dynamic behaviour of physical system is selected.

When the mathematical model of a physical system is solved for various input conditions, the result represents the dynamic response of the system. The mathematical model of a system is *linear*, if it obeys the *principle of superposition and homogeneity*. This principle implies that if a system model has responses $y_1(t)$ and $y_2(t)$ to any two inputs $x_1(t)$ and $x_2(t)$ respectively, then the system response to the linear combination of these inputs

$$\alpha_1 x_1(t) + \alpha_2 x_2(t)$$

is given by the linear combination of the individual outputs, i.e.,

$$\alpha_1 y_1(t) + \alpha_2 y_2(t)$$

where α_1 and α_2 are constants.

Mathematical models of most physical systems are characterized by differential equations. A mathematical model is linear, if the differential equation describing it has coefficients, which are either functions only of the independent variable or are constants. If the coefficients of the describing differential equations are functions of time (the independent variable), then the mathematical model is *linear time-varying*. On the other hand, if the coefficients of the describing differential equations are constants, the model is *linear time-invariant*.

The differential equations describing a linear time-invariant system can be reshaped into different forms for the convenience of analysis. For example, for transient response or frequency response analysis of single-input-single-output linear systems, *the transfer function representation* (to be discussed later in this chapter) forms a useful model. On the other hand, when a system has multiple inputs and outputs, the *vector-matrix notation* (discussed in Chapter 12) may be more convenient. The mathematical model of a system having been obtained, the available mathematical tools can then be utilized for analysis or synthesis of the system.

Powerful mathematical tools like the Fourier and Laplace transforms are available for use in linear systems. Unfortunately no physical system in nature is perfectly linear. Therefore certain assumptions must always be made to get a linear model which, as pointed out earlier, is a compromise between the simplicity of the mathematical model and the accuracy of results obtained from it. However, it may not always be possible to obtain a valid linear model, for example, in the presence of a strong nonlinearity or in presence of distributive effects which cannot be represented by lumped parameters.

A commonly adopted approach for handling a new problem is: first build a simplified model, linear as far as possible, by ignoring certain nonlinearities and other physical properties which may be present in the system and thereby get an approximate idea of the dynamic response of the system; a more complete model is then built for more complete analysis.

2.2 Differential Equations of Physical Systems

This section presents the method of obtaining differential equations of physical systems by utilizing the physical laws of the process. Depending upon the system, various well known physical laws like Newton's laws, Kirchhoff's laws, etc., will be used to build mathematical models.

MECHANICAL SYSTEMS

In the analysis of mechanical systems, it is convenient to make use of three idealized elements: the mass, the spring and the damper. These elements represent three essential phenomena which occur in various ways in mechanical systems.

Mechanics of Translation

The three ideal elements for translational mechanical systems are shown in Fig. 2.1 with their relevant properties and conventions.

The ideal mass element represents a particle of mass which is the lumped approximation of the mass of a body concentrated at the centre of mass. The symbol chosen to represent the ideal mass element shows two connection points or *terminals*, one is free and has associated with it the motional variable v, the other terminal represents a reference (commonly earth) with respect to which the motional variable of the free terminal is measured.

The concept of elastic deformation of a body is symbolized by the ideal element shown as a helical spring.

The friction exists in physical systems whenever mechanical surfaces are operated in sliding contact. The friction encountered in physical systems may be of many types:

(i) *Coulomb friction force*: The force of sliding friction between dry surfaces. This force is substantially constant.

(ii) *Viscous friction force*: The force of friction between moving surfaces separated by viscous fluid or the force between a solid body and a fluid medium. This force is approximately linearly proportional to velocity over a certain limited velocity range.

(iii) *Stiction*: The force required to initiate motion between two contacting surfaces (which is obviously more than the force required to maintain them in relative motion).

In most physical situations of interest, the viscous friction predominates. The relation given in Fig. 2.1 is based on this assumption.

The friction force acts in a direction opposite to that of velocity. However, it should be realised that friction is not always undesirable in physical

(1) The mass element

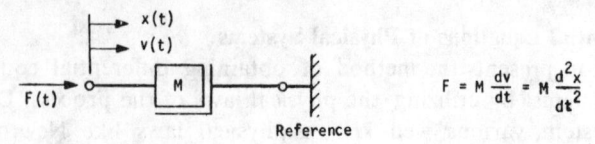

$$F = M\frac{dv}{dt} = M\frac{d^2x}{dt^2}$$

(2) The spring element

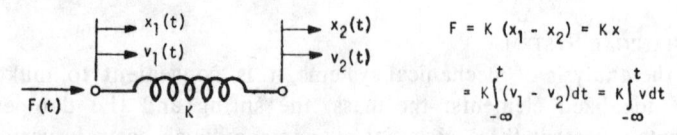

$$F = K(x_1 - x_2) = Kx$$
$$= K\int_{-\infty}^{t}(v_1 - v_2)dt = K\int_{-\infty}^{t}vdt$$

(3) The damper element

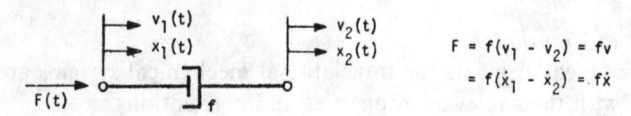

$$F = f(v_1 - v_2) = fv$$
$$= f(\dot{x}_1 - \dot{x}_2) = f\dot{x}$$

x(m), v(m/sec), M(kg), F(newton), K(newton/m), f(newton per m/sec)

(4) The inertia element

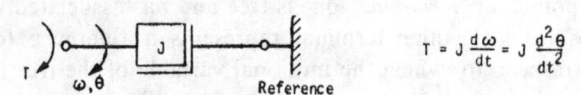

$$T = J\frac{d\omega}{dt} = J\frac{d^2\theta}{dt^2}$$

(5) The torsional spring element

$$T = K(\theta_1 - \theta_2) = K\theta$$
$$= K\int_{-\infty}^{t}(\omega_1 - \omega_2)dt = K\int_{-\infty}^{t}\omega dt$$

(6) The damper element

$$T = f(\omega_1 - \omega_2) = f\omega$$
$$= f(\dot{\theta}_1 - \dot{\theta}_2) = f\dot{\theta}$$

θ(rad), ω(rad/sec), J(kg-m^2), T(newton-m)
K(newton-m/rad), f(newton-m per rad/sec)

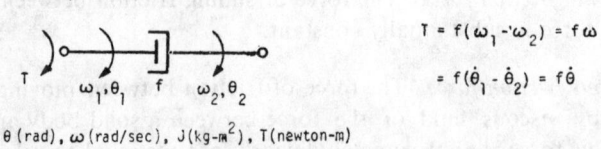

Figure 2.1 Ideal elements for mechanical systems.

systems. Sometimes it may even be necessary to introduce friction intentionally to improve the dynamic response of the system (discussed in Chapter 5). Friction may be introduced intentionally in a system by use of a *dashpot* shown in Fig. 2.2. It consists of a piston and oil filled cylinder with a narrow annular passage between piston and cylinder. Any relative motion between piston and cylinder is resisted by oil with a friction force (fv).

Mathematical Models of Physical Systems 17

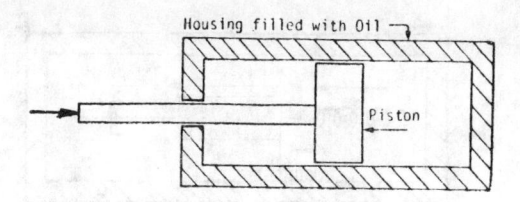

Figure 2.2 Dashpot construction.

Let us consider now the mechanical system shown in Fig. 2.3a. It is simply a mass M attached to a spring (stiffness K) and a dashpot (viscous friction coefficient f) on which the force F operates. Displacement x is positive in the direction shown. The zero position is taken to be at the point where the spring and mass are in static equilibrium.*

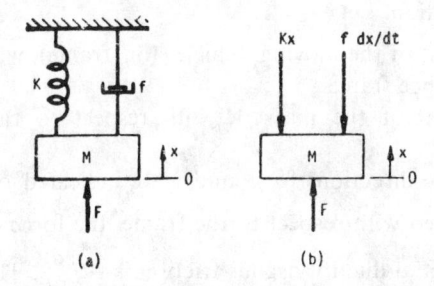

Figure 2.3 (a) A mass-spring-dashpot system; (b) Free-body diagram.

The systematic way of analyzing such a system is to draw a free-body diagram** as shown in Fig. 2.3b. Then by applying Newton's law of motion to the free-body diagram, the force equation can be written as

$$F - f\frac{dx}{dt} - Kx = M\frac{d^2x}{dt^2}$$

or

$$F = M\frac{d^2x}{dt^2} + f\frac{dx}{dt} + Kx \quad (2.1)$$

Equation (2.1) is a linear, constant coefficient differential equation of second-order.

MECHANICAL ACCELEROMETER

In its simplest form, an accelerometer consists of a spring-mass-dashpot system shown in Fig. 2.4. The frame of the accelerometer is attached to the moving vehicle.

*Note that the gravitational effect is eliminated by this choice of zero position.
**In Example 2.1, we shall see that there is one to one correspondence between free-body diagram approach and nodal method of analysis.

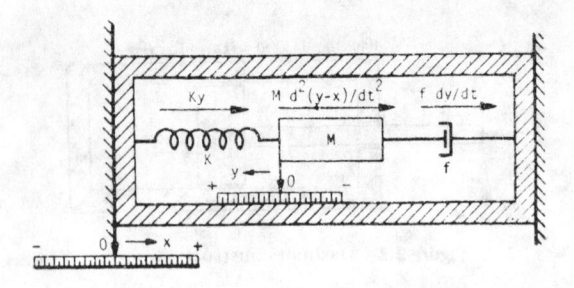

Figure 2.4 Simplified diagram of an accelerometer.

Whenever the moving vehicle and hence the frame of the accelerometer is accelerated, the spring deflects until it produces enough force to accelerate the mass at the same rate as the frame. The deflection of the spring which may be measured by a linear-motion potentiometer is a direct measure of acceleration. Let

x = displacement of the moving vehicle (or frame) with respect to a fixed reference frame.

y = displacement of the mass M with respect to the accelerometer frame.

The positive directions for x and y are indicated on the diagram.

Since y is measured with respect to the frame, the force on the mass due to spring is $-Ky$ and due to viscous friction is $-f\dfrac{dy}{dt}$. The motion of the mass with respect to the fixed reference frame in the positive direction of y is $(y-x)$.

The force equation for the system becomes

$$M \frac{d^2(y-x)}{dt^2} + f \frac{dy}{dt} + Ky = 0$$

or
$$M \frac{d^2y}{dt^2} + f \frac{dy}{dt} + Ky = M \frac{d^2x}{dt^2} = Ma \qquad (2.2)$$

where a is the input acceleration.

If a constant acceleration is applied to the accelerometer, the output displacement y becomes constant under steady-state as the derivatives of y become zero, i.e.,

$$Ma = Ky$$

or
$$a = \left(\frac{K}{M}\right) y$$

The steady-state displacement y is thus a measure of the constant input acceleration. This instrument can also be used for displacement measurements as explained later in section 2.3.

Fixed-axis Rotation

Mechanical systems involving fixed-axis rotation occur in the study of machinery of many types and are very important. The modelling procedure is very close to that used in translation. In these systems, the variables of interest are the torque and angular velocity (or displacement). The three basic components for rotational systems are: moment of inertia, torsional spring and viscous friction.

The three ideal rotational elements with their relevant properties and conventions are shown in Fig. 2.1.

Let us consider now, the rotational mechanical system shown in Fig. 2.5a which consists of a rotatable disc of moment of inertia J and a shaft of stiffness K. The disc rotates in a viscous medium with viscous friction coefficient f.

Let T be the applied torque which tends to rotate the disc. The free-body diagram is shown in Fig. 2.5b.

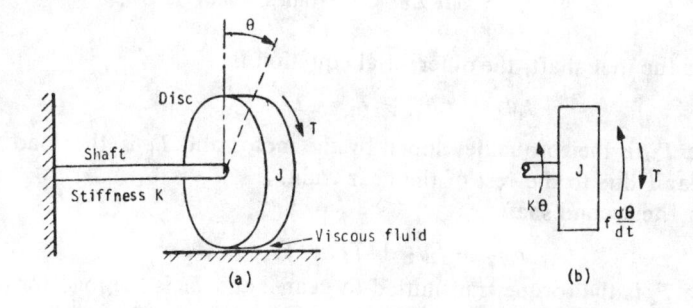

Figure 2.5 (a) Rotational mechanical system; (b) Free-body diagram.

The torque equation obtained from the free-body diagram is

$$T - f\frac{d\theta}{dt} - K\theta = J\frac{d^2\theta}{dt^2}$$

or

$$T = J\frac{d^2\theta}{dt^2} + f\frac{d\theta}{dt} + K\theta \tag{2.3}$$

Equation (2.3) is a linear constant coefficient differential equation describing the dynamcs of the system shown in Fig. 2.5a.

GEAR TRAINS

Gear trains are used in control systems to attain the mechanical matching of motor to load. Usually a servomotor operates at high speed but low torque. To drive a load with high torque and low speed by such a motor, the torque magnification and speed reduction are achieved by gear trains. Thus in mechanical systems gear trains act as matching devices like transformers in electrical systems.

Figure 2.6 shows a motor driving a load through a gear train which consists of two gears coupled together. The gear with N_1 teeth is called the primary gear (analogous to primary winding of a transformer) and gear with N_2 teeth is called the secondary gear.

Angular displacements of shafts 1 and 2 are denoted by θ_1 and θ_2 respectively. The moment of inertia and viscous friction of motor and gear 1 are denoted by J_1 and f_1 and those of gear 2 and load are denoted by J_2 and f_2 respectively.

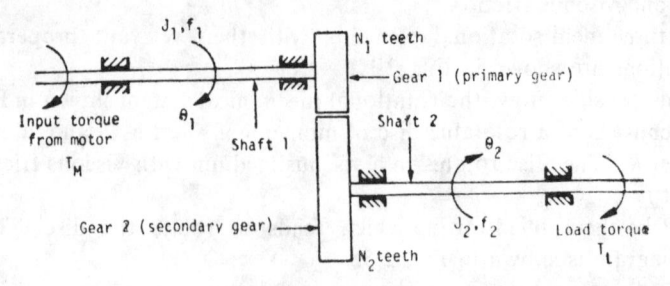

Figure 2.6 Gear train system.

For the first shaft, the differential equation is

$$J_1\ddot{\theta}_1 + f_1\dot{\theta}_1 + T_1 = T_M \tag{2.4}$$

where T_M is the torque developed by the motor and T_1 is the load torque on gear 1 due to the rest of the gear train.

For the second shaft

$$J_2\ddot{\theta}_2 + f_2\dot{\theta}_2 + T_L = T_2 \tag{2.5}$$

where T_2 is the torque transmitted to gear 2 and T_L is the load torque.

Let r_1 be the radius of gear 1 and r_2 be that of gear 2. Since the linear distance travelled along the surface of each gear is same, $\theta_1 r_1 = \theta_2 r_2$. The number of teeth on gear surface being proportional to gear radius, we obtain

$$\frac{\theta_2}{\theta_1} = \frac{N_1}{N_2} \tag{2.6}$$

Here the stiffness of the shafts of the gear train is assumed to be infinite. In an ideal case of no loss in power transfer, the work done by gear 1 is equal to that of gear 2. Therefore,

$$T_1\theta_1 = T_2\theta_2 \tag{2.7}$$

Combining eqns. (2.6) and (2.7) we have

$$\frac{T_1}{T_2} = \frac{\theta_2}{\theta_1} = \frac{N_1}{N_2} \tag{2.8}$$

Differentiating eqn. (2.8) twice, we have the following relation for speed and acceleration.

$$\frac{\ddot{\theta}_2}{\ddot{\theta}_1} = \frac{\dot{\theta}_2}{\dot{\theta}_1} = \frac{N_1}{N_2} \tag{2.9}$$

Thus if $N_1/N_2 < 1$, from eqns. (2.8) and (2.9) it is found that the gear train reduces the speed and magnifies the torque.

Eliminating T_1 and T_2 from eqns. (2.4) and (2.5) with the help of eqn. (2.8), we obtain

$$J_1\ddot{\theta}_1 + f_1\dot{\theta}_1 + \frac{N_1}{N_2}(J_2\ddot{\theta}_2 + f_2\dot{\theta}_2 + T_L) = T_M \qquad (2.10)$$

Elimination of θ_2 from eqn. (2.10) with the help of eqn. (2.9) yields,

$$\left[J_1 + \left(\frac{N_1}{N_2}\right)^2 J_2\right]\ddot{\theta}_1 + \left[f_1 + \left(\frac{N_1}{N_2}\right)^2 f_2\right]\dot{\theta}_1 + \left(\frac{N_1}{N_2}\right) T_L = T_M \qquad (2.11)$$

Thus the equivalent moment of inertia and viscous friction of gear train referred to shaft 1 are

$$J_{1eq} = J_1 + \left(\frac{N_1}{N_2}\right)^2 J_2; \quad f_{1eq} = f_1 + \left(\frac{N_1}{N_2}\right)^2 f_2$$

In terms of equivalent moment of inertia and friction, eqn. (2.11) may be written as

$$J_{1eq}\ddot{\theta}_1 + f_{1eq}\dot{\theta}_1 + \left(\frac{N_1}{N_2}\right) T_L = T_M$$

Here $(N_1/N_2) T_L$ is the load torque referred to shaft 1.

Similarly, expressing θ_1 in terms of θ_2 in eqn. (2.10) with the help of eqn. (2.9), the equivalent moment of inertia and viscous friction of gear train referred to load shaft are

$$J_{2eq} = J_2 + \left(\frac{N_2}{N_1}\right)^2 J_1; \quad f_{2eq} = f_2 + \left(\frac{N_2}{N_1}\right)^2 f_1$$

Torque equation referred to the load shaft may then be expressed as

$$J_{2eq}\ddot{\theta}_2 + f_{2eq}\dot{\theta}_2 + T_L = \left(\frac{N_2}{N_1}\right) T_M$$

ELECTRICAL SYSTEMS

The resistor, inductor and capacitor are the three basic elements of electrical circuits. These circuits are analyzed by the application of Kirchhoff's voltage and current laws.

Let us analyze the *L-R-C* series circuit shown in Fig. 2.7 by using Kirchhoff's voltage law. The governing equations of the system are

$$L\frac{di}{dt} + Ri + \frac{1}{C}\int_{\infty}^{t} i\, dt = e \qquad (2.12)$$

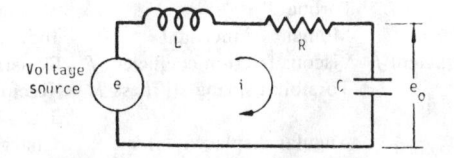

Figure 2.7 *L-R-C* series circuit.

$$\frac{1}{C} \int_{-\infty}^{t} i\, dt = e_0 \qquad (2.13)$$

In terms of electric charge $q = \int i\, dt$, eqn. (2.12) becomes

$$L \frac{d^2q}{dt^2} + R \frac{dq}{dt} + \frac{1}{C} q = e \qquad (2.14)$$

Similarly, using Kirchhoff's current law, we obtain the following equations for L-R-C parallel circuit shown in Fig. 2.8.

$$\frac{1}{L} \int_{-\infty}^{t} e\, dt + \frac{e}{R} + C \frac{de}{dt} = i \qquad (2.15)$$

In terms of magnetic flux linkage $\phi = \int e\, dt$, eqn. (2.15) may be written as

$$C \frac{d^2\phi}{dt^2} + \frac{1}{R} \frac{d\phi}{dt} + \frac{1}{L} \phi = i \qquad (2.16)$$

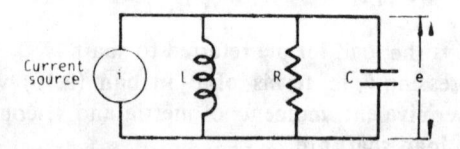

Figure 2.8 L-R-C parallel circuit.

ANALOGOUS SYSTEMS

Comparing eqn. (2.1) for the mechanical translational system shown in Fig. 2.3a or eqn. (2.3) for the mechanical rotational system shown in Fig. 2.5a and eqn. (2.14) for the electrical system shown in Fig. 2.7, it is seen that they are of identical form. Such systems whose differential equations are of identical form are called *analogous systems*. The force F (torque T) and voltage e are the analogous variables here. This is called the Force (Torque)-Voltage analogy. A list of analogous variables in this analogy is given in Table 2.1.

Table 2.1 ANALOGOUS QUANTITIES IN FORCE (TORQUE)-VOLTAGE ANALOGY

Mechanical translational systems	Mechanical rotational systems	Electrical systems
Force F	Torque T	Voltage e
Mass M	Moment of inertia J	Inductance L
Viscous friction coefficient f	Viscous friction coefficient f	Resistance R
Spring stiffness K	Torsional spring stiffness K	Reciprocal of capacitance $1/C$
Displacement x	Angular displacement θ	Charge q
Velocity $\dot{x}$	Angular velocity $\dot{\theta}$	Current i

Similarly eqns. (2.1) and (2.3) referred above and eqn. (2.16) for the electrical system shown in Fig. 2.8 are also identical. In this case force F (torque T) and current i are the analogous variables. This is called the Force (Torque)-Current analogy. A list of analogous quantities in this analogy is given in Table 2.2.

Table 2.2 ANALOGOUS QUANTITIES IN FORCE (TORQUE)-CURRENT ANALOGY

Mechanical translational systems	Mechanical rotational systems	Electrical systems
Force F	Torque T	Current i
Mass M	Moment of inertia J	Capacitance C
Viscous friction coefficient f	Viscous friction coefficient f	Reciprocal of resistance $1/R$
Spring stiffness K	Torsional spring stiffness K	Reciprocal of inductance $1/L$
Displacement x	Angular displacement θ	Magnetic flux linkage ϕ
Velocity $\dot{x}$	Angular velocity $\dot{\theta}$	Voltage e

The concept of analogous systems is a useful technique for the study of various systems like electrical, mechanical, thermal, liquid-level, etc. If the solution of one system is obtained, it can be extended to all other systems analogous to it. Generally it is convenient to study a non-electrical system in terms of its electrical analog as electrical systems are more easily amenable to experimental study.

THERMAL SYSTEMS

The basic requirement for the representation of thermal systems by linear models is that the temperature of the medium be uniform which is generally not the case. Thus for precise analysis a distributed parameter model must be used. Here, however, in order to simplify the analysis, uniformity of temperature is assumed and thereby the system is represented by a lumped parameter model.

Consider the simple thermal system shown in Fig. 2.9. Assume that the tank is insulated to eliminate heat loss to the surrounding air, there is no heat storage in the insulation and liquid in the tank is kept at uniform temperature by perfect mixing with the help of a stirrer. Thus a single temperature may be used to describe the thermal state of the entire liquid. (If complete mixing is not present, there is a complex temperature distribution throughout the liquid and the problem becomes one of the distributed parameters, requiring the use of partial differential equations.) Assume that the steady-state temperature of the inflowing liquid is θ_i and that of the outflowing liquid is θ. The steady-state heat input rate from the heater is H. The liquid flow rate is of course assumed constant.

Let ΔH (joules/min) be a small increase in the heat input rate from its steady-state value. This increase in heat input rate will result in increase of the heat outflow rate by an amount ΔH_1 and a heat storage rate of the liquid in the tank by an amount ΔH_2. Consequently the temperature of the liquid in the tank and therefore of the outflowing liquid rises by $\Delta\theta$ (°C). Since the insulation has been regarded as perfect, the increase in outflow rate is only due to the rise in temperature of the outflowing liquid and is given by

$$\Delta H_1 = Qs\Delta\theta$$

where Q = steady liquid flow rate in kg/min; and s = specific heat of the liquid in joules/kg/°-C.

The above relationship can be written in the form

$$\Delta H_1 = \Delta\theta/R \tag{2.17}$$

where $R = 1/Qs$, is defined as the *thermal resistance* and has the units of °C/joules/min.

The rate of heat storage in the tank is given by

$$\Delta H_2 = Ms\frac{d(\Delta\theta)}{dt}$$

where M = mass of liquid in the tank in kg; and $\dfrac{d(\Delta\theta)}{dt}$ = rate of rise of temperature in the tank.

The above equation can be expressed in the form

$$\Delta H_2 = C\frac{d(\Delta\theta)}{dt} \tag{2.18}$$

where $C = Ms$, is defined as the *thermal capacitance* and has the units of joules/°C. For the system of Fig. 2.9, the heat flow balance equation is

$$\Delta H = \Delta H_1 + \Delta H_2$$
$$= \frac{\Delta\theta}{R} + C\frac{d(\Delta\theta)}{dt}$$

or
$$RC\frac{d(\Delta\theta)}{dt} + \Delta\theta = R(\Delta H) \tag{2.19}$$

Equation (2.19) describes the dynamics of the thermal system with the assumption that the temperature of the inflowing liquid is constant.

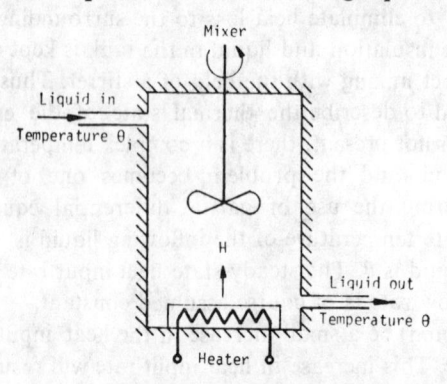

Figure 2.9 Thermal system.

In practice, the temperature of the inflowing liquid fluctuates. Thus along with a heat input signal from the heater, there is an additional signal due to change in the temperature of the inflowing liquid which is known as the *disturbance signal*.

Let $\Delta\theta_i$ be change in the temperature of the inflowing liquid from its steady-state value. Now in addition to the change in heat input from the heater, there is a change in heat carried by the inflowing liquid. The heat flow equation, therefore becomes

$$\Delta H + \frac{\Delta\theta_i}{R} = \frac{\Delta\theta}{R} + C\frac{d}{dt}(\Delta\theta)$$

or
$$RC\frac{d}{dt}(\Delta\theta) + \Delta\theta = \Delta\theta_i + R(\Delta H) \qquad (2.20)$$

Let us now relax the assumption that the tank insulation is perfect. As the liquid temperature increases by $\Delta\theta$, the rate of heat flow through the tank walls to the ambient medium increases by

$$\Delta H_3 = \frac{\Delta\theta}{R_t}$$

where R_t is the thermal resistance of the tank walls. Equation (2.20) is then modified to

$$\Delta H + \frac{\Delta\theta_i}{R} = \left(\frac{\Delta\theta}{R} + \frac{\Delta\theta}{R_t}\right) + C\frac{d}{dt}(\Delta\theta)$$

or
$$R'C\frac{d}{dt}(\Delta\theta) + \Delta\theta = \left(\frac{R'}{R}\right)\Delta\theta_i + R'(\Delta H)$$

where $R' = \dfrac{RR_t}{R + R_t}$ = effective thermal resistance due to liquid outflow and tank walls (it is a parallel combination of R and R_t).

It is still being assumed above that there is no heat storage in the tank walls. Relaxing this assumption will simply add to the thermal capacitance C.

Fluid Systems

The dynamics of the fluid systems can be represented by ordinary linear differential equations only if the fluid is incompressible and fluid flow is laminar. Industrial processes often involve fluid flow through connecting pipes and tanks where the flow is usually turbulent resulting in nonlinear equations describing the system.

Velocity of sound is a key parameter in fluid flow to determine the compressibility property. If the fluid velocity in much less than the velocity of sound, compressibility effects are usually small. As the velocity of sound in liquids is about 1500 m/sec, compressibility* effects are rarely of import-

---

*The tendency of so-called incompressible fluids to compress slightly under pressure is called *fluid compliance*. This type of effect is accounted for in hydraulic pumps and motors discussed in section 4.5.

ance in liquids and the treatment of compressibility is generally restricted to gases, where the velocity of sound is about 350 m/sec.

Another important fluid property is the type of fluid flow—laminar or turbulent. Laminar flow is characterized by smooth motion of one laminar of fluid past another, while turbulent flow is characterized by anirregular and nearly random motion superimposed on the main motion of fluid. The transition from laminar to turbulent flow was first investigated by Osborne Reynolds, who after experimentation found that for pipe flow the transition conditions could be correlated by a dimensionless group which is now known as *Reynolds number*, Re.

From his experiments, Reynolds found that pipe flow will be laminar for Re less than 2,000 and turbulent for Re greater than 3,000. When Re is between 2,000 and 3,000, the type of flow is unpredictable-and often changes back and forth between the laminar and turbulent states because of flow disturbances and pipe vibrations.

The pressure drop across a pipe section is given by (for proof, refer Shames[1])

$$P = \frac{128l\mu}{\pi D^4} Q \ ; \quad \text{for laminar flow} \tag{2.21a}$$

$$= RQ$$

$$P = \frac{8K_t \rho l}{\pi^2 D^5} Q^2 \ ; \quad \text{for turbulent flow} \tag{2.21b}$$

$$= K_T Q^2$$

where $l =$ length of pipe section (m); $D =$ diameter of pipe (m); $\mu =$ viscosity (newton-sec/m^2); $Q =$ volumetric flow rate (cub-m/sec); $K_t =$ a constant (to be determined experimentally); and $\rho =$ mass density (kg/m^3).

Equation (2.21a) representing laminar flow is linear, i.e.,

$$P = RQ \tag{2.22a}$$

where $R = \dfrac{128l\mu}{\pi D^4} \left(\dfrac{\text{newton/m}^2}{\text{cub-m/sec}} \right)$ is the *fluid resistance*.

Equation (2.21b) representing turbulent flow is nonlinear, i.e.,

$$P = K_T Q^2$$

where $$K_T = \frac{8K_t \rho l}{\pi^2 D^5}.$$

This equation can be linearized about the operating point (P_0, Q_0) by techniques discussed at length in section 4.2. At the operating point,

$$P_0 = K_T Q_0^2$$

Expanding the turbulent flow equation given above in Taylor series about the operating point and retaining first-order term only, we have

$$P = P_0 + \frac{dP}{dQ}\bigg|_{(P_0, Q_0)} (Q - Q_0)$$

It follows that
$$P - P_0 = 2K_T Q_0 (Q - Q_0)$$
or
$$\Delta P = R \Delta Q \quad (2.22b)$$
where $R = 2K_T Q_0$ is the *turbulent flow resistance*.

Equation (2.22b) relates the incremental fluid flow to incremental pressure around the operating point in the case of turbulent flow. This is illustrated by Fig. 2.10.

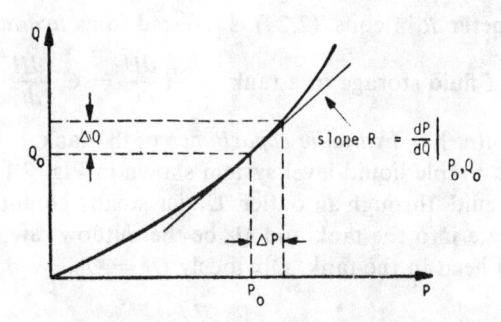

Figure 2.10 Pressure versus flow rate for turbulent flow.

Large pipes even when long offer small resistance while short devices that contain some contrictions (orifices, nozzles, valves, etc.) offer large resistance to fluid flow. For these dissipation devices, head loss
$$P = \frac{8K\rho}{\pi^2 D^4} Q^2$$
where K is a constant. Experimentally determined values of K for various dissipation devices can be found in handbooks. This equation is analogous to eqn. (2.21b) and can be linearized about the operating point to obtain resistance offered by a dissipation device.

The other ideal element used in modelling fluid systems is the *fluid capacitance*. Consider a tank with cross-sectional area $= A(m^2)$.

The rate of fluid storage in the tank $= A \dfrac{dH}{dt} = \dfrac{A}{\rho g} \dfrac{dP}{dt}$

$$= C \frac{dP}{dt} \quad (2.23)$$

where $H =$ fluid head in the tank (m); $P = \rho g H$ (newton/m^2) $=$ pressure at tank bottom; and $C = \dfrac{A}{\rho g} \left(\dfrac{\text{cub-m}}{\text{newton/m}^2} \right) = $ *capacitance* of the tank.

Liquid-level Systems

In terms of head $H(m)$, the fluid pressure is given by
$$P = \rho g H$$

The pressure-flow rate relations given by eqns. (2.22a) and (2.22b) may be expressed as the following head-flow rate relations:

$$H = RQ; \text{ for laminar flow} \qquad (2.24a)$$

where
$$R = \frac{128l\mu}{\pi D^2 \rho g}$$

$$\Delta H = R\Delta Q; \text{ for turbulent flow} \qquad (2.24b)$$

where
$$R = \frac{2K_T Q_0}{\rho g}$$

The parameter R in eqns. (2.24) is referred to as *hydraulic resistance*.

The rate of fluid storage in a tank $= A\dfrac{dH}{dt} = C\dfrac{dH}{dt}$

where $C = A(m^2) =$ *hydraulic capacitance* of the tank.

Consider a simple liquid-level system shown in Fig. 2.11 where a tank is supplying liquid through an outlet. Under steady conditions, let Q_i be the liquid flow rate into the tank and Q_0 be the outflow rate, while H_0 is the steady liquid head in the tank. Obviously $Q_i = Q_0$.

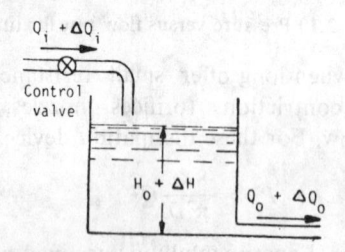

Figure 2.11 Liquid-level system.

Let ΔQ_i be a small increase in the liquid inflow rate from its steady-state value. This increase in liquid inflow rate causes increase of head of the liquid in the tank by ΔH, resulting in increase of liquid outflow rate by

$$\Delta Q_0 = \Delta H/R$$

The system dynamics is described by the liquid flow rate balance equation:

Rate of liquid storage in the tank = rate of liquid inflow − rate of liquid outflow

Therefore

$$C\frac{d(\Delta H)}{dt} = \Delta Q_i - \frac{\Delta H}{R}$$

or

$$RC\frac{d(\Delta H)}{dt} + \Delta H = R(\Delta Q_i) \qquad (2.25)$$

where C is the capacitance of the tank and R is the total resistance offered by the tank outlet and pipe.

Pneumatic Systems

We shall assume in our discussion that velocities of gases are a small fraction of the velocity of sound, which is true in a number of engineering applications. With this assumption, we treat pneumatic flow also as nearly incompressible. Therefore, the results presented earlier are directly applicable to this class of pneumatic systems.

Consider a simple pneumatic system shown in Fig. 2.12. A pneumatic source is supplying air to the pressure vessel through a pipe line.

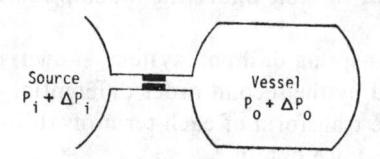

Figure 2.12 Simple pneumatic system.

Let us define:

P_i = air pressure of the source at steady-state (newton/m²).
P_0 = air pressure in the vessel at steady-state (newton/m²).
ΔP_i = small change in air pressure of the source from its steady-state value.
ΔP_0 = small change in air pressure of the vessel from its steady-state value.

System dynamics is described by the equation:

Rate of gas storage in vessel = rate of gas inflow

or

$$C \frac{d(\Delta P_0)}{dt} = \frac{\Delta P}{R} = \frac{\Delta P_i - \Delta P_0}{R}$$

Table 2.3 ANALOGOUS QUANTITIES

Electrical systems	Thermal systems	Liquid-level systems	Pneumatic systems
Charge, coulombs	Heat flow, joules	Liquid flow, cub-m	Air flow, cub-m
Current, amps	Heat flow rate, joules/min	Liquid flow rate, cub-m/min	Air flow rate, cub-m/min
Voltage, volts	Temperature, °C	Head, m	Pressure, newton/m²
Resistance, ohms	Resistance, °C / joules/min.	Resistance, m / cub-m/min	Resistance, newton/m² / cub-m/min
Capacitance, farads	Capacitance, joules/°C	Capacitance, cub-m/m	Capacitance, cub-m / newton/m²

or

$$RC \frac{d(\Delta P_0)}{dt} + \Delta P_0 = \Delta P_i \qquad (2.26)$$

Table 2.3 summarizes the variables and parameters of the thermal, liquid-level and pneumatic systems which are analogous to those of electrical systems.

2.3 Transfer Functions

The transfer function of a linear time-invariant system is defined to be the ratio of the Laplace transform of the output variable to the Laplace transform of the input variable under the assumption that all initial conditions are zero.

Consider the mass-spring-dashpot system shown in Fig. 2.3a, whose dynamics is described by the second-order differential equation (2.1).

Taking the Laplace transform of each term of this equation (assuming zero initial conditions), we obtain

$$F(s) = Ms^2 X(s) + fsX(s) + KX(s)$$

Then the transfer function is

$$G(s) = \frac{X(s)}{F(s)} = \frac{1}{Ms^2 + fs + K} \qquad (2.27)$$

The highest power of the complex variable s in the denominator of the transfer function determines the *order of the system*. The mass-spring-dashpot system under consideration is thus a second-order system, a fact which is already recognized from its differential equation.

The transfer function of the L-R-C circuit shown in Fig. 2.7 is similarly obtained by taking the Laplace transform of eqns. (2.12) and (2.13), with zero initial conditions. The resulting equations are

$$sLI(s) + RI(s) + \frac{1}{s} \frac{I(s)}{C} = E(s)$$

$$\frac{1}{s} \frac{I(s)}{C} = E_0(s)$$

If e is assumed to be the input variable and e_0 the output variable, the transfer function of the system is

$$\frac{E_0(s)}{E(s)} = \frac{1}{LCs^2 + RCs + 1} \qquad (2.28)$$

Equations (2.27) and (2.28) reveal that the transfer function is an expression in s-domain, relating the output and input of the linear time-invariant system in terms of the system parameters and is independent of the input. It describes the input-output behaviour of the system and does not give any information concerning the internal structure of the system. Thus, when the transfer function of a physical system is determined, the system can be represented by a *block*, which is a short-hand pictorial representation of the cause and effect relationship between input and output of the system. Functional operation of a system can be more readily visualized by examination of a block diagram rather than by examination of the equations describing the physical system. Therefore, when working with a linear time-invariant system, we can think of a system or its sub-systems simply as interconnected blocks with each block described by a transfer function.

From eqn. (2.19), the transfer function of the thermal system shown in Fig. 2.9 is

$$\frac{\Delta\theta(s)}{\Delta H(s)} = \frac{R}{RCs + 1} \tag{2.29}$$

The block diagram representation of the system is shown in Fig. 2.13a. When this system is subjected to a disturbance, the dynamics is described by eqn. (2.20). Taking the Laplace transform of this equation, we get

$$(RCs + 1)\Delta\theta(s) = \Delta\theta_i(s) + R\Delta H(s) \tag{2.30}$$

The corresponding block diagram representation is given in Fig. 2.13b.

(a)

(b)

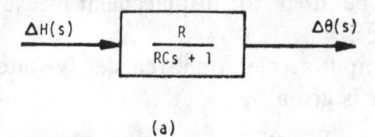

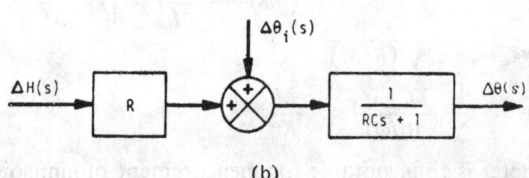

Figure 2.13 Block diagram of the thermal system shown in Fig. 2.9.

SINUSOIDAL TRANSFER FUNCTIONS

The steady-state response of a control system to a sinusoidal input is obtained by replacing s with $j\omega$ in the transfer function of the system.

Transfer function of the mechanical accelerometer shown in Fig. 2.4, obtained from eqn. (2.2), is

$$\frac{Y(s)}{X(s)} = \frac{Ms^2}{Ms^2 + fs + K}$$

$$= \frac{s^2}{s^2 + \frac{f}{M}s + \frac{K}{M}}$$

Its sinusoidal transfer function becomes

$$\frac{Y(j\omega)}{X(j\omega)} = \frac{(j\omega)^2}{(j\omega)^2 + \frac{f}{M}(j\omega) + \frac{K}{M}} \tag{2.31}$$

Equation (2.31) represents the behaviour of the accelerometer when used as a device to measure sinusoidally varying displacement. If the frequency of the sinusoidal input signal $X(j\omega)$ is very low, i.e., $\omega \ll \omega_n = \sqrt{(K/M)}$, then the transfer function given by eqn. (2.31) may be approximated by

$$\frac{Y(j\omega)}{X(j\omega)} \approx \frac{-\omega^2}{K/M}$$

The output signal is very weak for values of frequency $\omega \ll \omega_n$. Weak output signal coupled with the fact that some inherent noise may always be present in the system, makes the displacement measurement by the accelerometer in the low frequency range as quite unreliable.

For $\omega \gg \omega_n$, the transfer function given by eqn. (2.31) may be approximated by

$$\frac{Y(j\omega)}{X(j\omega)} \approx 1$$

Thus, at very high frequencies the accelerometer output follows the sinusoidal displacement input. For this range of frequencies the basic accelerometer system can be used for displacement measurement particularly in seismographic studies.

For a sinusoidal input acceleration, the steady-state sinusoidal response of the accelerometer is given by

$$\frac{Y(j\omega)}{A(j\omega)} = \frac{1}{(j\omega)^2 + \frac{f}{M}(j\omega) + \frac{K}{M}}$$

As long as $\omega \ll \omega_n = \sqrt{(K/M)}$,

$$\frac{Y(j\omega)}{A(j\omega)} \approx \frac{M}{K}$$

The accelerometer is thus suitable for measurement of sinusoidally varying acceleration from zero frequency (constant acceleration) to a frequency which depends upon the choice of ω_n for the accelerometer. The sinusoidal behaviour of this type of transfer functions will be studied in greater details in Chapter 8.

PROCEDURE FOR DERIVING TRANSFER FUNCTIONS

The following assumptions are made in deriving transfer functions of physical systems.

1. It is assumed that there is no loading, i.e., no power is drawn at the output of the system. If the system has more than one nonloading elements in tandem, then the transfer function of each element can be determined independently and the overall transfer function of the physical system is determined by multiplying the individual transfer functions. In case of systems consisting of elements which load each other, the overall transfer function should be derived by basic analysis without regard to the individual transfer functions.

Mathematical Models of Physical Systems

2. The system should be approximated by a linear lumped constant parameters model by making suitable assumptions.

To illustrate the point (1) above, let us consider two identical RC circuits connected in cascade so that the output from the first circuit is fed as input to the second as shown in Fig. 2.14.

The describing equations for this system are

$$\frac{1}{C}\int_{-\infty}^{t}(i_1-i_2)dt + Ri_1 = e_i \tag{2.32}$$

$$\frac{1}{C}\int_{-\infty}^{t}(i_2-i_1)dt + Ri_2 = -\frac{1}{C}\int_{-\infty}^{t}i_2 dt = -e_0 \tag{2.33}$$

Taking the Laplace transforms of eqns. (2.32) and (2.33), assuming zero initial conditions, we obtain

$$\frac{1}{sC}\left[I_1(s)-I_2(s)\right] + RI_1(s) = E_i(s)$$

$$\frac{1}{sC}\left[I_2(s)-I_1(s)\right] + RI_2(s) = -\frac{1}{sC}I_2(s) = -E_0(s)$$

The transfer function obtained by eliminating $I_1(s)$ and $I_2(s)$ from the above equations is

$$\frac{E_0(s)}{E_i(s)} = \frac{1}{\tau^2 s^2 + 3\tau s + 1} \tag{2.34}$$

where $\tau = RC$.

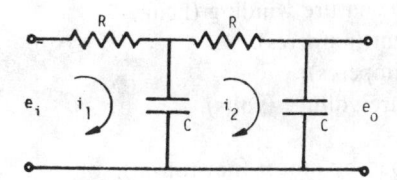

Figure 2.14 RC circuits in cascade.

The transfer function of each of the individual RC circuits is $1/(1+s\tau)$. From eqn. (2.34) it is seen that overall transfer function of the two RC circuits connected in cascade is not equal to $[1/(\tau s + 1)][1/(\tau s + 1)]$ but instead it is $1/(\tau^2 s^2 + 3\tau s + 1)$.

This difference is explained by the fact that while deriving the transfer function of a single RC circuit, it is assumed that the output is unloaded. However, when the input of second circuit is obtained from the output of first, a certain amount of energy is drawn from the first circuit and hence its original transfer function is no longer valid. The degree to which the overall transfer function is modified from the product of individual transfer functions depends upon the amount of loading.

As an example to illustrate the point (2) above, let us derive the transfer function of a d.c. servomotor. In servo applications, a d.c. motor is required to produce rapid accelerations from standstill. Therefore the physical requirements of such a motor are low inertia and high starting torque. Low inertia is attained with reduced armature diameter with a consequent increase in armature length such that the desired power output is achieved. Thus, except for minor differences in constructional features, a d.c. servomotor is essentially an ordinary d.c. motor.

In control systems, the d.c. motors are used in two different control modes—armature-control mode with fixed field and field-control mode with fixed armature current.

Armature-control

Consider the armature-controlled d.c. motor shown in Fig. 2.15.

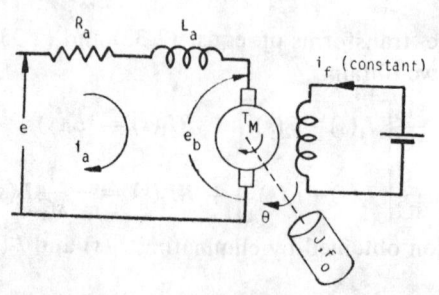

Figure 2.15 Armature-controlled d.c. motor.

In this system,
R_a = resistance of armature (ohms).
L_a = inductance of armature winding (henrys).
i_a = armature current (amperes).
i_f = field current (amperes).
e = applied armature voltage (volts).
e_b = back emf (volts).
T_M = torque developed by motor (newton-m).
θ = angular displacement of motor-shaft (rad).
J = equivalent moment of inertia of motor and load referred to motor shaft (kg-m²).
f_0 = equivalent viscous friction coefficient of motor and load referred to motor shaft $\left(\dfrac{\text{newton-m}}{\text{rad/sec}}\right)$.

In servo applications, the d.c. motors are generally used in the linear range of the magnetization curve. Therefore, the air gap flux ϕ is proportional to the field current, i.e.,

$$\phi = K_f i_f \tag{2.35}$$

where K_f is a constant.

The torque T_M developed by the motor is proportional to the product of the armature current and air gap flux, i.e.,

$$T_M = K_1 K_f i_f i_a \tag{2.36}$$

where K_1 is a constant.

In the armature-controlled d.c. motor, the field current is kept constant, so that eqn. (2.36) can be written as

$$T_M = K_T i_a \tag{2.37}$$

where K_T is known as the motor torque constant.

The motor back emf being proportional to speed is given as

$$e_b = K_b \frac{d\theta}{dt} \tag{2.38}$$

where K_b is the back emf constant.

The differential equation of the armature circuit is

$$L_a \frac{di_a}{dt} + R_a i_a + e_b = e \tag{2.39}$$

The torque equation is

$$J \frac{d^2\theta}{dt^2} + f_0 \frac{d\theta}{dt} = T_M = K_T i_a \tag{2.40}$$

Taking the Laplace transform of eqns. (2.38) to (2.40), assuming zero initial conditions, we get

$$E_b(s) = K_b s \theta(s) \tag{2.41}$$

$$(L_a s + R_a) I_a(s) = E(s) - E_b(s) \tag{2.42}$$

$$(Js^2 + f_0 s) \theta(s) = T_M(s) = K_T I_a(s) \tag{2.43}$$

From eqns. (2.41) to (2.43), the transfer function of the system is obtained as

$$G(s) = \frac{\theta(s)}{E(s)} = \frac{K_T}{s[(R_a + sL_a)(Js + f_0) + K_T K_b]} \tag{2.44}$$

The block diagram representation of eqn. (2.42) is shown in Fig. 2.16a where the circular block representing the differencing action is known as the *summing point*. Equation (2.43) is represented by a block shown in Fig. 2.16b.

Figure 2.16c represents eqn. (2.41) where a signal is taken off from a *take off* point and fed to the feedback block. Figure 2.16d is the complete block diagram of the system under consideration, obtained by connecting the block diagrams shown in Figs. 2.16a, b and c.

However, it should be noted that the block diagram of the system under consideration can be directly obtained from the physical system of Fig. 2.15 by using the transfer functions of simple electrical and mechanical networks derived already. The voltage applied to the armature circuit is $E(s)$ which is opposed by the back emf $E_b(s)$. The net voltage $(E - E_b)$ acts on a linear circuit comprised of resistance and inductance in series, having the transfer function $1/(sL_a + R_a)$. The result is an armature current $I_a(s)$. For fixed field, the torque developed by the motor is $K_T I_a(s)$. This torque rotates the load at a speed $\theta(s)$ against the moment of inertia J and viscous friction with coefficient f_0 [the transfer function is $1/(Js + f_0)$]. The back emf signal

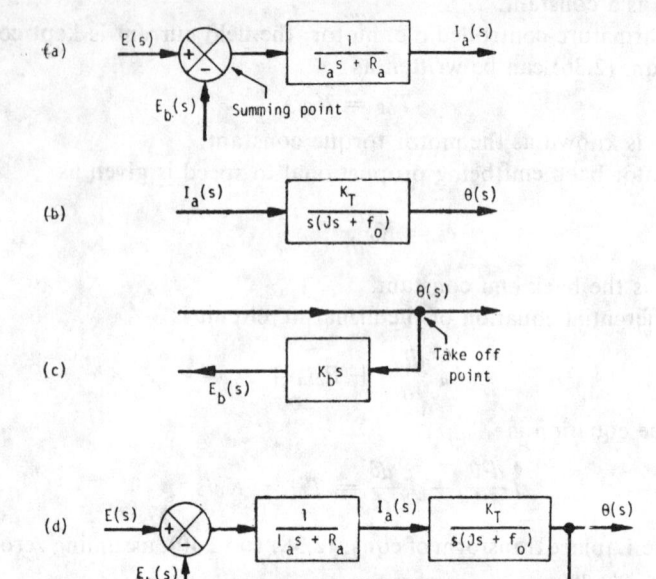

Figure 2.16 Block diagram of armature-controlled d.c. motor.

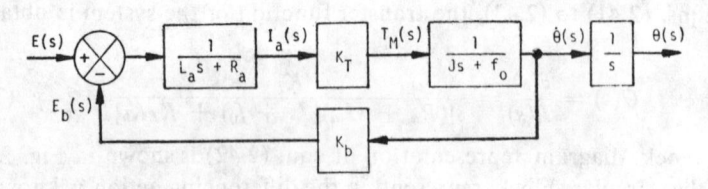

Figure 2.17 Block diagram of armature-controlled d.c. motor

$E_b = K_b \theta(s)$ is taken off from the shaft speed and fedback negatively to the summing point. The angle signal $\theta(s)$ is obtained by integrating (i.e., $1/s$) the speed $\dot\theta(s)$. This results in the block diagram of Fig. 2.17, which is equivalent to that of Fig. 2.16 as can be seen by shifting the take off point from $\dot\theta(s)$ to $\theta(s)$.

The armature circuit inductance L_a is usually negligible. Therefore from eqn. (2.44), the transfer function of the armature controlled motor simplifies to

$$\frac{\theta(s)}{E(s)} = \frac{K_T/R_a}{Js^2 + s(f_0 + K_T K_b/R_a)} \quad (2.45)$$

The term $(f_0 + K_T K_b/K_a)$ indicates that the back emf of the motor effectively increases the viscous friction of the system. Let

$$f = f_0 + K_T K_b/R_a$$

be the effective viscous friction coefficient. Then from eqn. (2.45)

$$\frac{\theta(s)}{E(s)} = \frac{K_T/R_a}{s(Js+f)} \quad (2.46)$$

The transfer function given by eqn. (2.46) may be written in the form

$$\frac{\theta(s)}{E(s)} = \frac{K_m}{s(s\tau_m + 1)} \quad (2.47)$$

where $K_m = K_T/R_a f$ = motor gain constant, and $\tau_m = J/f$ = motor time constant.

The motor torque and back emf constants K_T, K_b are interrelated. Their relationship is deduced below. In metric units, K_b is in volts/rad/sec and K_T is in newton-m/amp.

Electrical power converted to mechanical form = $e_b i_a = K_b \dot{\theta} i_a$ watts
Power at shaft (in mechanical form) = $T\dot{\theta} = K_T i_a \dot{\theta}$ watts
At steady speed these two powers balance. Hence

$$K_b \dot{\theta} i_a = K_T i_a \dot{\theta}$$

or $\quad K_b = K_T$ (in MKS units)

This result can be used to advantage in practice as K_b can be measured more easily and with greater accuracy than K_T.

Field-control

A field-controlled d.c. motor is shown in Fig. 2.18a.

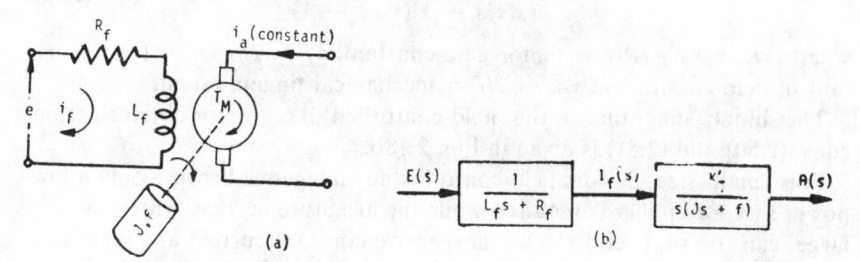

Figure 2.18 (a) Field-controlled d.c. motor;
(b) Block diagram of field-controlled motor.

In this system,

R_f = field winding resistance (ohms).
L_f = field winding inductance (henrys).
e = field control voltage (volts).
i_f = field current (amperes).
T_M = torque developed by motor (newton-m).
J = equivalent moment of inertia of motor and load referred to motor shaft (kg-m²).
f = equivalent viscous friction coefficient of motor and load referred to motor shaft $\left(\dfrac{\text{newton-m}}{\text{rad/sec}}\right)$.
θ = angular displacement of motor shaft (rad).

In the field-controlled motor, the armature current is fed from a constant current source. Therefore, from eqn. (2.36)

$$T_M = K_1 K_f i_f i_a = K_T' i_f$$

where K_T' is a constant.

The equation for the field circuit is

$$L_f \frac{di_f}{dt} + R_f i_f = e \qquad (2.48)$$

The torque equation is

$$J \frac{d^2\theta}{dt^2} + f \frac{d\theta}{dt} = T_M = K_T' i_f \qquad (2.49)$$

Taking the Laplace transform of eqns. (2.48) and (2.49), assuming zero initial conditions, we get

$$(L_f s + R_f) I_f(s) = E(s) \qquad (2.50)$$

$$(Js^2 + fs) \theta(s) = T_M(s) = K_T' I_f(s) \qquad (2.51)$$

From the above equations, the transfer function of the motor is obtained as

$$\frac{\theta(s)}{E(s)} = \frac{K_T'}{s(L_f s + R_f)(Js + f)}$$

$$= \frac{K_m}{s(\tau_f s + 1)(\tau_{me} s + 1)} \qquad (2.52)$$

where $K_m = K_T'/R_f f$ = motor gain constant; $\tau_f = L_f/R_f$ = time constant of field circuit; and $\tau_{me} = J/f$ = mechanical time constant.

The block diagram of the field-controlled d.c. motor obtained from eqns. (2.50) and (2.51) is given in Fig. 2.18b.

For small size motors field-control is advantageous because only a low power servo amplifier is required while the armature current which is not large can be supplied from an inexpensive constant current amplifier. For large size motors it is on the whole cheaper to use armature-control scheme. Further in armature-controlled motor, back emf contributes additional damping over and above that provided by load friction.

2.4 Block Diagram Algebra

As introduced earlier, the input-output behaviour of a linear system or element of a linear system is given by its transfer function

$$G(s) = C(s)/R(s)$$

where $R(s)$ = Laplace transform of the input variable; and $C(s)$ = Laplace transform of the output variable.

A convenient graphical representation of this behaviour is the *block diagram* as shown in Fig. 2.19a wherein the signal into the block represents the input $R(s)$ and the signal out of the block represents the output $C(s)$, while the block itself stands for the transfer function $G(s)$. The flow of information (signal) is unidirectional from the input to the output with the output being equal to the input multiplied by the transfer function of the block. A complex system comprising of several non-loading elements is represented by the interconnection of the blocks for individual elements. The blocks are connected by lines with arrows indicating the unidirectional flow of information from the output of one block to the input of the other. In addition to this, *summing* or *differencing* of signals is indicated by the symbols shown in Fig. 2.19b, while the *take-off* point of a signal is represented by Fig. 2.19c.

Block diagrams of some of the control systems turn out to be very complex such that the evaluation of their performance requires simplification (or reduction) of block diagrams which is carried out by block diagram rearrangements. Some of the important block diagram rearrangements are discussed in this section.

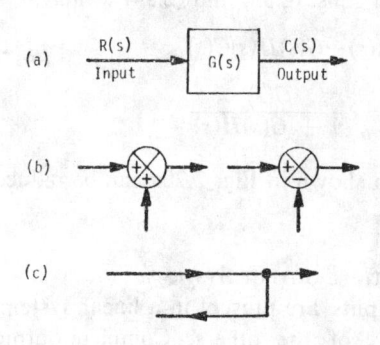

Figure 2.19.

BLOCK DIAGRAM OF A CLOSED-LOOP SYSTEM

Figure 2.20a shows the block diagram of a negative feedback system. With reference to this figure, the terminology used in block diagrams of control systems is given below.

$R(s)$ = reference input.
$C(s)$ = output signal or controlled variable.
$B(s)$ = feedback signal.
$E(s)$ = actuating signal.
$G(s) = C(s)/E(s)$ = forward path transfer function.
$H(s)$ = transfer function of the feedback elements.
$G(s)H(s) = B(s)/E(s)$ = loop transfer function.
$T(s) = C(s)/R(s)$ = closed-loop transfer function.

From Fig. 2.20a we have

$$C(s) = G(s)E(s) \tag{2.53}$$
$$E(s) = R(s) - B(s)$$
$$= R(s) - H(s)C(s) \tag{2.54}$$

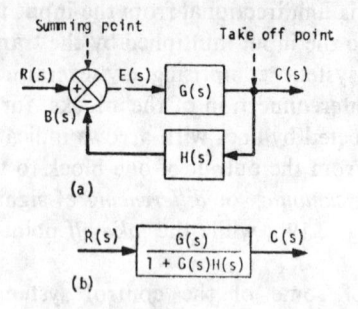

Figure 2.20 (a) Block diagram of a closed-loop system;
(b) Reduction of block diagram shown in Fig. 2.20a.

Eliminating $E(s)$ from eqns. (2.53) and (2.54) we have

$$C(s) = G(s)R(s) - G(s)H(s)C(s)$$

or

$$\frac{C(s)}{R(s)} = T(s) = \frac{G(s)}{1 + G(s)H(s)} \tag{2.55}$$

Therefore, the system shown in Fig. 2.20a can be reduced to a single block shown in Fig. 2.20b.

Multiple-Input-Multiple-Output Systems

When multiple inputs are present in a linear system, each input can be treated independently of the others. Complete output of the system can then be obtained by superposition, i.e., outputs corresponding to each input alone are added together.

Consider a two-input linear system shown in Fig. 2.21a. The response to the reference input can be obtained by assuming $U(s) = 0$. The corresponding block diagram shown in Fig. 2.21b gives

$C_R(s)$ = output due to $R(s)$ acting alone

$$= \frac{G_1(s)\,G_2(s)}{1 + G_1(s)\,G_2(s)H(s)} R(s) \tag{2.56}$$

Similarly the response to the input $U(s)$ is obtained by assuming $R(s) = 0$. The block diagram corresponding to this case is shown in Fig. 2.21c, which gives

$C_U(s)$ = output due to $U(s)$ acting alone

$$= \frac{G_2(s)}{1 + G_1(s)\,G_2(s)\,H(s)} U(s) \tag{2.57}$$

The response to the simultaneous application of $R(s)$ and $U(s)$ can be obtained by adding the two individual responses.

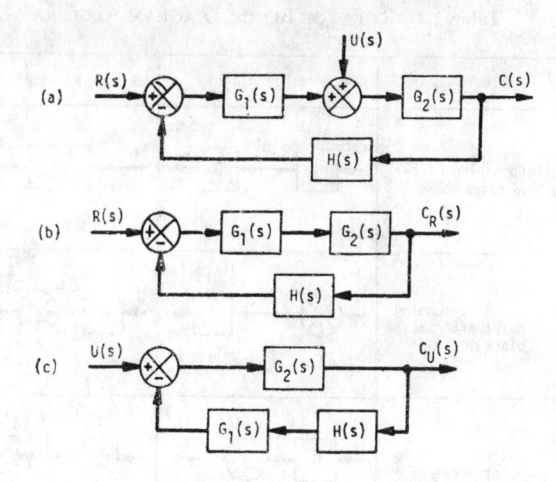

Figure 2.21 Block diagrams of two-input system.

Adding eqns. (2.56) and (2.57), we get

$$C(s) = C_R(s) + C_U(s)$$
$$= \frac{G_2(s)}{1 + G_1(s)G_2(s)H(s)} [G_1(s)R(s) + U(s)] \quad (2.58)$$

In case of multiple-input-multiple-output system shown in Fig. 2.22 (r inputs and m outputs), the i-th output $C_i(s)$ is given by the principle of superposition as

$$C_i(s) = \sum_{j=1}^{r} G_{ij}(s) R_j(s); \; i = 1, 2, \ldots, m$$

where $R_j(s)$ is the j-th input and $G_{ij}(s)$ is the transfer function between the i-th output and j-th input with all other inputs reduced to zero.

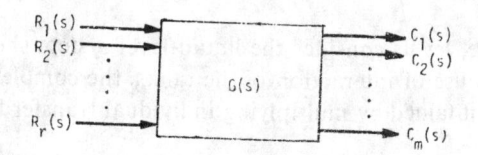

Figure 2.22 Multiple-input-multiple-output system.

BLOCK DIAGRAM REDUCTION

As indicated earlier, a complex block diagram configuration can be simplified by certain rearrangements of block diagrams using the rules of block diagram algebra. Some of the important rules are given in Table 2.4. All these rules are derived by simple algebraic manipulations of the equations representing the blocks.

Table 2.4 RULES OF BLOCK DIAGRAM ALGEBRA

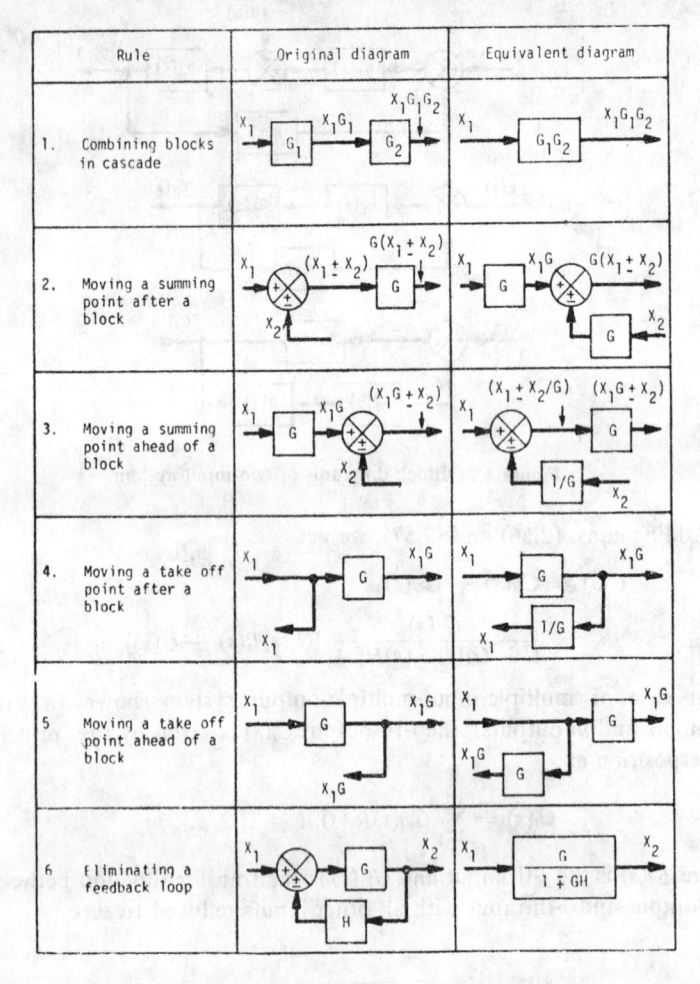

Rule	Original diagram	Equivalent diagram
1. Combining blocks in cascade	$x_1 \to [G_1] \to x_1 G_1 \to [G_2] \to x_1 G_1 G_2$	$x_1 \to [G_1 G_2] \to x_1 G_1 G_2$
2. Moving a summing point after a block	$x_1 \to \oplus \xrightarrow{(x_1 \pm x_2)} [G] \to G(x_1 \pm x_2)$, with x_2 into summer	$x_1 \to [G] \xrightarrow{x_1 G} \oplus \to G(x_1 \pm x_2)$, with $x_2 \to [G]$ into summer
3. Moving a summing point ahead of a block	$x_1 \to [G] \xrightarrow{x_1 G} \oplus \to (x_1 G \pm x_2)$, with x_2 into summer	$x_1 \to \oplus \xrightarrow{(x_1 \pm x_2/G)} [G] \to (x_1 G \pm x_2)$, with $x_2 \to [1/G]$ into summer
4. Moving a take off point after a block	$x_1 \to [G] \to x_1 G$, with x_1 taken off before block	$x_1 \to [G] \to x_1 G$, with x_1 obtained via $[1/G]$ after block
5. Moving a take off point ahead of a block	$x_1 \to [G] \to x_1 G$, with $x_1 G$ taken off after block	$x_1 \to [G] \to x_1 G$, with $x_1 G$ obtained via $[G]$ before block
6. Eliminating a feedback loop	$x_1 \to \oplus \to [G] \to x_2$, feedback $[H]$	$x_1 \to \left[\dfrac{G}{1 \mp GH}\right] \to x_2$

As an example, let us consider the liquid-level system shown in Fig. 2.23 (note that because of interaction of the tanks, the complete transfer function cannot be obtained by multiplying individual transfer functions of the tanks).

In this system, a tank having liquid capacitance C_1 is supplying liquid through a pipe of resistance R_1 to another tank of liquid capacitance C_2,

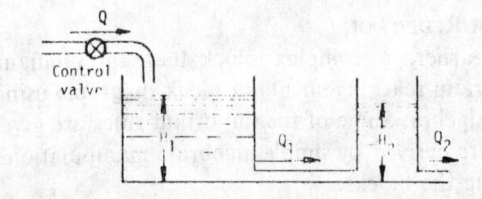

Figure 2.23 Liquid-level system.

which delivers this liquid through a pipe of resistance R_2. The steady-state outflow rates of tank 1 and that of tank 2 are Q_1 and Q_2 and heads are H_1 and H_2 respectively.

Let ΔQ be a small deviation in the inflow rate Q. This results in

ΔH_1 = small deviation of the head of tank 1 from its steady-state value.
ΔH_2 = small deviation of the head of tank 2 from its steady-state value.
ΔQ_1 = small deviation of the outflow rate of tank 1 from its steady-state value.
ΔQ_2 = small deviation of the outflow rate of tank 2 from its steady-state value.

The flow balance equation for tank 1 is

$$\Delta Q = \Delta Q_1 + C_1 \frac{d}{dt}(\Delta H_1)$$

Similarly for tank 2

$$\Delta Q_1 = \Delta Q_2 + C_2 \frac{d}{dt}(\Delta H_2)$$

where

$$\Delta Q_1 = \frac{\Delta H_1 - \Delta H_2}{R_1}$$

and

$$\Delta Q_2 = \frac{\Delta H_2}{R_2}$$

Taking the Laplace transform of the above equations we get

$$\Delta Q(s) - \Delta Q_1(s) = sC_1 \Delta H_1(s) \qquad (2.59)$$

$$\Delta Q_1(s) - \Delta Q_2(s) = sC_2 \Delta H_2(s) \qquad (2.60)$$

$$\Delta Q_1(s) = \frac{\Delta H_1(s) - \Delta H_2(s)}{R_1} \qquad (2.61)$$

$$\Delta Q_2(s) = \frac{\Delta H_2(s)}{R_2} \qquad (2.62)$$

The block diagrams corresponding to eqns. (2.59)-(2.62) are given in Figs. 2.24a-d. Connecting the block diagrams of Figs. 2.24a and b gives the block diagram for tank 1, which is shown in Fig. 2.24e. Similarly connecting the block diagrams of Figs. 2.24c and d gives the block diagram for tank 2 which is shown in Fig. 2.24f. Connecting the block diagrams of Figs. 2.24e and f, gives the overall block diagram of the system as shown in Fig. 2.24g. This block diagram is reduced in steps given below.

(i) In Fig. 2.24g shift the take off point T_1 after the block with transfer function $1/R_2$ (rule 4 of Table 2.4). This results in the block diagram of Fig. 2.24h.

(ii) Minor feedback loop enclosed in dotted line is now reduced to a single block by rules 1 and 6 of Table 2.4 resulting in Fig. 2.24i.

(iii) Shift the take off point T_2 to the right of the block with transfer function $1/(R_2 C_2 s + 1)$ resulting in Fig. 2.24j.

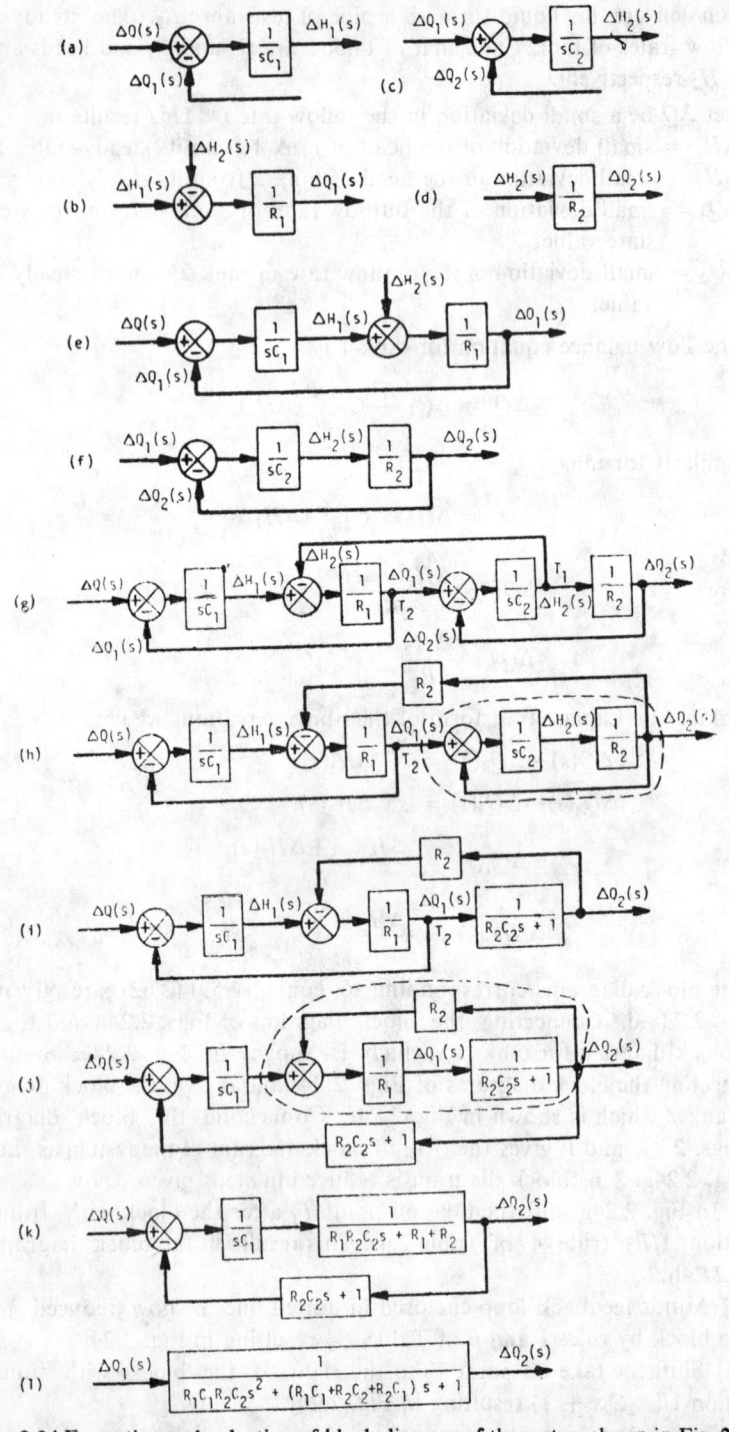

Figure 2.24 Formation and reduction of block diagram of the system shown in Fig. 2.23.

(iv) Reduce the encircled feedback loop giving Fig. 2.24k.
(v) Reduce Fig. 2.24k to the single block of Fig. 2.24l which gives the overall transfer function of the system.

2.5 Signal Flow Graphs

Block diagrams are very useful for representing control systems, but for complicated systems, the block diagram reduction process is tedious and time consuming. An alternate approach is that of *signal flow graphs* developed by S.J. Mason, which does not require any reduction process because of availability of a flow graph gain formula which relates the input and output system variables.

A signal flow graph is a graphical representation of the relationships between the variables of a set of linear algebraic equations. It consists of a network in which *nodes* representing each of the system variables are connected by *directed branches*. The closed-loop system whose block diagram is shown in Fig. 2.20a has the signal flow representation given in Fig. 2.25a. The formulation of this signal flow graph is explained through the various signal flow terms defined below.

(1) *Node.* It represents a system variable which is equal to the sum of all incoming signals at the node. Outgoing signals from the node do not affect the value of the node variable. For example, R, E and C are nodes in Fig. 2.25a.

(2) *Branch.* A signal travels along a branch from one node to another in the direction indicated by the branch arrow and in the process gets multiplied by the *gain* or *transmittance* of the branch. For example, the signal reaching the node C from the node E is given by GE where G is the branch transmittance and the branch is directed from the node E to the node C in Fig. 2.25a.

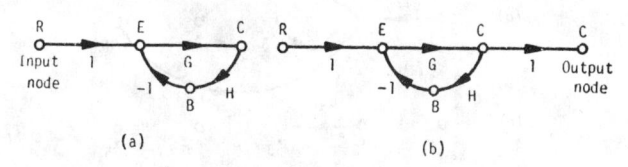

Figure 2.25 Signal flow graph of a closed-loop system.

(3) *Input node or source.* It is a node with only outgoing branches. For example, R is an input node in Fig. 2.25a.

(4) *Output node or sink.* It is a node with only incoming branches. However, this condition is not always met. An additional branch with unit gain may be introduced in order to meet this specified condition. For example, the node C in Fig. 2.25a has one outgoing branch but after introducing an additional branch with unit transmittance as shown in Fig. 2.25b the node becomes an output node.

(5) *Path.* It is the traversal of connected branches in the direction of the branch arrows such that no node is traversed more than once.

(6) *Forward path.* It is a path from the input node to the output node. For example, R-E-C is a forward path in Fig. 2.25a.

(7) *Loop.* It is a path which originates and terminates at the same node. For example, E-C-B-E is a loop in Fig. 2.25a.

(8) *Non-touching loops.* Loops are said to be non-touching if they do not possess any common node.

(9) *Forward path gain.* It is the product of branch gains encountered in traversing a forward path. For example, forward path gain of the path R-E-C in Fig. 2.25a is G.

(10) *Loop gain.* It is the product of the branch gains encountered in traversing the loop. For example, loop gain of the loop E-C-B-E in Fig. 2.25a is $-GH$.

CONSTRUCTION OF SIGNAL FLOW GRAPHS

The signal flow graph of a system is constructed from its describing equations. To outline the procedure, let us consider a system described by the following set of equations:

$$\begin{aligned} x_2 &= a_{12}x_1 + a_{32}x_3 + a_{42}x_4 + a_{52}x_5 \\ x_3 &= a_{23}x_2 \\ x_4 &= a_{34}x_3 + a_{44}x_4 \\ x_5 &= a_{35}x_3 + a_{45}x_4 \end{aligned} \qquad (2.63)$$

where x_1 is the input variable and x_5 is the output variable.

The signal flow graph for this system is constructed as shown in Fig. 2.26. First the nodes are located as shown in Fig. 2.26a. The first equation in

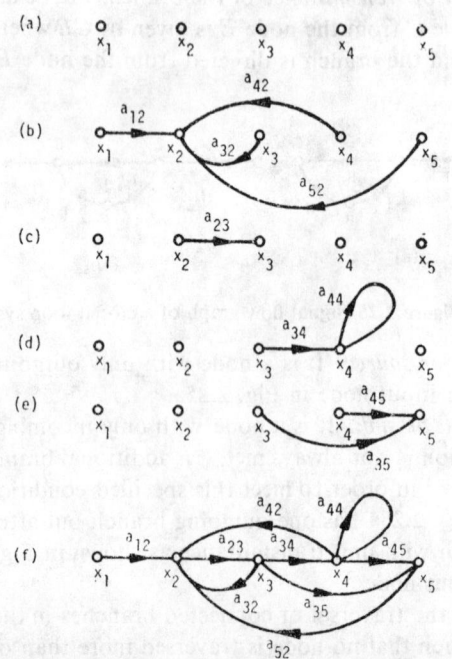

Figure 2.26 Construction of signal flow graph for eqns. (2.63).

(2.63) states that x_2 is equal to sum of four signals* and its signal flow graph is shown in Fig. 2.26b. Similarly, the signal flow graphs for the remaining three equations in (2.63) are constructed as shown in Figs. 2.26c, d and e respectively giving the complete signal flow graph of Fig. 2.26f.

The overall gain from input to output may be obtained by Mason's gain formula.

MASON'S GAIN FORMULA [2]

The relationship between an input variable and an output variable of a signal flow graph is given by the net gain between the input and output nodes and is known as the overall gain of the system. Mason's gain formula for the determination of the overall system gain is given by:

$$T = \frac{1}{\Delta} \sum_K P_K \Delta_K \qquad (2.64)$$

where P_K = path gain of K-th forward path; Δ = determinant of the graph = 1 − (sum of loop gains of all individual loops) + (sum of gain products of all possible combinations of two non-touching loops) − (sum of gain products of all possible combinations of three non-touching loops) + ..., i.e.,

$$\Delta = 1 - \sum_m P_{m1} + \sum_m P_{m2} - \sum_m P_{m3} + \ldots \qquad (2.65)$$

where P_{mr} = gain product of m-th possible combination of r non-touching loops; Δ_K = the value of Δ for that part of the graph not touching the K-th forward path; and T = overall gain of the system.

Let us illustrate the use of Mason's formula by finding the overall gain of the signal flow graph shown in Fig. 2.26. The following conclusions are drawn by inspection of this signal flow graph.

1. There are two forward paths with path gains

$P_1 = a_{12}a_{23}a_{34}a_{45}$ Fig. 2.27a

$P_2 = a_{12}a_{23}a_{35}$ Fig. 2.27b

2. There are five individual loops with loop gains

$P_{11} = a_{23}a_{32}$ Fig. 2.27c
$P_{21} = a_{23}a_{34}a_{42}$ Fig. 2.27d
$P_{31} = a_{44}$ Fig. 2.27e
$P_{41} = a_{23}a_{34}a_{45}a_{52}$ Fig. 2.27f
$P_{51} = a_{23}a_{35}a_{52}$ Fig. 2.27g

3. There are two possible combinations of two non-touching loops with loop gain products

$P_{12} = a_{23}a_{32}a_{44}$ Fig. 2.27h
$P_{22} = a_{23}a_{35}a_{52}a_{44}$ Fig. 2.27i

*a_{ij} is the transmittance of the branch directed from node x_i to node x_j as per the notation used here.

4. There are no combinations of three non-touching loops, four non-touching loops, etc. Therefore

$$P_{m3} = P_{m4} = \ldots = 0$$

Hence from eqn. (2.65)

$$\Delta = 1 - (a_{23}a_{32} + a_{23}a_{34}a_{42} + a_{44} + a_{23}a_{34}a_{45}a_{52} + a_{23}a_{35}a_{52})$$
$$+ (a_{23}a_{32}a_{44} + a_{23}a_{35}a_{52}a_{44})$$

5. First forward path is in touch with all the loops. Therefore, $\Delta_1 = 1$. The second forward path is not in touch with one loop (Fig. 2.27j). Therefore, $\Delta_2 = 1 - a_{44}$.

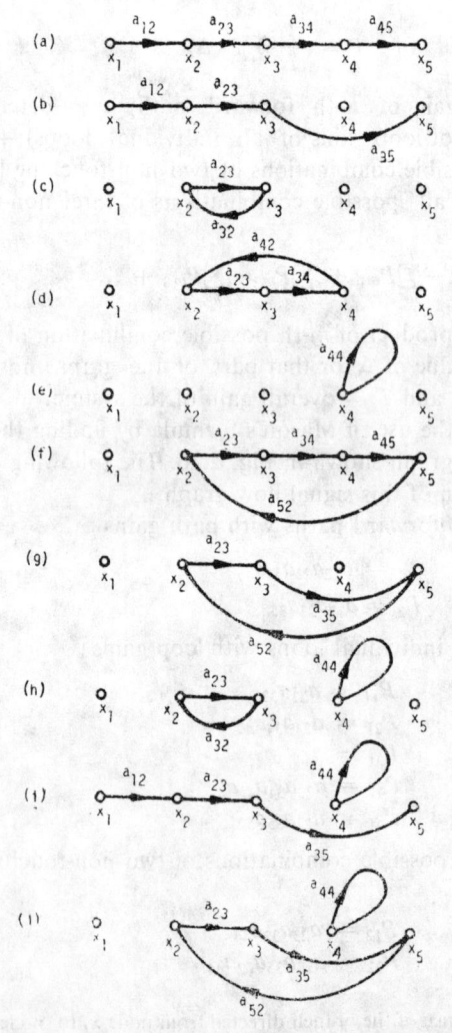

Figure 2.27 Application of Mason's formula to the signal flow graph shown in Fig. 2.26.

From eqn. (2.64), the overall gain

$$T = \frac{x_5}{x_1} = \frac{P_1 \Delta_1 + P_2 \Delta_2}{\Delta}$$

$$= \frac{a_{12}a_{23}a_{34}a_{45} + a_{12}a_{23}a_{35}(1-a_{44})}{1 - a_{23}a_{32} - a_{23}a_{34}a_{42} - a_{44} - a_{23}a_{34}a_{45}a_{52} - a_{23}a_{35}a_{52} + a_{23}a_{32}a_{44} + a_{23}a_{35}a_{52}a_4}$$

APPLICATION OF SIGNAL FLOW GRAPHS TO CONTROL SYSTEMS

There are countless useful applications of signal flow graphs to control systems. Let us discuss as an application, the evaluation of the transfer function of the electromechanical system shown in Fig. 2.28. The objective of this system is to move the load at a desired speed.

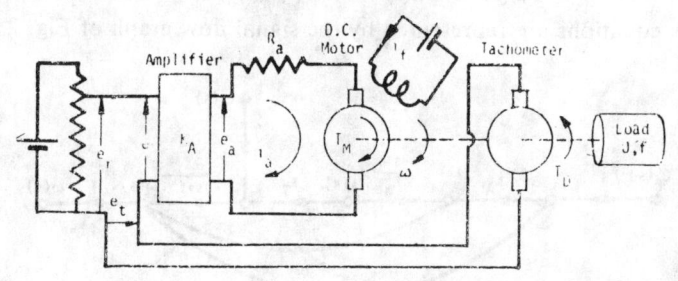

Figure 2.28 A speed control system.

The d.c. tachometer gives a voltage e_t proportional to the output speed ω. This voltage is subtracted from the reference voltage e_r creating the difference signal e which after amplification is used to control i_a of the motor such that the motor acquires the desired speed ω_0.

A d.c. tachometer is just a conventional d.c. generator with permanent magnet excitation. Its open circuit voltage is given by

$$e_t = K_1 \phi \omega$$

As ϕ is held constant (permanent magnet excitation), the generated voltage is directly proportional to the speed ω. The above equation can then be written as

$$e_t = K_t \omega \qquad (2.66)$$

where K_t (volts/rad/sec) is known as the tachometer constant.

The voltage at the armature terminals of the motor is given by

$$e_a = K_A e = K_A(e_r - e_t) \qquad (2.67)$$

For the armature circuit

$$R_a i_a + K_b \omega = e_a \qquad (2.68)$$

where K_b is the back emf constant of the motor.

For constant field current, the torque developed by the motor from eqn. (2.37) is

$$T_M = K_T i_a$$

Let the load consist of moment of inertia J, viscous friction with coefficient f and a load (disturbance) torque T_D. Then the torque equation becomes

$$T_D + J\frac{d\omega}{dt} + f\omega = T_M = K_T i_a \tag{2.69}$$

Taking the Laplace transform of eqns. (2.66) to (2.69), we get

$$E_t(s) = K_t \omega(s)$$
$$E_a(s) = K_A[E_r(s) - E_t(s)]$$
$$R_a I_a(s) + K_b \omega(s) = E_a(s)$$
$$K_T I_a(s) = T_M(s) = (Js+f)\omega(s) + T_D(s)$$

or
$$T_M(s) - T_D(s) = (Js+f)\omega(s)$$

These equations are represented by the signal flow graph of Fig. 2.29.

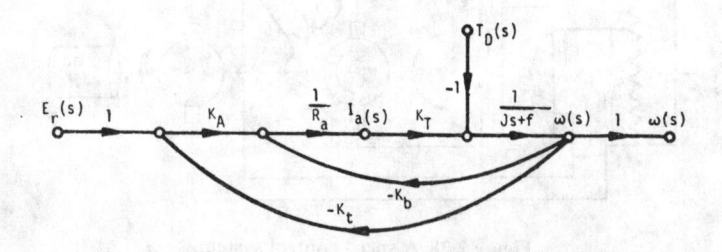

Figure 2.29 Signal flow graph of system shown in Fig. 2.28.

Consider first the case with zero disturbance torque. By inspection of the signal flow graph, with $T_D(s) = 0$, it is found that:

1. There is only one forward path with path gain

$$P_1 = \frac{K_A K_T}{R_a(Js+f)}$$

2. There are two individual loops with loop gains

$$P_{11} = \frac{-K_T K_b}{R_a(Js+f)}$$

$$P_{21} = \frac{-K_A K_T K_t}{R_a(Js+f)}$$

3. There are no combinations of two non-touching loops, three non-touching loops, etc. Therefore

$$P_{m2} = P_{m3} = \ldots = 0$$

Hence from eqn. (2.65)

$$\Delta = 1 - \left[-\frac{K_T K_b}{R_a(Js+f)} - \frac{K_A K_T K_t}{R_a(Js+f)} \right] = 1 + \frac{K_T K_b + K_A K_T K_t}{R_a(Js+f)}$$

4. The forward path is in touch with both the loops. Therefore

$$\Delta_1 = 1$$

From eqn. (2.64), the overall gain is

$$T(s) = \frac{\omega(s)}{E_r(s)} = \frac{P_1\Delta_1}{\Delta} = \frac{K_A K_T}{R_a(Js+f) + K_T K_b + K_A K_T K_t} \quad (2.70)$$

With $K_t = 0$, the system is reduced to open-loop with the transfer function

$$G(s) = \frac{K_A K_T}{R_a(Js+f) + K_b K_T} = \frac{K}{(\tau s + 1)} \quad (2.71)$$

where $\quad K = \dfrac{K_A K_T}{R_a f + K_T K_b}; \tau = \dfrac{R_a J}{R_a f + K_T K_b}$

From eqn. (2.70), the closed-loop transfer function of the system is given by

$$T(s) = \frac{K/\tau}{s + \left(\dfrac{1 + KK_t}{\tau}\right)} \quad (2.72)$$

When the disturbance torque $T_D(s)$ is present, the only change in the graph is the additional input $T_D(s)$.

Applying Mason's gain formula to the graph, the following transfer function is obtained between output speed and disturbance torque with zero reference voltage, i.e., $E_r(s) = 0$.

$$\left.\frac{\omega(s)}{T_D(s)}\right|_{E_r(s)=0} = \frac{\omega_D(s)}{T_D(s)} = \frac{-1}{Js + f + \dfrac{K_T}{R_a}(K_A K_t + K_b)} \quad (2.73)$$

When there is no feedback ($K_t = 0$), eqn. (2.73) modifies to

$$\left.\frac{\omega(s)}{T_D(s)}\right|_{E_r(s)=0} = \frac{\omega_D(s)}{T_D(s)} = \frac{-1}{Js + f + \dfrac{K_T K_b}{R_a}} \quad (2.74)$$

2.6 Illustrative Examples

Example 2.1: Consider the mechanical system shown in Fig. 2.30a. A force $F(t)$ is applied to mass M_2. The free-body diagrams for the two masses are shown in Fig. 2.30b. From this figure, we have the following differential equations describing the dynamics of the system

$$F(t) - f_2(\dot{y}_2 - \dot{y}_1) - K_2(y_2 - y_1) = M_2 \ddot{y}_2$$
$$f_2(\dot{y}_2 - \dot{y}_1) + K_2(y_2 - y_1) - f_1 \dot{y}_1 - K_1 y_1 = M_1 \ddot{y}_1$$

Rearranging we get

$$M_2 \ddot{y}_2 + f_2(\dot{y}_2 - \dot{y}_1) + K_2(y_2 - y_1) = F(t) \quad (2.75)$$
$$M_1 \ddot{y}_1 + f_1 \dot{y}_1 - f_2(\dot{y}_2 - \dot{y}_1) + K_1 y_1 - K_2(y_2 - y_1) = 0 \quad (2.76)$$

These are two simultaneous second-order linear differential equations. Manipulation of these equations will result in a single differential equation (fourth-order) relating the response y_2 (or y_1) to input $F(t)$.

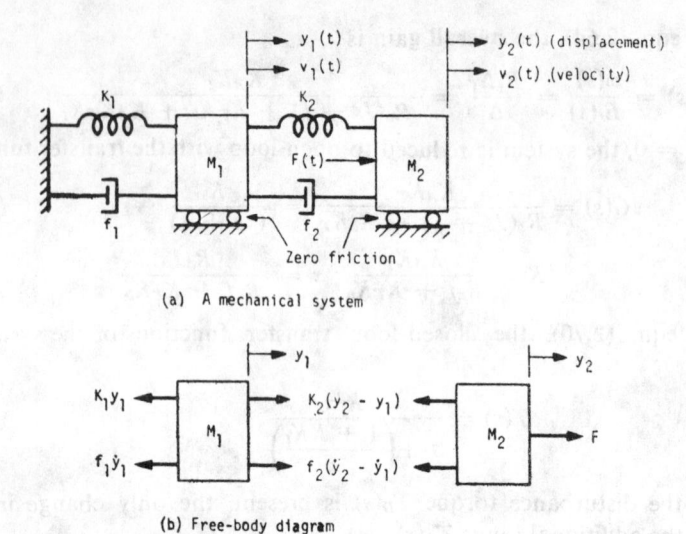

Figure 2.30 (a) A mechanical system; (b) Free-body diagram.

A spring-mass-damper system may be schematically represented as a network by showing the inertial reference frame as the second terminal of every mass (or inertia) element. As an example, the mechanical system of Fig. 2.30a is redrawn in Fig. 2.31 which may be referred to as the *mechanical network*. Analogous electrical circuit based on force-current analogy (Table 2.2) is shown in Fig. 2.32. A look at Fig. 2.32 (analog of Fig. 2.30a) and Fig. 2.31 reveals that they are alike topologically.

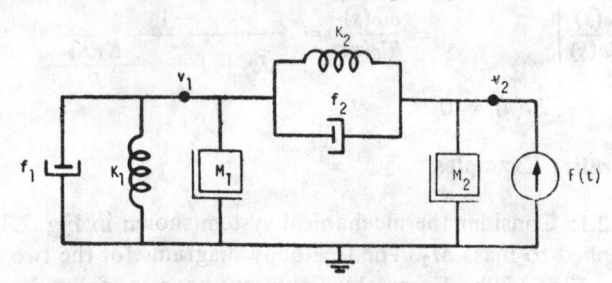

Figure 2.31 Mechanical network for the system of Fig. 2.30.

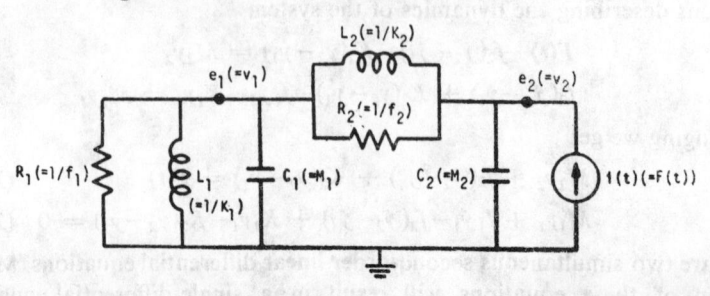

Figure 2.32 Electrical analog for the system of Fig. 2.30a.

The dynamical equations of the system [eqns. (2.75)-(2.76)] could also be obtained by writing nodal equations for the electrical network of Fig. 2.32 or for the mechanical network of Fig. 2.31 (with force and velocity analogous to current and voltage respectively) since the two are alike topologically. The result is:

$$f_1 v_1 + K_1 \int_{-\infty}^{t} v_1 dt + M_1 v_1 + K_2 \int_{-\infty}^{t} (v_1 - v_2) dt + f_2(v_1 - v_2) = 0$$

$$M_2 v_2 + K_2 \int_{-\infty}^{t} (v_2 - v_1) dt + f_2(v_2 - v_1) = F(t)$$

The result is same as obtained earlier (with $y = \int_{-\infty}^{t} v \, dt$, $\dot{y} = v$ and $\ddot{y} = \dot{v}$) in eqns. (2.75) and (2.76) using the free-body diagram approach.

Example 2.2: Consider a salt mixing tank shown in Fig. 2.33. A solution of salt in water at a concentration C_f (moles* of salt/m³ of solution) is mixed with pure water to obtain an outflow stream with salt concentration C_0. The water flow rate is assumed fixed at Q_w and the solution flow rate may be varied to achieve the desired concentration C_0 (also see Problem 3.7). Volumetric hold-up of the tank is V, which is held constant. Let us assume that stirring causes perfect mixing so that composition of the liquid in the tank is uniform throughout.

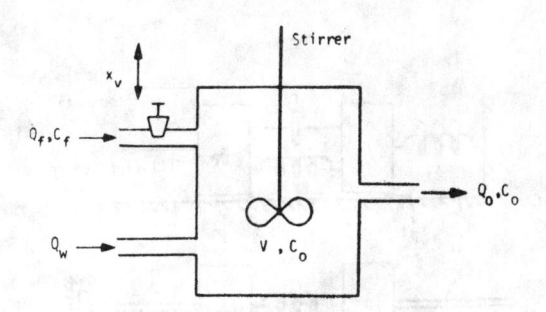

Figure 2.33 A salt mixing tank.

For this system,

$$Q_f = K_v x_v; \quad K_v \text{ is valve coefficient}$$
$$Q_0 = Q_w + Q_f$$

The rate of salt inflow in the tank

$$m_i = Q_f C_f; \quad \left(\frac{m^3}{sec} \cdot \frac{moles}{m^3} = moles/sec \right)$$

---

*A mole of a substance is defined as the amount of substance whose mass numerically equals its molecular weight. For example, a gram-mole of helium would have a mass of 4.003 g (molecular weight of helium ≃ 4.003).

The rate of salt outflow from the tank

$$m_0 = Q_0 C_0; \text{ (moles/sec)}$$

The rate of salt accumulation in the tank

$$m_a = \frac{d}{dt}[VC_0(t)] = V\frac{dC_0}{dt}$$

where $VC_0(t)$ is the salt hold-up of the tank at time t.
From the law of conservation of mass, we have

$$m_i = m_a + m_0$$

or

$$Q_f C_f = V\frac{dC_0}{dt} + Q_0 C_0$$

or

$$\tau \frac{dC_0}{dt} + C_0 = Kx_v$$

where $K = (C_f K_v)/Q_0$ and $\tau = V/Q_0$ is the tank hold-up time.
The transfer function of the system is

$$\frac{C_0(s)}{X_v(s)} = \frac{K}{s\tau + 1}$$

Problems

2.1 Obtain the transfer functions of the mechanical systems shown in Figs. P-2.1a and b.

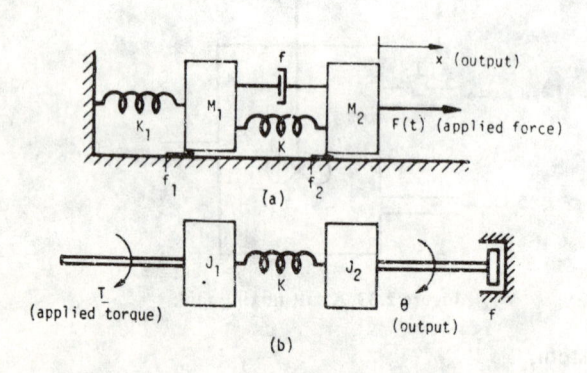

Figure P-2.1.

2.2 Write the differential equations governing the behaviour of the mechanical system shown in Fig. P-2.2. Also obtain an analogous electrical circuit based on force-voltage analogy.

2.3 Write the differential equations for the mechanical system shown in Fig. P-2.3. Also obtain an analogous electrical circuit based on force-current analogy.

2.4 Find the transfer function $X(s)/E(s)$ for the electromechanical system shown in Fig. P-2.4.
(*Hint*: For a simplified analysis, assume that the coil has a back emf $e_b = k_1 \, dx/dt$ and the coil current i produces a force $F_c = k_2 i$ on the mass M)

Figure P-2.2.

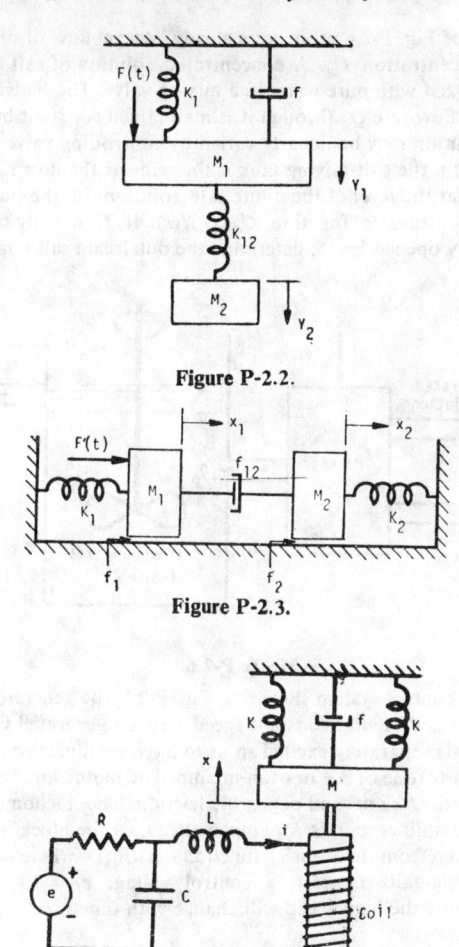

Figure P-2.3.

Figure P-2.4.

2.5 Figure P-2.5 shows a thermometer plugged into a bath of temperature θ_i. Obtain the transfer function $\theta(s)/\theta_i(s)$ of the thermometer and its electrical analogue. (The thermometer may be considered to have a thermal capacitance C which stores heat and a thermal resistance R which limits the heat flow.) How the temperature indication of the thermometer will vary with time after the thermometer is suddenly plugged in ?

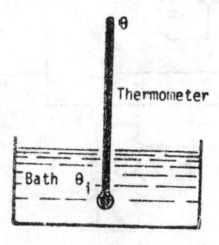

Figure P-2.5.

56 Control Systems Engineering

2.6 The scheme of Fig. P-2.6 produces a steady stream flow of dilute salt solution with controlled concentration C_o. A concentrated solution of salt with concentration C_i is continuously mixed with pure water in a mixing valve. The valve characteristic is such that the total flow rate Q_o through it is maintained constant but the inflow Q_i of concentrated salt solution may be linearly varied by controlling valve stem position x. The outflow rate from the salt mixing tank is the same as the flow rate into it from the mixing valve, such that the level of the dilute salt solution in the tank is maintained constant. Obtain the transfer function $C_o(s)/X(s)$. If from fully closed position, the valve stem is suddenly opened by x_o, determine the outstream salt concentration C_o as a function of time.

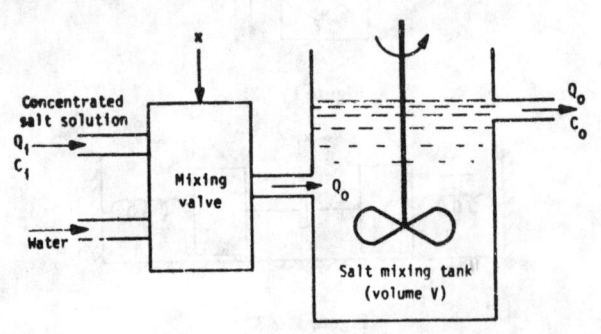

Figure P-2.6.

2.7 In the speed control system shown in Fig. P-2.7, the generator field time constant is negligible. It is driven at constant speed giving a generated voltage of K_g volts/field amp. The motor is separately excited so as to have a counter emf of K_b volts per rad/sec. It produces a torque of K_T newton-m/amp. The motor and its load have a combined moment of inertia J kg-m² and negligible friction. The tachometer has K_t volts per rad/sec and the amplifier gain is K_A amps/volt. Draw the block diagram of this system and determine therefrom the transfer function $\omega(s)/E_i(s)$, where ω is the load speed.

With the system originally at rest, a control voltage $e_i = 100$ volts is suddenly applied. Determine how the load speed will change with time.
Given:

$J = 6$ kg-m² $K_A = 4$ amp/volt $K_T = 1.5$ newton-m/amp
$K_g = 50$ volts/amp $R_a = 1$ ohm
$K_t = 0.2$ volts per rad/sec

(*Hint*: $K_b = K_T$ in MKS units)

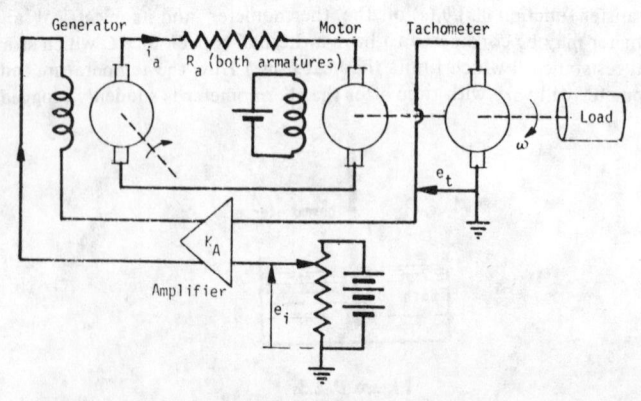

Figure P-2.7.

Mathematical Models of Physical Systems

2.8 Consider the positional servomechanism shown in Fig. P-2.8. Assume that the input to the system is the reference shaft position θ_R and the system output is the load shaft position θ_L. Draw the block diagram of the system indicating the transfer function of each block. Simplify the block diagram to obtain $\theta_L(s)/\theta_R(s)$ for the closed-loop system and also when the loop is open (in opening the loop, the lead from the output potentiometer driven by θ_C is disconnected and grounded). The parameters of the system are given below:

Sensitivity of error detector,	$K_P = 10$ volt/rad
Gain of d.c. amplifier,	$K_A = 50$ volts/volt
Motor field resistance,	$R_f = 100$ ohms
Motor field inductance,	$L_f = 20$ henrys
Motor torque constant,	$K_T = 10$ newton-m/amp
Moment of inertia of load,	$J_L = 250$ kg-m²
Coefficient of viscous friction of load,	$f_L = 2,500$ newton-m per rad/sec
Motor to load gear ratio,	$(\dot{\theta}_L/\dot{\theta}_M) = 1/50$
Load to potentiometer gear ratio,	$(\dot{\theta}_C/\dot{\theta}_L) = 1$

Motor inertia and friction are negligible.

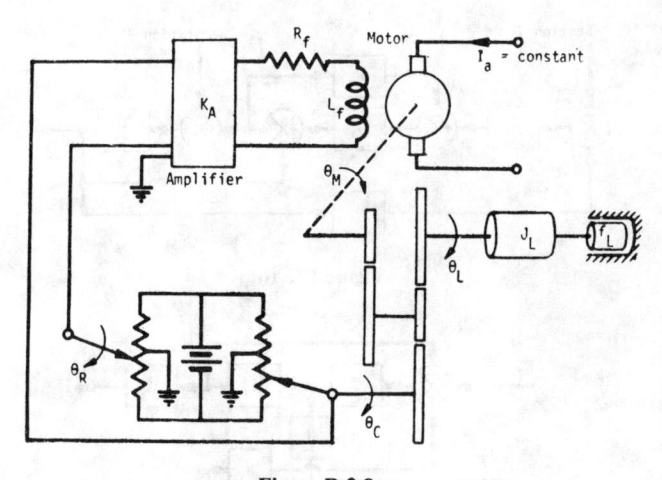

Figure P-2.8.

2.9 Using block diagram reduction techniques, find the closed-loop transfer functions of the systems whose block diagrams are given in Figs. P-2.9a and b.

2.10 For the system represented by the block diagram shown in Fig. P-2.10, evaluate the closed-loop transfer functions, when the input R is (i) at station I, (ii) at station II.

2.11 From the block diagram shown in Fig. P-2.11, determine C_1/R_1 and C_2/R_1 (assuming $R_2 = 0$).

2.12 Figure P-2.12 shows a schematic diagram of liquid-level system. The flow of liquid Q_i into the tank is controlled by the pressure P of the incoming liquid and valve opening v_x (note that this is a more realistic model than the one shown in Fig. 2.11) through a nonlinear relationship

$$Q_i = f(P, v_x)$$

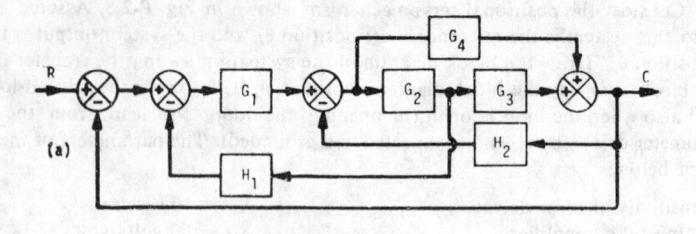

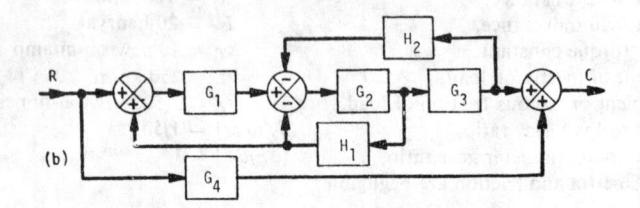

Figure P-2.9.

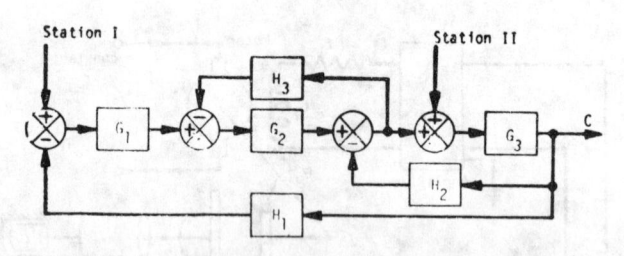

Figure P-2.10.

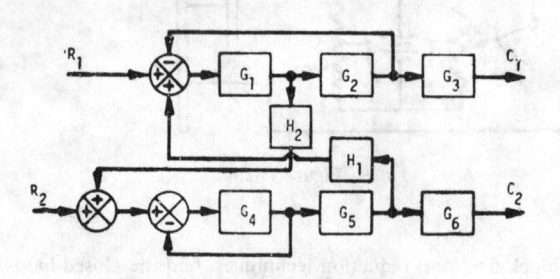

Figure P-2.11.

Linearized liquid-level model about the operating point $(P_0, Q_i = Q_0, H_0)$ is given as $\Delta Q_i = K_1 \Delta P + K_2 \Delta v_x$. Draw the signal flow graph and obtain therefrom the transfer function $\left.\dfrac{\Delta Q_0(s)}{\Delta v_x(s)}\right|_{\Delta Q_D = 0}$ with pressure remaining constant.

(The tank and output pipe may be considered to have liquid capacitance C and flow resistance R respectively.)

Mathematical Models of Physical Systems

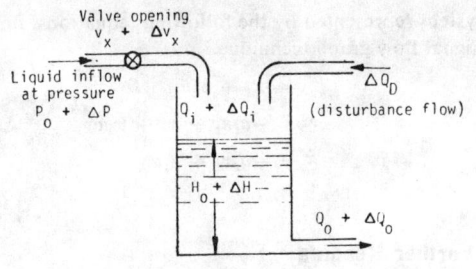

Figure P-2.12.

2.13 Draw a signal flow graph and evaluate the closed-loop transfer function of a system whose block diagram is given in Fig. P-2.13.

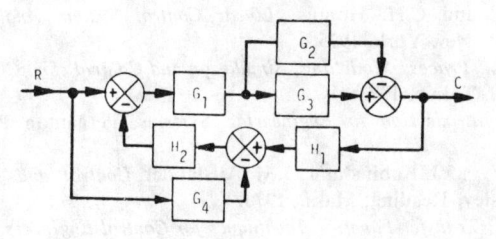

Figure P-2.13.

2.14 Obtain the overall transfer function C/R from the signal flow graph shown in Fig. P-2.14.

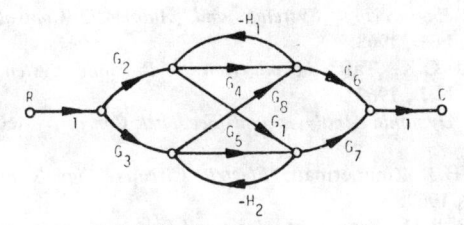

Figure P-2.14.

2.15 Figure P-2.15 gives the signal flow graph of a system with two inputs and two outputs. Find expressions for the outputs C_1 and C_2. Also determine the condition that makes C_1 independent of R_2 and C_2 independent of R_1.

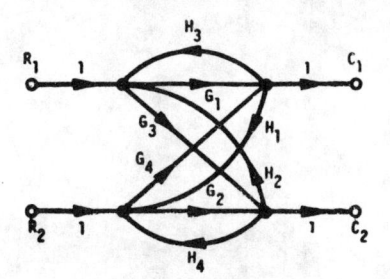

Figure P-2.15

2.16 For the system represented by the following equations, find the transfer function $X(s)/U(s)$ by signal flow graph technique.

$$x = x_1 + \beta_3 u$$
$$\dot{x}_1 = -a_1 x_1 + x_2 + \beta_2 u$$
$$\dot{x}_2 = -a_2 x_1 + \beta_1 u$$

References and Further Reading

1. Shames, I.H., *Mechanics of Fluids*, McGraw-Hill, New York, 1962.
2. Younger, D., "A simple derivation of Mason's gain formula", *Proc. IEEE*, 1963, 51, 7, 1043-1044.
3. D'Azzo, J.J. and C.H. Houpis, *Linear Control System Analysis and Design*, McGraw-Hill, New York, 1975.
4. Luyben, W.L., *Process Modelling, Simulation and Control for Chemical Engineers*, McGraw-Hill, Tokyo, 1973.
5. Seely, S., *An Introduction to Engineering Systems*, Pergamon Press, New York, 1972.
6. Takahashi, Y., M.J. Rabins and D.M. Auslander, *Control and Dynamic Systems*, Addison-Wesley, Reading, Mass., 1970.
7. Towill, D.R., *Transfer Function Techniques for Control Engineers*, Iliffe Books Ltd., London, 1970.
8. Perkins, W.R. and J.B. Cruz Jr., *Engineering of Dynamic Systems*, John Wiley, New York, 1969.
9. Crandall, S.H., D.C. Karnopp et al., *Dynamics of Mechanical and Electromechanical Systems*, McGraw-Hill, New York, 1968.
10. Dransfield, P., *Engineering Systems and Automatic Control*, Prentice-Hall, Englewood Cliffs, N.J., 1968.
11. Reswick, J.B. and C.K. Taft, *Introduction to Dynamic Systems*, Prentice-Hall, Englewood Cliffs, N.J., 1967.
12. Doebelin, E.O., *Dynamic Analysis and Feedback Control*, McGraw-Hill, New York, 1962.
13. Mason, S.J. and H.J. Zimmerman, *Electric Circuits, Signals and Systems*, John Wiley, New York, 1960.
14. Thaler, G.J. and R.G. Brown, *Analysis and Design of Feedback Control Systems*, McGraw-Hill, New York, 1960.

3. Feedback Characteristics of Control Systems

3.1 Feedback and Non-feedback Systems

Feedback systems play an important role in modern engineering practice because they have the possibility of being adopted to perform their assigned tasks automatically. A *non-feedback (open-loop) system* represented by the block diagram and signal flow graph in Fig. 3.1a, is activated by a single signal at the input (for single-input systems). There is no provision within this system for supervision of the output and no mechanism is provided to correct (or compensate) the system behaviour for any lack of proper performance of system components. On the other hand, a *feedback (closed-loop) system* represented by the block diagram and signal flow graph in Fig. 3.1b is driven by two signals (more signals could be employed), one the input signal and the other, a signal called the feedback signal derived from the output of the system. The feedback signal gives this system the capability to act as a self-correcting mechanism as explained below.

The output signal c is measured by a sensor $H(s)$, which produces a feedback signal b. The comparator compares the feedback signal b with the input (command) signal r generating the actuating signal e, which is a measure of discrepancy between r and b. The actuating signal is applied

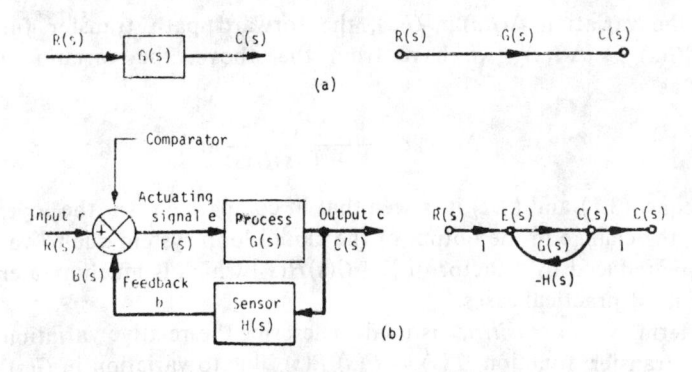

Figure 3.1 (a) A non-feedback (open-loop) system;
(b) A feedback (closed-loop) system.

to the process $G(s)$ so as to influence the output c in a manner which tends to reduce the error e.

Feedback as a means of automatic regulation and control is, in fact, inherent in nature and can be noticed in many physical, biological and soft systems. For example, the body temperature of any living being is automatically regulated through a process which is essentially a feedback process, only it is far more complex than the diagram of Fig. 3.1b.

3.2 Reduction of Parameter Variations by Use of Feedback

One of the primary purposes of using feedback in control systems is to reduce the sensitivity of the system to parameter variations. The parameters of a system may vary with age, with changing environment (e.g., ambient temperature), etc. Conceptually, *sensitivity* is a measure of the effectiveness of feedback in reducing the influence of these variations on system performance.

Let us define sensitivity on a quantitative basis. In the open-loop case

$$C(s) = G(s)R(s)$$

Suppose due to parameter variations, $G(s)$ changes to $[G(s) + \Delta G(s)]$ where $|G(s)| \gg |\Delta G(s)|$. The output of the open-loop system then changes to

$$C(s) + \Delta C(s) = [G(s) + \Delta G(s)]R(s)$$

or
$$\Delta C(s) = \Delta G(s) R(s) \qquad (3.1)$$

Similarly, in the closed-loop case, the output

$$C(s) = \frac{G(s)}{1 + G(s)H(s)} R(s)$$

changes to

$$C(s) + \Delta C(s) = \frac{G(s) + \Delta G(s)}{1 + G(s)H(s) + \Delta G(s)H(s)} R(s)$$

due to the variation $\Delta G(s)$ in $G(s)$, the forward path transfer function. Since $|G(s)| \gg |\Delta G(s)|$, we have from the above, the variation in the output as

$$\Delta C(s) \approx \frac{\Delta G(s)}{1 + G(s)H(s)} R(s) \qquad (3.2)$$

From eqns. (3.1) and (3.2) it is seen that in comparison to the open-loop system, the change in the output of the closed-loop system due to variation in $G(s)$ is reduced by a factor of $[1 + G(s)H(s)]$ which is much greater than unit in most practical cases.

The term *system sensitivity* is used to describe the relative variation in the overall transfer function $T(s) = C(s)/R(s)$ due to variation in $G(s)$ and is defined below:

$$\text{Sensitivity} = \frac{\text{percentage change in } T(s)}{\text{percentage change in } G(s)}$$

For small incremental variation in $G(s)$, the sensitivity is written in the quantitative form as

$$S_G^T = \frac{\partial T/T}{\partial G/G}$$

where S_G^T denotes the sensitivity of T with respect to G.

In accordance with the above definition, the sensitivity of the closed-loop system is

$$S_G^T = \frac{\partial T}{\partial G} \times \frac{G}{T} = \frac{(1+GH)-GH}{(1+GH)^2} \times \frac{G}{G/(1+GH)}$$

$$= \frac{1}{1+GH} \quad (3.3)$$

Similarly, the sensitivity of the open-loop system is

$$S_G^T = \frac{\partial T}{\partial G} \times \frac{G}{T} = 1 \quad \text{(in this case } T = G\text{)}$$

Thus, the sensitivity of a closed-loop system with respect to variation in G is reduced by a factor $(1+GH)$ as compared to that of an open-loop system.

The sensitivity of T with respect to H, the feedback sensor, is given as

$$S_H^T = \frac{\partial T}{\partial H} \times \frac{H}{T} = G\left[\frac{-G}{(1+GH)^2}\right]\frac{H}{G/(1+GH)}$$

$$= \frac{-GH}{1+GH}$$

The above equation shows that for large values of GH, sensitivity of the feedback system with respect to H approaches unity. Thus, we see that the changes in H directly affect the system output. Therefore, it is important to use feedback elements which do not vary with environmental changes or can be maintained constant.

The use of feedback in reducing sensitivity to parameter variations is an important advantage of feedback control systems. To have a highly accurate open-loop system, the components of $G(s)$ must be selected to meet the specifications rigidly in order to fulfil the overall goals of the system. On the other hand, in a closed-loop system $G(s)$ may be less rigidly specified, since the effects of parameter variations are mitigated by the use of feedback. However, a closed-loop system requires careful selection of the components of the feedback sensor $H(s)$. Since $G(s)$ is made up of power elements and $H(s)$ is made up of measuring elements which operate at low power levels, the selection of accurate $H(s)$ is far less costly than that of $G(s)$ to meet the exact specifications.

The price for improvement in sensitivity by use of feedback is paid in terms of *loss of system gain*. The open-loop system has a gain $G(s)$, while the gain of the closed-loop system is $G(s)/[1 + G(s) H(s)]$. Hence by use of feedback, the system gain is reduced by the same factor as by which the sensitivity of the system to parameter variations is reduced. Sufficient open-loop gain can, however, be easily built into a system so that we can afford to lose some gain to achieve improvement in sensitivity.

As an example of control of system sensitivity, let us consider the speed control system of Fig. 2.28 which may be operated in open-or closed-loop mode. The signal flow graph of this system is given in Fig. 2.29. The reduced signal flow graph of this system with $T_D = 0$, is drawn in Fig. 3.2, where

$$K = \frac{K_A K_T}{R_{af} + K_T K_b} \; ; \; \tau = \frac{R_a J}{R_{af} + K_T K_b}$$

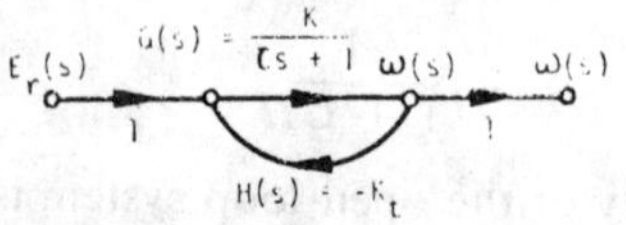

Figure 3.2 Reduced signal flow graph ($T_D = 0$) obtained from Fig. 2.29.

The sensitivity of the open-loop mode of operation to variation in the constant K is unity, while the corresponding sensitivity of the closed-loop mode is evaluated below.

From the signal flow graph of Fig. 3.2

$$T(s) = \frac{K}{\tau s + (1 + KK_t)}$$

$$S_K^T = \frac{\partial T}{\partial K} \times \frac{K}{T} = \frac{s + \frac{1}{\tau}}{s + \left(\frac{1 + KK_t}{\tau}\right)} \quad (3.4)$$

The expression (3.4) can also be obtained by substituting $G(s) = K/(\tau s + 1)$ and $H(s) = K_t$ in eqn. (3.3).

For a typical application of this system, we might have $1/\tau = 0.1$ and $(1 + KK_t)/\tau = 10$. Therefore from eqn. (3.4) we obtain

$$S_K^T = \frac{s + 0.1}{s + 10}$$

It follows from above that the sensitivity is a function of s and must be evaluated over the complete frequency band within which input has significant components. Our interest is to determine the upper limit for the sensitivity function $|S_K^T|$ over the frequency band and the frequency at which the maximum value occurs.

At a particular frequency, e.g., $s = j\omega = j1$, the magnitude of the sensitivity is approximately:

$$|S_K^T| = 0.1$$

Thus the sensitivity of the closed-loop speed control system at this frequency is reduced by a factor of ten compared to that of the open-loop case.

Sensitivity studies in the frequency domain will be taken up in Chapter 9.

3.3 Control over System Dynamics by Use of Feedback

Let us consider an elementary system shown in Fig. 3.3.

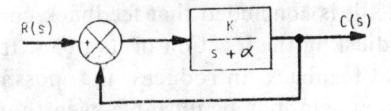

Figure 3.3 A simple feedback system.

The open-loop transfer function of the system is

$$G(s) = \frac{K}{s + \alpha}$$

which has a real pole* at $s = -\alpha$ in the s-plane. Let us evaluate the response of the system to a unit impulse** input [for unit impulse $R(s) = 1$]. From Fig. 3.3, the output for the non-feedback system is given by

$$C(s) = \frac{K}{s + \alpha}$$

and for the feedback system by

$$(s) = \frac{K}{s + \alpha + K}.$$

Taking the inverse Laplace transform of the above equations we get

$$c(t) = Ke^{-\alpha t} \quad \text{(for non-feedback system)} \quad (3.5)$$

$$= Ke^{-(K+\alpha)t} \quad \text{(for feedback system)} \quad (3.6)$$

These responses are plotted in Fig. 3.4.

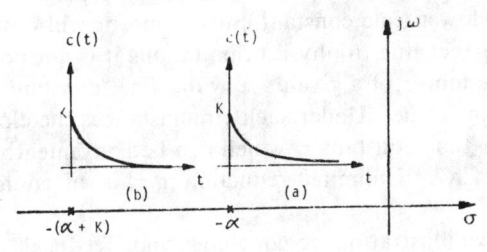

Figure 3.4 Impulse response of system shown in Fig. 3.3.

For the non-feedback system with the pole located at $s = -\alpha$, the impulse response is shown in Fig. 3.4a. The nature of the response is an exponential decay with a time constant of $\tau = 1/\alpha$. For positive values of K, the effect of the feedback is to shift the pole negatively to $s = -(\alpha + K)$

*The pole of a function of complex variable s is the value at which the function becomes infinite.
**Discussed in detail in Chapter 5.

and so the time constant reduces to $1/(K + \alpha)$. This implies that as K increases the system dynamics continuously becomes faster, i.e., the transient response decays more quickly as shown in Fig. 3.4b.

From this example, it is concluded that feedback controls the dynamics of the system by adjusting the location of its poles. It is, however, important to note here that feedback introduces the possibility of instability, that is, a closed-loop system may be unstable even though the open-loop is stable. The question of stability of control systems is treated in details in Chapter 6.

Consider once again the speed control system of Fig. 2.28. Let the system be subjected to a step* input for which $E_r(s) = A/s$, where A is a constant. The output response of the system obtained by reference to the signal flow graph of Fig. 3.2 or directly from eqns. (2.71) and (2.72) is given by

$$\omega(s) = \frac{KA/\tau}{s\left(s + \dfrac{1}{\tau}\right)} \quad \text{(for open-loop operation, i.e., } K_t = 0\text{)}$$

$$= \frac{KA/\tau}{s\left(s + \dfrac{1 + KK_t}{\tau}\right)} \quad \text{(for closed-loop operation)}$$

Taking the inverse Laplace transform of the above equations, we get

$$\omega(t) = KA(1 - e^{-t/\tau}) \quad \text{(for open-loop operation)} \tag{3.7}$$

$$= \frac{KA}{1 + KK_t}(1 - e^{-t/\tau_c}) \text{ (for closed-loop operation)} \tag{3.8}$$

where τ_c (closed-loop time constant) $= \tau/(1 + KK_t)$.

It is seen from above that if the open-loop time constant τ is large, the transient response is poor and one choice is to replace the motor by another one with a lower time constant. Such a motor will obviously be more expensive and further due to physical limitations it is not possible to design and manufacture motor of a given size with time constant lower than a certain minimum value. Under such circumstances the closed-loop mode provides a lower time constant τ_c which can be conveniently adjusted by a suitable choice of KK_t. Unlimited reduction in τ_c is of course not practicable.

From the above illustration we conclude that feedback is a powerful technique for control of system dynamics.

3.4 Control of the Effects of Disturbance Signals by Use of Feedback

Figure 3.5 shows the signal flow graph of a closed-loop system with the disturbance signal T_D in the forward path.

The ratio of the output $C(s)$ to the disturbance signal $T_D(s)$, when $R(s) = 0$, is obtained by applying the signal flow gain formula to the graph of Fig. 3.5, and is given by

*Discussed in detail in Chapter 5.

Feedback Characteristics of Control Systems

$$\frac{C_D(s)}{T_D(s)} = \frac{-G_2(s)}{1 + G_1(s)G_2(s)H(s)} \tag{3.9}$$

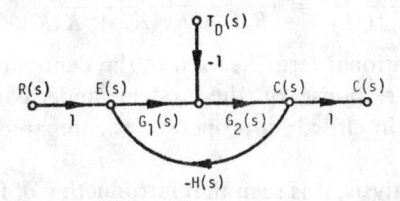

Figure 3.5 A closed-loop system with a disturbance signal.

If $|G_1G_2H(s)| \gg 1$ over the working range of s, then from eqn. (3.9)

$$\frac{C_D(s)}{T_D(s)} \approx \frac{-1}{G_1(s)H(s)}$$

Therefore, it is seen that if $G_1(s)$ is made sufficiently large, the effect of disturbance can be decreased by feedback.

Let us discuss the effect of load (disturbance) torque in the control system shown in Fig. 2.28. Assume that the load torque is a step signal for which $T_D(s) = A/s$ where A is a constant.

With reference to the signal flow graph of Fig. 2.29 or directly from eqns. (2.73) and (2.74), the change in speed due to load torque alone $[E_r(s) = 0]$ is given by

$$\omega_D(s) = \frac{-A}{s\left(Js + f + \dfrac{K_T K_b}{R_a}\right)} \tag{3.10}$$

(for open-loop operation, $K_t = 0$)

$$= \frac{-A}{s\left[Js + f + \dfrac{K_T}{R_a}(K_A K_t + K_b)\right]} \tag{3.11}$$

(for closed-loop operation)

The steady-state error in speed due to load is given by the *final value theorem* as

$$e_{ss} = \lim_{t \to \infty} \omega_D(t) = \lim_{s \to 0} s\omega_D(s) \tag{3.12}$$

Using eqns. (3.10), (3.11) and (3.12), the steady-state speed errors in open- and closed-loop cases are obtained as

$$e_{ss}(OL) = \frac{-AR_a}{R_a f + K_T K_b} \tag{3.13}$$

and

$$e_{ss}(CL) = \frac{-AR_a}{R_a f + K_T(K_b + K_A K_t)} \tag{3.14}$$

The ratio of steady-state error in output speed due to load torque; in open-loop and closed-loop cases obtained from eqns. (3.13) and (3.14), is

$$\frac{e_{ss}(CL)}{e_{ss}(OL)} = \frac{R_{af} + K_T K_b}{R_{af} + K_T(K_b + K_A K_t)}$$

Because of the additional term $K_T K_A K_t$ in the denominator, the effect of disturbance on the response of the system under consideration can be considerably reduced in closed-loop operation compared to that of the open-loop.

From the above analysis, it is seen that introduction of feedback decreases the effects of disturbances and noise signals in the forward path of the feedback loop. Feedback is introduced by a set of additional elements, called the measurement sensor H which may itself generate some noise. Let us evaluate the effect of this noise on the system performance.

Figure 3.6 shows the signal flow graph of a system with the noise signal $N(s)$ in the feedback path. Using the gain formula for the signal flow graph, the following result is obtained.

$$\left.\frac{C(s)}{N(s)}\right|_{R(s) = 0} = \frac{C_n(s)}{N(s)} = \frac{-G_1(s) G_2(s) H_2(s)}{1 + G_1(s) G_2(s) H_1(s) H_2(s)}$$

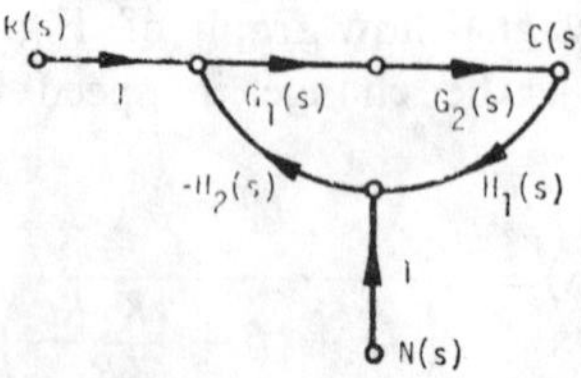

Figure 3.6 Closed-loop system with measurement noise.

For large values of loop gain $(|G_1 G_2 H_1 H_2(s)| \gg 1)$, the above equation reduces to

$$\frac{C_n(s)}{N(s)} \approx - \frac{1}{H_1(s)}$$

Therefore, the effect of noise on output is

$$C_n(s) = - \frac{N(s)}{H_1(s)} \tag{3.15}$$

Thus, for optimum performance of the system, the measurement sensor should be designed such that $H_1(s)$ is maximum, which is equivalent to maximizing the signal-to-noise ratio of the sensor.

The design specifications of the feedback sensor are far more stringent than those of the forward path transfer function. The feedback sensor must have low parameter variations as these are directly reflected in system response (the sensitivity $S_H^T \approx -1$). Further the signal-to-noise ratio for the sensor must be high as explained above. Usually it is possible to design and construct the sensor with such stringent specifications and at reasonable cost because the feedback elements operate at low power level.

To conclude, the use of feedback has the advantages of reducing sensitivity, improving transient response and minimizing the effects of disturbance signals in control systems. On the other hand, the use of feedback increases the number of components of the system, thereby increasing its complexity. Further it reduces the gain of the system and also introduces the possibility of instability. However, in most cases the advantages outweigh the disadvantages and therefore the feedback systems are commonly employed in practice.

3.5 Regenerative Feedback

The preceding material in this chapter has emphasized a negative or degenerative type of feedback. In regenerative feedback, the output is fedback with positive sign as shown in Fig. 3.7.

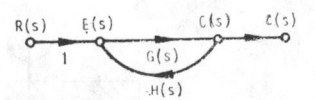

Figure 3.7 A regenerative feedback control system.

In this case, the transfer function is given by

$$\frac{C(s)}{R(s)} = \frac{G(s)}{1-G(s)H(s)} \quad (3.16)$$

There is a negative sign in the denominator of eqn. (3.16) which indicates the possibility of denominator becoming equal to zero thereby giving an infinite output for a finite input which is the condition of instability.

The regenerative feedback is sometimes used for increasing the loop gain of feedback systems. Figure 3.8 shows a feedback system with an inner loop having regenerative feedback. This signal flow graph reduces to a single loop graph whose loop gain is

$$\frac{-G(s)H(s)}{1-G_f(s)}$$

If $G_f(s)$ is selected to be nearly unity, the loop gain becomes very high and the closed-loop transfer function approximates to

$$\frac{C(s)}{R(s)} = \frac{G(s)}{1 - G_f(s)+G(s)H(s)} \approx \frac{1}{H(s)}$$

Figure 3.8 Increasing loop gain by regenerative feedback.

Thus due to high loop gain provided by the inner regenerative feedback loop, the closed-loop transfer function becomes insensitive to $G(s)$

3.6 Illustrative Examples

The effect of feedback on the performance of control systems is further illustrated with the help of following examples.

Example 3.1: Consider the temperature control system of Fig. 3.9 which is set up to produce a steady stream flow of hot liquid at a controlled temperature. The temperature of the outflowing liquid is regulated automatically by means of a feedback sensor (say a *thermocouple*) which produces an output voltage e_t proportional to the temperature of the outflowing liquid. This voltage is subtracted from the reference voltage e_r to generate the error signal, $e = (e_r - e_t)$, which in turn regulates the current i_c through the heater element (and therefore the rate of heat input to the liquid) by means of silicon controlled rectifiers (SCRs) connected in full-wave operation with suitable logic circuitry.

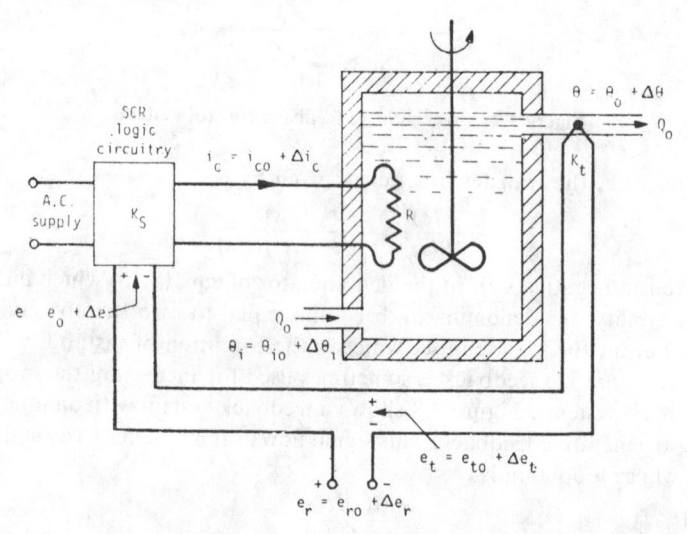

Figure 3.9 A temperature control system.

To reduce the complexity of this problem, certain simplifying assumptions given below are made at this stage.

(i) The liquid inflow and outflow rates for the tank are equal so that the liquid level in the tank is maintained constant during the operation.

(ii) The liquid in the tank is well-stirred so that its state can be described by the temperature θ of the outflowing liquid.

(iii) The tank is well-lagged so that the heat loss through its walls is negligible. Also the heat storage capacity of the tank walls is negligible.

(iv) The operation of the SCR circuit is linear, i.e., $i_c = K_S e$, where K_S is the circuit gain in amps/volt.

The heat flow balance equation is

Rate of heat generated by the heater = rate of heat storage in the tank
+ rate of heat removed by the outflowing liquid

Mathematically this can be expressed as (in consistent units)

$$i_c^2 R = Mc\frac{d\theta}{dt} + Q_0 \rho c\,(\theta - \theta_i)$$

$$= C\frac{d\theta}{dt} + \frac{1}{R_t}(\theta - \theta_i)$$

where M = mass of the liquid in the tank; c = specific heat of the liquid; ρ = density of the liquid; Q_0 = volume flow rate of the liquid; θ_i = temperature of the inflowing liquid; R = resistance of the heater element; $C = Mc$ = thermal capacitance of the liquid in the tank; and $R_t = \dfrac{1}{Q_0 \rho c}$ = thermal resistance of the heat transfer process.

Substituting all the variables in terms of their steady plus incremental values, the above equation can be written as

$$K_S^2(e_0^2 + 2e_0\Delta e)R = C d\Delta\theta/dt + (\theta_0 - \theta_{i0})/R_t + (\Delta\theta - \Delta\theta_i)/R_t \quad (3.17)$$

wherein it has been assumed that $(\Delta e)^2 \approx 0$. Under steady operation, i.e., with incremental values put as zero, we have

$$K_S^2 e_0^2 = (\theta_0 - \theta_{i0})/R_t \quad (3.18)$$

Subtracting eqn. (3.18) from eqn. (3.17), we have the describing equation in terms of the incremental values about the operating point as

$$2K_S^2 e_0 R\Delta e = C(d\Delta\theta/dt) + (\Delta\theta - \Delta\theta_i)/R_t \quad (3.19)$$

The incremental error is given by

$$\Delta e = \Delta e_r - \Delta e_t \quad (3.20)$$

Now

$$\Delta e_t = K_t \Delta\theta \quad (3.21)$$

K_t being the constant of the temperature sensor.

Taking the Laplace transform of eqns. (3.19), (3.20) and (3.21) and reorganizing, we get

$$\Delta\theta(s) = \frac{K\Delta E(s)}{\tau s + 1} + \frac{\Delta\theta_i(s)}{\tau s + 1} \quad (3.22)$$

$$\Delta E(s) = \Delta E_r(s) - \Delta E_t(s) \quad (3.23)$$

$$\Delta E_t(s) = K_t \Delta\theta(s) \quad (3.24)$$

where

$$K = 2K_S^2 e_0 R R_t; \quad \tau = R_t C$$

From eqns. (3.22), (3.23) and (3.24) we can draw the block diagram of the system as shown in Fig. 3.10 where the open-loop transfer function is

$$G(s) = \frac{K}{\tau s + 1}$$

and $\Delta\theta_i(s)$ is the change in the temperature of the inflowing liquid which can be regarded as a disturbance input entering the system through $1/(\tau s + 1)$.

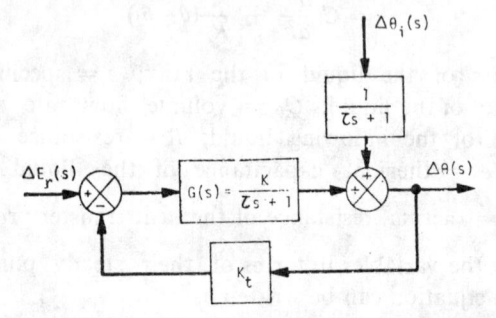

Figure 3.10 Block diagram of the system shown in Fig. 3.9.

Assuming the disturbance signal $\Delta\theta_i$ to be zero, the steady change in the temperature of the outflowing liquid caused by an unwanted step change ΔE_r in the reference voltage is given by

$$\Delta\theta(t)\Big|_{t\to\infty} = s\Big|_{s\to 0} \left(\frac{\Delta E_r}{s}\right) \frac{K}{\tau s + 1 + KK_t}$$

$$= \frac{\Delta E_r K}{1 + KK_t} \quad \text{(for closed-loop)}$$

$$= \Delta E_r K \quad \text{(for open-loop; } K_t = 0\text{)}$$

It is easily observed from above that the steady change in the temperature of the outflowing liquid caused by an unwanted change in reference voltage is reduced by the factor $1/(1 + KK_t)$ in the closed-loop compared to the open-loop case.

Similarly, if the reference input is held fixed, i.e., $\Delta e_r = 0$, a step change $\Delta\theta_i$ in the temperature of the inflowing liquid, causes a steady change in the output temperature of

$$\Delta\theta(t)\Big|_{t\to\infty} = s\Big|_{s\to 0} \left(\frac{\Delta\theta_i}{s}\right) \frac{1}{\tau s + 1 + KK_t}$$

$$= \frac{\Delta\theta_i}{1 + KK_t} \quad \text{(closed-loop case)}$$

$$= \Delta\theta_i \quad \text{(open-loop; } K_t = 0\text{)}$$

Thus, the change in temperature of the outflowing liquid caused by a change in temperature of the inflowing liquid can be reduced to any prescribed value by suitable choice of the loop gain KK_t.

Feedback Characteristics of Control Systems

If desired, the heat loss through the tank walls can be accounted for by introducing the term

$$Ah(\Delta\theta - \Delta\theta_e)$$

on the right hand side of eqn. (3.19), where h is the heat transfer constant and A is the surface area of the tank. This results in the additional disturbance signal $\Delta\theta_e$ caused by a change in the environmental temperature θ_{e0}.

Example 3.2: Consider the feedback control system shown in Fig. 3.11. The normal value of process parameter K is 1. Let us evaluate the sensitivity of transfer function $T(s) = C(s)/R(s)$ to variations in parameter K.

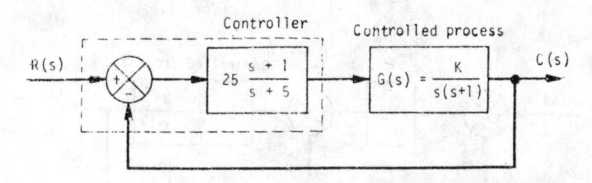

Figure 3.11 A single-loop configuration for $T(s) = \dfrac{25}{s^2+5s+25}$

From Fig. 3.11,

$$T(s) = \frac{C(s)}{R(s)} = \frac{25K}{s^2 + 5s + 25K}$$

$$= \cfrac{1}{1 + \cfrac{s^2 + 5s}{25K}}$$

Therefore

$$S_K^T = \frac{\partial T}{\partial K} \times \frac{K}{T} = \frac{s(s+5)}{s^2 + 5s + 25K}$$

Since the normal value of K is 1, we have

$$S_K^T = \frac{s(s+5)}{s^2 + 5s + 25} \qquad (3.25)$$

S_K^T may be evaluated at various values of frequency (discussed in Chapter 9). At a particular frequency $\omega = 5$, the magnitude of sensitivity is approximately $|S_K^T| = 1.41$.

In a typical design problem, the process transfer function $G(s)$ is given. We select $T(s)$ which would meet the given specifications. $T(s)$ may be achieved by different feedback structures. Sensitivity function then allows a quantitative comparison of these structures.

For example, the transfer function $T(s)$ of system of Fig. 3.11 may be realized by a two-loop structure of Fig. 3.12 (normal value of $K = 1$). For this structure,

$$T(s) = \frac{25K}{s^2 + (1 + 4K)s + 25K}$$

$$= \frac{25K}{s(s+1) + K(4s+25)}$$

$$S_K^T = \frac{s(s+1)}{s(s+1) + K(4s+25)}$$

$$= \frac{s(s+1)}{s^2 + 5s + 25} \quad \text{(for } K = 1\text{)} \tag{3.26}$$

The magnitude of sensitivity at $\omega = 5$ is $|S_K^T| \approx 1$.
This shows the superiority of the two-loop system of Fig. 3.12 over a single-loop system of Fig. 3.11. Thus the sensitivity function gives a firm basis for comparison of alternative designs.

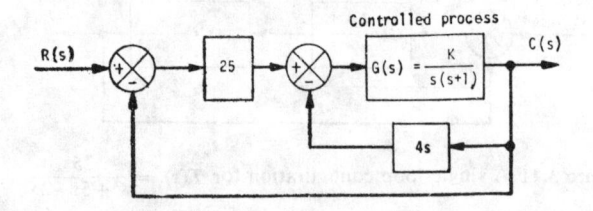

Figure 3.12 A two-loop configuration for $T(s) = \dfrac{25}{s^2+5s+25}$

Problems

3.1 A simple voltage regulator is shown in Fig. P-3.1. A potentiometer is used at the output terminals of the generator to give feedback voltage KV_0 where K is a constant ($K \leqslant 1$). The potentiometer resistance is high enough that it may be assumed to draw negligible current. The amplifier has a gain of 20 volts/volt. The generator gain is 50 volts/field amp. Reference voltage $v_r = 50$ volts.

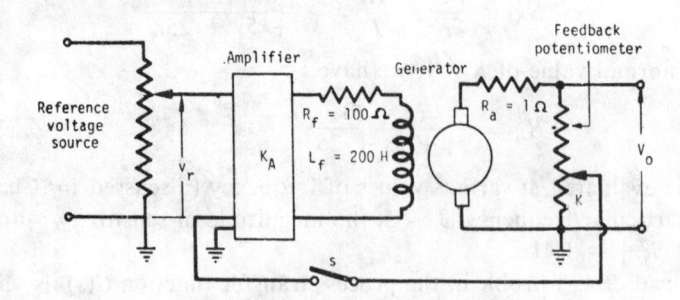

Figure P-3.1.

(a) Draw a block diagram of the system when the generator is supplying a load current. Indicate the transfer function of each block.

(b) The system is operated closed-loop ('s' closed). Determine the value of K in order to give a steady no-load generator terminal voltage of 250 volts. What is the change in terminal voltage caused by a steady load current of 30 amps? What reference voltage would be required to restore the generator terminal voltage of 250 V?

(c) The system is operated open-loop ('s' open), determine the reference voltage needed to obtain a steady no-load voltage of 250V. What would be the change in terminal voltage for a load of 30 amps?

(d) Compare the changes in terminal voltage in parts (b) and (c) and comment upon the effect of feedback in countering the changes in terminal voltage caused by load current.

3.2 The block diagram of a position control system is shown in Fig. P-3.2. Determine the sensitivity of closed-loop transfer function T with respect to G and H, the forward path and feedback path transfer functions respectively for $\omega = 1$ rad/sec.

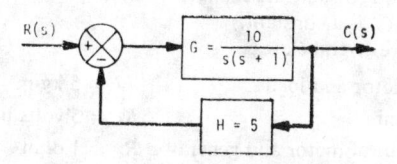

Figure P-3.2.

3.3 A servo system is represented by the signal flow graph shown in Fig. P-3.3. The variable T is the torque and E is the error. Determine:
(a) the overall transmission if $K_1 = 1$, $K_2 = 5$ and $K_3 = 5$;
(b) the sensitivity of the system to changes in K_1 for $\omega = 0$.

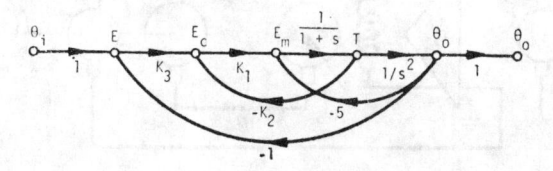

Figure P-3.3.

3.4 The field of a d.c. servomotor is separately excited by means of a d.c. amplifier of gain $K_A = 90$ (see Fig. P-3.4). The field has an inductance of 2 henrys and a resistance of 50 ohms. Calculate the field time constant.

A voltage proportional to the field current is now fedback negatively to the amplifier input. Determine the value of the feedback constant K to reduce the field time constant to 4 milliseconds.

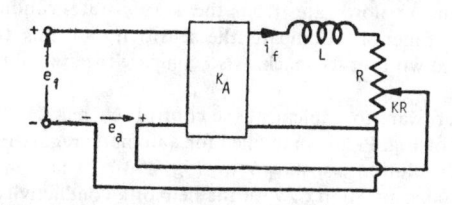

Figure P-3.4.

3.5 For the speed control system shown in Fig. P-3.5, assume that
(i) the reference and feedback tachometers are identical;

(ii) generator field time constant is negligible and its generated voltage is K^a volts/amp;
(iii) friction of motor and mechanical load is negligible.

(a) Find the time variation of output speed (ω_0) for a sudden reference input of 10 rad/sec. Find also the steady-state output speed.

(b) If the feedback loop is opened and gain K_A adjusted to give the same steady-state speed as in the case of the closed-loop, determine how the output speed varies with time and compare the speed of response in the two cases.

(c) Compare the sensitivity of ω_0 to changes in amplifier gain K_A and generator speed ω_g, with and without feedback.

(*Hint*: The generator gain constant K_g changes in direct proportion to generator speed, i.e., $K_g = K_g' \omega_g$, where K_g' is a constant.)

The system constants are given below :

Moment of inertia of motor and load	$J = 5$ kg-m²
Motor back emf constant	$K_b = 5$ volts per rad/sec
Total armature resistance of motor and generator	$R_a = 1$ ohm
Generator gain constant	$K_g = 50$ volts/amp
Amplifier gain	$K_A = 5$ amps/volt
Tachometer constant	$K_t = 0.5$ volts per rad/sec

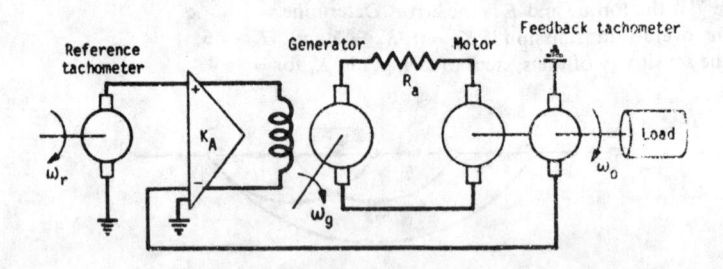

Figure P-3.5.

3.6. Figure P-3.6 shows a control system for supplying a steady flow of oil. The volume flow rate Q_0 of the oil from the tank is measured and the supply rate Q of oil to the tank is increased by an amount $K(Q - Q_0)$, when the outflow differs from the desired flow Q. Under steady conditions $Q_0 = \bar{Q}$ and the corresponding head in the tank is H. The tank and output pipe may be considered to have a liquid capacitance C and flow resistance R.

For small deviations in flow rate from the steady-state condition, determine the closed-loop transfer function. Compare the sensitivity of this transfer function to changes in R with and without feedback. Also compare the system time constants in the two cases.

Suggest suitable hardware to implement the control scheme.

3.7 The scheme of Fig. P-2.6 is modified for automatic regulation of the concentration of outstream salt solution as shown in Fig. P-3.7. The controller continuously monitors the output concentration C_0 with the help of a conductivity cell located in the exit pipe and compares it with the desired concentration C_d to generate the error signal $(C_d - C_0)$. The controller manipulates the valve stem position according to the linear law, $x = K_c(C_d - C_0)$.

(i) Draw the signal flow graph of the system and obtain therefrom the transfer function $C_0(s)/C_d(s)$.

Feedback Characteristics of Control Systems

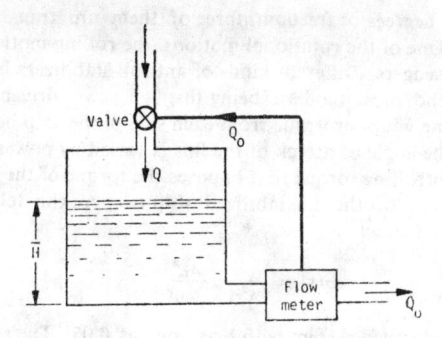

Figure P-3.6.

(ii) Compare the open-loop and closed-loop systems of Fig. P-2.6 and Fig. P-3.7 respectively for steady-state error in outstream salt concentration for unit-step disturbance in the inflow rate Q_i of the concentrated salt solution.

(iii) The system described above is operating around the point defined by (C_d^0, C_0^0, Q_i^0). Compare the steady-state error caused in the outstream salt concentration to unit-step disturbance (ΔC_i) in the concentration of the inflowing concentrated salt solution for the open-loop case of Fig. P-2.6 and the closed-loop case of Fig. P-3.7.

[*Hint* : For part (iii): Change in salt (mass) inflow rate is

$$(C_i^0 + \Delta C_i)(Q_i^0 + \Delta Q_i) - C_i^0 Q_i^0 \approx C_i^0[\Delta Q_i + (Q_i^0/C_i^0)\Delta C_i]$$

on the assumption that $\Delta C_i \Delta Q_i \approx 0$. The disturbance ΔC_i in C_i can thus be equivalently regarded as a disturbance $(Q_i^0/C_i^0)\Delta C_i$ in Q_i. Now

$$V\frac{d(\Delta C_0)}{dt} + Q_0(\Delta C_0) = C_i^0[\Delta Q_i + \underbrace{(Q_i^0/C_i^0)\Delta C_i}_{\text{Disturbance signal}}]$$

$$\Delta Q_i = K\Delta x; \quad \Delta x = K_c(-\Delta C_0) \text{ since } (\Delta C_d = 0)]$$

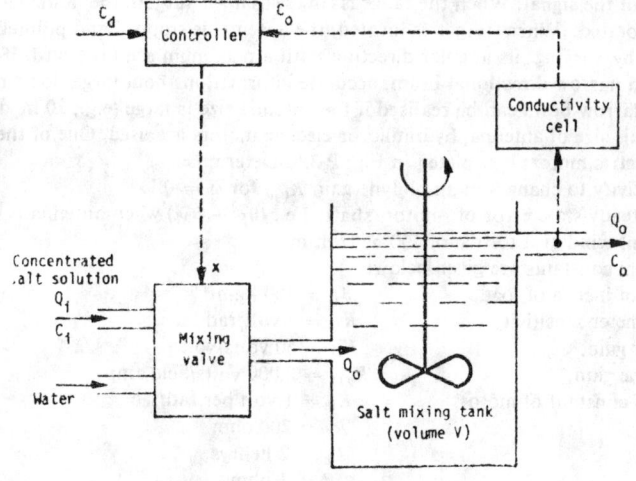

Figure P-3.7.

3.8 A ship has six degrees of freedom, three of them are translational motions and three are rotational. One of the rotational motions, the rolling motion contributes much discomfort to the passengers. Different kinds of antiroll stabilizers have been employed in ships, probably the most modern being through power driven fins. The fins which look like small airplane wings protrude from each side of the ship below the water line. Either the area or the angle of attack of the fins is varied by power actuators in such a way as to produce controlling torque that opposes the torque of the sea.

The rolling dynamics of the unstabilized ship is approximately represented by the transfer function given below:

$$G(s) = \frac{\omega_n^2}{s^2 + 2\zeta\omega_n s + \omega_n^2}$$

This is a highly underdamped system with ζ as low as 0.05. The roll sensor, senses the roll and provides a signal to the power actuator which drives the fins to produce controlling torque $T_f(s)$. The block diagram of the system is shown in Fig. P-3.8.

(a) Determine the sensitivity of the system to changes in K_f (for $\omega = 0$).
(b) Discuss the effect of feedback in stabilizing the ship against rolling motion.

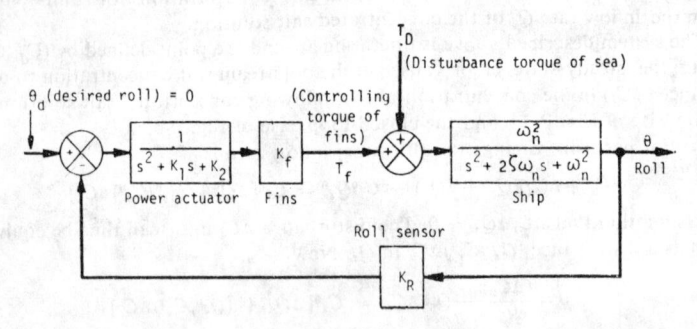

Figure P-3.8.

3.9 In a radar system, an electromagnetic pulse is radiated from an antenna into space. An echo pulse is received back if a conducting surface such as an airplane appears in the path of the signal. When the radar is in search of target, the antenna is continuously rotated. When target is located, the antenna is stopped and pointed towards the target by varying its angular direction until a maximum echo is heard. If energy is radiated in a narrow directional beam, accurate information about target location can be obtained. Narrow beam can be realised if the antenna size is large (e.g., 20 m diameter). To drive this size of antenna, hydraulic or electric motors are used. One of the schemes utilizing electric motors is depicted in Fig. P-3.9. Determine:

(a) Sensitivity to changes in amplidyne gain K_{am} for $\omega = 0.1$.
(b) The steady-state error of motor shaft, i.e., $(\theta_R - \theta_M)$ when antenna is subjected to a constant wind gust torque of 100 newton-m.

The system constants are given below:
Moment of inertia of load, $J_L = 400$ kg-m²
Potentiometer sensitivity, $K_P = 1$ volt/rad
Amplifier gain, $K_A = 50$ volts/volt
Amplidyne gain, $K_{am} = 2,000$ volts/field amp
Back emf constant of motor, $K_b = 1$ volt per rad/sec
    $R_c = 200$ ohm
    $L_c = 2$ henrys
    $R = 1$ ohm
    $N_1 = 40 = N_3$
    $N_2 = 800$

Feedback Characteristics of Control Systems

Load friction, motor inertia and friction, and quadrature axis time constant of amplidyne are assumed to be negligible.

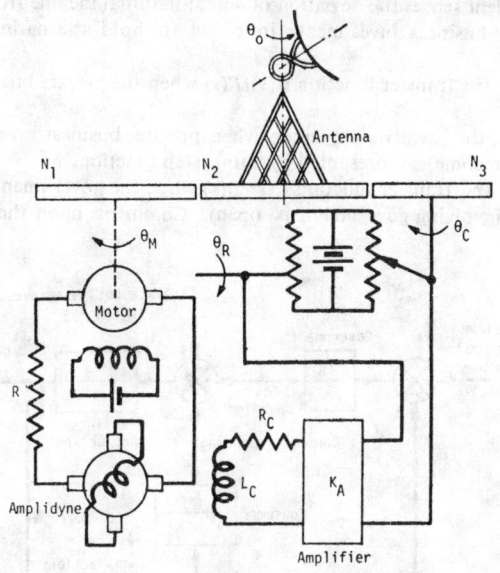

Figure P-3.9.

3.10 The liquid-level system shown in Fig. P-2.12 is modified for the automatic regulation of liquid head H (this means indirectly regulating the liquid outflow Q_0) by use of a float level sensor and a lever mechanism to adjust the opening of the valve as shown in Fig. P-3.10. For a change ΔH of head the lever-mechanism adjusts the valve opening by $\Delta v_x (\Delta v_x = - K_x \Delta H)$.

(i) Draw the signal flow graph of the system and obtain therefrom the transfer functions

$$\left.\frac{\Delta H(s)}{\Delta P(s)}\right|_{\Delta Q_D = 0} \text{ and } \left.\frac{\Delta Q_0(s)}{\Delta P(s)}\right|_{\Delta Q_D = 0}$$

(ii) Compare the open-loop and closed-loop systems of Fig. P-2.12 and Fig. P-3.10 respectively for the steady-state error in liquid outflow to the unit-step disturbance ΔQ_D.

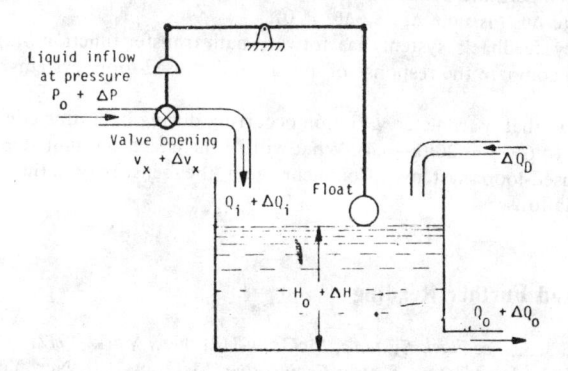

Figure P-3.10.

3.11 National economy control may be represented by the block diagram of Fig. P-3.11. Private business investment fluctuates and represents a disturbance to the system. Government senses the deviation of actual national income from the desired one and changes the business investments in order to hold the national income within tolerable limits.

(a) Determine the transfer function $C(s)/R(s)$ when the private business investment is zero.

(b) Determine the steady-state error when private business investment is zero and desired national income is represented by a unit step function.

(c) Determine the transfer function $C(s)/D(s)$ when the government does not act as a controller (loop involving government is open). Comment upon the stability of this system.

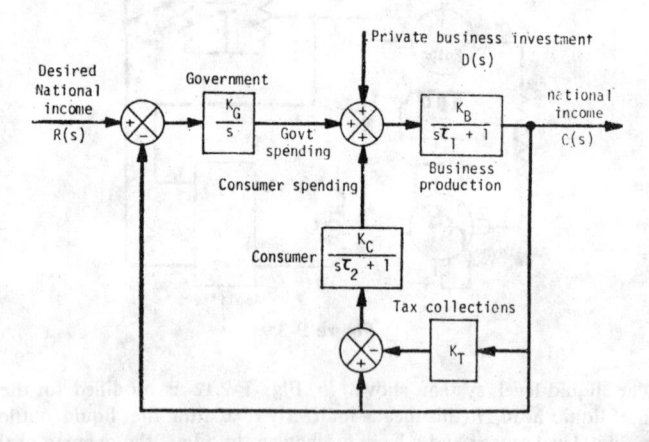

Figure P-3.11. [From William A. Lynch and John G. Truxal, *Signals and Systems in Electrical Engineering*, 1962, McGraw-Hill, New York. Reproduced with permission.]

3.12 (a) From the block diagram of Fig. 3.10 for the temperature control system of Fig. 3.9, find the transfer function $\Delta\theta(s)/\Delta E_r(s)$ when $\Delta\theta_i = 0$ for open- and closed-loop conditions. Determine also the sensitivity of these transfer functions to change in K under steady d.c. conditions, i.e., $s = 0$.

(b) Modify the block diagram of Fig. 3.10 to include the effect of the disturbance signal $\Delta\theta_e$ caused by changes in the environmental temperature θ_{e0}. Find the steady change in the temperature of the outflowing liquid for open- and closed-loop cases due to a step change $\Delta\theta_e$ (assume $\Delta e_r = \Delta\theta_i = 0$).

3.13 A unity feedback system has forward path transfer function $G(s) = 20/(s+1)$. Determine and compare the response of open- and closed-loop systems for unit-step input.

Suppose now that parameter variation occurring during operating conditions causes $G(s)$ to modify to $G'(s) = 20/(s+0.4)$. What will be the effect on unit-step response of open- and closed-loop systems? Comment upon the sensitivity of the two systems to parameter variations.

References and Further Reading

1. Cruz, Jr., J.B., *Feedback Systems*, McGraw-Hill, New York, 1972.
2. Truxal, J.G., *Introductory System Engineering*, McGraw-Hill, New York, 1972.

3. Barbe, E.C., *Linear Control Systems*, International Textbook Co., Scranton, Pennsylvania, 1963.
4. Horowitz, I.M., *Synthesis of Feedback Systems*, Academic Press, New York, 1963.
5. Taylor, P.L., *Servomechanisms*, Longmans, London, 1960.

4. Control Systems and Components

4.1 Introduction

A closed-loop control system can be represented by the general block diagram shown in Fig. 4.1. Such a system is composed of three basic elements; the feedback element, controller and controlled system.

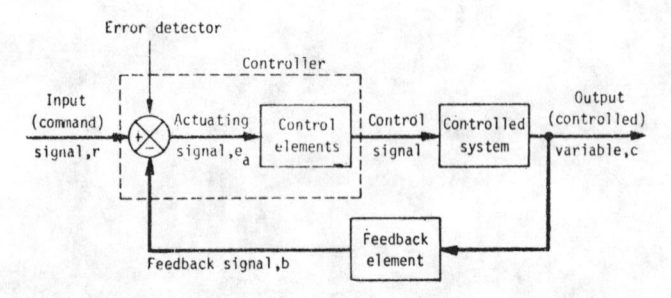

Figure 4.1 General block diagram of a closed-loop control system.

The *feedback element* is a device which converts the *output* (*controlled*) *variable c* into another suitable variable, the feedback signal *b* which then is compared with the input (command) signal.

The *controller* consists of an *error detector* and *control elements*. The error detector compares the feedback signal obtained from the plant output with the *input* (*command*) *signal* and determines therefrom the deviation known as the *actuating signal*. The actuating signal is usually at low power level. It is suitably manipulated by the control elements to produce a *control signal*. The manipulation may involve amplification, generation of a suitable function of the actuating signal and a power stage. The power stage in control elements is essential so that the control signal can drive the *controlled system* (a plant or process) to produce the desired output variable. Large power amplification may be involved in the plant or process being controlled.

This chapter is devoted to a general study of control system components, the emphasis being on those commonly used in servo-systems.

4.2 Linear Approximation of Nonlinear Systems

Practically none of the physical systems in nature are perfectly linear. Since most powerful mathematical tools are available for linear analysis, it is desirable to make linearizing assumptions whenever a compromise can be obtained between the simplicity of analysis and accuracy of results. As the components to be discussed in this chapter have somewhat nonlinear behaviour, it is worthwhile discussing a general linearizing technique before entering into a detailed discussion of components. In fact the technique has already been employed in the previous chapters where some of the components introduced were essentially nonlinear. The intuitive basis of linearization is that a smooth curve differs very little from its tangent line so long as the variable does not wander far from the point of tangency. Thus if the region of operation is restricted to a narrow range, a nonlinear system can be treated approximately as a linear system. This kind of approximation appears to be valid in most control systems since their purpose is to keep the controlled variable very close to the desired value. However, if a system is required to follow a varying desired value, one can analyze the system by linearizing it at several points along the curve.

Consider a general element with input $x(t)$ and output $y(t)$ whose input-output relationship $y = f(x)$ is shown graphically in Fig. 4.2. The relationship may be *nonlinear* but is assumed to be *continuous*.

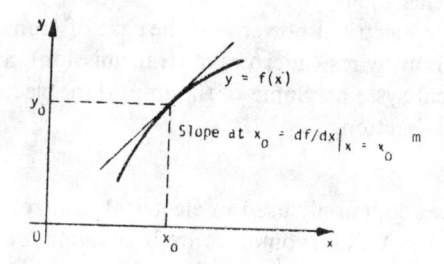

Figure 4.2 Linearization of a nonlinear element.

Expansion of the equation $y = f(x)$ into a Taylor's series about the normal operating point (x_0, y_0) gives

$$y = f(x) = f(x_0) + \frac{df}{dx}\bigg|_{x=x_0} \frac{x - x_0}{1!} + \frac{d^2f}{dx^2}\bigg|_{x=x_0} \frac{(x - x_0)^2}{2!} + \cdots \quad (4.1)$$

If the variation $(x-x_0)$ of the input about the normal operating point is small, higher than first-order terms in $(x-x_0)$ can be neglected yielding a linear approximation,

$$y = y_0 + m(x - x_0) \quad (4.2)$$

where $m = \dfrac{df}{dx}\bigg|_{x=x_0}$ is the slope at the operating point.

Equation (4.2) may be rewritten as

$$y - y_0 = m(x - x_0)$$

or

$$\Delta y = m\Delta x$$

If the output variable y depends on several input variables $x_1, x_2, ..., x_n$, i.e., $y = f(x_1, x_2,...,x_n)$, then the linear approximation for y can be obtained similarly by expanding this equation into Taylor's series about the operating point $(x_{10}, x_{20},...,x_{n0}, y_0)$ and neglecting all terms of second and higher derivatives. This approximation gives

$$y = f(x_{10}, x_{20}, ..., x_{n0})$$
$$+ \frac{\partial f}{\partial x_1}\bigg|_{\substack{x_1 = x_{10} \\ \cdots \\ x_n = x_{n0}}} (x_1 - x_{10}) + \frac{\partial f}{\partial x_2}\bigg|_{\substack{x_1 = x_{10} \\ \cdots \\ x_n = x_{n0}}} (x_2 - x_{20}) + ... + \frac{\partial f}{\partial x_n}\bigg|_{\substack{x_1 = x_{10} \\ \cdots \\ x_n = x_{n0}}} (x_n - x_{n0})$$

This technique is employed in several examples in the following sections.

4.3 Electrical Systems

The availability of electrical power and the ease of transmitting electrical signals with or without wires (microwave transmission) are the desirable features of electrical systems. Some of the important electrical components are discussed in this section.

SERVOMOTORS

The power devices commonly used in electrical control systems are a.c. and d.c.* servomotors. A.C. servomotors are best suited for low power applications. They are rugged, light in weight and have no brush contacts as is the case with d.c. servomotors. The error signal derived from synchros (to be discussed later in this section) can be amplified by a.c. amplifiers to produce a control signal for servomotors.

A.C. Servomotors

An a.c. servomotor is basically a two-phase induction motor except for certain special design features. A two-phase induction motor consists of two stator windings** oriented 90° electrical apart in space and excited by a.c. voltages which differ in time-phase by 90°. Figure 4.3 shows the schematic diagram for balanced operation of the motor, i.e., voltages of equal rms magnitude and 90° phase difference are applied to the two stator phases, thus making their respective fields 90° apart in both time and space, result-

---

*d.c. servomotors have been discussed in Chapter 2.
**A two pole construction is commonly used.

Control Systems and Components

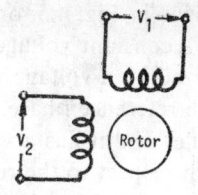

Figure 4.3 Schematic diagram of a two-phase induction motor.

ing in a magnetic field of constant magnitude rotating at synchronous speed. The direction of rotation depends upon phase relationship of voltages V_1 and V_2. As the field sweeps over the rotor, voltages are induced in it producing current in the short-circuited rotor. The rotating magnetic field interacts with these currents producing a torque on the rotor in the direction of field rotation. The general shape of the torque-speed characteristics of a two-phase induction motor is shown in Fig. 4.4.

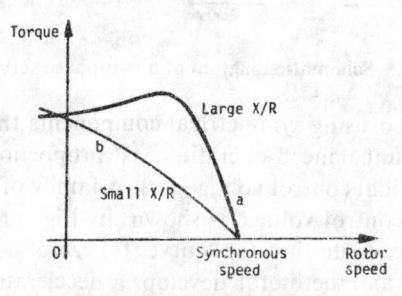

Figure 4.4 Torque-speed characteristics of induction motor.

It is seen from this figure that the shape of the characteristic depends upon the ratio of the rotor reactance X to the rotor resistance R. In normal induction motors, X/R ratio is generally kept high so as to obtain the maximum torque close to the operating region which is usually around 5% slip.

A two-phase servomotor differs in two ways from a normal induction motor.

1. The rotor of the servomotor is built with high resistance so that its X/R ratio is small and the torque-speed characteristic, as shown by the curve b of Fig. 4.4, is nearly linear in contrast to the highly nonlinear characteristic with large X/R. It will be shown later that if a conventional induction motor with high X/R ratio is used for servo applications, then because of the positive slope for part of the characteristic, the system using such a motor becomes unstable.

The rotor construction is usually squirrel cage or drag-cup type. The diameter of the rotor is kept small in order to reduce inertia and thus to obtain good accelerating characteristics. Drag-cup construction is used for very low inertia applications.

2. In servo applications, the voltages applied to the two stator windings

are seldom balanced. As shown in Fig. 4.5, one of the phases known as the *reference phase* is excited by a constant voltage and the other phase, known as the *control phase* is energized by a voltage which is 90° out of phase with respect to the voltage of the reference phase. The control phase voltage is supplied from a servo amplifier and it has a variable magnitude and polarity ($\pm 90°$ phase angle with respect to the reference phase). The direction of rotation of the motor reverses as the polarity of the control phase signal changes sign.

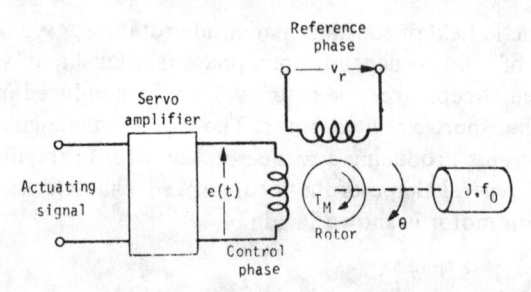

Figure 4.5 Schematic diagram of a two-phase servomotor.

It can be proved using symmetrical components that starting torque of servomotor under unbalanced operation, is proportional to E, the rms value of the sinusoidal control voltage $e(t)$. A family of torque-speed curves with variable rms control voltage is shown in Fig. 4.6a. All these curves have negative slope. Note that the curve for zero control voltage goes through the origin and the motor develops a decelerating torque.

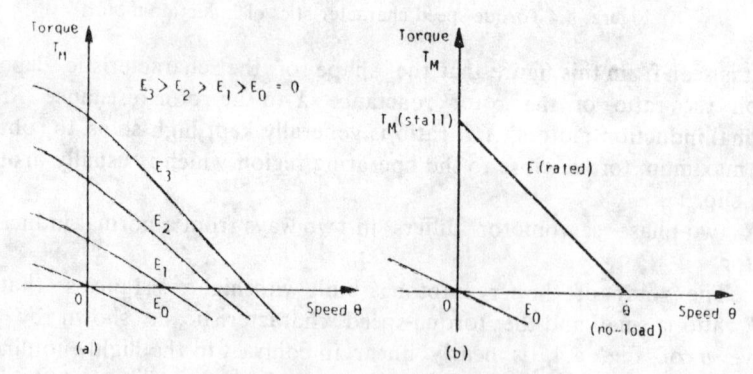

Figure 4.6 Servomotor characteristics.

As seen from Fig. 4.6a, the torque-speed curves are still somewhat nonlinear. However, in the low-speed region, the curves are nearly linear and equidistant, i.e., the torque varies linearly with speed as well as with control voltage. Since a servomotor seldom operates at high speeds, these curves can be linearized about the operating point.

The torque generated by the motor is a function of both the speed $\dot{\theta}$ and rms control voltage E, i.e., $T_M = f(\dot{\theta}, E)$. Expanding this equation into

Taylor's series about the normal operating point (T_{M0}, E_0, θ_0) and dropping off all the terms of second- and higher-order derivatives, we get

$$T_M = T_{M0} + \left.\frac{\partial T_M}{\partial E}\right|_{\substack{E=E_0 \\ \dot\theta=\dot\theta_0}} (E - E_0) + \left.\frac{\partial T_M}{\partial \dot\theta}\right|_{\substack{E=E_0 \\ \dot\theta=\dot\theta_0}} (\dot\theta - \dot\theta_0) \quad (4.3)$$

or $\quad T_M - T_{M0} = K(E - E_0) - f(\dot\theta - \dot\theta_0)$

or $\quad \Delta T_M = K\Delta E - f\Delta\dot\theta \qquad (4.4)$

where $\quad K = \left.\dfrac{\partial T_M}{\partial E}\right|_{\substack{E=E_0 \\ \dot\theta=\dot\theta_0}} \quad$ and $\quad f = -\left.\dfrac{\partial T_M}{\partial \dot\theta}\right|_{\substack{E=E_0 \\ \dot\theta=\dot\theta_0}} \quad$ are constants.

If the load consists of inertia J and viscous friction f_0, the torque equation, expressed in incremental notation, becomes

$$\Delta T_M = J\Delta\ddot\theta + f_0\Delta\dot\theta = K\Delta E - f\Delta\dot\theta$$

The above equation is valid even when ΔE varies with time so long as this variation is slow compared to sin $\omega_c t$; in frequency domain, it means that ω (cut-off) $\ll \omega_c$, where ω is the frequency of the signal ΔE, and ω_c is the carrier frequency. Control systems using such devices are called *carrier control systems*.

Taking the Laplace transform of above equation, we have

$$(Js^2 + f_0 s)\Delta\theta(s) = K\Delta E(s) - fs\Delta\theta(s)$$

The incremental motor transfer function about an operating point is then given by

$$G_M(s) = \frac{\Delta\theta(s)}{\Delta E(s)} = \frac{K}{Js^2 + (f_0 + f)s}$$

$$= \frac{K_m}{s(\tau_m s + 1)} \qquad (4.5)$$

where $K_m = \dfrac{K}{f_0 + f} = $ motor gain constant; and $\tau_m = \dfrac{J}{f_0 + f} = $ motor time constant.

From eqn. (4.5) it is seen that the negative slope of the torque-speed characteristic of the servomotor contributes the viscous friction term f to the load friction f_0. If the slope

$$\left.\frac{\partial T_M}{\partial \dot\theta}\right|_{\substack{E=E_0 \\ \dot\theta=\dot\theta_0}} = -f_0$$

is positive (as in part of torque-speed characteristic of a normal induction motor), then the effective viscous friction ($f_0 + f$) may become negative. The control system using such a motor may have negative damping and may therefore be unstable. This is one of the reasons why the conventional induction motors are not used as servomotors.

When a.c. servomotor is used in a position control system, the operating point is ($E_0 = 0$, $\dot{\theta}_0 = 0$), so that

$$\Delta\theta = \theta \text{ and } \Delta E = E$$

Therefore

$$G(s) = \frac{\theta(s)}{E(s)} = \frac{K_m}{s(\tau_m s + 1)}$$

The motor constants K and f can be easily estimated from two well-known tests—no-load and stall torque tests. The no-load speed ($T_M = 0$) and stall torque ($\dot{\theta} = 0$, i.e., motor rotor held stationary) at rated voltage applied to control phase are plotted in Fig. 4.6b. The straight line joining $\dot{\theta}$(no-load speed) and T_M (stall torque) points approximately represents the torque-speed characteristic of a servomotor. Therefore,

$$K = \frac{\text{Stall torque (rated voltage)}}{\text{Rated control phase voltage}} \qquad (4.6a)$$

$$f = \frac{\text{Stall torque (rated voltage)}}{\text{No-load speed (rated voltage)}} \qquad (4.6b)$$

It is further observed from Fig. 4.6a that the slope of the torque-speed characteristic reduces as control phase voltage decreases. In motors used in practice, the torque-speed slope in low speed region is nearly one-half the slope at rated voltage. Therefore, for servomotors used in position control systems,

$$f = \frac{1}{2} \frac{\text{Stall torque (rated voltage)}}{\text{No-load speed (rated voltage)}} \qquad (4.7)$$

approximately.

A.C. TACHOMETER

The principle of operation of an *a.c. tachometer* or *drag-cup generator* can be readily understood by referring to the schematic diagram of Fig. 4.7. As in the case of an a.c. servomotor, the two stator field coils are mounted at right angles to each other, i.e., in space quadrature. The tachomotor rotor is merely a thin aluminium cup that rotates in an air gap between a fixed magnetic structure. This light inertia rotor of highly conducting material provides a uniformly short-circuited secondary conductor.

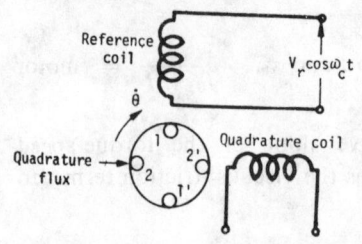

Figure 4.7 A.C. tachometer.

An approximate analysis of the a.c. tachometer is presented below.
Let the voltage applied to the reference coil be $V_r \cos \omega_c t$. It produces a reference flux of $\phi_r \sin \omega_c t$, as the voltage drop in resistance and reactance of the reference coil is of negligible order. Let the rotor speed be

$$\dot{\theta}(t) = \dot{\theta}_m \cos a\omega_c t$$

For ease of analysis, the rotor can be replaced by equivalent imaginary conductor pairs (1, 1') and (2, 2').

Speed voltage induced in the imaginary conductor pair (1, 1')

$$= k_1(\phi_r \sin \omega_c t)(\dot{\theta}_m \cos a\omega_c t)$$

Assuming the rotor reactance to be negligible, the current in (1, 1') will be proportional to the induced emf causing a quadrature flux of

$$\phi_q = k_2 \dot{\theta}_m \cos a\omega_c t \sin \omega_c t$$

An emf is induced in the quadrature coil due to time rate of change of ϕ_q and is given by

$$e_q = k_3 \dot{\theta}_m \frac{d}{dt}[\cos a\omega_c t \sin \omega_c t]$$

as $\dot{\theta}_m$ is constant. Differentiating we get

$$e_q = k_3 \dot{\theta}_m [-a\omega_c \sin a\omega_c t \sin \omega_c t + \omega_c \cos a\omega_c t \cos \omega_c t]$$

If $a \ll 1$, i.e., the frequency of sinusoidally varying speed $\dot{\theta}(t)$ is much less than the carrier frequency ω_c,

$$e_q \approx K_t \dot{\theta}_m \cos a\omega_c t \cos \omega_c t$$

or $\qquad e_q = K_t \dot{\theta}(t) \cos \omega_c t = v_t(t) \cos \omega_c t$

The emf induced in the quadrature coil is thus proportional to rotor speed and is in phase with the voltage applied to the reference coil. The modulated signal (emf) available at the terminals of the quadrature coil can be expressed as

$$v_t(t) = K_t \dot{\theta}(t) \qquad (4.8)$$

where K_t is the tachometer constant. This essentially is the tachometer action.

It may be noted that the transformer emf induced in the imaginary conductor pair (2, 2') produces a current and flux which does not link the quadrature coil. The current in (2, 2') merely causes a reactive voltage drop in the reference coil which has been neglected in the analysis.

AMPLIDYNE

When a control signal is required to have a power level higher than the capability of linear electronic amplifiers, *rotating amplifier* is an ideal choice*. An ordinary d.c. generator, in fact, is such an amplifier where a field voltage at a low power level ($v_f i_f$) controls a large armature power ($v_a i_a$). Larger power amplifications are obtained for a given machine frame size by use of the general class of machines called *cross field machines*. An

---
*Thyristor amplifiers are becoming increasingly popular for this purpose. This topic is of course beyond the scope of this book.

amplidyne is such a cross field machine which is specially suited for control applications. As we shall see presently, this device is equivalent to accommodating two stages of power amplification in a single machine frame.

A schematic diagram of an amplidyne driving an armature-controlled d.c. motor is shown in Fig. 4.8. The salient constructional features and principle of operation are discussed below.

An amplidyne is usually a two-pole machine, though its stator structure looks like four poles as it has to accommodate two field windings spaced 90°. The control field winding is placed so as to produce a flux ϕ_d in the *direct axis*. The field axis at 90° to it called the *quadrature axis*, accommodates the series quadrature winding. The armature is driven by a prime-mover at constant speed. It is wound like a two-pole d.c. armature but it carries two sets of brushes on the commutator—one set in the direct axis and the other at 90° to it in the quadrature axis. Like in a d.c. machine, flux in any axis, say the direct axis, causes a voltage to be induced in the armature which appears across the quadrature axis brushes. Similarly a quadrature axis flux causes a voltage across the direct axis brushes. Current flowing in any one set of armature brushes produces a flux in that particular brush axis on account of armature reaction.

Under steady d.c. conditions, control field voltage causes a constant field current to flow which establishes the control flux ϕ_d in the direct axis as shown in Fig. 4.8. The flux ϕ_d induces a voltage e_q across the brushes qq. The quadrature brushes are short-circuited through the series quadrature winding. A current i_q is thus established in quadrature axis and it produces its own flux ϕ_q. The purpose of the series quadrature winding is to increase the effective turns producing the flux ϕ_q so that a reduced quadrature axis current is sufficient to produce a given ϕ_q. Keeping down i_q reduces the commutation difficulties which are intense in this type of machine.

The quadrature axis flux in turn produces a voltage e_d across the direct axis brushes. As the current i_a is drawn from the direct axis brushes to drive the d.c. motor, a reaction flux ϕ_a is set up in the direct axis opposing the control flux ϕ_d. Since the degenerative action is undesirable as it tends

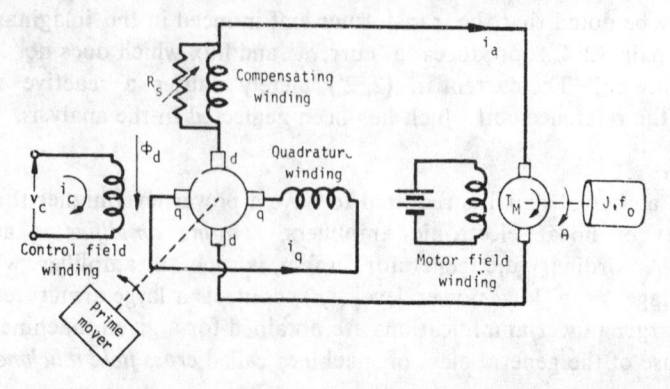

Figure 4.8 An amplidyne with d.c. motor load.

to reduce e_d, a compensating winding is provided in the direct axis and is connected in series with the amplidyne output leads so as to carry the current i_a. The compensating winding is so connected as to produce a flux which neutralizes the effect of ϕ_a. Near 100% compensation can be achieved by adjusting the resistor R_s connected in shunt across the compensating winding.

The transfer function $E_d(s)/V_c(s)$ between open-circuit voltage and control voltage is derived below, based on the assumption that there is no coupling between the direct axis and quadrature axis windings and that the saturation of the magnetic circuit is negligible.

For the control winding circuit,

$$\frac{I_c(s)}{V_c(s)} = \frac{1}{R_c(1 + s\tau_c)}$$

where $\tau_c = L_c/R_c$ = control winding time constant.

The control winding current i_c generates a flux ϕ_d which induces the voltage e_q such that

$$e_q = K_q i_c$$

where K_q is a constant.

Now, for the quadrature axis circuit

$$\frac{I_q(s)}{E_q(s)} = \frac{1}{R_q(1 + s\tau_q)}$$

where $\tau_q = L_q/R_q$ = quadrature-axis time constant.

Finally, the amplidyne open-circuit output voltage e_d is proportional to the current i_q, i.e.,

$$e_d = K_d i_q$$

where K_d is a constant.

The block diagram of the amplidyne is shown in Fig. 4.9a.

Usually the amplidyne output voltage is connected to the armature winding of an armature-controlled d.c. motor which drives some mechanical load as shown in Fig. 4.8. Let us derive the transfer function of the amplidyne together with the d.c. motor assuming 100% compensation so

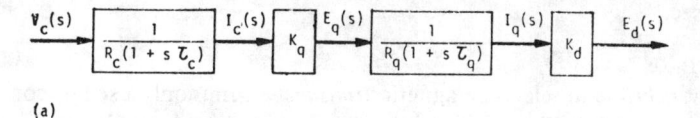

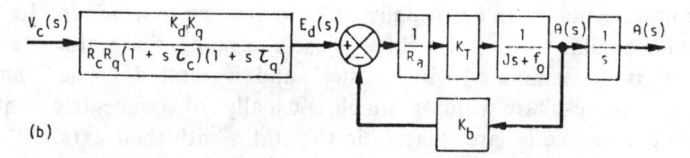

Figure 4.9 (a) Block diagram of amplidyne without load;
(b) Block diagram of the system shown in Fig. 4.8.

that the motor armature current has no effect on the control flux ϕ_d and that the direct axis inductance is negligible. Let

J = equivalent moment of inertia of motor and load referred to motor shaft

f_0 = combined viscous friction coefficient of motor and load referred to motor shaft

R_a = total resistance of amplidyne armature, compensating winding and motor armature

K_T = motor torque constant

K_b = motor back emf constant

The block diagram of the amplidyne and motor combination obtained from Fig. 4.9a and Fig. 2.17 is given in Fig. 4.9b. From this figure, the transfer function of the overall system is obtained as

$$\frac{\theta(s)}{V_c(s)} = \frac{K_d K_q}{R_c R_q (1 + s\tau_c)(1 + s\tau_q)} \times \frac{K_T}{s[R_a(Js + f_0) + K_T K_b]}$$

$$= \frac{K}{s(1 + s\tau_c)(1 + s\tau_q)(1 + s\tau_m)} \qquad (4.9)$$

where $K = \dfrac{K_T K_d K_q}{R_c R_q R_a f}$; $\tau_m = J/f$ = motor time constant; and $f = f_0 + \dfrac{K_T K_b}{R_a}$

= effective viscous friction coefficient.

From the above analysis it is seen that the amplidyne can be considered as a two-stage power amplifier. The first stage amplification is between control winding and quadrature axis circuit and the second stage between quadrature axis circuit and direct axis circuit.

Power amplification factor of first stage = $\dfrac{e_q i_q}{v_c i_c}$

Power amplification factor of second stage = $\dfrac{e_d i_a}{e_q i_q}$

Usually the power amplification factor of the first stage is more than that of the second stage, approximate values being 200 and 50, respectively. The overall power amplification factor is of the order of 10^4. The effective time constant of the amplidyne is usually small and is of the order of 0.05 sec.

SYNCHROS

A *synchro* is an electromagnetic *transducer* commonly used to convert an angular position of a shaft into an electric signal. It is commercially known as a *selsyn* or an *autosyn*.

The basic synchro unit is usually called a *synchro transmitter*. Its construction is similar to that of a three-phase alternator. The stator (stationary member) is of laminated silicon steel and is slotted to accommodate a balanced three-phase winding which is usually of concentric coil type (three identical coils are placed in the stator with their axis 120° apart) and is Y-connected. The rotor is of dumb-bell construction and is wound with a concentric coil. An a.c. voltage is applied to the rotor winding through slip rings. The constructional features and schematic diagram of a synchro transmitter are shown in Figs. 4.10 and 4.11, respectively.

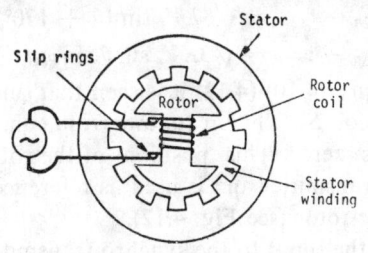

Figure 4.10 Constructional features of synchro transmitter.

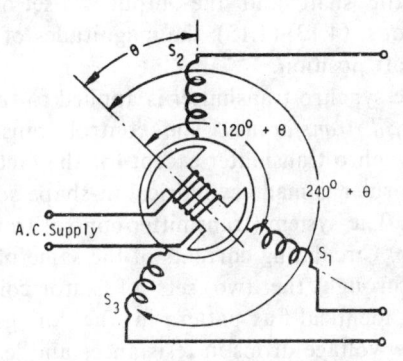

Figure 4.11 Schematic diagram of synchro transmitter.

Let an a.c. voltage
$$v_r(t) = V_r \sin \omega_c t$$
be applied to the rotor of the synchro transmitter as shown in Fig. 4.11. This voltage causes a flow of magnetizing current in the rotor coil which produces a sinusoidally time varying flux directed along its axis and distributed nearly sinusoidally in the air gap along the stator periphery. Because of transformer action, voltages are induced in each of the stator coils. As the air gap flux is sinusoidally distributed, the flux linking any stator coil is proportional to the cosine of the angle between the rotor and stator coil axes and so is the voltage induced in each stator coil. The stator coil voltages are of course in time phase with each other. Thus we see that the synchro transmitter acts like a single-phase transformer in which the rotor coil is the primary and the stator coils form the three secondaries.

Let v_{s1n}, v_{s2n}, and v_{s3n}, respectively be the voltages induced in the stator coils S_1, S_2 and S_3 with respect to the neutral. Then, for the rotor position of the synchro transmitter shown in Fig. 4.11 where the rotor axis makes an angle θ with the axis of the stator coil S_2

$$v_{s1n} = KV_r \sin \omega_c t \cos (\theta + 120°) \tag{4.10}$$

$$v_{s2n} = KV_r \sin \omega_c t \cos \theta \tag{4.11}$$

$$v_{s3n} = KV_r \sin \omega_c t \cos[(\theta + 240°) \tag{4.12}$$

The three terminal voltages of the stator are

$$v_{s_1s_2} = v_{s_1n} - v_{s_2n} = \sqrt{3}KV_r \sin(\theta + 240°) \sin \omega_c t \quad (4.13)$$

$$v_{s_2s_3} = v_{s_2n} - v_{s_3n} = \sqrt{3}KV_r \sin(\theta + 120°) \sin \omega_c t \quad (4.14)$$

$$v_{s_3s_1} = v_{s_3n} - v_{s_1n} = \sqrt{3}KV_r \sin \theta \sin \omega_c t \quad (4.15)$$

When $\theta = 0$, from eqns. (4.10)-(4.12) it is seen that maximum voltage is induced in the stator coil S_2, while it follows from eqn. (4.15) that the terminal voltage $v_{s_3s_1}$ is zero. This position of the rotor is defined as the *electrical zero* of the transmitter and is used as reference for specifying the angular position of the rotor (see Fig. 4.12).

Thus it is seen that the input to the synchro transmitter is the angular position of its rotor shaft and the output is a set of three single-phase voltages given by eqns. (4.13)-(4.15); the magnitudes of these voltages are functions of the shaft position.

The output of the synchro transmitter is applied to the stator windings of a *synchro control transformer*. The control transformer is similar in construction to a synchro transmitter except for the fact that the rotor of the control transformer is made cylindrical in shape so that the air gap is practically uniform. The system (transmitter-control transformer pair) acts as an error detector. Circulating currents of the same phase but of different magnitudes flow through the two sets of stator coils. The result is the establishment of an identical flux pattern in the air gap of the control transformer as the voltage drops in resistances and leakage reactances of the two sets of stator coils are usually small. The control transformer flux axis thus being in the same position as that of the synchro transmitter rotor, the voltage induced in the control transformer rotor is proportional to the cosine of the angle between the two rotors and is given by

$$e(t) = K'V_r \cos \phi \sin \omega_c t \quad (4.16)$$

where ϕ is the angular displacement between the two rotors. When $\phi = 90°$, i.e., the two rotors are at right angles, then the voltage induced in the control transformer rotor is zero. This position is known as the *electrical zero* position of the control transformer. In Fig. 4.12, the transmitter and control transformer rotors are shown in their respective electrical zero positions. Let the rotor of the transmitter rotate through an angle θ in the

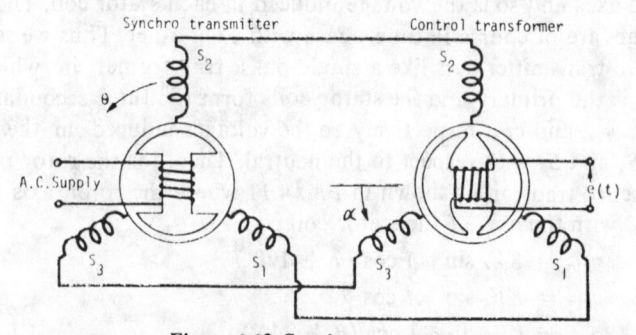

Figure 4.12 Synchro error detector.

direction indicated and let the control transformer rotor rotate in the same direction through an angle α resulting in a net angular separation of $\phi = (90° - \theta + \alpha)$ between the two rotors. From eqn. (4.16), the voltage at the rotor terminals of the control transformer is then

$$e(t) = K'V_r \sin(\theta - \alpha) \sin \omega_c t \tag{4.17}$$

For small angular displacement between the two rotor positions,

$$e(t) = K'V_r(\theta - \alpha) \sin \omega_c t \tag{4.18}$$

The synchro transmitter-control transformer pair thus acts as an error detector giving a voltage signal at the rotor terminals of the control transformer proportional to the angular difference between the transmitter and control transformer shaft positions. Equation (4.18), though derived for constant $(\theta - \alpha)$, is valid for varying conditions as well, so long as the rate of angle change is small enough for the speed voltages induced in the device to be negligible.

Equation (4.18) is represented graphically in Fig. 4.13 for an arbitrary time variation of $(\theta - \alpha)$.

We see from this diagram that the output of synchro error detector is a modulated signal, the modulating wave has the information regarding the lack of correspondence between the two rotor positions and the carrier wave is the a.c. input to the rotor of the synchro transmitter. This type of modulation is known as *suppressed-carrier modulation*. From eqn. (4.18)

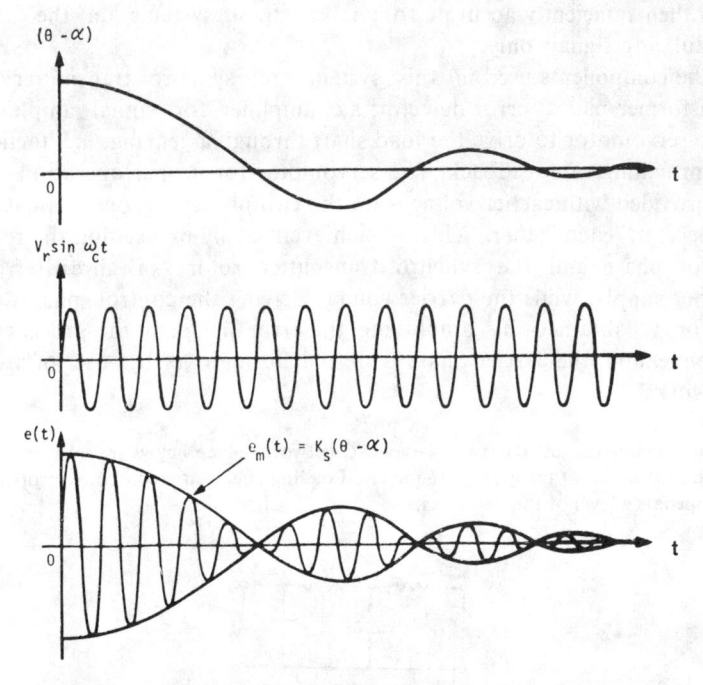

Figure 4.13 Typical wave forms of synchro error detector.

the modulating signal representing the discrepancy between the two shaft positions is

$$e_m(t) = K_s(\theta - \alpha)$$

where K_s is known as *sensitivity* of error detector and has the units of volts (rms)/rad angular difference of the shafts of the synchro pair.

As pointed out earlier the rotor of the control transformer is made cylindrical in shape so that the air gap is practically uniform. This is essential for a control transformer, since its rotor terminals are usually connected to an amplifier, therefore the change in the rotor output impedance with rotation of the shaft must be minimized. Another distinguishing feature is that the stator winding of the control transformer has a higher impedance per phase. This feature permits several control transformers to be fed from a single transmitter.

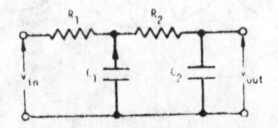

A.C. Position Control System

Consider the system shown in Fig. 4.14 in which the position of the mechanical load is controlled in accordance with the position of the reference shaft. This system employs a.c. components and all the signals other than the input and output shaft positions are suppressed-carrier modulated signals. Such systems are known as *carrier control systems* and are designed so that the signal cut-off frequency is much less than the carrier frequency. It is then sufficiently accurate to analyze these systems on the basis of modulating signals only.

The components used in this system are: synchro transmitter-control transformer pair as error detector; a.c. amplifier for signal amplification; a.c. servomotor to drive the load shaft through a gearing; a.c. tachometer for providing rate feedback. The servomotor for proper operation has to be provided with carrier voltages on the two phases 90° out of phase with respect to each other. This is achieved easily by exciting the reference motor phase and the synchro transmitter rotor coil directly from the carrier supply, while the carrier voltage driving the control phase of the motor is obtained by amplifying the error signal. In the process of this amplification, the carrier phase is shifted through 90° by use of two *RC* networks*.

---

*The parameters of the two *L*-type *RC* networks given below are selected to give a phase shift of 90° at the carrier frequency. Loading effects are reduced by proper choice of impedance level of the networks.

Control Systems and Components 97

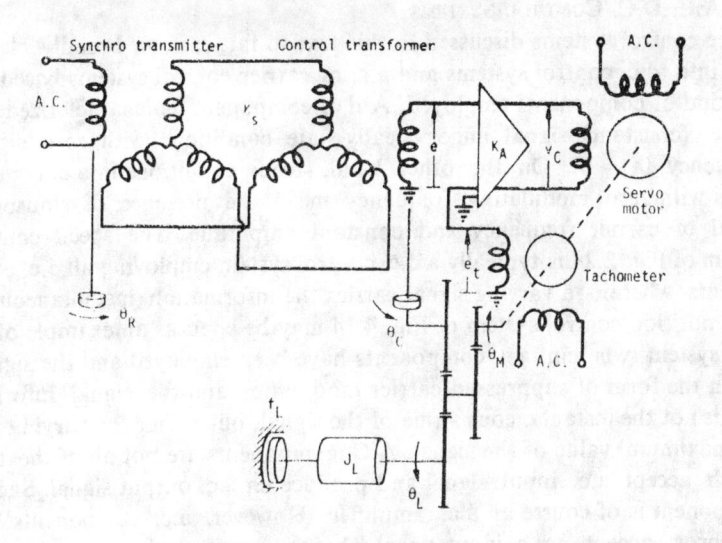

Figure 4.14 A.C. position control system.

Linearized transfer functions derived earlier in this chapter are valid for the various components of this system as the signals are always small, it being a position tracking system. The operating point of the a.c. servomotor is ($v_C = 0$, $\theta_M = 0$) as the motor is stationary under steady conditions. Referring to the component transfer functions, the signal flow graph of Fig. 4.15 can be at once drawn. The overall transfer function of the system obtained therefrom is

$$\frac{\theta_C(s)}{\theta_R(s)} = \frac{K_m K_A K_s n}{\tau_m s^2 + (1 + K_m K_A K_t)s + K_m K_A K_s n} \quad (4.19)$$

where K_m = motor gain constant in rad/volt; τ_m = motor time constant in sec; K_s = synchro sensitivity in volts/rad; K_t = tachometer constant in volts/rad/sec; K_A = amplifier gain in volts/volt; and n (gear ratio) = θ_C/θ_M = θ_L/θ_M.

The tachometer feedback is of course employed negatively and as will be seen in Chapter 5, it improves the damping factor of the closed-loop system.

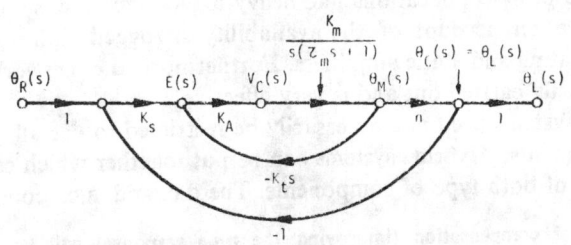

Figure 4.15 Signal flow graph of the system shown in Fig. 4.14.

THE A.C.-D.C. CONTROL SYSTEMS

The control systems discussed in this text so far, can be broadly classified into d.c. control systems and a.c. or carrier control systems based on the kind of components employed. A d.c. component is characterized by a d.c. (constant) signal under steady-state conditions with zero signal frequency ($s = 0$). On the other hand, steady conditions in a.c. components with zero modulating frequency mean the presence of sinusoidal signal of carrier frequency and constant amplitude. The speed control system of Fig. 2.28 is typically a d.c. control system employing all d.c. components wherein a varying signal carries the information instantaneously. The position control system of Fig. 4.14 may be cited as an example of an a.c. system wherein a.c. components have been employed and the signals are in the form of suppressed-carrier modulation and the signal information is not the instantaneous value of the signal, but rather the varying rms (or maximum) value of the signal. A.C. components are not all of the type which accept a.c. input signal and produce an a.c. output signal. Such a component is of course an a.c. amplifier. However, a.c. components like synchros, accept a.d.c. input signal (shaft angular position), are provided with a carrier supply and produce a carrier modulated signal; while components like a.c. servomotors accept carrier modulated input signal at the control phase, are provided with carrier supply at the reference phase and produce a demodulated (d.c.) output signal in the form of an angular position or velocity.

The choice between d.c. and a.c. systems is chiefly based on economic considerations, ease of operations and maintenance and considerations of size and weight particularly for air-borne systems. One of the advantages of a.c. systems is the use of an a.c. amplifier which is inherently stable unlike the d.c. amplifier which suffers from the possibility of drift. Another major advantage of a.c. systems is the smaller frame size for a.c. components particularly for high carrier frequency. Even though network compensation* is not very convenient, tachometer feedback compensation can be very effective in a.c. systems. On account of weight and size considerations, a.c. systems are preferred for aircraft and missile systems and on account of low-cost a.c. amplifier required for such systems, they are also preferred for instrumentation and low power ground applications; e.g., the servo multipliers used in analog computers are basically a.c. position control systems.

For large power applications like heavy drives, etc., d.c. systems are an ideal choice on account of the availability of rugged high power amplification rotating and static amplifiers. Furthermore, d.c. network compensation is easily carried out and is very effective.

Control systems need not necessarily be restricted to use all d.c. or all a.c. components. Hybrid systems can be put together which combine the advantages of both type of components. The d.c. and a.c. components in

*The idea of compensation (improving the time response) will be discussed in Chapter 10.

a system have to be coupled together through modulators or demodulators so that they receive and put out the kind of signal desired.

4.4 Stepper Motor

A stepper motor is an electromechanical device which actuates a train of step angular (or linear) movements in response to a train of input pulses on one to one basis—one step actuation for each pulse input. A stepper motor is the actuator element of incremental motion control systems—computer peripherals like printers, tape drives, capstan drives etc., machine tool and process control systems.

The two most widely used types of stepper motors are:
1. Variable reluctance motor
2. Permanent magnet motor

Here we shall describe the basic operational features of a variable reluctance motor.

A variable reluctance stepper motor consists of a single or several stacks of stators and rotors—stators have a common frame and rotors have a common shaft as shown in the longitudinal cross-sectional view of Fig. 4.16

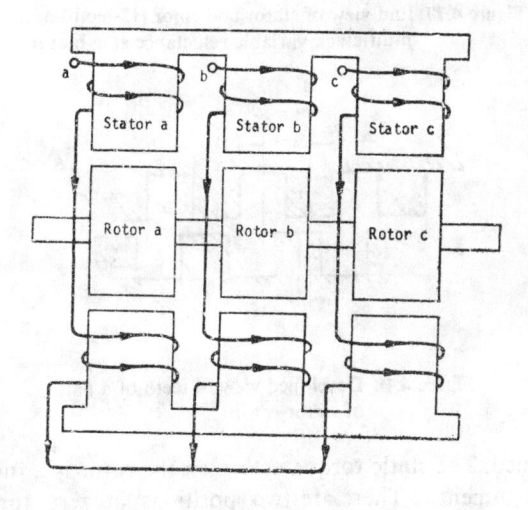

Figure 4.16 Longitudinal cross-sectional view of 3-stack variable reluctance motor.

for a 3-stack motor. Both stators and rotors have toothed structure as shown in the end view of Fig. 4.17. The stator and rotor teeth are of same size and therefore can be aligned as shown in this figure. The stators are pulse excited, while the rotors are unexcited.

Consider a particular stator and rotor set shown in the developed diagram of Fig. 4.18. As the stator is excited, the rotor is pulled into the nearest minimum reluctance position—the position where stator and rotor

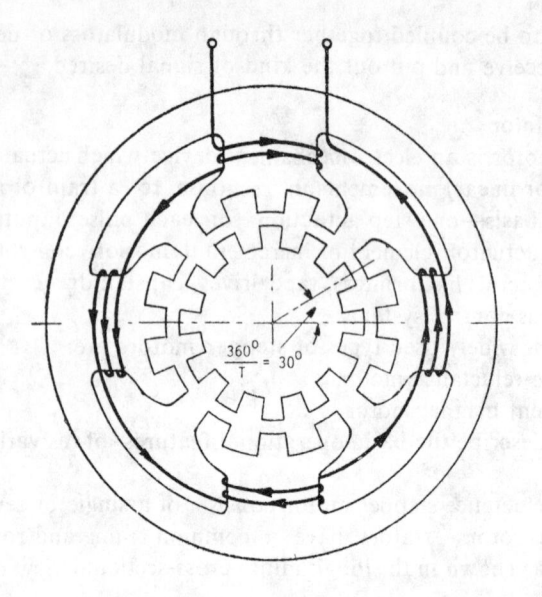

Figure 4.17 End view of stator and rotor (12-teeth) of a multistack variable reluctance step motor.

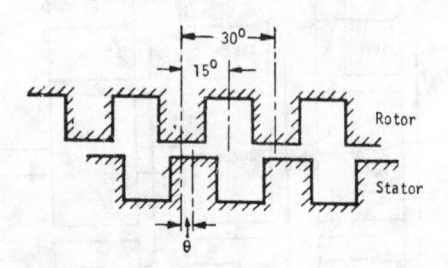

Figure 4.18 Developed view of teeth of a pair of stator-rotor.

teeth are aligned. The static torque acting on the rotor is a function of the angular misalignment θ. There are two positions of zero torque: $\theta = 0$, rotor and stator teeth aligned and $\theta = 360°/(2 \times T) = 180°/T$ (T = number of rotor teeth), rotor teeth aligned with stator slots. The shape of the static torque-angle curve of one stack of a stepper motor is shown in Fig. 4.19. Teeth aligned position ($\theta = 0$) is a stable position, i.e., slight disturbance from this position in either direction brings the rotor back to it. Tooth-slot aligned position ($\theta = 180°/T$) is unstable, i.e., slight disturbance from this position in either direction makes the rotor move away from it. The rotor thus locks into stator in position $\theta = 0$ (or multiple of $360°/T$). The dynamic torque-speed characteristic will differ from this due to speed emf induced in stator coils.

Control Systems and Components

While the teeth on all the rotors are perfectly aligned, stator teeth of various stacks differ by an angular displacement of

$$\alpha = \frac{360°}{nT} \qquad (4.20)$$

where n = number of stacks.

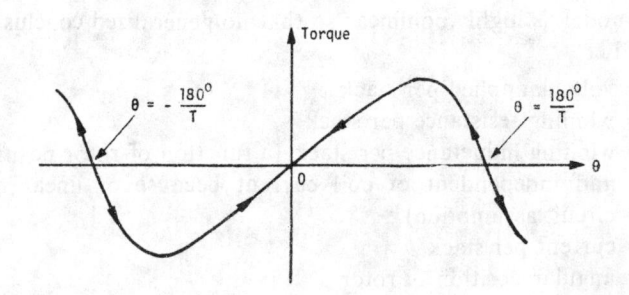

Figure 4.19 Static torque-angle curve of a stepper motor stack.

Figure 4.20 shows the developed diagram of a 3-stack stepper motor with rotor in such a position that stack c rotor teeth are aligned with its stator. Here

$$\alpha = \frac{360°}{3 \times 12} = 10° \text{ (number of rotor teeth} = 12)$$

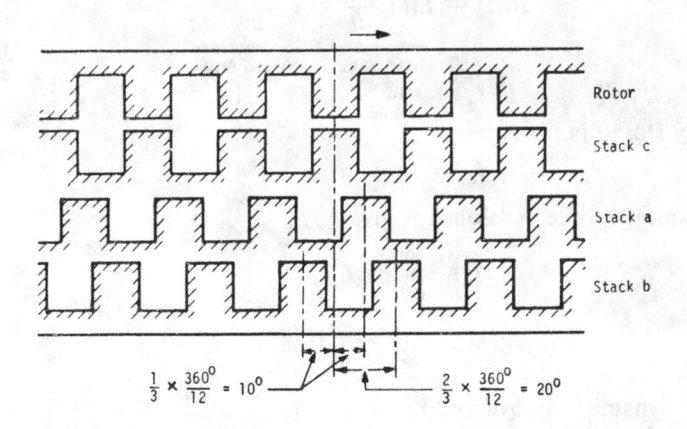

Figure 4.20 Developed view of rotor and stator stacks of a 3-stack stepper motor.

In a multiple stack rotor, number of phases equals number of stacks. If phase a stator is pulse excited, the rotor will move by 10° in the direction indicated. If instead phase b is excited, the rotor will move by 10° opposite to the direction indicated. Pulse train with sequence *abcab* will make the rotor go through incremental motion in indicated direction, while sequence *bacba* will make it move in opposite direction. Directional control is only possible with three or more phases.

Using basic principles of electromechanical energy conversion, a simplified analysis of one stack of the stepper motor is presented here. It is assumed that the magnetic circuit is linear (unsaturated). Even then the resulting model is highly nonlinear so that no generalized conclusions can be drawn. Let

$e(t)$ = voltage applied per stack
R = winding resistance per stack
$L(\theta)$ = winding inductance per stack (a function of rotor position only and independent of coil current because of linear magnetic circuit assumption)
$i(t)$ = current per stack
$\theta(t)$ = angular position of rotor

Kirchoff's mesh equation for stator winding is

$$e(t) = Ri(t) + \frac{d\lambda}{dt}$$

where λ = flux linkages of stator winding = $iL(\theta)$.
Therefore,

$$e(t) = Ri(t) + L(\theta)\frac{di}{dt} + i\frac{dL(\theta)}{dt}$$

$$= Ri(t) + \underbrace{L(\theta)\frac{di}{dt}}_{\text{Transformer emf}} + \underbrace{i\frac{dL(\theta)}{d\theta}\frac{d\theta}{dt}}_{\text{Speed emf}} \quad (4.21)$$

Energy stored in air gap is

$$W = \tfrac{1}{2}L(\theta)i^2(t) \quad (4.22)$$

Mechanical torque developed is given by

$$T = \frac{\partial}{\partial \theta}W(i, \theta)$$

$$= \frac{1}{2}i^2(t)\frac{dL(\theta)}{d\theta} \quad (4.23)$$

Rotor dynamics is governed by

$$T = J\frac{d^2\theta}{dt^2} + f\frac{d\theta}{dt} \quad (4.24)$$

In a toothed structure, reluctance and therefore winding inductance varies cosinusoidally (even function) as function of θ over and above an average value, i.e.,

$$L(\theta) = L_1 + L_2 \cos T\theta \quad (4.25)$$

Substituting in eqn. (4.23),

$$T = -\frac{1}{2}L_2 Ti^2(t) \sin T\theta$$

$$= -Ki^2(t) \sin T\theta \qquad (4.26)$$

This indeed is the reluctance torque and has sinusoidal form compared to the torque-angle curve of Fig. 4.19. Notice the similarity of the overall shape of the curve to a sinusoidal wave.

Equations (4.21), (4.26) and (4.24) govern the dynamic behaviour of one stack of a stepper motor under application of $e(t)$, a pulse wave shape. Assuming that mechanical and electrical transients are over in the time intervening between pulses, single-stack analysis would suffice. Equations being highly nonlinear, approximate linear model is not feasible. Solution must be obtained on an analog or digital computer.

Intervening time between the application of two pulses must be long enough to allow for the motor rotor to lock in position otherwise the motor may skip a step which is undesirable. Step skipping may also occur if rotor oscillation amplitude about the locking position is too large.

USE OF STEPPER MOTORS IN CONTROL SYSTEMS

There are two distinctly different ways of using stepper motors in control systems. One is the open-loop mode and other is the closed-loop mode.

As has been discussed earlier, the stepper motor is a digital device whose output in shaft angular displacement is completely determined by the number of input pulses. Consequently, there is no need for a feedback device to determine the position of motor shaft and therefore, of the load connected to the motor shaft. This means that an open-loop step servo system can be designed to yield the same accuracy as a closed-loop analog system. This leads to simplified system design. Figure 4.21 illustrates the use of stepper motor in open-loop mode.

Figure 4.21 Use of stepper motor in open-loop mode.

In the closed-loop (positional feedback) mode, the motor is used like a conventional servo motor. A signal from the output is fedback and is used to operate a gate controlling the pulses from a pulse generator (Fig. 4.22).

Since stepper motors are essentially digital actuators, there is no need to use digital-to-analog and analog-to-digital converters in constructing digital control systems (refer Chapter 11).

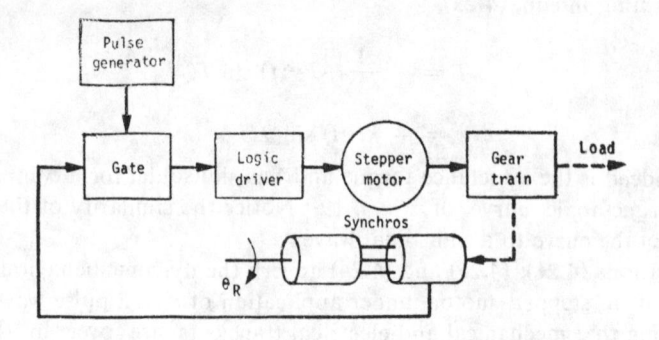

Figure 4.22 Use of stepper motor in closed-loop mode.

4.5 Hydraulic Systems

Hydraulically operated components are frequently used in hydraulic feedback systems and in combined electromechanical-hydraulic systems. In these elements, power is transmitted through the action of fluid flow under pressure. The fluid used is relatively incompressible such as petroleum-base oils or certain non-inflammable synthetic fluids.

The main advantage of the hydraulic systems lies in the hydraulic motor which can be made much smaller in physical size than an electric motor for the same power output. In addition to this, hydraulic components are rapidly acting and more rugged compared to the corresponding electrical components. On the other hand, hydraulic systems have the inherent problems of leak and of sealing them against foreign particles, operating noise and the tendency to become sluggish at low temperatures because of increased viscosity of the fluid. Furthermore, hydraulic lines are not as flexible as electric cables.

Common hydraulic control applications are power steering and brakes in automobiles, the steering mechanisms of large ships, the control of large machine tools, etc.

The hydraulic output devices used in control systems are generally of two types—those intended to produce rotary motion and those whose output is translational. The first type of devices are known as hydraulic motors and the second ones are known as hydraulic linear actuators.

Some of the important hydraulic components are discussed in this section.

HYDRAULIC PUMPS AND MOTORS

A device frequently used in control systems giving large output torque and short response time is the *hydraulic transmission* shown in Fig. 4.23. It consists of a variable stroke hydraulic pump and a fixed stroke hydraulic motor. Control of the motor is exercised by varying the amount of oil delivered by the pump. This is carried out by mechanically changing the pump stroke. Like in a d.c. generator and motor, there is no essential difference between hydraulic pump and motor. In a pump, the input is mechanical power (torque at a certain speed) and output, hydraulic power (flow at a certain pressure) and in a motor, the input is hydraulic and output mechanical.

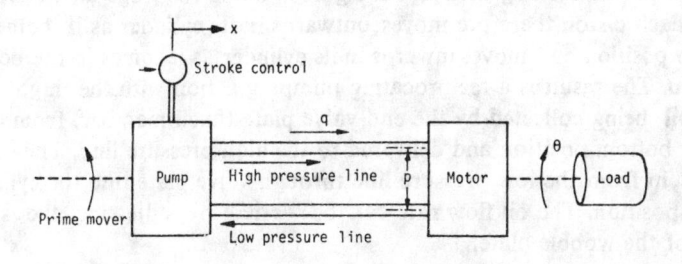

Figure 4.23 Hydraulic transmission.

Figure 4.24 shows the constructional features of hydraulic pump and motor. The pistons in hydraulic pump bear against the stationary wobble plate and are carried round by the cylinder block which is made to rotate by the prime mover. (In some designs, the cylinder block is kept stationary and the wobble plate rotated.) First consider that the wobble plate of the pump is in the neutral position. Now as the shaft is driven, the

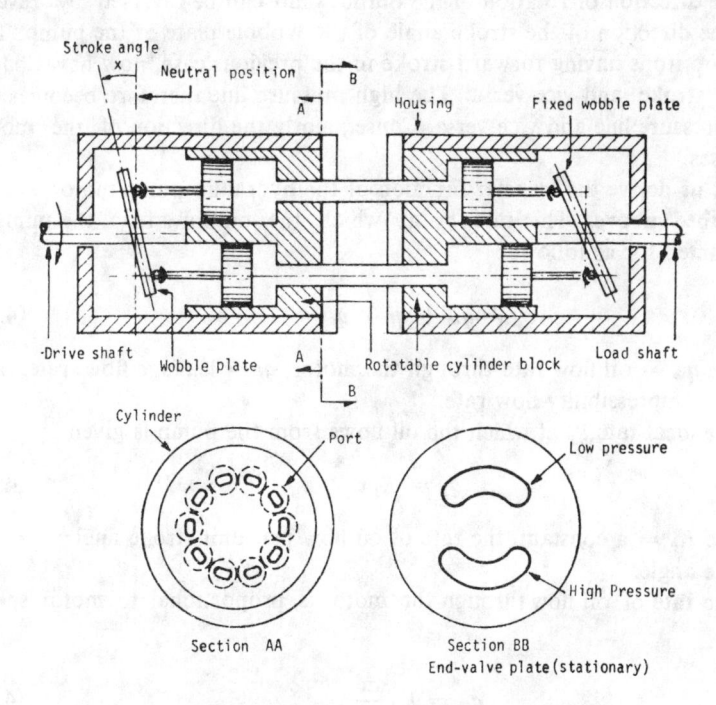

Figure 4.24 Constructional details of hydraulic transmission system.

cylinder block with pistons rotates, but there is no displacement of the pistons into the cylinders and hence no pumping action takes place. Let the wobble plate be now tilted (Fig. 4.24). As the cylinder block is now rotated, the pistons rotate along with and are guided along the ridge of the wobble plate. Each piston therefore moves outwards in its cylinder as it comes in the top position and moves inwards in its cylinder as it comes in the bottom position. The result is a reciprocating pumping action with the high pressure oil being collected by the end-valve plate (fixed member) from cylinders in bottom position and delivered to the high pressure line. The oil is drawn in from the low pressure line through valve plate into the cylinders in top position. The oil flow rate can be varied by adjusting the stroke angle of the wobble plate.

From the pump, the oil is forced under pressure into the high pressure inlet of the end-valve plate of the motor. It exerts a force against the piston that may be resolved into a component normal to the fixed wobble plate and other in the plane of the wobble plate. The normal force is opposed by bearings but the other force causes a torque around the output shaft of the hydraulic motor; thus the output shaft velocity is approximately proportional to flow of oil and hence stroke angle of the wobble plate of the pump. The pump constantly pushes oil into the high pressure line while it constantly receives oil from the low pressure line. Thus oil is merely circulated from pump to motor and back again.

The direction of rotation of the output shaft can be reversed by reversing the direction of the stroke angle of the wobble plate of the pump. The pump pistons having forward stroke in the previous case, now have backward stroke and vice-versa. The high pressure line therefore becomes the low pressure line and vice-versa. Consequently the direction of the motor reverses.

Let us derive the transfer function of the hydraulic pump-motor system described above. The rate q_p at which the oil flows from the pump is accounted for as follows :

$$q_p = q_m + q_l + q_c \qquad (4.27)$$

where q_m = oil flow rate through the motor; q_l = leakage flow rate; and q_c = compressibility flow rate.

The ideal rate q_p at which the oil flows from the pump is given by

$$q_p = K_p x \qquad (4.28)$$

where K_p = a constant, the rate of oil flow per unit stroke angle; x = the stroke angle.

The rate of oil flow through the motor is proportional to motor speed, i.e.,

$$q_m = K_m \frac{d\theta}{dt} \qquad (4.29)$$

Control Systems and Components

where θ = angle through which motor turns; and K_m = motor displacement constant.

All the oil from the pump does not flow through the motor in the proper channels. Due to back pressure in the motor, which is built up by the hydraulic flow to overcome the resistance to free movement offered by load on motor shaft, a portion of the ideal flow from the pump leaks back past the pistons of motor and pump. It is usually assumed that the leakage flow is proportional to motor pressure, i.e.,

$$q_l = K_l p \tag{4.30}$$

where K_l = constant; p = pressure drop across the motor = pressure difference between high and low pressure lines.

The back pressure built up by the motor not only causes leakage flow in the motor and pump as well, but also causes the oil in the lines to compress. Compressibility of the oil may usually be considered negligible except when the oil is at very high pressure and when there is a large quantity of oil in the system (refer foot-note on p. 25).

Volume compressibility flow is essentially proportional to pressure and therefore the rate of compressibility flow is proportional to the rate of change of pressure, i.e.,

$$q_c = K_c \frac{dp}{dt} \tag{4.31}$$

where K_c = coefficient of compressibility.

Substituting the values of various flow rates in eqn. (4.27), the flow rate equation can be written as

$$K_p x = K_m \frac{d\theta}{dt} + K_l p + K_c \frac{dp}{dt} \tag{4.32}$$

The torque T_M developed by the motor is proportional to the pressure drop and balances the load torque. If the load is assumed to consist of moment of inertia J and viscous friction with coefficient f, the torque equation is

$$T_M = K_T p = J \frac{d^2\theta}{dt^2} + f \frac{d\theta}{dt} \tag{4.33}$$

where K_T = motor torque constant.

If the motor losses are included as part of load, the hydraulic input power $(q_m p)$ must equal the mechanical output power $[T_M(d\theta/dt)]$, i.e.,

$$q_m p = T_M \frac{d\theta}{dt} \tag{4.34}$$

or

$$K_m \frac{d\theta}{dt} p = K_T p \frac{d\theta}{dt}$$

or

$$K_m = K_T \text{ (in MKS units)} \tag{4.35}$$

Equation (4.33), now becomes

$$K_m p = J\ddot{\theta} + f\dot{\theta} \tag{4.36}$$

From eqns. (4.32) and (4.36), we get

$$K_p x = \frac{K_c J}{K_m}\ddot{\theta} + \left(\frac{K_c f}{K_m} + \frac{K_l J}{K_m}\right)\ddot{\theta} + \left(\frac{K_{lf}}{K_m} + K_m\right)\dot{\theta} \qquad (4.37)$$

Taking the Laplace transform of eqn. (4.37), we get

$$G(s) = \frac{\theta(s)}{X(s)} = \frac{K_p}{s\left[\frac{K_c J}{K_m}s^2 + \left(\frac{K_c f}{K_m} + \frac{K_l J}{K_m}\right)s + \left(K_m + \frac{K_{lf}}{K_m}\right)\right]} \qquad (4.38)$$

Normally $K_c \ll K_m$ and therefore eqn. (4.38) may be simplified as

$$G(s) = \frac{\theta(s)}{X(s)} = \frac{K_p}{s\left[\frac{K_l J}{K_m}s + \left(K_m + \frac{K_{lf}}{K_m}\right)\right]}$$

$$= \frac{K}{s(\tau s + 1)} \qquad (4.39)$$

where
$$K = \frac{K_p}{K_m + \frac{K_{lf}}{K_m}}; \quad \tau = \frac{J K_l}{K_m^2 + K_{lf}}$$

It is observed that the transfer function of hydraulic transmission is of the same form as that of an electric motor given by eqn. (4.5). Further, the simplified transfer function of eqn. (4.39) can be obtained directly by neglecting the compressibility flow in eqn. (4.27).

HYDRAULIC VALVES

In place of the variable displacement pump, one frequently used device is a constant pressure source and a valve to control the flow of oil to the hydraulic motor. This system has the advantage that the movable parts of the valve can be made much lighter than those of the stroke-control mechanism in the pump, as a result of which, the time constants are reduced and the system becomes fast acting. The disadvantage of this system is that the valve has nonlinear characteristics and introduces problems of its own.

Figure 4.25 shows a three-way spool valve controlling the oil flow to the hydraulic motor. When the spool is in the neutral position ($x = 0$), the oil flow is blocked completely. As the spool is moved downwards (x positive), the upper pipe line to the motor gets connected to the high pressure source and the lower pipe line to the sump causing the motor to rotate in a particular direction. If the spool is moved upward (x negative), the lower pipe line to the motor gets connected to the high pressure source and the upper pipe line to the sump. The direction of oil flow in the motor and hence direction of its rotation is reversed.

Because turbulent flow exists at a sharp-edged orifice such as a valve, the flow equation is (refer section 2.2)

$$q_1 = Kx\sqrt{P_0 - P_1}; \quad x > 0$$

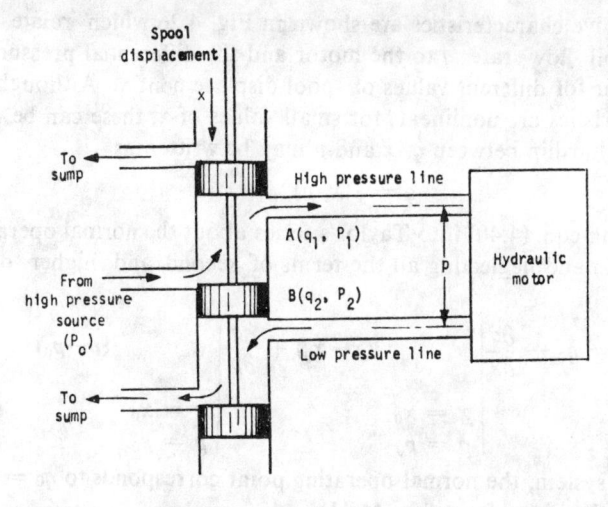

Figure 4.25 Hydraulic valve.

where q_1 is the rate of flow through port A and K is flow coefficient which is a function of valve and fluid properties.

In addition, it follows that

$$q_2 = Kx\sqrt{P_2}; \quad x > 0$$

where q_2 is the flow rate at port B.

Since $q_1 = q_2 = q$, we have from above equations

$$q = \frac{Kx}{\sqrt{2}}\sqrt{P_0 - p}; \quad x > 0$$

where $p = P_1 - P_2$.

For $x < 0$, the relationship between p, q and x may be obtained in a similar manner.

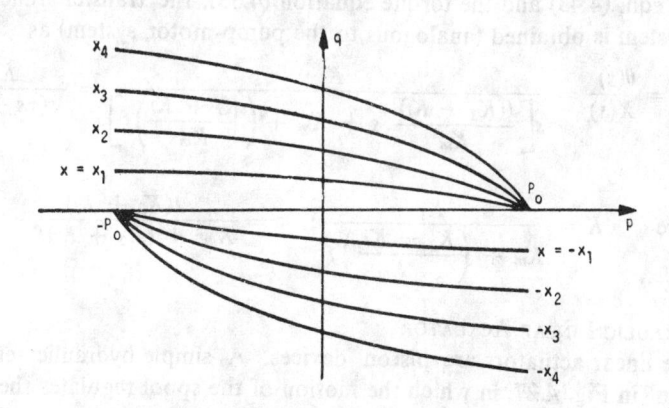

Figure 4.26 Typical valve characteristics.

The valve characteristics are shown in Fig. 4.26 which relate the volumetric oil flow rate q to the motor and the differential pressure p across the motor for different values of spool displacement x. Although the valve characteristics are nonlinear, for small values of x, these can be linearized. The relationship between q, x and p may be written as

$$q = f(x, p) \tag{4.40}$$

Expanding eqn. (4.40) into Taylor's series about the normal operating point (q_0, x_0, p_0) and neglecting all the terms of second and higher derivatives, we get

$$q = q_0 + \left.\frac{\partial q}{\partial x}\right|_{\substack{x=x_0 \\ p=p_0}} (x-x_0) + \left.\frac{\partial q}{\partial p}\right|_{\substack{x=x_0 \\ p=p_0}} (p-p_0) \tag{4.41}$$

For this system, the normal operating point corresponds to $q_0 = 0$, $p_0 = 0$, $x_0 = 0$, therefore, from eqn. (4.41)

$$q = K_1 x - K_2 p \tag{4.42}$$

where $$K_1 = \left.\frac{\partial q}{\partial x}\right|_{\substack{x=0 \\ p=0}} \quad ; \quad K_2 = -\left.\frac{\partial q}{\partial p}\right|_{\substack{x=0 \\ p=0}}$$

Equation (4.42) gives a linearized relationship among q, x and p.

The response equation of the valve-motor combination can now be written down from eqn. (4.32) of the pump-motor system by replacing its left-hand side by $q = K_1 x - K_2 p$, the inflow rate to the pipe-line-motor combination. Furthermore the compressibility is neglected right at this stage. Thus

$$q = K_1 x - K_2 p = K_m \dot{\theta} + K_l p \tag{4.43}$$

From eqn. (4.43) and the torque equation (4.33), the transfer function of the system is obtained (analogous to the pump-motor system) as

$$G(s) = \frac{\theta(s)}{X(s)} = \frac{K_1}{s\left[\frac{J(K_2 + K_l)}{K_m} s + K_m + \left(\frac{K_2 + K_l}{K_m}\right)f\right]} = \frac{K}{s(\tau s + 1)} \tag{4.44}$$

where $$K = \frac{K_1}{K_m + \left(\frac{K_2 + K_l}{K_m}\right)f} \quad ; \quad \tau = \frac{J(K_2 + K_l)}{K_m^2 + (K_2 + K_l)f}.$$

HYDRAULIC LINEAR ACTUATOR

The linear actuators are piston devices. A simple hydraulic actuator is shown in Fig. 4.27, in which the motion of the spool regulates the flow of oil to either side of the power cylinder. When the spool moves to the right,

Control Systems and Components

the oil from the high pressure source enters into the power cylinder, on the left of the power piston. This creates a differential pressure across the piston which causes the power piston to move to the right, pushing the oil in front of it to the sump. The oil is pressurized by a pump and is recirculated in the system. The load rigidly coupled to the piston moves a distance y from its reference position in response to the displacement x of the valve spool from its neutral position.

Let us first carry out a *simplified analysis* of the linear actuator. The relationship between the volumetric oil flow rate q into the power piston and the differential pressure p across the piston for small values of spool displacement x, has already been given in eqn. (4.42) and can be rewritten in the form given below:

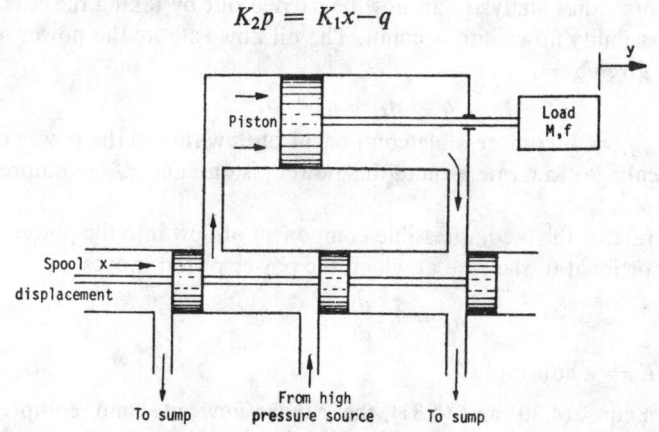

$$K_2 p = K_1 x - q$$

Figure 4.27 Hydraulic linear actuator.

Assuming leakage and compressibility flows to be negligible, the rate of oil flow into the piston is proportional to the rate at which the piston moves, i.e.,

$$q = A \frac{dy}{dt} \quad (4.45)$$

where A is the area of the piston.

The force on the piston is (Ap) which moves the load consisting of mass M and viscous friction with coefficient f, according to the differential equation

$$Ap = M\ddot{y} + f\dot{y} \quad (4.46)$$

From eqns. (4.42), (4.45) and (4.46) we can draw the signal flow graph given in Fig. 4.28.

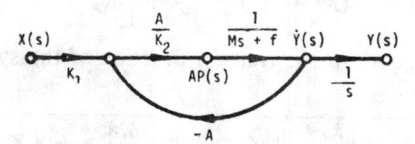

Figure 4.28 Signal flow graph of hydraulic actuator.

The transfer function, as obtained from the signal flow graph, is

$$\frac{Y(s)}{X(s)} = \frac{A(K_1/K_2)}{s\left[Ms + \left(f + \frac{A^2}{K_2}\right)\right]} \qquad (4.47)$$

$$= \frac{K}{s(\tau s + 1)}$$

where $K = \dfrac{AK_1}{A^2 + fK_2}$; $\tau = \dfrac{MK_2}{A^2 + fK_2}$.

It is found that the transfer function of the linear actuator is similar to that of hydraulic and electric motors and contains in it the term $1/s$ which accounts for the integrating action of the device.

A more exact analysis can now be carried out by taking the leakage and compressibility flows into account. The oil flow rate in the power cylinder is then given by

$$q = q_p + q_l + q_c \qquad (4.48)$$

where q_p = incompressible component of flow rate to the power cylinder; q_l = leakage flow rate around the power piston; and q_c = compressibility flow rate.

The rate of the incompressible component of flow into the power cylinder is proportional to the rate at which the power piston moves, i.e.,

$$q_p = K_p \frac{dy}{dt} \qquad (4.49)$$

where K_p = a constant.

From eqns. (4.30) and (4.31), the leakage flow rate and compressibility flow rate are given as

$$q_l = K_l p \qquad (4.50)$$

$$q_c = K_c \frac{dp}{dt} \qquad (4.51)$$

Substituting the various flow rates in eqn. (4.48), we have

$$q = K_1 x - K_2 p = K_p \frac{dy}{dt} + K_l p + K_c \frac{dp}{dt}$$

or $\qquad K_p \dfrac{dy}{dt} + K_c \dfrac{dp}{dt} + (K_l + K_2)\, p = K_1 x \qquad (4.52)$

The differential equation governing the load remains the same as in eqn. (4.46). Substituting for p from eqn. (4.46) in (4.52), we obtain

$$\frac{K_c M}{A}\dddot{y} + \left[\frac{K_c f}{A} + (K_l + K_2)\frac{M}{A}\right]\ddot{y} + \left[K_p + (K_l + K_2)\frac{f}{A}\right]\dot{y} = K_1 x$$

From the above equation, the transfer function of the system is

$$\frac{Y(s)}{X(s)} = \frac{K_1}{s\left[\dfrac{K_c M}{A}s^2 + \left\{\dfrac{K_c f}{A} + (K_l + K_2)\dfrac{M}{A}\right\}s + \left\{\dot{K}_p + (K_l + K_2)\dfrac{f}{A}\right\}\right]}$$

As previously stated, K_c is very small, therefore $K_c M/A$ and $K_c f/A$ can be neglected. The transfer function then reduces to

$$\frac{Y(s)}{X(s)} = \frac{K}{s(\tau s + 1)} \tag{4.53}$$

where $\quad K = \dfrac{K_1}{K_p + (K_l + K_2)f/A}; \quad \tau = \dfrac{(K_l + K_2)M/A}{K_p + (K_l + K_2)f/A}.$

With $K_l = 0$ and $K_p = A$, this transfer function reduces to the simplified case transfer function given in eqn. (4.47).

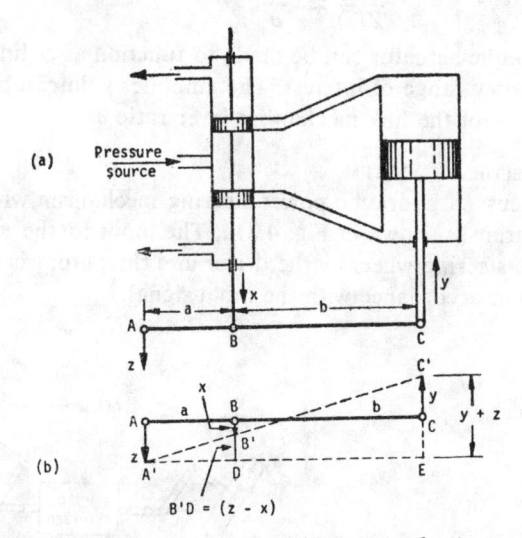

Figure 4.29 Hydraulic actuator with negative feedback.

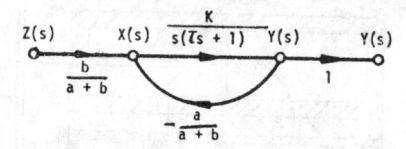

Figure 4.30 Signal flow graph of system shown in Fig. 4.29a.

Hydraulic linear actuator can also be modified to act as hydraulic amplifier (proportional controller) by providing a negative feedback through a link mechanism as shown in Fig. 4.29a. For small movements x, y and z can be regarded as linear. From the link geometry in Fig. 4.29b,

$$\frac{C'E}{A'E} = \frac{B'D}{A'D}$$

or

$$\frac{y+z}{a+b} = \frac{z-x}{a}$$

or

$$x = \frac{b}{a+b} z - \frac{a}{a+b} y \tag{4.54}$$

The transfer function $Y(s)/X(s)$ of the actuator already derived, is given in eqn. (4.47). Combining eqns. (4.47) and (4.54) we draw the signal flow graph of Fig. 4.30. The overall transfer function is obtained therefrom as

$$\frac{Y(s)}{Z(s)} = \frac{bK}{(a+b)s(\tau s + 1) + Ka} \tag{4.55}$$

In the normal frequency range of hydraulic control systems

$$|(a+b)s(\tau s + 1)| \ll Ka.$$

Therefore,

$$\frac{Y(s)}{Z(s)} \approx \frac{b}{a} \tag{4.56}$$

Thus the hydraulic actuator can be made to function as a linear amplifier over the frequency range of interest. The amplifier gain can be adjusted by a suitable choice of the link mechanism lever ratio b/a.

HYDRAULIC FEEDBACK SYSTEM

Let us discuss a hydraulic power steering mechanism whose simplified schematic diagram is shown in Fig. 4.31a. The input to the system is the rotation of the steering wheel by the driver and the output is positioning of the car wheels in accordance with the input signal.

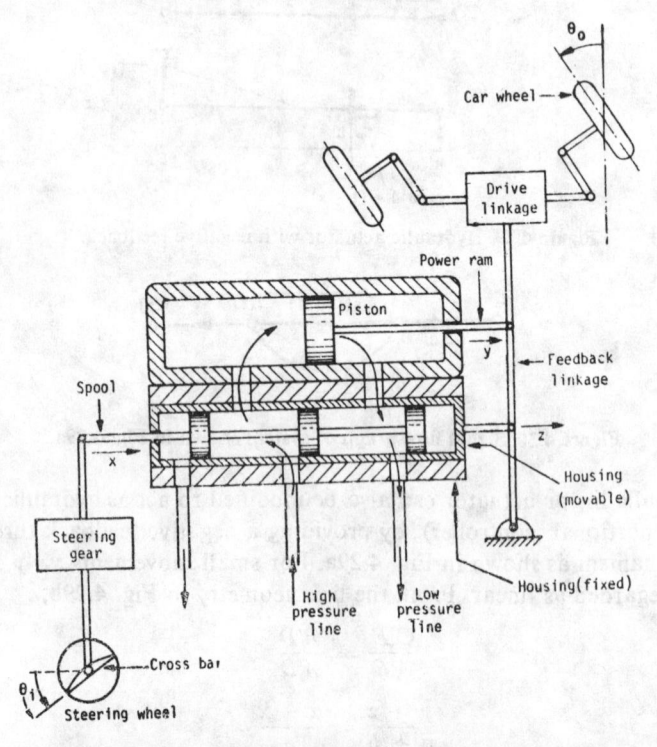

Figure 4.31 (a) Power steering mechanism. [From V. Del Toro and S. Parker, *Principles of Control Systems Engineering*, 1960, McGraw-Hill, New York. Reproduced with permission.]

When the steering wheel is in the zero position, i.e., the cross bar is horizontal, the wheels are directed parallel to the longitudinal axis of the car. For this condition, the spool is in the neutral position and the oil supply to the power cylinder is cut-off. When the steering wheel is turned anticlockwise through an angle θ_i, the spool is made to move towards right by an amount x with the help of the gear mechanism. The high pressure oil enters on the left hand side of the power cylinder causing the power piston and hence the power ram to move towards right by an amount y. Through a proper drive linkage, a torque is applied to the wheels causing the desired displacement θ_0 of the wheels.

A rigid linkage bar connects the power ram and the moveable valve housing. When the power ram moves towards right, the linkage moves as shown in Fig. 4.31b. It is seen from this figure that movement of the power ram towards right causes a movement of the movable valve housing in such a direction as to seal off the high pressure side. The system then operates with a fixed θ_0 for a given input x under steady conditions.

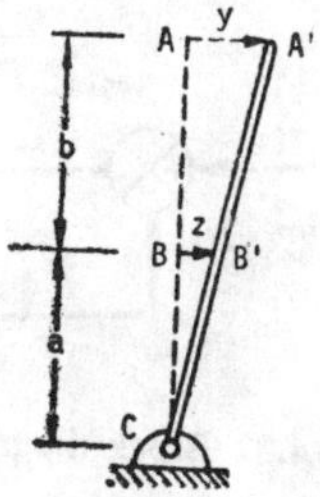

The geometry of the linkage is shown in Fig. 4.31b, from which we obtain for small movements,

Figure 4.31 (b) Geometry of feedback linkage for small values of movement y.

$$\frac{AA'}{AC} = \frac{AA'}{AB + BC} = \frac{BB'}{BC}$$

or
$$\frac{y}{a+b} = \frac{z}{a} \qquad (4.57)$$

Therefore
$$Z(s) = \frac{a}{a+b} Y(s) \qquad (4.58)$$

Assuming the compressibility and leakage flows to be negligible the flow rate of oil into power cylinder due to spool movement is [refer eqn. (4.42)] given by

$$q = K_1 e - K_2 p$$

where $e = (x - z)$ is the net displacement of spool relative to the valve housing.

Therefore
$$Q(s) = K_1 E(s) - K_2 P(s) \qquad (4.59)$$
and
$$E(s) = X(s) - Z(s) \qquad (4.60)$$

The flow rate of oil into the power cylinder is proportional to the rate at which the power piston displaces the oil, i.e.,

$$Q(s) = K_p s Y(s) \qquad (4.61)$$

Assuming the load on power piston to consist of mass M and viscous friction with coefficient f, the force equation is given by

$$AP(s) = Ms^2Y(s) + fsY(s) \tag{4.62}$$

where A is the area of the piston.

From eqns. (4.59) and (4.62), we get

$$Q(s) = K_1 E(s) - K_2 \frac{Ms^2 + fs}{A} Y(s) \tag{4.63}$$

The block diagram of the power steering mechanism obtained from eqns. (4.58)-(4.63) is shown in Fig. 4.32.

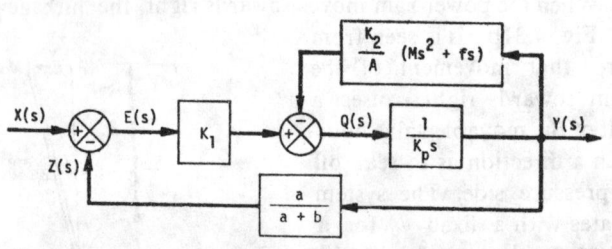

Figure 4.32 Block diagram of power steering mechanism shown in Fig. 4.31a.

4.6 Pneumatic Systems

As different from hydraulic fluid, air medium (or other gases in special situations) is used in pneumatic control systems. Air medium has the advantage of being non-inflammable and having almost negligible viscosity compared to the high viscosity of hydraulic fluids, which also varies considerably with temperature causing a marked effect on the performance of control systems.

On the other hand, the high incompressibility of a hydraulic fluid causes the force wave to travel faster and therefore the hydraulic systems have a shorter response time; while in pneumatic systems there is a considerable amount of compressibility flow so that such systems are characterized by longer time delays.

Pneumatic systems find considerable application in the process control field. These are sometimes used in guided missiles and aircraft systems. Some of the important pneumatic components are discussed in this section.

PNEUMATIC BELLOWS

It consists of a hollow chamber with thin metallic walls. The side walls of bellows are corrugated, while the input and output surfaces are flat as shown in Fig. 4.33. The action of the bellows is similar to that of a spring. An increase in the pressure within the bellows results in an increase in the separation between the input and output surfaces.

The force acting to separate two surfaces $= (\Delta P)A$, where $A =$ area of each flat surface; and $\Delta P =$ differential pressure (internal pressure minus external pressure).

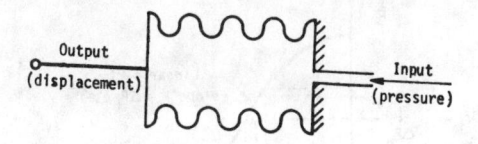

Figure 4.33 Pneumatic bellows.

The force opposing the separation $= K(\Delta x)$, where $K =$ stiffness of the bellows; and $\Delta x =$ displacement of movable surface from the reference. Therefore, in equilibrium

$$K(\Delta x) = (\Delta P)A$$

or
$$\frac{\Delta X(s)}{\Delta P(s)} = \frac{A}{K} \qquad (4.64)$$

This is the transfer function of the bellows.

PNEUMATIC FLAPPER VALVE

This is an important component of many pneumatic systems. The power source for this device is the supply of air at constant pressure. A schematic diagram of the valve is shown in Fig. 4.34.

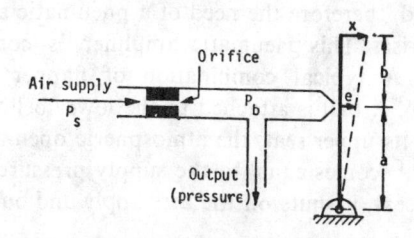

Figure 4.34 A flapper valve.

Pressurized air is fed through the orifice and is ejected from the nozzle towards the flapper. The flapper is positioned against the nozzle opening and the nozzle back pressure P_b is controlled by the nozzle-flapper distance e. As the flapper approaches the nozzle, the resistance to the flow of air through the nozzle increases with the result that the nozzle back pressure increases. If the nozzle is completely closed by the flapper, the nozzle back pressure P_b becomes equal to the supply pressure P_s. If the flapper is moved away from the nozzle so that nozzle flapper distance is large, then there is practically no restriction to flow and the nozzle back pressure takes on a minimum value close to the ambient pressure.

Thus the flapper valve converts small changes in the position of the flapper into large changes in the back pressure. A typical curve relating the nozzle back pressure P_b to the nozzle flapper distance e is shown in

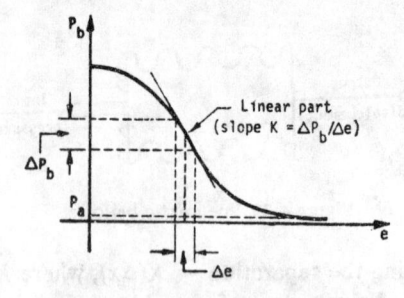

Figure 4.35 Flapper valve characteristic.

Fig. 4.35. The linear part of the curve is utilized in the valve operation. For this operating region, the transfer function of the valve is

$$\frac{\Delta P_b(s)}{\Delta X(s)} = \left(\frac{a}{a+b}\right) K \qquad (4.65)$$

where $K < 0$ is the slope of the linear part of the curve and $e = [a/(a+b)]x$ is the nozzle-flapper distance as shown in Fig. 4.34.

PNEUMATIC RELAY

For the flapper valve of the type discussed above, it is necessary to restrict the range of flapper displacement to very small value so that the linearity of operation is maintained. Because of this, the change in the output pressure is very small and therefore the need of a pneumatic amplifier in cascade with this device arises. This pneumatic amplifier is commonly known as pneumatic relay. A typical combination of flapper valve and relay is shown in Fig. 4.36. A ball is attached to the lower bellows surface. When the ball rests on its upper seat, the atmospheric opening is closed and the output pressure P becomes equal to the supply pressure P_s. When the ball rests on its lower seat, it shuts off the air supply and output pressure drops

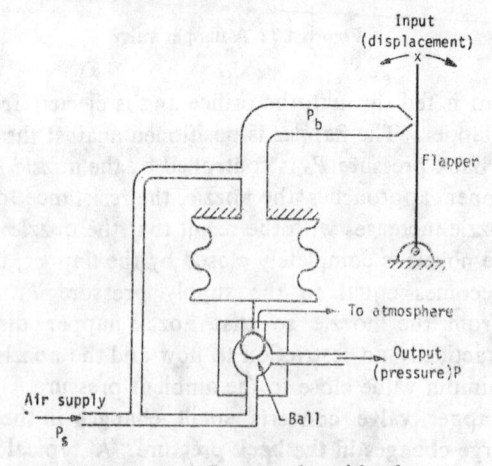

Figure 4.36 A flapper valve with relay.

to ambient pressure. The output pressure can thus be made to vary from ambient to full supply pressure.

The movement of the flapper away from the nozzle causes the nozzle back pressure P_b to decrease, thus the bellows contracts, moving the ball upwards. The atmospheric opening is partially closed and the output pressure increases. When the flapper moves towards the nozzle, the opposite action takes place.

It is thus seen that an increase in separation between nozzle and flapper results in a decrease in output pressure in the case of the flapper valve and an increase in output pressure when the flapper valve is used with the pneumatic relay. The transfer function of the combination is therefore

$$\frac{\Delta P(s)}{\Delta X(s)} = \frac{a}{a+b} K \qquad (4.66)$$

where $K > 0$.

PNEUMATIC ACTUATOR

The majority of the modern pneumatic control systems require translatory output motion. This motion is achieved by pneumatic actuators responding to changes in input air pressure. A schematic diagram of the pneumatic actuator is shown in Fig. 4.37. Assume that the actuator is used to position a load consisting of a spring with stiffness K, viscous friction with coefficient f and mass M. For a small variation ΔP in the input pressure, the force acting on the diaphragm is $A(\Delta P)$ where A is the area of the diaphragm. If Δy is the displacement of the actuator stem because of this force, then the force balance equation is

$$A(\Delta P) = M\Delta \ddot{y} + f\Delta \dot{y} + K\Delta y$$

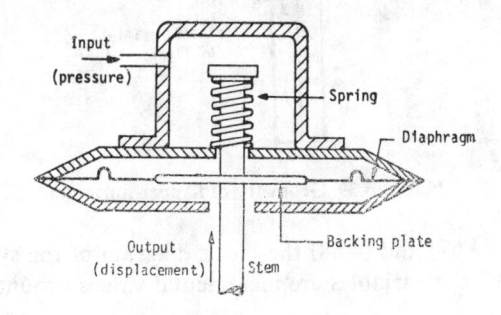

Figure 4.37 Pneumatic actuator.

Therefore the transfer function of the actuator is

$$\frac{\Delta Y(s)}{\Delta P(s)} = \frac{A}{Ms^2 + fs + K} \qquad (4.67)$$

It may be noted that the stiffness and mass of the diaphragm and backing plate, are considered to be negligible.

Pneumatic Position Control System

The pneumatic position control system shown in Fig. 4.38 is used to position a fluid flow valve. The principle of operation of this system is that an input signal is applied to the flapper corresponding to the required opening of the plug valve. The nozzle back pressure operates the actuator and the actuator stem moves to give the required opening of the plug valve.

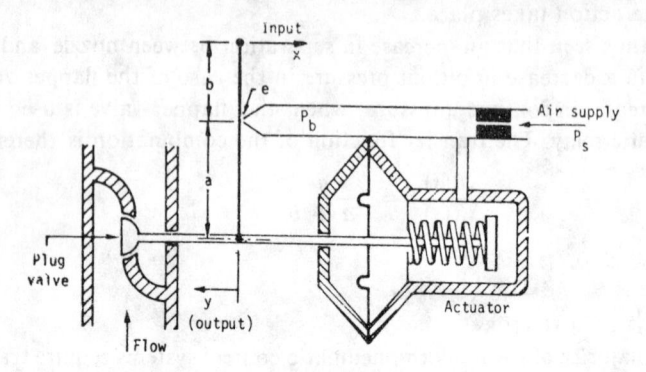

Figure 4.38 Pneumatic position control system.

The geometry of the flapper for small input and output displacements is shown in Fig. 4.39. From this figure, we have

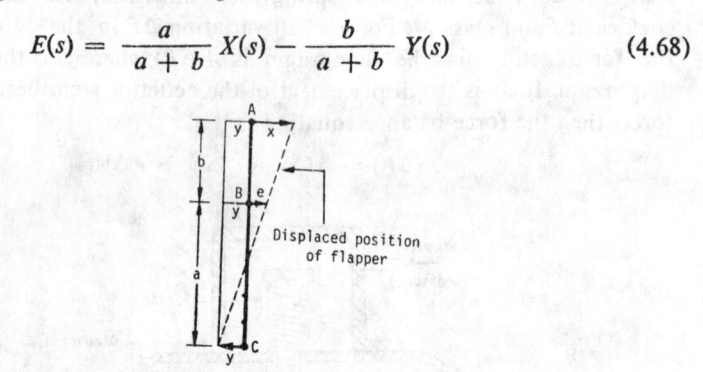

$$E(s) = \frac{a}{a+b} X(s) - \frac{b}{a+b} Y(s) \qquad (4.68)$$

Figure 4.39 Geometry of flapper movement.

Using eqns. (4.67) and (4.68) the block diagram of the system is drawn in Fig. 4.40. All the variables are incremental values around the operating point.

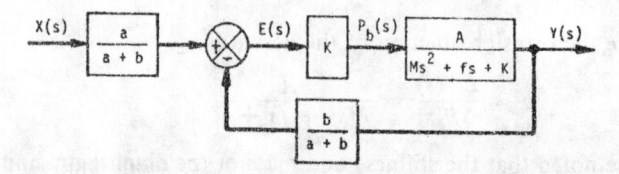

Figure 4.40 Block diagram of the system shown in Fig. 4.38.

4.7 Gyroscopes—Inertial Navigation

Gyroscopes (gyros) are devices used for measuring or indicating inclination of frames in inertial navigation systems. Their action is based on two physical principles enunciated below.

1. A spinning wheel in the absence of external torques acting on it, maintains the direction of its spin axis in space. Such a spinning wheel is known as a *free gyro*.

2. When a torque is applied to an axis inclined to the spin axis of a wheel, the wheel rotates in the plane of the two axes (i.e., about an axis at 90° to both the spin and input torque axis) in a direction tending to align the spin axis with the input torque axis. This rotation is termed as *precession*.

CONSTRUCTIONAL FEATURES

The constructional features of a gyroscope are illustrated in the skeleton diagram of Fig. 4.41. The spinning wheel rotates in bearings located in the inner gimbal (linkages). The inner gimbal is free to rotate in the bearing placed in the outer gimbal which in turn can rotate in the frame to which the gyro is attached. This is a free gyro.

The gyro wheel is made to spin by means of a synchronous motor whose rotor is mounted on the axis of the spinning wheel.

The position of spin axis relative to the earth determines the basic type of gyro. When the axis is in horizontal position, the gyroscope is referred to as *positional* or *directional gyro* and when the axis is in vertical position, it is called *vertical gyro*. Figure 4.41 illustrates a directional gyro wherein the wheel spins about the horizontal axis while the outer gimbal is free to rotate in the bearing located in the vehicle frame. Figure 4.42 gives the schematics of a vertical gyro.

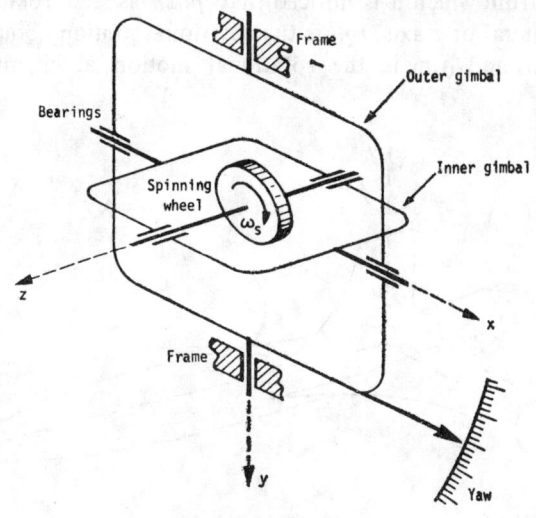

Figure 4.41 Gyroscope.

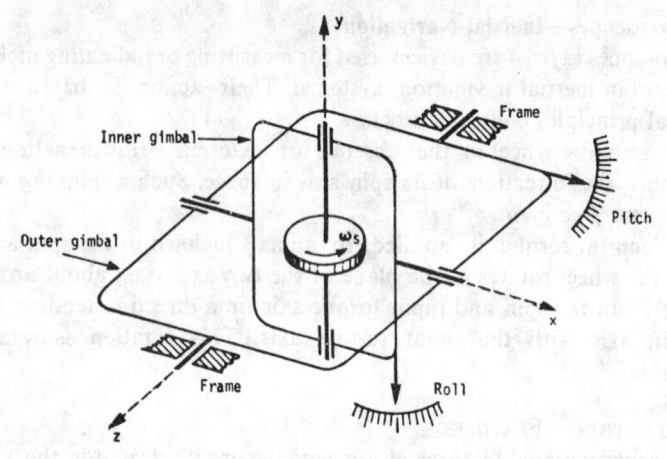

Figure 4.42 Vertical gyroscope.

In actual practice, the spin axis tends to drift slowly away from its initial direction because of the torques acting on wheels due to friction in gimbal bearings. To compensate for the drift and for earth's rotation, the reference direction of spin axis has to be adjusted. This is accomplished by means of two torque motors—one with rotor on the inner gimbal axis and stator fastened to outer gimbal and the other with rotor on outer gimbal axis and stator fastened to frame.

The reference direction of a vertical gyro is automatically controlled by means of a more accurate primary reference such as the average position of a vertical pendulum by employing the error signal for control purposes.

Vertical and directional gyros are used to measure or indicate the three rotational motions of a vehicle. These rotational motions are illustrated in Fig. 4.43 from which it is noticed that *pitch* is the rotational motion about the lateral or z-axis, *roll* is the rotational motion about the longitudinal or x-axis and *yaw* is the rotational motion about the normal or y-axis.

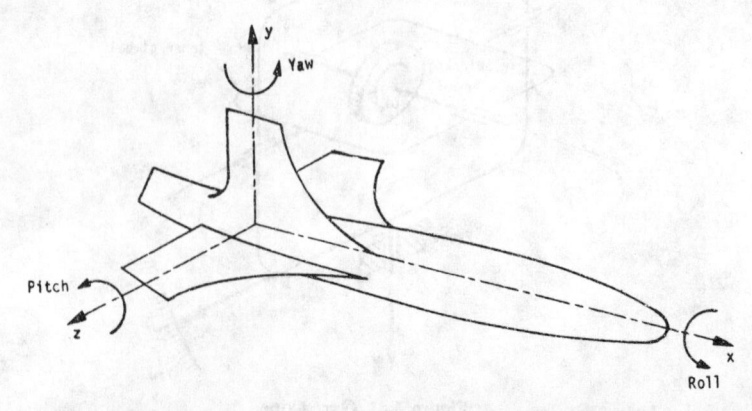

Figure 4.43 Pitch, roll, and yaw motions of a vehicle.

A vertical gyro fastened to a vehicle measures pitch and roll. The pitch and roll scales as shown in Fig. 4.42 are fixed to the vehicle frame. As the vehicle pitches, the frame and pitch scale rotate about the z-axis so that the pointer indicates the pitch. The pitch motion causes the roll scale to move at right angles to the roll pointer so the roll reading does not change. Similarly, as the vehicle rolls about the x-axis the roll scale rotates and the roll pointer indicates the roll.

A directional gyro fastened to a vehicle measures yaw as shown in Fig. 4.41. As the vehicle yaws, the yaw scale rotates about the y-axis and the pointer gives the indication of yaw magnitude.

If an electrical signal is desired, pitch, roll and yaw scales could be replaced by synchros.

Gyro Dynamics

Consider a wheel spinning with velocity ω_s initially in alignment with the x-axis as shown in Fig. 4.44. The momentum about the spin axis is

$$H = J_s \omega_s$$

where J_s = moment of inertia of the wheel about the spin axis.

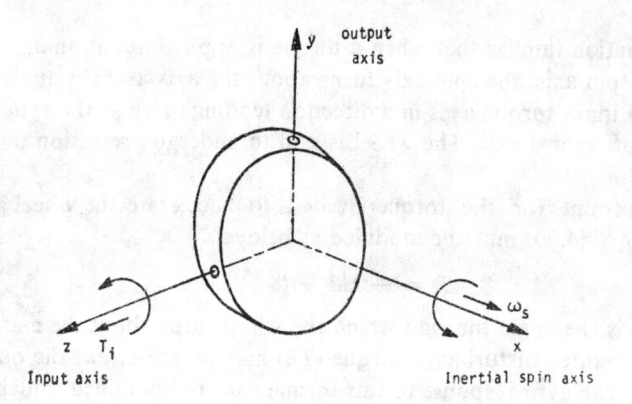

Figure 4.44.

Angular velocity, momentum and torque being vectors, are represented along the axis of rotation, their direction being along the thumb of the right hand with fingers of the hand curled in the direction of rotation or torque.

Inertias of the wheel about the other two axes are:

I = moment of inertia about the z-axis (input axis).

J = moment of inertia about the y-axis (output axis).

Torque T_i applied about the input axis in the direction indicated causes a change in the momentum of the wheel by an amount ΔH in the direction of T_i in time Δt (Fig. 4.45a);

$$\Delta H = T_i \Delta t \qquad (4.69)$$

In this time the spin axis turns through an angle ($-\Delta\theta$) which for small change can be approximated as

$$\tan(-\Delta\theta) \approx -\Delta\theta = \frac{\Delta H}{H} = \frac{T_i}{H}\Delta t$$

or

$$\frac{d\theta}{dt} = \dot\theta = -\frac{T_i}{H}$$

or

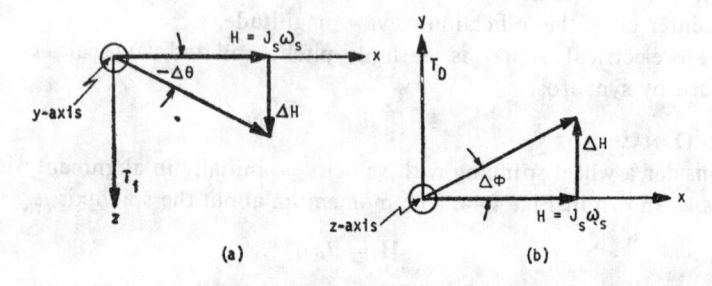

$$T_i = -H\dot\theta \qquad (4.70)$$

Figure 4.45.

This equation implies that when a torque is applied about an axis inclined to the spin axis, the spin axis turns about the axis at right angles to both spin and input torque axes in a direction tending to align the spin axis with the input torque axis. The wheel is said to undergo precession under input torque.

To account for the torque needed to accelerate the wheel about the z-axis, eqn. (4.70) must be modified as below:

$$T_i = -H\dot\theta + I\ddot\phi \qquad (4.71)$$

where ϕ is the angle through which the wheel turns about the z-axis.

An unwanted disturbance torque (T_D) may be present at the output axis (y-axis). The gyro response to this torque can be obtained independently as long as the angular movements of the spin axis are small. Figure 4.45b shows the response to disturbance torque. Now

$$\Delta H = T_D \Delta t$$

$$\tan \Delta\phi \approx \Delta\phi = \frac{\Delta H}{H} = \frac{T_D}{H}\Delta t$$

or

$$\frac{d\phi}{dt} = \dot\phi = \frac{T_D}{H}$$

or

$$T_D = H\dot\phi \qquad (4.72)$$

In presence of moment of inertia J, eqn. (4.72) modifies to

$$T_D = H\dot\phi + J\ddot\theta \qquad (4.73)$$

where θ is the angle through which the wheel turns about the y-axis.

The damping and spring torques about the input axis are negligible compared to applied torque. These torques about the output axis must, however, be accounted for. Equation (4.73) in presence of damping and spring restoring torques gets modified to

$$T_D = H\dot{\phi} + J\ddot{\theta} + f\dot{\theta} + K\theta \qquad (4.74)$$

where f is the damping torque coefficient and K is spring constant.

Taking the Laplace transforms of eqns (4.71) and (4.74), assuming zero initial conditions, we get

$$T_i(s) = -sH\theta(s) + s^2 I\phi(s) \qquad (4.75)$$
$$T_D(s) = sH\phi(s) + (Js^2 + fs + K)\theta(s) \qquad (4.76)$$

Solving eqns. (4.75) and (4.76), we have

$$\theta(s) = \frac{sT_D(s) - (H/I)T_i(s)}{s(Js^2 + fs + K + H^2/I)} \qquad (4.77)$$

$$\phi(s) = \frac{s(H/I)\,T_D(s) + ((Js^2 + fs + K)/I)\,T_i(s)}{s^2(Js^2 + fs + K + H^2/I)} \qquad (4.78)$$

In the absence of a disturbing torque, the gyro response to input torque is given by

$$\frac{\theta(s)}{T_i(s)} = \frac{-H/I}{s(Js^2 + fs + K + H^2/I)} \qquad (4.79)$$

THE FREE GYRO

If there are no restraining forces in any direction on the gyro, the gyro wheel remains fixed in space so that the gyro measures the angular position of the vehicle with respect to the gyro as a reference (see Figs. 4.41 and 4.42). This type of gyro provides an inertial frame of relatively low accuracy.

Setting J, f, K and T_D equal to zero in eqn. (4.77), we get

$$s\theta(s) = -\frac{T_i(s)}{H}$$

or

$$\frac{d\theta}{dt} = -\frac{T_i}{H} \qquad (4.80)$$

Similarly setting J, f, K and T_i equal to zero in eqn. (4.78), we get

$$\frac{d\phi}{dt} = \frac{T_D}{H} \qquad (4.81)$$

When used in this mode drift rates of the order of a fraction of a degree per minute are quite common.

THE RATE GYRO

For the purpose of providing rate feedback from roll, pitch or yaw to damp out vehicle oscillations about these axes, rate gyros are employed. These have low accuracy.

To understand the operation of a rate gyro, set $T_D = 0$ in eqn. (4.76). If a large spring constant is provided at the output axis (i.e., large K), eqn. (4.76) can be approximated as

$$s\phi(s) = -\frac{K}{H}\theta(s)$$

or
$$\theta = -\frac{H}{K}\frac{d\phi}{dt} \qquad (4.82)$$

Thus a signal proportional to $d\phi/dt$ is obtained resulting in a rate gyro.

With reference to Fig. 4.41, a helical spring is provided between outer gimbal and the frame to convert it into a rate gyro.

THE RESTRAINED (INTEGRATING) GYRO

If only one axis is free to move, the gyro has a single degree of freedom with respect to the frame. In eqn. (4.76) setting $T_D = 0$ and providing a large f and neglecting K, we have

$$\theta(s) = -\frac{H}{f(\tau s + 1)}\phi(s)$$

where $\tau = J/f$, is negligible for large f.

Therefore
$$\theta = -\frac{H}{f}\phi \qquad (4.83)$$

Compared to eqn. (4.82), θ is now proportional to integral of the rate of ϕ. Hence this gyro is also called integrating rate gyro or simply 'integrating gyro'. The integrating gyro is very rugged and extremely accurate.

With reference to Fig. 4.41, the outer gimbal is fixed to the frame to convert it into an integrating gyro. As the frame (and outer gimbal) rotates about the y-axis, the wheel precession is caused about the x-axis, which means that the inner gimbal rotates about the x-axis. In actual construction, the inner gimbal takes the form of a cylinder filled with inert gas and hermatically sealed. The outer case is filled with viscous fluid whose density is such that the inner gimbal remains suspended. Thus the load on jewelled bearings is negligibly small; thereby the effect of drift is reduced.

Problems

4.1 A two-phase servomotor has rated voltage applied to its reference winding. The torque-speed characteristic of the motor with 115 volts, 50 c/s applied to control winding is shown in Fig P-4.1. The moment of inertia of the motor is 1×10^{-5} kg-m^2 and friction

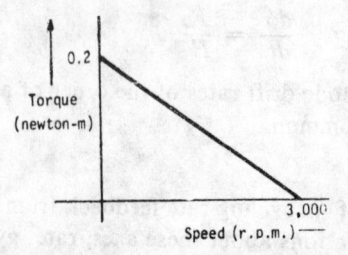

Figure P-4.1.

is negligible. Find the transfer function connecting the shaft position θ with the control winding voltage v_c.

4.2 The schematic diagram of a servo system is shown in Fig. P-4.2. The two-phase servomotor develops a torque in accordance with the equation

where
$$T_M = -K_1 \dot{\theta}_M + K_2 v_C$$
$$K_1 = 0.5 \times 10^{-3} \text{ newton-m per rad/sec}$$
$$K_2 = 2 \times 10^{-3} \text{ newton-m/volt}$$

Draw the block diagram of the system indicating the transfer function of each block. Calculate the value of amplifier gain K_A such that the system has a steady-state error $(\theta_R - \theta_C)$ of 0.1 deg for an input velocity of 1 rad/sec. What will be the control field voltage under this condition? Given:

Load inertia,	$J_L = 2.5$ kg-m²
Coefficient of load friction,	$f_L = 250$ newton-m per rad/sec
Motor to load gear ratio,	$(\dot{\theta}_L/\dot{\theta}_M) = 1/50$
Motor to synchro gear ratio,	$(\dot{\theta}_C/\dot{\theta}_M) = 1/1$
Sensitivity of synchros,	$K_s = 100$ volts/rad

The motor inertia and friction are negligible.

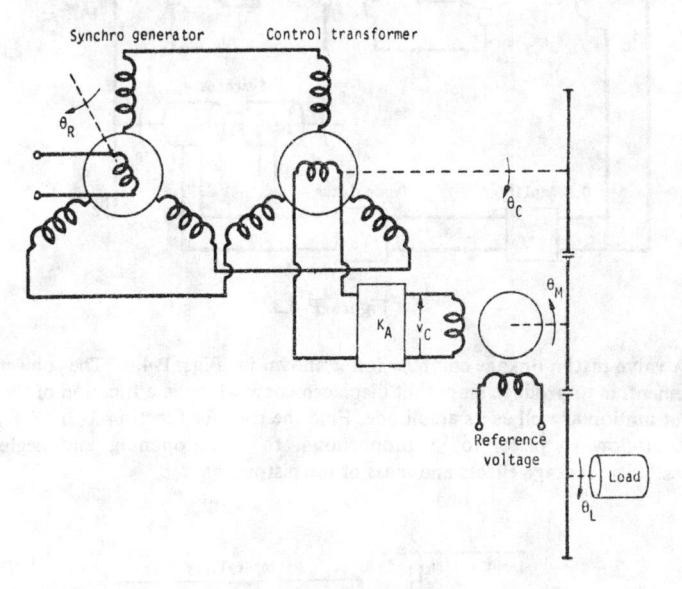

Figure P-4.2.

4.3 An a.c.–d.c. servo system is shown in Fig. P-4.3. The transfer function of the demodulator is given as K_d d.c. volts/a.c. volt. The sensitivity of the synchro error-detector is K_s volts/rad; amplidyne gain is K_{am} volts/field amp; gain of d.c. generator is K_g volts/field amp. The d.c. motor is separately excited and has a counter emf of K_b volts per rad/sec and a torque constant of K_T newton-m/amp. Motor inertia and friction are negligible. Draw the block diagram of the system indicating the transfer function of each block. Obtain $\theta_C(s)/\theta_R(s)$ and also find the steady-state error for a step input of 1 rad/sec to the system.

The system parameters are given below:

$K_s = 30$ volts/rad $\qquad K_g = 100$ volts/field amp
$K_d = 4$ d.c. volts/a.c. volt $\qquad R = 1$ ohm

$K_A = 5$ volts/volt
$R_C = 200$ ohms
$L_C = 2$ henrys
$K_{am} = 1,000$ volts/field amp
$R_f = 100$ ohms
$L_f = 2$ henrys

$K_b = 1$ volt per rad/sec
$J_L = 0.5$ kg-m^2
$f_L = 1$ newton-m per rad/sec
$N_1/N_2 = 1$

Assume the quadrature axis time constant of the amplidyne to be of negligible order.

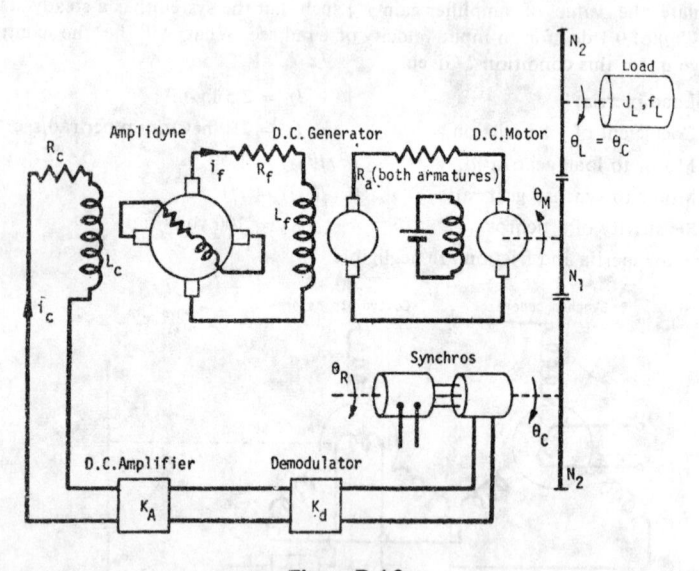

Figure P-4.3.

4.4 A valve piston linkage combination is shown in Fig. P-4.4. The object of this arrangement is to produce an output displacement x which is a function of the speed of the input motion as well as its amplitude. Find the transfer function $X(s)/Y(s)$ assuming rate of oil-flow to piston to be proportional to valve opening and neglecting oil compressibility, leakage effects and mass of the piston.

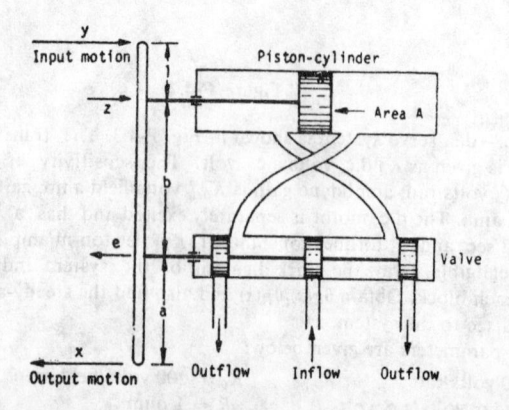

Figure P-4.4.

Control Systems and Components

4.5 A hydraulic servo system used to control the transverse feed of a machine tool is shown in Fig. P-4.5. Each angular position of the cam corresponds to a desired reference position x_r such that $y_r = K_r x_r$. The load on the piston is that due to tool reactive force and may be assumed to consist of mass M, friction f and spring with constant K.

Draw the block diagram and obtain the transfer function $Y_c(s)/X_r(s)$, assuming that:
Rate of oil flow to the piston = $K_1 \times$ valve opening
Leakage flow across the piston = $K_2 \times$ pressure across the piston
The compressibility effect is negligible.

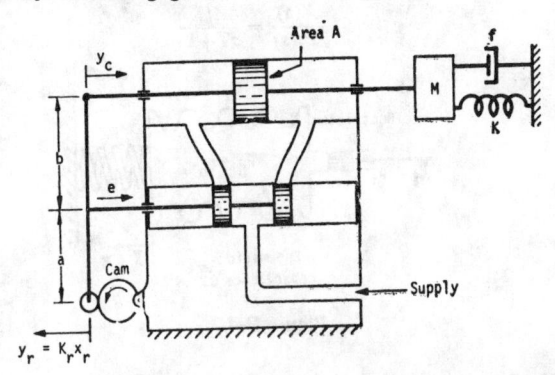

Figure P-4.5 [From F. Raven, *Automatic Control Engineering*, 2nd ed, 1968, McGraw-Hill New York. Reproduced with permission.]

4.6 The electrohydraulic position control system shown in Fig. P-4.6 positions a mass M with negligible friction. Assume that the rate of oil flow to the piston is $q = K_x x - K_p p$, where x is the control valve opening and p is the differential pressure. The mass of piston, oil compressibility and leakage are assumed negligible. Draw the signal flow graph of the system and obtain therefrom the transfer function $Y(s)/X_i(s)$. How the transfer function will modify if the mass M is also negligible such that the differential pressure p needed to move the piston is nearly zero?

The system constants are given below:

Mass, $\quad M = 1{,}000$ kg
Control valve constants, $\quad K_x = 200$ cm³/sec/cm of valve opening

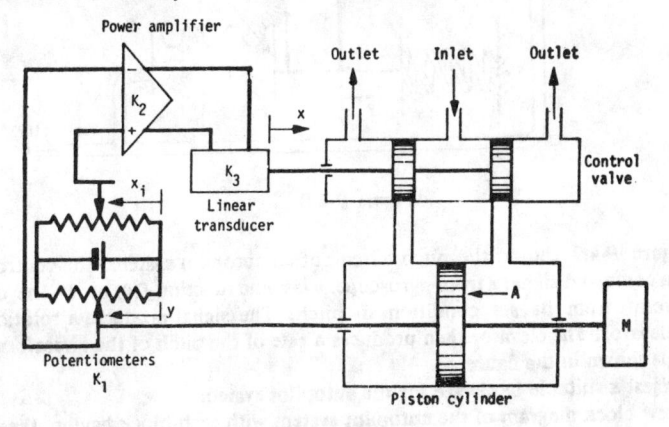

Figure P-4.6.

Potentiometer sensitivity, $K_p = 0.5$ cm³/sec per gm-wt/cm²
Power amplifier gain, $K_1 = 1$ volt/cm
Linear transducer constant, $K_2 = 500$ mA/volt
Piston area, $K_3 = 0.1$ cm/mA
$A = 100$ cm²

4.7 Figure P-4.7 shows a bellows with a restrictor. Show that the transfer function of the system is

$$\frac{X(s)}{P_i(s)} = \frac{A/K}{sT+1}$$

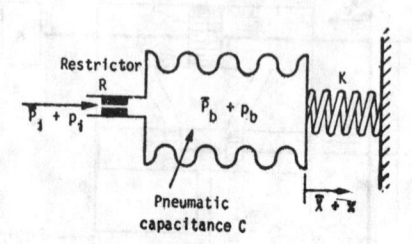

Figure P-4.7.

where $T = RC$; $R =$ pneumatic resistance of restrictor; $C =$ pneumatic capacitance of bellows; and $A =$ cross-sectional area of base of bellows.

4.8 A position balance pneumatic controller is shown diagrammatically in Fig. P-4.8. Determine the transfer function of the system relating the controlled pressure p_c to the movement e, in terms of lever dimensions a and b and the restrictor calibration $T = RC$.

(The supply pressure P_s is assumed to be constant and controlled pressure $p_c = K_p x$, where K_p is a constant.)

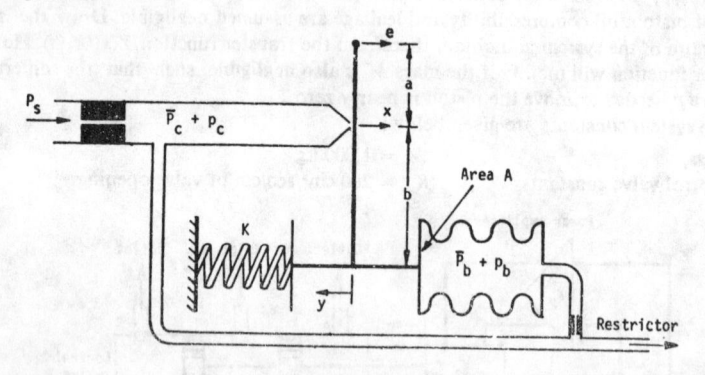

Figure P-4.8.

4.9 Figure P-4.9 shows the pitch circuit of an autopilot system. The control system receives a command signal v from gyroscope. v is some function $f(\theta)$ of θ—the deviation of the aircraft from its set condition in pitch. The signal produces a rotation of the aircraft elevator. The elevator then produces a rate of the pitch of the aircraft which is fedback as shown in the figure.

(a) Suggest a suitable hardware for the autopilot system.

(b) Draw block diagram of the autopilot system with each block having the transfer function of the corresponding component in the pitch circuit.

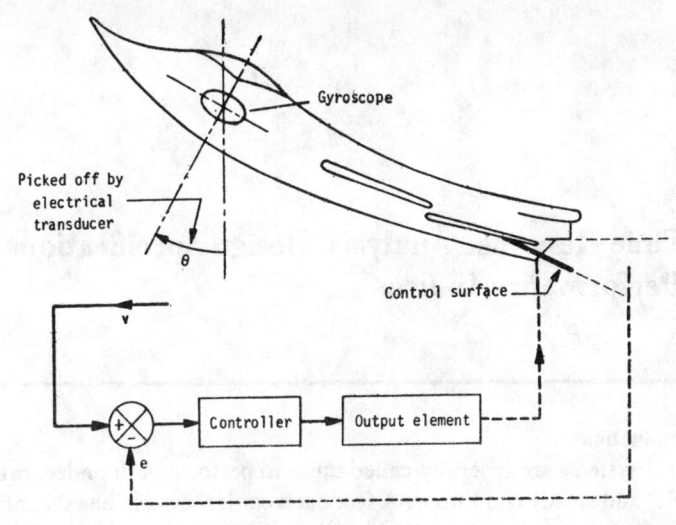

Figure P-4.9.

References and Further Reading

1. Miller, R.W., *Servomechanisms Devices and Fundamentals*, Reston Publishing Co., Reston, Virginia, 1977.
2. Kuo, B.C., *Theory and Applications of Step Motors*, West Publishing Company, St. Paul, Minn., 1974.
3. Baeck, H.S., *Practical Servomechanism Design*, McGraw-Hill, New York, 1968.
4. Chubb, B.A., *Modern Analytical Design of Instrument Servomechanisms*, Addison-Wesley, Reading, Mass., 1967.
5. Canfield, E.B., *Electromechanical Control Systems and Devices*, John Wiley, New York, 1965.
6. Ivey, K.A., *A-C Carrier Control Systems*, John Wiley, New York, 1964.
7. Savant, Jr., C.J., *Control System Design*, McGraw-Hill, New York, 1964.
8. Bukstein, E.D., *Basic Servomechanisms*, Holt, Rinehert and Winston, New York, 1963.
9. Johnson, E.R., *Servomechanisms*, Prentice-Hall, Englewood Cliffs, N.J., 1963.
10. Fitzgerald, A.E., and C. Kingsley, Jr., *Electric Machinery*, McGraw-Hill, New York, 1961.
11. Raven, F., *Automatic Control Engineering*, McGraw-Hill, New York, 1961.
12. Ahrendt, W.R. and C.J. Savant, Jr., *Servomechanism Practice*, McGraw-Hill, New York, 1960.
13. Taylor, P.L., *Servomechanisms*, Longmans, London, 1960.
14. Gibson, J.E. and F.B. Tuteur, *Control System Components*, McGraw-Hill, New York, 1958.

5. Time Response Analysis, Design Specifications and Performance Indices

5.1 Introduction

Control systems are generally called upon to perform both under transient (dynamic) and steady conditions. A feedback control system has the inherent capability that its parameters can be adjusted to alter both its transient and steady-state behaviour. In order to analyze the transient and steady-state behaviour (these two together are referred to as time response) of control systems, the first step always is to obtain a mathematical model of the system. For any specific input signal a complete time response expression can then be obtained through the Laplace transform inversion (or through convolution integral in case the input is such that we cannot obtain its Laplace transform). This expression yields the steady-state behaviour of the system with time tending to infinity. In case of simple deterministic signals, steady-state response can also be obtained directly without obtaining the time response expression by use of the final value theorem.

Before proceeding with the time response analysis of a control system, it is necessary to test the stability of the system (the concept of stability shall be explained in Chapter 6). System stability can be tested through indirect tests without actually obtaining the transient response. In case the system happens to be unstable, we need not proceed with its transient response analysis.

Usually the input signals to control systems are not known fully ahead of time. In most cases these signals may be random in nature, e.g., in a radar tracking system, the position and speed of the target to be tracked may vary in a random fashion. It is thus difficult to express the actual input signals mathematically by simple equations. The characteristics of actual signals which severely strain a control system are a sudden shock, a sudden change, a constant velocity and a constant acceleration. System dynamic behaviour for analysis and design is therefore judged and compared under application of standard test signals—an impulse, a step, a constant velocity (a ramp input) and constant acceleration (a parabolic input). Another standard test signal of great importance is a sinusoidal signal. Steady-state response to a sinusoidal test signal yields a great deal of information about the system particularly because the actual inputs can generally be recognized to contain a band of frequencies from zero onwards (feedback control systems are low-pass filters). Because of its importance sinusoidal steady-state response of control systems will be dealt in considerable detail in Chapters 8 and 9. The response of a control system to the above mentioned standard test signals has normally a good correlation with its response to actual inputs.

The nature of the transient response is revealed by any of the test signals mentioned above as this nature is dependent upon system poles only and not on the type of input. It is therefore sufficient to analyze the transient response to one of the standard test signals—a step is generally used for this purpose. Steady-state response is then examined with respect to this particular test signal as well as other test signals, the advantage being that the time consuming transient analysis need not be carried for all the test signals, while the steady-state response can be quickly determined by the final value theorem as pointed out already.

The time response performance of a control system is measured by computing several time response performance indices as well as steady-state accuracy. These indices give a quantitative method to compare the performance of alternative system configurations or to adjust the parameters of a given system. As a given parameter is varied, various performance indices may change in a conflicting manner. The best parameter choice would thus be the best compromise solution. Certain of the performance indices may be specified as upper or lower bounds in a design.

5.2 Standard Test Signals

STEP SIGNAL

The *step* is a signal whose value changes from one level (usually zero) to another level A in zero time. The mathematical representation of the step function is

$$r(t) = Au(t)$$

where
$$u(t) = 1; \quad t > 0 \quad \quad (5.1)$$
$$= 0; \quad t < 0$$

In the Laplace transform form,

$$R(s) = A/s$$

The graphical representation of a step signal is shown in Fig. 5.1a.

RAMP SIGNAL

The *ramp* is a signal which starts at a value of zero and increases linearly with time. Mathematically,

$$r(t) = At; \quad t > 0$$
$$= 0; \quad t < 0 \quad \quad (5.2)$$

In the Laplace transfrom form,
$$R(s) = A/s^2$$
The graphical representation of a ramp signal is shown in Fig. 5.1b. From eqns. (5.1) and (5.2), it is seen that a ramp signal is integral of a step signal.

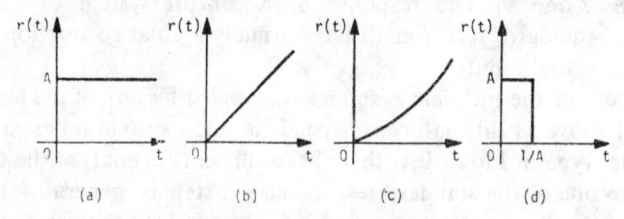

Figure 5.1 Standard test signals.

PARABOLIC SIGNAL

The mathematical representation of this signal is
$$r(t) = At^2/2; \quad t > 0$$
$$= 0; \quad t < 0 \qquad (5.3)$$
In the Laplace transform form,
$$R(s) = A/s^3$$
The graphical representation of a parabolic signal is shown in Fig. 5.1c. From eqns. (5.2) and (5.3), it is seen that a parabolic signal is integral of a ramp signal.

IMPULSE SIGNAL

A unit-impulse is defined as a signal which has zero value everywhere except at $t = 0$, where its magnitude is infinite. It is generally called the δ-function and has the following property:
$$\delta(t) = 0; \quad t \neq 0$$
$$\int_{-\epsilon}^{+\epsilon} \delta(t)dt = 1$$
where ϵ tends to zero.

Since a perfect impulse cannot be achieved in practice, it is usually approximated by a pulse of small width but unit area as shown in Fig. 5.1d.

Mathematically, an impulse function is the derivative of a step function, i.e.,
$$\delta(t) = \dot{u}(t)$$
The Laplace transform of a unit-impulse is
$$\mathcal{L}\delta(t) = 1 = R(s)$$
The impulse response of a system with transfer function $C(s)/R(s) = G(s)$, is given by
$$C(s) = G(s)R(s)$$
$$= G(s)$$

or
$$c(t) = \mathcal{L}^{-1}G(s) = g(t) \tag{5.4}$$

Thus the impulse response of a system, indicated by $g(t)$, is the inverse Laplace transform of its transfer function. The system's impulse response $g(t)$ is sometimes referred to as *weighting function* of the system.

5.3 Time Response of First-order Systems

Let us consider a system shown in Fig. 5.2 which mathematically represents the pneumatic system of Fig. 2.12 whose dynamics is described by eqn. (2.26).

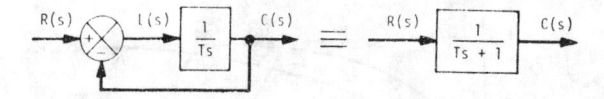

Figure 5.2 Block diagram of a first-order system.

The transfer function of the system from Fig. 5.2 is
$$\frac{C(s)}{R(s)} = \frac{1}{Ts+1} \tag{5.5}$$

In the following sections we shall analyze the system response to unit-step and unit-ramp inputs assuming zero initial conditions.

RESPONSE TO THE UNIT-STEP INPUT

For the unit-step input $[R(s) = 1/s]$, from eqn. (5.5), the output response is given by
$$C(s) = \frac{1}{s(Ts+1)}$$
$$= \frac{1}{s} - \frac{T}{Ts+1}$$

Taking the inverse Laplace transform, we get
$$c(t) = 1 - e^{-t/T}$$

which is plotted in Fig. 5.3.

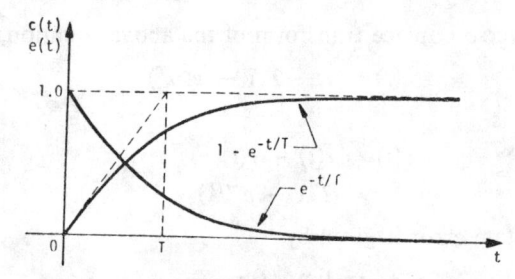

Figure 5.3 Unit-step response of a first-order system.

It is seen that the output rises exponentially from zero value to the final value of unity. The initial slope of the curve at $t = 0$ is given by

$$\left.\frac{dc}{dt}\right|_{t=0} = \left.\frac{1}{T} e^{-t/T}\right|_{t=0} = \frac{1}{T}$$

where T is known as the *time constant* of the system.

The time constant is indicative of how fast the system tends to reach the final value. The speed of response can be quantitatively defined as the time for the output to become a particular percentage of its final value. A large time constant corresponds to a sluggish system and a small time constant corresponds to a fast response as shown in Fig. 5.4.

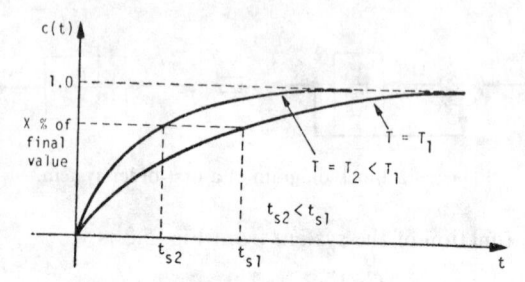

Figure 5.4 Effect of time constant on system response.

Consider now, the error response of the system which is given by
$$e(t) = r(t) - c(t) = e^{-t/T}$$
which is also plotted in Fig. 5.3. The steady-state error is given by
$$e_{ss} = \lim_{t \to \infty} e(t) = 0$$
Thus this system tracks the unit-step input with zero steady-state error.

RESPONSE TO THE UNIT-RAMP INPUT

From eqn. (5.5), the output response for the unit-ramp input $[R(s) = 1/s^2]$ is given by
$$C(s) = \frac{1}{s^2(Ts+1)}$$
$$= \frac{1}{s^2} - \frac{T}{s} + \frac{T^2}{Ts+1}$$

Taking the inverse Laplace transform of the above equation, we get
$$c(t) = t - T(1 - e^{-t/T})$$

The error signal is
$$e(t) = r(t) - c(t)$$
$$= T(1 - e^{-t/T})$$

and the steady-state error is given by
$$e_{ss} = \lim_{t \to \infty} e(t) = T$$

Thus the first-order system under consideration will track the unit-ramp input with a steady-state error T, which is equal to the time constant of the system, as shown in Fig. 5.5. Reducing the system time constant therefore not only improves its speed of response but also reduces its steady-state error to a ramp input.

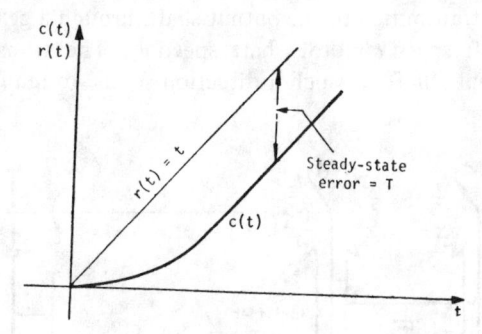

Figure 5.5 Unit-ramp response of a first-order system.

If we examine the derivative of $c(t)$, i.e.,

$$\dot{c}(t) = 1 - e^{-t/T}$$

we find that it is identical to the system response to the unit-step input. The transient response to the ramp input signal thus yields no additional information about the speed of response of the system. We therefore need examine only the steady-state error to the ramp input which can be obtained directly by the final value theorem as given below:

$$\begin{aligned} e_{ss} &= \lim_{t \to \infty} e(t) = \lim_{s \to 0} s\,E(s) \\ &= \lim_{s \to 0} s[R(s) - C(s)] \\ &= \lim_{s \to 0} s\left[\frac{1}{s^2} - \frac{1}{s^2(Ts+1)}\right] \\ &= T \end{aligned}$$

This avoids the need to obtain the inverse Laplace transform resulting in a considerable labour saving in higher-order systems.

5.4 Time Response of Second-order Systems

Consider the servomechanism shown in Fig. 5.6, which controls the position of a mechanical load in accordance with the position of the reference shaft. The two potentiometers convert the input and output positions into proportional electrical signals, which are in turn compared and an error signal equal to the difference of the two appears at the leads coming from potentiometer wiper arms.

The error signal (voltage) is

$$v_e = K_P(r - c)$$

where r = reference shaft position in rad; c = output shaft position in rad; and K_P = potentiometer sensitivity in volts/rad.

The error signal is amplified by a factor K_A by the amplifier and is applied to the armature circuit of the d.c. motor whose field winding is excited with a constant voltage. If any error exists, the motor develops a torque which is transmitted to the output shaft through a gear train of ratio n (n = load shaft speed c/motor shaft speed θ). The transmitted torque rotates the output shaft in such a direction so as to reduce the error to zero.

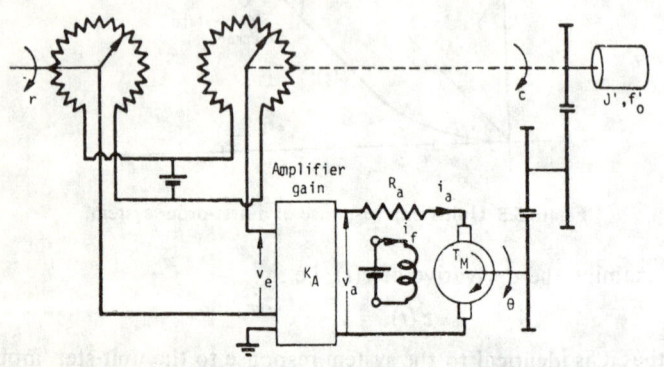

Figure 5.6 Schematic diagram of position servomechanism.

The block diagram of the system is shown in Fig. 5.7a. Here J and f_0 are the equivalent inertia and friction at the motor shaft.

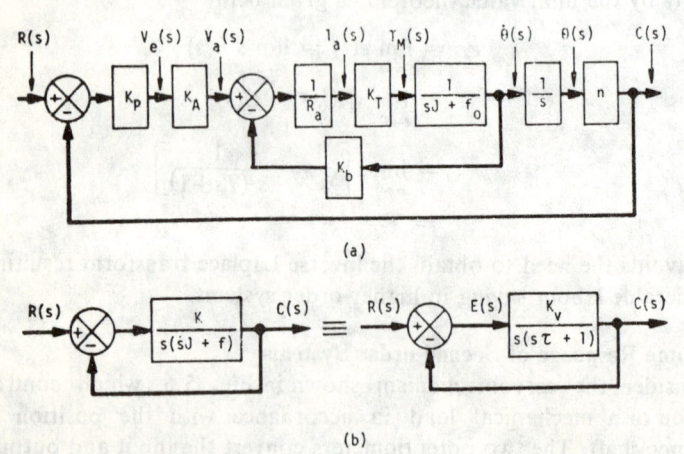

Figure 5.7 (a) Block diagram of the system shown in Fig. 5.6; (b) Simplified block diagram of the system.

The inner loop can be reduced to give the motor transfer function as

$$\frac{\theta(s)}{V_a(s)} = \frac{K_T/R_a}{s(sJ+f)}$$

where $f = f_0 + \dfrac{K_T K_b}{R_a}$.

Time Response Analysis

The block diagram can now be simplified to the form of Fig. 5.7b, where

$$K = K_P K_A K_T \frac{n}{R_a}$$

The forward transfer function can be written in the time constant form as

$$G(s) = \frac{K_v}{s(\tau s + 1)}$$

where $K_v = K/f$, $\tau = J/f$.

Figure 5.7b is in fact a standard second order system involving one forward path integration. As shall be explained later in this chapter, this configuration belongs to a general class of systems called type-1.

Response of Second-order System to the Unit-step

The analysis, now onwards, is not restricted to the particular system used as an example above but is valid for a general second-order type-1 system. From Fig. 5.7b, the overall transfer function of the system is

$$\frac{C(s)}{R(s)} = \frac{K_v}{\tau s^2 + s + K_v} \tag{5.6}$$

$$= \frac{K_v/\tau}{s^2 + \frac{1}{\tau}s + \frac{K_v}{\tau}}$$

It can also be written in the standard form

$$\frac{C(s)}{R(s)} = \frac{\omega_n^2}{s^2 + 2\zeta\omega_n s + \omega_n^2} = \frac{p(s)}{q(s)} \tag{5.7}$$

where $\zeta = $ *damping factor* (or *damping ratio*) $= \dfrac{1}{2\sqrt{(K_v \tau)}} = \dfrac{f}{2\sqrt{(KJ)}}$; and $\omega_n = $ *undamped natural frequency* $= \sqrt{(K_v/\tau)} = \sqrt{(K/J)}$.

The time response of any system is characterized by the roots of the denominator polynomial $q(s)$, which in fact are the poles of the transfer function. The denominator polynomial $q(s)$ is therefore called the *characteristic polynomial* and

$$q(s) = 0 \tag{5.8}$$

is called the *characteristic equation*. The characteristic equation of the system under consideration is

$$s^2 + 2\zeta\omega_n s + \omega_n^2 = 0$$

The roots of this characteristic equation are given by

$$(s^2 + 2\zeta\omega_n s + \omega_n^2) = (s - s_1)(s - s_2)$$

For $\zeta < 1$,

$$s_1, s_2 = -\zeta\omega_n \pm j\omega_n\sqrt{(1-\zeta^2)} = -\zeta\omega_n \pm j\omega_d$$

where $\omega_d = \omega_n\sqrt{(1-\zeta^2)}$, is called the *damped natural frequency*.

Control systems are normally designed with damping factor $\zeta < 1$ (this is explained in detail later in this chapter) and therefore only this case will be investigated in detail here.

From eqn. (5.7), for the unit-step input, the output response is given by

$$C(s) = \frac{\omega_n^2}{s[s+\zeta\omega_n - j\omega_n\sqrt{(1-\zeta^2)}][s+\zeta\omega_n+j\omega_n\sqrt{(1-\zeta^2)}]} \quad (5.9)$$

The Laplace inverse of eqn. (5.9) is obtained, by the method of residues (refer Appendix I) as

$$c(t) = \frac{\omega_n^2}{s^2+2\zeta\omega_n s+\omega_n^2}\bigg|_{s=0}$$

$$+2Re\left[\frac{\omega_n^2}{s[s+\zeta\omega_n - j\omega_n\sqrt{(1-\zeta^2)}]}\bigg|_{s=-\zeta\omega_n - j\omega_n\sqrt{(1-\zeta^2)}} e^{-\zeta\omega_n t} e^{-j\omega_n\sqrt{(1-\zeta^2)}t}\right]$$
(5.10)

$$= 1 - \frac{e^{-\zeta\omega_n t}}{\sqrt{(1-\zeta^2)}} \sin\left[\omega_n\sqrt{(1-\zeta^2)}t + \tan^{-1}\frac{\sqrt{(1-\zeta^2)}}{\zeta}\right] \quad (5.11)$$

The steady-state value of $c(t)$ is given as

$$c_{ss} = \lim_{t\to\infty} c(t) = 1$$

The time response of an underdamped ($\zeta < 1$) second-order system is plotted from eqn. (5.11) in Fig. 5.8. It is damped sinusoid. The response reaches a steady-state value of $c_{ss} = 1$, i.e., the steady-state error of this system approaches zero. The time response for various values of ζ plotted against normalized time $\omega_n t$ is shown in Fig. 5.9. The system breaks into continuous oscillations for $\zeta = 0$, as can be seen from eqn. (5.11). The mathematical expression for the time response in this case is given by

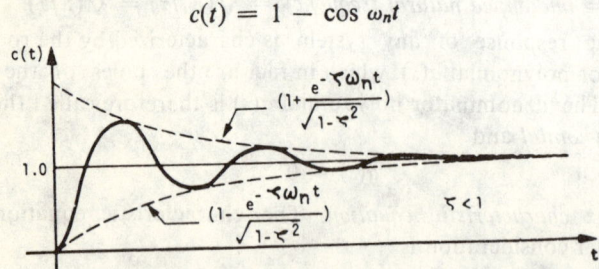

$$c(t) = 1 - \cos \omega_n t$$

Figure 5.8 Unit-step response of underdamped second-order system.

As ζ is increased, the response becomes progressively less oscillatory till it becomes *critically damped* (just non-oscillatory) for $\zeta = 1$ and becomes *overdamped* for $\zeta > 1$. The response expressions for these two cases are not of great significance in control systems and are therefore not presented here.

Figure 5.10 shows the locus of the poles of the second-order system discussed above with ω_n held constant and ζ varying from 0 to ∞. As ζ

Time Response Analysis

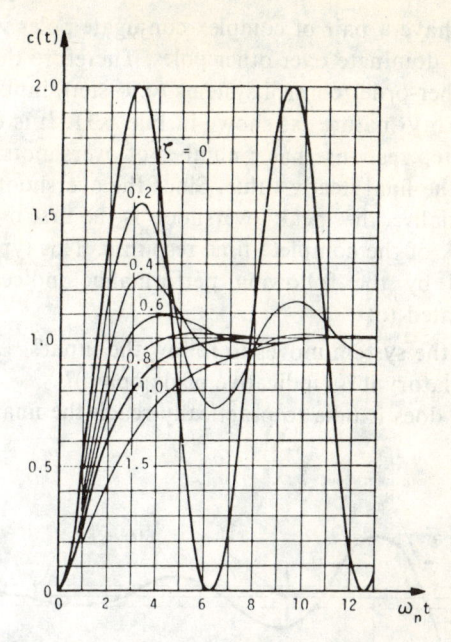

Figure 5.9 Unit-step response curves of second-order system.

increases the poles move away from the imaginary axis along a circular path of radius ω_n meeting at the point $\sigma = -\omega_n$ and then separating and travelling along the real axis, one towards zero and the other towards infinity. For $0 < \zeta < 1$, the poles are complex conjugate pair making an angle of $\theta = \cos^{-1} \zeta$ with the negative real axis.

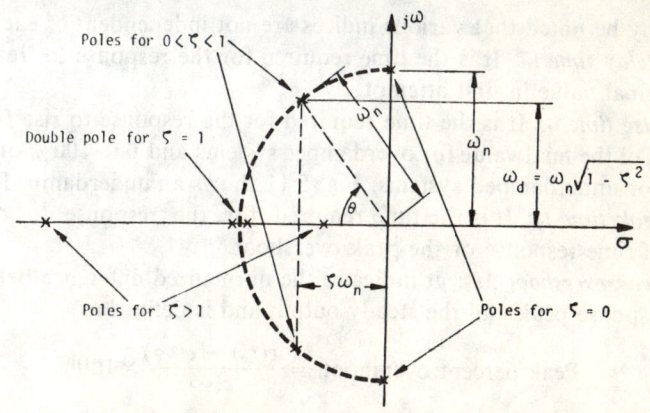

Figure 5.10 Pole locations for a second-order system.

TIME RESPONSE SPECIFICATIONS

As pointed out earlier, control systems are generally designed with damping less than one, i.e., oscillatory step response. Higher-order control systems usually have a pair of complex conjugate poles with damping less than one which dominate over other poles. Therefore the time response of second- and higher-order control systems to a step input is generally of damped oscillatory nature as shown in Fig. 5.11. It is observed from this figure that the step response has a number of overshoots and undershoots with respect to the final steady value. Since the overshoots and undershoots decay exponentially, the peak overshoot is the first overshoot and is the same as the peak of the complete time response. This type of step response is characterized by the following performance indices. The indices are qualitatively related to:

(i) How fast the system moves to follow the input?
(ii) How oscillatory it is (indicative of damping)?
(iii) How long does it take to practically reach the final value?

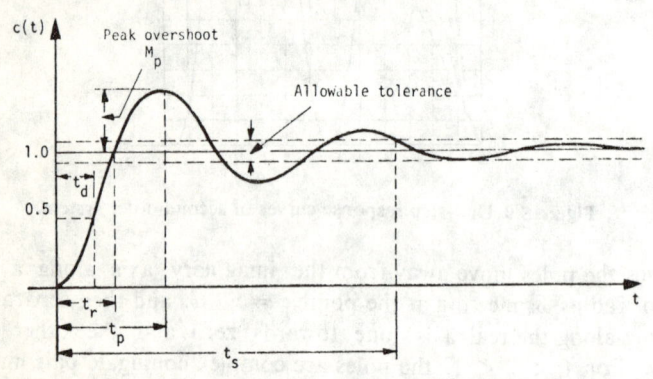

Figure 5.11 Time response specifications.

It may be noted that various indices are not independent of each other.

1. *Delay time t_d*: It is the time required for the response to reach 50% of the final value in first attempt.

2. *Rise time t_r*: It is the time required for the response to rise from 10% to 90% of the final value for overdamped systems and 0 to 100% of the final value for underdamped systems. Fig. 5.11 shows an underdamped case.

3. *Peak time t_p*: It is the time required for the response to reach the peak of time response or the peak overshoot.

4. *Peak overshoot M_p*: It indicates the normalized difference between the time response peak and the steady output and is defined as

$$\text{Peak percent overshoot} = \frac{c(t_p) - c(\infty)}{c(\infty)} \times 100\%$$

In most control systems (except type-0, defined later in this chapter) the steady output for step input is the same as the input. For example, for the second order system of Fig. 5.7b, $c(\infty) = c_{ss} = 1$, i.e., the same as the input.

5 *Settling time t_s*: It is the time required for the response to reach and stay within a specified tolerance band (usually 2% or 5%) of its final value.

6. *Steady-state error e_{ss}*: It indicates the error between the actual output and desired output as t tends to infinity, i.e.,

$$e_{ss} = \lim_{t \to \infty} [r(t) - c(t)]$$

From Fig. 5.11 it is seen that by specifying t_d, t_r, t_p, M_p, t_s and e_{ss}, the shape of the unit-step time response curve is virtually fixed. It should however, be noted that these specifications are mutually dependent and must be specified in a consistent manner.

TIME RESPONSE SPECIFICATIONS OF SECOND-ORDER SYSTEMS

Let us obtain the expressions for the rise time, peak time, peak overshoot and settling time for a second-order system whose block diagram is shown in Fig. 5.7b and whose dynamics for the underdamped case is described by eqn. (5.11).

1. *Rise time t_r*: From eqn. (5.11) the rise time t_r is obtained when $c(t)$ reaches unity for the first time, i.e.,

$$c(t_r) = 1 - \frac{e^{-\zeta\omega_n t_r}}{\sqrt{(1-\zeta^2)}} \sin\left[\omega_n\sqrt{(1-\zeta^2)}t_r + \tan^{-1}\frac{\sqrt{(1-\zeta^2)}}{\zeta}\right] = 1$$

From the above equation, we get

$$t_r = \frac{\pi - \tan^{-1}\frac{\sqrt{(1-\zeta^2)}}{\zeta}}{\omega_n\sqrt{(1-\zeta^2)}} \tag{5.12}$$

For $0 < \zeta < 1$,

$$0 < \tan^{-1}\frac{\sqrt{(1-\zeta^2)}}{\zeta} < \frac{\pi}{2} \tag{5.13}$$

2. *Peak time t_p*: Differentiating eqn. (5.11) with respect to t and equating to zero, we get

$$\frac{dc(t)}{dt} = \frac{\zeta\omega_n}{\sqrt{(1-\zeta^2)}} e^{-\zeta\omega_n t} \sin\left[\omega_n\sqrt{(1-\zeta^2)}t + \tan^{-1}\frac{\sqrt{(1-\zeta^2)}}{\zeta}\right]$$
$$- \frac{e^{-\zeta\omega_n t}}{\sqrt{(1-\zeta^2)}} \omega_n\sqrt{(1-\zeta^2)} \cos\left[\omega_n\sqrt{(1-\zeta^2)}t + \tan^{-1}\frac{\sqrt{(1-\zeta^2)}}{\zeta}\right] = 0 \tag{5.14}$$

With reference to Fig. 5.12, eqn. (5.14) may be written as

$$\sin[\omega_n\sqrt{(1-\zeta^2)}t + \phi]\cos\phi - \cos[\omega_n\sqrt{(1-\zeta^2)}t + \phi]\sin\phi = 0$$

or

$$\sin[\omega_n\sqrt{(1-\zeta^2)}t] = 0 \tag{5.15}$$

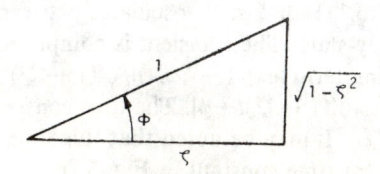

Figure 5.12.

Therefore, the time to various peaks is given by
$$\omega_n\sqrt{(1-\zeta^2)}t = 0, \pi, 2\pi, 3\pi, \ldots$$
Since the peak time corresponds to the first overshoot,
$$t_p = \frac{\pi}{\omega_n\sqrt{(1-\zeta^2)}} \tag{5.16}$$

The first undershoot will occur at $t = 2\pi/\omega_n\sqrt{(1-\zeta^2)}$, the second overshoot at $t = 3\pi/\omega_n\sqrt{(1-\zeta^2)}$ and so on. A plot of normalized peak time $\omega_n t_p$ versus ζ is given in Fig. 5.13.

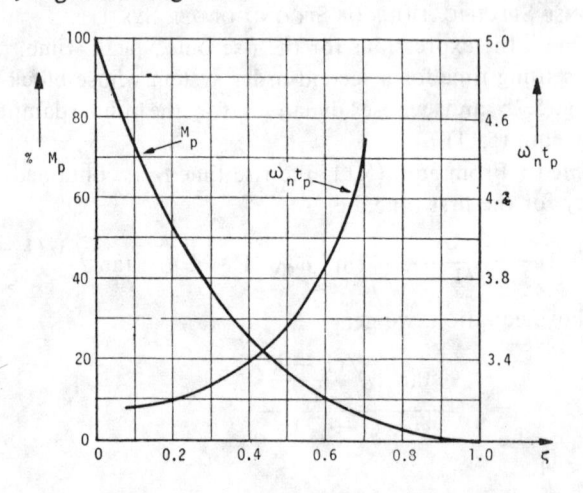

Figure 5.13 M_p and $\omega_n t_p$ versus ζ for a second-order system.

3. *Peak overshoot M_p*: From eqn. (5.11) and Fig. 5.11,
$$M_p = c(t_p) - 1$$
$$= -\frac{e^{-\zeta\omega_n t_p}}{\sqrt{(1-\zeta^2)}}\sin\left[\omega_n\sqrt{(1-\zeta^2)}t_p + \tan^{-1}\frac{\sqrt{(1-\zeta^2)}}{\zeta}\right]$$
$$= e^{-\pi\zeta/\sqrt{(1-\zeta^2)}}$$

Therefore the peak percent overshoot
$$= 100\ e^{-\pi\zeta/\sqrt{(1-\zeta^2)}}\ \% \tag{5.17}$$

As seen from Fig. 5.13, the peak overshoot is a monotonically decreasing function of damping ζ and is independent of ω_n. It is therefore an excellent measure of system damping.

4. *Settling time t_s*: From Fig. 5.8, it is observed that the time response $c(t)$ given by eqn. (5.11) for $\zeta < 1$, oscillates between a pair of envelopes before reaching steady-state. The transient is comprised of a product of an exponentially decaying term $[\exp(-\zeta\omega_n t)]/\sqrt{(1-\zeta^2)}$ and a sinusoidally oscillating term $\sin[\omega_n\sqrt{(1-\zeta^2)}t + \phi]$. The time constant of the exponential envelopes is $T = 1/\zeta\omega_n$. It may be noted that this time constant is equal to 2τ where τ is the motor time constant in Fig. 5.7b.

Figure 5.14 shows the unit-step response versus normalized settling time $\omega_n t_s$ for various values of ζ as well as the tolerance band. It is observed from this figure that as the damping ζ is reduced from the value of one (corresponding to the critical damping case), the normalized settling time decreases monotonically till the first overshoot just touches the upper limit of the tolerance band. As the damping is decreased further, $\omega_n t_s$ suddenly jumps up (i.e., it has a discontinuity) and then increases slightly with further decrease in damping. The plot of $\omega_n t_s$ versus ζ is shown in Fig. 5.15. The plot has another discontinuity corresponding to the first undershoot touching the lower limit of the tolerance band and similar discontinuities occur corresponding to each overshoot and undershoot, respectively touching the upper and lower limits of the tolerance band.

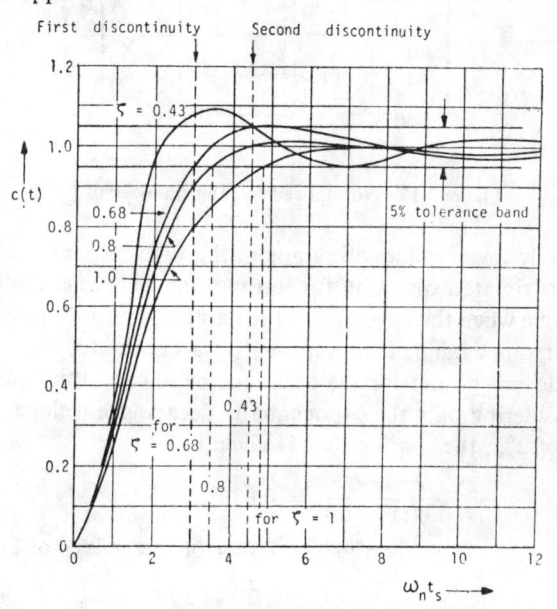

Figure 5.14 Settling time for various values of ζ.

The normalized settling time $\omega_n t_s$ is the least at the first discontinuity ($\zeta = 0.76$ for 2% tolerance band and $\zeta = 0.68$ for 5% tolerance band). This fact justifies why the control systems are normally designed to be underdamped. This argument is further strengthened by the observation that rise time reduces monotonically as the damping is decreased.

In fact many control systems are designed to have even lower damping. Justification of this lies in the fact that almost all practical systems possess some coulomb friction and other nonlinearities (backlash, binding in gears and linkages, etc.). The presence of these nonlinearities tends to introduce steady-state error. As these nonlinearities do not appear in the linear mathematical model, therefore in order to compensate for error introduced by them, the linear system is designed with somewhat higher gain and hence lower damping since $\zeta = 1/2\sqrt{(K_v \tau)}$.

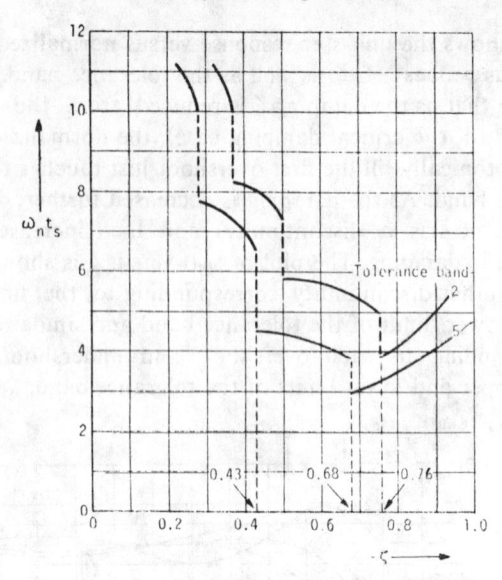

Figure 5.15 Normalized settling time versus ζ.

For relatively lower values of damping, the time response is oscillatory and an approximate expression for settling time can be easily obtained by finding the time when the envelope of the damped sinusoid shown in Fig. 5.8 enters the tolerance band. It is sufficiently accurate to assume that the overshoots do not go outside the tolerance band after this time.

Thus, considering only the exponentially decaying envelope for a tolerance band of 2%, the settling time is given by

$$\frac{e^{-\zeta \omega_n t_s}}{\sqrt{(1-\zeta^2)}} = 0.02$$

or
$$e^{-\zeta \omega_n t_s} \approx 0.02 \text{ (for low values of } \zeta\text{)}$$

or
$$t_s \approx \frac{4}{\zeta \omega_n} = 4T \tag{5.18}$$

where T is the time constant of the exponential term [in fact $T = 2\tau$, where τ is the time constant of the open-loop transfer function $G(s)$ in Fig. 5.7b]. Approximate value of t_s is $3/\zeta\omega_n$ or $3T$ for 5% tolerance band.

It is seen that the settling time is inversely proportional to ω_n.

5. *Steady-state error e_{ss}*: From eqn. (5.11), for unit-step input

$$e_{ss} = \lim_{t \to \infty} [1 - c(t)] = 0$$

Thus the type of second-order system under consideration has zero steady-state error to unit-step input.

From eqn. (5.7), the response to a unit-ramp input $[r(t) = t; R(s) = 1/s^2]$, is given by

$$c(t) = \mathcal{L}^{-1}\left[\frac{\omega_n^2}{s^2(s^2 + 2\zeta\omega_n s + \omega_n^2)}\right]$$

$$= t - \frac{2\zeta}{\omega_n} + \frac{e^{-\zeta\omega_n t}}{\omega_n\sqrt{(1-\zeta^2)}} \sin[\omega_n\sqrt{(1-\zeta^2)}t + \phi]$$

$$\text{for } \zeta < 1$$

$c(t)$ for various values of ζ is plotted in Fig. 5.16. The steady-state error for unit-ramp input is given by

$$e_{ss} = \lim_{t \to \infty} [t - c(t)]$$

$$= \frac{2\zeta}{\omega_n} = \frac{1}{K_v}$$

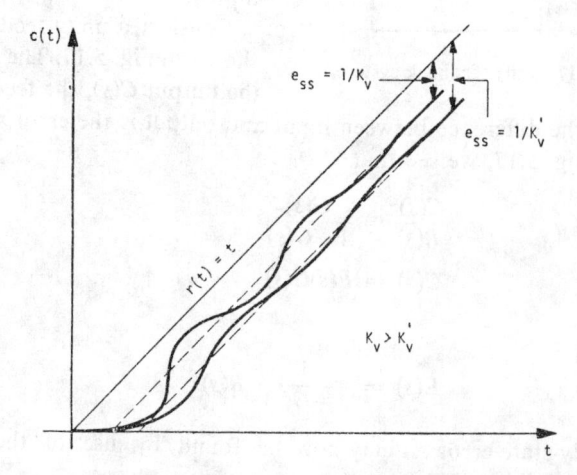

Figure 5.16 Unit-ramp response of second-order system.

From Fig. 5.16, it is observed that the nature of the transient response to the ramp input is similar to that of the step input, i.e., it is damped oscillatory and yields no new information about the transient response of the system. As pointed out earlier, it is therefore sufficient to test the transient response of the system for a step input only. Ramp input response, of course, gives new information about its steady-state behaviour which may be evaluated directly by the final value theorem.

For the system under consideration,

$$e_{ss} = \lim_{s \to 0} s \left[\frac{1}{s^2} - C(s) \right]$$

Substituting for $C(s)$ from eqn. (5.7), we get

$$e_{ss} = 2\zeta/\omega_n = 1/K_v \tag{5.19}$$

The result will be further illustrated in the following section on steady-state errors.

5.5 Steady-state Errors and Error Constants

Steady-state errors constitute an extremely important aspect of system performance, for it would be meaningless to design for dynamic accuracy if the steady output differed substantially from the desired value for one reason or another. The steady-state error is a measure of system accuracy. These errors arise from the nature of inputs, type of system and from non-linearities of system components such as static friction, backlash, etc. These are generally aggravated by amplifier drifts, aging or deterioration.

As discussed in the introduction to this chapter, the steady-state performance of a stable control system is generally judged by its steady-state error to step, ramp and parabolic inputs.

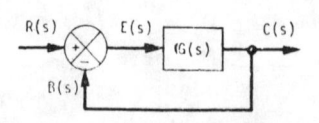

Figure 5.17 Unity feedback system.

Consider a unity feedback system shown in Fig. 5.17. The input is $R(s)$, the output $C(s)$, the feedback signal $B(s)$ and the difference between input and output is the error signal $E(s)$.

From Fig. 5.17, we see that

$$\frac{C(s)}{R(s)} = \frac{G(s)}{1+G(s)}$$

$$C(s) = E(s)G(s)$$

Therefore

$$E(s) = \frac{1}{1+G(s)} R(s) \qquad (5.20)$$

The steady-state error e_{ss} may now be found by use of the final value theorem as follows.

$$e_{ss} = \lim_{t \to \infty} e(t) = \lim_{s \to 0} sE(s) = \lim_{s \to 0} \frac{sR(s)}{1+G(s)} \qquad (5.21)$$

Equation (5.21) shows that the steady-state error depends upon the input $R(s)$ and the forward transfer function $G(s)$. The expression for steady-state errors for various types of standard test signals are derived below:

1. UNIT-STEP INPUT

$$\text{Input } r(t) = u(t)$$

$$R(s) = 1/s$$

From eqn. (5.21)

$$e_{ss} = \lim_{s \to 0} \frac{1}{1+G(s)} = \frac{1}{1+G(0)} = \frac{1}{1+K_p} \qquad (5.22)$$

where $K_p = G(0)$ is defined as the *position error constant*.

2. UNIT-RAMP (VELOCITY) INPUT

Input $\qquad r(t) = t \qquad$ or $\qquad \dot{r}(t) = 1$

$$R(s) = 1/s^2$$

From eqn. (5.21)
$$e_{ss} = \lim_{s \to 0} \frac{1}{s + sG(s)} = \lim_{s \to 0} \frac{1}{sG(s)} = \frac{1}{K_v} \quad (5.23)$$
where $K_v = \lim_{s \to 0} sG(s)$ is defined as the *velocity error constant*.

3. UNIT-PARABOLIC (ACCELERATION) INPUT

Input $r(t) = t^2/2$ or $\ddot{r}(t) = 1$
$R(s) = 1/s^3$

From eqn. (5.21)
$$e_{ss} = \lim_{s \to 0} \frac{1}{s^2 + s^2 G(s)} = \lim_{s \to 0} \frac{1}{s^2 G(s)} = \frac{1}{K_a} \quad (5.24)$$
where $K_a = \lim_{s \to 0} s^2 G(s)$ is defined as the *acceleration error constant*.

TYPES OF FEEDBACK CONTROL SYSTEMS

The open-loop transfer function of a unity feedback system can be written in two standard forms—the *time-constant form* and the *pole-zero form*. In these two forms, $G(s)$ is as given below.

$$G(s) = \frac{K(T_{z1}s+1)(T_{z2}s+1)\ldots}{s^n(T_{p1}s+1)(T_{p2}s+1)\ldots} \quad \text{(time-constant form)} \quad (5.25)$$

$$= \frac{K'(s+z_1)(s+z_2)\ldots}{s^n(s+p_1)(s+p_2)\ldots} \quad \text{(pole-zero form)} \quad (5.26)$$

The gains in the two forms are related by

$$K = K' \frac{\prod_i z_i}{\prod_j p_j} \quad (5.27)$$

With the gain relation of eqn. (5.27) for the two forms of $G(s)$, it is sufficient to obtain steady-state errors in terms of the gains of any one of the forms. We shall use the time constant form in the discussions below.

Equation (5.25) involves the term s^n in the denominator which corresponds to number of integrations in the system. As s tends to zero, this term dominates in determining the steady-state error. Control systems are therefore classified in accordance with the number of integrations in the open-loop transfer function $G(s)$ as described below:

1. *Type-0 System*

If $n = 0$, the steady-state errors to various standard inputs, obtained from eqns. (5.22), (5.23), (5.24) and (5.25) are

$$e_{ss} \text{ (position)} = \frac{1}{1+G(0)} = \frac{1}{1+K} = \frac{1}{1+K_p} \quad (5.28)$$

$$e_{ss} \text{ (velocity)} = \lim_{s \to 0} \frac{1}{sG(s)} = \infty$$

$$e_{ss}(\text{acceleration}) = \lim_{s \to 0} \frac{1}{s^2 G(s)} = \infty$$

Thus a system with $n = 0$ or no integration in $G(s)$ has a constant position error, infinite velocity and acceleration errors. The position error constant is given by the open-loop gain of the transfer function in the time-constant form.

2. *Type-1 System*

If $n = 1$, the steady-state errors to various standard inputs are

$$e_{ss} \text{ (position)} = \frac{1}{1+G(0)} = \frac{1}{1+\infty} = 0$$

$$e_{ss} \text{ (velocity)} = \lim_{s \to 0} \frac{1}{sG(s)} = \frac{1}{K} = \frac{1}{K_v} \qquad (5.29)$$

$$e_{ss} \text{ (acceleration)} = \lim_{s \to 0} \frac{1}{s^2 G(s)} = \infty$$

Thus a system with $n = 1$ or with one integration in $G(s)$ has zero position error, a constant velocity error and an infinite acceleration error at steady-state.

3. *Type-2 System*

If $n = 2$, the steady-state errors to various standard inputs are

$$e_{ss} \text{ (position)} = \frac{1}{1+G(0)} = 0$$

$$e_{ss} \text{ (velocity)} = \lim_{s \to 0} \frac{1}{sG(s)} = 0$$

$$e_{ss} \text{ (acceleration)} = \lim_{s \to 0} \frac{1}{s^2 G(s)} = \frac{1}{K} = \frac{1}{K_a} \qquad (5.30)$$

Thus a system with $n = 2$ or two integrations in $G(s)$ has a zero position error, zero velocity error and a constant acceleration error at steady-state.

Steady-state errors for various inputs and systems are summarized in Table 5.1

Table 5.1 STEADY-STATE ERROR FOR VARIOUS INPUTS AND SYSTEMS

Type of input	Steady-state error		
	Type-0 system	Type-1 system	Type-2 system
Unit-step	$1/(1+K_p)$	0	0
Unit-ramp	∞	$1/K_v$	0
Unit-parabolic	∞	∞	$1/K_a$
	$K_p = \lim_{s \to 0} G(s)$	$K_v = \lim_{s \to 0} sG(s)$	$K_a = \lim_{s \to 0} s^2 G(s)$

For nonunity feedback systems (Fig. 5.18) the difference between the input signal $R(s)$ and feedback signal $B(s)$ is the actuating error signal $E_a(s)$ which is given by [refer eqn. (5.20)]

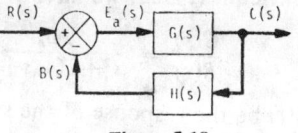

Figure 5.18.

$$E_a(s) = \frac{1}{1+G(s)H(s)} R(s) \qquad (5.31a)$$

Therefore, the steady-state actuating error is

$$e_{ass} = \lim_{s \to 0} \frac{sR(s)}{1+G(s)H(s)} \qquad (5.31b)$$

The error constants for nonunity feedback systems may be obtained by replacing $G(s)$ by $G(s)H(s)$ in Table 5.1.

The error constants K_p, K_v and K_a describe the ability of a system to reduce or eliminate steady-state errors. As the type of system becomes higher (i.e., increasing number of integrations), progressively more steady-state errors are eliminated. Although it appears that there is no limit to the number of integrations, types-0, -1 and -2 are the most commonly employed systems in practice. Systems of type higher than 2, i.e., with more than two integrations are not employed in practice because of two reasons:

(i) These are more difficult to stabilize (this point will be discussed in detail in Chapter 9).

(ii) The dynamic errors for such systems tend to be larger than those for types-0, -1, and -2, although their steady-state performance is desirable.

One of the disadvantages of error constants is that they do not give information on the steady-state error when inputs are other than the three basic types—step, ramp and parabolic. Another difficulty is that the error constants fail to indicate the exact manner in which error function changes with time. The dynamic error may be evaluated using the dynamic error coefficients—the concept generalized to include inputs of almost any arbitrary function of time (refer Problem 5.13).

5.6 Effect of Adding a Zero to a System

The expressions for time response specifications established in Section 5.4 are valid only for the second-order closed-loop transfer function of the form given in eqn. (5.7). Let a zero at $s = -z$ be added to this transfer function. Then we have

$$\frac{C(s)}{R(s)} = \frac{(s+z)(\omega_n^2/z)}{s^2+2\zeta\omega_n s+\omega_n^2} \qquad (5.32a)$$

Note that multiplication term in the numerator of above expression has been adjusted so that *steady-state gain* $\frac{C}{R}(0)$ of the system is unity. This gives the steady-state value of output $c_{ss} = 1$ when input is unit step. Thus the system will track the step input with zero steady-state error.

From eqn. (5.32a) we have

$$\frac{C(s)}{R(s)} = \frac{\omega_n^2}{s^2+2\zeta\omega_n s+\omega_n^2} + \frac{s}{z}\left(\frac{\omega_n^2}{s^2+2\zeta\omega_n s+\omega_n^2}\right) \quad (5.32b)$$

Let $c_z(t)$ be the response of the system with a zero at $s = -z$. Then, from eqn. (5.32b), we have

$$c_z(t) = c(t) + \frac{1}{z}\frac{d}{dt}c(t) \quad (5.32c)$$

where $c(t)$ is the response given by eqn. (5.11).

The effect of added derivative term may be seen by examining Fig. 5.19 where a case for a typical value of ζ (less than one) is considered. We see from this figure that the effect of the zero is to contribute a pronounced early peak to the system's response whereby the peak overshoot may increase appreciably. From eqn. (5.32c) and Fig. 5.19 it is seen that the smaller the value of z, i.e., the closer the zero to origin, the more pronounced is the peaking phenomenon. On account of this fact, *the zeros on the real axis near the origin are generally avoided in design.* However, in a sluggish system the artful introduction of a zero at the proper position can improve the transient response.

We further observe from eqn. (5.32c) that as z increases, i.e., the zero moves farther into the left half of s-plane, its effect becomes less pronounced. For sufficiently large values of z, the effect of zero on transient response may become negligible.

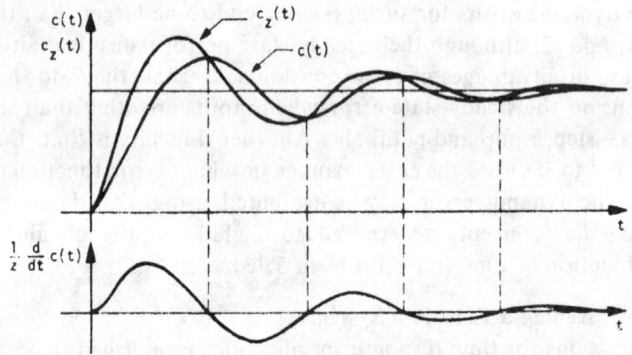

Figure 5.19 Effect of an additional zero on unit-step response of a second-order system.

5.7 Design Specifications of Second-order Systems

A control system is generally required to meet three time response specifications: steady-state accuracy (specified in terms of permissible error e_{ss}), damping factor ζ (or peak overshoot to step input, M_p) and settling time t_s. If the rise time t_r is also specified, it should be consistent with the specification of t_s as both these depend upon ζ and ω_n. Steady-state accuracy requirement is met by a suitable choice of K_p, K_v or K_a depending upon the type of the system. As explained earlier, the damping factor ζ sufficiently less than one is preferred in control systems. For most control systems ζ in the range of 0.7-0.28 (or peak overshoot of 5-40%) is considered acceptable rather desirable from a practical point of view. For this range of ζ, the closed-loop pole locations are restricted to the shaded region of the s-plane as shown in Fig. 5.20.

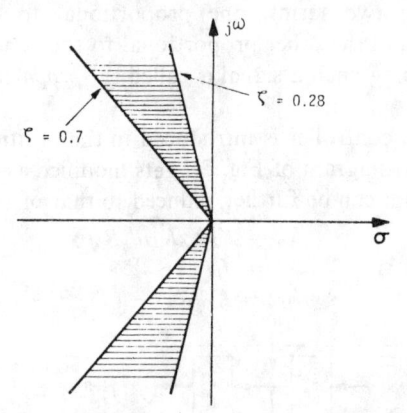

Figure 5.20 Desirable region of pole locations for a second-order system.

Let us now examine the expressions for e_{ss}, ω_n, ζ and t_s for a type-1 second-order system, already derived in this chapter and reproduced below.

$$\zeta = \frac{1}{2\sqrt{(K_v\tau)}}; \quad \omega_n = \sqrt{(K_v/\tau)} \qquad (5.33)$$

$$t_s \approx 4/\zeta\omega_n \text{ (for 2\% tolerance band)} \qquad (5.34)$$

$$e_{ss} = \frac{2\zeta}{\omega_n} = \frac{1}{K_v} \quad \text{(for unit-ramp input)} \qquad (5.35)$$

It is obvious from the above equations that only two of the three specifications, i.e., e_{ss}, ζ, t_s can be met exactly by a suitable choice of K_v and τ provided both of these are variable. The third specification, if consistent, can then be met as an upper or lower bound. However, for the practical example of the second-order system of Fig. 5.6, the time constant τ (motor time constant) is generally fixed. Thus K_v or the open-loop gain is the only adjustable parameter of the system (this can be adjusted by varying the amplifier gain). This second-order system can then meet only one of the specifications exactly. Usually this specification is on the allowable steady-state error. In most systems when the open-loop gain is adjusted to meet the steady-state specified accuracy, the ζ reduces below the specified value impairing the system's dynamic reponse.

We conclude from the above discussion that even to meet two independent specifications a second-order system requires to be modified. This modification termed as *compensation* should allow for high open-loop gains to meet the specified steady-state accuracy and yet preserve a satisfactory dynamic performance. Some of the practical modification schemes are discussed below.

DERIVATIVE ERROR COMPENSATION

A system is said to possess derivative error compensation when the generation of its output depends in some way on the rate of change of actuating signal. For the system shown in Fig. 5.6, this type of compensation is easily introduced by using an amplifier which provides an output signal containing two terms, one proportional to the derivative of the actuating signal and the other proportional to the actuating signal itself. A controller producing such a signal is called a *proportional plus derivative controller*.

When such a controller is introduced in the position control system of Fig. 5.6, the block diagram of Fig. 5.7 gets modified as shown in Fig. 5.21a. This block diagram can be further reduced to that of Fig. 5.21b, where

$$K_v = K_P K_A K_T n/(R_a f)$$
$$\tau = J/f$$
$$K_D' = K_D/K_A$$

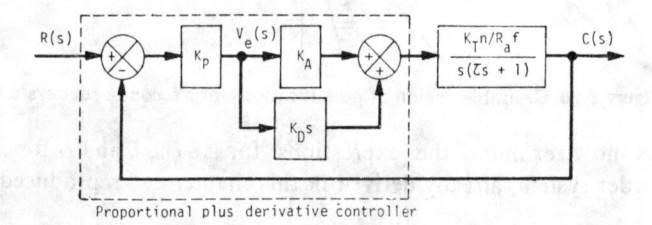

Figure 5.21(a) Proportional plus derivative controller for the system shown in Fig. 5.6.

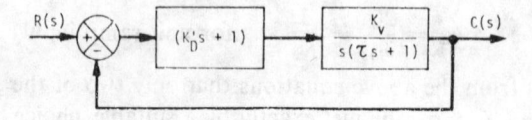

Figure 5.21(b) Second-order system with error derivative control.

From Fig. 5.21b the closed-loop transfer function of the system is given by

$$T(s) = \frac{\frac{K_v}{\tau}(K_D's+1)}{s^2 + \left(\frac{1+K_v K_D}{\tau}\right)s + \frac{K_v}{\tau}}$$

Accordingly, the characteristic equation is modified from that of eqn. (5.8) to the following:

$$s^2 + \left(\frac{1+K_v K_D}{\tau}\right)s + \frac{K_v}{\tau} = 0$$

The damping and natural frequency of the compensated system are given by

$$\omega_n' = \sqrt{(K_v/\tau)} = \omega_n \qquad (5.36)$$

$$\zeta' = \frac{1+K_v K_D'}{2\sqrt{(K_v \tau)}} = \zeta + \frac{K_D'}{2}\sqrt{(K_v/\tau)} \qquad (5.37)$$

We observe that compared to the uncompensated system, for the same K_v, the natural frequency of the compensated system remains unchanged, while its damping is increased by $(K_D'/2)\sqrt{(K_v/\tau)}$. If the steady-state error e_{ss} to velocity input is specified (this is generally the case), the system K_v gets fixed (as $e_{ss} = 1/K_v$) and hence the amplifier gain K_A. The closed-loop damping ζ' can now be raised to the desired level by a suitable choice of K_D from eqn (5.37). $K_D = K_D' K_A$ gives the constant of the derivative term for the amplifier.

An additional advantage of the scheme is that as the damping increases due to compensation with ω_n remaining fixed, the system settling time reduces (as $t_s = 4/\zeta\omega_n$).

DERIVATIVE OUTPUT COMPENSATION

A system is said to possess a derivative output compensation when the generation of its output depends in some way upon the rate at which the controlled variable is changing. For a servo system, a common way of obtaining this compensation is by means of tachometer feedback.

To obtain this type of compensation for the position control system of Fig. 5.6, a d.c. tachometer is coupled to the motor and its output voltage v_t which is proportional to speed (derivative of position), i.e., $v_t = K_t \dot{\theta}$, is fedback negatively to the amplifier input. The system with this modification is resketched in Fig. 5.22a. The new block diagram in which the inner loop due to motor back emf has already been simplified is presented in Fig. 5.22b.

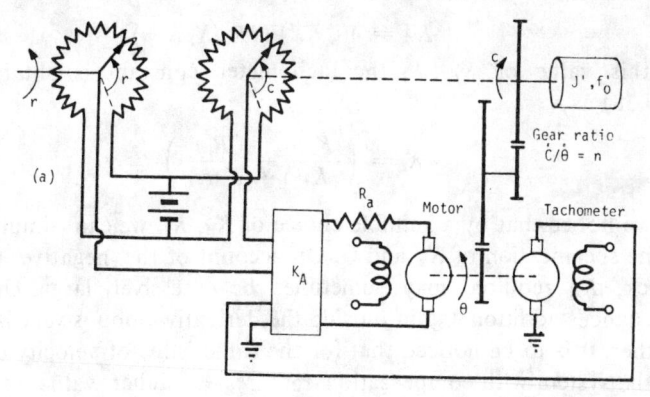

Figure 5.22 (a) Second-order position control system with tachometer (derivative) feedback.

Due to the tachometer feedback, the gain of the inner loop and the effective time constant get reduced. Simplifying the inner loop, we get the block diagram of Fig. 5.22c, from which we can write

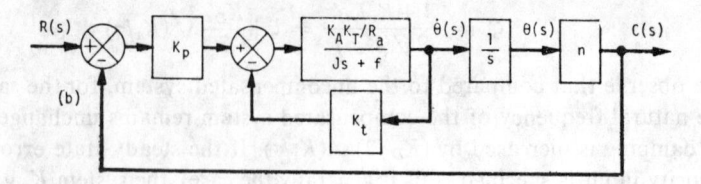

Figure 5.22(b) Block diagram of Fig. 5.22a.

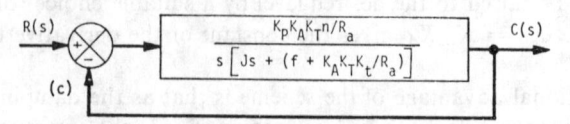

Figure 5.22(c) Reduced block diagram of Fig. 5.22b.

$$K_v' = \frac{K_P K_A K_T n}{R_a f + K_A K_T K_t} \tag{5.38}$$

$$\tau' = \frac{J R_a}{R_a f + K_A K_T K_t} \tag{5.39}$$

The natural frequency and damping of the closed-loop can then be expressed as

$$\omega_n' = \sqrt{(K_v'/\tau')} = \sqrt{(K_P K_A K_T n/J R_a)} \tag{5.40}$$
$$\zeta' = 1/2\sqrt{(K_v'\tau')} = (R_a f + K_A K_T K_t)/2\sqrt{(K_P K_A K_T n J R_a)} \tag{5.41}$$

For a specified K_v' and ζ', we can write from eqns. (5.38) and (5.41)

$$\zeta' K_v' = \tfrac{1}{2}\sqrt{(K_P K_A K_T n/J R_a)}$$

from which K_A is determined as

$$K_A = 4(\zeta' K_v')^2 (J R_a/K_P K_T n)$$

Using this value of K_A, K_t the tachometer constant is obtained from eqn. (5.38) as

$$K_t = \left(\frac{K_P n}{K_v'} - \frac{R_a f}{K_A K_T} \right)$$

Thus, we notice that by a suitable choice of K_A, K_t, we can simultaneously meet the specification of K_v' and ζ'. On account of the negative derivative feedback, K_A required may sometimes be excessively large. Under such circumstances additional gain outside the derivative loop is very helpful.

Further, it is to be noticed that for the same value of velocity error constant, the system with compensation requires a higher value of K_A and hence it will have a higher natural frequency (eqn. 5.40). Compensation thus increases both the system damping and natural frequency resulting in reduced settling time.

INTEGRAL ERROR COMPENSATION

In an integral error compensation scheme, the output response depends in some manner upon the integral of the actuating signal. This type of compensation is introduced by using a controller which produces an output signal consisting of two terms, one proportional to the actuating signal and the other proportional to its integral. Such a controller is called *proportional plus integral controller*.

The block diagram of the system of Fig. 5.6 with proportional plus integral compensation is shown in Fig. 5.23a, while its simplified block diagram is given in Fig. 5.23b where

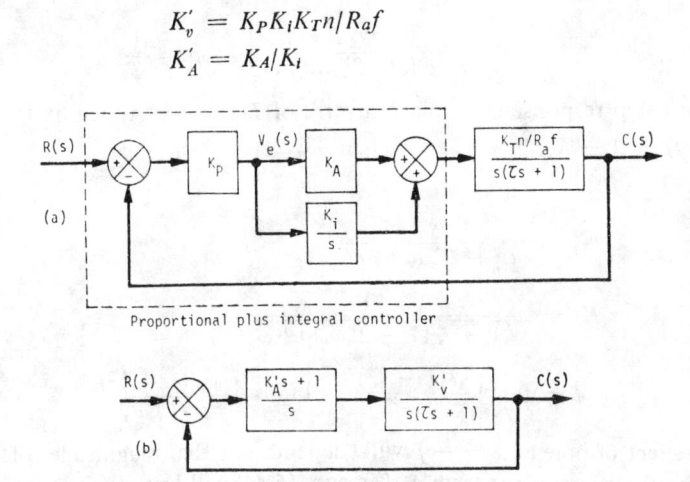

$$K'_v = K_P K_i K_T n / R_a f$$
$$K'_A = K_A / K_i$$

Figure 5.23 Proportional plus integral controller for the system shown in Fig. 5.6.

The closed-loop transfer function is easily obtained as

$$\frac{C}{R}(s) = \frac{K'_v (K'_A s+1)}{(\tau s^3 + s^2 + K'_v K_A s + K'_v)} \tag{5.42}$$

We find that the integral error compensation changes a second-order system into a third-order one. The effect of compensation on system dynamics therefore cannot be visualized as easily as in the previous two types of compensation schemes. The techniques to study the dynamics of higher-order systems will be taken up later in Chapters 7 and 9. A significant contribution of integral error compensation to the system steady-state performance is, however, obvious as the additional integration in the forward path changes the system from type-1 to type-2 and that the error of velocity input is eliminated or considerably reduced as the practical integration may not be perfect. Thus, we can use integral error compensation to meet high accuracy requirements. However as we shall see in Chapter 6 on stability that a third-order characteristic equation introduces a distinct possibility of system instability.

5.8 Design Considerations for Higher-order Systems

The results presented in the previous section hold for second-order systems. These results could be extended to higher-order systems with a *dominant pair of complex poles*, i.e., as long as the real part of these poles is much less than those of the other poles, the time response of higher-order systems can be approximated by that obtained by neglecting all the system poles other than the dominant poles.

For example, let us consider a third-order system with transfer function

$$\frac{C(s)}{R(s)} = T(s) = \frac{\omega_n^2}{(1+sT)(s^2+2\zeta\omega_n s+\omega_n^2)}$$

$$= \frac{\omega_n^2 p}{(s+p)(s^2+2\zeta\omega_n s+\omega_n^2)} \; ; \; p = \frac{1}{T} \quad (5.43)$$

The unit-step response is (Laplace inverse of $T(s)/s$; see Appendix I)

$$c_p(t) = 1 - K_1 e^{-pt} + K_2 e^{-\zeta\omega_n t} \sin[\omega_n \sqrt{(1-\zeta^2)} t - \phi] \quad (5.44)$$

where

$$K_1 = \frac{\omega_n^2}{\omega_n^2 - 2\zeta\omega_n p + p^2}$$

$$K_2 = \frac{1}{\sqrt{1-\zeta^2(1-2\zeta\omega_n/p+\omega_n^2/p^2)}}$$

$$\phi = \tan^{-1}\frac{\sqrt{(1-\zeta^2)}}{-\zeta} + \tan^{-1}\frac{\omega_n\sqrt{(1-\zeta^2)}}{p-\zeta\omega_n}$$

The effect of pole at $s = -p$ will depend on the magnitude of p. If $p \ll \zeta\omega_n$, the effect of the term e^{-pt} in eqn. (5.44) will last much longer than that of the term whose magnitude is governed by $\exp(-\zeta\omega_n t)$. Further, from eqn. (5.44) we observe that

$$\lim_{p \to 0} K_1 = 1; \quad \lim_{p \to 0} K_2 = 0$$

Therefore as the pole at $s = -p$ moves closer to the origin, the response $c_p(t)$ in eqn. (5.44) is dominated by this pole and the effect of the two complex poles at $-\zeta\omega_n \pm j\omega_n\sqrt{(1-\zeta^2)}$ diminishes.

On the other hand, if $\zeta\omega_n \ll p$, the effect of the term e^{-pt} in eqn. (5.44) will be over much more quickly than the term whose magnitude is governed by $\exp(-\zeta\omega_n t)$. Again, from eqn. (5.44) we observe that

$$\lim_{p \to \infty} K_1 = 0; \quad \lim_{p \to \infty} K_2 = \frac{1}{\sqrt{(1-\zeta^2)}}$$

Therefore, as the pole at $s = -p$ moves far away from the origin, the response $c_p(t)$ in eqn. (5.44) is dominated by the complex poles at

$$-\zeta\omega_n \pm j\omega_n\sqrt{(1-\zeta^2)},$$

Figure 5.24 shows unit-step response of a third-order system of eqn. (5.43) with $\zeta = 0.3$. It is observed that for $p/\zeta\omega_n = 6$, the response of third-order system is, for all practical purposes, the response $c(t)$ of the system with the transfer function

$$T(s) = \frac{\omega_n^2}{s^2+2\zeta\omega_n s+\omega_n^2}$$

Time Response Analysis

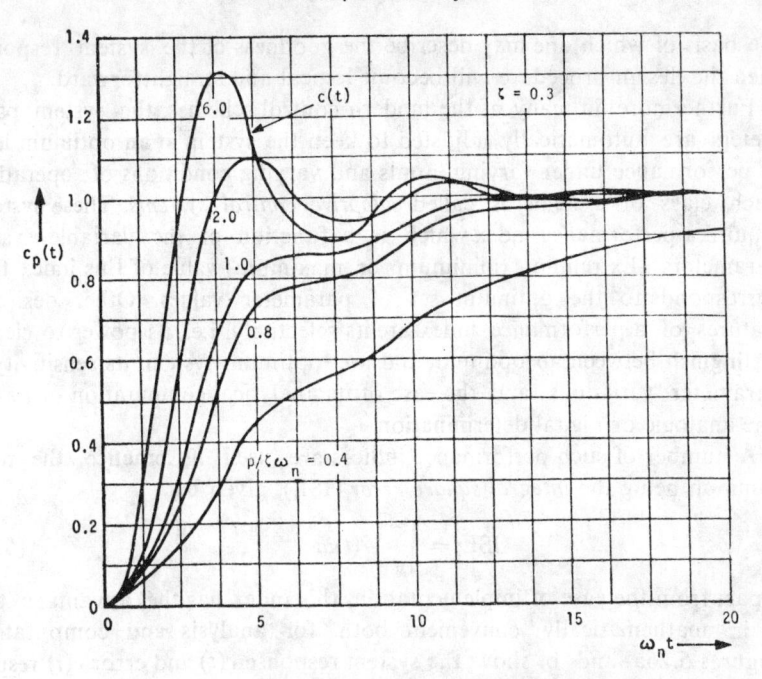

Figure 5.24 Unit-step response of a third-order system.

The relative location of dominant poles of the third-order system is shown in Fig. 5.25.

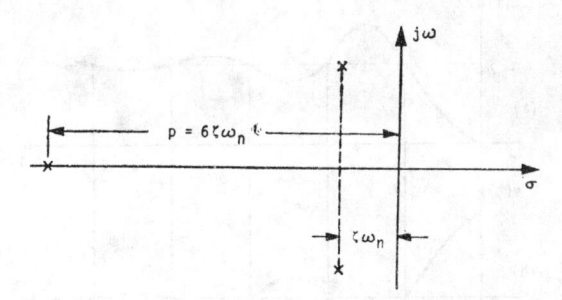

Figure 5.25 Location of dominant poles of the third-order system.

5.9 Performance Indices

As discussed already, the design of a control system is an attempt to meet a set of specifications which define the overall performance of the system in terms of certain measurable quantities. A number of performance measures have been introduced so far in respect of dynamic response to step input (ζ, M_p, t_r, t_p, t_s, etc.) and the steady-state error, e_{ss}, to both step and higher-order inputs. These measures have to be satisfied simultaneously in design and hence the design necessarily becomes a trial and error procedure. If, however, a single performance index could be established on the basis of which one may describe the goodness of the system response, then the design procedure will become logical and straightforward.

Furthermore, in many of the modern control schemes, the system parameters are automatically adjusted to keep the system at an optimum level of performance under varying inputs and varying conditions of operation. Such class of systems is called *adaptive control systems*. These systems require a performance index which is a function of the variable system parameters. Extremum (minimum or maximum) value of this index then corresponds to the optimum set of parameter values. Other desirable features of a performance index are its selectivity, i.e., its power to clearly distinguish between an optimum and non-optimum system, its sensitivity to parameter variations and the ease of its analytical computation or its on-line analogic or digital determination.

A number of such performance indices are used in practice, the most common being the *integral square error* (ISE), given by

$$\text{ISE} = \int_0^\infty e^2(t)dt \tag{5.47}$$

Apart from the ease of implementation, this index has the advantage that it is mathematically convenient both for analysis and computation. Figures 5.26a and b show the system response $c(t)$ and error $e(t)$ respectively to unit-step input. The square error is shown in Fig. 5.26c and its integral in Fig. 5.26d. It is obvious that ISE converges to a limit as $t \to \infty$. Minimization of ISE by adjusting system parameters is a good compromise

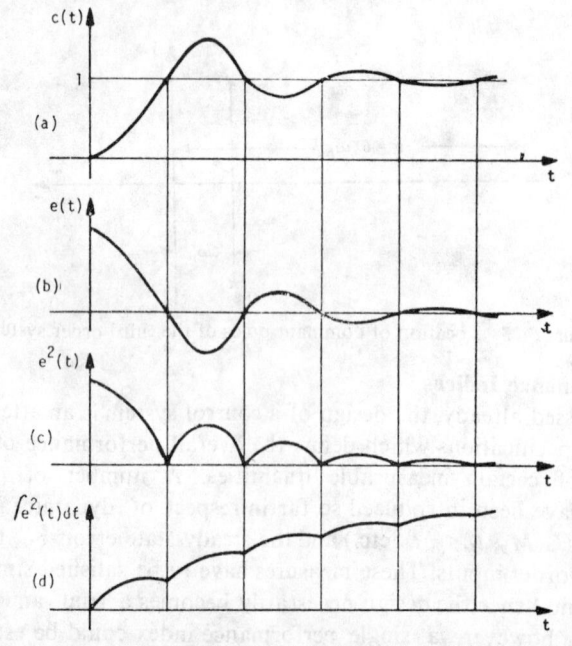

Figure 5.26 Calculation of integral square error.

Time Response Analysis

between reduction of rise time to limit the effect of large initial error, reduction of peak overshoot and reduction of settling time to limit the effect of small error lasting for a long time.

Consider now the second-order system discussed previously in this chapter. From eqn. (5 11), the error response to unit-step is given by

$$e(t) = \frac{e^{-\zeta\omega_n t}}{\sqrt{(1-\zeta^2)}} \sin\left[\omega_n\sqrt{(1-\zeta^2)}t + \tan^{-1}\frac{\sqrt{(1-\zeta^2)}}{\zeta}\right]$$

Therefore

$$\text{ISE} = \int_0^\infty e^2(t)dt = \frac{1}{\omega_n}\int_0^\infty e^2(t_n)dt_n$$

where $t_n = \omega_n t$, i.e., the normalized time.

Simplification of the above expression leads to the following result:

$$\text{ISE}^* = \frac{1}{2\omega_n}\left(\frac{1}{2\zeta} + 2\zeta\right) \tag{5.48}$$

For a fixed choice of ω_n, ISE given by eqn. (5.48) is minimized for $\zeta = 0.5$. This result corroborates the previously elaborated intuitive design criterion that the damping factor of a control system should be sufficiently less than 1.

For higher-order systems, ISE must be computed numerically. Since it is not practicable to integrate up to infinity, the limit ∞ is replaced by T which is chosen sufficiently large so that $e(t)$ for $t > T$ is negligible. Usually T is estimated to be settling time t_s or multiple of settling time, i.e.,

$$\text{ISE} = \int_0^{T=nt_s} e^2(t)dt$$

*Details of the derivation are given below:

$$e(t_n) = \frac{e^{-\zeta t_n}}{\sqrt{(1-\zeta^2)}} \sin[\sqrt{(1-\zeta^2)}t_n + \phi]$$

where $\phi = \tan^{-1}\frac{\sqrt{(1-\zeta^2)}}{\zeta}$

$$e^2(t_n) = \frac{e^{-2\zeta t_n}}{(1-\zeta^2)} \sin^2[\sqrt{(1-\zeta^2)}t_n + \phi]$$

$$= \frac{e^{-2\zeta t_n}}{2(1-\zeta^2)} [1 - \cos 2\{\sqrt{(1-\zeta^2)}t_n + \phi\}]$$

$$= \frac{e^{-2\zeta t_n}}{2(1-\zeta^2)} [1 - (2\zeta^2 - 1)\cos 2\sqrt{(1-\zeta^2)}t_n + 2\zeta\sqrt{(1-\zeta^2)}\sin 2\sqrt{(1-\zeta^2)}t_n]$$

$$\frac{1}{\omega_n}\int_0^\infty e^2(t_n)dt = \frac{1}{2\omega_n(1-\zeta^2)} \int_0^\infty e^{-2\zeta t_n}[1 - (2\zeta^2 - 1)\cos 2\sqrt{(1-\zeta^2)}t_n + 2\zeta\sqrt{(1-\zeta^2)}\sin 2\sqrt{(1-\zeta^2)}t_n]dt_n$$

$$= \frac{1}{2\omega_n}\left(\frac{1}{2\zeta} + 2\zeta\right)$$

Another easily instrumented performance index is the *integral of the absolute magnitude of error* (IAE), which is written as

$$\text{IAE} = \int_0^T |e(t)|\, dt$$

In order to reduce the weighting of the large initial error and to penalize the small errors occurring later in the response more heavily, the following indices are proposed.

Integral time-absolute error, $\quad \text{ITAE} = \int_0^T t|e(t)|\, dt$

Integral time-square error, $\quad \text{ITSE*} = \int_0^T t e^2(t)\, dt$

Figure 5.27 shows the comparison of the different performance indices for the second-order system having transfer function $C(s)/R(s) = 1/(s^2+2s+1)$.

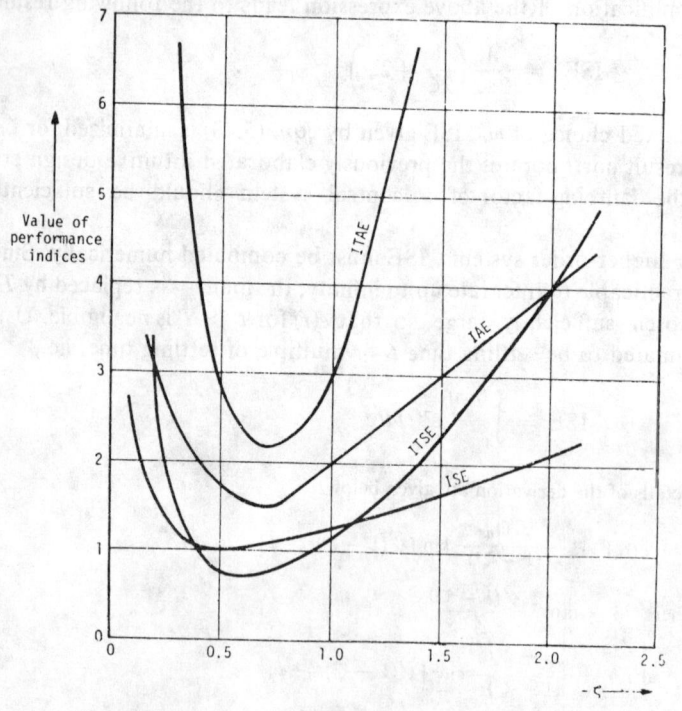

Figure 5.27 Performance indices for a second-order system.

*ITSE for the second-order system discussed in this chapter is obtained as below:

$$\text{ITSE} = \int_0^\infty t e^2(t)\, dt = \frac{1}{\omega_n^2} \int_0^\infty t_n e^2(t_n)\, dt_n$$

$$= \frac{1}{2\omega_n^2(1-\zeta^2)} \int_0^\infty t_n e^{-2\zeta t_n}[1 - (2\zeta^2 - 1) \cos 2\sqrt{(1-\zeta^2)}\,t_n$$
$$+ 2\zeta\sqrt{(1-\zeta^2)} \sin 2\sqrt{(1-\zeta^2)}\,t_n]\, dt_n$$

Integrating each part we get

$$\text{ITSE} = \frac{1}{2\omega_n^2(1-\zeta^2)} \left[\frac{1}{4\zeta^2} - (2\zeta^2 - 1)\frac{(2\zeta^2 - 1)}{4} + 2\zeta\sqrt{(1-\zeta^2)}\, \frac{1}{2}\, \zeta\sqrt{(1-\zeta^2)} \right]$$

$$= \frac{1}{2\omega_n^2}\left(\frac{1}{4\zeta^2} + 2\zeta^2 \right)$$

For fixed ω_n, ITSE is minimised for $\zeta = 1/(8)^{1/4} = 0.60$

Inspection of these curves reveals that ISE is not very sensitive to parameter variations since the curve is rather flat near the point where the performance index is minimum (i.e., $\zeta = 0.5$ as already stated). Therefore, the selectivity of this performance index is poor. It, however, has the advantage of being easy to deal with mathematically. The IAE performance index (IAE is minimized for $\zeta = 0.7$ for the second-order system under consideration) gives slightly better selectivity than ISE. The ITAE performance index (ITAE is minimized for $\zeta = 0.707$ for the second-order system under consideration) produces smaller overshoots and oscillations than the IAE and ISE indices. In addition, it is the most sensitive of the three, i.e., it has the best selectivity. The ITSE index is somewhat less sensitive but is not comfortable computationally.

In practice, relatively insensitive criterion may be more useful and based on this logic, ISE may be the most desirable performance index in some practical applications and is considered further in Chapter 10. Where selectivity is more important, the ITAE performance index is a better choice.

Optimal Control Systems

A control system is optimum when the selected performance index is minimized. The optimum value of the system parameters depends directly on the definition of optimality.

A paper by Graham and Lathrop [1] created a great deal of interest in system optimization using ITAE criterion. They suggested the form that the system transfer function should take, for various order systems, in order to achieve zero steady-state step and ramp error systems and minimize ITAE.

In the case of the zero steady-state step error systems, the general closed-loop transfer function is

$$T(s) = \frac{C(s)}{R(s)} = \frac{a_n}{s^n + a_1 s^{n-1} + \cdots + a_{n-1}s + a_n} \quad (5.49)$$

Table 5.2 shows the optimum forms of the closed-loop transfer functions based on the ITAE criterion. The procedure used to produce this table was to vary each coefficient in eqn. (5.49) until the ITAE value became a minimum. Then the successive coefficients were varied in sequence to minimize the ITAE value. These standard forms of transfer functions provide a quick and simple method for synthesizing an optimum dynamic response (other standard forms based on different performance indices are also available).

In the case of zero steady-state ramp error systems, the general closed-loop transfer function is

$$T(s) = \frac{C(s)}{R(s)} = \frac{a_{n-1}s + a_n}{s^n + a_1 s^{n-1} + \cdots + a_{n-1}s + a_n} \quad (5.50)$$

Table 5.3 gives the optimum forms of the closed-loop transfer functions based on the ITAE criterion (refer Problem 5.18).

Table 5.2 OPTIMUM FORMS OF THE CLOSED-LOOP TRANSFER FUNCTIONS BASED ON THE ITAE CRITERION (ZERO STEADY-STATE STEP ERROR SYSTEMS)

$$\frac{C(s)}{R(s)} = \frac{a_n}{s^n + a_1 s^{n-1} + \ldots + a_{n-1}s + a_n}$$

$$s + \omega_n$$

$$s^2 + 1.4\omega_n s + \omega_n^2$$

$$s^3 + 1.75\omega_n s^2 + 2.15\omega_n^2 s + \omega_n^3$$

$$s^4 + 2.1\omega_n s^3 + 3.4\omega_n^2 s^2 + 2.7\omega_n^3 s + \omega_n^4$$

$$s^5 + 2.8\omega_n s^4 + 5.0\omega_n^2 s^3 + 5.5\omega_n^3 s^2 + 3.4\omega_n^4 s + \omega_n^5$$

$$s^6 + 3.25\omega_n s^5 + 6.60\omega_n^2 s^4 + 8.60\omega_n^3 s^3 + 7.45\omega_n^4 s^2 + 3.95\omega_n^5 s + \omega_n^6$$

Table 5.3 OPTIMUM FORMS OF THE CLOSED-LOOP TRANSFER FUNCTIONS BASED ON THE ITAE CRITERION (ZERO STEADY-STATE RAMP ERROR SYSTEMS)

$$\frac{C(s)}{R(s)} = \frac{a_{n-1}s + a_n}{s^n + a_1 s^{n-1} + \ldots + a_{n-1}s + a_n}$$

$$s^2 + 3.2\omega_n s + \omega_n^2$$

$$s^3 + 1.75\omega_n s^2 + 3.25\omega_n^2 s + \omega_n^3$$

$$s^4 + 2.41\omega_n s^3 + 4.93\omega_n^2 s^2 + 5.14\omega_n^3 s + \omega_n^4$$

$$s^5 + 2.19\omega_n s^4 + 6.50\omega_n^2 s^3 + 6.30\omega_n^3 s^2 + 5.24\omega_n^4 s + \omega_n^5$$

$$s^6 + 6.12\omega_n s^5 + 13.42\omega_n^2 s^4 + 17.16\omega_n^3 s^3 + 14.14\omega_n^4 s^2 + 6.76\omega_n^5 s + \omega_n^6$$

5.10 Illustrative Examples

The design concepts introduced in this chapter are further illustrated with the help of following examples.

Example 5.1: Reconsider the servomechanism shown in Fig. 5.6. The parameters of the system are given as follows:

Sensitivity of error detector, $K_P = 1$ volt/rad
Gain of dc amplifier, K_A (variable)
Resistance of armature of motor, $R_a = 5$ ohms
Equivalent inertia at the motor shaft, $J = 4 \times 10^{-3}$ kg-m^2

Equivalent friction at the motor shaft, $f_0 = 2\times 10^{-3}$ newton-m per rad/sec
Torque constant of motor, $K_T = 1$ newton-m/amp
Gear ratio, $n = 1/10$

The value of the back emf constant K_b is not given in the above list of parameters. However, a definite relationship exists between K_b and K_T (refer Section 2.3). In MKS units $K_b = K_T$.

The open-loop transfer function of the system in the pole-zero form is (refer Fig. 5.7a)

$$G(s) = \frac{K}{s(sJ+f)}$$

where

$$f = f_0 + \frac{K_T K_b}{R_a} = 202\times 10^{-3}$$

$$J = 4\times 10^{-3}$$

$$K = K_P K_A K_T \frac{n}{R_a} = \frac{K_A}{50}$$

The open-loop transfer function of the system in the time-constant form is (refer Fig. 5.7b)

$$G(s) = \frac{K_v}{s(s\tau+1)}$$

where

$$K_v = K/f = \frac{10^3 K_A}{50\times 202}$$

$$\tau = J/f = 4/202$$

The closed-loop transfer function of the system is

$$\frac{C(s)}{R(s)} = \frac{5K_A}{s^2+50.5s+5K_A}$$

The natural undamped frequency of the system is

$$\omega_n = \sqrt{(5K_A)} \tag{5.51}$$

The damping ratio is

$$\zeta = 25.25/\omega_n \tag{5.52}$$

The system under consideration is a type-1 system. Therefore, from Table 5.1 we find that the steady-state error to unit step input is zero and steady-state error to unit-ramp input is

$$e_{ss} = \frac{1}{K_v} = \frac{50\times 202}{10^3 K_A} = \frac{1.01}{K_A} \tag{5.53}$$

Proportional Control

Let us study the effects on the time response when the value of the gain K_A is varied. From eqns. (5.51) (5.53), we note that an increase in K_A increases the natural undamped frequency ω_n but decreases the damping ratio ζ. Therefore, if we choose to improve the steady-state accuracy of the system by increasing the forward gain (eqn. (5.53)), the transient response becomes more oscillatory.

Consider for example, that we desire

$$e_{ss} \text{ (to unit ramp input)} = 0.001$$

Therefore,

$$K_A = 1,010$$
$$\omega_n = 71$$
$$\zeta = \frac{25.25}{71} = 0.355$$

This tells us that with $K_A = 1,010$, the system becomes quite oscillatory and therefore is not a desirable one.

Proportional plus Derivative Control

Consider that the positional control system of Fig. 5.6 is modified by replacing the dc amplifier with proportional plus derivative control so that the open-loop transfer function now is (Fig. 5.21b)

$$G(s) = \frac{K_v\left(\frac{K_D}{K_A}s+1\right)}{s(\tau s+1)}$$

The characteristic equation for the system is

$$1+G(s) = 0$$

or

$$s^2+(50.5+5K_D)s+5K_A = 0$$

Notice that the derivative control has the effect of increasing the coefficient of s term. It means that the damping of the system is increased. If K_A is fixed by considerations of steady-state accuracy, we can control parameter K_D to achieve desired transient response.

Suppose that keeping $K_A = 1,010$, we desire a damping ratio of 0.6. We can now achieve it by adjusting the derivative constant (K_D) as below:

$$50.5+5K_D = 2\times0.6\times71$$

or

$$K_D = 6.94$$

Proportional plus Integral Control

Let us consider the position control system of Fig. 5.6 again. With proportional plus integral control, the open-loop transfer function of the system becomes (Fig. 5.23)

$$G(s) = \frac{5(K_As+K_i)}{s^2(s+50.5)}$$

The characteristic equation is

$$s^3+50.5s^2+5K_As+5K_i = 0 \qquad (5.54)$$

Time Response Analysis 167

The effect of K_i on the transient behaviour of the system may be investigated by plotting the roots of eqn. (5.54) as a function of K_i for various values of K_A (refer Section 7.4). The steady-state error to ramp input is of course zero for the system with integral control.

Example 5.2: A turntable with a moment of inertia of 10 kg-m² is used with a proportional error controller in a unity negative feedback system. The controller develops a torque of 60 newton-m per radian of misalignment. The viscous friction is such that the damping factor is 0.3.

(a) Draw a signal flow graph of the system and determine therefrom $\frac{\theta_0}{\theta_i}(s)$ and $\frac{\theta_e}{\theta_i}(s)$, where θ_i, θ_e and θ_0 are respectively the input, error and output signals.

(b) Determine the steady-state tracking error for a constant velocity input of 0.04 rad/sec.

(c) Explain what modification could be made to the system to eliminate error to a ramp input as in part (b).

Solution:
(a) Signal flow graph is drawn in Fig. 5.28.

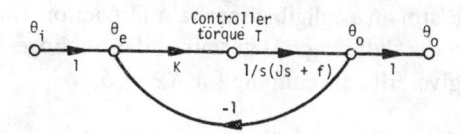

Figure 5.28.

From the signal flow graph

$$\frac{\theta_0}{\theta_i}(s) = \frac{K}{Js^2+fs+K} = \frac{K/J}{s^2+(f/J)s+K/J}$$

$$K/J = \omega_n^2 = 60/10 = 6$$

$$f/J = 2\zeta\omega_n = 2\times 0.3 \times \sqrt{6} = 0.6\sqrt{6}$$

$$\therefore \quad \frac{\theta_0}{\theta_i}(s) = \frac{6}{s^2+0.6\sqrt{(6)}s+6}$$

Now from the signal flow graph

$$\frac{\theta_e}{\theta_i}(s) = \frac{s(Js+f)}{Js^2+fs+K} = \frac{s(s+0.6\sqrt{6})}{s^2+0.6\sqrt{(6)}s+6}$$

(b) $$\theta_i(s) = \frac{0.04}{s^2}$$

$$\lim_{t\to\infty} \theta_e(t) = \lim_{s\to 0}\left[s\times \frac{0.04}{s^2}\times \frac{s(s+0.6\sqrt{6})}{s^2+0.6\sqrt{(6)}s+6}\right]$$

$$= 0.0098 \text{ rad}$$

(c) Introduce an integral error term in the controller such that the controller transfer function becomes

$$\left(K+\frac{K_I}{s}\right) = \frac{Ks+K_I}{s}$$

Since this introduces one more integration in the forward path, the system becomes type-2 and hence the steady-state error to ramp input is reduced to zero.

Example 5.3: In a d.c. position control servomechanism the load is driven by a motor supplied with constant armature current. The motor field current is supplied from a d.c. amplifier, the input to which is the difference between the voltages obtained from input and output potentiometers.

The load and motor together have a moment of inertia of $J = 0.4$ kg-m^2 and the viscous friction is $f = 2$ newton-m/rad/sec. Each potentiometer constant is $K_P = 0.6$ V/rad. The motor develops a torque of $K_T = 2$ newton-m per amp of field current. The field time constant is negligible.

(a) Make a sketch of the system showing how the hardware is connected.

(b) Derive the equation of motion of the system and find the value of the amplifier gain K_A (in amperes output per volt input) to give a natural frequency of 10 rad/sec.

(c) A tachogenerator of negligible inertia and friction is connected in the system to improve the damping. Determine the tachogenerator constant (in V/rad/sec) to give critical damping for $K_A = 5$.

Solution:

(a) Figure 5.29 gives the sketch of the position control servomechanism. The switch 'S' can be closed to provide the tachogenerator feedback.

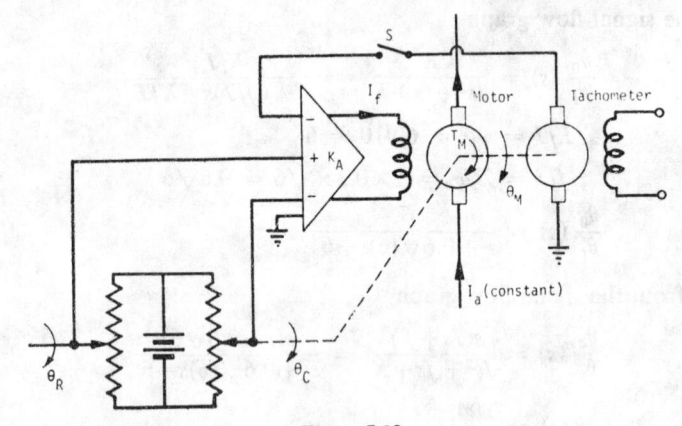

Figure 5.29.

(b) Figure 5.30 gives the signal flow graph of the system with switch 'S' open, i.e., tachogenerator not in the loop.

$$\frac{\theta_C(s)}{\theta_R(s)} = \frac{\theta_M(s)}{\theta_R(s)} = \frac{K_P K_A K_T}{s(Js+f)+K_P K_A K_T}$$

Time Response Analysis

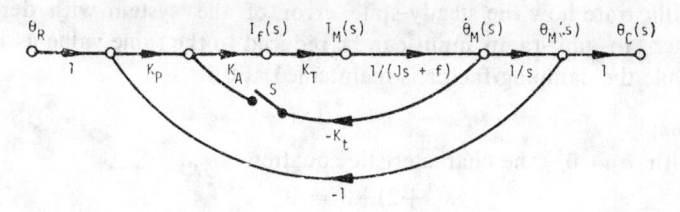

Figure 5.30.

The governing differential equation can be written from the transfer function as

$$J\ddot{\theta}_M(t) + f\dot{\theta}_M(t) + K_P K_A K_T \theta_M(t) = K_P K_A K_T \theta_R(t)$$

The system characteristic equation is

$$s^2 + (f/J)s + K_P K_A K_T/J = 0$$

$$\therefore \quad \omega_n^2 = \frac{K_P K_A K_T}{J}$$

or

$$10^2 = \frac{0.6 \times K_A \times 2}{0.4}$$

or

$$K_A = 33.3 \text{ amp/volt}$$

(c) With tachogenerator (switch 'S' closed)

$$\frac{\theta_M}{\theta_R}(s) = \frac{K_P K_A K_T}{Js^2 + (f + K_A K_t K_T)s + K_P K_A K_T}$$

$$2\zeta\omega_n = (f + K_A K_t K_T)/J; \quad \omega_n = \sqrt{(K_P K_A K_T/J)}$$

$$\therefore \quad \zeta = \frac{1}{2} \frac{f + K_A K_t K_T}{\sqrt{(J K_P K_A K_T)}}$$

or

$$1 = \frac{1}{2} \frac{2 + 5 \times 2 \times K_t}{\sqrt{(0.4 \times 0.6 \times 5 \times 2)}} \quad \text{or} \quad K_t = 0.11 \text{ V/rad/sec.}$$

Example 5.4: The system illustrated in Fig. 5.31 is a unity feedback control system with a minor feedback loop (output derivative feedback).

(a) In the absence of derivative feedback ($a = 0$), determine the damping factor and natural frequency. Also determine the steady-state error resulting from a unit-ramp input.

(b) Determine the derivative feedback constant a which will increase the damping factor of the system to 0.7. What is the steady-state error to unit-ramp input with this setting of the derivative feedback constant?

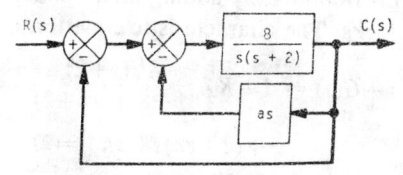

Figure 5.31.

(c) Illustrate how the steady-state error of the system with derivative feedback to unit-ramp input can be reduced to the same value as in part (a), while the damping factor is maintained at 0.7.

Solution:
(a) With $a = 0$, the characteristic equation is
$$s(s+2)+8 = 0$$
or
$$s^2+2s+8 = 0$$
$$\omega_n = \sqrt{8} = 2\sqrt{2} \text{ rad/sec}$$
$$2\zeta\omega_n = 2$$
$$\therefore \quad \zeta = \frac{1}{2\sqrt{2}} = 0.353$$

System $K_v = 8/2 = 4$
$$\therefore \quad e_{ss} \text{ (to unit-ramp)} = 1/4 = 0.25$$

(b) With derivative feedback, the characteristic equation is
$$1+G(s) = 1+\frac{\frac{8}{s(s+2)}}{1+\frac{8as}{s(s+2)}} = 0$$
or
$$s^2+(2+8a)s+8 = 0$$
$$\therefore \quad 2\zeta\omega_n = 2+8a$$
$$2\times 0.7 \times 2\sqrt{2} = 2+8a$$
$$a = 0.245$$

System $K_v = 8/(2+8a)$
$$\therefore \quad e_{ss} = (2+8a)/8 = 0.495$$

(c) Let the gain of 8 in the forward loop be adjusted to a higher value K_A. The new characteristic equation is
$$s^2+(2+aK_A)s+K_A = 0$$
$$\therefore \quad 2\zeta\omega_n = 2+aK_A$$
$$2\times 0.7\sqrt{K_A} = 2+aK_A \quad (5.55)$$

System $K_v = K_A/(2+aK_A)$
$$\therefore \quad e_{ss} = (2+aK_A)/K_A = 0.25 \quad (5.56)$$

Solving eqns. (5.55) and (5.56) we obtain
$$K_A = 31.36; \quad a = 0.186$$

Alternative solution is obtained by adding an amplifier of gain K_A between the two summing blocks. The characteristic equation now becomes
$$1+G(s) = 1+K_A \frac{8/s(s+2)}{1+8as/s(s+2)} = 0$$
or
$$s^2+(2+8a)s+8K_A = 0$$
$$\therefore \quad 2\zeta\omega_n = 2+8a$$

$$2\times 0.7\times 2\sqrt{2}\sqrt{K_A} = 2+8a \tag{5.57}$$

System $K_v = 8K_A/(2+8a)$

$$\therefore \qquad e_{ss} = (2+8a)/8K_A = 0.25 \tag{5.58}$$

Solving eqns. (5.57) and (5.58) we obtain

$$K_A = 3.92;\ a = 0.73$$

Note: The second solution requires a smaller gain but a separate amplification stage.

Example 5.5: Consider the liquid-level system shown in Fig. 5.32. The pump controls the liquid head h by supplying liquid at a rate Q m³/sec to the tank of cross-sectional area 1 m². We shall assume that the flow rate Q is proportional to the error in liquid level (desired level − actual level). Under these assumptions, the system equations are:

(i) $Q = h$ or $\dfrac{H(s)}{Q(s)} = \dfrac{1}{s}$

(ii) $Q = Ke$; e = error in liquid level, K = gain constant

The block diagram representation of the system is given in Fig. 5.33. Let us pose the problem of computing the value of K that minimizes the ISE for unit-step input.

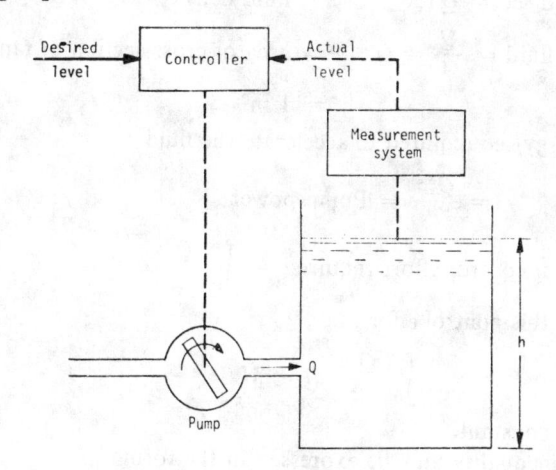

Figure 5.32 A liquid-level system.

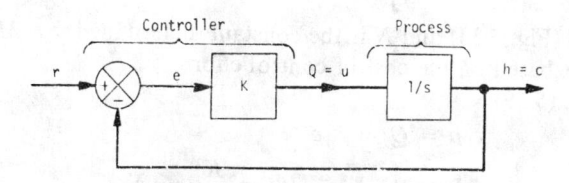

Figure 5.33 Block diagram representation of Fig. 5.32.

From Fig. 5.33 we have
$$\frac{E(s)}{R(s)} = \frac{s}{s+K}$$
For unit-step input
$$E(s) = \frac{1}{s+K}$$
Therefore
$$e(t) = e^{-Kt} \tag{5.59}$$
$$\text{ISE} = J_e = \int_0^\infty e^2(t)dt$$
$$= 1/2K \tag{5.60}$$

Obviously, the minimum value of J_e is obtained as $K \to \infty$. This is an impractical solution resulting in excessive strain on physical components of the system. Increasing the gain means in effect increasing the pump size.

Sound engineering judgement tells us that we must include in our performance index the 'cost' of the control effort. We can do this in many ways. One of the ways is to modify the performance index J_e so as to include the cost of control effort as a performance measure. For the system under consideration, this cost is proportional to power rating of pump which depends upon kinetic energy required to accelerate the fluid.

Mass of fluid/sec $= Q\rho$; $\quad \rho =$ fluid density

Velocity of fluid $= \dfrac{Q}{A} = Q$; $A =$ area of cross-section of tank
$$= 1 \text{ m}^2$$

Kinetic energy/sec required to accelerate the fluid
$$= \frac{Q^3\rho}{2} = \text{Pump power}$$

Therefore, total control effort required $= \displaystyle\int_0^\infty \frac{Q^3\rho}{2} dt$

We may limit this control effort, i.e.,
$$\int_0^\infty \frac{Q^3\rho}{2} dt \leqslant M$$
where M is a constant.

The above inequality may be expressed in the form
$$J_u = \int_0^\infty u^3 dt \leqslant N \tag{5.61}$$
where $u = Q$ (Fig. 5.33) and N is the constant determined by M and ρ. J_u is the index based on the cost of control effort.

As per eqn. (5.59)
$$u = Q = K e^{-Kt}$$
$$\therefore \quad J_u = \int_0^\infty K^3 e^{-3Kt} dt = \frac{K^2}{3} \leqslant N$$

Now J_e given by (5.60) is minimized when K has the largest permissible value, i.e.,

$$\frac{K^2}{3} = N$$

or
$$K = \sqrt{3N} \qquad (5.62)$$

Hence
$$J_e(\min) = \frac{1}{2K} = \frac{1}{2\sqrt{3N}} \qquad (5.63)$$

Example 5.6: A unity feedback position control system has a forward path transfer function

$$G(s) = K/s$$

For unit-step input, compute the value of K that minimizes ISE.

Solution:

For the system under consideration

$$\frac{E(s)}{R(s)} = \frac{1}{1+G(s)} = \frac{s}{s+K}$$

For unit-step input

$$E(s) = \frac{1}{s+K}$$

Therefore,
$$e(t) = e^{-Kt} \qquad (5.64)$$

$$\text{ISE} = \int_0^\infty e^2(t)dt = \frac{1}{2K} \qquad (5.65)$$

Obviously, the minimum value of ISE is obtained as $K \to \infty$. This is an impractical solution resulting in excessive strain on physical components of the system. A more practical solution results by limiting the second derivative of the output which amounts to limiting the maximum torque of the system.

Let the constraint be

$$|\ddot{c}| \leqslant M \qquad (5.66)$$

where c is the output and M is a constant.

For the system considered here, the output is

$$C(s) = \frac{K}{s(s+K)}$$

$$c(t) = 1 - e^{-Kt}$$

Therefore,
$$\ddot{c}(t) = -K^2 e^{-Kt}$$

$$|\ddot{c}(t)|_{\max} = K^2 \leqslant M \qquad (5.67)$$

The maximum value of K satisfying the constraint (5.66) will give the minimum value of ISE. Thus the optimum value of K obtained from eqn. (5.67) is

$$K = \sqrt{M}$$

Therefore, minimum ISE $= 1/2\sqrt{M}$

Example 5.7: For the system of Example 5.6, compute the value of K that minimizes the following performance index.

$$J = \int_0^\infty (e^2 + \lambda \ddot{e}) dt \,; \lambda \text{ is a positive constant}$$

What is the minimum value of J?

Solution:
Before solving this problem we note the following:

$$e(t) = 1 - c(t)$$
or
$$\ddot{e}(t) = -\ddot{c}(t)$$

The output acceleration could thus be limited by a single performance index of the type given in this problem.

From eqn. (5.64)

$$\ddot{e} = K^2 e^{-Kt}$$

Therefore,

$$J = \int_0^\infty (e^{-2Kt} + \lambda K^4 e^{-2Kt}) dt$$

$$= (1 + \lambda K^4)/2K$$

The minimum value of J is obtained when

$$\frac{\partial J}{\partial K} = -\frac{1}{2K^2} + \frac{3}{2}\lambda K^2 = 0$$

or
$$K = (3\lambda)^{-1/4}$$

Note:
$$\frac{\partial^2 J}{\partial K^2} = \frac{1}{K^3} + 3\lambda K > 0$$

The minimum value of J is $\dfrac{2}{3}(3\lambda)^{1/4}$.

The constant λ is called the *weighting factor*. When $\lambda = 0$, there is no constraint imposed on the output acceleration. When λ is large, it means more importance is given to the constraint on output acceleration compared to the performance of the system. A suitable value of λ is chosen so that the relative importance of the system performance is contrasted with the importance of the limit on output acceleration.

Figure 5.34 gives a plot of the performance index versus K for various values of λ.

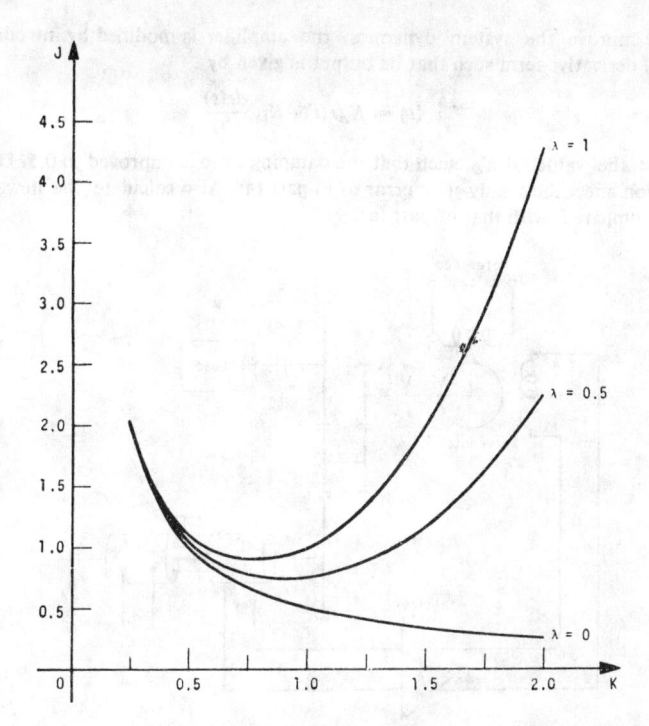

Figure 5.34 Performance index vs. gain K for Example 5.7.

Problems

5.1 A servomechanism is used to control the angular position θ_0 of a mass through a command signal θ_i. The moment of inertia of moving parts referred to the load shaft is 200 kg-m^2 and the motor torque at the load is 6.88×10^4 newton-m per rad of error. The damping torque coefficient referred to the load shaft is 5×10^3 newton-m per rad/sec.

(a) Find the time response of the servomechanism to a step input of 1 rad and determine the frequency of transient oscillation, the time to rise to the peak overshoot and the value of the peak overshoot.

(b) Determine the steady-state error when the command signal is a constant angular velocity of 1 revolution/min.

(c) Determine the steady-state error which exists when a steady torque of 1,200 newton-m is applied at load shaft.

5.2 In the position control system shown in Fig. P-5.2, the sensitivity of the synchro error detector is 1 volt/deg. The transfer function of the motor is found to be

$$\frac{\theta_M(s)}{V_C(s)} = \frac{K_m}{s(s\tau_m + 1)}$$

where $K_m = 15$ rad/sec/volt; $\tau_m = 0.15$ sec.

The gear ratios are given as

$$\theta_C/\theta_L = 1; \quad \theta_L/\theta_M = 1/50$$

(a) If the input shaft is driven at a constant speed of π rad/sec, determine the value of the amplifier gain such that the steady-state deviation between input and output positions is less than 5 deg. For this value of amplifier gain, determine the damping ratio and settling time of the system.

(b) To improve the system dynamics, the amplifier is modified by introducing an additional derivative term such that its output is given by

$$v_C(t) = K_A e(t) + K_D \frac{de(t)}{dt}$$

Determine the value of K_D such that the damping ratio is improved to 0.5. Does this modification affect the steady-state error as in part (a)? Also calculate the new settling time and compare it with that of part (a).

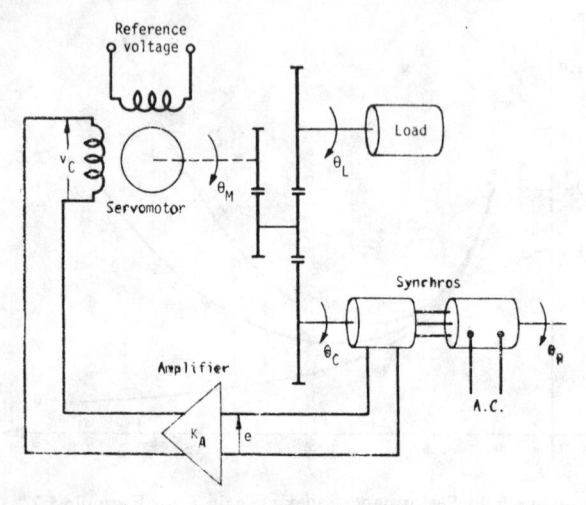

Figure P-5.2.

5.3 A d.c. position control servomechanism comprises of a split-field d.c. servomotor with constant armature current, potentiometer error detector, a d.c. amplifier and a tachogenerator coupled to motor shaft. A fraction K of tachogenerator output is fed-back to produce stabilizing effect. The following particulars refer to the system:

Moment of inertia of motor, J_M	$= 2 \times 10^{-3}$ kg-m²
Moment of inertia of load, J_L	$= 5$ kg-m²
Motor to load gear ratio, $\dot{\theta}_L/\dot{\theta}_M$	$= 1/50$
Load to potentiometer gear ratio, $\dot{\theta}_L/\dot{\theta}_C$	$= 1$
Motor torque constant, K_T	$= 2$ newton-m/amp
Tachogenerator constant, K_t	$= 0.2$ volt/rad/sec
Sensitivity of error detector, K_P	$= 0.6$ volt/rad
Amplifier gain	$= K_A$ amps/volt

Motor and load frictions and motor field time constant are assumed to be negligible.
(a) Make a sketch of the system, showing how the hardware is connected.
(b) Determine the transfer function of the system.
(c) Determine the amplifier gain required and the fraction of the tachogenerator voltage fedback to give an undamped natural frequency of 4 rad/sec and damping of 0.8.

5.4 The schematic diagram of a servomechanism is given in Fig. P-5.4. The system constants are as follows:

Synchro sensitivity, K_s	$= 1$ volt/rad
Amplifier gain, K_A	$= 20$ volt/volt
Motor torque constant, $K_T =$	10^{-5} newton-m/volt

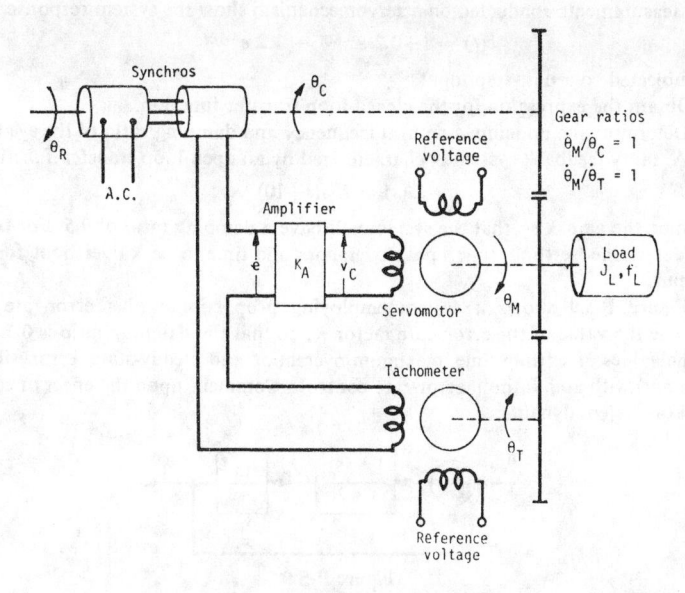

Figure P-5.4

Load inertia, J_L $= 1.5 \times 10^{-5}$ kg-m²
Viscous friction, f_L $= 1 \times 10^{-5}$ newton-m/rad/sec
Tachometer constant, K_t $= 0.2$ volt/rad/sec

Motor inertia and friction are assumed to be negligible.

(a) Find the value of the damping ratio assuming that the tachometer is disconnected. Determine also the steady-state error corresponding to an input velocity of 1 rad/sec.

(b) Determine the value of damping ratio when the tachometer is included as part of the system.

(c) The tachometer is now removed and the amplifier is replaced by another which gives an output voltage of

$$v_C = K_A e + K_A \int e\, dt$$

Compare the steady-state behaviour of the system with that of part (a).

5.5 The open-loop transfer function of a unity feedback system is given by

$$G(s) = K/s(Ts+1)$$

where K and T are positive constants.

By what factor should the amplifier gain be reduced so that the peak overshoot of unit-step response of the system is reduced from 75% to 25%?

5.6 The following tests are conducted on the position control system of Fig. 5.6 (with gear ratio $n = 1/10$)

(1) Unit-step response is recorded and M_p measured therefrom has a value of 25%.
(2) A constant velocity input of 1 rad/sec produces a steady-state error of 0.04 rad.
(3) With the reference potentiometer shaft held fixed, a torque of 10 newton-m applied to the output shaft produces a steady-state error of 0.01 rad.

From the above data, determine:

(a) The natural frequency ω_n.
(b) The moment of inertia J referred to the motor shaft.
(c) The coefficient of viscous friction f referred to the motor shaft.

5.7 Measurements conducted on a servomechanism show the system response to be
$$c(t) = 1 + 0.2\, e^{-60t} - 1.2\, e^{-10t}$$
when subjected to a unit-step input.
(a) Obtain the expression for the closed-loop transfer function.
(b) Determine the undamped natural frequency and damping ratio of the system.

5.8 A unity feedback system is characterized by an open-loop transfer function
$$G(s) = K/s(s+10)$$
Determine the gain K so that the system will have a damping ratio of 0.5. For this value of K determine settling time, peak overshoot and time to peak overshoot for a unit-step input.

5.9 Figure P-5.9 shows a system employing proportional plus error-rate control. Determine the value of the error-rate factor K_e so that the damping ratio is 0.5. Determine the values of settling time, maximum overshoot and steady-state error (for unit-ramp input) with and without error-rate control. Comment upon the effect of error-rate control on system dynamics.

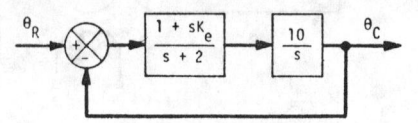

Figure P-5.9.

5.10 A feedback system employing output-rate damping is shown in Fig. P-5.10.
(a) In the absence of derivative feedback ($K_0 = 0$), determine the damping factor and natural frequency of the system. What is the steady-state error resulting from unit-ramp input?
(b) Determine the derivative feedback constant K_0, which will increase the damping factor of the system to 0.6. What is the steady-state error to unit-ramp input with this setting of the derivative feedback constant?
(c) Illustrate how the steady-state error of the system with derivative feedback to unit-ramp input can be reduced to same value as in part (a), while the damping factor is maintained at 0.6.

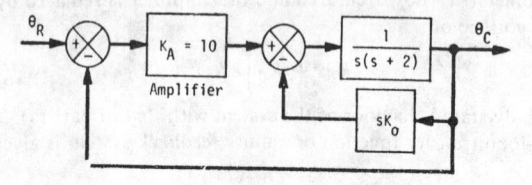

Figure P-5.10.

5.11 A unity feedback system is characterized by the open-loop transfer function
$$G(s) = 1/s(0.5s+1)(0.2s+1)$$
Determine the steady-state errors for unit-step, unit-ramp and unit-acceleration input. Also determine the damping ratio and natural frequency of the dominant roots.

5.12 The open-loop transfer function of a servo system with unity feedback is
$$G(s) = 10/s(0.1s+1)$$
Evaluate the static error constants (K_p, K_v, K_a) for the system. Obtain the steady-state error of the system when subjected to an input given by the polynomial
$$r(t) = a_0 + a_1 t + \frac{a_2}{2} t^2$$

5.13 For Problem 5.12, evaluate the dynamic error using the dynamic error coefficients.

Hint: For a unity feedback system

$$\frac{E(s)}{R(s)} = \frac{1}{1+G(s)} = \frac{1}{K_1} + \frac{1}{K_2}s + \frac{1}{K_3}s^2 + \cdots$$

Coefficients $K_1, K_2, K_3, \ldots$ are defined to be dynamic-error coefficients.

$$E(s) = \frac{1}{K_1}R(s) + \frac{1}{K_2}sR(s) + \frac{1}{K_3}s^2R(s) + \cdots$$

$$e(t) = \frac{1}{K_1}r(t) + \frac{1}{K_2}\dot{r}(t) + \frac{1}{K_3}\ddot{r}(t) + \cdots$$

5.14 A machine tool is required to cut a 30° circular arc of 1 cm radius. The tool moves at a constant feed velocity of 0.1 cm/sec parallel to the x-axis, as shown in Fig. P-5.14. A unity feedback servomechanism with open-loop transfer function $G(s) = 10/s(s+1)$ drives the tool in y-direction. Estimate the error when $x = 0.3$.

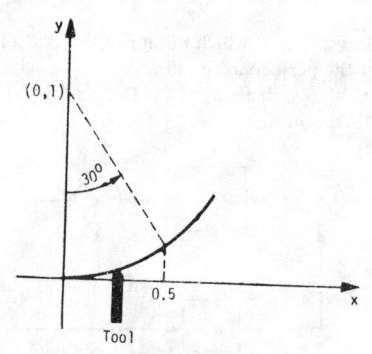

Figure P-5.14.

Hint: The equation of the circular arc is

$$x^2 + (y-1)^2 = 1$$

which gives

$$y = 1 - \sqrt{(1-x^2)}, \text{ for } 0 < x < 0.5$$

Since $x = 0.1t$,

$$y = 1 - \sqrt{[1-(0.1t)^2]}$$

This is the input to the unity feedback system.

5.15 Consider a unity feedback system with the closed-loop transfer function

$$\frac{C(s)}{R(s)} = \frac{\omega_n^2}{s^2 + 2\omega_n s + \omega_n^2}$$

For unit-step input, compute the following performance indices:

1. $\text{IAE} = \int_0^\infty |e(t)|\, dt$

2. $\text{ITAE} = \int_0^\infty t\,|e(t)|\, dt$

Calculate the values of ω_n which minimize IAE and ITAE respectively. Are these values of ω_n practical? If not, then choose suitable values of ω_n and determine the corresponding performance indices.

Hint: (i) Since damping factor $\zeta = 1$, the system does not overshoot. Therefore

$$|e(t)| = e(t)$$

(ii) From Appendix I, $\quad \int_0^\infty e(t)\,dt = \lim_{s \to 0} E(s);$

$$\int_0^\infty t e(t)\,dt = \lim_{s \to 0}\left[-\frac{d}{ds}E(s)\right]$$

5.16 A unity feedback system has the forward path transfer function

$$G(s) = K/s(s+\alpha)$$

(i) Taking K as constant, determine the value of α which minimizes ISE.
(ii) Taking α as constant, determine the value of K which minimizes ISE.

5.17 A unity feedback system has an open-loop transfer function

$$G(s) = 1/s(s+2\zeta)$$

For unit-step input, compute the following:

(a) $\text{ISE} = \int_0^\infty e^2(t)\,dt;$ (b) $\text{ITSE} = \int_0^\infty t\,e^2(t)\,dt$

Calculate the optimal values of $\zeta > 0$ which minimize ISE and ITSE respectively. What are the minimum values of the performance indices?

5.18 Consider the system shown in Fig. P-5.18. Determine the values of K_A, K and α which will satisfy the ITAE criterion.

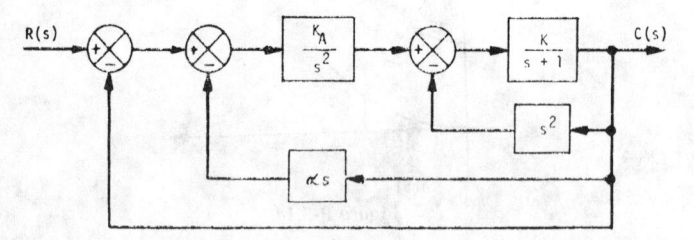

Figure P-5.18.

References and Further Reading

1. Grahm, D. and R.C. Lathrop, "The Synthesis of Optimum Response: Criteria and Standard Forms," *Trans. AIEE*, Part 2, **72**, Nov. 1953, pp. 273-288.
2. Kuo, B.C., *Automatic Control Systems*, (3rd Ed.), Prentice-Hall, Englewood Cliffs, N.J., 1975.
3. Thaler, G.J., *Design of Feedback Systems*, Dowden, Hutchinson and Ross, Stroudsburg, Pennsylvania, 1973.
4. Shinners, S.M., *Modern Control System—Theory and Application*, Addison-Wesley, Reading, Mass., 1972.
5. Gupta, S.C. and L. Hasdorff, *Fundamentals of Automatic Control*, John Wiley, New York, 1970.
6. Ogata, K., *Modern Control Engineering*, Prentice-Hall, Englewood Cliffs, N.J., 1970.
7. Baeck, H.S., *Practical Servomechanism Design*, McGraw-Hill, New York, 1968.
8. Chubb, B.A., *Modern Analytical Design of Instrument Servomechanisms*, Addison-Wesley, Reading, Mass., 1967.

6. Concepts of Stability and Algebraic Criteria

6.1 The Concept of Stability

Roughly speaking, stability in a system implies that small changes in the system input, in initial conditions or in system parameters, do not result in large changes in system output. Stability is a very important characteristic of the transient performance of a system. Almost every working system is designed to be stable. Within the boundaries of parameter variations permitted by stability considerations, we can then seek to improve the system performance.

A linear time-invariant system is stable if the following two notions of system stability are satisfied:

(i) When the system is excited by a bounded input, the output is bounded.
(ii) In the absence of the input, the output tends towards zero (the equilibrium state of the system) irrespective of initial conditions. (This stability concept is known as *asymptotic stability*.)

The second notion of stability generally concerns a free system relative to its transient behaviour. The first notion concerns a system under the influence of an input. Clearly, if a system is subjected to an unbounded input and produces an unbounded response, nothing can be said about its stability. But if it is subjected to a bounded input and produces an unbounded response, it is by definition, unstable. Actually the output of an unstable system may increase to a certain extent and then the system may break down or become nonlinear after the output exceeds a certain magnitude, so that the linear mathematical model no longer applies.

As we shall see below, the two notions of stability defined above are essentially equivalent in linear time-invariant systems. Simple and powerful tools are available to determine the stability of such systems. For nonlinear systems, because of the possible existence of multiple equilibrium states and other anomalies, the concept of stability is difficult even to define, so that there is no clearcut correspondence between the two notions of stability defined above. For a free stable nonlinear system, there is no guarantee that output will be bounded whenever input is bounded. Also if the output is bounded for a particular bounded input, it may not be bounded for other bounded inputs. Many of the important results obtained thus far, concern the stability of the nonlinear systems in the sense of the second notion above, i.e., when the system has no input. We shall discuss some of these results in Chapter 14. Here in this chapter, we are concerned with the stability determination of linear time-invariant systems.

Let us observe the physical implication of the two notions of stability defined earlier, by considering a single-input, single-output system with transfer function

$$\frac{C(s)}{R(s)} = G(s) = \frac{b_0 s^m + b_1 s^{m-1} + \ldots + b_m}{a_0 s^n + a_1 s^{n-1} + \ldots + a_n}; \quad m < n \quad (6.1)$$

With initial conditions assumed zero, the output of the system is given by

$$c(t) = \mathcal{L}^{-1}[G(s)R(s)]$$

Therefore (see Appendix I)

$$c(t) = \int_0^\infty g(\tau) r(t - \tau) d\tau$$

where $g(t) = \mathcal{L}^{-1}(G(s))$ is the *impulse response* of the system (eqn. (5.4)).

Taking the absolute value on both sides we get

$$|c(t)| = \left| \int_0^\infty g(\tau) r(t - \tau) d\tau \right|$$

Since the absolute value of integral is not greater than the integral of the absolute value of the integrand,

$$|c(t)| \leqslant \int_0^\infty |g(\tau) r(t - \tau)| d\tau$$

$$\leqslant \int_0^\infty |g(\tau)| |r(t - \tau)| d\tau \quad (6.2)$$

The first notion of stability is satisfied if for every bounded input ($|r(t)| \leqslant M_1 < \infty$), the output is bounded ($|c(t)| \leqslant M_2 < \infty$). From (6.2), we have for bounded input, the bounded output condition as

$$|c(t)| \leqslant M_1 \int_0^\infty |g(\tau)| d\tau \leqslant M_2$$

Thus the first notion of stability is satisfied if the impulse response $g(t)$ is absolutely integrable, i.e., $\int_0^\infty |g(\tau)| d\tau$ is finite (area under the absolute-value curve of the impulse response $g(t)$ evaluated from $t = 0$ to $t = \infty$ must be finite).

The nature of $g(t)$ is dependent on the poles of the transfer function $G(s)$ which are the roots of the characteristic equation. These roots may be both real and complex conjugate and may have multiplicity of various orders. The nature of response terms contributed by all possible types of roots are given in Table 6.1 and have been illustrated in Fig. 6.1.

Certain observations are easily made from the study of Table 6.1. All the roots which have nonzero real parts [cases (i), (ii), (iii) and (iv)], contribute response terms with a multiplying factor of $e^{\sigma t}$. If $\sigma < 0$ (i.e., the roots have negative real parts), the response terms vanish as $t \to \infty$ and if $\sigma > 0$ (i.e., the roots have positive real parts), the response terms increase without bound. Roots on the $j\omega$-axis with multiplicity two or higher [cases (vi) and (viii)] also contribute terms which increase without bound as $t \to \infty$.

Table 6.1 RESPONSE TERMS CONTRIBUTED BY VARIOUS TYPES OF ROOTS

Type of roots	Nature of response terms contributed	
(i) Single root at $s = \sigma$	$Ae^{\sigma t}$	(Fig. 6.1a)
(ii) Roots of multiplicity k at $s=\sigma$	$(A_1+A_2 t+\ldots+A_k t^{k-1})e^{\sigma t}$	
(iii) Complex conjugate root pair at $s=\sigma \pm j\omega$	$Ae^{\sigma t} \sin(\omega t+\beta)$	(Fig. 6.1b)
(iv) Complex conjugate root pairs of multiplicity k at $s=\sigma \pm j\omega$	$[A_1 \sin(\omega t+\beta_1)+A_2 t \sin(\omega t+\beta_2)+$ $\ldots+A_k t^{k-1} \sin(\omega t+\beta_k)]e^{\sigma t}$	
(v) Single complex conjugate root pair on the $j\omega$-axis (i.e., at $s=\pm j\omega$)	$A \sin(\omega t+\beta)$	(Fig. 6.1c)
(vi) Complex conjugate root pair of multiplicity k on the $j\omega$-axis	$A_1 \sin(\omega t+\beta_1)+A_2 t \sin(\omega t+\beta_2)+$ $\ldots+A_k t^{k-1} \sin(\omega t+\beta_k)$	(Fig. 6.1c)
(vii) Single root at origin (i.e., at $s=0$)	A	(Fig. 6.1d)
(viii) Roots of multiplicity k at origin	$(A_1+A_2 t+\ldots+A_k t^{k-1})$	(Fig. 6.1d)

Single root at origin [case (vii)] or non-multiple root pairs [case (v)] on the $j\omega$-axis contribute response terms which are constant amplitude or constant amplitude oscillation. These observations lead us to the following general conclusions regarding system stability (refer Fig. 6.2).

(1) If all the roots of the characteristic equation have negative real parts, then the impulse response is bounded and eventually decreases to zero. Therefore, $\int_0^\infty |g(\tau)| \, d\tau$ is finite and the system is *bounded-input, bounded-output* stable.

(2) If any root of the characteristic equation has a positive real part, $g(t)$ is unbounded and $\int_0^\infty |g(\tau)| \, d\tau$ is infinite. The system is therefore unstable.

(3) If the characteristic equation has repeated roots on the $j\omega$-axis, $g(t)$ is unbounded and $\int_0^\infty |g(\tau)| \, d\tau$ is infinite. The system is therefore unstable.

(4) If one or more nonrepeated roots of the characteristic equation are on the $j\omega$-axis, then $g(t)$ is bounded but $\int_0^\infty |g(\tau)| \, d\tau$ is infinite. The system is therefore unstable.

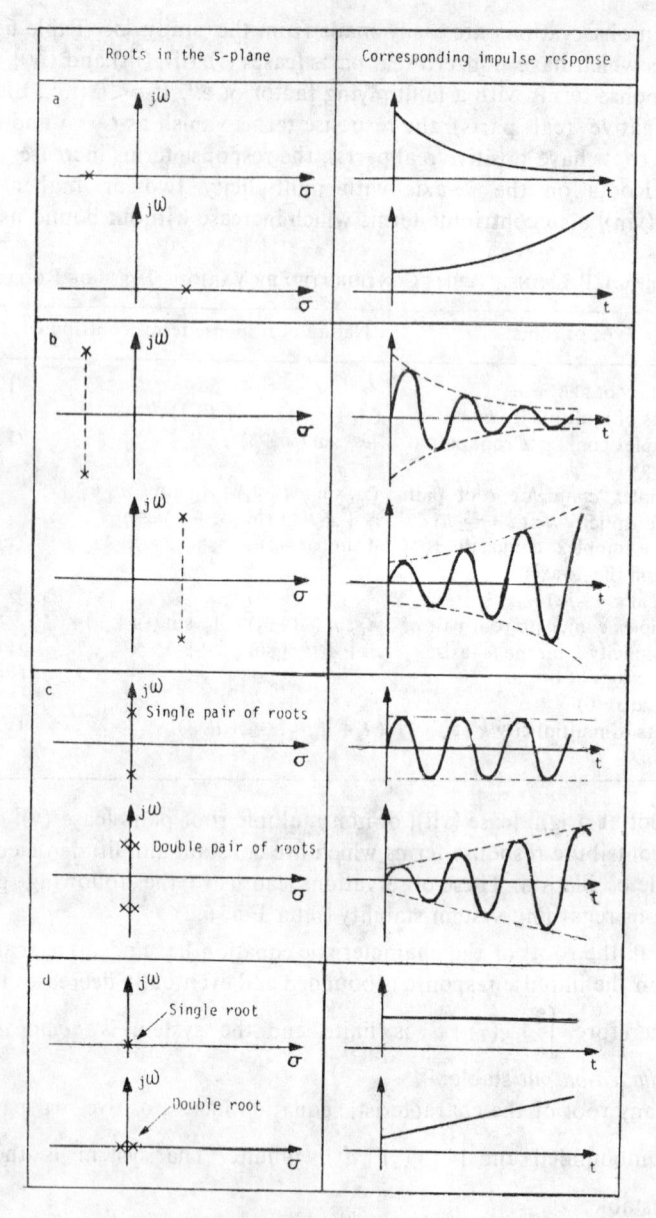

Figure 6.1 Response terms contributed by various types of roots.

The response to initial conditions is not evident from the model of eqn. (6.1) since the transfer function of a system is derived with the assumption of zero initial conditions. However, as we shall see in Chapter 14, the conditions for bounded-input, bounded-output stability also satisfy the requirement of asymptotic stability of the zero-input response of linear time-invariant systems.

Concepts of Stability and Algebraic Criteria

There are a few exceptions to the foregoing definition of stability. When a unit-step input ($R(s) = 1/s$) is applied to a perfect integrator ($G(s) = 1/s$), the output is unbounded. However, an integrator is a useful system. From Fig. 6.1 we observe that for nonrepeated poles of $G(s)$ on the $j\omega$-axis, the response is bounded unless input has a pole, matching one of the system poles on the $j\omega$-axis. The zero-input response in such cases is bounded but non-asymptotic ($c(t)$ does not tend to zero as $t \to \infty$). Depending upon the amplitude of the ultimate response, such a system may be treated as acceptable or nonacceptable. This situation is generally referred to as the case of *marginally* or *limitedly* stable system.

This suggests that in our concern for complete generality, we should not be misled by the complexity of occasional special cases. For instance, if $m = n$ in eqn. (6.1) we have (transfer functions with $m > n$ are not physically realizable),

$$G(s) = \frac{b_0 s^n + b_1 s^{n-1} + \ldots + b_n}{a_0 s^n + a_1 s^{n-1} + \ldots + a_n} = \frac{p(s)}{q(s)}$$

$$= \underbrace{\frac{b_0}{a_0}}_{\text{I}} + \underbrace{\frac{p'(s)}{q(s)}}_{\text{II}}$$

Assume that all the roots of $q(s)$ are in the left half of s-plane. Part I of the transfer function can be treated separately; it corresponds to the output component which depends directly on the input. Whether we call such a system stable or unstable, is again a matter of personal preference.

In a vast majority* of practical systems, the following statements on stability are quite useful (Fig. 6.2):

(i) If all the roots of the characteristic equation have negative real parts, the system is *stable*.
(ii) If any root of the characteristic equation has a positive real part or if there is a repeated root on the $j\omega$-axis, the system is *unstable*.
(iii) If the condition (i) is satisfied except for the presence of one or more nonrepeated roots on the $j\omega$-axis, the system is *limitedly stable*.

In further subdivision of the concept of stability, a linear system is characterized as:

(i) *Absolutely stable* with respect to a parameter of the system if it is stable for all values of this parameter.
(ii) *Conditionally stable* with respect to a parameter, if the system is stable for only certain bounded ranges of values of this parameter.

---

*In calculation of $G(s)$ of a system for a particular output, there may be poles of $G(s)$ which are cancelled by the numerator factors in $G(s)$. Such a transfer function does not exhibit all the modes of the system. However, in vast majority of situations, transfer functions exhibit all natural frequencies; system stability may therefore be determined by examination of poles of the transfer function.

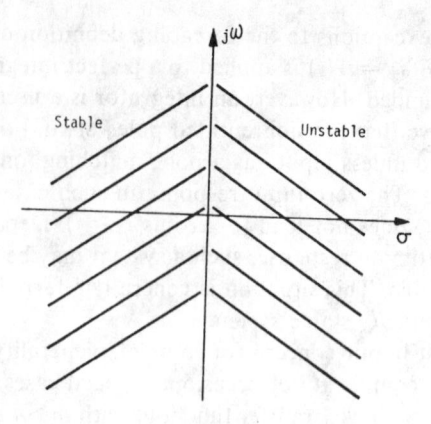

Figure 6.2 Regions of root-locations for stable, unstable and limitedly stable systems.

It follows from the above discussion that stability can be established by determining the roots of the characteristic equation. Unfortunately, no general formula in algebraic form is available to determine the roots of the characteristic equations of higher than second-order. Though various numerical methods exist for root determination of a characteristic equation, these are quite cumbersome even for third- and fourth-order systems.

However, simple graphical and algebraic criteria have been developed which permit the study of stability of a system without the need of actually determining the roots of its characteristic equation. These criteria answer the question, whether a system be stable or not, in 'yes' or 'no' form.

RELATIVE STABILITY

In pratical systems, it is not sufficient to know that the system is stable but a stable system must meet the specifications on *relative stability* which is a quantitative measure of how fast the transients die out in the system.

Relative stability may be measured by relative settling times of each root or pair of roots. It has been shown in the preceding chapter that the settling time of a pair of complex conjugate poles is inversely proportional to the real part (negative) of the roots. This result is equally valid for real roots. As a root (or a pair of roots) moves farther away from the imaginary axis as shown in Fig. 6.3, the relative stability of the system improves.

6.2 Necessary Conditions for Stability

Certain conclusions regarding the stability of a system can be drawn by merely inspecting the coefficients of its characteristic equation in polynomial form. In the following sections, we shall show that a necessary (but not sufficient) condition for stability of a linear system is that all the coefficients of its characteristic equation $q(s) = 0$, be real and have the same sign. Furthermore, none of the coefficients should be zero.

Consider the characteristic equation

$$q(s) = a_0 s^n + a_1 s^{n-1} + \ldots + a_{n-1} s + a_n = 0; \quad a_0 > 0 \qquad (6.3)$$

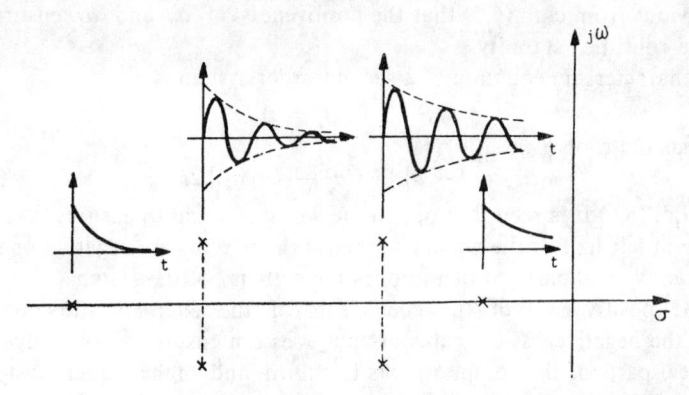

Figure 6.3 Relative stability for various root-locations in the s-plane.

It is to be noted that there is no loss of generality in assuming $a_0 > 0$. In case $a_0 < 0$, it can be made positive by multiplying the characteristic equation by -1 throughout.

Equation (6.3) may be written in the factored form as

$$q(s) = a_0 \Pi(s+\sigma_k)\Pi[(s+\sigma_l)^2+\omega_l^2] \qquad (6.4)$$

For the system to be stable, the roots should have negative real parts, which is satisfied if in eqn. (6.4) all σ_k and σ_l are positive real. It means that all the factors of eqn. (6.4) have positive terms only. As these factors are multiplied together to get the characteristic equation in polynomial form, all the coefficients of the resulting polynomial must work out to be positive. However, if one or more roots have positive real parts, the coefficients of the characteristic equation may or may not be all positive and hence the above conclusion.

Furthermore, it is to be noted that none of the coefficients can be zero or negative unless one (or more than one) of the following occurs:

(1) one or more roots have positive real parts;
(2) a root (or roots) at origin, i.e., $\sigma_k = 0$ and hence $a_n = 0$;
(3) $\sigma_l = 0$ for some l, which implies the presence of roots on the $j\omega$-axis.

We therefore conclude that the absence or negativeness of any of the coefficients of the characteristic equation (with $a_0 > 0$) indicates that the system is either unstable or at most limitedly stable.

From the foregoing discussions, let us prove the following propositions:

1. The positiveness of the coefficients of characteristic equation is necessary as well as sufficient condition for stability of systems of first- and second-order.

The characteristic equation of a first-order system is

$$a_0 s + a_1 = 0$$

which has a single root

$$s_1 = -a_1/a_0 \qquad (6.5)$$

It is obvious from eqn. (6.5) that the positiveness of a_0 and a_1 ensures a negative root, i.e., stability.

The characteristic equation of a second-order system is
$$a_0 s^2 + a_1 s + a_2 = 0$$
which has the roots
$$s_1, s_2 = [-a_1 \pm \sqrt{(a_1^2 - 4a_0 a_2)}]/2a_0 \qquad (6.6)$$
From eqn. (6.6) it is seen that positiveness of a_0, a_1 and a_2 ensures that the roots lie in left half of the s-plane (either both the roots are negative or they have negative real parts) which implies the stability of the system.

2. The positiveness of the coefficients of the characteristic equation ensures the negativeness of real roots but does not ensure the negativeness of the real parts of the complex roots for third- and higher-order systems. Therefore it cannot be a sufficient condition for stability of third- and higher-order systems.

Consider a third-order system with characteristic equation
$$s^3 + s^2 + 2s + 8 = 0 \qquad (6.7)$$
Equation (6.7) may be put in the factored form as
$$(s+2)\left(s - 0.5 + j\frac{\sqrt{15}}{2}\right)\left(s - 0.5 - j\frac{\sqrt{15}}{2}\right) = 0$$
We see that the real part of the complex roots is positive indicating instability of the system even though all the coefficients of the characteristic equation (6.7) are positive.

Therefore, if the characteristic equation of a system is of degree higher than second, the possibility of its instability cannot be excluded even when all the coefficients of its characteristic equation are positive. The first step in analyzing the stability of the system is to examine its characteristic equation. If some of the coefficients are zero or negative it can be concluded that the system is not stable. On the other hand, if all the coefficients of the characteristic equation are positive (or negative), the possibility of stability of system exists and one should proceed further to examine the sufficient conditions of stability.

A. Hurwitz and E.J. Routh independently published the method of investigating the sufficient conditions of stability of a system. The Hurwitz Criterion is in terms of determinants and the Routh Criterion is in terms of array formulation, which is more convenient to handle. We first discuss the Hurwitz Criterion.† The Routh Criterion which is derivable from that of Hurwitz, is then presented.

6.3 Hurwitz Stability Criterion

The characteristic equation of an nth order system is
$$q(s) = a_0 s^n + a_1 s^{n-1} + \ldots + a_{n-1} s + a_n = 0$$
For the stability of this system, it is necessary and sufficient that the n determinants formed from the coefficients $a_0, a_1, \ldots, a_n$ of the characteristic equation

†The proof will be given in Chapter 14.

be positive, where these determinants are taken as the principal minors of the following arrangement (called the Hurwitz determinant):

$$\begin{vmatrix} a_1 & a_0 & 0 & 0 & 0 & 0..0 & 0 \\ a_3 & a_2 & a_1 & a_0 & 0 & 0...0 & 0 \\ a_5 & a_4 & a_3 & a_2 & a_1 & a_0...0 & 0 \\ \vdots & \vdots & \vdots & \vdots & \vdots & \vdots & \vdots \\ a_{2n-1} & a_{2n-2} & a_{2n-3} & . & . & . & a_{n+1} & a_n \end{vmatrix} \quad (6.8)$$

where the coefficients with indices larger than n or with negative indices are replaced by zeros.

In other words, the necessary and sufficient conditions for stability are,

$$\Delta_1 = a_1 > 0;$$

$$\Delta_2 = \begin{vmatrix} a_1 & a_0 \\ a_3 & a_2 \end{vmatrix} > 0,$$

$$\Delta_3 = \begin{vmatrix} a_1 & a_0 & 0 \\ a_3 & a_2 & a_1 \\ a_5 & a_4 & a_3 \end{vmatrix} > 0; \ldots$$

$$\Delta_n = \text{entire arrangement of (6.8)} > 0.$$

When $\Delta_{n-1} = 0$, the system is limitedly stable.

Example 6.1: Consider a fourth-order system with the characteristic equation

$$s^4 + 8s^3 + 18s^2 + 16s + 5 = 0$$

The Hurwitz arrangement is given below:

$$\begin{array}{c|cccc} a_1 & 8 & 1 & 0 & 0 \\ a_3 & 16 & 18 & 8 & 1 \\ a_5 & 0 & 5 & 16 & 18 \\ a_7 & 0 & 0 & 0 & 5 \end{array}$$

Therefore,

$$\Delta_1 = 8 > 0$$

$$\Delta_2 = \begin{vmatrix} 8 & 1 \\ 16 & 18 \end{vmatrix} = 128 > 0$$

$$\Delta_3 = \begin{vmatrix} 8 & 1 & 0 \\ 16 & 18 & 8 \\ 0 & 5 & 16 \end{vmatrix} = 1{,}728 > 0$$

$$\Delta_4 = \begin{vmatrix} 8 & 1 & 0 & 0 \\ 16 & 18 & 8 & 1 \\ 0 & 5 & 16 & 18 \\ 0 & 0 & 0 & 5 \end{vmatrix} = 5\Delta_3 > 0$$

Therefore the system under consideration is stable.

6.4 Routh Stability Criterion

This criterion is based on ordering the coefficients of the characteristic equation into an array, called the Routh array as given below:

$$q(s) = a_0 s^n + a_1 s^{n-1} + a_2 s^{n-2} + \ldots + a_{n-1} s + a_n = 0$$

ROUTH ARRAY

$$\begin{array}{c|cccccc} s^n & a_0 & a_2 & a_4 & a_6 & \cdots & \\ s^{n-1} & a_1 & a_3 & a_5 & & & \\ s^{n-2} & b_1 & b_2 & b_3 & & & \\ s^{n-3} & c_1 & c_2 & & & & \\ s^{n-4} & d_1 & d_2 & & & & \\ \vdots & \vdots & \vdots & & & & \\ s^2 & e_1 & a_n & & & & \\ s^1 & f_1 & & & & & \\ s^0 & a_n & & & & & \end{array}$$

(6.9)

The coefficients $b_1, b_2, \ldots$, are evaluated as follows:

$$b_1 = (a_1 a_2 - a_0 a_3)/a_1;$$
$$b_2 = (a_1 a_4 - a_0 a_5)/a_1; \ldots$$

This process is continued till we get a zero as the last coefficient in the third row. In a similar way, the coefficients of 4th, 5th,..., nth and $(n+1)$th rows are evaluated, e.g.,

$$c_1 = (b_1 a_3 - a_1 b_2)/b_1;$$
$$c_2 = (b_1 a_5 - a_1 b_3)/b_1; \ldots$$

and

$$d_1 = (c_1 b_2 - b_1 c_2)/c_1;$$
$$d_2 = (c_1 b_3 - b_1 c_3)/c_1; \ldots$$

It is to be noted here that in the process of generating the Routh array, the missing terms are regarded as zeros. Also all the elements of any row can be divided by a positive constant during the process to simplify the computational work.

The Routh stability criterion is stated as below.

Concepts of Stability and Algebraic Criteria   191

For a system to be stable, it is necessary and sufficient that each term of first column of Routh array [as given in eqn. (6.9)] of its characteristic equation be positive if $a_0 > 0$. If this condition is not met, the system is unstable and number of sign changes of the terms of the first column of the Routh array corresponds to the number of roots of the characteristic equation in the right half of the s-plane.

The Routh criterion stated above and the Hurwitz criterion are equivalent, as is shown below.

Elements of first column of the Routh array can be interpreted in terms of Hurwitz determinants as follows:

$$b_1 = \frac{a_1 a_2 - a_0 a_3}{a_1} = \frac{\begin{vmatrix} a_1 & a_0 \\ a_3 & a_2 \end{vmatrix}}{a_1} = \frac{\Delta_2}{\Delta_1}$$

Similarly $c_1 = \Delta_3/\Delta_2;$
$d_1 = \Delta_4/\Delta_3;\ldots$

Therefore, the condition of positiveness of the Hurwitz determinants corresponds to the condition of positiveness of the elements of the first column of the Routh array.

Example 6.2: Consider the fourth-order system of Example 6.1 with the characteristic equation

$$s^4 + 8s^3 + 18s^2 + 16s + 5 = 0$$

The Routh array for this system is given below:

s^4	1	18	5
s^3	8	16	0 (for the missing term)
s^2	$\frac{8 \times 18 - 1 \times 16}{8} = 16$	$\frac{8 \times 5 - 1 \times 0}{8} = 5$	
s^1	$\frac{16 \times 16 - 8 \times 5}{16} = 13.5$	0	
s^0	5		

The elements of the first column are all positive and hence the system is stable.

Example 6.3: Consider the following characteristic equation:

$$3s^4 + 10s^3 + 5s^2 + 5s + 2 = 0$$

The Routh array is given below:

s^4	3	5	2
s^3	10	5	
s^3	2	1	
s^2	$\frac{2 \times 5 - 3 \times 1}{2} = 3.5$	$\frac{2 \times 2 - 0 \times 3}{2} = 2$	
s^1	$\frac{3.5 \times 1 - 2 \times 2}{3.5} = -\frac{0.5}{3.5}$		
s^0	2		

It may be noted that in order to simplify computational work, the s^3-row in the formation of the Routh array has been modified by dividing it by 5 throughout. The modified s^3-row is then used to complete the process of array formation.

Examining the first column of the Routh array, it is found that there are two changes in sign (from 3.5 to $-0.5/3.5$ and from $-0.5/3.5$ to 2). Therefore the system under consideration is unstable having two poles in the right half of the s-plane.

It is to be noted that the Routh criterion gives only the number of roots in the right half of the s-plane. It gives no information as regards the values of the roots and also does not distinguish between real and complex roots.

SPECIAL CASES

Occasionally, in applying the Routh stability criterion, certain difficulties arise causing the breakdown of the Routh's test. The difficulties encountered are generally of the following types.

Difficulty 1: When the first term in any row of the Routh array is zero while rest of the row has at least one nonzero term. Because of this zero term, the terms in the next row become infinite and Routh's test breaks down. The following methods can be used to overcome this difficulty:

(a) Substitute a small positive number ϵ for the zero and proceed to evaluate the rest of the Routh array. Then examine the signs of the first column of Routh array by letting $\epsilon \to 0$.

(b) Modify the original characteristic equation by replacing* s by $1/z$. Apply the Routh's test on the modified equation in terms of z. The number of z-roots with positive real parts are the same as the number of s-roots with positive real parts. This method works in most but not all cases.

The following example illustrates these methods.

Example 6.4: Consider the characteristic equation

$$s^5 + s^4 + 2s^3 + 2s^2 + 3s + 5 = 0$$

The Routh array is

s^5	1	2	3
s^4	1	2	5
s^3	ϵ	-2	
s^2	$\dfrac{2\epsilon + 2}{\epsilon}$	5	
s^1	$\dfrac{-4\epsilon - 4 - 5\epsilon^2}{2\epsilon + 2} \to -2$		
s_0	5		

*This transformation maps the left half of the s-plane into the left half of the z plane and the right half of the s-plane into the right half of the z-plane.

From the Routh array, it is seen that first element in the third row is 0. This is replaced by ϵ, a small positive number. The first element in the 4th row is now $(2\epsilon+2)/\epsilon$ which has a positive sign as $\epsilon \to 0$. (In the Routh stability criterion, we are interested only in the signs of the terms in the first column and not in their magnitudes.) The first term of the fifth row is $(-4\epsilon - 4 - 5\epsilon^2)/(2\epsilon+2)$, which has a limiting value of -2 as $\epsilon \to 0$. Examining the terms in the first column of the Routh array, it is found that there are two changes in sign and hence the system is unstable having two poles in the right half s-plane.

Consider now the second method of overcoming the difficulty caused by a zero term in the first column of the Routh array.

Replacing s by $1/z$ in the characteristic equation and rearranging, we get

$$5z^5+3z^4+2z^3+2z^2+z+1 = 0$$

The Routh array for this equation is

z^5	5	2	1
z^4	3	2	1
z^3	$-4/3$	$-2/3$	
z^2	$1/2$	1	
z^1	2		
z^0	1		

There are two changes of sign in the first column of the Routh array, which tell us that there are two z-roots in the right half z-plane. Therefore, the number of s-roots in the right half s-plane is also two.

Difficulty 2: *When all the elements in any one row of the Routh array are zero.*

This condition indicates that there are symmetrically located roots in the s-plane (pair of real roots with opposite signs and/or pair* of conjugate roots on the imaginary axis and/or complex conjugate roots forming quadrates in the s-plane). The polynomial whose coefficients are the elements of the row just above the row of zeros in the Routh array is called an *auxiliary polynomial*. This polynomial gives the number and location of root pairs of the characteristic equation which are symmetrically located in the s-plane. The order of the auxiliary polynomial is always even.

Because of a zero row in the array, the Routh's test breaks down. This situation is overcome by replacing the row of zeros in the Routh array by a row of coefficients of the polynomial generated by taking the first derivative of the auxiliary polynomial. The following example illustrates the procedure.

Example 6.5: Consider a sixth-order system with the characteristic equation

$$s^6+2s^5+8s^4+12s^3+20s^2+16s+16 = 0$$

*There may be more than one such pairs.

The Routh array is

$$
\begin{array}{c|cccc}
s^6 & 1 & 8 & 20 & 16 \\
s^5 & 2 & 12 & 16 & \\
s^5 & 1 & 6 & 8 & \\
s^4 & 2 & 12 & 16 & \\
s^4 & 1 & 6 & 8 & \\
s^3 & 0 & 0 & &
\end{array}
$$

Since the terms in the s^3-row are all zero, the Routh's test breaks down. Now the auxiliary polynomial is formed from the coefficients of the s^4-row, which is given by

$$A(s) = s^4 + 6s^2 + 8$$

The derivative of the polynomial with respect to s is

$$\frac{d}{ds}A(s) = 4s^3 + 12s$$

The zeros in the s^3-row are now replaced by the coefficients 4 and 12. The Routh array then becomes

$$
\begin{array}{c|cccc}
s^6 & 1 & 8 & 20 & 16 \\
s^5 & 1 & 6 & 8 & \\
s^4 & 1 & 6 & 8 & \\
s^3 & 4 & 12 & & \\
s^3 & 1 & & & \\
s^2 & 3 & & & \\
s^1 & 1/3 & & & \\
s^0 & 8 & & &
\end{array}
$$

We see that there is no change of sign in the first column of the new array.

By solving for the roots of auxiliary polynomial

$$s^4 + 6s^2 + 8 = 0$$

we find that the roots are

$$s = \pm j\sqrt{2} \text{ and } s = \pm j2$$

These two pairs of roots are also the roots of the original characteristic equation. Since there is no sign change in the new array formed with the help of the auxiliary polynomial, we conclude that no root of characteristic equation has positive real part. Therefore the system under consideration is limitedly stable.

APPLICATION OF THE ROUTH STABILITY CRITERION TO
LINEAR FEEDBACK SYSTEMS

The Routh stability criterion is frequently used for the determination of the condition of stability of linear feedback control systems. Consider the closed-loop feedback system shown in Fig. 6.4. Let us determine the range of K for which the system is stable.

Concepts of Stability and Algebraic Criteria

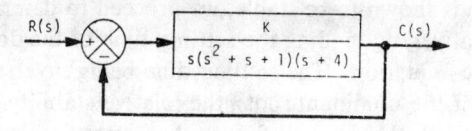

Figure 6.4.

The closed-loop transfer function of the system is
$$\frac{C(s)}{R(s)} = \frac{K}{s(s^2+s+1)(s+4)+K}$$
Therefore, the characteristic equation is
$$s(s^2+s+1)(s+4)+K = 0$$
or
$$s^4+5s^3+5s^2+4s+K = 0$$
The Routh array for this equation is

$$\begin{array}{c|ccc}
s^4 & 1 & 5 & K \\
s^3 & 5 & 4 & \\
s^2 & 21/5 & K & \\
s^1 & \left(\frac{84}{5} - 5K\right)/\frac{21}{5} & & \\
s^0 & K & &
\end{array}$$

Since for a stable system, the signs of elements of the first column of the Routh array should be all positive, the condition of system stability requires that
$$K > 0$$
and
$$(84/5 - 5K) > 0$$
Therefore for stability, K should lie in the range
$$84/25 > K > 0$$
When $K = 84/25$, there will be a zero at the first entry in the fourth row of the Routh array. As explained earlier, this corresponds to the presence of a pair of symmetrical roots which, as shown below, are pure imaginary in this case. Therefore $K = 84/25$, will cause sustained self-oscillations in the closed-loop system.

For $K = 84/25$, the auxiliary polynomial, given by the coefficients of the third row, is
$$(21/5)s^2+84/25 = 0$$
which gives the roots as
$$s = \pm j\sqrt{(4/5)} = \pm j\omega_0$$
Hence the frequency of sustained self-oscillations at
$$K = 84/25 \text{ is } \sqrt{(4/5)} \text{ rad/sec}$$

6.5 Relative Stability Analysis

Once a system is shown to be stable, we proceed to determine its relative stability quantitatively by finding the settling time of the dominant roots of its characteristic equation. The settling time being inversely proportional to the real part of the dominant roots, the relative stability can be specified by requiring that all the roots of the characteristic equation be more negative than a certain value, i.e., all the roots must lie to the left of the line $s = -\sigma_1 (\sigma_1 > 0)$. The characteristic equation of the system under study is then modified by shifting the origin of the s-plane to $s = -\sigma_1$, i.e., by the substitution,

$$s = z - \sigma_1$$

as illustrated in Fig. 6.5. If the new characteristic equation in z satisfies the Routh criterion, it implies that all the roots of the original characteristic equation are more negative than $-\sigma_1$.

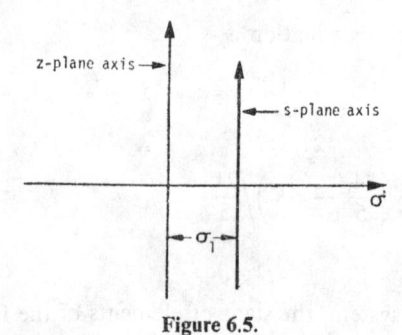

Figure 6.5.

Example 6.6: Consider a third-order system with the characteristic equation

$$s^3 + 7s^2 + 25s + 39 = 0$$

which by the Routh's test can be shown to have all its roots in the left half of the s-plane. Let us check if all the roots of this equation have real parts more negative than -1.

Shift the origin to $s = -1$ by the substitution

$$s = z - 1$$

in the characteristic equation. The characteristic equation in the new variable z is

$$z^3 + 4z^2 + 14z + 20 = 0$$

Forming the Routh array, we have

z^3	1	14
z^2	4	20
z^1	9	
z^0	20	

As the signs of all the elements of the first column of the Routh array are positive, the roots of the characteristic equation in z lie in the left half

z-plane, which implies that all the roots of the original characteristic equation in s lie to the left of $s = -1$ in the s-plane.

6.6 More on the Routh Stability Criterion

Some additional useful information regarding the Routh stability criterion which supplements what is already presented in this Chapter, is given below:

1. During the formation of a Routh array, if at any stage a row (not comprising all zeros) having one or more zero elements and/or one or more negative elements appears, the completion of Routh array will always result in sign changes in the first column, i.e., the characteristic polynomial has one or more roots in the right half s-plane. Therefore, for stability determination in such cases, we need not proceed with array formation beyond this stage except when we are interested in determining the number of right half s-plane roots.

2. If the first element in any row of a Routh array happens to be zero, while the rest of the row has at least one non-zero element, the system is unstable as per (1) above. In proceeding with array formation for determining the number of right half s-plane roots, this zero element is replaced by a small number ϵ. The limit $\epsilon \to 0$ is then taken to determine the changes in algebraic signs of the first column terms yielding the information regarding the number of right half s-plane roots.

This method, however, fails if the characteristic polynomial has roots on the imaginary axis [1, 2]. In order to establish a modified procedure for such cases, it is necessary to examine the rationale of the method.

If the polynomial under test had slightly different coefficients, the troublesome first-column zero would instead be some small non-zero number. As long as the polynomial has no roots on the imaginary axis, sufficiently small perturbations of its coefficients will not alter the number of right half s-plane roots. Rather than to actually perturb the polynomial coefficients, the effect of such perturbation is indicated by replacing the zero term with ϵ. Therefore for a polynomial with no imaginary-axis roots, $\epsilon \to 0$ from positive or negative side would give identical results for the number of right half s-plane roots. ϵ is generally taken to be positive for convenience.

For the case where the characteristic equation has imaginary-axis roots, replacement of the first-column zero element by ϵ would cause the imaginary-axis roots to move into either left or right half of the s-plane, with the result that the roots of characteristic polynomial in the right half s-plane can not be correctly determined.

Therefore to apply the ϵ-limiting method for a characteristic polynomial with the imaginary-axis roots, one must first extract these roots before the test is applied on the remainder polynomial having no imaginary-axis roots.

For example, consider the polynomial

$$q(s) = s^6 + s^5 + 3s^4 + 3s^3 + 3s^2 + 2s + 1$$

The Routh array is

$$\begin{array}{c|cccc}
s^6 & 1 & 3 & 3 & 1 \\
s^5 & 1 & 3 & 2 & \\
s^4 & \varepsilon & 1 & 1 & \\
s^3 & \dfrac{3\epsilon - 1}{\varepsilon} & \dfrac{2\epsilon - 1}{\epsilon} & & \\
s^2 & \dfrac{-2\epsilon^2 + 4\epsilon - 1}{3\epsilon - 1} & 1 & & \\
s^1 & \dfrac{4\epsilon^2 - \epsilon}{2\epsilon^2 - 4\epsilon + 1} & & & \\
s^0 & 1 & & &
\end{array}$$

As $\epsilon \to 0$, the elements of s^1 row tend to zero. This indicates that there are symmetrically located roots in the s-plane. We therefore need to examine the auxiliary polynomial to find out the possibility of the imaginary-axis roots. If no such roots exist, the usual procedure of replacing the all-zero row by coefficients of the derivative of the auxiliary polynomial is adopted. If the imaginary-axis roots are found to exist, the original polynomial is divided out by the auxiliary polynomial and the test is performed on the remainder polynomial.

For the example under consideration, the auxiliary equation is (let $\epsilon \to 0$ in s^2-row)

$$s^2 + 1 = 0$$

yielding two roots on the imaginary axis. Dividing the original polynomial $q(s)$ by (s^2+1), we get

$$q'(s) = s^4 + s^3 + 2s^2 + 2s + 1$$

The Routh array for this polynomial is

$$\begin{array}{c|ccc}
s^4 & 1 & 2 & 1 \\
s^3 & 1 & 2 & \\
s^2 & \varepsilon & 1 & \\
s^1 & \dfrac{2\varepsilon - 1}{\epsilon} & & \\
s^0 & 1 & &
\end{array}$$

As $\epsilon \to 0$, there are two sign changes in the first-column elements. This indicates that there are two roots in the right half s-plane.

Problems

6.1 Find the roots of the characteristic equation for systems whose open-loop transfer functions are given below. Locate the roots in the s-plane and indicate the stability of each system.

(a) $G(s)H(s) = \dfrac{1}{(s+2)(s+4)}$

(b) $G(s)H(s) = \dfrac{5(s+3)}{s(s+3)(s+8)}$

(c) $G(s)H(s) = \dfrac{9}{s^2(s+2)}$

Note: Refer Appendix III.

6.2 The characteristic equation of a servo system is given by

$$a_0 s^4 + a_1 s^3 + a_2 s^2 + a_3 s + a_4 = 0$$

Determine the conditions which must be satisfied by the coefficients of the characteristic equation for the system to be stable.

6.3 By means of the Routh criterion, determine the stability of the systems represented by the following characteristic equations. For systems found to be unstable, determine the number of roots of the characteristic equation in the right half s-plane.

(a) $s^4 + 2s^3 + 8s^2 + 4s + 3 = 0$
(b) $s^4 + 2s^3 + s^2 + 4s + 2 = 0$
(c) $s^5 + s^4 + 3s^3 + 9s^2 + 16s + 10 = 0$
(d) $s^6 + 3s^5 + 5s^4 + 9s^3 + 8s^2 + 6s + 4 = 0$

6.4 The characteristic equations for certain feedback control systems are given below. In each case, determine the range of values of k for the system to be stable.

(a) $s^3 + 3ks^2 + (k+2)s + 4 = 0$
(b) $s^4 + 4s^3 + 13s^2 + 36s + k = 0$
(c) $s^4 + 20ks^3 + 5s^2 + 10s + 15 = 0$

6.5 The open-loop transfer function of a unity feedback control system is given by

$$G(s) = \dfrac{K}{(s+2)(s+4)(s^2+6s+25)}$$

By applying the Routh criterion, discuss the stability of the closed-loop system as a function of K. Determine the values of K which will cause sustained oscillations in the closed-loop system. What are the corresponding oscillation frequencies?

6.6 A system oscillates with frequency ω, if it has poles at $s = \pm j\omega$ and no poles in the right half s-plane. Determine values of K and a, so that the system shown in Fig. P-6.6 oscillates at a frequency 2 rad/sec.

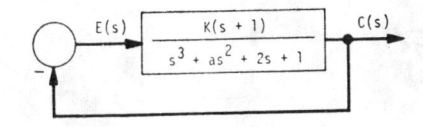

Figure P-6.6.

6.7 A unity feedback control system is characterized by the open-loop transfer function

$$G(s) = \dfrac{K(s+13)}{s(s+3)(s+7)}$$

(a) Using the Routh criterion, calculate the range of values of K for the system to be stable.
(b) Check if for $K = 1$, all the roots of the characteristic equation of the above system have damping factor greater than 0.5.

Note: Part (b) requires actual determination of the roots.

6.8 Determine whether the largest time constant of the characteristic equation given below is greater than, less than, or equal to 1.0 sec.

$$s^3+4s^2+6s+4 = 0$$

6.9 A feedback system has an open-loop transfer function of

$$G(s)H(s) = \frac{Ke^{-s}}{s(s^2+5s+9)}$$

Determine by use of the Routh criterion, the maximum value of K for the closed-loop system to be stable.
(*Hint* : For low frequencies $e^{-s} \approx (1 - s)$).

6.10 Determine the range of values of $K(K > 0)$ such that the characteristic equation

$$s^3+3(K+1)s^2+(7K+5)s+(4K+7) = 0$$

has roots more negative than $s = -1$.

References and Further Reading

1. Chang, T.S. and C.T. Chen, "On the Routh-Hurwitz Criterion," *IEEE Trans. Auto. Contr.*, 1974, **19**, 3, 250-251.
2. Hostetter, G.H., "Additional Comments on 'on the Routh-Hurwitz Criterion'," 1975, **20**, 2, 296-297.
3. Jury, E.I., *Inners and Stability of Dynamic Systems*, John Wiley, New York, 1974.
4. Chen, C.F. and I.J. Haas, *Elements of Control Systems Analysis*, Prentice-Hall, Englewood Cliffs, N.J., 1968.
5. Hsu, J.C. and A.U. Meyer, *Modern Control Principles and Applications*, McGraw-Hill, New York, 1968.
6. Ku, Y.H., *Analysis and Control of Linear Systems*, International Text Book Co., Scranton, Pennsylvania, 1962.

7. The Root Locus Technique

7.1 Introduction

It has been discussed earlier that the possibility of unstable operation is inherent in all feedback control systems because of the very nature of the feedback itself. An unstable system, obviously cannot perform the control task required of it. Therefore, while analyzing a given system, the very first investigation that needs to be made is, whether the system is stable. However, the determination of stability of a system is necessary but not sufficient, for a stable system with low damping is still undesirable. In an analysis problem one must, therefore, proceed to determine not only the absolute stability of a system but also its relative stability (peak overshoot, settling time, etc.). As discussed in the preceding chapter, the relative stability is directly related to the location of the closed-loop poles of a system. The Routh's criterion gives a satisfactory answer to the question of stability but its adoption to determine the relative stability is tedious and requires trial and error procedure even in the analysis problem.

Consider now a design problem in which the designer is required to achieve the desired performance for a system by adjusting the location of its closed-loop poles in the s-plane by varying one or more system parameters. The Routh's criterion, obviously does not help much in such problems. For determining the location of the closed-loop poles, one may resort to the classical techniques of factoring the characteristic polynomial and determining its roots, since the closed-loop poles are the roots of the characteristic equation. This technique is very laborious when the degree of the characteristic polynomial is three or higher. Furthermore, repeated calculations are required as a system parameter is varied for adjustments.

A simple technique, known as the *root locus technique*, for finding the roots of the characteristic equation, introduced by W.R. Evans, is extensively used in control engineering practice. This technique provides a graphical method of plotting the locus of the roots in the s-plane as a given system parameter is varied over the complete range of values (maybe from zero to infinity). The roots corresponding to a particular value of the system parameter can then be located on the locus or the value of the parameter for a desired root location can be determined from the locus. The root locus is a powerful technique as it brings into focus the complete dynamic response of the system and further, being a graphical technique, an approximate root locus sketch can be made quickly and the designer can easily visualize the effects of varying various system parameters on root locations. The root locus also provides a measure of sensitivity of roots to the variation in the parameter being considered. It may further be pointed out here that the root locus technique is applicable for single as well as multiple-loop systems.

In this chapter, we shall discuss the concepts underlying the root locus technique, the general procedure for sketching root locus plots, and the analysis of feedback systems by use of this technique. The use of this technique for design problems will be considered in Chapter 10.

7.2 The Root Locus Concept

To understand the concepts underlying the root locus technique, consider the simple second-order system shown in Figure 7.1. The open-loop transfer function of this system is

$$G(s) = \frac{K}{s(s+a)}$$

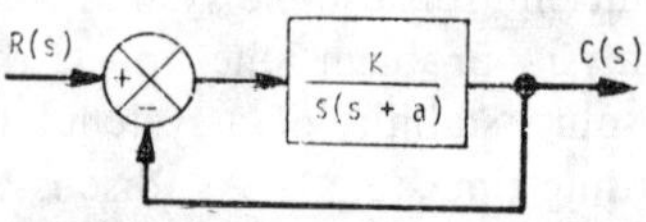

where K and a are constants.

Figure 7.1 A second-order system.

The open-loop transfer function has two poles—one at origin $s = 0$ and the other at $s = -a$. The closed-loop transfer function of this system is

$$\frac{C(s)}{R(s)} = \frac{K}{s^2 + as + K} \tag{7.1}$$

From eqn. (7.1), the characteristic equation of the system is

$$s^2 + as + K = 0 \tag{7.2}$$

The second-order system under consideration is always stable for positive values of a and K but its dynamic behaviour is controlled by the roots of eqn. (7.2) and hence in turn by the magnitudes of a and K, since the roots are given by

$$s_1, s_2 = -\frac{a}{2} \pm \sqrt{\left[\left(\frac{a}{2}\right)^2 - K\right]} \tag{7.3}$$

From eqn. (7.3), it is seen that as any of the system parameters (a or K) varies, the roots of the characteristic equation change. Let us consider the commonly occurring case of variable gain K, while the parameter a is held fixed. As K is varied from zero to infinity, the two roots (s_1, s_2) describe loci in the s-plane. Root locations for various ranges of values of K are:

(1) $0 \leqslant K < a^2/4$, the roots are real and distinct.

When $K = 0$, the two roots are $s_1 = 0$, $s_2 = -a$, i.e., they coincide with the open-loop poles of the system.

(2) $K = a^2/4$, the roots are real and equal in value, i.e., $s_1 = s_2 = -a/2$.

(3) $a^2/4 < K < \infty$, the roots are complex conjugate with real part $= -a/2$, i.e., unvarying real part.

The root locus with varying K is plotted in Fig. 7.2. These loci give the following information about the system behaviour.

1. The root locus plot has two branches starting at the two open-loop poles ($s = 0$ and $s = -a$) for $K = 0$.
2. As K is increased from 0 to $a^2/4$, the roots move towards the point $(-a/2, 0)$ from opposite directions. Both the roots lie on the negative real axis which corresponds to an overdamped system. The two roots meet at $s = -a/2$ for $K = a^2/4$. This corresponds to a critically-damped system. As K is increased further ($K > a^2/4$), the roots break away from the real axis, become complex conjugate and since the real part of both the roots remains fixed at $-a/2$, the roots move along the line $\sigma = -a/2$ and the system becomes underdamped.
3. For $K > a^2/4$, the real parts of the roots are fixed, therefore the settling time is nearly constant.

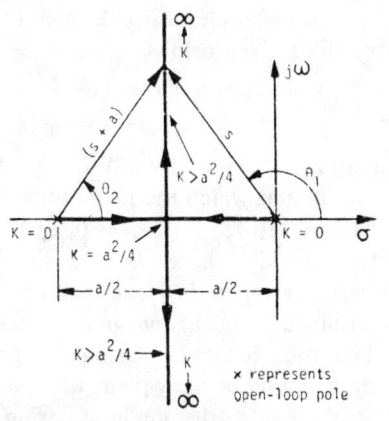

Figure 7.2 Root loci of $s^2 + as + K = 0$ as a function of K.

The root locus shown in Fig. 7.2 has been drawn by the direct solution of the characteristic equation. This procedure becomes highly tedious for higher-order systems. Evans developed a graphical technique by use of which the root locus for third- and higher-order systems can be constructed as easily as for a second-order system. The characteristic equation of any system is given by

$$\Delta(s) = 0 \qquad (7.4)$$

where $\Delta(s)$ is the determinant of the signal flow graph of the system and is given by eqn. (2.65), which is reproduced below:

$$\Delta(s) = 1 - \sum_m P_{m1} + \sum_m P_{m2} - \sum_m P_{m3} + \ldots$$

where P_{mr} = gain product of mth possible combination of r non-touching loops of the graph. Thus, the characteristic equation can always be written in the form

$$1 + P(s) = 0 \qquad (7.5)$$

For the single-loop system shown in Fig. 7.3

$$P(s) = G(s)H(s)$$

where $G(s)H(s)$ is the open-loop transfer function in block diagram terminology or loop transmittance in signal flow graph terminology.

From eqn. (7.5) it is seen that the roots of the characteristic equation (i.e., the closed-loop poles of the system) occur only for those values of s where

$$P(s) = -1 \qquad (7.6)$$

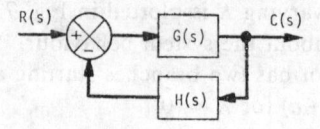

Figure 7.3 Single-loop feedback system.

Since s is a complex variable, eqn. (7.6) is converted into the two Evans conditions given below:

$$|P(s)| = 1 \tag{7.7}$$

$$\angle P(s) = \pm 180°(2q+1); \quad q = 0, 1, 2,... \tag{7.8}$$

Equations (7.7) and (7.8) imply that the roots of $1+P(s) = 0$ are those values of s at which the magnitude of $P(s)$ equals 1 and the angle of $P(s)$ equals $\pm 180°(2q+1)$; $q = 0, 1, 2,...$ A plot of the points in the complex plane satisfying the *angle criterion* of eqn. (7.8) is the root locus. The value of gain corresponding to a root, i.e., a point on the root locus, can be determined from the *magnitude criterion* of eqn. (7.7).

The root locus can be quickly drawn by checking the angle criterion at various points of the s-plane using a protractor. A simple device called the *spirule** checks the angle criterion and at the same time gives the system gain at a root location. The root locus technique is thus much easier to apply as compared to the direct solution of the characteristic equation.

Consider for example the feedback system of Fig. 7.3 with

$$G(s) = K\frac{(s+b)}{s(s+a)} \tag{7.9}$$

$$H(s) = 1$$

Its characteristic equation is

$$1 + G(s) = 0 \tag{7.10}$$

The open-loop transfer function $G(s)$ has two poles at $s = 0, -a$ and a zero at $s = -b$ as shown in Fig. 7.4. The angle criterion for the root locus is

$$\angle \left[\frac{(s+b)}{s(s+a)}\right] = \pm 180°(2q+1); \quad q = 0, 1, 2,... \tag{7.11}$$

With $s = (\sigma + j\omega)$

$$\angle \left[\frac{(s+b)}{s(s+a)}\right] = \angle \left[\frac{(\sigma+j\omega+b)}{(\sigma+j\omega)(\sigma+j\omega+a)}\right]$$

Therefore $\tan^{-1}\left(\frac{\omega}{\sigma+b}\right) - \tan^{-1}\left(\frac{\omega}{\sigma}\right) - \tan^{-1}\left(\frac{\omega}{\sigma+a}\right) = -\pi$

or $\tan^{-1}\left(\frac{\omega}{\sigma}\right) + \tan^{-1}\left(\frac{\omega}{\sigma+a}\right) = \pi + \tan^{-1}\left(\frac{\omega}{\sigma+b}\right)$

---

*This is a device invented by Evans for drawing the root locus speedily. It is a combination of a protector and a rule with a logarithmic spiral drawn on it. Both the protector and the rule can be rotated and the combination is very helpful for complex number multiplication in polar form.

Taking tan on both sides, we have

$$\frac{\dfrac{\omega}{\sigma}+\dfrac{\omega}{\sigma+a}}{1-\dfrac{\omega^2}{\sigma(\sigma+a)}} = \frac{\dfrac{\omega}{\sigma+b}+\tan\pi}{1-(\tan\pi)\left(\dfrac{\omega}{\sigma+b}\right)}$$

Simplifying we get

$$(\sigma+b)^2+\omega^2 = b^2 - ab \tag{7.12}$$

Equation (7.12) gives the root locus in the s-plane. It is a circle centred at $(-b, 0)$, i.e., at the zero of the open-loop transfer function and of radius $\sqrt{(b^2 - ab)}$. The locus is drawn in Fig. 7.4. The sections of the real axis which lie on the root locus and the direction of arrows for increasing K are easily determined by the rules given in the next section.

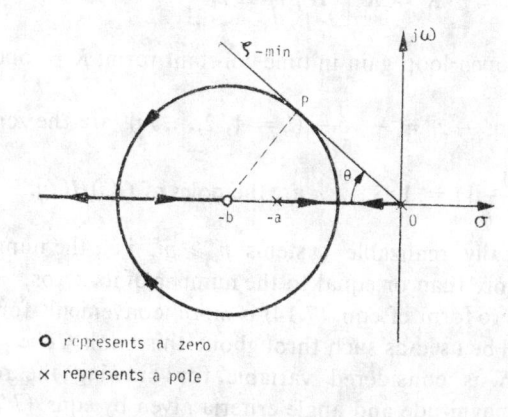

○ represents a zero
× represents a pole

Figure 7.4 Root locus plot of $1+K(s+b)/s(s+a) = 0$.

As K is varied, the least-damped (minimum damping factor) complex conjugate poles are obtained by drawing OP tangential to the circular locus. By geometry

$$OP = \sqrt{[b^2 - (b^2 - ab)]} = \sqrt{(ab)}$$

$$\zeta_{min} = \cos\theta = \sqrt{(ab)}/b = \sqrt{(a/b)}$$

The magnitude criterion can be used to determine the value of K for any particular location of roots.

7.3 Construction of Root Loci

Consider the basic feedback system of Fig. 7.3. Its characteristic equation is

$$1+G(s)H(s) = 0$$

The open-loop transfer function $G(s)H(s)$ is generally known in the factored form as it is obtained by modelling the transfer function of individual components comprising the system. Therefore $G(s)H(s)$ [or $P(s)$ in case of multiple-loop systems] can generally be expressed in either of the factored forms given below:

$$1+G(s)H(s) = 1 + \frac{K' \prod_{i=1}^{m}(\tau_{zi}s+1)}{\prod_{j=1}^{n}(\tau_{pj}s+1)} = 0 \text{ (time-constant form)} \quad (7.13)$$

$$= 1 + \frac{K \prod_{i=1}^{m}(s+z_i)}{\prod_{j=1}^{n}(s+p_j)} = 0 \text{ (pole-zero form)} \quad (7.14)$$

These forms are interrelated by the following expression

$$K = K' \left[\prod_{j=1}^{n} p_j \bigg/ \prod_{i=1}^{m} z_i \right]$$

where K' = open-loop gain in time-constant form; K = open-loop gain in pole-zero form; $-z_i = -\dfrac{1}{\tau_{zi}}$ ($i = 1, 2, ..., m$) are the zeros of $G(s)H(s)$;

$-p_j = -\dfrac{1}{\tau_{pj}}$ [$j = 1, 2, ..., n$ are the poles of $G(s)H(s)$].

In all physically realizable systems $n \geqslant m$, i.e., the number of poles of $G(s)H(s)$ is more than or equal to the number of its zeros.

The pole-zero form of eqn. (7.14) is more convenient for drawing root locus and will be used as such throughout this book. The open-loop gain parameter* K is considered variable for drawing the root locus of the system. The magnitude and angle criteria given by eqns. (7.7) and (7.8) can now be expressed as

$$\frac{K \prod_{i=1}^{m} |s+z_i|}{\prod_{j=1}^{n} |s+p_j|} = 1 \quad (7.15)$$

$$\sum_{i=1}^{m} \angle(s+z_i) - \sum_{j=1}^{n} \angle(s+p_j) = \pm(2q+1)180°; \quad q = 0, 1, 2,... \quad (7.16)$$

The points on root locus can be determined by satisfying the angle criterion through trial and error procedure. The poles and zeros of $G(s)H(s)$ are located on the s-plane on a graph sheet and a trial point s_0 is selected. The lines joining each of the poles $(-p_j)$ and zeros $(-z_i)$ to s_0 represent the phasors (s_0+p_j) and (s_0+z_i) respectively as shown in Fig. 7.5. The angles of these phasors can be measured by placing the protector or spirule at the pole and zero locations and the angle values are substituted in the angle

---

*If the variable parameter is other than the open-loop gain, the characteristic equation can be rearranged such that the parameter of interest appears as a multiplying factor in eqn. (7.14). This technique is illustrated in Sec. 7.4.

criterion of eqn. (7.16). The trial point s_0 is suitably shifted till the angle criterion is satisfied to a desired degree of accuracy. A more convenient method of reading the angles is to place the protector or spirule at the trial point and to read the angles in anticlockwise direction with reference to the dotted line shown in the figure. The dotted line is drawn parallel to the real axis and to the left of the point s_0. A number of points on the root locus are determined in this manner and a smooth curve drawn through these points gives the root locus with variable K. The value of K for a particular root location s_0 can be obtained from the magnitude criterion of eqn. (7.15) by substituting the phasor magnitudes read to scale from phasor lengths. It is to be noted that the same scale must be employed for both the real and imaginary axes of the complex s-plane.

From the magnitude criterion of eqn. (7.15)

$$K = \frac{\prod_{j=1}^{n} |s_0 + p_j|}{\prod_{i=1}^{m} |s_0 + z_i|}$$

With reference to Fig. 7.5, we can write

$$K = \frac{\text{Product of phasor lengths* from } s_0 \text{ to open-loop poles}}{\text{Product of phasor lengths* from } s_0 \text{ to open-loop zeros}}$$

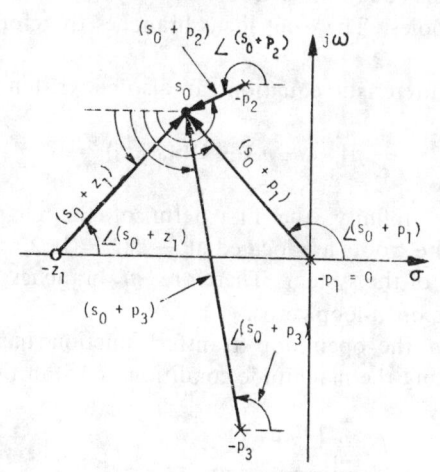

Figure 7.5 Determining a point on root locus.
×represents a pole; O represents a zero.

In order to help reduce the tedium of the trial and error procedure, certain rules have been developed for making a quick approximate sketch of the root locus. This approximate sketch provides a guide for selection of a trial point such that more accurate root locus can be obtained by a few trials. In fact, considerable time in root locus drawing is saved by this rough sketch.

*Read to scale.

Further, the approximate root locus sketch, as obtained by the rules discussed below, is very useful in visualizing the effects of variation of system gain K, the effects of shifting pole-zero locations and of bringing in a new set of poles and zeros.

CONSTRUCTION RULES

Rule 1: *The root locus is symmetrical about the real axis (σ-axis).*

We know that the roots of the characteristic equation are either real or complex conjugate or combinations of both. Therefore their locus must be symmetrical about the σ-axis of the s-plane.

Rule 2: *As K increases from zero to infinity, each branch of the root locus originates from an open-loop pole with $K = 0$ and terminates either on an open-loop zero or on infinity with $K = \infty$. The number of branches terminating on infinity equals the number of open-loop poles minus zeros.*

The characteristic eqn. (7.14) can be written as

$$\prod_{j=1}^{n} (s+p_j) + K \prod_{i=1}^{m} (s+z_i) = 0$$

When $K = 0$, this equation has roots at $-p_j$ ($j = 1, 2, ..., n$) which are the open-loop poles. The root locus branches therefore start at the open-loop poles.

The same characteristic equation can also be written as

$$\frac{1}{K} \prod_{j=1}^{n} (s+p_j) + \prod_{i=1}^{m} (s+z_i) = 0$$

As K tends to infinity, the first term of the characteristic equation vanishes and the roots are located at $-z_i$ ($i = 1, 2, ..., m$) which are the open-loop zeros of the system. Therefore m branches of the root locus terminate on the open-loop zeros.

In case $m < n$, the open-loop transfer function has $(n-m)$ zeros at infinity. Examining the magnitude condition (7.15) in the form

$$\frac{\prod_{i=1}^{m} |(s+z_i)|}{\prod_{j=1}^{n} |(s+p_j)|} = \frac{1}{K}$$

we find that this is satisfied by $s \to \infty e^{j\phi}$ as $K \to \infty$. Therefore, $(n-m)$ branches of the root locus terminate on infinity.

Rule 3: *A point on the real axis lies on the locus if the number of open-loop poles plus zeros on the real axis to the right of this point is odd.*

Consider the open-loop pole-zero configuration shown in Fig. 7.6. Let us examine any point s_0 on the real axis. As this point is joined by phasors to all the open-loop poles and zeros, it is easily seen that (i) the poles and zeros on the real axis to the right of this point contribute an angle of 180° each, (ii) the poles and zeros to the left of this point contribute an angle of 0° each, and (iii) the net angle contribution of a complex conjugate pole or zero pair is always zero. Therefore

$$\angle[G(s)H(s)] = (m_r - n_r)180°$$
$$= \pm(2q+1)180°; \quad q = 0, 1, 2, \dots$$

where m_r = number of zeros on the right of s_0; and n_r = number of poles on the right of s_0.

We therefore see that for a point on the real axis, the angle criterion is only met if $(n_r - m_r)$ or $(n_r + m_r)$ is odd and hence the rule. If this rule is satisfied at any point on the real axis, it continues to be satisfied as the point is moved on either side unless the point crosses a real axis pole or zero. By use of this additional fact, the real axis can be divided into segments on-locus and not-on-locus, the dividing points being the real open-loop poles and zeros. The on-locus segments of the real axis must alternate.

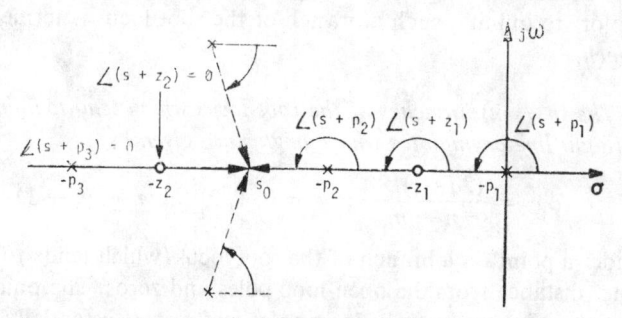

Figure 7.6 Angle contributions for a point on the real axis.

Example 7.1: Consider the system shown in Fig. 7.7 with the open-loop transfer function:

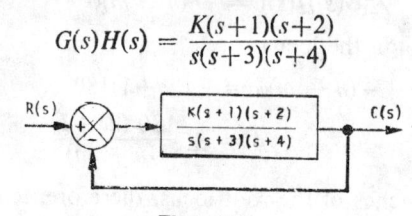

$$G(s)H(s) = \frac{K(s+1)(s+2)}{s(s+3)(s+4)}$$

Figure 7.7.

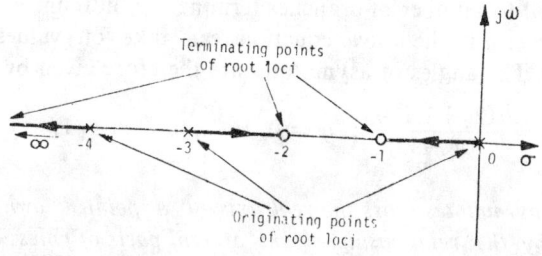

Figure 7.8 Root locus plot for the system shown in Fig. 7.7.

The open-loop pole-zero-configuration is shown in Fig. 7.8.

From the rules described so far, the following information concerning the root locus plot is easily obtained.

1. There are three branches of the root locus as there are three open-loop poles. These branches start with $K = 0$ at each of the poles, $s = 0$, -3, -4.

2. As K increases, the branches leave the open-loop poles and seek open-loop zeros, two branches terminate on the two open-loop zeros ($s = -1$, -2) and one terminates on infinity.

3. The real axis segments between 0 and -1, -2 and -3, -4 and $-\infty$, lie on the root locus (shown by thick lines in Fig. 7.8). This fact is verified by rule 3, e.g., for any point on the real axis between -2 and -3, $(n_r + m_r) = 1 + 2 = 3$ is odd and therefore this segment is on the root locus.

If a root locus branch moves along the real axis from an open-loop pole to zero or to infinity, such a branch of the root locus is termed as a *real-root branch*.

Rule 4: *The $(n - m)$ branches of the root locus which tend to infinity, do so along straight line asymptotes whose angles are given by*

$$\phi_A = \frac{(2q+1)180°}{n-m}; \quad q = 0, 1, 2, ..., (n-m-1).$$

Consider a point on a branch of the root locus (which tends to infinity) at a remote distance from the open-loop poles and zeros. The phasors drawn from such a point to the open-loop poles and zeros essentially make the same angle (say ϕ) to each pole and zero. Therefore, for the point under consideration

$$\angle[G(s)H(s)] = -(n-m)\phi$$

For this point to lie on the locus

$$-(n-m)\phi = \pm(2q+1)180°$$

or
$$\phi = \pm \frac{180°(2q+1)}{(n-m)} \qquad (7.17)$$

The $(n - m)$ branches of the root locus, therefore, tend to infinity along asymptotes having angles given by eqn. (7.17). The number of such asymptotes is equal to number of branches terminating on infinity, i.e., $(n - m)$. Thus the integer q in the above equation can take on values from 0 to $(n - m - 1)$. The angles of asymptotes are therefore given by

$$\phi_A = \frac{(2q+1)180°}{(n-m)}; \quad q = 0, 1, ..., (n-m-1) \qquad (7.18)$$

Rule 5: *The asymptotes cross the real axis at a point known as centroid, determined by the relationship: (sum of real parts of poles — sum of real parts of zeros)/(number of poles — number of zeros).*

From eqn. (7.14), the open-loop transfer function is written in the form

$$G(s)H(s) = \frac{K(s+z_1)(s+z_2)\ldots(s+z_m)}{(s+p_1)(s+p_2)\ldots(s+p_n)}; \quad m \leq n$$

$$= \frac{K[s^m + (\sum_{i=1}^{m} z_i)s^{m-1} + \ldots + (\prod_{i=1}^{m} z_i)]}{[s^n + (\sum_{j=1}^{n} p_j)s^{n-1} + \ldots + (\prod_{j=1}^{n} p_j)]}$$

Dividing the denominator by the numerator, we get

$$G(s)H(s) = \frac{K}{[s^{n-m} + (\sum_{j=1}^{n} p_j - \sum_{i=1}^{m} z_i)s^{n-m-1} + \ldots]} \quad (7.19)$$

As s tends to infinity, the terms with higher powers of s dominate. Therefore, eqn. (7.19) can be approximated as

$$G(s)H(s)\bigg|_{s\to\infty} \approx \frac{K}{[s^{n-m} + (\sum_{j=1}^{n} p_j - \sum_{i=1}^{m} z_i)s^{n-m-1}]} \quad (7.20)$$

Consider now the following function

$$P(s) = \frac{1}{(s+\sigma_A)^{n-m}}$$

which when expanded, gives

$$P(s) = \frac{1}{[s^{n-m} + (n-m)\sigma_A s^{n-m-1} + \ldots]} \quad (7.21)$$

The characteristic equation $1 + P(s) = 0$ has $(n-m)$ root locus branches which are straight lines* passing through the point $s = -\sigma_A$ on the real axis and having angles $[(2q+1)180°]/(n-m)$, $q = 0, 1, 2, \ldots, (n-m-1)$. If σ_A is selected such that

*$P(s) = 1/(s+\sigma_A)^{n-m}$ has no zeros and a real pole of multiplicity $(n-m)$ at $s = -\sigma_A$. Consider any point s_0 on a straight line drawn from $-\sigma_A$ at an angle ϕ (see diagram). Then

$$\angle P(s)\bigg|_{s=s_0} = -(n-m)\phi$$

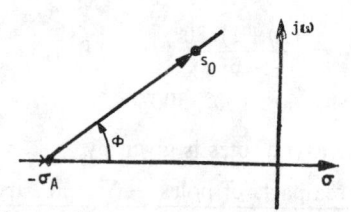

For this point to be on the root locus

$$-(n-m)\phi = -(2q+1)180°$$

or

$$\phi = \frac{(2q+1)180°}{(n-m)}; \quad q = 0, 1, 2, \ldots, (n-m-1)$$

Thus the root locus of $1 + P(s) = 0$ is a set of $(n-m)$ lines drawn from $s = -\sigma_A$ at angles ϕ given by the above expression.

$$(n - m)\sigma_A = \sum_{j=1}^{n} p_j - \sum_{i=1}^{m} z_i \qquad (7.22)$$

the function $G(s)H(s)$ behaves in the same manner as $P(s)$ for values of s approaching infinity, because the first two higher-order terms of their denominators are identical. In other words, the general function $G(s)H(s)$ approaches the particular function $P(s)$ as s approaches infinity. Therefore, the branches of the root locus of $1+G(s)H(s) = 0$ which tend to infinity, approach the straight line root locus branches of $1+P(s) = 0$. Hence the straight line root loci of $1+P(s) = 0$ act as asymptotes to the $(n - m)$ root locus branches of $1+G(s)H(s) = 0$ which tend to infinity.

From eqn. (7.22), the *centroid* of the asymptotes is given by

$$-\sigma_A = \frac{\sum_{j=1}^{n}(-p_j) - \sum_{i=1}^{m}(-z_i)}{n - m} \qquad (7.23)$$

Because all the complex poles and zeros occur in conjugate pairs, σ_A is always a real quantity. Therefore eqn. (7.23) may be reduced to the following form:

$$-\sigma_A = \frac{\sum \text{real parts of poles} - \sum \text{real parts of zeros}}{\text{number of poles} - \text{number of zeros}} \qquad (7.24)$$

Example 7.2: Consider a feedback system with the characteristic equation

$$1+K\frac{1}{s(s+1)(s+2)} = 0$$

The open-loop pole-zero configuration of this system is shown in Fig. 7.9. From the rules described so far, the following information concerning the root locus is obtained.

1. There are three branches of the root locus.
2. The branches of the root locus originate with $K = 0$ from the open-loop poles $s = 0, -1$ and -2.
3. Since there is no open-loop zero in the finite region, all the three branches terminate on infinity, along the asymptotes whose angles with the real axis are given by

$$\phi_A = \frac{(2q+1)180°}{3}; \quad q = 0, 1, 2$$

$$= 60°, 180°, 300°$$

4. The centroid of the asymptotes is given by

$$-\sigma_A = \frac{\sum \text{real parts of poles} - \sum \text{real parts of zeros}}{\text{number of poles} - \text{number of zeros}}$$

$$-\sigma_A = \frac{-1-2}{3} = -1$$

5. The segments of the real axis between 0 and -1, -2 and $-\infty$, lie on the root locus.

The Root Locus Technique   213

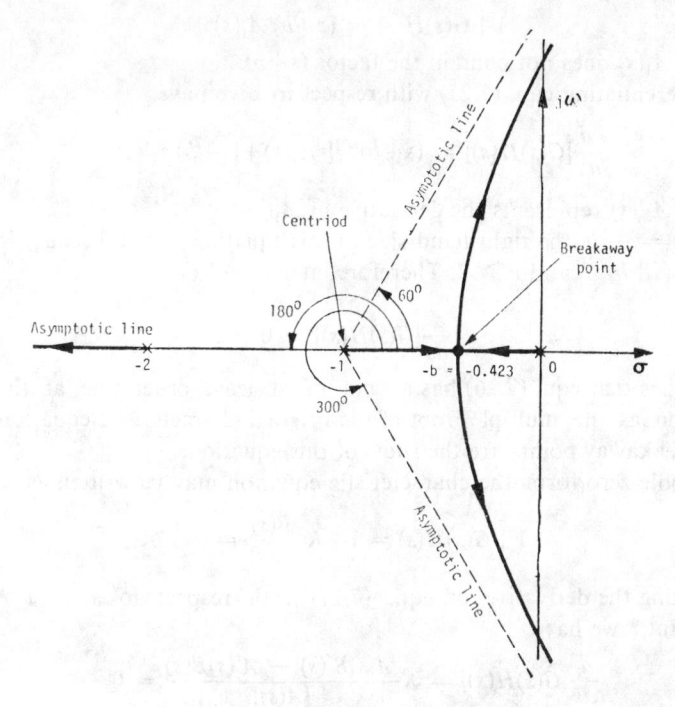

Figure 7.9 Root locus plot of the equation
$1+K/s(s+1)(s+2) = 0$.

From Fig. 7.9, it is seen that out of the three branches of the root locus, one is a real-root branch which originates from $s = -2$ and terminates on $-\infty$. The other two branches originate from $s = 0$ and $s = -1$, and move on the real axis approaching each other as K is increased. These two branches must therefore meet on the real axis. The characteristic equation has a double root at such a point. As the gain K is further increased, the root locus branches break away from the real axis to give a complex conjugate pair of roots. The point which represents a double root is known as *breakaway point* (determination of the breakaway point is discussed in the next rule). The branches which represent complex roots are known as *complex-root branches*.

In Fig. 7.9, the two real-root branches originating from the open-loop poles $s = 0$ and $s = -1$ approach each other, breakaway at the point $-b$ and then one branch moves to infinity along 60° asymptote and the other moves to infinity along 300° asymptote. The third branch being a real-root branch coincides with the 180° asymptote.

Rule 6: *The breakaway points (points at which multiple roots of the characteristic equation occur) of the root locus are the solutions of $dK/ds = 0$.*

Assume that the characteristic equation $1+G(s)H(s) = 0$ has a multiple root at $s = -b$ of multiplicity r. Then

$$1+G(s)H(s) = (s+b)^r A_1(s) \qquad (7.25)$$

where $A_1(s)$ does not contain the factor $(s+b)$.

Differentiating eqn. (7.25) with respect to s we have

$$\frac{d}{ds}[G(s)H(s)] = (s+b)^{r-1}[rA_1(s)+(s+b)A_1'(s)]$$

where $A_1'(s)$ represents the derivative of $A_1(s)$.

At $s = -b$, the right-hand side of this equation is zero because it has a factor $(s+b)^{r-1}$ and $r \geqslant 2$. Therefore, at $s = -b$

$$\frac{d}{ds}[G(s)H(s)] = 0 \qquad (7.26)$$

It implies that eqn. (7.26) has a root of at least order one at the same location as the multiple root of the original characteristic equation. Thus the breakaway points are the roots of this equation.

In pole-zero form, the characteristic equation may be written as

$$1+G(s)H(s) = 1+K\frac{B(s)}{A(s)} = 0 \qquad (7.27)$$

Taking the derivative of eqn. (7.27) with respect to s with K as a constant,* we have

$$\frac{d}{ds}[G(s)H(s)] = K\frac{A(s)B'(s) - A'(s)B(s)}{[A(s)]^2} = 0$$

As per eqn. (7.26) the roots of $\frac{d}{ds}[G(s)H(s)] = 0$ are the breakaway points, therefore the breakaway points are also given by the roots of

$$A(s)B'(s) - A'(s)B(s) = 0 \qquad (7.28)$$

From eqn. (7.27) we can write

$$K = -A(s)/B(s)$$

Differentiating K with respect to s, we get

$$\frac{dK}{ds} = \frac{A(s)B'(s) - A'(s)B(s)}{[B(s)]^2} \qquad (7.29)$$

At the breakaway points, the numerator of eqn. (7.29) is zero as per eqn. (7.28). Therefore, the breakaway points of the original characteristic equation are determined from the roots of

$$\frac{dK}{ds} = 0 \qquad (7.30)$$

Equation (7.26) or eqn. (7.30) could be used to determine the breakaway points.

In general, a breakaway point may involve two or more than two branches. If a breakaway point is obtained when r branches of the root

---

*This value of K pertains to the multiple roots of the characteristic equation.

locus come together and meet at a point, then that point represents the rth order root of the characteristic equation. Since the characteristic equation can have real as well as complex multiple roots, its root locus can have real as well as complex breakaway points, but because of conjugate symmetry of the root loci, the breakaway points must either be on the real axis or occur in complex conjugate pairs.

From the above discussions, we conclude that the breakaway points are the roots of the equation $dK/ds = 0$. It should, however, be noted that not all the roots of equation $dK/ds = 0$ correspond to the actual breakaway points. The actual breakaway points are those roots of the equation at which the root locus angle criterion is met.

Breakaway Points on the Real Axis

We discuss below different methods of evaluating breakaway points on the real axis.

Method 1: An analytical approach. This method requires the determination of the roots of the equation $dK/ds = 0$ to evaluate the breakaway points. The practical efficacy of the method is limited to third-order systems as in higher-order cases the determination of the roots of $dK/ds = 0$ in itself is a time consuming job. The method is illustrated through the example below.

Example 7.3: In the root locus plot of equation $1 + K/s(s+1)(s+2) = 0$ shown in Fig. 7.9 we found that there is a breakaway point on the real axis between 0 and -1 as the two real-root branches are oppositely directed on this segment.

From the characteristic equation, the gain K is given by

$$K = -s(s+1)(s+2) \qquad (7.31)$$
$$= -(s^3 + 3s^2 + 2s)$$

Differentiating eqn. (7.31), we get

$$\frac{dK}{ds} = -(3s^2 + 6s + 2)$$

The roots of the equation $dK/ds = 0$, are

$$s_{1,2} = \frac{-6 \pm \sqrt{(36-24)}}{6}$$
$$= -0.423, -1.577$$

Since the breakaway point must lie between 0 and -1, it is clear that $s = -0.423$ corresponds to the actual breakaway point.

Method 2: A graphical approach. It is a more practical method for determining the breakaway points. Along the real axis, the condition $dK/ds = 0$ implies that the gain K is extremized with respect to the real variable $s = \sigma$. A breakaway on the real axis may occur in two ways. First, a breakaway point may result from two real-root branches moving towards each other as K is increased. After the breakaway point these branches become complex root branches. In this case K is increasing along the real axis from either side to the breakaway point. The condition $dK/ds = 0$, therefore results in maximization of K at the breakaway point. Secondly, a real axis breakaway point may occur with complex-root branches moving towards the real axis and meeting at the breakaway point. These branches then become real-root branches and move in opposite direction along the real axis as K is further increased. In this case K decreases as breakaway point is approached from either side of the real axis. Therefore, the condition $dK/ds = 0$ results in minimization of K at the breakaway point.

As an example of the first case consider the third-order system whose root locus plot is drawn in Fig. 7.9. Figure 7.10 shows plot of K for this system for various values of s between 0 and -1. The maximum value of K is 0.385 at $s = -0.423$. Thus the breakaway point occurs at $s = -0.423$ with $K = 0.385$.

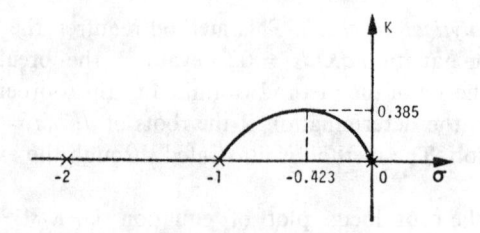

Figure 7.10 Graphical evaluation of a real axis breakaway point.

An example of the second case is shown in the root locus plot of $1+K(s+2)/(s^2+2s+2) = 0$ in Fig. 7.11.

Another graphical method of finding a real axis breakaway point is to locate, through use of angle criterion, a few points (two or three such points will do) on the complex-root branches close to the real axis. A smooth curve drawn through these points intersects the real axis at the breakaway point.

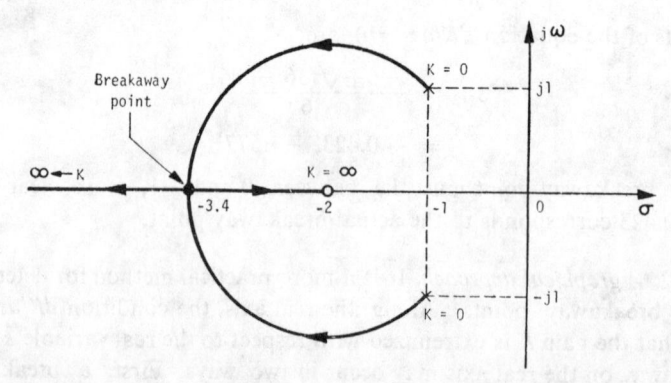

Figure 7.11 Root locus plot of $1+K(s+2)/(s^2+2s+2) = 0$.

Complex Breakaway Points

The application of eqn. (7.30) for evaluation of complex breakaway points is illustrated in the following example.

Example 7.4: The open-loop transfer function of a feedback system is

$$G(s)H(s) = \frac{K}{s(s+4)(s^2+4s+20)}$$

The open-loop pole-zero configuration is shown in Fig. 7.12. From the rules described so far, the following information concerning the root locus plot is obtained.

1. There are 4 branches of root locus originating at $s = 0$, $s = -4$, $s = -2+j4$ and $s = -2 -j4$ respectively.

2. Since there is no open-loop zero in the finite region, all the four branches terminate on infinity along the asymptotes whose angles with the real axis are

$$\phi_A = \frac{(2q+1)\ 180°}{4}; \quad q = 0, 1, 2, 3$$

$$= 45°, 135°, 225°, 315°$$

3. The centroid of the asymptotes is given by

$$-\sigma_A = \frac{\sum \text{real parts of poles} - \sum \text{real parts of zeros}}{\text{number of poles} - \text{number of zeros}}$$

$$= (-4 - 2 - 2)/4 = -2$$

4. The points between 0 and -4 on the real axis lie on the root locus.

5. From the characteristic equation we have

$$K = -s(s+4)(s^2+4s+20)$$
$$= -(s^4+8s^3+36s^2+80s)$$

Therefore

$$\frac{dK}{ds} = -(4s^3+24s^2+72s+80) = 0$$

the roots* of which are found to be at $s = -2$ and $s = -2\pm j2.45$.

Therefore, there is one breakaway point on the real axis at $s = -2$ and two complex conjugate breakaway points at $s = -2\pm j2.45$. The root locus plot is sketched in Fig. 7.12.

Breakaway Directions of Root Locus Branches

The root locus branches must approach or leave the breakaway point on the real axis at an angle of $\pm 180°/r$, where r is the number of root locus branches approaching or leaving the point.

The above statement can easily be verified by considering the root locus of Fig. 7.9. It is seen that two root locus branches approach the breakaway point and therefore according to the above statement, the root locus branches must leave the real axis breakaway point at an angle of $\pm 90°$.

*See Appendix III.

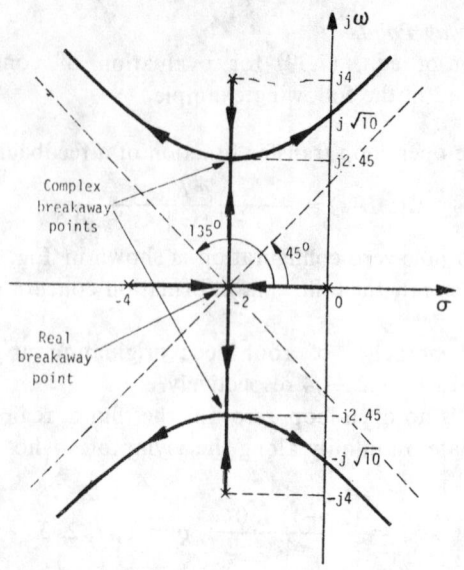

Figure 7.12 Root locus plot with complex breakaway points.

Take a point at an angle of 90° to the real axis and very close to the breakaway point. It can be shown by the angle criterion that the point lies on the root locus.

Rule 7: The angle of departure from an open-loop pole is given by

$$\phi_p = \pm 180°(2q+1) + \phi; \quad q = 0, 1, 2, \ldots \quad (7.32)$$

where ϕ is the net angle contribution, at this pole, of all other open-loop poles and zeros.

Similarly the angle of arrival at an open-loop zero is given by

$$\phi_z = \pm 180°(2q+1) - \phi; \quad q = 0, 1, 2, \ldots$$

where ϕ means the net angle contribution at the zero under consideration of all other open-loop poles and zeros.

It easily follows from this rule that the angle of departure from a real open-loop pole or the angle of arrival at a real open-loop zero is always 0° or 180°. The angle of departure or arrival need therefore be calculated only for complex poles and zeros.

Consider the open-loop pole-zero configuration shown in Fig. 7.13. Around a pole p (complex), select a point s on the root locus branch starting from p and located very close to it. The net angle contribution of all other poles and zeros at this point is

$$\phi = \theta_4 - (\theta_1 + \theta_2 + \theta_3 + \theta_5)$$

The angle contribution of the pole p at s is ϕ_p. In the limit as the point s on the root locus approaches p, ϕ_p equals the angle of departure of the root locus from the pole p. By the angle criterion, we have

$$\phi - \phi_p = \pm 180°(2q+1); \quad q = 0, 1, 2, \ldots$$

or
$$\phi_p = \pm 180°(2q+1)+\phi; \quad q = 0, 1, 2,...$$

It may be noted that in the limit as the point s approaches p, ϕ becomes the net angle contribution of all other open-loop poles and zeros at the pole p.

The rule for the angle of arrival at a zero follows similarly.

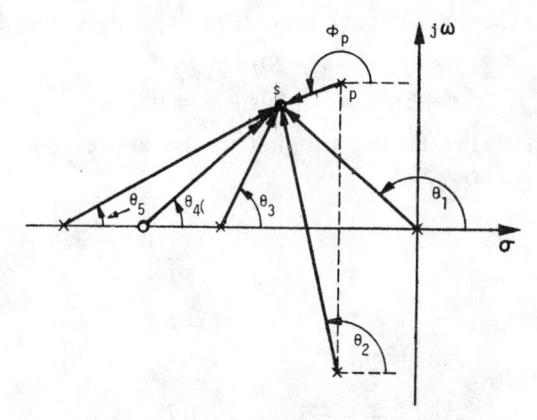

Figure 7.13 Determination of the angle of departure of root locus.

Example 7.5: Let us evaluate the angle of departure of the root locus branch from the pole at $s = -1+j1$ in Fig. 7.11. The open-loop pole-zero configuration of this figure is reproduced in Fig. 7.14.

From Fig. 7.14 and eqn. (7.32), the angle of departure from the pole $(-1+j1)$ is given by

$$\phi_p = \pm(2q+1)180° + (\theta_2 - \theta_1)$$
$$= 180° + (45° - 90°) = +135°$$

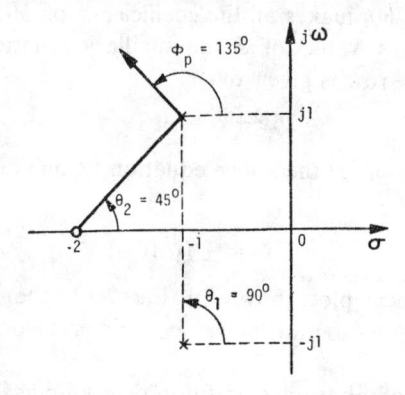

Figure 7.14 Determining the angle of departure of root locus branch in Fig. 7.11.

Rule 8: *The intersection of root locus branches with the imaginary axis can be determined by use of the Routh criterion.*

This rule is simply an application of the Routh criterion discussed in Chapter 6. To illustrate the rule, let us reconsider Example 7.4. The open-loop transfer function of the system considered in this example is

$$G(s)H(s) = \frac{K}{s(s+4)(s^2+4s+20)}$$

Therefore, the characteristic equation of the system is given by

$$s(s+4)(s^2+4s+20)+K = 0$$

or

$$s^4+8s^3+36s^2+80s+K = 0$$

Application of the Routh criterion to the above equation gives the following Routh array:

s^4	1	36	K
s^3	8	80	
s^3	1	10	
s^2	26	K	
s^1	$\dfrac{260-K}{26}$	0	
s^0	K		

For all the roots of the characteristic equation to lie to the left of the imaginary axis, the following conditions should be satisfied:

$$K > 0; \quad (260-K)/26 > 0$$

Therefore, the critical value of K (which corresponds to location of roots on the $j\omega$-axis) is given by

$$(260-K)/26 = 0 \quad \text{or} \quad K = 260$$

The value of $K = 260$, makes all the coefficients of s^1-row of the Routh array zero. For this value of K the auxiliary equation formed from the coefficients of the s^2-row is given by

$$26s^2+K = 0$$

For $K = 260$, the roots of the above equation lie on the $j\omega$-axis and are given by

$$s = \pm j\sqrt{10}$$

Thus for the root locus plot shown in Fig. 7.12, the branches intersect the $j\omega$-axis at $s = \pm j\sqrt{10}$ and the value of K corresponding to these roots is 260.

The rules described above are useful in determining the general configuration of root locus and are summarized in Table 7.1 for easy reference. While each of these rules has already been illustrated through examples, the comprehensive example given below further illustrates the use of all the rules in a single example.

The Root Locus Technique   221

Table 7.1 Rules for Construction of Root Loci of $1+G(s)H(s) = 0$ where the Open-loop Transfer Function $G(s)H(s)$ is Known in Pole-zero form with $n =$ Number of Open-loop Poles, $m =$ Number of Open-loop Zeros

S. No.	Rule
1.	The root locus is symmetrical about the real axis.
2.	Each branch of the root locus originates from an open-loop pole at $K = 0$ and terminates at $K = \infty$, either on an open-loop zero or on infinity. The number of branches of the root locus terminating on infinity is equal to $(n - m)$, i.e., the number of open-loop poles minus the number of zeros.
3.	Segments of the real axis having an odd number of real axis open-loop poles plus zeros to their right are parts of the root locus.
4.	The $(n - m)$ root locus branches that tend to infinity, do so along straight line asymptotes making angles with the real axis given by $$\phi_A = \frac{180°(2q+1)}{(n-m)}; \quad q = 0, 1, 2, \ldots, (n-m-1)$$
5.	The point of intersection of the asymptotes with the real axis is at $s = -\sigma_A$ where $$-\sigma_A = \frac{\Sigma \text{ real parts of poles} - \Sigma \text{ real parts of zeros}}{n - m}$$
6.	The breakaway points of the root locus are determined from the roots of the equation $dK/ds = 0$. r branches of root locus which meet at a point, break away at an angle of $\pm 180°/r$.
7.	The angle of departure from an open-loop pole is given by $$\phi_p = \pm 180°(2q+1) + \phi; \quad q = 0, 1, 2, \ldots$$ where ϕ is the net contribution at the pole of all other open-loop poles and zeros Similarly the angle of arrival at an open-loop zero is given by $$\phi_z = \pm 180°(2q+1) - \phi; \quad q = 0, 1, 2, \ldots$$ where ϕ is the net angle contribution at the zero of all other open-loop poles and zeros.
8.	The intersection of root locus branches with the imaginary axis can be determined by use of the Routh criterion.
9.	The open-loop gain K in pole-zero form at any point s_0 on the root locus is given by $$K = \frac{\prod_{j=1}^{n} |s_0+p_j|}{\prod_{i=1}^{m} |s_0+z_i|}$$ $$= \frac{\text{Product of phasor lengths* from } s_0 \text{ to open-loop poles}}{\text{Product of phasor lengths * from } s_0 \text{ to open-loop zeros}}$$

*Read to scale.

Example 7.6: A feedback control system has an open-loop transfer function

$$G(s)H(s) = \frac{K}{s(s+3)(s^2+2s+2)}$$

Find the root locus as K is varied from 0 to ∞.

Solution

The open-loop poles are located at $s = 0, -3, (-1+j1)$ and $(-1-j1)$ while there are no finite open-loop zeros. The pole-zero configuration is shown in Fig. 7.15.

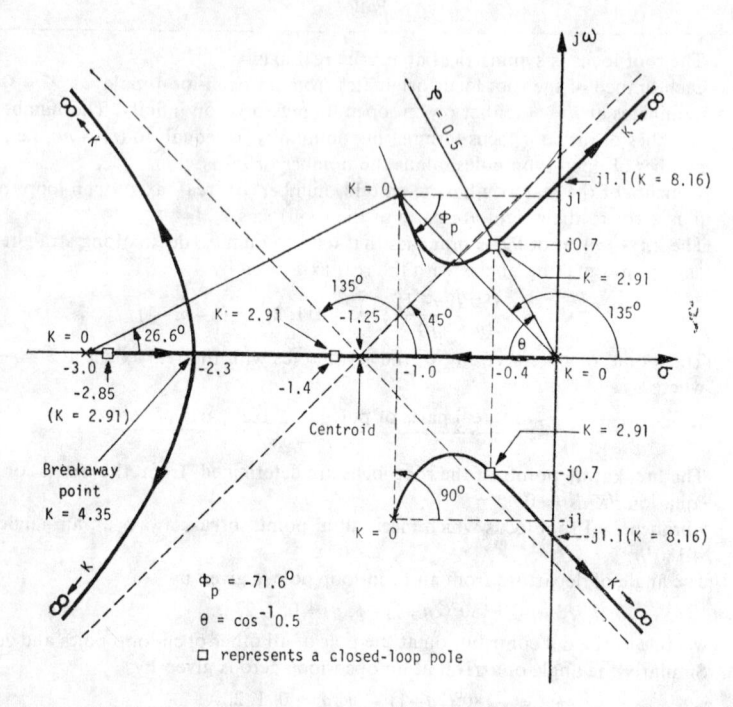

Figure 7.15 Root locus plot of $1 + K/s(s+3)(s^2+2s+2) = 0$.

Rule 2 tells us that the four branches of the root locus originate at $K=0$ from the four open-loop poles and terminate at $K = \infty$ on infinity, since there are no finite zeros.

Rule 3 tells us that the root locus exists on the real axis for $-3 \leqslant s \leqslant 0$, shown by thick line on the real axis in Fig. 7.15.

Rule 4 tells us that the four branches tend to infinity along asymptotes whose angles with the real axis are

$$\phi_A = \frac{(2q+1)180°}{4-0}; \quad q = 0, 1, 2, 3$$

$$= 45°, 135°, 225°, 315°$$

Rule 5 tells us that the centroid is given by

$$-\sigma_A = \frac{-3-1-1}{4} = -1.25$$

The asymptotes are shown by dotted lines in Fig. 7.15.

Rule 6 is used below to determine the breakaway points:

From the characteristic equation of the system

$$K = -s(s+3)(s^2+2s+2) = -(s^4+5s^3+8s^2+6s)$$

$$\frac{dK}{ds} = -4(s^3+3.75s^2+4s+1.5) = 0$$

The roots of this equation which give the possible breakaway points of the root locus are $s = -2.3;\ -0.725 \pm j0.365$. A breakaway point must occur at $s = -2.3$ as this part of the real axis is on the root locus and the two root locus branches starting from $s = 0$ and $s = -3$ are approaching each other. It can be checked that $s = -0.725 \pm j0.365$ are not the breakaway points as the angle criterion is not met at these points. The rule further tells us that the two branches break away at an angle of $\pm 90°$.

The value of K at the breakaway point is evaluated as

$$K = \{\,|s\|s+3\|s+1+j1\|s+1-j1\,\}_{s=-2.3}$$
$$= A \cdot B \cdot C \cdot D = 2.3 \times 0.7 \times 1.68 \times 1.68 = 4.54$$

where A, B, C, D are the lengths of the phasors drawn from the open-loop poles to the point $s = -2.3$.

Rule 7 tells us that the root locus branch leaves the pole at $s = (-1+j1)$ at angle ϕ_p given by

$$\phi_p = 180°(2q+1) + [-135° - 90° - 26.6°]$$
$$\therefore \quad \phi_p = -71.6°$$

Rule 8 tells us to check by use of the Routh criterion if the two branches originating from $s = -1+j1$ and $s = -1-j1$ will intersect the imaginary axis. The system characteristic equation is

$$s(s+3)(s^2+2s+2)+K = 0$$

or

$$s^4+5s^3+8s^2+6s+K = 0$$

ROUTH ARRAY

s^4	1	8	K
s^3	5	6	
s^2	34/5	K	
s^1	$\dfrac{(204/5 - 5K)}{34/5}$	0	
s^0	K	0	

Examination of the elements in the first column of the Routh array reveals that the above mentioned root locus branches will intersect the imaginary axis at a value of K given by

$$(204/5) - 5K = 0$$

from which $K = 8.16$

The auxiliary equation, formed from the coefficients of the s^2-row when $K = 8.16$, is

$$(34/5)s^2 + 8.16 = 0$$

from which
$$s = \pm j1.1$$
Therefore, purely imaginary closed-loop poles of the system are located at $s = \pm j1.1$ as shown in Fig. 7.15.

With the information obtained through the use of these rules, the root locus is sketched in Fig. 7.15 from where it is seen that for $K > 8.16$, the system has two closed-loop poles in the right half of the s-plane and is thus unstable.

DETERMINATION OF ROOTS FROM ROOT LOCUS

Two problems of practical significance which can be worked with the root locus technique are explained below.

(i) *Determination of roots for a specified open-loop gain.* Along a particular root locus branch, a region is determined by trial and error such that the values of the open-loop gain (calculated by the magnitude criterion) at various points of the region are close to specified value. Further, trial and error will then yield the root location. A better organized graphical procedure is to plot open-loop gain versus the real part of the points along the locus. From this graph the real part of the root is read off for the specified open-loop gain, which then is used to locate the root on the locus.

The above procedure is repeated for each root locus branch.

(ii) *Determination of the open-loop gain for a specified damping of the dominant roots.* A damping line (making an angle $\theta = \cos^{-1} \zeta$ with the negative real axis) is drawn for the specified damping. In the region where the rough root locus sketch intersects the ζ-line, a trial and error procedure is adopted along the ζ-line to determine the point of intersection accurately. This point is then the desired root, at which the open-loop gain is computed. Roots along other branches can then be obtained for this open-loop gain using the procedure outlined in (i) above.

As an example consider the root locus plot in Fig. 7.15. Let the specified damping of the dominant roots be 0.5. A ζ-line is drawn in the second quadrant at an angle of $\theta = \cos^{-1} 0.5$ with the negative real axis. Adopting the trial and error procedure of (ii) above, it is found that the point $s = -0.4 + j0.7$ which lies on the ζ-line, satisfies the angle criterion. Therefore the dominant roots of the system are $s_{1,2} = -0.4 \pm j0.7$. The value of the open-loop gain K at this root is

K = product of distances from poles to this point
$= 0.84 \times 1.86 \times 2.74 \times 0.68 = 2.91$

The other two roots of the characteristic equation are obtained from the branches originating at the poles at $s = 0$ and $s = -3$. From Fig. 7.15, it is seen that at the breakaway point, the value of K is 4.35. Therefore the points corresponding to $K = 2.91$ must lie on the real-root portions of these branches. Using the trial and error procedure given in (i) above, it is found that the points $s = -1.4$ and $s = -2.85$ have open-loop gain $K \approx 2.91$. Thus the closed-loop transfer function of the system under consideration is

$$T(s) = \frac{C(s)}{R(s)} = \frac{2.91}{(s+0.4+j0.7)(s+0.4-j0.7)(s+1.4)(s+2.85)}$$

SOME TYPICAL ROOT LOCUS PLOTS

Apart from the root locus plots presented so far, Table 7.2 gives additional root locus plots of some of the commonly occurring pole-zero configurations.

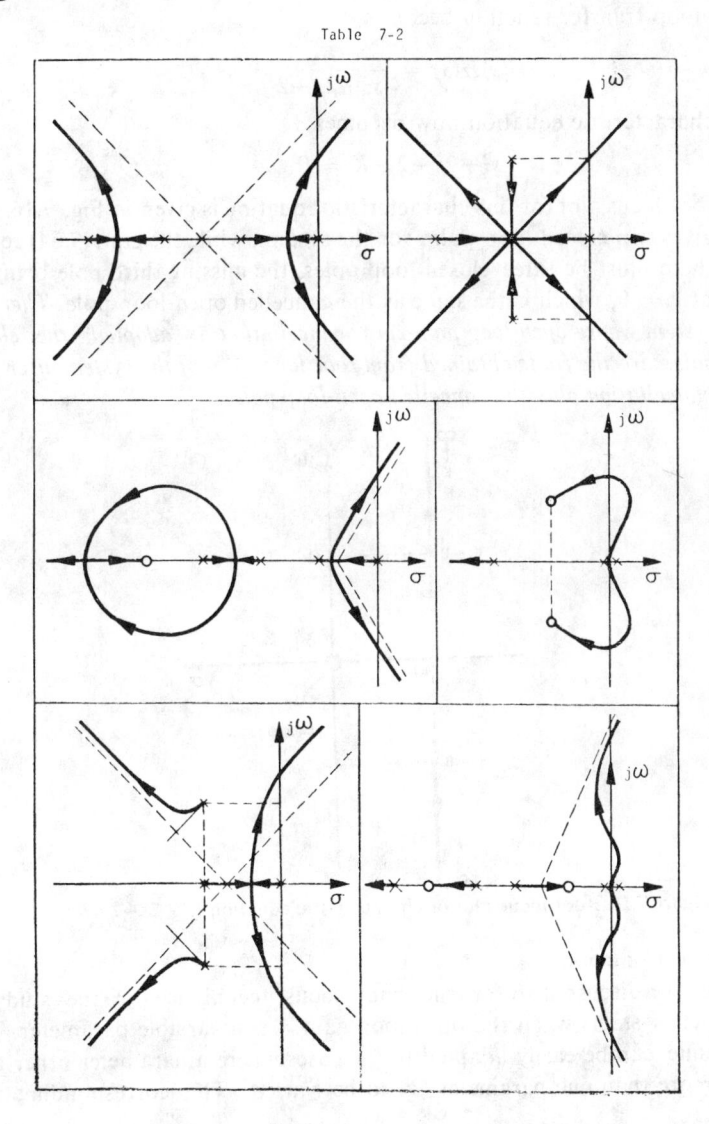

Table 7-2

Cancellation of Poles and Zeros

Consider the open-loop transfer function

$$G(s)H(s) = \frac{K(s+z)}{s(s^2+2s+2)} \tag{7.33}$$

Its characteristic equation is

$$s^3+2s^2+(2+K)s+Kz = 0 \tag{7.34}$$

When $z=0$, the numerator and denominator of eqn. (7.33) have a common factor s such that the open-loop pole at $s = 0$ and the open-loop zero at $s = 0$ cancel each other. After cancelling the common factor, the resulting open-loop transfer function becomes

$$G(s)H(s) = \frac{K}{s^2+2s+2}$$

The characteristic equation now becomes

$$s^2+2s+2+K = 0$$

The root locus plot for this characteristic equation is given in Fig. 7.16. This plot gives two closed-loop poles for the system, while the eqn. (7.34) reveals that there must be three closed-loop poles, the missing third pole being the pole at $s = 0$, which is the same as the cancelled open-loop pole. *Therefore in a system where open-loop pole-zero cancellation is adopted, the closed-loop poles are the roots obtained from root locus plot of the system after pole-zero cancellation plus the cancelled open-loop poles.*

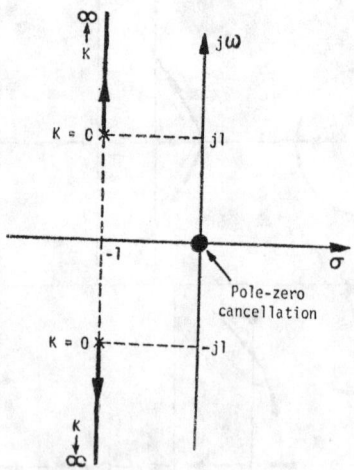

Figure 7.16 Root locus plot of characteristic equation $(s^2+2s+2+K) = 0$.

7.4 Root Contours

We have discussed so far the root locus technique for the study of feedback systems when the open-loop gain K is a variable parameter. This technique can be easily adapted to the cases where a parameter other than K or more than one parameter are to be varied. The corresponding root locus plots are called *root contours* which can be constructed following the rules already enunciated in the previous section.

The basic equation for construction of root locus as already defined in eqn. (7.14) is reproduced below

$$1 + G(s)H(s) = 1 + \frac{K \prod_{i=1}^{m}(s+z_i)}{\prod_{j=1}^{n}(s+p_j)}$$

$$= 1 + K\left(\frac{s^m + b_1 s^{m-1} + \ldots + b_{m-1}s + b_m}{s^n + a_1 s^{n-1} + \ldots + a_{n-1}s + a_n}\right) \quad (7.35)$$

where K, the open-loop gain is a variable parameter. If the variable parameter is other than K (say the distance α of an open-loop pole from the $j\omega$-axis), then the first step in construction of root locus is to manipulate the characteristic equation in the form of eqn. (7.35) above, with α in place of the gain parameter K. Auxiliary root locus plot may be needed to bring eqn. (7.35) in the factored pole-zero form.

To illustrate the construction of root contours, consider the system shown in Fig. 7.1, whose characteristic equation is

$$s^2 + as + K = 0 \quad (7.36)$$

The root contours are to be plotted by allowing both a and K to vary. This can be carried out conveniently by rewriting the characteristic equation as

$$1 + a\left(\frac{s}{s^2 + K}\right) = 0 \quad (7.37)$$

such that a appears as a gain parameter. Equation (7.37) is in the form that its root contours can be drawn for various values of K by allowing a to vary from 0 to ∞. The following information is derived from the root locus construction rules.

The root contours of eqn. (7.37) originate from poles at $s = \pm j\sqrt{K}$ and terminate on zero at $s = 0$ and on $s = -\infty$. From eqn. (7.37)

$$a = -(s^2 + K)/s$$

Therefore the roots of

$$da/ds = -(s^2 - K)/s^2 = 0$$

give the breakaway points at $s = \pm\sqrt{K}$. Figure 7.17 shows that the breakaway point must lie between $s = 0$ and $s = -\infty$. Therefore $s = -\sqrt{K}$ corresponds to the breakaway point. The root contours* for various values of K with a varying from 0 to ∞ are shown in Fig. 7.17.

*In this case it can be shown that the complex-root branches are circular.

$$\angle \frac{\sigma + j\omega}{(\sigma + j\omega + j\sqrt{K})(\sigma + j\omega - j\sqrt{K})} = 180°(2q+1)$$

or $\tan^{-1}(\omega/\sigma) - [\tan^{-1}(\omega + \sqrt{K})/\sigma + \tan^{-1}(\omega - \sqrt{K})/\sigma] = 180°$

Transposing $\tan^{-1}(\omega/\sigma)$ to the right and taking tan on both sides we get,

$$\frac{(\omega + \sqrt{K})/\sigma + (\omega - \sqrt{K})/\sigma}{1 - (\omega^2 - K)/\sigma^2} = \omega/\sigma$$

which on simplification yields, $\sigma^2 + \omega^2 = K$, i.e., the complex-root branches are parts of a circle centred at (0, 0) and of radius $\sqrt{K}$.

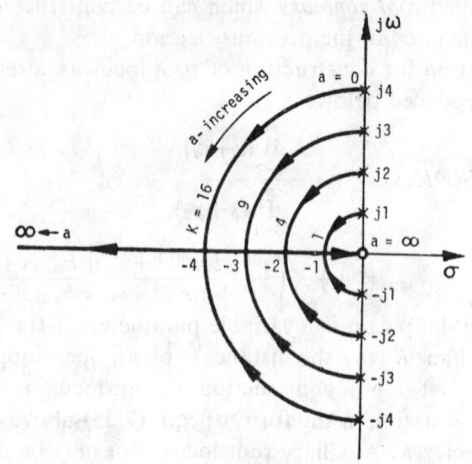

Figure 7.17 Root contours of the system shown in Fig. 7.1.

Example 7.7: Consider the feedback control system with an open-loop transfer function

$$G(s)H(s) = \frac{K}{s(s+1)(s+\alpha)}$$

in which the open-loop gain and the pole $s = -\alpha$ are both regarded as variable.

The characteristic equation of the system is

$$1 + \frac{K}{s(s+1)(s+\alpha)} = 0$$

or
$$s^2(s+1) + \alpha(s+1)s + K = 0 \qquad (7.38)$$

It can be manipulated into the form below where α appears as a root locus parameter

$$1 + \frac{\alpha s(s+1)}{s^2(s+1) + K} = 0 \qquad (7.39)$$

Though eqn. (7.39) is in the form that root locus with respect to the parameter α can be drawn, however, we cannot proceed with it without determining its open loop poles from the reduced characteristic equation

$$s^2(s+1) + K = 0$$

where K is a variable. The reduced characteristic equation can be obtained from the complete characteristic eqn. (7.38) by putting $\alpha = 0$. The reduced characteristic equation can be rewritten as

$$1 + \frac{K}{s^2(s+1)} = 0 \qquad (7.40)$$

The root locus of the reduced characteristic equation with K as a variable parameter is plotted in Fig. 7.18a. The three roots of eqn. (7.40) for a particular value of K contribute the open-loop poles of eqn. (7.39).

The Root Locus Technique

The root contours for various values of K with varying α are drawn in Fig. 7.18b. The value of α at which the root contours will cross the $j\omega$-axis and the system will become unstable is obtained by application of the Routh criterion to the characteristic equation (7.38) as given below:

$$s^2(s+1)+\alpha s(s+1)+K = 0$$
or
$$s^3+(\alpha+1)s^2+\alpha s+K = 0$$

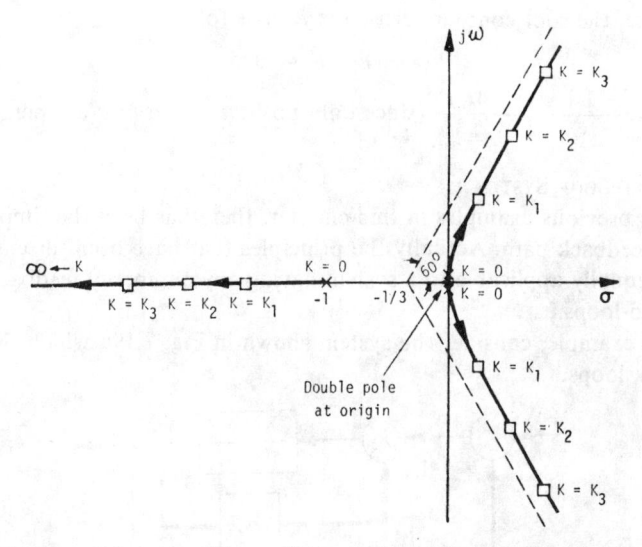

Figure 7.18(a) Root locus plot of the auxiliary characteristic equation $s^2(s+1)+K = 0$.

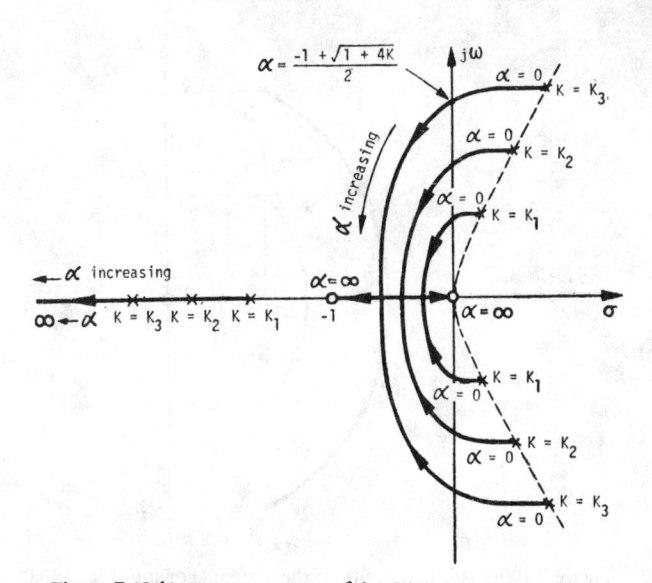

Figure 7.18(b) Root contours of $1+K/s(s+1)(s+\alpha) = 0$.

The Routh array is

s^3	1	α
s^2	$(\alpha+1)$	K
s^1	$\dfrac{\alpha(\alpha+1)-K}{(\alpha+1)}$	0
s^0	K	

Therefore, the root contours cross the $j\omega$-axis for

$$\alpha(\alpha+1)-K=0$$

or $\alpha = \dfrac{-1+\sqrt{(1+4K)}}{2}$ (since only positive values of α are permitted).

MULTIPLE-LOOP SYSTEM

In the previous examples in this chapter, there has been the implication of one feedback path. Actually the principles that have been discussed so far are equally applicable to a system having any number of feedback paths or closed-loops.

As an example, consider the system shown in Fig. 7.19a which has two feedback loops.

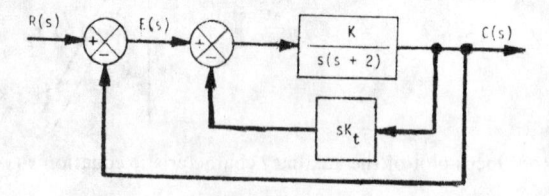

Figure 7.19(a) A feedback system with two loops.

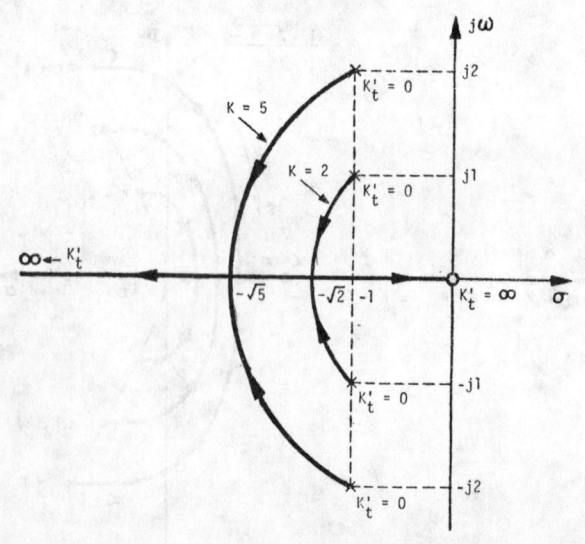

Figure 7.19(b) Root contours of the system shown in Fig. 7.19a.

The open-loop transfer function of the system is derived as

$$G(s) = \frac{C(s)}{E(s)} = \frac{K/s(s+2)}{1+sKK_t/s(s+2)} = \frac{K}{s(s+2)+sKK_t}$$

Therefore the characteristic equation of the system is

$$s(s+2)+sKK_t+K = 0 \tag{7.41}$$

Equation (7.41) may be rearranged as

$$1 + \frac{K_t K s}{s(s+2)+K} = 0$$

or

$$1 + \frac{K_t K s}{[(s+1+j\sqrt{(K-1)}][(s+1-j\sqrt{(K-1)}]} = 0$$

The root contours plotted for various values of K with $K_t' = KK_t$ varying from zero to infinity are shown in Fig. 7.19b.

7.5 Systems with Transportation Lag

All the feedback systems discussed so far are represented by linear, lumped parameter mathematical models (transfer functions consisting of the ratios of algebraic polynomials). This is valid so long as the time taken for energy transmission is negligible, i.e., the output begins to appear immediately on application of input. This is not quite true of transmission channel-lines, pipes, belt conveyors, etc. In such cases a definite time elapses after application of the input before the output begins to appear. This type of pure time lag is known as *transportation lag* or *dead time*. Linear lumped parameter model is not valid under such situations unless the pure time lag is negligible compared to other lags in the system. For example, the transmission line (pipe) between the hydraulic pump and motor causes a time lag in transportation of oil from pump to motor and vice versa. Similarly the slow rate of transmission of heat energy by conduction or convection introduces serious transportation lags in process control systems.

The transfer function corresponding to a transportation lag can be easily determined. Consider a component whose output is the same as the input but delayed by T sec. Mathematically this can be expressed as

$$c_0(t) = c_i(t - T)$$

Taking the Laplace transform of the above equation we get

$$C_0(s) = C_i(s)e^{-sT}$$

or $\quad C_0(s)/C_i(s) = e^{-sT} \tag{7.42}$

Thus, the transfer function for a constant transportation lag is given by e^{s-T} (a non-minimum phase function).

Let us now analyze the system shown in Fig. 7.20 with the help of the root locus technique. The system shows an arrangement of controlling the thickness of a steel plate produced by a rolling mill. A voltage signal corresponding to the desired thickness is the reference input to the system. A thickness gauge provides a feedback voltage signal proportional to the actual thickness of the plate. The error voltage actuates the motor which positions the rolls.

The open-loop transfer function of the system is given by

$$G(s)H(s) = \frac{K_1}{s(\tau_m s + 1)} \quad (7.43)$$

where for the sake of simplicity, we assume a single dominant time constant in the loop, probably that of the motor.

In the transfer function given by eqn. (7.43), we have ignored the fact that a finite time must elapse before a change in thickness of steel plates between rolls reaches the point of measurement at the gauge.

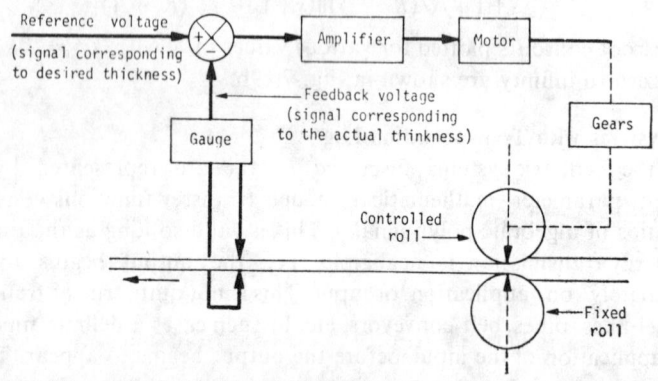

Figure 7.20 A feedback system with transportation lag.

If this delay in transportation of signal is assumed to be T sec, then the open-loop transfer function of the system becomes

$$G(s)H(s) = K_1 e^{-sT}/s(\tau_m s + 1)$$

The above equation may be rearranged in the following form

$$G(s)H(s) = K e^{-sT}/s(s + \alpha) \quad (7.44)$$

Plotting the root locus for a system with transportation lag is considerably more complicated than for a system without it. However, if the transportation lag is small, $e^{-sT} \simeq (1 - sT)$. Eqn. (7.44) may now be rearranged in the following form:

$$G(s)H(s) = -K\left(s - \frac{1}{T}\right)\bigg/s(s + \alpha)$$

The characteristic equation becomes

$$1 - K\left(s - \frac{1}{T}\right)\bigg/s(s + \alpha) = 1 - P(s) = 0 \quad (7.45)$$

Comparing the above equation* with eqn. (7.5), we observe that the angle criterion given by eqn. (7.8) is now modified as

$$\angle P(s) = \pm 180°(2q); \quad q = 0, 1, 2, \ldots$$

*Such cases also occur in positive feedback systems.

This leads to following modifications in Rules 3, 4 and 7 of Table 7.1.
(i) Replace 'odd number' by 'even number'.
(ii) Replace '180°(2q+1)' by '180°(2q)'.

The root locus plot for eqn. (7.45) with $\alpha = 2$ and $T = 1$ sec is shown in Fig. 7.21. This plot generally agrees with the exact plot in the region of small values of s where the dominant roots are located.

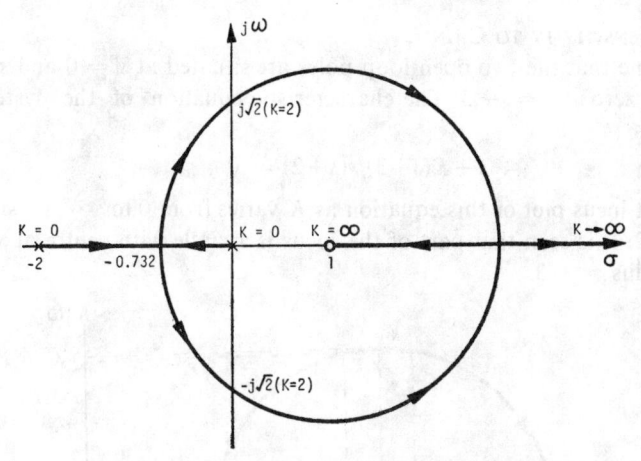

Figure 7.21 Root locus plot of $1+Ke^{-s}/s(s+2)=0$ where e^{-s} is approximated as $(1-s)$.

7.6 Sensitivity of the Roots of the Characteristic Equation

Earlier in Chapter 3 the sensitivity of the system transfer function $T(s)$ to variation of a parameter has been defined as

$$S_K^T = \frac{d(\ln T)}{d(\ln K)} = \frac{\partial T/T}{\partial K/K}$$

where K is the parameter of interest. Because of the importance of root locus as a design technique, it is useful to define the sensitivity of the roots of the characteristic equation (i.e., the closed-loop poles of the system) to variation of a parameter K. This is defined as

$$S_K^{-r_k} = \frac{\partial(-r_k)}{\partial(\ln K)} = \frac{\partial(-r_k)}{\partial K/K} \tag{7.46}$$

where $-r_k$ is a root of the characteristic equation. In order that the system dynamic response be relatively unaffected by parameter changes, the system must be so designed that the roots of the characteristic equation and in particular the dominant roots are made insensitive to parameter changes, i.e., the root sensitivity $S_K^{-r_k}$ is made less than a specified value.

Example 7.8: Consider a feedback control system with the characteristic equation

$$1 + \frac{K(s+z)}{s(s+p)} = 0 \tag{7.47}$$

In this case, the parameters of interest are the gain K, the zero $s = -z$ and the pole $s = -p$. An unwanted change in any of these parameters will cause the roots of the characteristic equation to shift. The amount and nature of this shift is determined by the root sensitivity to each of these parameters. Let us consider the root sensitivity to these parameters one by one.

ROOT SENSITIVITY TO GAIN K

Assume that the two open-loop poles are situated at $s = 0$ and $s = -2$ and the zero at $s = -3$. The characteristic equation of the system then becomes

$$1 + K(s+3)/s(s+2) = 0 \qquad (7.48)$$

The root locus plot of this equation as K varies from 0 to ∞ is shown in Fig. 7.22. It is seen that part of the locus is a circle with centre at $s = -3$ and radius $= \sqrt{3}$.

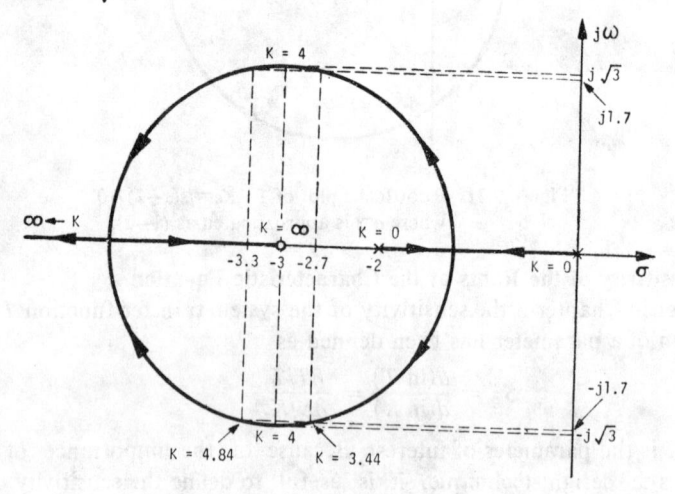

Figure 7.22 Root locus plot of eqn. (7.48).

Let the nominal system gain be $K = K_0 = 4$. This results in two complex roots, $s_1 = -r_1 = -3 + j\sqrt{3}$ and $s_2 = -r_2 = -3 - j\sqrt{3} = \bar{r}_1$. Since the roots are complex conjugate, the root sensitivity for $-r_1$ is the conjugate of the root sensitivity for $-r_2$. Therefore, we need to evaluate only the root sensitivity of one of the roots, say $-r_1$.

Assume that because of changes in gain K, the root $-r_1$ gets shifted to a new location $s = -3.3 + j1.7$ on the root locus, of course. Then

$$\Delta(-r_1) = (-3.3 + j1.7) - (-3 + j1.73)$$
$$= -0.3 - j0.03$$

From Fig. 7.22, the value of gain K for this new location of the root (obtained by the magnitude criterion) is 4.84. Therefore

$$\Delta K = K - K_0 = 0.84$$

$$\Delta K/K = 0.84/4 = 0.21$$

From eqn. (7.46), the sensitivity of the root $-r_1$ is

$$S_{K+}^{(-r_1)} = \frac{\Delta(-r_1)}{\Delta K/K} = \frac{-0.3 - j0.03}{0.21} = 1.43 \angle 186°$$

for positive changes of gain.

Considering the new root location at $s = -2.7+j1.7$, the gain corresponding to this location obtained from Fig. 7.22 is 3.44. Therefore

$$\Delta(-r_1) = (-2.7+j1.7) - (-3+j1.73) = 0.3 - j0.03$$

$$\Delta K = 0.54$$

$$\Delta K/K = 0.54/4 = 0.135$$

The sensitivity of the root $-r_1$ is now given by

$$S_{K-}^{(-r_1)} = (0.3 - j0.03)/(0.135) = 2.22 \angle -6°$$

for negative changes in gain.

It can easily be verified that as the percentage change in gain $\Delta K/K$ decreases, the sensitivity measures $S_{K+}^{(-r_1)}$ and $S_{K-}^{(-r_1)}$ approach equality in magnitude and a difference in angle of 180°. Thus for very small changes in gain $\Delta K/K \to 0$, the sensitivity measures are related as

$$|S_{K+}^{(-r_1)}| = |S_{K-}^{(-r_1)}| \qquad (7.49)$$

$$\angle S_{K+}^{(-r_1)} = 180° + \angle S_{K-}^{(-r_1)} \qquad (7.50)$$

Usually the desired root sensitivity measure is for small changes in the parameter. Therefore we need to evaluate only one sensitivity measure, say for positive changes in parameter; other sensitivity measure (for negative changes of parameter) can be obtained by use of eqns. (7.49) and (7.50).

ROOT SENSITIVITY FOR SMALL PARAMETER CHANGES

Once the roots have been determined for nominal parameter values, the root sensitivity for any particular root for small parameter changes can be computed analytically from the knowledge of the open-loop poles and zeros using the root sensitivity expressions derived below.

It has been shown towards the end of Section 7.4 that the characteristic equation can always be manipulated such that the parameter K of interest appears in the form

$$1 + K \frac{\prod_{i=1}^{m}(s+z_i)}{\prod_{j=1}^{n}(s+p_j)} = 1 + K \frac{B(s)}{A(s)} = 0 \qquad (7.51)$$

where $B(s) = \prod_{i=1}^{m}(s+z_i)$ and $A(s) = \prod_{j=1}^{n}(s+p_j)$ are polynomials of degree m and n respectively with the coefficient of their highest degree term in s as unity.

The characteristic eq. (7.51) can be rewritten as

$$q(s) = A(s) + K B(s) = K_1 \prod_{k=1}^{n}(s+r_k) = 0 \tag{7.52}$$

where $-r_k (k = 1, 2, ..., n)$ are the roots of the characteristic equation. In eqn. (7.52) K_1 is given by

$$K_1 = 1 \quad \text{if } n > m$$
$$= 1+K \quad \text{if } n = m$$

Taking logarithm of each side of eqn. (7.52), we get

$$\ln q(s) = \ln K_1 + \sum_{k=1}^{n} \ln (s+r_k) \tag{7.53}$$

Differentiating eqns. (7.52) and (7.53) with respect to the parameter K, we have

$$\frac{\partial q(s)/\partial K}{q(s)} = \frac{B(s)}{q(s)} = K_1^{-1} \frac{\partial K_1}{\partial K} - \sum_{k=1}^{n} \frac{\partial(-r_k)/\partial K}{(s+r_k)} \tag{7.54}$$

Each side of eqn. (7.54) must have the same residue at any pole. Taking the residue of both the sides at the pole $s = -r_k$, we have

$$\frac{\partial(-r_k)}{\partial K} = -(s+r_k) \frac{B(s)}{q(s)}\bigg|_{s=-r_k} = -\frac{B(-r_k)}{q'(-r_k)} \tag{7.55}$$

where $\quad q'(-r_k) = \dfrac{dq(s)}{ds}\bigg|_{s=-r_k}$

In taking the residue above it has been assumed that the root $-r_k$ is a non-repeated root of the characteristic equation.

It follows from eqn. (7.51) that

$$B(-r_k) = -\frac{A(-r_k)}{K}$$

Therefore we can write eqn. (7.55) as

$$S_K^{-r_k} = \frac{\partial(-r_k)}{\partial K/K} = \frac{A(-r_k)}{q'(-r_k)} \tag{7.56}$$

The sensitivity of the root $s = -r_k$ can thus be computed from the expression* given in eqn. (7.56).

The transfer function sensitivity and sensitivities of the roots of the characteristic equation can be correlated as follows.

The transfer function can be expressed as

$$T(s) = \frac{K_2 \prod_{i=1}^{m'}(s+z_i)}{\prod_{j=1}^{n}(s+r_k)} \tag{7.57}$$

---

*The derivation of this expression is from Horowitz [1]. Reproduced with permission.

where the zeros* of $T(s)$ are contributed by the zeros of $G(s)$ and the poles of $H(s)$ for a single-loop system [i.e., m' = number of zeros of $G(s)$ plus number of poles of $H(s)$]. Taking logarithm of eqn. (7.57) and differentiating it with respect to $\ln K$, K being the parameter of interest, we get

$$\frac{d(\ln T)}{d(\ln K)} = \frac{\partial(\ln K_2)}{\partial(\ln K)} + \sum_{k=1}^{n} \frac{\partial(-r_k)}{\partial(\ln K)} \frac{1}{(s+r_k)}$$

or

$$S_K^T = \frac{\partial(\ln K_2)}{\partial(\ln K)} + \sum_{k=1}^{n} S_K^{-r_k} \frac{1}{(s+r_k)} \qquad (7.58)$$

where it has been assumed that $\partial z_i/\partial(\ln K) = 0$, i.e., the zeros of $T(s)$ are independent of the parameter K.

The expression (7.58) can be used to compute the transfer function sensitivity from the root sensitivities for the parameter K. In the particular case where the system gain K_2 is independent of the parameter K, we have

$$S_K^T = \sum_{k=1}^{n} S_K^{-r_k} \frac{1}{(s+r_k)} \qquad (7.59)$$

Example 7.8 (continued): In Example 7.8 introduced earlier, the root sensitivity to each of the parameters K, p and z is computed below using the expression (7.56).

Root sensitivity to gain K. From the characteristic eqn. (7.48),

$$A(s) = s(s+2)$$
$$B(s) = (s+3)$$
$$q(s) = A(s) + KB(s)$$
$$= s(s+2) + 4(s+2)$$
$$\therefore \quad q'(s) = 2s+6$$

Using eqn. (7.56) the root sensitivity to changes in gain K with nominal gain $K = 4$ is computed below.

$$S_K^{-r_1} = \frac{A(-r_1)}{q'(-r_1)} = \frac{(-3+j\sqrt{3})(-1+j\sqrt{3})}{2(-3+j\sqrt{3}+3)} = -2$$

$$= 2\angle 180°$$

For a small change in the nominal gain (say $\Delta K = 0.4$), the new (shifted) root location can be calculated as

$$S_K^{-r_1} \approx \frac{\Delta(-r_1)}{\Delta K/K} \qquad (7.60)$$

---

*$T(s) = G(s)/[1+G(s)H(s)]$
Let $G(s) = N_G(s)/D_G(s)$ and $H(s) = N_H(s)/D_H(s)$

$\therefore \quad T(s) = \dfrac{N_G(s)D_H(s)}{D_G(s)D_H(s) + N_G(s)N_H(s)}$

The zeros of $T(s)$ thus are the roots of $N_G(s)D_H(s)$, i.e., the zeros of $G(s)$ and the poles of $H(s)$.

$$\therefore \quad \Delta(-r_1) = -2 \times 0.4/4 = -0.2$$

New root location $= (-r_1) + \Delta(-r_1) = (-3.2 + j\sqrt{3})$

If the gain decreases then $\Delta K = -0.4$. Therefore

$$\Delta(-r_1) = -2(-0.4)/4 = +0.2$$

The new root location $= (-r_1) + \Delta(-r_1)$
$$= (-2.8 + j\sqrt{3})$$

The expression for root sensitivity (7.56) is not valid for large changes in the parameter K, in which case the roots for changed value of K must be obtained from the root locus plot and the root sensitivity computed from eqn. (7.60).

Root sensitivity to open-loop zero $(s = -z)$. The characteristic equation can be rewritten with z as a variable parameter, i.e.,

$$1 + 4(s+z)/s(s+2) = 0$$
$$s^2 + 6s + 4z = 0$$

or
$$1 + z'/s(s+6) = 0$$

where
$$z' = 4z$$

The nominal value of $z' = 12$ for which the two roots are $-r_1, -r_2,$ $= -3 \pm j\sqrt{3}$.

Now

$$A(s) = s(s+6)$$
$$B(s) = 1$$
$$q(s) = s(s+6) + 12$$
$$q'(s) = 2s + 6$$

$$S_{z'}^{-r_1} = \frac{A(s)}{q'(s)}\bigg|_{s=-3+j\sqrt{3}} = (2\sqrt{3}) \angle 90°$$

Root sensitivity to open-loop pole $(s = -p)$. The characteristic equation is written below with p as a variable parameter:

$$1 + 4(s+3)/s(s+p) = 0$$

or
$$1 + ps/(s^2 + 4s + 12) = 0$$

For a nominal value of $p = 2$, the roots of the characteristic equation are $-r_1, -r_2 = -3 \pm j\sqrt{3}$. From the characteristic equation,

$$A(s) = s^2 + 4s + 12$$
$$B(s) = s$$
$$q(s) = (s^2 + 4s + 12) + 2s$$
$$q'(s) = 2s + 6$$

$$\therefore \quad S_p^{-r_1} = \frac{A(s)}{q'(s)}\bigg|_{s=-3+j\sqrt{3}} = 2 \angle -120°$$

The root sensitivity measure for parameter variations is useful for comparing several system designs. The use of this method as a design technique is somewhat limited because of repeated complex calculations and for lack of an obvious direction for adjusting the parameters to reduce or minimize sensitivity.

Problems

7.1 A unity feedback control system has an open-loop transfer function

$$G(s) = K/s(s^2+4s+13)$$

Sketch the root locus plot of the system by determining the following:
 (i) Centroid, number and angle of asymptotes.
 (ii) Angle of departure of root loci from the poles.
 (iii) Breakaway points if any.
 (iv) The value of K and the frequency at which the root loci cross the $j\omega$-axis.

7.2 Sketch the root locus plot of a unity feedback system with an open-loop transfer function

$$G(s) = K/s(s+2)(s+4)$$

Find the range of values of K for which the system has damped oscillatory response. What is the greatest value of K which can be used before continuous oscillations occur? Also determine the frequency of continuous oscillations.

Determine the value of K so that the dominant pair of complex poles of the system has a damping ratio of 0.5. Corresponding to this value of K, determine the closed-loop transfer function in factored form.

7.3 A unity feedback system has an open-loop transfer function

$$G(s) = K(s+1)/s(s-1)$$

Sketch the root locus plot with K as a variable parameter and show that the loci of complex roots are part of a circle with $(-1, 0)$ as centre and radius = $\sqrt{2}$.

Is the system stable for all values of K? If not, determine the range of K for stable system operation. Find also the marginal value of K which causes sustained oscillations and the frequency of these oscillations.

From the root locus plot, determine the value of K such that the resulting system has a settling time of 4 sec. What are the corresponding values of the roots?

7.4 The signal flow graph of a control system is shown in Fig. P-7.4. Comment upon the stability of this system when the switch S is open.

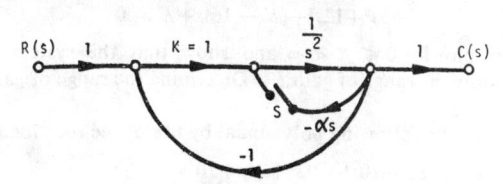

Figure P-7.4.

With the switch S closed, draw the root locus plot of the system with α as a varying parameter. Show that the complex-root branches are part of a circle. From the root locus plot, determine the value of α such that the resulting system has a damping ratio of 0.5. For this value of α, find the overall transfer function in factored form.

7.5 The block diagram of a control system is shown in Fig. P-7.5. Draw the root locus plot of the system with α as varying parameter.
 (a) Determine the steady-state error to the unit-ramp input, damping ratio and settling time for the system without derivative feedback, i.e.,

$$\alpha = 0.$$

(b) Discuss the effect of derivative feedback on transient as well as steady-state behaviour of the system assuming $\alpha = 0.2$.
(c) Determine the value of α for the system to be critically damped.

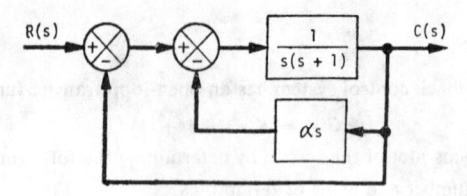

Figure P-7.5.

7.6 A unity feedback system has an open-loop transfer function

$$G(s) = K/s^2(s+2)$$

(a) By sketching a root locus plot, show that the system is unstable for all values of K.
(b) Add a zero at $s = -a$ ($0 \leqslant a < 2$) and show that the addition of zero stabilizes the system.
(c) If $a = 1$, sketch the root locus plot and determine approximately the value of K which gives the greatest damping ratio for the oscillatory mode. Find also the value of this damping ratio and the corresponding undamped natural frequency.

7.7 Open-loop transfer function of a unity feedback system is

$$G(s) = K/(s+2)^3$$

Sketch the root locus plot and determine the following:
(a) Static loop sensitivity for which the root locus crosses the $j\omega$-axis and the corresponding frequency of sustained oscillations.
(b) The position error constant corresponding to a damping ratio of 0.5. Also determine the peak overshoot, time to peak overshoot and settling time considering the effect of dominant poles only.

(*Note*: The static loop sensitivity is defined to be the gain in pole-zero form.)

7.8 The characteristic equation of a feedback control system is

$$s^4 + 3s^3 + 12s^2 + (K-16)s + K = 0$$

Sketch the root locus plot for $0 \leqslant K < \infty$ and show that the system is conditionally stable (stable for only a range of gain K). Determine the range of gain for which the system is stable.

7.9 Find the roots of the following polynomial by use of the root locus method.

$$3s^4 + 10s^3 + 21s^2 + 24s + 30 = 0$$

(*Hint*: See Appendix III.)

7.10 Figure P-7.10 shows an arrangement of controlling the thickness of steel plates. A signal proportional to the desired steel thickness is the reference input and a signal proportional to the actual thickness is the feedback signal. The error signal actuates the motor which positions the cutter. A definite time T sec elapses before the change in thickness of steel plates reaches the point of measurement. The open-loop transfer function may be assumed to be

$$G(s)H(s) = 100e^{-sT}/s(s+4)$$

If the steel plate moves at a constant velocity of 1 m/sec, find the maximum permissible distance d before sustained oscillations are produced.

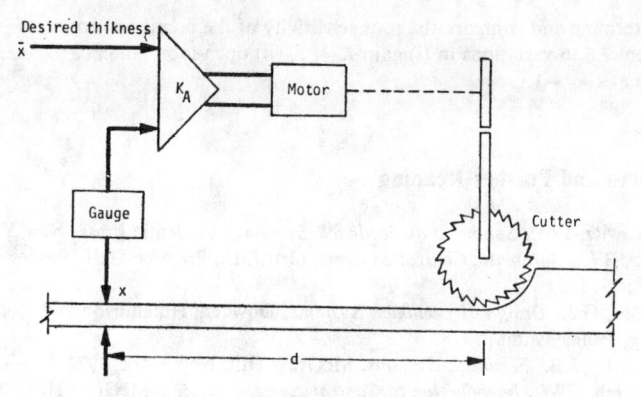

Figure P-7.10

7.11 The block diagram of Fig. P-7.11 shows a feedback system using a proportional plus integral controller.
(a) Draw the root contours of the system for $a = 0, 1, 2$ and 4.
(b) Find the range of values of a for which the system is stable.
(c) Determine the steady-state error for unit-parabolic input.
(d) Determine the damping ratio of the dominant roots for various values of $a(K_A = 5)$ and show that the system is very sensitive to the value of a.

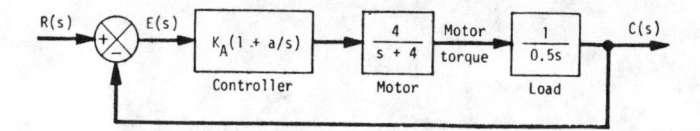

Figure P-7.11.

7.12 A unity feedback system has an open-loop transfer function

$$G(s) = 25/s(s+4)$$

(a) Determine the settling time and damping ratio of the system.
(b) The system dynamics is to be improved by cascading a compensating network having the transfer function

$$G_c(s) = (as+1)/(bs+1)$$

in the forward path as shown in Fig. P-7.12. By drawing root contours, discuss the effect of varying a and b on system dynamics. Also determine suitable values of a and b, so that the settling time is less than 1 sec and the dominant roots have damping ratio greater than 0.5.

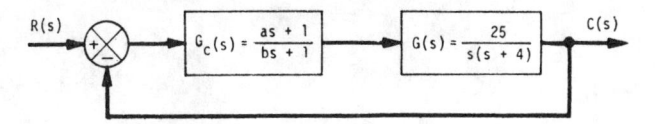

Figure P-7.12.

7.13 Determine the root sensitivity of the complex roots of the system of Problem 7.5 to variations in derivative feedback constant $\alpha = 0.2$.

7.14 Determine and compare the root sensitivity of the dominant roots of the system of Problem 7.6 to variations in (i) gain $K = 5$, (ii) open-loop pole at $s = -2$, (iii) open-loop zero at $s = -1$.

References and Further Reading

1. Horowitz, I.M., *Synthesis of Feedback Systems*, Academic Press, New York, 1963.
2. Kuo, B.C., *Automatic Control Systems*, (3rd Ed.), Prentice-Hall, Englewood Cliffs, N.J., 1975.
3. Thaler, G.J., *Design of Feedback Systems*, Dowden, Hutchinson and Ross, Stroudsburg, Pennsylvania, 1973.
4. Cruz, Jr., J.B., *Feedback Systems*, McGraw-Hill, New York, 1972.
5. Eveleigh, V.W., *Introduction to Control Systems Design*, McGraw-Hill, New York, 1972.
6. Shinners, S.M., *Modern Control System Theory and Application*, Addison-Wesley, Reading, Mass., 1972.
7. Gupta, S.C. and L. Hasdorff, *Fundamentals of Automatic Control*, John Wiley, New York, 1970.
8. Ogata, K., *Modern Control Engineering*, Prentice-Hall, Englewood Cliffs, N.J., 1970.
9. Dorf, R.C., *Modern Control Systems*, Addison-Wesley, Reading, Mass., 1967.
10. Truxal, J.G., *Automatic Feedback Control System Synthesis*, McGraw-Hill, New York, 1955.

8. Frequency Response Analysis

8.1 Introduction

Various standard test signals used to study the performance of control systems were discussed in Chapter 5. While the sinusoidal test signal was introduced there, the discussion on the same was postponed till this chapter which is fully devoted to it on account of its importance in control engineering. Consider a linear system with a sinusoidal input

$$r(t) = A \sin \omega t$$

Under steady-state, the system output as well as the signals at all other points in the system are sinusoidal. The steady-state output may be written as

$$c(t) = B \sin(\omega t + \phi)$$

The magnitude and phase relationship between the sinusoidal input and the steady-state output of a system is termed the *frequency response*. In linear time-invariant systems, the frequency response is independent of the amplitude and phase of the input signal.

The frequency response test on a system or a component is normally performed by keeping the amplitude A fixed and determining B and ϕ for a suitable range of frequencies. Signal generators and precise measuring instruments are readily available for various ranges of frequencies and amplitudes. The ease and accuracy of measurements are some of the advantages of the frequency response method. Wherever it is not possible to obtain the form of the transfer function of a system through analytical techniques, the necessary information to compute its transfer function can be extracted by performing the frequency response test on the system. The step response test can also be performed easily but the extraction of transfer function from the step response data is quite a laborious procedure. For systems with very large time constants, the frequency response test is cumbersome to perform as the time required for the output to reach steady-state for each frequency of the test signal is excessively long. Therefore, the frequency response test is not recommended for systems with very large time constants. Further the frequency response test obviously cannot be performed on non-interruptable systems. Under such circumstances a single shot test (step or impulse) is more convenient even though the computation of transfer function from it gets involved.

The design and parameter adjustment of the open-loop transfer function of a system for specified closed-loop performance is carried out somewhat more easily in frequency domain (i.e., through frequency response) than in time domain (i.e., through time response). Further, the effects of noise disturbance and parameter variations are relatively easy to visualize and assess through frequency response. If necessary, the transient response of a system can be obtained from its frequency response through the Fourier integral (discussed in the next section). This correlation is quite tedious to compute except for first and second-order systems. Usually for the sake of simplicity and ease of analysis, the correlation between frequency and time response of a second-order system is employed as an approximation for higher-order* systems as well. Through the use of this approximate correlation, a satisfactory transient response for higher-order systems can be achieved by the adjustment of their frequency response.

An interesting and revealing comparison of frequency and time domain approaches is based on the relative stability studies of feedback systems. The Routh criterion is a time domain approach which establishes with relative ease the stability of a system, but its adoption to determine the relative stability is involved and requires repeated application of the criterion. The root locus method is a very powerful time domain approach as it reveals not only stability but also the actual time response of the system, though it is somewhat laborious, being a technique of determining the roots of the characteristic equation. On the other hand, the Nyquist criterion (discussed later in Chapter 9) is a powerful frequency domain method of extracting the information regarding stability as well as relative stability of a system without the need to evaluate roots of the characteristic equation.

The frequency response is easily evaluated from the sinusoidal transfer function which can be obtained simply by replacing s with $j\omega$ in the system transfer function $T(s)$. The transfer function $T(j\omega)$ thus obtained, is a complex function of frequency and has both a magnitude and a phase angle. These characteristics are conveniently represented by graphical plots. The various graphical techniques to represent the sinusoidal transfer function $T(j\omega)$ are discussed in detail in this chapter.

8.2 Correlation between Time and Frequency Response

As discussed earlier, the correlation between time and frequency response has an explicit form only for systems of first- and second-order. Let us first discuss the explicit correlation for second-order systems.

---

*Such systems should approximately meet the condition for dominance of a pair of complex conjugate closed-loop poles.

Second-order Systems

A second-order system of the form shown in Fig. 8.1 has the transfer function

$$\frac{C(s)}{R(s)} = \frac{\omega_n^2}{s^2 + 2\zeta\omega_n s + \omega_n^2}$$

where ζ is the damping factor and ω_n is the undamped natural frequency of oscillations.

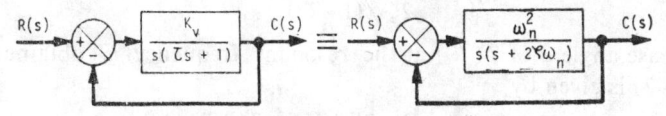

Figure 8.1 A second-order system.

The sinusoidal transfer function of the system is

$$\frac{C}{R}(j\omega) = T(j\omega) = \frac{\omega_n^2}{(j\omega)^2 + 2\zeta\omega_n(j\omega) + \omega_n^2}$$

$$= \frac{1}{(1 - u^2) + j2\zeta u} \quad (8.1)$$

where $u = \omega/\omega_n$ is the normalized driving signal frequency. From eqn. (8.1), we get

$$|T(j\omega)| = M = 1/\sqrt{[(1 - u^2)^2 + (2\zeta u)^2]} \quad (8.2)$$

and

$$\angle T(j\omega) = \phi = -\tan^{-1}[2\zeta u/(1 - u^2)] \quad (8.3)$$

The steady-state output of the system for a sinusoidal input of unit magnitude and variable frequency ω is given by

$$c(t) = \frac{1}{\sqrt{[(1 - u^2)^2 + (2\zeta u)^2]}} \sin\left(\omega t - \tan^{-1}\frac{2\zeta u}{1 - u^2}\right)$$

From eqns. (8.2) and (8.3), it is seen that when

$$u = 0, \quad M = 1 \quad \text{and} \quad \phi = 0$$

$$u = 1, \quad M = \frac{1}{2\zeta} \quad \text{and} \quad \phi = -\pi/2$$

$$u \to \infty, \quad M \to 0 \quad \text{and} \quad \phi \to -\pi$$

The magnitude and phase angle characteristics for normalized frequency u for certain values of ζ are shown in Figs. 8.2a and b respectively.

The frequency where M has a peak value is known as the *resonant frequency*. At this frequency, the slope of the magnitude curve is zero. Let ω_r be the resonant frequency and $u_r = (\omega_r/\omega_n)$ be the normalized resonant frequency. Then

$$\left.\frac{dM}{du}\right|_{u=u_r} = -\frac{1}{2}\frac{[-4(1 - u_r^2)u_r + 8\zeta^2 u_r]}{[(1 - u_r^2)^2 + (2\zeta u_r)^2]^{3/2}} = 0$$

which gives
$$4u_r^3 - 4u_r + 8\zeta^2 u_r = 0$$
or
$$u_r = \sqrt{(1 - 2\zeta^2)}$$
i.e.,
$$\omega_r = \omega_n \sqrt{(1 - 2\zeta^2)} \tag{8.4}$$

From eqn. (8.2), the maximum value of the magnitude known as the *resonant peak* is given by

$$M_r = \frac{1}{2\zeta\sqrt{(1-\zeta^2)}} \tag{8.5}$$

The phase angle ϕ of $T(j\omega)$ at the resonant frequency u_r obtained from eqn. (8.3) is given by

$$\phi_r = -\tan^{-1}[\sqrt{(1 - 2\zeta^2)}/\zeta]$$

From eqns. (8.4) and (8.5), it is seen that as ζ approaches zero, ω_r approaches ω_n and M_r approaches infinity. For $0 < \zeta \leqslant 1/\sqrt{2}$, the resonant frequency always has a value less than ω_n and the resonant peak has a value greater than 1.

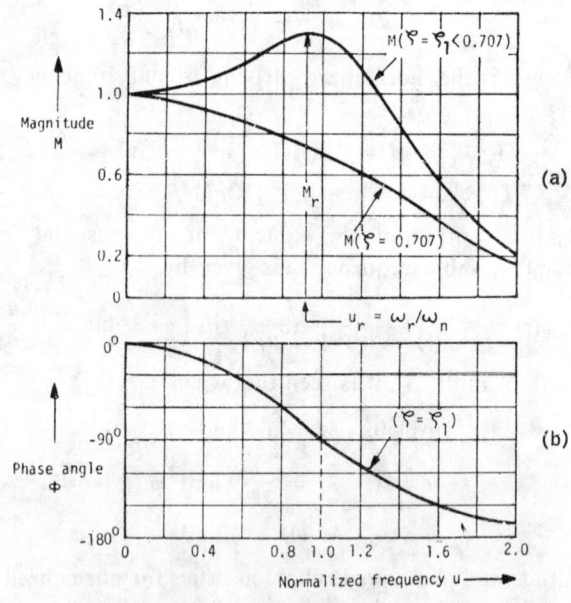

Figure 8.2 (a) Frequency response magnitude characteristics;
(b) Frequency response phase characteristic.

For $\zeta > 1/\sqrt{2}$, it is seen from eqn. (8.4) that dM/du, the slope of the magnitude curve does not become zero for any real value of ω. For this range of ζ, the magnitude M decreases monotonically from $M = 1$ at $u = 0$ with increasing u, as shown in Fig. 8.2. It therefore follows that for $\zeta > 1/\sqrt{2}$, there is no resonant peak as such and the greatest value of M equals one.

Frequency Response Analysis

We then find from eqns. (8.4) and (8.5) and the above discussion that for a second-order system, the resonant peak M_r of its frequency response is indicative of its damping factor ζ for $0 < \zeta \leqslant 1/\sqrt{2}$, and the resonant frequency ω_r of the frequency response is indicative of its natural frequency for a given ζ and hence indicative of its speed of response (as $t_s = 4/\zeta\omega_n$). M_r and ω_r of the frequency response could thus be used as performance indices for a second-order system.

Re-examining Fig. 8.2, we notice that for $\omega > \omega_r$, M decreases monotonically. The frequency at which M has a value* of $1/\sqrt{2}$ is of special significance and is called the *cut-off frequency* ω_c. The signal frequencies above cut-off are greatly attenuated in passing through a system.

For feedback control systems, the range of frequencies over which M is equal to or greater than $1/\sqrt{2}$ is defined as *bandwidth* ω_b. Control systems being low-pass filters (at zero frequency, $M = 1$), the bandwidth ω_b is equal to cut-off frequency ω_c. The definition of the bandwidth is depicted on typical frequency response of a feedback control system in Fig. 8.3a.

In general, the bandwidth of a control system indicates the noise-filtering characteristics of the system. Also, bandwidth gives a measure of the transient response properties as observed below.

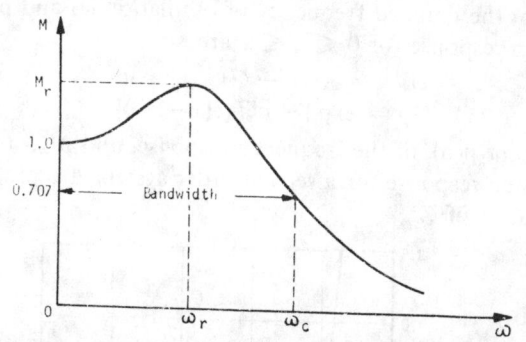

Figure 8.3(a) Typical magnification curve of a feedback control system.

The normalized bandwidth $u_b = \omega_b/\omega_n$ of the second-order system under consideration can be readily determined as follows:

$$M = \frac{1}{\sqrt{[(1 - u_b^2)^2 + (2\zeta u_b)^2]}} = \frac{1}{\sqrt{2}}$$

or
$$u_b^4 - 2(1 - 2\zeta^2)u_b^2 - 1 = 0$$

Solving for u_b we get

$$u_b = [1 - 2\zeta^2 + \sqrt{(2 - 4\zeta^2 + 4\zeta^4)}]^{1/2} \qquad (8.6)$$

As the bandwidth must be a positive real quantity, the negative sign in quadratic solution and the negative sign in taking the square-root have been discarded.

*This value corresponds to -3db point on the Bode plot of $T(j\omega)$ of the second-order system under consideration. Bode plots are dealt with in later part of this chapter.

We observe from eqn. (8.6) that the normalized bandwidth is a function of damping only; u_b versus ζ is plotted in Fig. 8.3b.

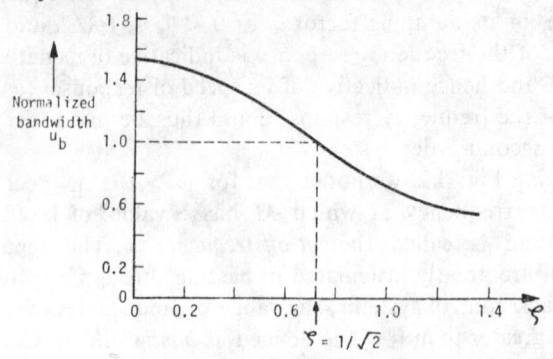

Figure 8.3(b) Bandwidth versus damping factor.

The denormalized bandwidth from eqn. (8.6) is given by
$$\omega_b = \omega_n[1 - 2\zeta^2 + \sqrt{(2 - 4\zeta^2 + 4\zeta^4)}]^{1/2} \tag{8.7}$$

Let us now consider the step response of a second-order system. The nature of this response has been discussed in detail in Chapter 5. The expressions for the damped frequency of oscillation ω_d and peak overshoot M_p of the step response for $0 \leqslant \zeta \leqslant 1$ are

$$\omega_d = \omega_n\sqrt{(1 - \zeta^2)} \tag{8.8}$$
$$M_p = \exp[-\pi\zeta/\sqrt{(1 - \zeta^2)}] \tag{8.9}$$

M_r, the resonant peak of the frequency response, and M_p, the peak overshoot of the step response for a second-order system, are plotted in Fig. 8.4 for various values of ζ.

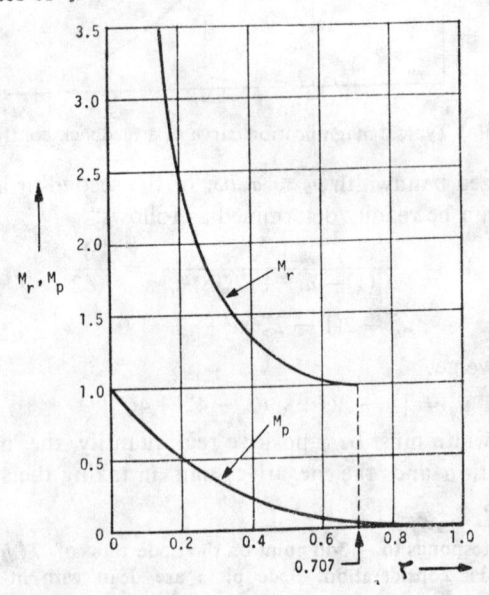

Figure 8.4 M_r, M_p versus ζ.

Frequency Response Analysis

The comparison of M_r and M_p plots shows that for $0 < \zeta < 1/\sqrt{2}$, the two performance indices are correlated as both are functions of the system damping factor ζ only. It means that a system with a given value of M_r of its frequency response, must exhibit a corresponding value of M_p if subjected to a step input. For $\zeta > 1/\sqrt{2}$, the resonant peak M_r does not exist and the correlation breaks down. This is not a serious problem as for this range of ζ, the step response oscillations are well-damped and M_p is hardly perceptible.

Similarly, the comparison of eqns. (8.4) and (8.8) reveals that there exists a definite correlation between the resonant frequency ω_r of frequency response and damped frequency of oscillations of the step response. The ratio of these two frequencies is

$$\omega_r/\omega_d = \sqrt{[(1 - 2\zeta^2)/(1 - \zeta^2)]}$$

which is a function of ζ and is plotted in Fig. 8.5.

It is further observed from eqn. (8.7) that the bandwidth, a frequency domain concept, is indicative of the undamped natural frequency of a system for a given ζ and therefore indicative of the speed of response ($t_s = 4/\zeta\omega_n$), a time domain concept.

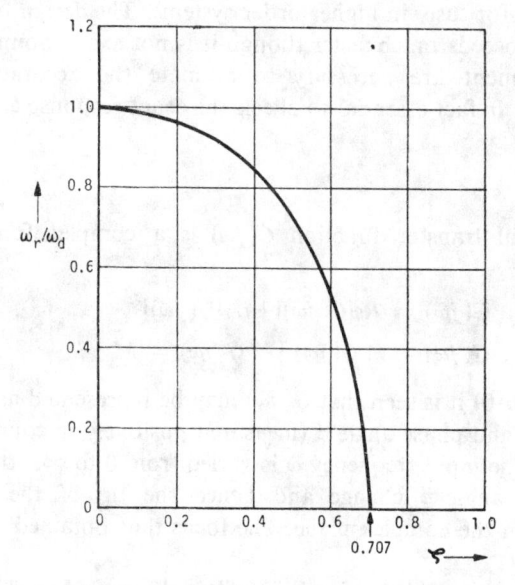

Figure 8.5 Correlation between ω_r and ω_d.

HIGHER-ORDER SYSTEMS

In general, the transient response and the frequency response of linear systems are related through the Fourier integral. This relationship forms an important basis of most design procedures and criteria. Usually the desired time domain behaviour is interpreted in terms of frequency response characteristics. The design is carried out in the frequency domain and the frequency response is then translated back into the time domain. The behaviour in the frequency domain of a given driving function $r(t)$ is given by the Fourier integral as (see Appendix I)

$$R(\omega) = \int_{-\infty}^{\infty} r(t)e^{-j\omega t}\,dt$$

Similarly, the frequency response $C(j\omega)$ may be translated back into the time domain by the inverse Fourier integral as

$$c(t) = \frac{1}{2\pi}\int_{-\infty}^{\infty} C(j\omega)e^{j\omega t}\,d\omega$$

However, the correlation between transient and frequency response through the Fourier integral is highly laborious to compute. It is therefore tempting to consider the possibility of using the second-order correlations for higher-order systems as well. Whenever, a higher-order system has a transfer function which is dominated by a pair of complex conjugate poles, it can be approximated by a second-order system whose poles are the dominant poles of the higher-order system. When such an approximation is possible, all the time and frequency domain correlations of a second-order system become valid for use in higher-order systems. The design based on these correlations proceeds much faster though it is not exact. Sound engineering skill and judgement are necessary to estimate the accuracy of such a procedure. It is in fact essential to check the exact response after the design is completed.

8.3 Polar Plots

The sinusoidal transfer function $G(j\omega)$ is a complex function and is given by

$$G(j\omega) = Re[G(j\omega)] + jI_m[G(j\omega)]$$

or $\qquad G(j\omega) = |G(j\omega)| \angle G(j\omega) = M \angle \phi \qquad (8.10)$

From eqn. (8.10) it is seen that $G(j\omega)$ may be represented as a phasor of magnitude M and phase angle ϕ (measured positively in counter clockwise direction). As the input frequency ω is varied from 0 to ∞, the magnitude M and phase angle ϕ change and hence the tip of the phasor $G(j\omega)$ traces a locus in the complex plane. The locus thus obtained is known as *polar plot.**

As an example, consider a simple RC filter shown in Fig. 8.6.
The transfer function of the system is

$$\frac{E_0}{E_i}(s) = G(s) = \frac{1/Cs}{R + 1/Cs} = \frac{1}{1 + Ts}$$

where $T = RC$.

*Other ways of graphically representing $G(j\omega)$ are discussed in later sections of this chapter.

Frequency Response Analysis

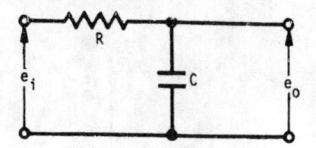

Figure 8.6 RC filter circuit.

Therefore the sinusoidal transfer function is

$$G(j\omega) = \frac{1}{1+j\omega T} \qquad (8.11)$$

$$= \frac{1}{\sqrt{(1+\omega^2 T^2)}} \angle -\tan^{-1}\omega T = M\angle\phi \qquad (8.12)$$

The polar plot of $G(j\omega)$ as in eqn. (8.12) is drawn in Fig. 8.7. When $\omega = 0$, $M = 1$ and $\phi = 0$. Therefore the phasor at $\omega = 0$ has unit length and lies along the positive real axis. As ω increases, M decreases and phase angle increases negatively. When $\omega = 1/T$, $M = 1/\sqrt{2}$ and $\phi = -45°$. As $\omega \to \infty$, M becomes zero and ϕ is $-90°$. This is represented by a phasor of zero length directed along the $-90°$ axis in the complex plane. In fact the locus of $G(j\omega)$ (polar plot) can be shown to be a semicircle.

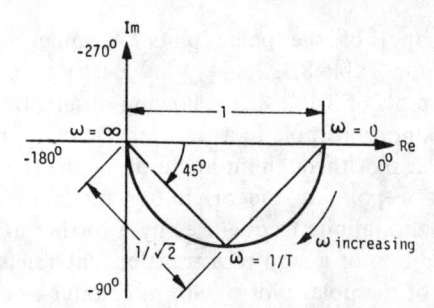

Figure 8.7 Polar plot of $1/(1+j\omega T)$.

Consider now the transfer function

$$G(j\omega) = \frac{1}{j\omega(1+j\omega T)} \qquad (8.13)$$

This transfer function may be rearranged as

$$G(j\omega) = \frac{-T}{1+\omega^2 T^2} - j\frac{1}{\omega(1+\omega^2 T^2)} \qquad (8.14)$$

From eqn. (8.14) we get

$$\lim_{\omega \to 0} G(j\omega) = -T - j\infty = \infty\angle -90°$$

$$\lim_{\omega \to \infty} G(j\omega) = -0 - j0 = 0\angle -180°$$

The general shape of the polar plot of this transfer function is shown in Fig. 8.8. The plot is asymptotic to the vertical line passing through the point $(-T, 0)$.

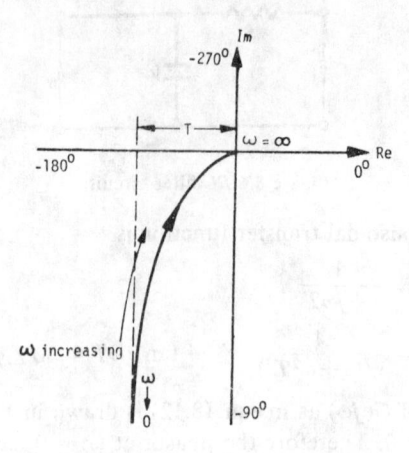

Figure 8.8 Polar plot of $1/j\omega(1+j\omega T)$.

The major advantage of the polar plot lies in stability study of systems. N. Nyquist (in 1932) related the stability of a system to the form of these plots. Because of his work, the polar plots are commonly referred to as *Nyquist plots*.

The general shapes of the polar plots of some important transfer functions are given in Table 8.1.

From the polar plots of Table 8.1, following observations are made:

(i) Addition of a nonzero pole to a transfer function results in further rotation of the polar plot through an angle of $-90°$ as $\omega \to \infty$.

(ii) Addition of a pole at the origin to a transfer function rotates the polar plot at zero and infinite frequencies by a further angle of $-90°$.

The effect of addition of a zero to a transfer function is to rotate the high frequency portion of the polar plot by $90°$ in counter-clockwise direction.

INVERSE POLAR PLOTS

The inverse polar plot of $G(j\omega)$ is a graph of $1/G(j\omega)$ as a function of ω. For example, for the *RC* filter shown in Fig. 8.6

$$\frac{1}{G(j\omega)} = G^{-1}(j\omega) = 1+j\omega T = \sqrt{(1+\omega^2 T^2)} \angle \tan^{-1}\omega T$$

The corresponding inverse polar plot is shown in Fig. 8.9.

It will be seen later in Chapter 9 that the inverse polar plots are useful in applying *M*-criterion and are also valuable in stability study of nonunity feedback systems.

8.4 Bode Plots

One of the most useful representations of a transfer function is a logarithmic plot which consists of two graphs, one giving the logarithm of $|G(j\omega)|$ and the other phase angle of $G(j\omega)$ both plotted against frequency in logarithmic scale. These plots are called *Bode plots* in honour of

H.W. Bode who did the basic work in this area. The transfer function $G(j\omega)$ is represented by

$$G(j\omega) = |G(j\omega)| e^{j\phi(\omega)}$$

Taking natural logarithm of both sides

$$\ln G(j\omega) = \ln |G(j\omega)| + j\phi(\omega) \qquad (8.15)$$

Table 8.1

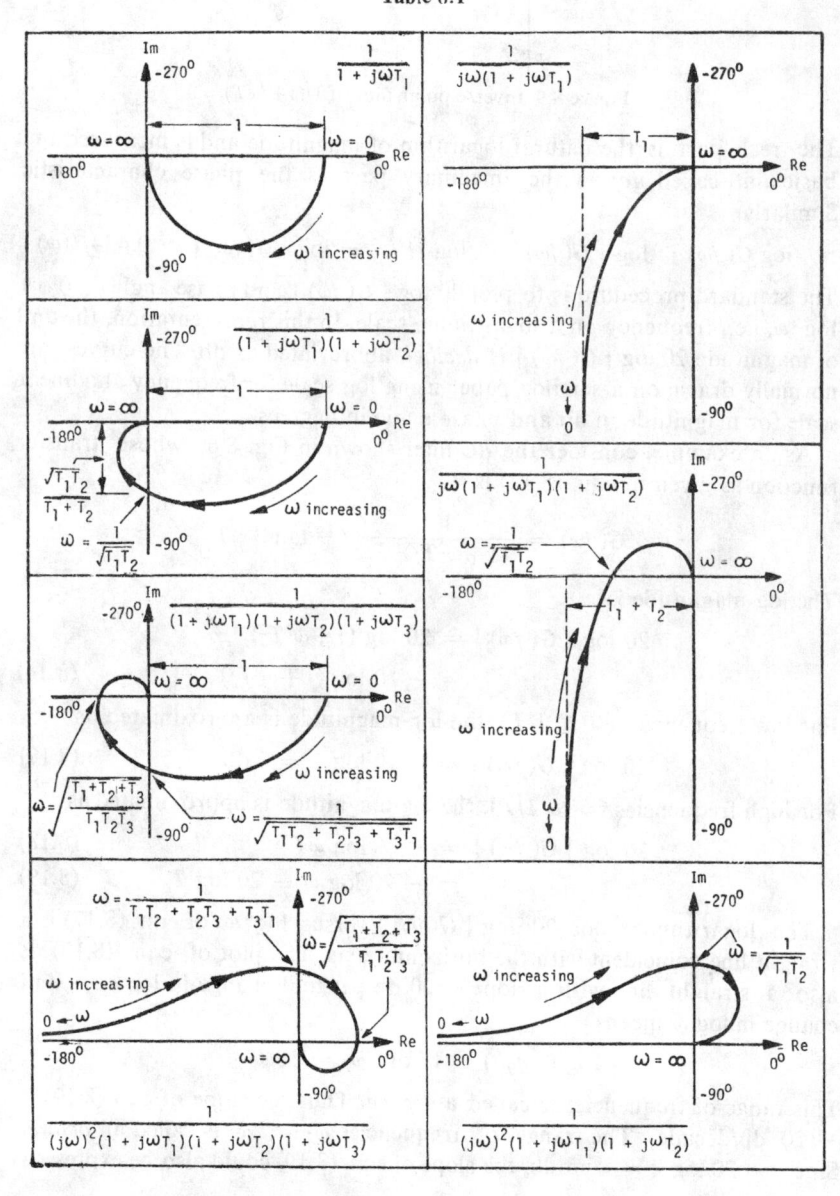

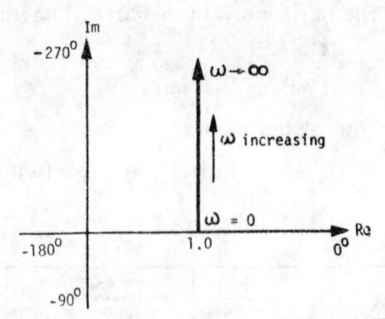

Figure 8.9 Inverse polar plot of $1/(1+j\omega T)$.

The real part is the natural logarithm of magnitude and is measured in a basic unit called *neper*; the imaginary part is the phase characteristic. Similarly,

$$\log G(j\omega) = \log |G(j\omega)| + \log e^{j\phi(\omega)} = \log |G(j\omega)| + 0.434 j\phi(\omega)$$

The standard procedure is to plot $20 \log |G(j\omega)|$ and phase angle $\phi(\omega)$ vs. $\log \omega$, i.e., frequency on a logarithmic scale. In this representation, the unit of magnitude $20 \log |G(j\omega)|$ is *decibel*, abbreviated as db. The curves are normally drawn on a semilog paper using log scale for frequency and linear scale for magnitude in db and phase angle in degrees.

As an example, consider the *RC* filter shown in Fig. 8.6, whose transfer function as given in eqn. (8.12) is

$$G(j\omega) = \frac{1}{(1+\omega^2 T^2)^{1/2}} \angle -\tan^{-1} \omega T$$

The log-magnitude is

$$20 \log |G(j\omega)| = 20 \log (1+\omega^2 T^2)^{-1/2}$$
$$= -10 \log (1+\omega^2 T^2) \quad (8.16)$$

For low frequencies ($\omega \ll 1/T$), the log-magnitude is approximated as

$$20 \log |G(j\omega)| = -10 \log 1 = 0 \text{ db} \quad (8.17)$$

For high frequencies ($\omega \gg 1/T$), the log-magnitude is approximated as

$$20 \log |G(j\omega)| = -20 \log \omega T \quad (8.18)$$
$$= -20 \log \omega - 20 \log T \quad (8.19)$$

The logarithmic plot $20 \log |G(j\omega)|$ versus $\log \omega$ of eqn. (8.17) is a straight line coincident with the horizontal axis. The plot of eqn. (8.19) is also a straight line with a slope -20 db per unit change in $\log \omega$. A unit change in $\log \omega$ means

$$\log (\omega_2/\omega_1) = 1 \quad \text{or} \quad \omega_2 = 10\omega_1$$

This range of frequencies is called a *decade*. Thus the slope of eqn (8.19) is -20 db/decade. The range of frequencies $\omega_2 = 2\omega_1$ is called an *octave*. Since $-20 \log 2 = -6$ db, the slope of eqn. (8.19) could also be expressed as -6 db/octave. Further at $\omega = 1/T$ the plot of eqn. (8.19) has a value of $20 \log 1 = 0$ db as shown in Fig. 8.10.

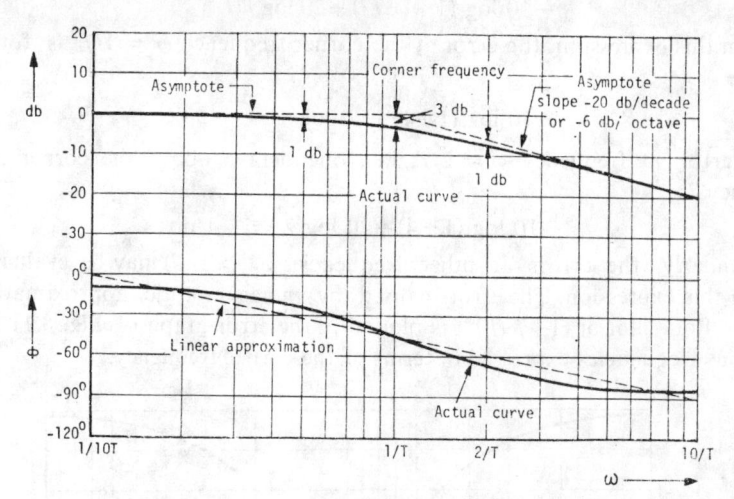

Figure 8.10 Bode plot of $(1+j\omega T)^{-1}$.

Though the straight line approximations of eqns. (8.17) and (8.19) hold good for $\omega \ll 1/T$ and $\omega \gg 1/T$ respectively, with some loss of accuracy these could be extended for frequencies $\omega \leqslant 1/T$ and $\omega \geqslant 1/T$. Therefore the log-magnitude versus log-frequency curve of $1/(1+j\omega T)$ can be approximated by two straight line asymptotes, one a straight line at 0 db for the frequency range $0 < \omega \leqslant 1/T$ and the other, a straight line with a slope -20 db/decade (or -6 db/octave) for the frequency range $1/T \leqslant \omega < \infty$. The frequency $\omega = 1/T$ at which the two asymptotes meet is called the *corner frequency* or the *break frequency*. The corner frequency divides the plot in two regions, a low frequency region and a high frequency region.

It is important to note that the log-magnitude plot of $(1+j\omega T)^{-1}$ shown in Fig. 8.10 is an asymptotic approximation of the actual plot. The actual plot can be obtained from it by applying corrections for the errors introduced by asymptotic approximation.

From eqns. (8.16) and (8.17), the error in log-magnitude for $0 < \omega \leqslant 1/T$ is given by

$$-10 \log (1+\omega^2 T^2) + 10 \log 1$$

Therefore, the error at the corner frequency $\omega = 1/T$ is

$$-10 \log (1+1) + 10 \log 1 = -3 \text{ db}$$

The error at frequency $(\omega = 1/2T)$ one octave below the corner frequency is

$$-10 \log (1+1/4) + 10 \log 1 = -1 \text{ db}$$

Similarly, the errors at other frequencies ($\omega < 1/T$) may also be evaluated from this expression.

For $1/T \leqslant \omega < \infty$, the error in log-magnitude given by eqns. (8.16) and (8.18) is

$$-10 \log (1+\omega^2 T^2) + 20 \log \omega T$$

From this expression, the error at the corner frequency $\omega = 1/T$ is found to be

$$-10 \log (1+1) + 20 \log 1 = -3 \text{ db}$$

The error at frequency $\omega = 2/T$, i.e., one octave above the corner frequency is

$$-10 \log (1+4) + 20 \log 2 = -1 \text{ db}$$

Similarly, the errors at other frequencies ($\omega > 1/T$) may be evaluated from this expression. The error caused by the asymptotic approximation of the Bode plot of $(1+j\omega T)^{-1}$ is plotted in the error-graph of Fig. 8.11 for various frequencies expressed in terms of the corner frequency.

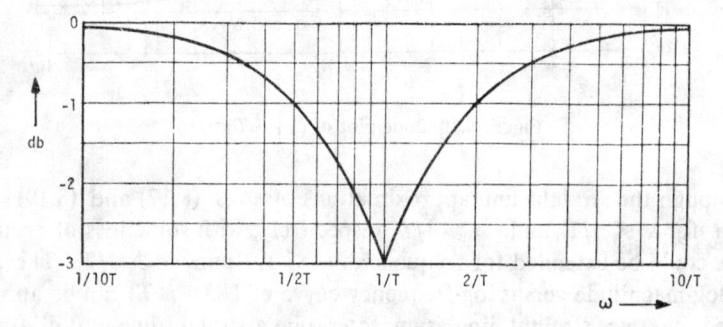

Figure 8.11 Error in log-magnitude versus frequency of $(1+j\omega T)^{-1}$.

In practice, a sufficiently accurate log-magnitude plot is obtained by correcting the asymptotic plot by -3 db at the corner frequency and by -1 db one octave below and one above the corner frequency, and then drawing a smooth curve through these three points approaching the low and high frequency asymptotes as shown in Fig. 8.10.

From eqn. (8.12), the phase angle ϕ of the factor $1/(1+j\omega T)$ is

$$\phi = -\tan^{-1} \omega T$$

At the corner frequency, the phase angle of this factor is

$$\phi = -\tan^{-1} (T/T) = -45°$$

At zero frequency, $\phi = 0$ and at infinity, it becomes $-90°$. Since the phase angle is given by inverse tangent function, the phase characteristic is skew symmetric about the inflection point $\phi = -45°$ as shown in Fig. 8.10.

Phase versus log-frequency plot can also be approximated by a straight line passing through $-45°$ at the corner frequency ($\omega = 1/T$), $0°$ at $\omega = 1/10T$ and $-90°$ at $\omega = 10/T$ as shown by the dotted line in Fig. 8.10. Such an approximation has a maximum error of about $6°$. Using linear approximations, the phase plot of $G(j\omega)$ can be readily obtained. However, in most analysis and design problems a more accurate phase plot is needed and therefore the linear approximation does not find much favour for use. Better accuracy and time saving can in fact be achieved by cutting out a template of the phase plot of $(1+j\omega T)^{-1}$ on a plastic or card-board piece for repeated use.

It is observed that compared to polar plots, Bode plots can be more quickly constructed. In these plots, both low and high frequency regions are brought into focus simultaneously. The main advantage of the Bode plots is the conversion of multiplicative factors into additive factors.

Consider a typical transfer function $G(j\omega)$ factored in the time-constant form*

$$G(j\omega) = \frac{K(1+j\omega T_a)(1+j\omega T_b)\ldots}{(j\omega)^r(1+j\omega T_1)(1+j\omega T_2)\ldots\left[1+2\zeta\left(j\frac{\omega}{\omega_n}\right)+\left(j\frac{\omega}{\omega_n}\right)^2\right]\ldots} \quad (8.20)$$

The transfer function $G(j\omega)$ has real zeros at $-1/T_a, -1/T_b\ldots$, a pole at the origin of multiplicity r, real poles at $-1/T_1, -1/T_2, \ldots$ and complex poles at $-\zeta\omega_n \pm j\omega_n\sqrt{(1-\zeta^2)},\ldots$.

If the transfer function has complex zeros, quadratic terms of the form given in denominator of eqn. (8.20) will appear in the numerator as well.

Note that the constant multiplier K is given by

$$K = \lim_{\omega \to 0} (j\omega)^r G(j\omega)$$

where r is the number of poles of $G(j\omega)$ at the origin, i.e., r is the system type number. For type-0, type-1 and type-2 systems, $K = K_p, K_v$ and K_a respectively.

From eqn. (8.20) the log-magnitude is given by

$20 \log |G(j\omega)| = 20 \log K + 20 \log |1+j\omega T_a| + 20 \log |1+j\omega T_b| +\ldots$
$\quad - 20r \log(\omega) - 20 \log |1+j\omega T_1| - 20 \log |1+j\omega T_2| \ldots$
$\quad - 20 \log |1+j2\zeta(\omega/\omega_n) - (\omega/\omega_n)^2| \ldots \quad (8.21)$

and the phase angle is given by

$\angle G(j\omega) = \tan^{-1}\omega T_a + \tan^{-1}\omega T_b +\ldots - r(90°) - \tan^{-1}\omega T_1$
$\quad - \tan^{-1}\omega T_2 - \ldots - \tan^{-1}\left(\dfrac{2\zeta\omega\omega_n}{\omega_n^2 - \omega^2}\right)\ldots \quad (8.22)$

From eqns. (8.21) and (8.22) it is seen that the Bode plots of $G(j\omega)$ may be obtained by adding the Bode plots of the factors of $G(j\omega)$. These factors are found to be of the forms given below:

1. Constant gain K
2. Poles at the origin $1/(j\omega)^r$
3. Pole on real axis $1/(1+j\omega T)$

*If the transfer function is given in pole-zero form as in eqn. (5.26), it should be rearranged in the time-constant form for construction of its Bode plot.

4. Zero on real axis $(1+j\omega T)$
5. Complex conjugate poles $1/[1+j2\zeta(\omega/\omega_n) - (\omega/\omega_n)^2]$
6. Complex conjugate zeros may also be present.

Let us now consider the plotting of each of these factors and the construction of the Bode plot for a given $G(j\omega)$ as sum of the plots of its individual factors.

FACTORS OF THE FORM $K/(j\omega)^r$

The log-magnitude of this factor is

$$20 \log \left|\frac{K}{(j\omega)^r}\right| = -20r \log \omega + 20 \log K \qquad (8.23)$$

and the phase is

$$\phi(\omega) = -90°r$$

With $\log \omega$ as abscissa, the plot of eqn. (8.23) is a straight line having a slope of $-20r$ db/decade and passing through $20 \log K$ db when $\log \omega = 0$, i.e., $\omega = 1$, as shown in Figure 8.12. Further, the plot has a value of 0 db at the frequency of

$$20r \log \omega = 20 \log K$$

or

$$\omega = (K)^{1/r}$$

The angle contribution of the factor K is zero and that of $1/(j\omega)^r$ is $-90°r$ at any frequency. The Bode magnitude plot for $r = 0, 1, 2$ and 3 are shown in Fig. 8.12. The phase diagram is simply a horizontal line at an angle of $-90°r$.

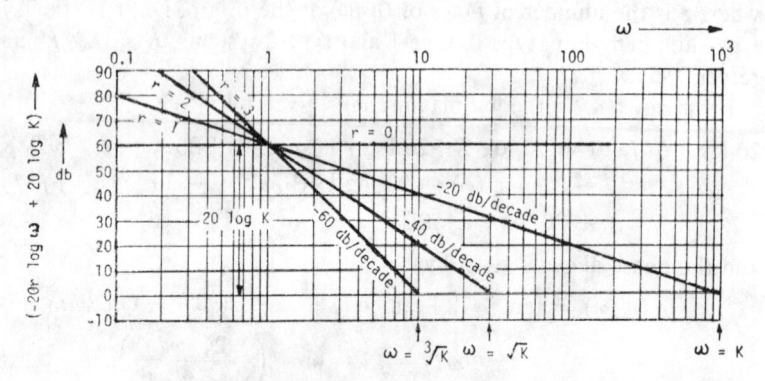

Figure 8.12 Log-magnitude plot of $K/(j\omega)^r$.

POLE OR ZERO ON THE REAL AXIS

The pole factor $1/(1+j\omega T)$ has already been considered in this section. The Bode plot of the zero factor $(1+j\omega T)$ can be drawn in the same manner as the pole factor but with a slope of $+20$ db/decade and a phase angle of $+\tan^{-1}\omega T$. The db-corrections can be read from Fig. 8.11 and are to be added to the asymptotic plot. The plot of a zero factor is shown in Fig. 8.13.

Frequency Response Analysis

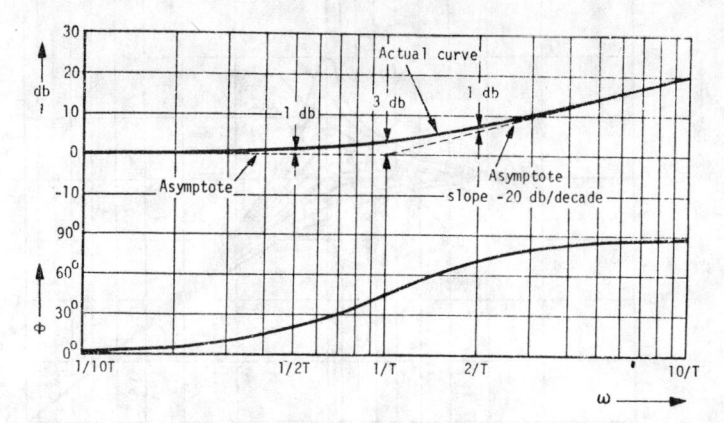

Figure 8.13 Bode plot of $(1+j\omega T)$.

COMPLEX CONJUGATE POLES

In normalized form, the quadratic factor for a pair of complex conjugate poles may be written as

$$\frac{1}{(1+j2\zeta u - u^2)}$$

where $u = \omega/\omega_n$ is the normalized frequency. The log-magnitude of this factor is

$$20 \log \left|\frac{1}{1+j2\zeta u - u^2}\right| = -20 \log [(1-u^2)^2+(2\zeta u)^2]^{1/2}$$
$$= -10 \log [(1-u^2)^2+4\zeta^2 u^2]$$

For $u \ll 1$, the log-magnitude is given by

$$20 \log \left|\frac{1}{1+j2\zeta u - u^2}\right| \approx -10 \log 1 = 0$$

and for $u \gg 1$, the log-magnitude is

$$20 \log \left|\frac{1}{1+j2\zeta u - u^2}\right| \approx -10 \log u^4 = -40 \log u$$

Therefore, the log-magnitude curve of the quadratic factor under consideration, consists of two straight line asymptotes, one horizontal line at 0 db for $u \ll 1$ and the other, a line with a slope -40 db/decade for $u \gg 1$. The two asymptotes meet on 0-db line at $u = 1$, i.e., $\omega = \omega_n$ which is the corner frequency of the plot. The asymptotic plot and actual plots are shown in Fig. 8.14.

The error between the actual magnitude and the asymptotic approximation is as given below:

For $0 < u \leqslant 1$, the error is

$$-10 \log [(1-u^2)^2+4\zeta^2 u^2]+10 \log 1$$

and for $1 < u \leqslant \infty$, the error is

$$-10 \log [(1-u^2)^2+4\zeta^2 u^2]+40 \log u$$

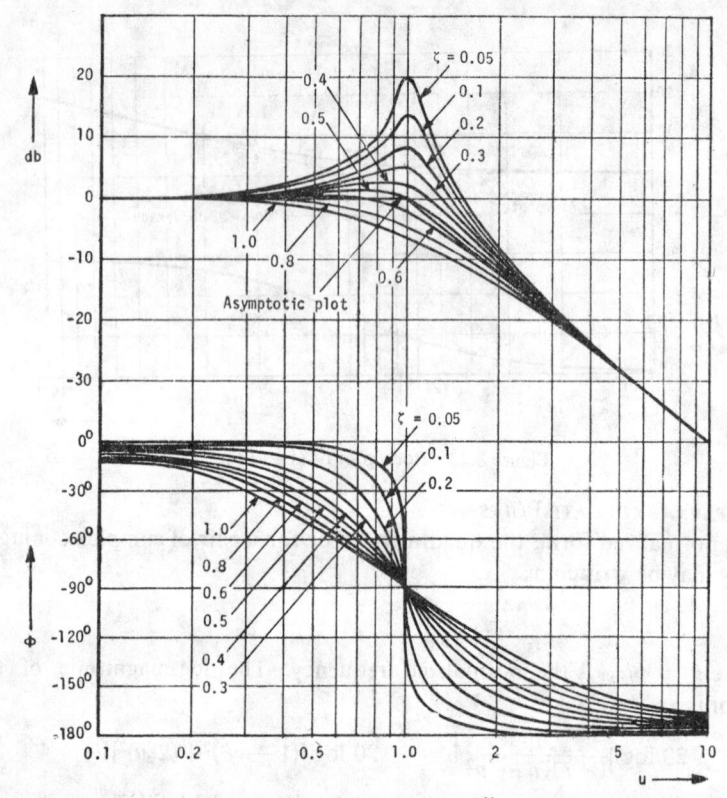

Figure 8.14 Bode plot of $1/(1+j2\zeta u - u^2)$.

From the above expressions, it is seen that the error is a function of ζ and u. The error versus u curves for different values of ζ are plotted in Fig. 8.1. The corrected log-magnitude curves for various values of ζ are shown in Fig. 8.14.

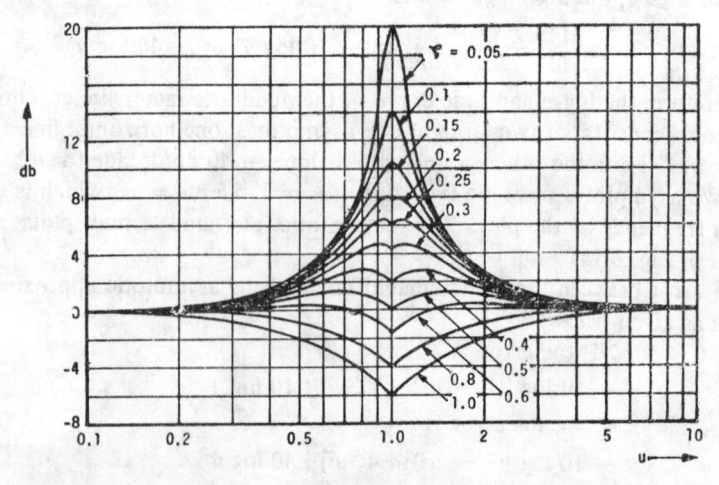

Figure 8.15 Error in db vs. frequency for asymptotic Bode plot of $1/(1+j2\zeta u - u^2)$.

Frequency Response Analysis

The phase angle of the quadratic factor $1/(1+j2\zeta u - u^2)$ is given by

$$\phi = -\tan^{-1}\left(\frac{2\zeta u}{1-u^2}\right) \tag{8.24}$$

We find that the phase angle is a function of both u and ζ. The phase angle plots for various values of ζ are given in Fig. 8.14. All these plots have a phase angle of $0°$ at $u = 0$, $-90°$ at $u = 1$ and $-180°$ at $u = \infty$. The curves become sharper in going from low frequency range to the high frequency range as ζ decreases, until for $\zeta = 0$ the curve jumps discontinuously from $0°$ down to $-180°$ at $u = 1$.

In the above discussion, the plots of factors $(j\omega)^r$, i.e., zeros at the origin and $[1+j2\zeta(\omega/\omega_n) - (\omega/\omega_n)^2]$, i.e., complex zeros, are not considered. The plots of these factors are similar to the plots of poles at the origin $1/(j\omega)^r$ and complex poles $1/[1+j2\zeta(\omega/\omega_n) - (\omega/\omega_n)^2]$ respectively, but with opposite signs.

GENERAL PROCEDURE FOR CONSTRUCTING BODE PLOTS

The following steps are generally involved in constructing the Bode plot for a given $G(j\omega)$.

1. Rewrite the sinusoidal transfer function in the time-constant form as given in eqn. (8.20).
2. Identify the corner frequencies associated with each factor of the transfer function.
3. Knowing the corner frequencies, draw the asymptotic magnitude plot. This plot consists of straight line segments with line slope changing at each corner frequency by $+20$ db/decade for a zero and -20 db/decade for a pole ($\pm 20m$ db/decade for a zero or pole of multiplicity m). For a complex conjugate zero or pole the slope changes by ± 40 db/decade ($\pm 40m$ db/decade for complex conjugate zero or pole of multiplicity m).
4. From the error graphs of Figs. 8.11 and 8.15, determine the corrections to be applied to the asymptotic plot.
5. Draw a smooth curve through the corrected points such that it is asymptotic to the line segments. This gives the actual log-magnitude plot.
6. Draw phase angle curve for each factor and add them algebraically to get the phase plot.

To illustrate the technique, let us draw the Bode plot for the transfer function

$$G(s) = \frac{64(s+2)}{s(s+0.5)(s^2+3.2s+64)}$$

The rearrangement of the transfer function in the time-constant form gives

$$G(s) = \frac{4(1+s/2)}{s(1+2s)(1+0.05s+s^2/64)}$$

Therefore, the sinusoidal transfer function in the time-constant form is given by

$$G(j\omega) = \frac{4(1+j\omega/2)}{j\omega(1+2j\omega)[1+j0.4(\omega/8)-(\omega/8)^2]} \qquad (8.25)$$

The factors of this transfer function in order of their occurrence as frequency increases, are

1. Constant gain, $K = 4$
2. Pole at origin, $1/j\omega$
3. Pole at $s = -0.5$; corner frequency $\omega_1 = 0.5$
4. Zero at $s = -2$; corner frequency $\omega_2 = 2$
5. Pair of complex conjugate poles with $\zeta = 0.2$ $\omega_n = 8$; corner frequency $\omega_3 = 8$

The pertinent characteristics of each factor of the transfer function are given in Table 8.2.

Table 8.2 Asymptotic Approximation Table for Construction of Bode Plot of

$$\frac{4(1+j\omega/2)}{j\omega(1+j2\omega)[1+j0.4(\omega/8)-(\omega/8)^2]}$$

Factor	Corner frequency	Asymptotic log-magnitude characteristic	Phase angle characteristic
$4/j\omega$	None	Straight line of constant slope -20 db/decade, passing through $20 \log 4 = 12$ db point at $\omega = 1$	Constant $-90°$
$1/(1+j2\omega)$	$\omega_1 = 0.5$	Straight line of 0 db for $\omega < \omega_1$, straight line of slope -20 db/decade for $\omega > \omega_1$	Phase angle varies, from 0 to $-90°$, angle at $\omega_1 = -45°$
$1+j0.5\omega$	$\omega_2 = 2$	Straight line of 0 db for $\omega < \omega_2$, straight line of slope $+20$ db/decade for $\omega > \omega_2$	Phase angle varies from 0 to $90°$, angle at $\omega_3 = 45°$
$\left[1+j0.4\left(\frac{\omega}{8}\right)-\left(\frac{\omega}{8}\right)^2\right]$	$\omega_3 = 1$ ($\zeta = 0.2$)	Straight line of 0 db for $\omega < \omega_3$, straight line of slope -40 db/decade for $\omega > \omega_3$	Phase angle varies from 0 to $-180°$, angle at $\omega_3 = -90°$

Asymptotic log-magnitude plot of the transfer function given in eqn. (8.25) is drawn in Fig. 8.16. The plot is obtained from Table 8.2 following the steps given below:

1. We start with the factor $(4/j\omega)$ corresponding to the pole at origin. Its log-magnitude plot is the asymptote '1', having a slope of -20 db/decade and passing through the point $20 \log 4 = 12$ db at $\omega = 1$.

2. Let us now add to the asymptote '1' the plot of the factor $1/(1+j2\omega)$ corresponding to the lowest corner frequency $\omega = \omega_1 = 0.5$. Since this

Frequency Response Analysis

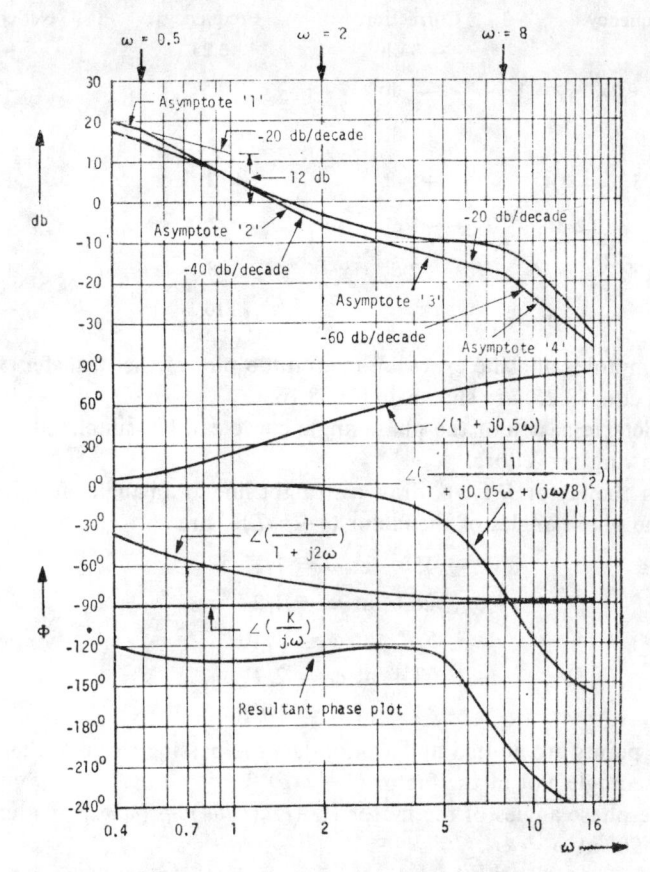

Figure 8.16 Bode plot of $\dfrac{4(1+j\omega/2)}{j\omega(1+j2\omega)[1+j0.4(\omega/8)-(\omega/8)^2]}$.

factor contributes zero db for $\omega \leqslant \omega_1 = 0.5$, the resultant plot up to $\omega = 0.5$ is the same as that of the asymptote '1'. For $\omega > \omega_1 = 0.5$ this factor contributes -20 db/decade such that the resultant plot of these two factors is the asymptote '2' of slope $(-20)+(-20) = -40$ db/decade.

3. Above $\omega_2 = 2$, the factor $(1+j0.5\omega)$ is effective. This gives rise to a straight line of slope $+20$ db/decade for $\omega > 2$, which when added to asymptote '2' results in asymptote '3' with a slope $(-40+20) = -20$ db/decade in the frequency band ω_2 to ω_3.

4. Above ω_3, the plot of $1/[1+j0.4(\omega/8)-(\omega/8)^2]$ is to be added. This factor is represented by a straight line of -40 db/decade for $\omega > \omega_3 = 8$, which when added to asymptote '3', results in asymptote '4' having a slope of $(-20)+(-40) = -60$ db/decade in the frequency band ω_3 to ∞.

To the asymptotic plot thus obtained, corrections are to be applied to get the actual plot. The following is the list of corrections obtained from Figs. (8.11) and (8.15).

Frequency	Correction	Frequency	Net correction
$\omega_1 = 0.5$	-3 db	0.25	-1 db
$\frac{1}{2}\omega_1 = 0.25, 2\omega_1 = 1$	-1 db	0.5	-3 db
$\omega_2 = 2$	$+3$ db	1	0 db
$\frac{1}{2}\omega_2 = 1, 2\omega_2 = 4$	$+1$ db	2	$+3$ db
$\omega_3 = 8, \zeta = 0.2$	$+8$ db	4	$+3$ db
$\frac{1}{2}\omega_3 = 4, 2\omega_3 = 16$	$+2$ db	8	$+8$ db
		16	$+2$ db

The asymptotes and the exact log-magnitude plot of the transfer function given in eqn. (8.25) are shown in Fig. 8.16.

The determination of the phase angle curve can be simplified by using the following procedure.

1. For the factor $K/(j\omega)^r$, draw a straight line at an angle of $-90°r$.
2. The phase angles of the factor $(1+j\omega T)^{\pm 1}$ are

$$\pm 45° \text{ at } \omega = 1/T$$
$$\pm 26.6° \text{ at } \omega = 1/2T$$
$$\pm 5.7° \text{ at } \omega = 1/10T$$
$$\pm 63.4° \text{ at } \omega = 2/T$$
$$\pm 84.3° \text{ at } \omega = 10/T$$

These points are plotted and a smooth curve passing through them gives the phase angle plot of the factor $(1+j\omega T)^{\pm 1}$.

3. The phase angles of the factor $[1+j2\zeta(\omega/\omega_n) - (\omega/\omega_n)^2]^{-1}$ are:

(a) $-90°$ at $\omega = \omega_n$.

(b) Corresponding to the value of ζ given in the factor, a few points of phase angles are read off from Fig. 8.14 or are calculated from the phase relation given in eqn. (8.24). These points are then located and joined to give a phase plot of the quadratic factor.

4. Once the phase curve of each factor of the transfer function has been drawn, the complete phase plot is obtained by adding the individual curves algebraically.

For the example under consideration, the phase angle contributions of the individual factors and the resultant phase angle plot are shown in Fig. 8.16.

8.5 All-pass and Minimum-phase Systems

The cases we have considered so far are termed as *minimum-phase transfer functions*, i.e., those with all poles and zeros in the left half of the s-plane.

Consider now the special class of transfer functions having a pole-zero pattern which is antisymmetric about the imaginary axis, i.e., for every pole in the left half plane, there is a zero in the mirror-image position. A common example of such a transfer function is

$$G(j\omega) = \frac{1 - j\omega T}{1 + j\omega T}$$

whose pole-zero configuration is shown in Fig. 8.17c. It has a magnitude of unity at all frequencies and a phase angle ($-2\tan^{-1}\omega T$) which varies from $0°$ to $-180°$ as ω is increased from 0 to ∞. The property of unit magnitude at all frequencies applies to all transfer functions with antisymmetric pole-zero pattern. Physical systems with this property, are called *all-pass systems*.

Consider next the case where the transfer function has poles in the left half s-plane and zeros in both the left- and right-half s-plane. Poles are not permitted to lie in the right-half s-plane because such a system would be unstable. Consider for example the transfer function

$$G_1(j\omega) = \frac{(1 - j\omega T)}{(1 + j\omega T_1)(1 + j\omega T_2)}$$

whose pole-zero pattern is shown in Fig. 8.17a. This transfer function may be rewritten as

$$G_1(j\omega) = \left[\frac{(1 + j\omega T)}{(1 + j\omega T_1)(1 + j\omega T_2)}\right]\left[\frac{(1 - j\omega T)}{(1 + j\omega T)}\right]$$

$$= G_2(j\omega)G(j\omega)$$

which now becomes the product of two transfer functions, one $G_2(j\omega)$ having no poles or zeros in the right half of s-plane (Fig. 8.17b) and the other, an all-pass transfer function, $G(j\omega)$ of Fig. 8.17c. It is evident that $G_1(j\omega)$ and $G_2(j\omega)$ have identical curves of magnitude versus frequency but their phase versus frequency curves are different as shown in Fig. 8.18, with $G_2(j\omega)$ having a smaller range of phase angle than $G_1(j\omega)$. It means

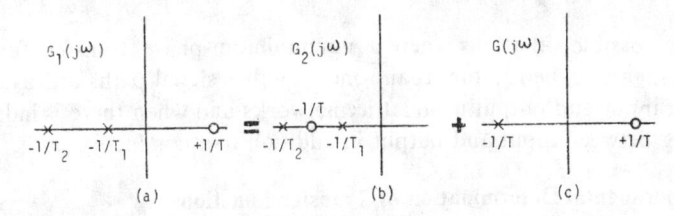

Figure 8.17 Pole-zero patterns for (a) nonminimum-phase function; (b) minimum-phase function; (c) all-pass function.

that for $G_1(j\omega)$, there is no unique relationship between magnitude and phase, as it is always possible to alter the phase curve without affecting the associated magnitude curve by the addition of an all-pass transfer function. A transfer function which has one or more zeros in the right half s-plane is known as *nonminimum-phase transfer function*.

In general, if a transfer function has any zeros in the right half s-plane, it is possible to extract them one by one by associating them with all-pass transfer functions in a manner shown in Fig. 8.17. Each time this is done the magnitude curve remains unaltered but the phase lag is reduced, until eventually we are left with a function which contains no zeros in the right-half s-plane. Such a transfer function has the least (minimum) phase angle range for a given magnitude curve and is called a *minimum-phase function*. It has a unique relationship between its phase and magnitude curves. Typical phase angle characteristics of minimum- and nonminimum-phase transfer functions are shown in Fig. 8.18. It will be seen in Chapter 10 that the larger the phase lags present in a system, the more complex are its stabilization problems. Therefore for control systems, elements with nonminimum-phase transfer functions are avoided as far as possible.

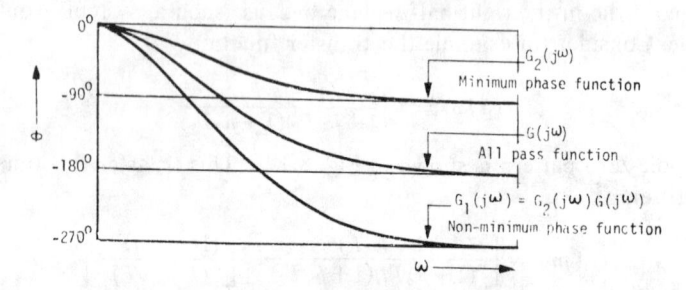

Figure 8.18 Phase angle characteristics of minimum-phase and nonminimum-phase functions.

A common example of a nonminimum-phase element is transportation lag (see Section 7.5) which has the transfer function

$$G(j\omega) = e^{-j\omega T}$$
$$= 1\angle -\omega T \text{ rad} = 1\angle -57.3\omega T \text{ deg}$$

The phase angle characteristics of transportation lag are shown in Figs. 8.19a and b.

Other possible situations where a nonminimum-phase transfer function can arise are, when more than one possible signal paths are available between input and output as in lattice networks and when there is inductive coupling between input and output in addition to a conductive path.

8.6 Experimental Determination of Transfer Functions

The Bode plots are of great value in situations where the transfer function of a given system is unknown. In such cases we proceed to obtain the frequency response data experimentally in the frequency range of interest. The system transfer function within a certain degree of accuracy can then be obtained by fitting an asymptotic log-magnitude plot to the experimental data as per the procedure outlined below.

1. The experimental data is used to plot the exact log-magnitude and phase angle versus frequency curves on a semilog graph sheet.

2. Asymptotes are then drawn on the log-magnitude curve keeping in view the fact that the slopes of these asymptotes must be multiples of ± 20 db/decade. The corner frequencies are so adjusted that the db value

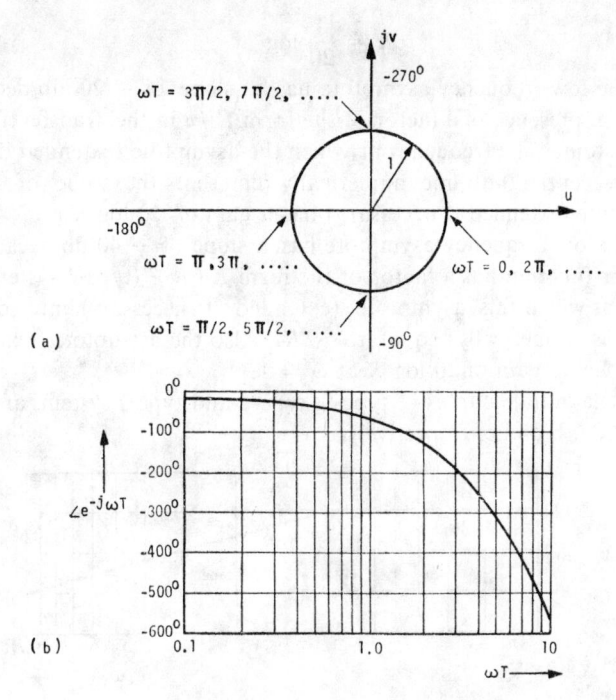

Figure 8.19 Phase angle characteristics of $e^{-j\omega T}$.

at the corner frequency on the asymptotic plot differs from the actual log-magnitude plot by an amount which is in close agreement with the db correction of the kind of factor revealed.

3. If the slope of asymptotic log-magnitude curve obtained in (2) above, changes by $-20\,m$ db/decade at $\omega = \omega_1$, it indicates that the factor $1/(1+j\omega/\omega_1)^m$ exists in the transfer function.

4. If the slope changes by $+20\,m$ db/decade at $\omega = \omega_2$, it indicates the presence of the factor $(1+j\omega/\omega_2)^m$ in the transfer function.

5. The change of slope by -40 db/decade at $\omega = \omega_3$ indicates that either a double pole (i.e., $m = 2$) or a pair of complex conjugate poles is present. If the error between the asymptotic and the actual curve is about -6 db, then a factor of the form $1/(1+j\omega/\omega_3)^2$ is present and if the error is positive then a quadratic factor of the form $1/[1+j2\zeta(\omega/\omega_3)+(j\omega/\omega_3)^2]$ is present. The corresponding value of ζ is obtained with the help of the error graph of Fig. 8.15.

6. In the low frequency range, the plot is dominated by a factor of the form $K/(j\omega)^r$.

In most practical systems r equals 0, 1 or 2. The value r is determined as follows:

(a) If the low frequency asymptote is a horizontal line at x db, it indicates that the transfer function represents a type-0 system with a gain K given by

$$20 \log K = x$$

or
$$K = \frac{1}{20} \log^{-1} x$$

(b) If the low frequency asymptote has a slope of -20 db/decade, it indicates the presence of a factor of the form $K/j\omega$ in the transfer function (type-1 system). The frequency at which the asymptote (extended if necessary) intersects the 0-db line numerically represents the value of K. Also the asymptote (extended if necessary) has a gain of $20 \log K$ at $\omega = 1$.

(c) If the low frequency asymptote has a slope of -40 db/decade, then the transfer function has a factor of the form $K/(j\omega)^2$ (type-2 system). The frequency at which this asymptote (extended if necessary) intersects the 0-db line is numerically equal to $\sqrt{K}$. Also the asymptote (extended if necessary) has a gain of $20 \log K$ at $\omega = 1$.

The log-magnitude curves of type-0, type-1 and type-2 systems are shown in Figs. 8.20a, b and c, respectively.

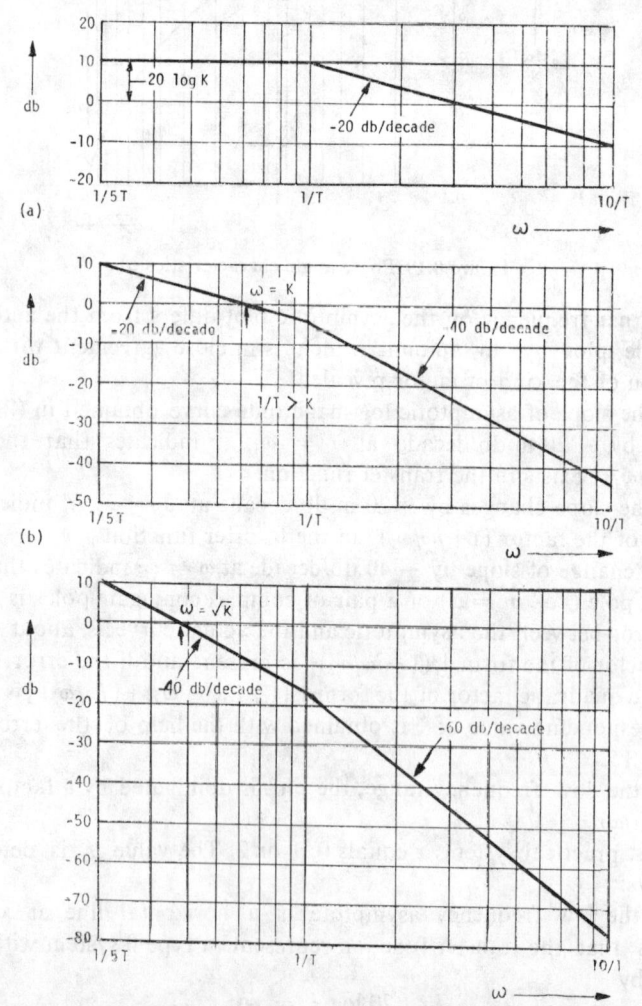

Figure 8.20 Log-magnitude curves of type-0, type-1, and type-2 systems.

Frequency Response Analysis

After obtaining the transfer function from the log-magnitude curve, the phase angle curve is constructed from it and is then compared with the one obtained experimentally. If the two curves are in fair agreement and if the phase angles of both the curves at very high frequencies tend to $-90°\,(q-p)$ where p and q are degrees of the numerator and denominator polynomials respectively of the transfer function, then the transfer function is of minimum-phase type. If the computed phase angle is 180° less negative than the one obtained experimentally, then the transfer function is of nonminimum-phase type and one of the zeros of the transfer function lies in the right half s-plane.

As an illustration, let us derive the transfer function of the system whose experimental log-magnitude and phase-angle curves are shown in Fig. 8.21. First of all, the asymptotes are drawn on the experimentally determined curve as shown. The low frequency asymptote has a slope of -20 db/decade and when extended, intersects the 0 db axis at $\omega = 5$. Therefore the asymptote is a plot of the factor $5/(j\omega)$. The corner frequencies are

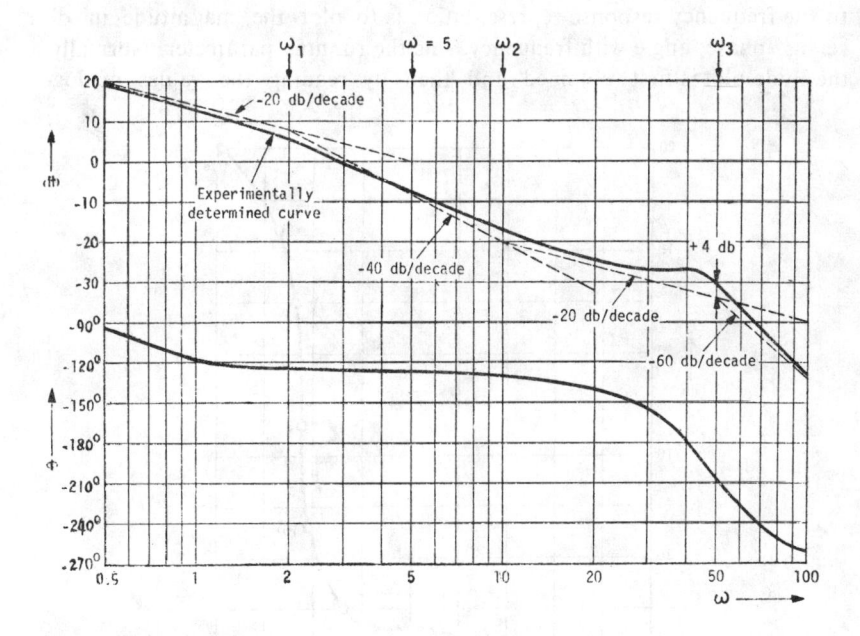

Figure 8.21 Experimentally obtained log-magnitude and phase characteristics.

found to be located at $\omega_1 = 2$, $\omega_2 = 10$ and $\omega_3 = 50$. At the first corner frequency, the slope of the curve changes by -20 db/decade and at the second corner frequency, it changes by $+20$ db/decade. Therefore the transfer function has the factors $1/(1+j\omega/2)$ and $1+j\omega/10$ corresponding to these corner frequencies. At $\omega = \omega_3$, the curve changes by a slope of -40 db/decade. At this frequency the error between actual and approximate plots is +4 db. Therefore, the transfer function has a quadratic factor

$$\frac{1}{1+j2\zeta(\omega/50)+(j\omega/50)^2}$$

where $\zeta = 0.3$, as obtained from the error graph of Fig. 8.15, corresponding to the error +4 db.

Thus the transfer function of the system becomes

$$G(j\omega) = \frac{5(1+j\omega/10)}{j\omega(1+j\omega/2)[1+j\,0.6(\omega/50)+(j\omega/50)^2]}$$

From the experimental phase angle curve shown in Fig. 8.21, it is seen that the phase angle at very high frequencies is $-270°$ which is equal to $-90°(q-p) = -90°(4-1)$. Therefore the Bode plot represents a minimum-phase transfer function.

8.7 Log-magnitude versus Phase Plots

In the previous sections, we considered polar plots and Bode plots as the graphical representations of frequency response. An alternative approach to the frequency response representation is to plot the magnitude in db versus phase angle with frequency ω as the running parameter. Normally, the Bode plot is first obtained and then by reading the values of log-

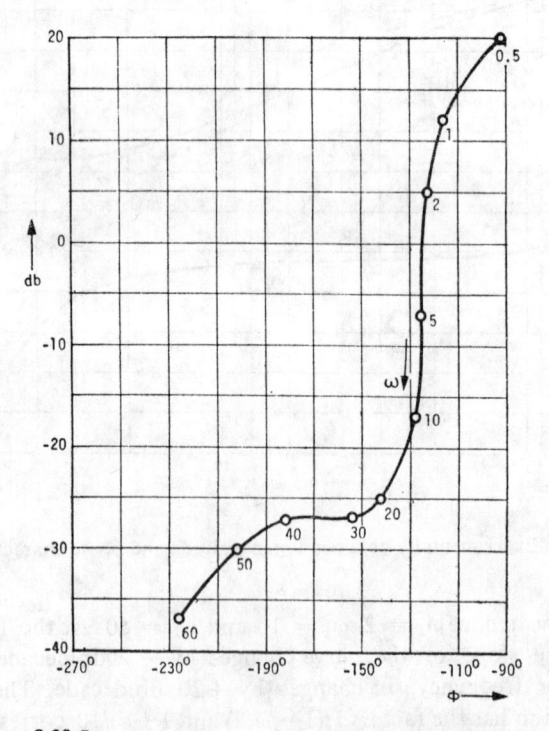

Figure 8.22 Log-magnitude vs. phase curve obtained from Fig. 8.21.

magnitude and phase angle at different frequencies from these plots, the log-magnitude vs. phase angle plot is constructed. The advantages of these plots are that the relative stability of closed-loop control systems can be determined quickly and the compensation can be carried out easily (discussed in Chapter 10).

The log-magnitude versus phase curve for the transfer function

$$G(j\omega) = \frac{5(1+j\omega/10)}{j\omega(1+j\omega/2)[1+j0.6(\omega/50)+(j\omega/50)^2]}$$

is shown in Fig. 8.22 which is obtained by utilizing the Bode plot of Fig. 8.21.

Problems

8.1 Sketch the polar plots of the transfer functions given below. Determine whether these plots cross the real axis. If so, determine the frequency at which the plots cross the real axis and the corresponding magnitude $|G(j\omega)|$.

(a) $G(s) = 1/(1+s)(1+2s)$

(b) $G(s) = 1/s(1+s)(1+2s)$

(c) $G(s) = 1/s^2(1+s)(1+2s)$

(d) $G(s) = \dfrac{(1+0.2s)(1+0.025s)}{s^3(1+0.005s)(1+0.001s)}$

8.2 Sketch the direct and inverse polar plots for a unity feedback system with open-loop transfer function

$$G(s) = 1/s(1+s)^2$$

Also find the frequency at which $|G(j\omega)| = 1$ and the corresponding phase angle $\angle G(j\omega)$.

8.3 Sketch the Bode plots showing the magnitude in decibels and phase angle in degrees as a function of log frequency for the transfer functions given below. Determine the gain cross-over frequency ω_c (the frequency at which the magnitude curve crosses the 0-db axis) in each case.

(a) $G(s) = \dfrac{10}{s(1+0.5s)(1+0.1s)}$

(b) $G(s) = \dfrac{75(1+0.2s)}{s(s^2+16s+100)}$

8.4 Sketch the Bode plots for the following transfer functions and determine in each case, the system gain K for the gain cross-over frequency ω_c to be 5 rad/sec.

(a) $G(s) = \dfrac{Ks^2}{(1+0.2s)(1+0.02s)}$

(b) $G(s) = \dfrac{Ke^{-0.1s}}{s(1+s)(1+0.1s)}$

8.5 The frequency response test datacertain of elements plotted on Bode diagrams and asymptotically approximated are shown in Fig. P-8.5. Find the transfer function of each element. (Elements are known to have minimum-phase characteristics.)

8.6 The following frequency response test data were obtained for the forward path elements of a unity feedback control system. Plot the data on a semilog graph paper and determine the transfer function of the forward path, using asymptotic approximation. What is the type of the system?

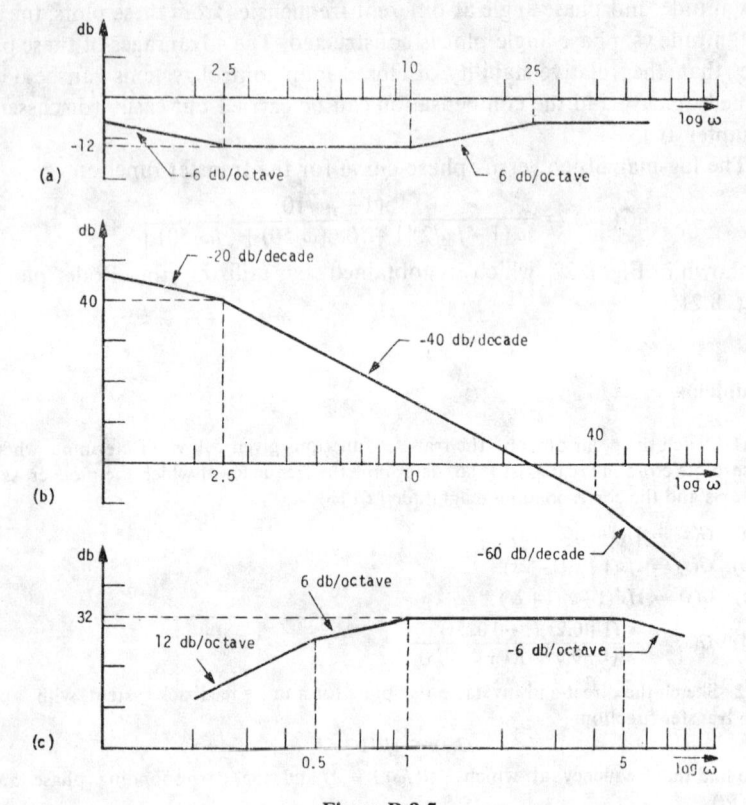

Figure P-8.5.

Gain (db)	34	28	24.6	14.2	8	1.5	−3.5	−7.2
Frequency (rad/sec)	0.1	0.2	0.3	0.7	1.0	1.5	2.0	2.5
Gain (db)		−12.5	−14.7	−16.0	−17.5	−17.5	−17.5	
Frequency (rad/sec)		4.0	5.0	6.0	9.0	20	35	

8.7 Consider the feedback system shown in Fig. P-8.7.

(a) Find the value of K and a to satisfy the following frequency domain specifications:

$$M_r = 1.04$$
$$\omega_r = 11.55 \text{ rad/sec}$$

(b) For the values of K and a determined in part (a), calculate the settling time and bandwidth of the system.

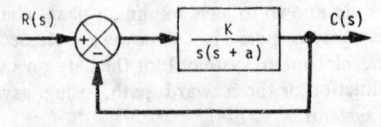

Figure P-8.7.

8.8 The closed-loop transfer function of a feedback system is given by

$$T(s) = \frac{1000}{(s+22.5)(s^2+2.45s+44.4)}$$

(a) Determine the resonance peak M_r and resonant frequency ω_r of the system by drawing the frequency response curve.

(b) What should be the values of damping ratio ζ and undamped natural frequency ω_n of an equivalent second-order system which will produce the same M_r and ω_r as determined in part (a)?

(c) Determine the bandwidth of the equivalent second-order system.

8.9 Unit-step response data of a second-order system is given below. Obtain the corresponding frequency response indices $(M_r, \omega_r, \omega_b)$ for the system.

t (sec)	0	0.05	0.10	0.15	0.20	0.25	0.30	0.35	0.40	0.45	0.50
$c(t)$	0	0.25	0.8	1.08	1.12	1.02	0.98	0.98	1.0	1.0	1.0

Further Reading

1. D'Azzo, J.J. and C.H. Houpis, *Linear Control System Analysis and Design*, McGraw-Hill, New York, 1975.
2. Thaler, G.J., *Design of Feedback Systems*, Dowden, Hutchinson and Ross, Stroudsburg, Pennsylvania, 1973.
3. Eveleigh, V.W., *Introduction to Control Systems Design*, McGraw-Hill, New York, 1972.
4. Chen, C.F. and I.J. Haas, *Elements of Control Systems Analysis*, Prentice-Hall, Englewood Cliffs, N.J., 1968.
5. Ku, Y.H., *Analysis and Control of Linear Systems*, International Text Book Co., Scranton, Pennsylvania, 1962.

9. Stability in Frequency Domain

9.1 Introduction

In earlier chapters, it has been shown that the stability of a system depends upon the location of the roots of its characteristic equation in the s-plane. For a system to be stable, the roots of the characteristic equation should not lie in the right half of the s-plane or on the imaginary axis, for such roots lead to instability or sustained oscillations of the system. The root locus technique discussed in Chapter 7 is an approach to determine the location of the roots in the s-plane, given the open-loop pole-zero configuration.

In this chapter, another equally important and useful technique, the *Nyquist stability criterion*, which relates the location of the roots of the characteristic equation to the open-loop frequency response of a system will be discussed. Unlike the root locus technique, the computation of closed-loop poles is not necessary in this case and the stability study can be carried out graphically from the open-loop frequency response. Therefore, experimentally determined open-loop frequency response can be used directly for the study of stability when the feedback path is closed.

The Nyquist stability criterion is based on a theorem of complex variables due to Cauchy, commonly known as *principle of argument*. In the following section, we shall discuss this theorem and then in the next section, we shall develop the Nyquist stability criterion.

9.2 Mathematical Preliminaries

Consider a function $q(s)$ that can be expressed as a quotient of two polynomials. Each polynomial may be assumed to be known in the form of product of linear factors as given below:

$$q(s) = \frac{(s-\alpha_1)(s-\alpha_2)\ldots(s-\alpha_m)}{(s-\beta_1)(s-\beta_2)\ldots(s-\beta_n)} \tag{9.1}$$

Let s be a complex variable, represented by $s = \sigma + j\omega$ on the complex s-plane. Then the function $q(s)$ is also complex (being a dependent variable) and may be defined as $q(s) = u + jv$ and represented on the complex $q(s)$-plane with coordinates u and v. Equation (9.1) indicates that for every point s in the s-plane at which $q(s)$ is analytic,† we can find a corresponding point $q(s)$ in the $q(s)$-plane. Alternatively, it can be stated that the function $q(s)$ maps the points in the s-plane into the $q(s)$-plane. Since any number of points of analycity in the s-plane, can be mapped into the $q(s)$-plane, it follows that for a contour in the s-plane which does not go through any singular point, there corresponds a contour in the $q(s)$-plane as shown in Fig. 9.1. The region to the right of a closed contour is considered *enclosed* by the contour when the contour is traversed in the clockwise direction*. Thus, the shaded area in Fig. 9.1a is enclosed by the closed contour.

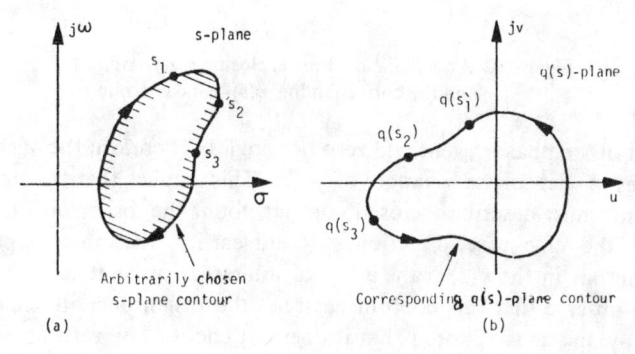

Figure 9.1 Arbitrarily chosen s-plane contour which does not go through singular points and the corresponding $q(s)$-plane contour.

As will be seen later in the next section that while developing the Nyquist criterion, we are not interested in the exact shape of the $q(s)$-plane contour. An important fact that concerns us is the encirclement of the origin by the $q(s)$-plane contour. To investigate this, consider an s-plane contour which encloses only one of the zeros of $q(s)$, say $s = \alpha_1$, while all the poles and remaining zeros are distributed in the s-plane outside the contour. As discussed above, for any non-singular point s on the s-plane contour, there corresponds a point $q(s)$ on the $q(s)$-plane contour. From eqn. (9.1), the point $q(s)$ is given by

$$|q(s)| = \frac{|(s-\alpha_1)| |(s-\alpha_2)|\ldots}{|(s-\beta_1)| |(s-\beta_2)|\ldots} \qquad (9.2)$$

$$\angle q(s) = \angle(s-\alpha_1) + \angle(s-\alpha_2) + \ldots - \angle(s-\beta_1) - \angle(s-\beta_2) - \ldots$$

From Fig. 9.2a it is found that as the point s follows the prescribed path (i.e., clockwise direction) on the s-plane contour, eventually returning to the starting point, the phasor $(s-\alpha_1)$ generates a net angle of -2π,

---

†A function $q(s)$ is analytic in the s-plane provided the function and all its derivatives exist. The points in the s-plane where the function (or its derivatives) does not exist, are called singular points. The poles of a function are singular points.
*This convention is opposite to that usually employed in complex variable theory but is equally applicable and is generally used in control system theory.

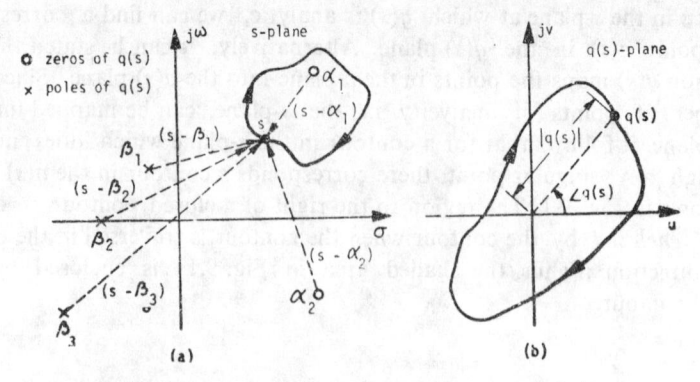

Figure 9.2 An s-plane contour enclosing a zero of $q(s)$ and the corresponding $q(s)$-plane contour.

while all other phasors generate zero net angles. Therefore the $q(s)$-phasor undergoes a net phase change of -2π. This implies that the tip of the $q(s)$-phasor must describe a closed contour about the origin of the $q(s)$-plane in the clockwise direction. As said earlier, the exact shape of the closed contour in the $q(s)$-plane is not of interest to us, but it is sufficient for us to observe that this contour encircles the origin once. If the contour in the s-plane is so chosen that it does not enclose any zero or pole, the corresponding contour in the $q(s)$-plane then will not encircle the origin. If the s-plane contour encloses two zeros, say at $s = \alpha_1$ and $s = \alpha_2$, the $q(s)$-plane contour encircles the origin twice in the clockwise direction as shown in Fig. 9.3. Generalizing, we can say that for each zero of $q(s)$ enclosed by the s-plane contour, the corresponding $q(s)$-plane contour encircles the origin once in the clockwise direction.

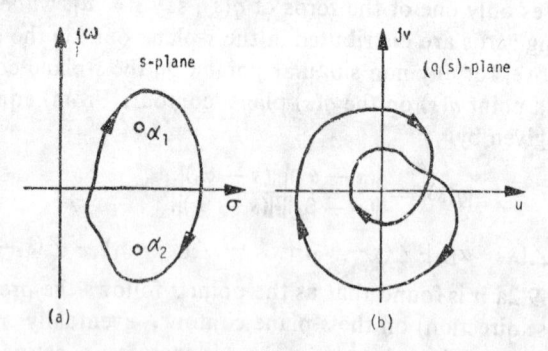

Figure 9.3 An s-plane contour enclosing two zeros of $q(s)$ and the corresponding $q(s)$-plane contour.

Consider now the enclosure of a pole of $q(s)$, say at $s = \beta_1$, by the s-plane contour. From an argument similar to the one used in conjunction with Fig. 9.2, it follows that the phasor $(s - \beta_1)$ generates an angle of -2π

as s traverses the prescribed path. Since $(s - \beta_1)$ is in the denominator of eqn. (9.1), the $q(s)$-plane contour experiences an angle change of $+2\pi$, which means one counter-clockwise encirclement of the origin. This argument holds for all other poles of $q(s)$.

Thus, if there are P poles and Z zeros of $q(s)$ enclosed by the s-plane contour, then the corresponding $q(s)$-plane contour must encircle the origin Z times in the clockwise direction and P times in the counter-clockwise direction, resulting in a net encirclement of the origin, $(P - Z)$ times in the counter-clockwise direction. For example, in case of 1 zero and 3 poles enclosed by the s-plane contour, the net encirclement of the origin by the $q(s)$-plane contour is $2\pi(3 - 1) = 4\pi$ rad, i.e., two counter-clockwise revolutions, as shown in Figs. 9.4a and b. This relation between the enclosure of poles and zeros of $q(s)$ by the s-plane contour and the encirclements of the origin by the $q(s)$-plane contour is commonly known as *principle of argument*.

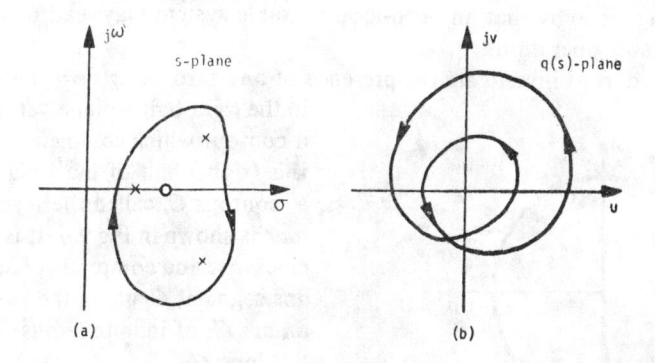

Figure 9.4 Mapping of the s-plane contour which encloses 1 zero and 3 poles.

9.3 Nyquist Stability Criterion

Consider the single-loop* feedback system shown in Fig. 9.5. The characteristic equation of the system is

$$q(s) = 1 + G(s)H(s) = 0 \tag{9.3}$$

The standard pole-zero form of the open-loop transfer function $G(s)H(s)$ considered in earlier chapters is

$$G(s)H(s) = K\frac{(s+z_1)(s+z_2)...(s+z_m)}{(s+p_1)(s+p_2)...(s+p_n)}; m \leqslant n \tag{9.4}$$

From eqns. (9.3) and (9.4) we obtain

$$q(s) = 1 + K\frac{(s+z_1)(s+z_2)...(s+z_m)}{(s+p_1)(s+p_2)...(s+p_n)} \tag{9.5}$$

---

*For the multiple-loop case $q(s) = 1 + P(s)$ as is shown by eqn. (7.5). With $P(s)$ expressed in pole-zero form, all the discussion and results that follow, apply for multiple-loop systems as well.

$$= \frac{(s+p_1)(s+p_2)...(s+p_n)+K(s+z_1)(s+z_2)...(s+z_m)}{(s+p_1)(s+p_2)...(s+p_n)}$$

$$= \frac{(s+z_1')(s+z_2')...(s+z_n')}{(s+p_1)(s+p_2)...(s+p_n)} \tag{9.6}$$

From eqn. (9.6) it is seen that the zeros of $q(s)$ at $-z_1', -z_2', ..., -z_n'$ are the roots of the characteristic equation and the poles of $q(s)$ at $-p_1, -p_2, ..., -p_n$ are same as the open-loop poles of the system. For the system to be stable, the roots of the characteristic equation and hence the zeros of $q(s)$ must lie in the left half of the s-plane. It is important to note that even if some of the open-loop poles lie in the right half s-plane, all the zeros of $q(s)$, i.e., the closed-loop poles may lie in the left half s-plane, meaning thereby that an open-loop unstable system may lead to a closed-loop stable operation.

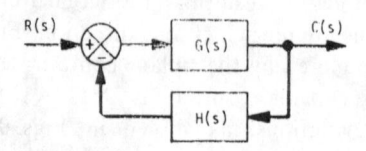

Figure 9.5 A feedback control system.

In order to investigate the presence of any zero of $q(s) = 1+G(s)H(s)$ in the right half s-plane, let us choose a contour which completely encloses this (right) half of the s-plane. Such a contour C, called the *Nyquist contour* is shown in Fig 9.6. It is directed clockwise and comprises of an infinite line segment C_1 along the $j\omega$-axis and an arc C_2 of infinite radius.

Along C_1,

$s = j\omega$ with s varying from $-j\infty$ to $+j\infty$

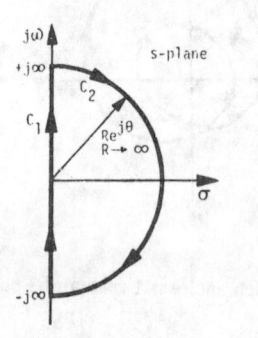

Figure 9.6 The Nyquist contour.

and along C_2,

$$s = \underset{R\to\infty}{Re^{j\theta}} \text{ with } \theta \text{ varying from } +\frac{\pi}{2} \text{ to } 0 \text{ to } -\frac{\pi}{2}$$

The Nyquist contour so defined encloses all the right half s-plane zeros and poles of $q(s) = 1+G(s)H(s)$. Let there be Z zeros and P poles of $q(s)$ in the right half s-plane. As s moves along the Nyquist contour in the s-plane, a closed contour Γ_q is traversed in the $q(s)$-plane which encloses the origin

$$N = P - Z \tag{9.7}$$

times in the counter-clockwise direction.

In order for the system to be stable, there should be no zeros of $q(s) = 1+G(s)H(s)$ in the right half s-plane, i.e.,

$$Z = 0$$

This condition is met if

$$N = P \tag{9.8}$$

that is, for a closed-loop system to be stable, the number of counter-clockwise encirclements of the origin of the $q(s)$-plane by the contour Γ_q should equal the number of the right half s-plane poles of $q(s)$ which are the poles of open-loop transfer function $G(s)H(s)$. As $G(s)H(s)$ is generally known in the factored form, the number of such poles (P) is easily ascertained by inspection.

In the special case (this is generally the case in most single-loop practical systems) of $P = 0$ (i.e., the open-loop stable system), the closed-loop system is stable if

$$N = P = 0 \tag{9.9}$$

which means that net encirclements, of the origin of the $q(s)$-plane by the Γ_q contour should be zero. It is easily observed that

$$G(s)H(s) = [1+G(s)H(s)] - 1.$$

It therefore follows that the contour Γ_{GH} of $G(s)H(s)$ corresponding to the Nyquist contour in the s-plane is the same as contour Γ_q of $1+G(s)H(s)$ drawn from the point $(-1+j0)$. Thus the encirclement of the origin by the contour Γ_q is equivalent to the encirclement of the point $(-1+j0)$ by the contour Γ_{GH} as shown in Fig. 9.7.

We can now state the Nyquist stability criterion as follows:

If the contour Γ_{GH} of the open-loop transfer function $G(s)H(s)$ corresponding to the Nyquist contour in the s-plane encircles the point $(-1+j0)$ in the counter-clockwise direction as many times as the number of right half s-plane poles of $G(s)H(s)$, the closed-loop system is stable.

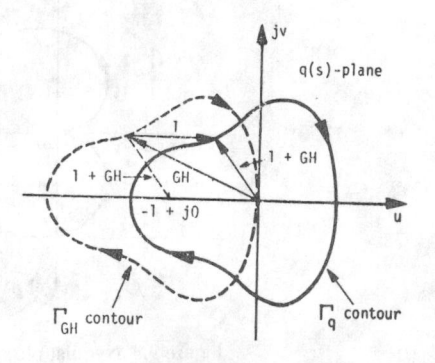

Figure 9.7.

In the commonly occurring case of the open-loop stable system, the closed-loop system is stable if the contour Γ_{GH} of $G(s)H(s)$ does not encircle $(-1+j0)$ point, i.e., the net encirclement is zero.

The mapping of the Nyquist contour into the contour Γ_{GH} is carried out as follows:

1. The mapping of the imaginary axis is carried out by substitution of $s = j\omega$ in $G(s)H(s)$. This converts the mapping function into a frequency function of $G(j\omega)H(j\omega)$.

2. In physical systems $(m \leqslant n)$, $\lim\limits_{\substack{s = Re^{j\theta} \\ R \to \infty}} G(s)H(s) = $ real constant (it is zero if $m < n$). Thus the infinite arc of the Nyquist contour maps into a point on the real axis.

The complete contour Γ_{GH} is thus the polar plot of $G(j\omega)H(j\omega)$ with ω varying from $-\infty$ to ∞. This is usually called the Nyquist plot or locus of $G(s)H(s)$. Further it is important to note that the Nyquist plot is symmetrical about the real axis since $G^*(j\omega)H^*(j\omega) = G(-j\omega)H(-j\omega)$.

Example 9.1: Consider a feedback system whose open-loop transfer function is given by

$$G(s)H(s) = \frac{K}{(T_1s+1)(T_2s+1)}$$

The plot† of $G(j\omega)H(j\omega)$ is shown in Fig. 9.8. The infinite semicircular arc of the Nyquist contour maps into origin. It is seen that the plot of $G(j\omega)H(j\omega)$ does not encircle the point $(-1+j0)$ for any positive values of K, T_1 and T_2. Therefore, the system is stable for all positive values of K, T_1 and T_2.

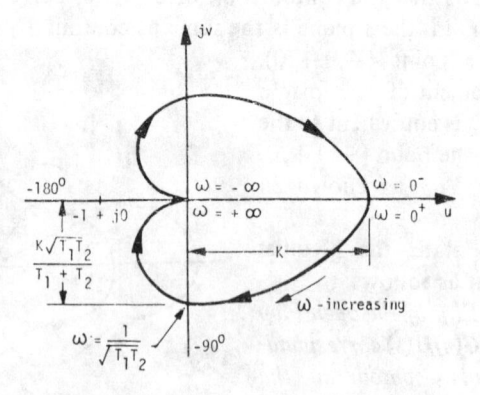

Figure 9.8 Nyquist plot of $K/(1+j\omega T_1)(1+j\omega T_2)$.

Example 9.2: Consider now, an open-loop unstable system with the transfer function

$$G(s)H(s) = \frac{s+2}{(s+1)(s-1)}$$

Let us determine whether the system is stable when the feedback path is closed.

From the transfer function of the open-loop system, it is observed that there is one open-loop pole in the right half s-plane. Therefore $P = 1$.

The $G(j\omega)H(j\omega)$-locus is sketched in Fig. 9.9 which indicates that the $(-1+j0)$ point is encircled by this locus once in the counter-clockwise direction. Therefore $N = 1 = P$. Thus $Z = 0$, i.e., there are no zeros of $1+G(s)H(s)$ in the right half s-plane and hence the closed-loop system is stable.

†Polar plot of this function is given in Table 8.1.

Stability in Frequency Domain

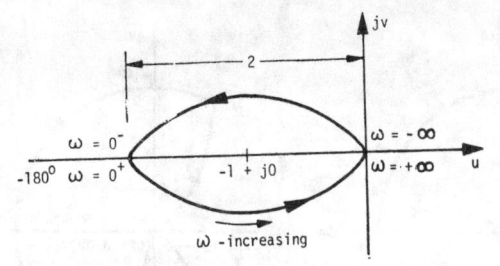

Figure 9.9 Nyquist plot of $(j\omega+2)/(j\omega+1)(j\omega-1)$.

OPEN-LOOP POLES ON THE $j\omega$-AXIS

If $G(s)H(s)$ and therefore $1+G(s)H(s)$ has any poles on the $j\omega$-axis, the Nyquist contour defined earlier (see Fig. 9.6) cannot be used as such since the s-plane contour should not pass through a singularity of $1+G(s)H(s)$. To study stability in such cases, the Nyquist contour must be modified so as to bypass any $j\omega$-axis pole. This is accomplished by indenting the Nyquist contour around the $j\omega$-axis poles along a semicircle of radius ϵ, where $\epsilon \to 0$ as shown in Fig. 9.10.

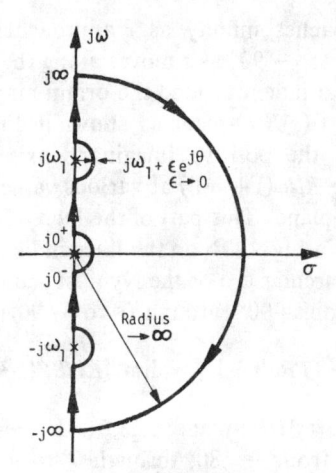

Figure 9.10 Indented Nyquist contour for $j\omega$-axis open-loop poles.

The following example illustrates the stability study in such cases.

Example 9.3: Consider a feedback system whose open-loop transfer function is given by

$$G(s)H(s) = \frac{K}{s(Ts+1)}$$

In this example, the open-loop system has a pole at the origin. The Nyquist contour must therefore be indented to bypass the origin as shown in Fig. 9.11a.

The mapping of the Nyquist contour may be carried out as follows:

1. The semicircular indent around the pole at origin represented by $s = \underset{\epsilon \to 0}{\epsilon}\, e^{j\theta}$ (θ varying from $-90°$ through $0°$ to $+90°$) maps into

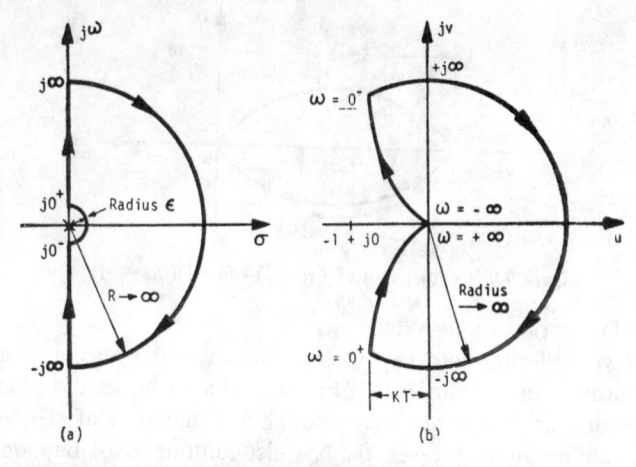

Figure 9.11 Nyquist contour and $G(s)H(s)$-locus for $G(s)H(s) = K/s(Ts+1)$.

$$\lim_{\epsilon \to 0} [K/(\epsilon e^{j\theta})(1+T\epsilon e^{j\theta})] = \lim_{\epsilon \to 0} (K/\epsilon e^{j\theta}) = \lim_{\epsilon \to 0} (K/\epsilon)e^{-j\theta}$$

The value K/ϵ approaches infinity as ϵ approaches zero and $-\theta$ varies from $+90°$ through $0°$ to $-90°$ as s moves along the semicircle. Thus the infinitesimal semicircular indent around the origin maps into a semicircular arc of infinite radius in $G(s)H(s)$-plane as shown in Fig. 9.11b.

2. The mapping of the positive imaginary axis ($\omega = 0^+$ to $+\infty$) is obtained by calculating $K/j\omega(1+j\omega T)$ at various values of ω and plotting them in the $G(s)H(s)$-plane. This part of the locus, for the problem under discussion, is sketched in Fig 9.11b on the lines indicated in Table 8.1.

3. The infinite semicircular arc of the Nyquist contour represented by $s = Re^{j\theta}$ (θ varying from $+90°$ through $0°$ to $-90°$) is mapped into
$R \to \infty$

$$\lim_{R \to \infty} [K/Re^{j\theta}(TRe^{j\theta}+1)] = \lim_{R \to \infty} [K/TR^2(e^{j2\theta})] = 0e^{-j2\theta}$$

i.e., the origin of the $G(s)H(s)$-plane. The $G(s)H(s)$-locus thus turns at the origin with zero radius from $-180°$ through $0°$ to $+180°$.

4. The mapping of the negative imaginary axis is the mirror image of that for the positive imaginary axis.

The complete Nyquist plot for $G(s)H(s) = K/s(Ts+1)$ is shown in Figure 9.11b.

In order to investigate the stability of this system, we first note that number of poles P of $G(s)H(s)$ in the right half s-plane is zero. Examination of Fig. 9.11b reveals that for all positive values of K and T, the locus of $G(j\omega)H(j\omega)$ does not encircle $(-1+j0)$ point. Therefore the system under consideration is always stable.

Example 9.4: Consider a system with an open-loop transfer function

$$G(s)H(s) = \frac{(4s+1)}{s^2(s+1)(2s+1)}$$

Stability in Frequency Domain

In this example, the open-loop system has a double pole at the origin. The Nyquist contour is therefore indented to bypass the origin as shown in Fig. 9.12a. The mapping of the Nyquist contour is obtained as follows:

1. Semicircular indent represented by $s = \lim_{\epsilon \to 0} \epsilon e^{j\theta}$ (where θ varies from $-90°$ through 0 to $90°$) is mapped into

$$\lim_{\epsilon \to 0} \left[\frac{4\epsilon e^{j\theta}+1}{\epsilon^2 e^{j2\theta}(\epsilon e^{j\theta}+1)(2\epsilon e^{j\theta}+1)} \right] = \lim_{\epsilon \to 0} \left(\frac{1}{\epsilon^2 e^{j2\theta}} \right) = \infty \, e^{-j2\theta}$$

$$= \infty \, (\angle 180° \to \angle 0° \to \angle -180°)$$

This part of the map is an infinite circle shown in Fig. 9.12b.

2. Along the $j\omega$-axis

$$G(j\omega)H(j\omega) = \frac{(1+j4\omega)}{(j\omega)^2(1+j\omega)(1+j2\omega)}$$

For various values of ω, $G(j\omega)H(j\omega)$ is calculated and plotted as shown in Fig. 9.12b.

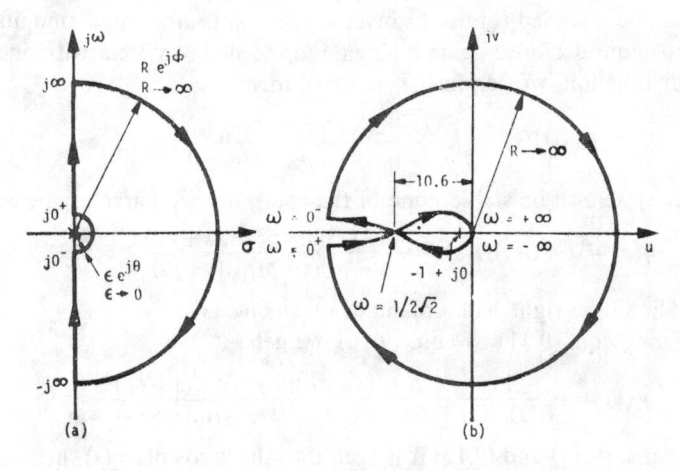

Figure 9.12 Nyquist contour and the corresponding mapping for $G(s)H(s) = (4s+1)/s^2(s+1)(2s+1)$.

The $G(j\omega)H(j\omega)$-locus intersects the real axis at a point where

$$\angle G(j\omega)H(j\omega) = -180°$$

or $-180° - \tan^{-1}\omega - \tan^{-1} 2\omega + \tan^{-1} 4\omega = -180°$

which gives

$$\omega = 1/2\sqrt{2} = 0.354 \text{ rad/sec}$$

Therefore $| G(j\omega)H(j\omega) |_{\omega=1/2\sqrt{2}} = 10.6$

Further $G(j\omega)H(j\omega) |_{\omega \to +j\infty} = 0 \angle -270°$, i.e., the map of $j\omega$-axis ends at $0 \angle -270°$ as $\omega \to +\infty$.

3. The infinite semicircle of the Nyquist contour represented by

$s = \lim_{R\to\infty} Re^{j\phi}$ (ϕ varies from $+90°$ through $0°$ to $-90°$) is mapped into

$$\lim_{R\to\infty} \frac{(1+4Re^{j\phi})}{R^2 e^{j2\phi}(1+Re^{j\phi})(1+2Re^{j\phi})} = \cdot 0 \, e^{-j3\phi}$$

$$= 0(\angle -270° \to \angle 0° \to \angle +270°)$$

The complete mapped plot corresponding to the Nyquist contour of Fig. 9.12a is shown in Fig. 9.12b. From this plot it is observed that $(-1+j0)$ point is encircled twice in the clockwise direction. Therefore $N = -2$. From the given transfer function it is seen that no pole of $G(s)H(s)$ lies in the right half s-plane, i.e., $P = 0$. Thus $-2 = 0 - Z$ or $Z = 2$. Hence two zeros of $q(s)$ lie in the right half s-plane from which we conclude that the system is unstable.

NYQUIST STABILITY CRITERION APPLIED TO INVERSE POLAR PLOTS

Occasionally, it is found more convenient* to work with the inverse function $1/G(j\omega)H(j\omega)$ rather than the direct function $G(j\omega)H(j\omega)$. In the following we shall see that the Nyquist stability criterion for direct polar plots can be extended for use to inverse polar plots after minor modification.

Let us consider once again a single-loop feedback system with open-loop transfer function (written in the standard form)

$$G(s)H(s) = K \frac{(s+z_1)(s+z_2)\dots(s+z_m)}{(s+p_1)(s+p_2)\dots(s+p_n)} ; \; m \leqslant n \quad (9.10)$$

For the system to be stable none of the roots of the characteristic equation

$$q(s) = 1 + G(s)H(s) = \frac{(s+z_1')(s+z_2')\dots(s+z_n')}{(s+p_1)(s+p_2)\dots(s+p_n)} \quad (9.11)$$

should lie in the right half s-plane or on the $j\omega$-axis.

Dividing eqn. (9.11) by eqn. (9.10), we get

$$q'(s) = \frac{1}{G(s)H(s)} + 1 = \frac{(s+z_1')(s+z_2')\dots(s+z_n')}{(s+z_1)(s+z_2)\dots(s+z_m)} \quad (9.12)$$

From eqns. (9.11) and (9.12) it is seen that the zeros of $q'(s)$ are same as zeros of $q(s)$, which are the roots of the characteristic equation. It is further noticed that poles of $q(s)$ are same as the poles of $G(s)H(s)$, while poles of $q'(s)$ are same as the poles of $1/G(s)H(s)$ or the zeros of $G(s)H(s)$.

It can be easily concluded from above that if $1/G(s)H(s)$ has P right half s-plane poles and the characteristic equation has Z right half s-plane zeros, the locus of $1/G(s)H(s)$ encircles the point $(-1+j0)$ N times in counterclockwise direction where

$$N = P - Z$$

Since stability implies absence of right half s-plane zeros of the characteristic equation, i.e., $Z = 0$, the Nyquist stability criterion for inverse polar plots immediately follows and can be stated as below:

---

*As shall be shown later, inverse polar plots are particularly useful for nonunity feedback systems.

Stability in Frequency Domain

If the Nyquist plot of $1/G(s)H(s)$, corresponding to the Nyquist contour in the s-plane, encircles the point $(-1+j0)$ as many times as are the number of right half s-plane poles of $1/G(s)H(s)$, the closed-loop system is stable. In the special case where $1/G(s)H(s)$ has no poles in the right half s-plane, the closed-loop system is stable provided the net encirclement of $(-1+j0)$ point by the Nyquist plot of $1/G(s)H(s)$ is zero.

Example 9.5: Consider a feedback system with an open-loop transfer function

$$G(s)H(s) = K/s(Ts+1)$$

The inverse polar plot of $G(s)H(s)$ corresponding to the s-plane Nyquist contour of Fig. 9.13a is obtained in steps below.

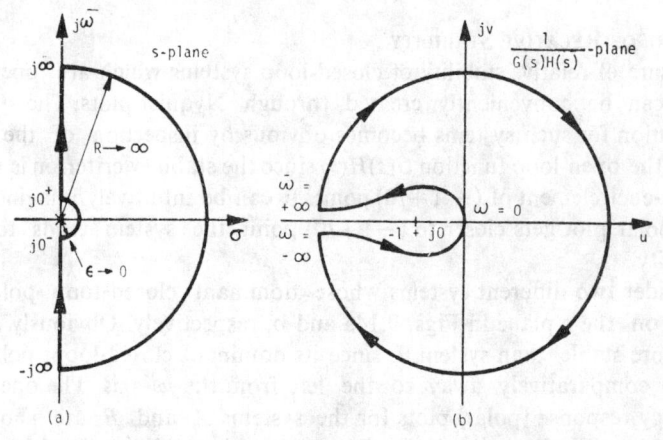

Figure 9.13 The Nyquist contour and the corresponding plot of $1/G(s)H(s) = s(sT+1)/K$.

1. The semicircular indent around the origin in the s-plane is represented by

$$s = \lim_{\epsilon \to 0} \epsilon e^{j\theta};\ \text{where}\ \theta\ \text{varies from}\ -90°\ \text{through}\ 0°\ \text{to}\ +90°.\ \text{It is}$$

mapped into $1/G(s)H(s)$-plane as

$$\lim_{\epsilon \to 0}\left[\frac{\epsilon e^{j\theta}(\epsilon e^{j\theta}T+1)}{K}\right] = \lim_{\epsilon \to 0}\frac{\epsilon}{K}e^{j\theta} = 0e^{j\theta}$$

2. Along the $j\omega$-axix $1/G(j\omega)H(j\omega) = j\omega(j\omega T+1)/K$.
3. The infinite semicircle in the s-plane represented by

$$s = \lim_{R \to \infty} Re^{j\theta};\ \theta\ \text{varies from}\ +90°\ \text{through}\ 0°\ \text{to}\ -90°$$

is mapped into the $1/G(s)H(s)$-plane as

$$\lim_{R \to \infty}\frac{e^{j\theta}(Re^{j\theta}+1)}{K} = \lim_{R \to \infty}\frac{R^2}{K}e^{j2\theta}$$

which is a circle of infinite radius with angle varying from $180°$ through $0°$ to $-180°$.

The inverse polar plot of $G(s)H(s)$ obtained from the above steps is shown in Fig. 9.13b. It is found that $(-1+j0)$ point is not encircled by $1/G(s)H(s)$-locus. Further since $1/G(s)H(s) = s(Ts+1)/K$ has no poles in the right half s-plane, the system is stable.

9.4 Assessment of Relative Stability Using Nyquist Criterion

In Chapter 7, the need to determine the relative stability of a system, in addition to its absolute stability was clearly established. In this section the use of Nyquist plot shall be extended to extract information regarding relative stability, thus eliminating the need to determine the location of the roots of characteristic equation. The Nyquist plot thus becomes a powerful designer's tool which provides information regarding both stability and relative stability of a feedback system.

MEASURE OF RELATIVE STABILITY

Measure of relative stability of closed-loop systems which are open-loop stable can be conveniently created through Nyquist plots. The stability information for such systems becomes obvious by inspection of the polar plot of the open-loop function $G(s)H(s)$ since the stability criterion is merely the non-encirclement of $(-1+j0)$ point. It can be intuitively imagined that as the polar plot gets closer to $(-1+j0)$ point, the system tends towards instability.

Consider two different systems whose dominant closed-loop poles are shown on the s-plane in Figs. 9.14a and b, respectively. Obviously system A is 'more stable' than system B, since its dominant closed-loop poles are located comparatively away to the left from the $j\omega$-axis. The open-loop frequency response (polar) plots for the systems A and B are shown in Figs. 9.14c and d, respectively. The comparison of the closed-loop pole

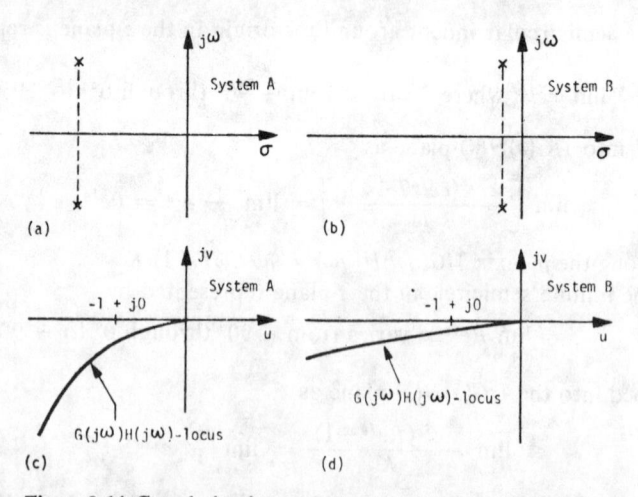

Figure 9.14 Correlation between the closed-loop s-plane root locations and open-loop frequency response curves.

Stability in Frequency Domain

locations of these two systems with their corresponding polar plots reveals that as a polar plot moves closer to $(-1+j0)$ point, the system closed-loop poles move closer to the $j\omega$-axis and hence the system becomes relatively less stable and vice versa.

Figure 9.15 shows a typical $G(j\omega)H(j\omega)$-locus which crosses the negative real axis at a frequency $\omega = \omega_2$ with an intercept of a. Let a unit circle centred at origin [obviously it passes through the point $(-1+j0)$] intersect the $G(j\omega)H(j\omega)$-locus at a frequency $\omega = \omega_1$ and let the phasor $G(j\omega_1) H(j\omega_1)$ make an angle of ϕ with the negative real axis measured positively in counter-clockwise direction. It is immediately observed that as $G(j\omega)H(j\omega)$-locus approaches $(-1+j0)$ point, the relative stability reduces. Simultaneously, the value of a approaches unity and that of ϕ tends to zero. The relative stability could thus be measured in terms of the intercept a or the angle ϕ. These concepts are used to define *gain margin* and *phase margin* as practical measures of relative stability. It must be stressed here that gain margin and phase margin concepts are applicable to open-loop stable systems only. Vast majority of practical systems of course lie in this category.

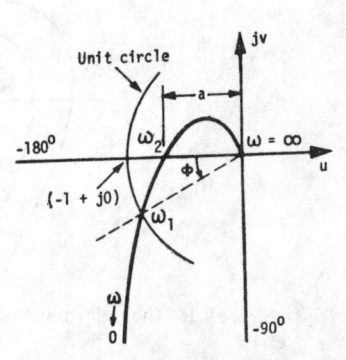

Figure 9.15 A typical $G(j\omega)H(j\omega)$-locus.

Gain Margin and Phase Margin

Consider a feedback system with the following open-loop transfer function:

$$G(j\omega)H(j\omega) = \frac{K}{j\omega(j\omega T_1+1)(j\omega T_2+1)} \quad (9.13)$$

Let us investigate the stability of this system for various values of K. The polar plots of $G(j\omega)H(j\omega)$ for various values of K are shown in Fig. 9.16. The point of intersection of the polar plot with the negative real axis can be determined by setting the imaginary part of $G(j\omega)H(j\omega)$ equal to zero. From eqn. (9.13)

$$G(j\omega)H(j\omega) = u+jv = \frac{-K(T_1+T_2) - jK(1/\omega)(1 - \omega^2 T_1 T_2)}{1+\omega^2(T_1^2+T_2^2)+\omega^4 T_1^2 T_2^2} \quad (9.14)$$

Let ω_2 be the frequency at the point of intersection. Then from eqn. (9.14) we have

$$v = \frac{-K(1/\omega_2)(1 - \omega_2^2 T_1 T_2)}{1+\omega_2^2(T_1^2+T_2^2)+\omega_2^4 T_1^2 T_2^2} = 0$$

which gives

$$\omega_2 = 1/\sqrt{(T_1 T_2)} \quad (9.15)$$

From eqn. (9.14) the magnitude of the real part of $G(j\omega)H(j\omega)$ at the frequency ω_2 is given by

$$u = \frac{-K(T_1+T_2)}{1+\omega_2^2(T_1^2+T_2^2)+\omega_2^4 T_1^2 T_2^2}$$

$$= -\frac{KT_1T_2}{(T_1+T_2)} \qquad (9.16)$$

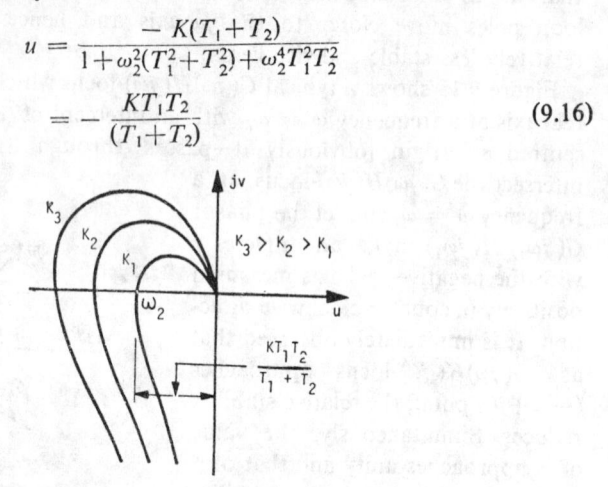

Figure 9.16 The polar plots of $G(j\omega)H(j\omega)$ for various values of gain K.

From Fig. 9.16 it is seen that for the system to be stable

$$K\frac{T_1T_2}{T_1+T_2} < 1$$

or

$$K < (T_1+T_2)/T_1T_2$$

When the gain K is less than $(T_1+T_2)/T_1T_2$, the $(-1+j0)$ point is not encircled by $G(j\omega)H(j\omega)$-plot and the system is stable. When K is equal to $(T_1+T_2)/T_1T_2$, the plot passes through the point $(-1+j0)$ which indicates that the system has roots on the $j\omega$-axis. When K is further increased so as to be greater than $(T_1+T_2)/T_1T_2$, the $G(j\omega)H(j\omega)$-plot encircles the $(-1+j0)$ point and hence the system is unstable. The margin between the actual gain and critical gain causing sustained oscillations, is a measure of relative stability and is called the gain margin which is defined below.

Gain margin. It is the factor by which the system gain can be increased to drive it to the verge of instability.

In Fig. 9.15, it is seen that at $\omega = \omega_2$, the phase angle $\angle G(j\omega)H(j\omega)$ is $180°$ and $|G(j\omega)H(j\omega)|$ is a. If the gain of the system is increased by a factor $1/a$, then $|G(j\omega)H(j\omega)|_{\omega=\omega_2}$ becomes $a(1/a) = 1$ and hence the $G(j\omega)H(j\omega)$-plot will pass through $(-1+j0)$ point, driving the system to the verge of instability. Therefore, the gain margin (GM) may be defined as the reciprocal of the gain at the frequency at which the phase angle becomes $180°$. The frequency at which the phase angle is $180°$ is called the *phase cross-over frequency*.

With reference to Fig. 9.15, we have

$$\text{GM} = 1/a$$

where $a = |G(j\omega)H(j\omega)|_{\omega=\omega_2}$.

In decibels the increase in gain for $G(j\omega)H(j\omega)$-plot to pass through $(-1+j0)$ is given by

$$GM = -20 \log a \text{ db}$$

Since a is less than 1 for stable systems, $\log a$ is negative and hence GM is positive.

In the example considered earlier in this section if $T_1 = 1$, $T_2 = 0.5$, then for $K = 0.75$, the gain margin is given by

$$GM = \left[\frac{KT_1T_2}{T_1+T_2}\right]^{-1} = 4$$

In decibels, the gain margin is given by

$$GM = 20 \log 4 = 12 \text{ db}$$

This value of gain margin indicates that the system gain may be increased by a factor of 4 before the stability limit is reached.

Phase margin. The frequency at which $|G(j\omega)H(j\omega)| = 1$ is called the *gain cross-over frequency*. It is given by the intersection of the $G(j\omega)H(j\omega)$-plot and a unit circle centred at the origin as shown in Fig. 9.15. At this frequency, the phase angle $\angle G(j\omega_1)H(j\omega_1)$ is equal to $(-180°+\phi)$. If an additional phase-lag equal to ϕ is introduced at the gain cross-over frequency, the phase angle $\angle G(j\omega_1)H(j\omega_1)$ will become $-180°$, while the magnitude remains unity. The $G(j\omega)H(j\omega)$-plot will then pass through $(-1+j0)$ point, driving the system to the verge of instability. This additional phase-lag ϕ is known as the phase margin (PM).

The phase margin is thus defined as the amount of additional phase-lag at the gain cross-over frequency required to bring the system to the verge of instability. From Fig. 9.15 it is seen that the phase margin is measured positively in counter-clockwise direction from the negative real axis. The *phase margin is always positive for stable feedback systems.*

The value of phase margin for any system can be computed from

$$\text{Phase margin } \phi = \angle G(j\omega)H(j\omega)|_{\omega=\omega_1} + 180° \qquad (9.17)$$

where the angle at ω_1, the gain cross frequency, is measured negatively.

Gain margin (GM) and phase margin (PM) are frequently used for frequency response specifications by designers. It is important to note once again that these measures of stability are valid for open-loop stable systems only. A large gain margin or a large phase margin indicates a very stable feedback system but usually a very sluggish one. A GM close to unity or a PM close to zero corresponds to a highly oscillatory system. Usually a GM of about 6 db or a PM of 30-35° results in a reasonably good degree of relative stability. In most practical systems a good gain margin automatically guarantees a good phase margin and vice versa. However, the cases where the specification on one does not necessarily satisfy the other, also exist as shown in Figs. 9.17 and 9.18.

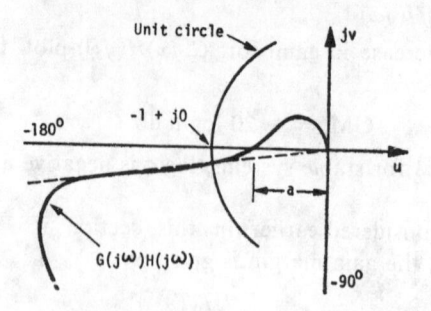

Figure 9.17 Polar plot of a system with good GM and poor PM.

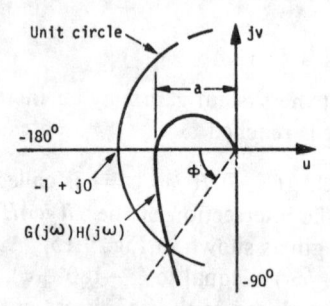

Figure 9.18 Polar plot of a system with good PM and poor GM.

Figure 9.19 Polar plot of a second-order system with GM always equal to infinity while PM reduces continuously with increasing gain.

In a second-order system with $G(j\omega)H(j\omega) = K/j\omega(j\omega T+1)$ whose polar plot is shown in Fig. 9.19, GM always remains fixed at infinite value as the plot always reaches the real axis at the origin, while the PM reduces continuously with increasing system gain. In this case, as shall be seen below, PM is the correct measure of relative stability.

CORRELATION BETWEEN PHASE MARGIN AND DAMPING FACTOR

Consider a unity feedback second-order system with an open-loop transfer function

$$G(s)H(s) = \frac{K}{s(\tau s+1)} = \frac{\omega_n^2}{s(s+2\zeta\omega_n)}$$

where $\omega_n = \sqrt{(K/\tau)}$ and $2\zeta\omega_n = 1/\tau$.

Replacing s by $j\omega$ for obtaining the polar plot, we have

$$G(j\omega)H(j\omega) = \frac{\omega_n^2}{j\omega(j\omega+2\zeta\omega_n)} \qquad (9.18)$$

At the gain cross-over frequency $\omega = \omega_1$, the magnitude $|G(j\omega)H(j\omega)| = 1$. Therefore from eqn. (9.18), we have

$$\frac{\omega_n^2}{\omega_1\sqrt{(\omega_1^2+4\zeta^2\omega_n^2)}} = 1$$

or
$$(\omega_1^2)^2 + 4\zeta^2\omega_n^2(\omega_1)^2 - \omega_n^4 = 0$$

which yields
$$(\omega_1/\omega_n)^2 = \sqrt{(4\zeta^4+1)} - 2\zeta^2$$

The phase margin of this system is given by

$$\phi = -90° - \tan^{-1}\left(\frac{\omega_1}{2\zeta\omega_n}\right) + 180$$

$$= 90° - \tan^{-1}\left[\frac{1}{2\zeta}\{(4\zeta^4+1)^{1/2} - 2\zeta^2\}^{1/2}\right]$$

$$= \tan^{-1}\left[2\zeta\left\{\frac{1}{(4\zeta^4+1)^{1/2} - 2\zeta^2}\right\}^{1/2}\right] \qquad (9.19)$$

Equation (9.19) gives a relationship between ζ and ϕ for an underdamped second-order system. The ζ-ϕ relationship is plotted in Fig. 9.20. In the range $\zeta \leqslant 0.7$, a reasonably good linear approximation for ζ-ϕ relationship is given below and is shown by the dotted line in Fig. 9.20.

$$\zeta \approx 0.01\phi \qquad (9.20)$$

where ϕ is in degrees.

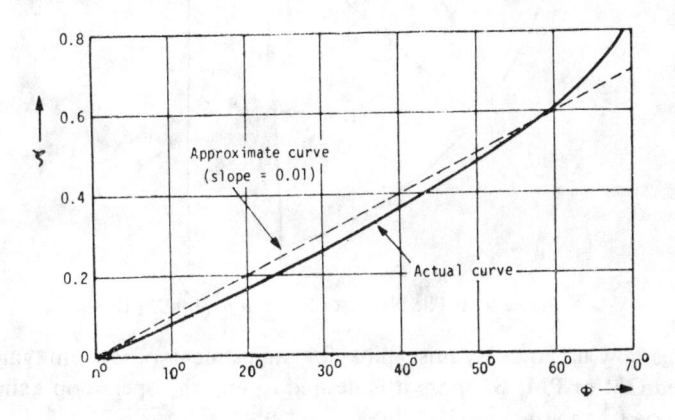

Figure 9.20 Plot of ζ versus ϕ for a second-order system.

Equation (9.19) and its approximation of eqn. (9.20) hold good for second-order systems, but can be used for higher-order systems as well, provided the transient response of the system is primarily contributed by a pair of dominant underdamped roots.

COMPUTATION OF GAIN MARGIN AND PHASE MARGIN

GM and PM may be computed by the use of various plots—direct polar plot, inverse polar plot, Bode plot, log-magnitude versus phase angle plot. In relatively simple cases GM and PM may be computed directly.

Example 9.6: Consider a unity feedback system having an open-loop transfer function

$$G(j\omega) = \frac{K}{j\omega(j0.2\omega+1)(j0.05\omega+1)}$$

For $K = 1$

$$|G(j\omega)| = \frac{1}{\omega\sqrt{[1+(0.2\omega)^2]}\sqrt{[1+(0.05\omega)^2]}}$$

and $\quad \angle G(j\omega) = -90° - \tan^{-1} 0.2\omega - \tan^{-1} 0.05\omega$

The Nyquist plot of $G(j\omega)$, for $K = 1$, is shown in Fig. 9.21. From this plot it is found that

$$GM = 20 \log (1/0.04) = 20 \log 25 = 28 \text{ db}$$
$$PM = 76°$$

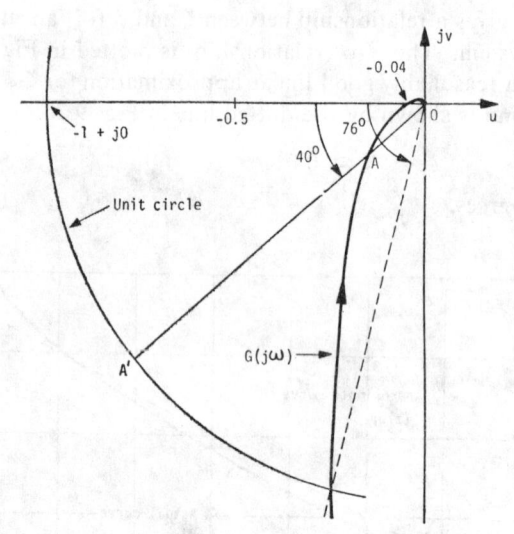

Figure 9.21 The Nyquist plot of $G(j\omega)$ for $K=1$.

Let us now use the Nyquist plot for adjustment of system gain for specified GM or PM. Suppose it is desired to find the open-loop gain for
(i) a GM of 20 db
(ii) a PM of 40°

(i) For a GM of 20 db, the Nyquist plot should intersect the real axis at $-a$ where

$$20 \log (1/a) = 20$$
or $\quad a = 0.1$

This is achieved if the system gain is increased by a factor of $0.1/0.04 = 2.5$. Thus $K = 2.5$.

(ii) A PM of 40° is obtained if the system gain is increased such that point A is shifted to location A' in Fig. 9.21. This is achieved if the system gain is increased by a factor of $0A'/0A = 1/0.191 = 5.24$. Thus $K = 5.24$.

Stability in Frequency Domain

The above example can also be solved by computation without the need of drawing Nyquist plot (this is easily possible for first- and second-order systems and for those third-order systems which have one open-loop pole at origin and no open-loop zeros).

(i) As calculated already, for a GM of 20 db,
$$a = 0.1$$
The Nyquist plot intersects the real axis at a point where
$$G(j\omega) = \frac{K}{j\omega(j0.2\omega+1)(j0.05\omega+1)} = \frac{K}{-0.25\omega^2+j\omega(1-0.01\omega^2)} \quad (9.21)$$
is real. Setting the imaginary part of eqn. (9.21) equal to zero, we have
$$\omega = \omega_2 = 10 \text{ rad/sec}$$
Now
$$|G(j\omega)|_{\omega=\omega_2} = \frac{K}{0.25(10)^2} = a = 0.1$$
which gives
$$K = 2.5$$

(ii) Let $\omega = \omega_1$ be the gain cross-over frequency. Then for a PM of 40°
$$-90° - \tan^{-1} 0.2\omega_1 - \tan^{-1} 0.05\omega_1 + 180° = 40°$$
or
$$\tan^{-1} 0.2\omega_1 + \tan^{-1} 0.05\omega_1 = 50°$$
$$\frac{0.25\omega_1}{1-0.01\omega_1^2} = \tan 50° = 1.2$$
or
$$0.012\omega_1^2 + 0.25\omega_1 - 1.2 = 0$$
Solving for positive value of ω_1, we get
$$\omega_1 = 4 \text{ rad/sec}$$
Hence
$$|G(j\omega)|_{\omega=\omega_1} = \frac{K}{\omega_1\sqrt{[1+(0.2\omega_1)^2]}\sqrt{[1+(0.05\omega_1)^2]}} = 1$$
which gives
$$K = 5.2$$

Example 9.7: Let us determine the gain margin and phase margin of a unity feedback system having an open-loop transfer function
$$G(j\omega) = \frac{10}{j\omega(j0.1\omega+1)(j0.05\omega+1)}$$
by use of Bode plot.

The Bode plot of $G(j\omega)$ is shown in Fig. 9.22. From this figure, it is seen that GM = 12 db and PM 33°.

Let us now use the Bode plot for adjustment of the system gain for a specified GM or PM. Suppose it is desired to find the open-loop gain for (i) a GM of 20 db; and (ii) a PM of 24°.

(i) A GM of 20 db will be obtained if the log-magnitude plot in Fig. 9.22 is shifted downwards by $(20-12) = 8$ db. The system gain must therefore be reduced by -8 db or by a factor of 2.5. The corresponding open-loop gain is $10/2.5 = 4$.

(ii) From Fig. 9.22, it is observed that if the gain cross-over frequency is changed to 9.3 rad/sec, a PM of 24° is obtained. To change the gain cross-over frequency to 9.3 rad/sec, the log-magnitude curve should be raised upwards by 3.5 db or system gain should be raised upwards by 3.5 db or system gain should be increased by a factor of 1.5. Hence the open-loop gain for a PM of 24° is $10 \times 1.5 = 15$.

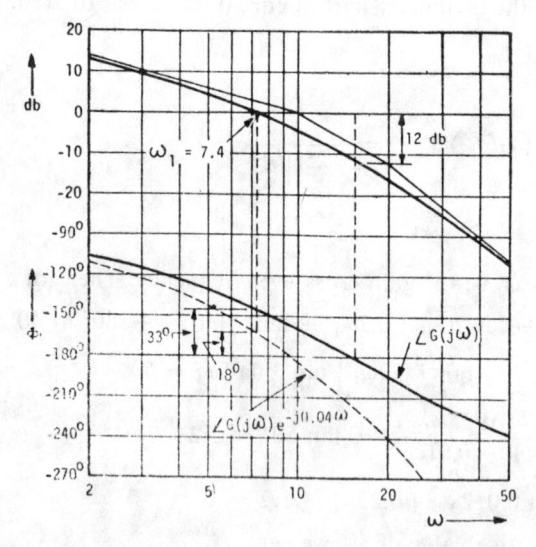

Figure 9.22 Bode plot of $G(j\omega) = 10/j\omega(j0.1\omega+1)(j0.05\omega+1)$.

Systems with Transportation Lag

As was discussed earlier in Chapters 7 and 8, certain elements in control systems are characterized by a dead time or transportation lag. For the block diagram shown in Fig. 9.23, the open-loop transfer function of the system is

$$G(s) = e^{-sT}G_1(s)$$

As $G(s)$ is a transcendental function, analytical approach to a system described by such an equation is difficult. One of the methods of handling a transcendental transfer function is to convert it into a rational function by

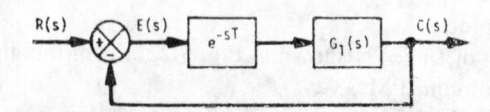

Figure 9.23 Simple system with transportation lag.

approximating the exponential term by a polynomial of s so that the usual techniques of analysis can be applied. The approximation suggested in Chapter 7 is

$$e^{-sT} \approx 1 - sT$$

This approximation works fairly well as long as the dead time T is small in comparison to the system time constant and in addition the input time function to the dead time element is smooth and continuous.

As was pointed out in Chapter 7, the control systems with dead time could be analyzed with root locus technique without the need to make any approximation but the method is quite complex and time consuming. On the other hand the frequency domain graphical methods provide a simple yet exact approach to handle this problem since the function $e^{-sT} \mid_{s=j\omega} = e^{-j\omega T}$ is readily interpreted in terms of either the Nyquist or Bode plot.

Assume that the transfer function $G_1(s)$ in Fig. 9.23 is given by

$$G_1(s) = 1/s(s+1) \qquad (9.22)$$

Then the open-loop transfer function of the system becomes

$$G(s) = e^{-sT}/s(s+1) \qquad (9.23)$$

The polar plot of e^{-sT} is shown in Fig. 8.19a, using which the Nyquist plot of $G(j\omega)$ for various values of T is drawn in Fig. 9.24. From this figure, it is observed that the effect of $e^{-j\omega T}$ term in eqn. (9.23) is simply to rotate each point of the $G_1(j\omega)$ plot by an angle of ωT rad in the clockwise direction. Relative stability of the system reduces as T increases. Sufficiently large values of T may drive the system to instability. The effect of e^{-sT} on stability can be determined by the Nyquist stability criterion as follows.

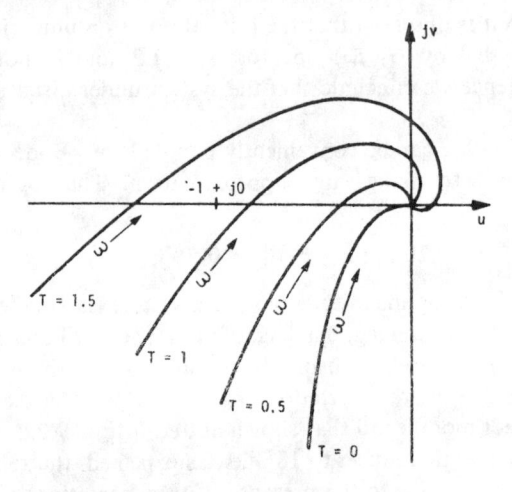

Figure 9.24 Nyquist plot of $G(s) = e^{-sT}/s(s+1)$.

The characteristic equation of the system shown in Fig. 9.23 is given by

$$1+G_1(s)e^{-sT} = 0$$

or
$$G_1(s) = \frac{1}{s(s+1)} = -e^{+sT} \qquad (9.24)$$

If eqn. (9.24) is satisfied at a particular frequency, the system will exhibit sustained oscillations. The corresponding condition for systems without transportation lag is

$$G_1(j\omega) = -1 \qquad (9.25)$$

Comparison of eqns. (9.24) and (9.25) reveals that the effect of e^{-sT} is simply to shift the critical stability point from $(-1+j0)$ to $(-e^{+j\omega T})$, i.e., the critical point now becomes a critical locus.

Example 9.8: Let us determine with the help of the Nyquist stability criterion, the maximum value of T for the stability of the closed-loop system of Fig. 9.23. From eqn. (9.24), the following condition is satisfied when the system is limitedly stable (i.e., it has sustained oscillations):

$$G_1(j\omega) = \frac{1}{j\omega(j\omega+1)} = -e^{+j\omega T}$$

The polar plots of $G_1(j\omega)$ and $-e^{+j\omega T}$ are shown in Fig. 9.25. These two plots intersect at the point A corresponding to $\omega = 0.75$ rad/sec on $G_1(j\omega)$-plot and $\omega T = 52°$ $(\pi/180) = 0.9$ rad/sec on $-e^{+j\omega T}$ plot. For the point A to pertain to the same frequency on both $G_1(j\omega)$ and $-e^{+j\omega T}$ plots, we have

$$0.75T = 0.9$$

or
$$T = 1.2 \text{ sec}$$

From Fig. 9.25 it is observed that the critical point on unit circle ($-e^{j\omega T}$-locus) is enclosed by $G_1(j\omega)$-plot for $T > 1.2$ and is not enclosed for $T < 1.2$ and hence we conclude that the system under discussion is stable if $T < 1.2$ sec.

Transportation lag can be conveniently handled on Bode plot as well without the need to make any approximation. The log-magnitude of transportation lag is

$$20 \log |e^{-j\omega T}| = 0$$

Thus the open-loop log-magnitude plot of a system is unaffected by the presence of transportation lag. The lag, of course, contributes a phase angle of $-(\omega T \times 180°)/\pi$, thereby causing the modification of the phase plot.

If in Example 9.8, a transportation lag e^{-sT} ($T = 0.04$ sec) is brought in, the phase plot modifies to that shown dotted in Fig. 9.22. For the given gain, the phase margin reduces to 18°, i.e., as expected, the relative stability of the system reduces due to the presence of transportation lag.

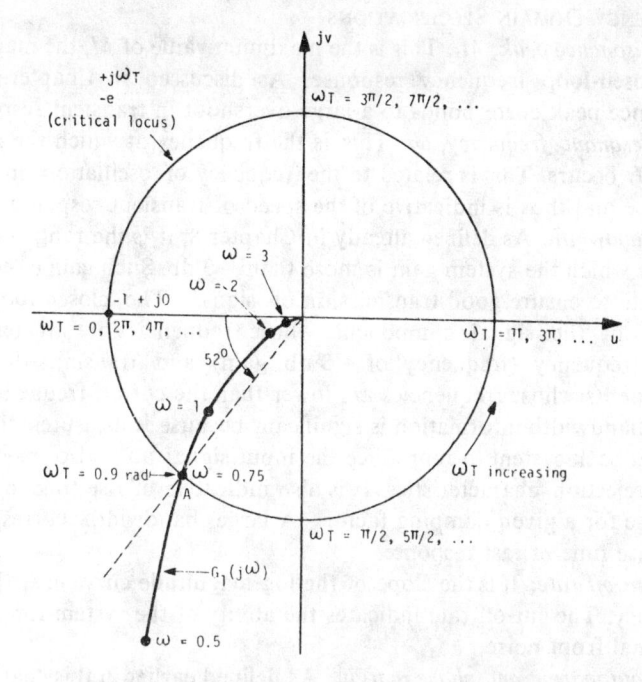

Figure 9.25 Critical locus and $G_1(j\omega)$-plot for Example 9.8.

The value of T for the system to be on verge of instability is obtained by setting the phase margin equal to zero, i.e.,

$$\angle G_1(j\omega) \bigg|_{\omega=\omega_1} - \frac{\omega_1 T \times 180°}{\pi} = -180°$$

where $G_1(j\omega)$ is the system without transportation lag and ω_1 is the gain crossover frequency.

For Example 9.8, we have from Fig. 9.22, $\omega_1 = 7.4$ rad/sec. Thus

$$-153° - \frac{7.4 T \times 180°}{\pi} = -180°$$

or
$$T = 0.063 \text{ sec}$$

9.5 Closed-loop Frequency Response

The study of closed-loop frequency response is very useful as it enables us to use the second-order correlations (discussed in Chapter 8) between frequency response and transient response to predict approximately the time response of feedback systems. With the help of these correlations, the time response specifications are first converted into a set of specifications in frequency domain. After design and compensation in frequency domain, the frequency response is translated back to give an approximate time response. Usually the specifications in frequency domain are given in the following terms.

Frequency Domain Specifications

1. *Resonance peak*, M_r. This is the maximum value of M, the magnitude of the closed-loop frequency response. As discussed in Chapter 8, a large resonance peak corresponds to a large overshoot in transient response.

2. *Resonant frequency*, ω_r. This is the frequency at which the resonance peak M_r occurs. This is related to the frequency of oscillation in the step response and thus is indicative of the speed of transient response.

3. *Bandwidth*. As defined already in Chapter 8, it is the range of frequencies for which the system gain is more than -3 db. Such gain is considered adequate to ensure good transmission of signal. The closed-loop system filters out the signal components whose frequencies are greater than the cut-off frequency (frequency of -3 db gain) and transmits those signal components whose frequencies are lower than the cut-off frequency.

The bandwidth information is significant because it measures the ability of a feedback system to reproduce the input signal and also measures its noise rejection characteristics. It is also indicative of rise time in transient response for a given damping factor. A large bandwidth corresponds to small rise time or fast response.

4. *Cut-off rate*. It is the slope of the log-magnitude curve near the cut-off frequency. The cut-off rate indicates the ability of the system to distinguish the signal from noise.

5. *Gain margin and phase margin*. As defined earlier in this chapter, these are the measures of relative stability and are related to the closeness of the closed-loop poles to the $j\omega$-axis. For a second-order system, the exact correlation between phase margin and ζ is given by eqn. (9.19).

From the frequency domain specifications, we find that the maximum value of M and the frequency at which it occurs are important figures of merit of a system. The frequency domain compensation methods (discussed in Chapter 10) are based upon the knowledge of these two factors. These factors can be evaluated from the closed-loop frequency response. However, this procedure, as is evident, is time consuming.

Fortunately, graphical techniques are available which help in determining the values of M_r and ω_r directly from the open-loop frequency response obviating the need of determining complete closed-loop frequency response. These graphical techniques require that constant-M contours be drawn on the complex plane. When these contours are used in conjunction with the open-loop frequency response, the values of M_r and ω_r are easily obtained.

In the following paragraphs, we shall show that the constant-M and constant-N ($N = \tan \alpha$, α being the phase angle of the frequency response) contours are circles.

Constant-M Circles

Consider any point $G(j\omega) = x+jy$, on the polar plot of $G(j\omega)$. The closed-loop frequency response is

$$T(j\omega) = \frac{C(j\omega)}{R(j\omega)} = \frac{G(j\omega)}{1+G(j\omega)} = \frac{x+jy}{1+x+jy} = Me^{j\alpha} \qquad (9.26)$$

From eqn. (9.26), the magnitude M is given by

$$M = \frac{|x+jy|}{|1+x+jy|} = \left[\frac{x^2+y^2}{(1+x)^2+y^2}\right]^{1/2}$$

or
$$M^2 = \frac{x^2+y^2}{(1+x)^2+y^2}$$

Rearranging this equation, we get

$$y^2 + \left[x + \frac{M^2}{M^2-1}\right]^2 = \frac{M^2}{(M^2-1)^2} \qquad (9.27)$$

Equation (9.27) is the equation of a circle with centre at

$$x_0 = -\frac{M^2}{M^2-1}; \quad y_0 = 0 \qquad (9.28)$$

and with radius

$$r_0 = \frac{M}{M^2-1} \qquad (9.29)$$

Using eqns. (9.28) and (9.29) constant-M circles for various values of M can be drawn. Consider Fig. 9.26 in which the $G(j\omega)$-plot and three constant-M circles are plotted. It is observed that M_2-circle is tangent to the $G(j\omega)$-plot. Therefore the maximum value of M is M_2, i.e., $M_r = M_2$.

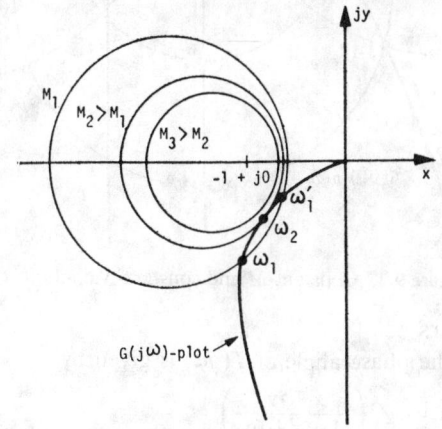

Figure 9.26 Constant-M circles and $G(j\omega)$-plot.

A family of M-circles for different values of M is constructed in Fig. 9.27. From eqns. (9.28) and (9.29) the following conclusions are easily drawn regarding M-circles.

(i) *For* $M > 1$. As M increases the radii of M-circles reduce monotonically and the centres located on negative real axis progressively shift towards $(-1+j0)$ till, for $M = \infty$, $r_0 = 0$ and $x_0 = -1$, i.e., the $M = \infty$ circle has a radius zero and is centred at $(-1+j0)$.

(ii) *For* $M = 1$. It follows that $r_0 = \infty$, $x_0 = -\infty$. Thus $M = 1$ circle is of infinite radius with centre at infinity, i.e., it is a straight line parallel to y-axis. Its intercept on x-axis is given by eqn. (9.27) as

$$x\Big|_{y=0} = \left[-\frac{M^2}{M^2-1} + \frac{M}{M^2-1}\right]\Big|_{M=1} = -\frac{1}{2}$$

(iii) *For* $M < 1$. As M decreases, the radii of M-circles reduce monotonically and centres located on positive real axis shift towards origin till, for $M = 0$, $r_0 = 0$ and $x_0 = 0$, i.e., the $M = 0$ circle is of radius zero centred at origin.

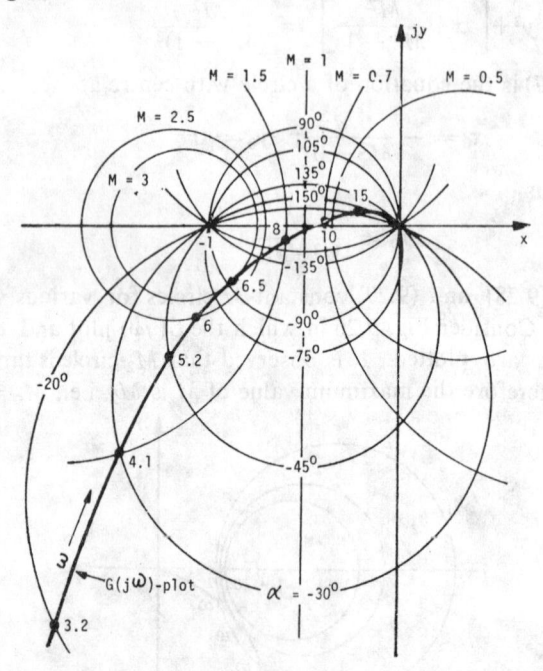

Figure 9.27 Constant-M and constant-N circles.

CONSTANT-N CIRCLES

From eqn. (9.26), the phase angle of $T(j\omega)$ is given by

$$\angle T(j\omega) = \alpha = \angle\left(\frac{x+jy}{1+x+jy}\right)$$
$$= \tan^{-1}(y/x) - \tan^{-1}[y/(1+x)] = \tan^{-1}[y/(x^2+x+y^2)]$$

Therefore

$$\tan \alpha = \frac{y}{x^2+x+y^2} = N \tag{9.30}$$

For a constant value of α, $N = \tan \alpha$ is also constant.

Rearranging eqn. (9.30), we get

$$\left(x+\frac{1}{2}\right)^2 + \left(y-\frac{1}{2N}\right)^2 = \frac{1}{2}\frac{N^2+1}{N^2} \tag{9.31}$$

This is the equation of a circle with centre at

$$x_0 = -1/2; \; y_0 = 1/2N \tag{9.32}$$

Stability in Frequency Domain

and with radius

$$r_0 = \frac{1}{2N}(N^2+1)^{1/2} \qquad (9.33)$$

For different values of α, a family of N-circles is also constructed in Fig. 9.27. Since eqn. (9.31) is satisfied for $x = 0, y = 0$ and for $x = -1, y = 0$, all the constant-N circles pass through the origin and $(-1+j0)$ point regardless of the value of N.

Example 9.9: Let us compute the closed-loop frequency response of a unity feedback system with open-loop transfer function

$$G(j\omega) = \frac{10}{j\omega(1+0.2j\omega)(1+0.05j\omega)}$$

by use of constant-M and constant-N circles.

Figure 9.27 shows $G(j\omega)$-plot superimposed on constant-M and -N circles. At various values of frequencies the magnitude M and phase angle α are read off. With the data thus obtained, the closed-loop frequency response curves are plotted in Fig. 9.28.

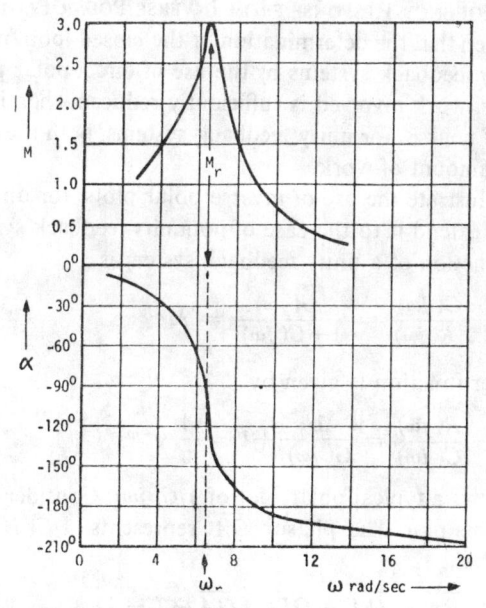

Figure 9.28 Closed-loop frequency response.

NONUNITY FEEDBACK SYSTEMS

For a nonunity feedback system, the closed-loop transfer function is given by

$$\frac{C(j\omega)}{R(j\omega)} = T(j\omega) = \frac{G(j\omega)}{1+G(j\omega)H(j\omega)} \qquad (9.34)$$

where $G(j\omega)$ is the forward path transfer function and $H(j\omega)$ is the feedback path transfer function.

Equation (9.34) may be written in the form

$$T(j\omega) = \frac{1}{H(j\omega)}\left[\frac{G(j\omega)H(j\omega)}{1+G(j\omega)H(j\omega)}\right]$$

$$= \frac{1}{H(j\omega)}\left[\frac{G_0(j\omega)}{1+G_0(j\omega)}\right] = \frac{1}{H(j\omega)}T_0(j\omega)$$

where $\quad G_0(j\omega) = G(j\omega)H(j\omega)$

and $\quad T_0(j\omega) = \dfrac{G_0(j\omega)}{1+G_0(j\omega)}$ \hfill (9.35)

Equation (9.35) is of the standard form of unity feedback systems. Constant-M and constant-N circles can be used in conjunction with $G_0(j\omega)$ to obtain $T_0(j\omega)$. $T(j\omega)$ can then be obtained by multiplying $T_0(j\omega)$ by $1/H(j\omega)$. This multiplication can be carried out easily by drawing Bode plots for $T_0(j\omega)$ and $H(\cdot j\omega)$ and then graphically subtracting the log-magnitude and phase angle of $H(j\omega)$ from that of $T_0(j\omega)$. The resulting Bode plots give the closed-loop frequency response $T(j\omega)$.

CLOSED-LOOP FREQUENCY RESPONSE FROM INVERSE POLAR PLOTS

We have just seen that the determination of the closed-loop frequency response of nonunity feedback systems by the use of direct polar plots is quite tedious. However, work involved is sufficiently reduced when inverse polar plots are used. Of course, for unity feedback systems, both the plots require almost the same amount of work.

We shall first illustrate the use of inverse polar plots for unity feedback systems and then extend it to the case of nonunity feedback systems.

The transfer function of a unity feedback system is

$$\frac{C(j\omega)}{R(j\omega)} = \frac{G(j\omega)}{1+G(j\omega)} = Me^{j\alpha}$$

Its inverse transfer function is given by

$$\frac{R(j\omega)}{C(j\omega)} = \frac{1}{G(j\omega)}+1 = \frac{1}{M}e^{-j\alpha} \quad (9.36)$$

Figure 9.29 shows a typical polar plot of $1/G(j\omega)$. Consider a point A of frequency ω_1 on this plot. The phasor $\overline{OA}$ represents $1/G(j\omega_1)$. We find from this figure that

$$\overline{PA} = \overline{OA} - \overline{OP} = \overline{OA} - (-1)$$

$$= \frac{1}{G(j\omega_1)}+1 = \frac{1}{M_1}e^{-j\alpha_1}$$

It follows from the above result that the directed line segment PA drawn from $(-1+j0)$ represents $e^{-j\alpha}/M$. Thus the contours of constant-M are circles with centre at $(-1+j0)$ point and with radii equal to $1/M$. The contours of constant values of $-\alpha$ are radial lines passing through $(-1+j0)$ point; $1/M$ circles and $-\alpha$ lines are shown in Fig. 9.30.

Stability in Frequency Domain

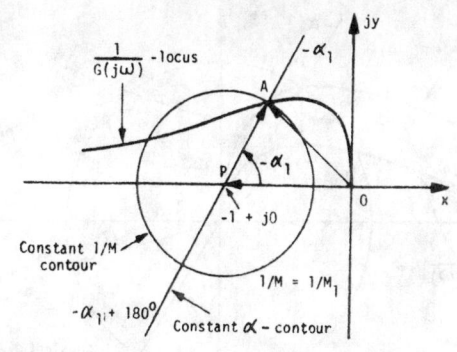

Figure 9.29 Determination of closed-loop frequency response.

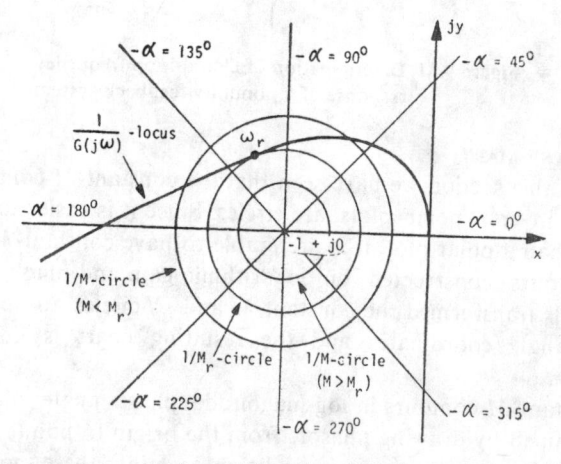

Figure 9.30 Determination of M_r and ω_r from inverse polar plots.

Figure 9.30 also shows a $1/G(j\omega)$-locus and $1/M$-circle which is tangent to it. If M is increased, then $1/M$-circle does not intersect the $1/G(j\omega)$-locus, while if M is decreased the $1/M$-circle will intersect the locus at two points. It therefore follows that the M of the tangent circle corresponds to the resonant peak M_r and the tangent frequency corresponds to the resonant frequency ω_r.

For nonunity feedback systems, the inverse transfer function is given by

$$\frac{R(j\omega)}{C(j\omega)} = \frac{1+G(j\omega)H(j\omega)}{G(j\omega)} = \frac{1}{G(j\omega)} + H(j\omega) = \frac{1}{M}e^{-j\alpha}$$

The $1/G(j\omega)$ and $H(j\omega)$-loci can be drawn separately and then added as shown in Fig. 9.31. Now the directed line segments from the origin to different points on $[1/G(j\omega)+H(j\omega)]$-locus give both the magnitude $1/M$ and phase angle $-\alpha$. Therefore the $1/M$-circles and α-contours for nonunity feedback systems will remain the same as for unity feedback systems but are drawn from the origin, as shown in Fig. 9.31.

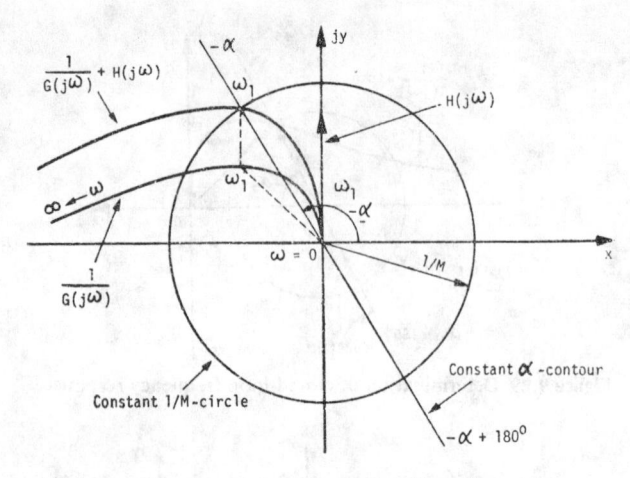

Figure 9.31 Determination of closed-loop frequency response of a nonunity feedback system.

The Nichols Chart

Earlier in this section we have seen that the constant-M contours on the direct and inverse polar plots are circles. Since it is easier to construct a Bode plot than a polar plot, it is preferable to have constant-M and constant-α contours constructed on logarithmic gain and phase coordinates. N.B. Nichols transformed the constant-M and -N circles to log-magnitude and phase angle coordinates and the resulting chart is known as the *Nichols chart*.

The constant-M contours in log-magnitude phase angle plane may be readily obtained by drawing phasors from the origin to points on constant-M circles in the direct polar plane and then locating these points in log-magnitude phase angle plane. When this is repeated for different values of M, we get a family of constant-M contours in log-magnitude phase angle plane. Similarly a family of constant-α contours may also be plotted in this plane. It is found that constant-M and constant-α contours repeat for every 360° interval and there is a symmetry at every 180° interval. Constant-M and constant-α contours for phase angles from 0° to $-210°$ are shown in Fig. 9.32.

The Nichols chart is very useful for determining the closed-loop frequency response from that of the open-loop. This is accomplished by superimposing the log-magnitude versus phase angle plot of $G(j\omega)$ on Nichols chart. The intersections of the log-magnitude versus phase angle plot and constant-M and -α contours give the magnitude M and phase angle α of the closed-loop frequency response at different frequency points.

Gain Adjustments

When a control system is found to be unstable or has poor transient response, the first step is to check if its performance can be modified by

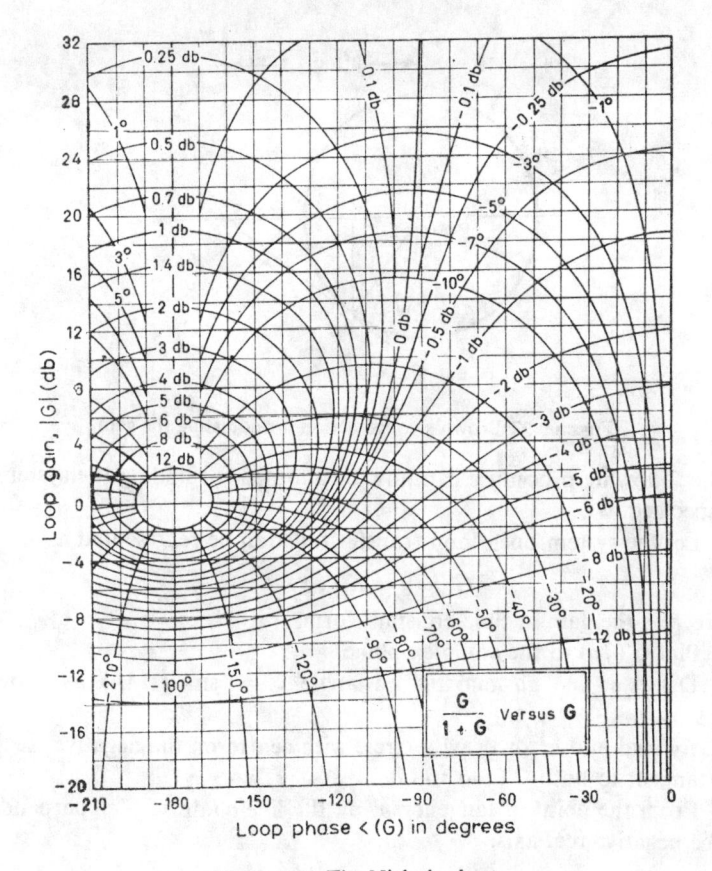

Figure 9.32 The Nichols chart.

adjustment of gain. This adjustment is usually based on a desirable value of M_r. The method of finding the gain K for a specified value of closed-loop M_r (where $M_r > 1$) is known as *Brown's construction*. To develop this construction, let us consider the geometry of a constant-M circle.

Gain adjustment by direct polar plots. In Fig. 9.33 the line OP is drawn from origin, tangent to the desired M_r-circle at point P. It makes an angle ψ with the negative real axis. It follows from the geometry of this figure that

$$\sin \psi = \frac{PA}{OA} = \frac{|M_r/(M_r^2 - 1)|}{|M_r^2/(M_r^2 - 1)|} = \frac{1}{M_r}$$

Also $\quad OC = OP \cos \psi$

or $\quad OC = \sqrt{[(OA)^2 - (AP)^2]}\sqrt{(1 - \sin^2 \psi)}$
$\qquad\quad = \sqrt{\{[M_r^2/(M_r^2 - 1)]^2 - [M_r/(M_r^2 - 1)]^2\}}\sqrt{(1 - 1/M_r^2)}$
$\qquad\quad = 1$

Thus we see that the point C coincides with $(-1 + j0)$ point in the complex plane.

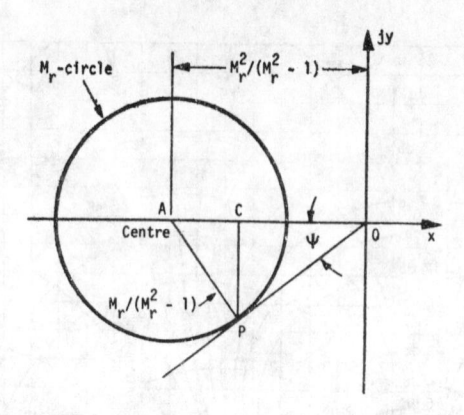

Figure 9.33 Brown's construction for adjusting the gain.

The following procedure can now be adopted for finding the system gain for specified M_r.

1. Let the system open-loop transfer function be represented as
$$G(j\omega) = KG'(j\omega)$$
where K is the gain in time-constant form.
2. Plot $G'(j\omega)$ in the complex plane.
3. Draw a line *od* making an angle $\psi = \sin^{-1}(1/M_r)$ as shown in Fig. 9.34.
4. By trial and error draw a circle with centre on the negative real axis and tangent to both $G'(j\omega)$ and *od*.
5. From the point of tangency p' on the line *od* draw $p'c'$ perpendicular to the negative real axis.

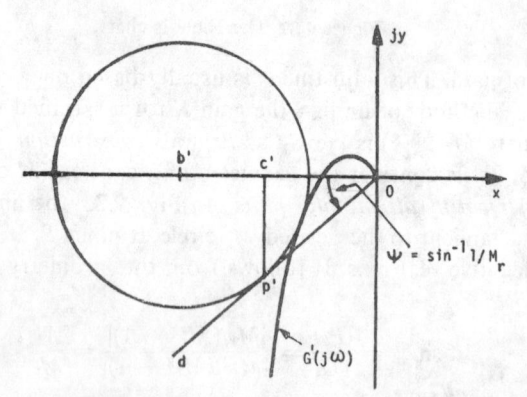

Figure 9.34 Construction for determination of K for specified M_r.

6. If all the phasors in the complex plane are now enlarged K times such that c' shifts to $(-1+j0)$, the enlarged complex plane plot will become $G(j\omega) = KG'(j\omega)$ and the circle will become the M_r-circle. Thus
$$K.oc' = 1$$

or
$$K \text{(for specified } M_r) = \frac{1}{oc'}$$

Gain adjustment by inverse polar plot. Figure 9.35 shows a $1/M$-circle and op is drawn tangent to the circle. As discussed previously the circle must be centred at $(-1+j0)$ for unity feedback systems. It is obvious from the geometry of this figure that

$$\psi = \sin^{-1}(1/M)$$

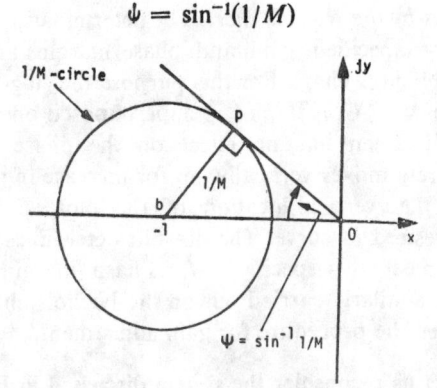

Figure 9.35 Geometry of $1/M$-circle.

This fact is used in the procedure given below for determining the system gain for a specified M_r.

1. Draw the complex plot of $1/G'(j\omega)$ where $G(j\omega) = KG'(j\omega)$. Such a plot is shown in Fig. 9.36.
2. Draw a line od in the second quadrant at an angle $\psi = \sin^{-1}(1/M_r)$ from the negative real axis.
3. By trial and error draw a circle centred on the negative real axis and tangent to both the line od and the plot $1/G'(j\omega)$. Let the centre of this circle be located at b'

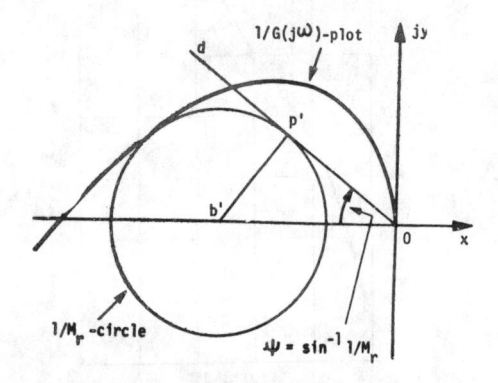

Figure 9.36 Gain adjustment by inverse polar plot.

4. When the gains for all the phasors are multiplied by $1/K$, the circle becomes the $1/M_r$-circle and the plot becomes $1/G(j\omega) = 1/KG'(j\omega)$ plot and the point b' shifts to $(-1+j0)$. Thus

$$\frac{1}{K} \cdot ob' = 1$$

or $$K = ob'$$

This gives us the method of obtaining K for specified M_r.

Gain adjustment by the Nichols chart. The determination of K for specified resonant peak or specified gain and phase margins are carried out very conveniently on Nichols chart. For this purpose the log-magnitude versus phase angle [db vs $\angle G(j\omega)$] plot is superimposed on the Nichols chart. Since the gain adjustment has no effect on the phase angle, the db vs $\angle G(j\omega)$ plot merely moves vertically up for increase in gain and down for decrease in gain. The vertical location of the plot is adjusted till it is tangent to the desired M-curve. The db-shift determines the adjustment in gain required to meet the specified M_r. Phase margin or gain margin adjustments are similarly carried out on the Nichols chart. The following example illustrates the procedure for gain adjustment.

Example 9.10: Let us reconsider the system discussed in Example 9.7. The system has an open-loop transfer function

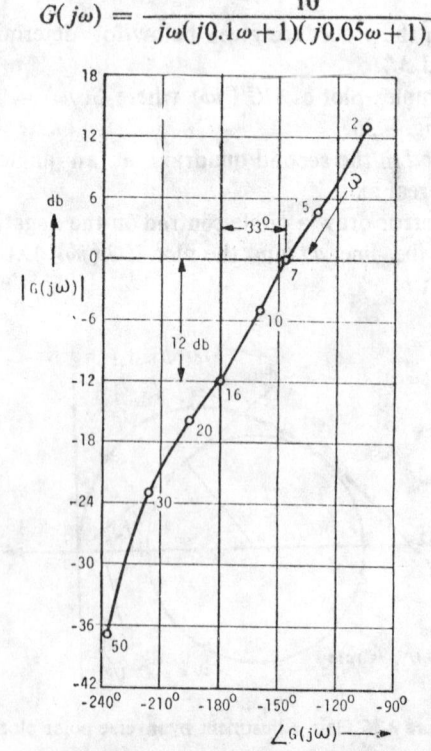

$$G(j\omega) = \frac{10}{j\omega(j0.1\omega+1)(j0.05\omega+1)}$$

Figure 9.37 Determination of GM and PM.

The db-$\angle G(j\omega)$ plot for this $G(j\omega)$ is shown in Fig. 9.37. (It is important to note that this plot is conveniently drawn by obtaining the log-magnitude and phase angle data for various values of frequency from the Bode plot of $G(j\omega)$ shown in Figure 9.22.) From Fig. 9.37, it is found that

$$GM = +12 \text{ db}; \quad PM = +33°$$

Gain adjustment for desired GM or PM. Suppose it is desired to find the open-loop gain for (i) a GM of 20 db, (ii) a PM of 24°.

(i) A GM of 20 db is obtained if the plot of Fig. 9.37 is shifted downwards by $(20-12) = 8$ db. The system gain is therefore changed by -8 db or decreased by a factor of 2.5.

(ii) A PM of 24° is obtained if the plot of Fig. 9.37 is raised upwards by 3.5 db or the system gain is increased by a factor of 1.5.

Closed-loop frequency response from the Nichols chart. The complete closed-loop frequency response of a system is easily obtained by superimposing the db-$\angle G(j\omega)$ plot on the Nichols chart as shown in Fig. 9.38. The intersections of this plot with M- and α-contours of the Nichols chart determine

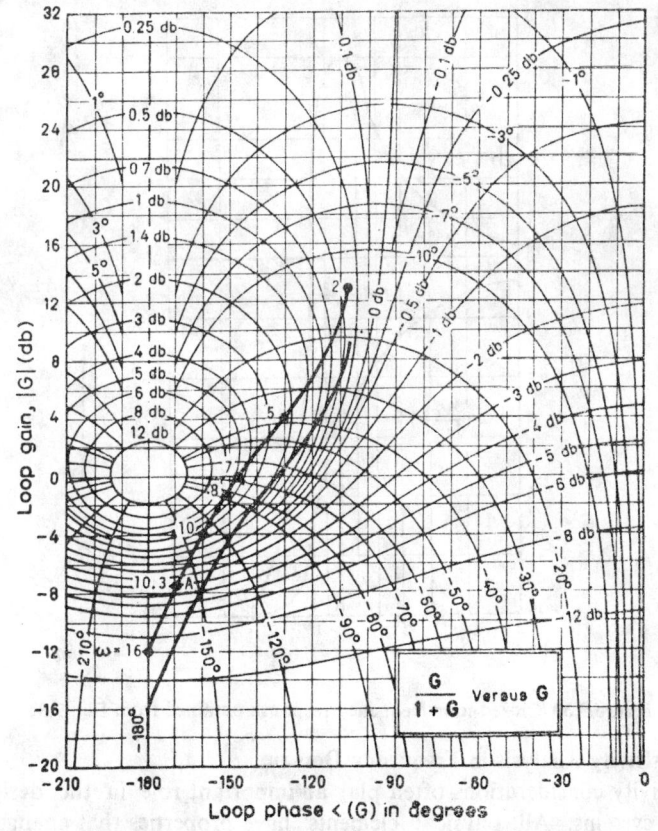

Figure 9.38 Log-magnitude vs phase angle plot superimposed on the Nichols chart.

the closed-loop M and α values for various frequencies from which the closed-loop frequency response of Fig. 9.39 is plotted.

From Fig. 9.38 it is seen that db-$\angle G(j\omega)$ plot is tangent to the 5 db M-contour at $\omega = 8$ rad/sec. Therefore for $K = 10$, $M_r = 5$ db and $\omega_r = 8$ rad/sec.

Bandwidth. Bandwidth is defined to be the frequency at which the closed-loop gain is -3 db. From Fig. 9.38 it is observed that db-$\angle G(j\omega)$ plot intersects the -3 db M-contour at the point A. The frequency parameter at this point can be easily determined by transferring the open-loop db data at A to the Bode plot of Fig. 9.22. We find this frequency to be 10.3 rad/sec therefore the bandwidth $\omega_b = 10.3$.

Gain adjustment for desired M_r. Finally, let us investigate the use of db-$\angle G(j\omega)$ plot to obtain the open-loop gain for desired M_r, say 2 db. We find from Fig. 9.38 that the db-$\angle G(j\omega)$ plot must be moved downwards by 3.5 db in order that it be tangent to the 2 db M-contour. Thus for $M_r = 2$ db, the open-loop gain should be decreased by 3.5 db or by a factor of 1.5 giving the required open-loop gain as $10/1.5 \approx 6.33$.

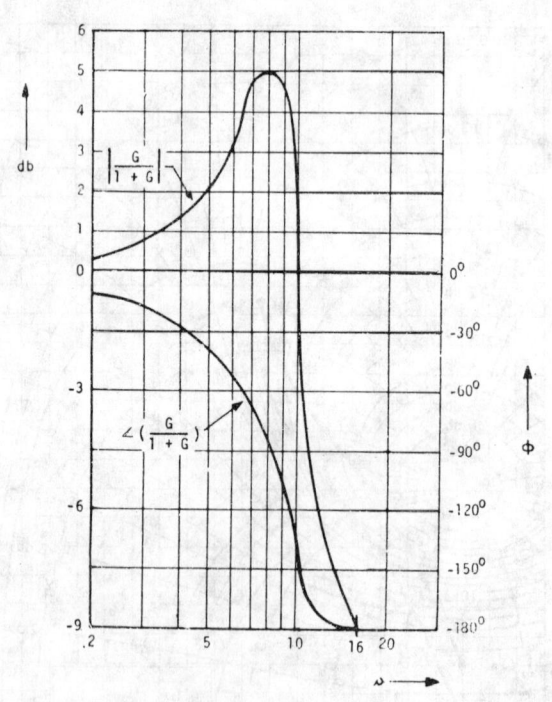

Figure 9.39 Closed-loop frequency response obtained from Fig. 9.38.

9.6 Sensitivity Analysis in Frequency Domain

Sensitivity considerations often play an important role in the design of control systems. All physical elements have properties that change with

Stability in Frequency Domain

environment and age. A good control system should be very insensitive to these parameter variations while being able to follow the command responsively.

The sensitivity of a system's closed-loop transfer function T with respect to the characteristic K of a given element is (refer Section 3.2)

$$S_K^T = \frac{\partial T/T}{\partial K/K}$$

Note that the sensitivity is a function of frequency. For example, the sensitivity of closed-loop transfer function of system of Fig. 3.11 to variations in the parameter K is (eqn. (3.25))

$$S_K^T(s) = \frac{s(s+5)}{s^2+5s+25}$$

Therefore

$$S_K^T(j\omega) = \frac{j\omega(j\omega+5)}{(j\omega)^2+5j\omega+25}$$

Figure 9.40 (curve (i)) gives a plot of $|S_K^T(j\omega)|$ vs ω. From this figure we observe that sensitivity becomes zero as ω approaches zero. The peak value of $|S_K^T(j\omega)|$ is reached at $\omega = 6$ rad/sec. This means that system of Fig. 3.11 is most sensitive to change in K at this frequency or more generally in this frequency range.

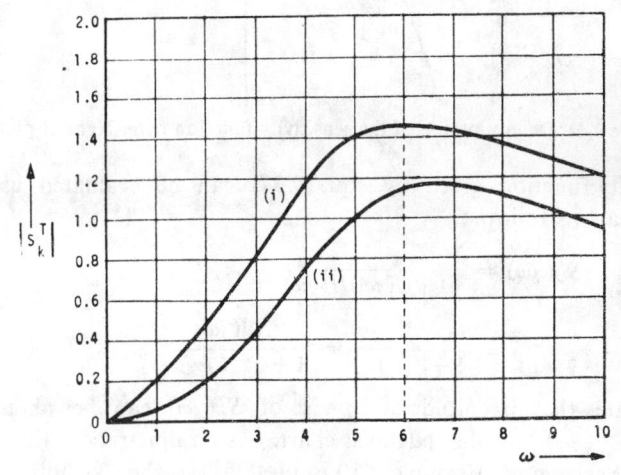

Figure 9.40 Sensitivity curves for the systems of Figs. 3.11 and 3.12.

For the two-loop configuration of Fig. 3.12, from eqn. (3.26) we have

$$S_K^T(j\omega) = \frac{j\omega(j\omega+1)}{(j\omega)^2+5j\omega+25}$$

The $|S_K^T(j\omega)|$ vs ω plot is shown in Fig. 9.40 (curve (ii)). The two sensitivity curves of this figure give an effective comparison of the two system configurations shown in Figs. 3.11 and 3.12.

Sensitivity of T to Variations in G

The sensitivity of closed-loop transfer function $T(s)$ to variations in open-loop transfer function $G(s)$ is given by the function (eqn. (3.3))

$$S_G^T(j\omega) = \frac{1}{1+G(j\omega)H(j\omega)} \tag{9.37}$$

The function $|S_G^T(j\omega)|$ may conveniently be evaluated from the Nyquist plot of $G(j\omega)H(j\omega)$. The phasor $1+G(j\omega)H(j\omega)$ is shown in the Nyquist plot of Fig. 9.41. At a frequency ω_1, $|1+G(j\omega_1)H(j\omega_1)|$ is K_1 and therefore $|S_G^T(j\omega_1)| = \frac{1}{K_1}$. The maximum sensitivity is given by the circle of radius K_m centred at $(-1+j0)$ which touches the $G(j\omega)H(j\omega)$ plot. The peak sensitivity is then $|S_G^T(j\omega)| = \frac{1}{K_m}$ at frequency ω_m.

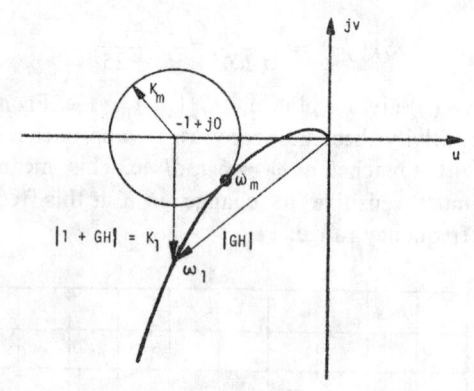

Figure 9.41 Determination of the sensitivity function from Nyquist plots.

Sensitivity function given by eqn. (9.37) can be evaluated using the Nichols chart as follows:

$$S_G^T(j\omega) = \frac{1}{1+G(j\omega)H(j\omega)}$$

$$= \frac{1}{1+G_0(j\omega)} = \frac{G_0^{-1}(j\omega)}{1+G_0^{-1}(j\omega)}$$

This indicates that magnitude and phase of $S_G^T(j\omega)$ can be obtained by plotting $G_0^{-1}(j\omega)$ on the Nichols chart. As an illustration, the function $G^{-1}(j\omega)$ for system of Example 9.10 is plotted on the Nichols chart* in Fig. 9.42. The intersection of $G^{-1}(j\omega)$ curve on the Nichols chart with the constant-M loci gives $|S_G^T(j\omega)|$ at the corresponding frequencies. From Fig. 9.42 it is observed that peak value of $|S_G^T(j\omega)|$ is 7 db at $\omega = 9$ rad/sec.

*Note that vertical coordinate of the Nichols chart is in decibels; $|G^{-1}(j\omega)|$ and $\angle G^{-1}(j\omega)$ can be obtained from $|G(j\omega)|$ and $\angle G(j\omega)$, since

$$db\,|G^{-1}(j\omega)| = -\,db\,|G(j\omega)|$$
$$\angle G^{-1}(j\omega) = -\angle G(j\omega)$$

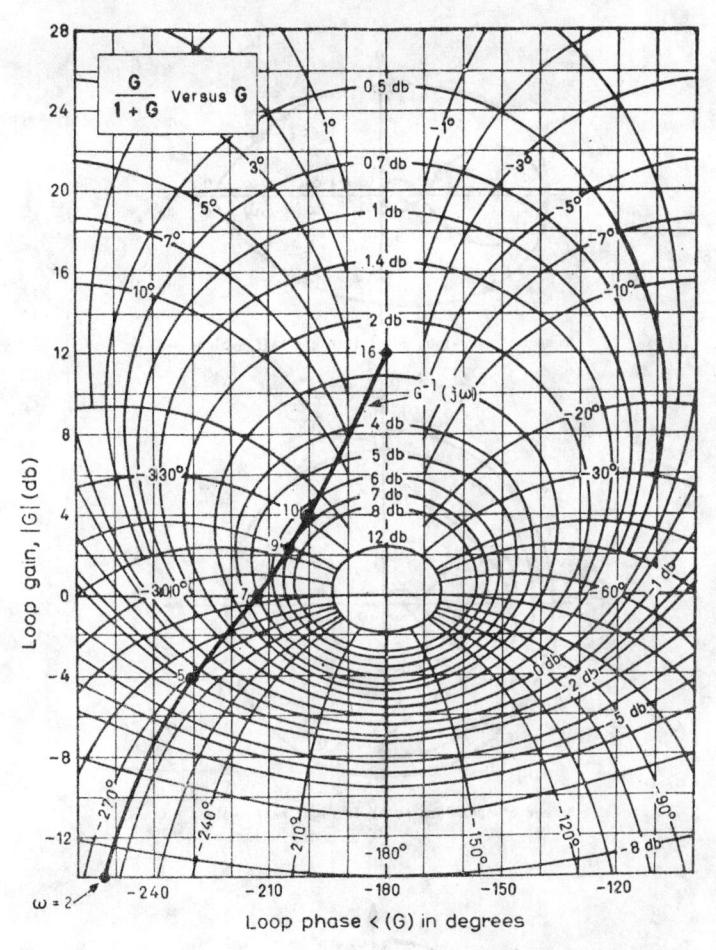

Figure 9.42 Determination of the sensitivity function from the Nichols chart.

Problems

9.1 By use of the Nyquist criterion, determine whether the closed-loop systems having the following open-loop transfer functions are stable or not. If not, how many closed-loop poles lie in the right half s-plane?

(a) $G(s)H(s) = \dfrac{1+4s}{s^2(1+s)(1+2s)}$

(b) $G(s)H(s) = \dfrac{1}{s(1+2s)(1+s)}$

(c) $G(s)H(s) = \dfrac{1}{s^2+100}$

9.2 Check the stability of systems whose Nyquist plots are shown in Fig. P-9.2.

9.3 Sketch the Nyquist plot for a system with the open-loop transfer function

$$G(s)H(s) = \frac{K(1+0.5s)(s+1)}{(1+10s)(s-1)}$$

Determine the range of values of K for which the system is stable.

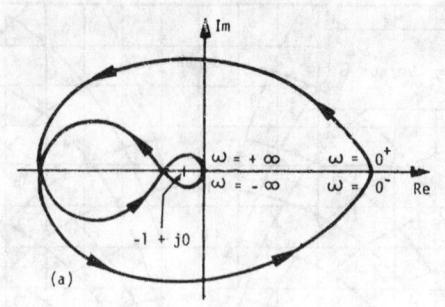

Figure P-9.2.

Plot (a): Number of open-loop poles in right half s-plane = 0

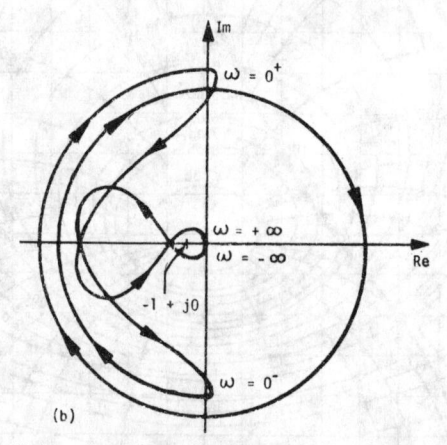

Plot (b): Number of open-loop poles in right half s-plane = 0

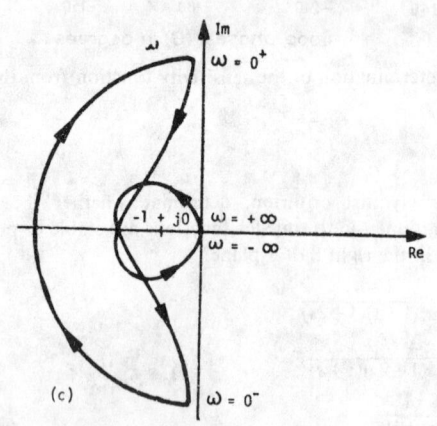

Plot (c): Number of open-loop poles in right half s-plane = 1

9.4 Consider the feedback system shown in Fig. P-9.4. Sketch the Nyquist plot for this system when $G_c(s) = 1$ and determine the maximum value of K for stability.

Stability in Frequency Domain

If $G_c(s)$ is modified to a controller having the transfer function $(1+1/s)$, what then is the maximum value of K for stability? Comment upon the result.

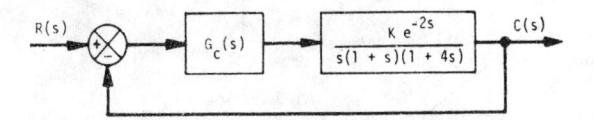

Figure P-9.4.

9.5 Consider a feedback system having the characteristic equation

$$1 + \frac{K}{(s+1)(s+1.5)(s+2)} = 0$$

It is desired that all the roots of the characteristic equation have real parts less than -1. Extend the Nyquist stability criterion to find the largest value of K, satisfying this condition.

[*Hint*: For the above said condition to be satisfied, Nyquist plot of $GH(-1+j\omega)$ should not encircle $(-1+j0)$ point.]

9.6 Sketch the Bode plot of a closed-loop system which has the open-loop transfer function

$$G(s)H(s) = \frac{2e^{-sT}}{s(1+s)(1+0.5s)}$$

Determine the maximum value of T for the system to be stable.

9.7 Sketch the Bode plot for a unity feedback system, characterized by the open-loop transfer function

$$G(s) = \frac{K(1+0.2s)(1+0.025s)}{s^3(1+0.001s)(1+0.005s)}$$

Show that the system is conditionally stable. Find the range of values of K for which the system is stable.

9.8. The open-loop transfer function of a unity feedback system is given by

$$G(s) = \frac{K}{s(T_1 s+1)(T_2 s+1)}$$

Derive an expression for gain K in terms of T_1, T_2 and specified gain margin G_m.

9.9. Phase margin for a second-order system is given by

$$\phi_{pm} = \tan^{-1} 2\zeta \left[\frac{1}{\sqrt{(4\zeta^4+1)} - 2\zeta^2} \right]^{1/2}$$

Write the approximate expression for ϕ_{pm} for low values of ζ.

Using the approximate expression, find the value of the gain K such that the system shown in Fig. P-9.9 has a phase margin of ϕ_s degrees.

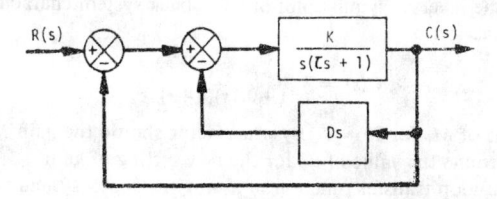

Figure P-9.9.

9.10 The straight line Bode plot of a feedback system is shown in Fig. P-9.10. Derive an expression for the value of ω_c to yield maximum phase margin, in terms of system constants ω_1, ω_2. Determine the maximum phase margin when $m = 1$.

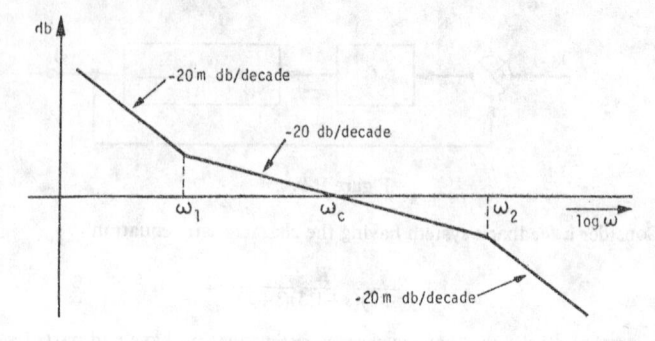

Figure P-9.10

9.11 The open-loop frequency response of a unity feedback system is given below:

ω	0.8	1.0	1.2	1.4	1.6
$Re[KG(j\omega)]$	−3.5	−2.9	−2.3	−2.0	−1.2
$Im[KG(j\omega)]$	−4.4	−3.2	−1.9	−1.2	−0.5

Determine the change in gain K required to make the resonant peak $M_r = 1.4$. For this value of gain, determine the phase margin of the system. Evaluate the value of damping ratio using M_r-criterion and phase margin criterion.

9.12 A unity feedback system has the following open-loop frequency response.

ω	2	3	4	5	6	8	10
$\lvert G(j\omega) \rvert$	7.5	4.8	3.15	2.25	1.70	1.00	0.64
$\angle G(j\omega)$	−118°	−130°	−140°	−150°	−157°	−170°	−180°

(a) Evaluate the gain margin and phase margin of the system.
(b) Determine the change in gain required so that the gain margin of the system is 20 db.
(c) Determine the change in gain required so that the phase margin of the system is 60 deg.

9.13 Sketch the inverse Nyquist plot of a feedback system characterized by the open-loop transfer function

$$G(s) = \frac{K}{s(1+0.1s)(1+s)}$$

Find the value of M_r for $K = 1$. By what factor should the gain K be changed so that M_r is 1.4? Determine the value of ω_r for the new setting of gain.

9.14 The open-loop transfer function of a unity feedback system is

$$G(s) = \frac{Ke^{-0.1s}}{s(1+0.1s)(1+s)}$$

Stability in Frequency Domain

By use of Bode plot and/or Nichols chart, determine the following:
(a) The value of K so that the gain margin of the system is 20 db.
(b) The value of K so that the phase margin of the system is 60 deg.
(c) The value of K so that resonant peak M_r of the system is 1 db. What are the corresponding values of ω_r and ω_b?
(d) The value of K so that the bandwidth ω_b of the system is 1.5 rad/sec.

9.15 The open-loop frequency response of a unity feedback system is given below:

ω	1.0	1.5	2.0	2.5	3.0	3.5	4.0
$\left\|\dfrac{1}{G(j\omega)}\right\|$	0.15	0.36	0.55	0.83	1.1	1.43	2.0
$-\angle G(j\omega)$	130°	140°	148°	153°	158°	160°	164°

Sketch the inverse Nyquist plot and determine values of M_r and ω_r.

The configuration of feedback system is now changed to include derivative feedback as shown in Fig. P-9.15. Determine graphically the resultant frequency response and therefrom determine values of M_r and ω_r. Comment upon the effect of derivative feedback on system stability.

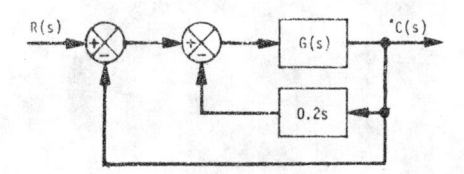

Figure P-9.15.

9.16 The block diagram of a position control system is shown in Fig. P-9.16. Determine the peak sensitivity of the closed-loop transfer function T with respect to G and H, the forward path and feedback path transfer functions respectively. Also determine the frequency at which the peak sensitivity occurs.

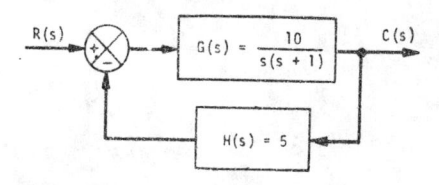

Figure P-9.16.

What will be the bandwidth of the system if it is designed to have

$$\left| S_G^T(j\omega) \right| < 1?$$

Further Reading

1. D'Azzo, J.J. and C.H. Houpis, *Linear Control System Analysis and Design*, McGraw-Hill, New York, 1975.
2. Kuo, B.C., *Automatic Control Systems* (3rd Ed.), Prentice-Hall, Englewood Cliffs, N.J., 1975.
3. Thaler, G.J., *Design of Feedback Systems*, Dowden, Hutchinson and Ross, Stroudsburg, Pennsylvania, 1973.
4. Eveleigh, V.W., *Introduction to Control Systems Design*, McGraw-Hill, New York, 1972.
5. Shinners, S.M., *Modern Control System Theory and Application*, Addison-Wesley, Reading, Mass., 1972.
6. Gupta, S.C. and L. Hasdorff, *Fundamentals of Automatic Control*, John Wiley, New York, 1970.
7. Ogata, K., *Modern Control Engineering*, Prentice-Hall, Englewood Cliffs, N.J., 1970.
8. Dorf, R.C., *Modern Control Systems*, Addison-Wesley, Reading, Mass., 1967.
9. Ku, Y.H., *Analysis and Control of Linear Systems*, International Text Book Co., Scranton, Pennsylvania, 1962.

10. Introduction to Design

10.1 The Design Problem

The design of automatic control systems is perhaps the most important function that the control engineer carries out. Systems analysis presented in the preceding chapters provides the necessary clue to design and construction of practical control systems. While some of the direct design methods can be abstracted from analysis, in most situations the design proceeds on a trial and error basis wherein analysis techniques are repeatedly applied.

Every control system designed for a specific application has to meet certain performance specifications. In Chapters 5 and 8, we discussed two methods of specifying the performance of a control system:

(i) By a set of specifications in time domain and/or in frequency domain such as peak overshoot, settling time, gain margin, phase margin, steady-state error, etc.

(ii) By optimality of a certain function, e.g., an integral function.

In addition to the performance specifications, some other constraints are always imposed on the control system design. Consider for example the tracking antenna control system discussed in Problem 3.9 (Fig. P-3.9). The objective of the system is to drive large size antenna. The first step in the design is the selection of an actuator tom ove the antenna. Depending upon the performance specifications, available power supply, space and economical limitations, etc., it could be a d.c. servomotor, a.c. servomotor or hydraulic servomotor. The size of the motor is determined by the inertia, velocity and acceleration ranges of the antenna. Further, since motor's rated speed is usually larger than the required load speed, a gear train of suitable gear ratio is also used. In system of Fig. P-3.9, armature controlled d.c. servomotor with amplidyne as power amplifier have been taken rather arbitrarily. This collection of devices (antenna, gear train, motor, amplidyne) is the *plant* of the control system.

From this discussion, it is evident that the choice of a plant is dictated not only by performance specifications but also by size, weight, available power supply, cost, etc. Therefore the plant chosen generally can not meet the performance specifications. Though the designer is free to choose a new plant, this is generally not done because of cost, availability and other constraints. However, some components of the chosen plant may be easily replaced, e.g., if an electronic amplifier is a component of a plant, its replacement is not a big problem because of the low-cost and wide-range of availability of such amplifiers. Merely by gain adjustment, it may be possible to meet the given specifications on performance of simple control systems. In such cases, gain adjustment seems to be the most direct and simple way of design. However, in most practical cases, the gain adjustment does not provide the desired result. As is usually the case, increasing the gain reduces the steady-state error but results in oscillatory transient response or, even, in instability. Under such circumstances, it is usually possible to introduce some kind of corrective subsystems to force the chosen plant to meet the given specifications. These subsystems are known as *compensators* and their job is to compensate for the deficiency in the performance of the plant. The design problem may therefore be stated as follows:

Given a plant and a set of specifications, design suitable compensators so that the overall system will meet the given specifications.

Approaches to the Design Problem

There are basically two approaches to the control system design problem.

1. We select the configuration of the overall system by introducing compensators and then choose the parameters of the compensators to meet the given specifications on performance.

2. For a given plant, we find an overall system that meets the given specifications and then compute the necessary compensators.

The first approach has been used at several places in the preceding chapters. In Chapter 5, we studied control system configurations wherein proportional, derivative and integral controllers were introduced to obtain the desired performance specified in terms of peak overshoot, settling time, steady-state error, etc., or in terms of optimality of a function. In Chapter 7, we discussed the root locus technique for parameter design, while in Chapter 9, we discussed the use of Nyquist and Bode plots for control systems design to achieve the desired performance specified in terms of resonant peak of closed-loop frequency response, bandwidth, gain margin, phase margin, etc.

The primary objective of this chapter is to discuss further the question of design and to present significant design and compensation methods for single-input single-output linear time-invariant systems. We shall assume here fixed system configuration and performance specified in terms of peak overshoot, settling time, gain margin, phase margin, steady-state error, etc., and we shall present the root locus method and frequency domain methods of design, frequently referred to as the *classical methods of design*.

The problem of design of parameters of compensators in a fixed system configuration will again be taken up in Chapter 13, wherein we shall assume that the performance specifications have been translated into the optimality of a certain function. This design technique is called the *parameter optimization*. The parameter optimization problem may be stated as follows: *Given a plant and a configuration of compensation, find the free parameters so that the resulting system is optimal with respect to some performance criteria.*

Chapter 13 will also deal with the second approach of design when the designer has complete freedom in using compensators. This approach deals with basically two problems:

(*i*) *Given a plant, find an overall system which is optimal with respect to given performance criteria.*

(*ii*) *Compute the compensators for the system obtained in step* (*i*).

In a control system, introduction of sampling, use of digital computer, etc., may be an additional requirement. Design of such systems is discussed in Chapters 11 and 13.

10.2 Preliminary Considerations of Classical Design

Proper selection of performance specifications is the most important step in control system design. The desired behaviour is specified in terms of transient response measures and steady-state error. The steady-state error is usually specified in terms of error constants for specific inputs while the transient response measures of relative stability and speed of response may be specified in time or frequency domain or even in both. In time domain, the measure of relative stability is damping factor ζ or peak overshoot M_p, while the speed of response is measured in terms of rise time, settling time or natural frequency. On the other hand, in frequency domain the measure of relative stability is resonant peak M_r or phase margin ϕ_{pm}, while the measure of speed of response is resonant frequency ω_r or bandwidth ω_b.

Once a set of performance specifications has been selected, the designer's next aim is to select a configuration for the overall system. The nature of compensation depends upon the given plant. The compensator may be an electrical, mechanical, hydraulic, pneumatic or other type of device or network. Usually, an electric network serves as compensator in many control systems. (Different types of electric networks used as compensation devices are discussed later in this chapter.) The compensator transfer function may be placed in cascade with the plant transfer function (*cascade or series compensation*) or in the feedback path (*feedback or parallel compensation*) as shown in Fig. 10.1.

Once the configuration of the overall system has been selected, the next step is to obtain the parameters of the compensator that satisfy the selected specifications in the best possible manner. Parameter search is of course restricted to the feasible domain. Practically most control systems require a trial and error parameter adjustment to achieve at least an acceptable performance if it is not possible to satisfy exactly all the performance specifications.

Control Systems Engineering

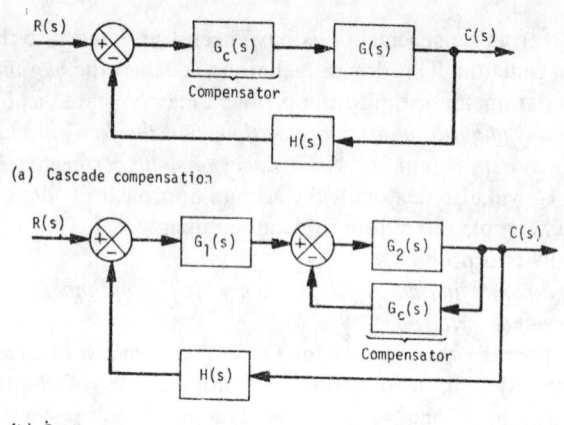

(a) Cascade compensation

(b) Feedback compensation

Figure 10.1 (a) Cascade compensation; and (b) feedback compensation.

A DESIGN EXAMPLE

Consider the system of Fig. 10.2 which has an open-loop transfer function.

$$G(s) = \frac{5K_A}{s(s/2 + 1)(s/6 + 1)} \tag{10.1}$$

From eqn. (10.1), K_v, the velocity error constant of the system is given by

$$K_v = \lim_{s \to 0} s\,G(s) = 5K_A \tag{10.2}$$

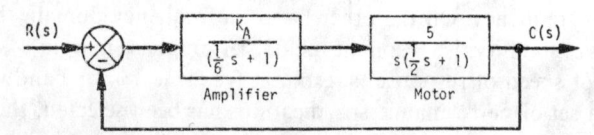

Figure 10.2 A position control system.

For $K_A = 1$ (i.e., $K_v = 5$), the steady-state error to unit velocity input ($e_{ss} = 1/K_v$) is 0.2, which may be assumed to be an acceptable value.

From eqn. (10.1), we have

$$G(s) = \frac{60K_A}{s(s+2)(s+6)} = \frac{K}{s(s+2)(s+6)} \tag{10.3}$$

The root locus plot of the uncompensated system appears in Fig. 10.3. Corresponding to a value $K = 60$ (i.e., $K_v = 5$) on the root locus plot, the dominant pair of roots is found to be

$$s_{1,2} = -0.3 \pm j2.8$$

For this root pair, the damping ratio ζ is 0.105 and the undamped natural frequency ω_n is 2.85. The settling time ($= 4/\zeta\omega_n$) for the response is therefore about 13.36 sec.

From the above investigation of the uncompensated system, it is seen that the system has poor relative stability and sluggish transient response.

Assume that the specifications on the transient performance are

$$\zeta = 0.6;$$

settling time < 4 sec.

A line corresponding to $\zeta = 0.6$ is shown in Fig. 10.3. The intersection of this line with root locus branch determines the value of K which yields the specified ζ. From Fig. 10.3 we find that

$K = 10.5[K_v = 10.5/(2 \times 6) = 0.875$, i.e., $K_A = 0.175]$;
$\omega_n = 1.26$;
settling time = 5.3 sec;
$e_{ss} = 1.14$.

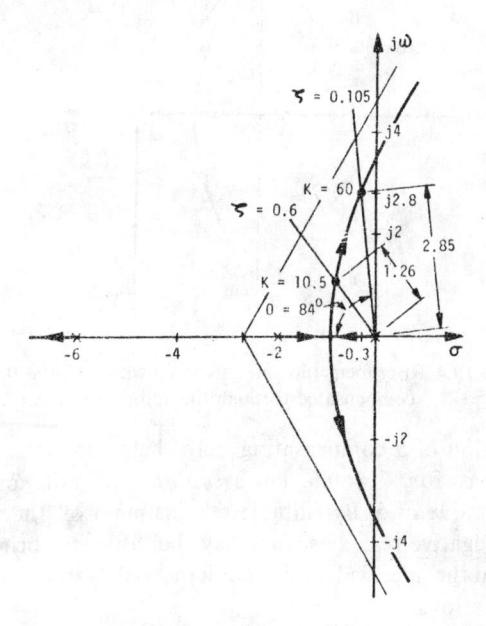

Figure 10.3 Root locus plot of system shown in Fig. 10.2.

Thus the reduction in gain satisfies the specification on ζ but the steady-state error increases well beyond the tolerable limit of 0.2 and further the settling time specifications are not fully satisfied. Therefore we conclude that our purpose is not served by mere gain adjustment.

In order to meet the specifications we need to increase ω_n keeping ζ constant (= 0.6) without sacrificing the steady-state performance (i.e., $K_v \geqslant 5$). To accomplish this goal let us consider the proposition of adding a compensating zero somewhere between the nonzero poles of the open-loop system.

The resulting root locus plot is shown in Fig. 10.4, from which we observe the following effects of the compensating zero.

1. All branches of the root locus now lie completely in the left half s-plane, so that K can be adjusted to any positive value without causing instability.

2. Complex-root branches of the root locus are now bent away from the $j\omega$-axis.

From Fig. 10.4, it is found that for $\zeta = 0.6$;
$K = 16[K_v = (16\times 3)/(2\times 6) = 4]$;
$\omega_n = 3.4$;
settling time $= 4/(0.6\times 3.4) = 1.96 < 4$ sec.

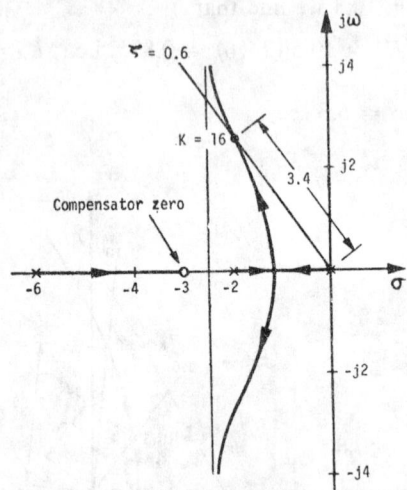

Figure 10.4 Root locus plot of system shown in Figure 10.2, compensated through the addition of a zero.

Thus the addition of a compensating zero helps to satisfy the settling time specifications for $\zeta = 0.6$. The system K_v still falls somewhat short of the specified value of 5. By suitable readjustment of the compensating zero along the negative real axis, one may be able to bring the steady-state error within the specified limit simultaneously satisfying the transient response specifications.

From Figs. 10.3 and 10.4 a further observation is made. Without the compensating zero, the real root in addition to the dominant complex-root pair is always located to the left of the open-loop pole at $s = -6$. On the other hand, in the presence of the compensating zero, this root moves over to the right of $s = -6$ and the dominance condition is therefore weakened. If this root (which is the closed-loop pole) is located close to the open-loop compensating zero, it will contribute a negligible term to the transient response such that the transient response continues to be dominated by the desired closed-loop poles.

The addition of a zero in the open-loop transfer function can be achieved by adding a compensator with the transfer function

$$G_c(s) = (s+z_c) = (s+1/\tau_c) \qquad (10.4)$$

in cascade with the forward transfer function. Since an isolated zero is not physically realizable, we must add a pole alongwith the compensating zero so as to achieve physical realizability. This pole must of course be added far away from the $j\omega$-axis such that it has relatively negligible effect on the root locus in the region where the two dominant complex closed-loop poles are to occur.

Figure 10.5 shows the root locus plot in the presence of both a compensating zero and pole. The root locus near the dominant closed-loop poles gets somewhat modified by the presence of the compensating pole. The effect on the transient response of the system is not very pronounced. For $\zeta = 0.6$, we obtain from root locus

$K = 67.5 [K_v = (67.5 \times 3)/(2 \times 6 \times 10) = 1.7]$;
$\omega_n = 2$;
settling time $= 4/(2 \times 0.6) = 3.33 < 4$ sec.

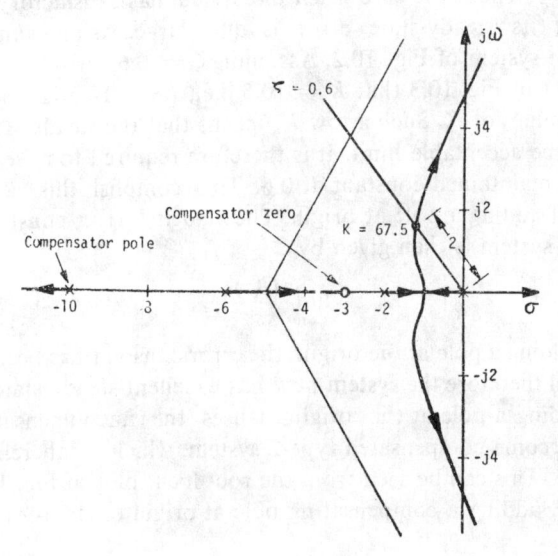

Figure 10.5 Root locus plot of system shown in Figure 10.2 with compensating pole and zero.

By adjusting the compensating zero and pole, it may be possible to raise the K_v to the specified value and yet satisfying the transient response specifications.

We see from the above discussion that the compensating pole must be located to the left of the compensating zero. The transfer function of such a compensator then becomes

$$G_c(s) = \frac{(s+z_c)}{(s+p_c)} = \frac{(s+1/\tau)}{(s+1/\alpha\tau)}; \quad \alpha = \frac{z_c}{p_c} < 1, \tau > 0 \qquad (10.5)$$

Note that $\alpha < 1$ ensures that the pole is located to the left of the zero.

The compensator having a transfer function of the form given in eqn. (10.5) is known as a *lead compensator*. The block diagram of the system with lead compensator is shown in Fig. 10.6.

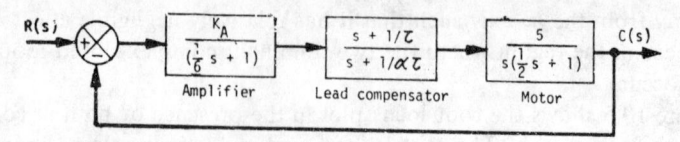

Figure 10.6 Position control system with lead compensator.

From the foregoing discussion we conclude that a *lead compensator speeds up the transient response and increases the margin of stability of a system. It also helps to increase the system error constant though to a limited extent.*

Let us now consider the case when the system has satisfactory transient response but its steady-state error is quite large. As an example, let us reconsider the system of Fig. 10.2. Assuming $\zeta = 0.6$, we find from the root locus plot of Fig. 10.3 that $K = 10.5$ [i.e., $K_v = 10.5/(2 \times 6) = 0.875$] yields this value of ζ. Such a low K_v means that the steady-state error is well beyond the acceptable limit. It is therefore required to raise the system K_v, while ζ is maintained constant at 0.6. To accomplish this, consider the proposition of adding a pole at origin. The velocity error constant of the compensated system is then given by

$$K_v = \lim_{s \to 0} \frac{sG(s)}{s} \to \infty$$

Thus by adding a pole at the origin, the error constant has been increased to infinity and therefore the system now has excellent steady-state response. However, adding a pole at the origin causes the uncompensated type-1 system to become compensated type-2 system, which is inherently closed-loop unstable. This can be seen from the root locus plot of Fig. 10.7, which is obtained by adding a compensating pole at origin to the root locus plot of Fig. 10.3.

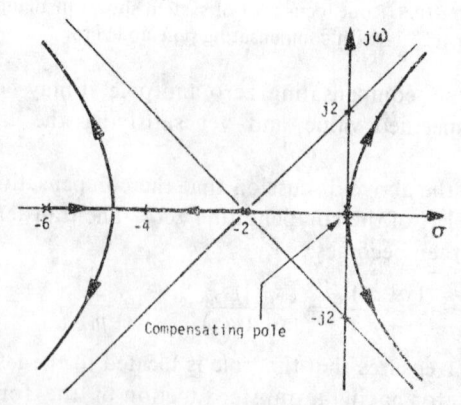

Figure 10.7 Root locus plot of system shown in Figure 10.2 with compensating pole at origin.

Introduction to Design

It is to be noted that addition of a compensating pole at origin to a type-0 system would not change the transient response so drastically as in the case of a type-1 system, still it does degrade the transient response.

This difficulty is remedied by adding a compensating zero very close to the pole at origin and in the left half of the s-plane. The compensated system then continues to be type-2, while the additional zero approximately cancels the effect of the compensating pole on the root locus. The compensated system thus has $K_v = \infty$, while its transient response is not significantly different from that of the uncompensated system.

Figure 10.8 shows the root locus plot of the compensated system with the compensating pole at origin and compensating zero at -0.2. Due to the presence of the compensating zero, an additional real closed-loop pole appears. For $\zeta = 0.6$ which corresponds to $K = 12$, this particular closed-loop pole is located at $s = -0.25$. It contributes an additional term in the transient response which decays slowly but is very small in magnitude on account of its proximity to the open-loop zero. This proximity must be ensured in the compensator design failing which the slowly decaying term will increase the settling time of the system.

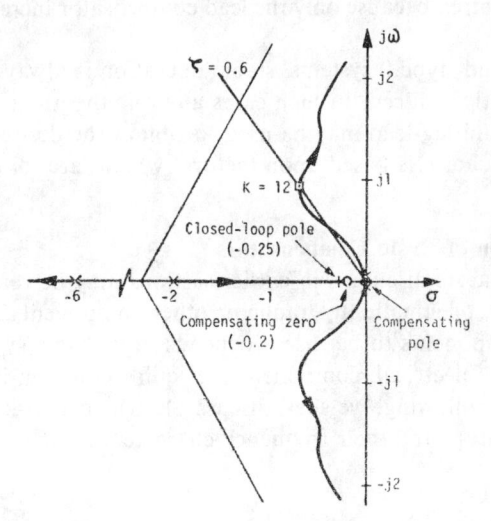

Figure 10.8 Root locus plot of system shown in Figure 10.2 with compensating pole and zero.

The addition of the compensating pole and zero as discussed above can be achieved by a cascade compensator with the transfer function

$$G_c(s) = \frac{(s+z_c)}{s} \qquad (10.6)$$

In actual practice, the considerations of physical realizability require that the pole at origin be shifted on the real axis slightly to the left of origin. The transfer function of the compensator, then becomes

$$G_c(s) = \frac{(s+z_c)}{(s+p_c)}; \quad \frac{z_c}{p_c} = \beta > 1 \qquad (10.7)$$

$\beta > 1$ ensures that pole is to the right of zero, i.e., nearer the origin than zero.

The compensator having a transfer function of the form given in eqn. (10.7) is called a *lag compensator*.

From the foregoing discussion we conclude that a *lag compensator improves the steady-state behaviour of a system, while nearly preserving its transient response.*

When both the transient and steady-state responses require improvement, a *lag-lead compensator* is required. This is basically a lag and a lead compensator connected in series.

SELECTION OF A COMPENSATOR

In general, there are two situations in which compensation is required. In the first case, the system is absolutely unstable and the compensation is required to stabilize it as well as to achieve a specified performance. In the second case, the system is stable but the compensation is required to obtain the desired performance. The systems which are type-2 or higher, are usually absolutely unstable. For these types of systems, clearly lead compensator is required because only the lead compensator increases the margin of stability.

In type-1 and type-0 systems, stable operation is always possible if the gain is sufficiently reduced. In such cases any of the three compensators, viz., lag, lead and lag-lead may be used to obtain the desired performance. The particular choice is based upon factors which are discussed later in this chapter.

10.3 Realization of Basic Compensators

The compensators discussed in earlier sections may be realized by electrical, mechanical, pneumatic, hydraulic or other components. The choice of the type of components to be used depends upon the system structure. Realization by electrical components is quite common in many control systems. In the following, we shall discuss electric network realization of basic compensators and their frequency characteristics.

LEAD COMPENSATOR

The s-plane representation of the lead compensator is shown in Fig. 10.9. It has a zero at $s = -1/\tau$ and a pole at $s = -1/\alpha\tau$ with the zero closer to the origin than the pole. The general form of the lead compensator is

$$G_c(s) = \frac{(s+z_c)}{(s+p_c)} = \frac{(s+1/\tau)}{(s+1/\alpha\tau)}; \quad \alpha = z_c/p_c < 1, \tau > 0 \quad (10.8)$$

$$= \alpha\left(\frac{\tau s+1}{\alpha\tau s+1}\right) \quad (10.9)$$

The lead compensator with the transfer function (10.8) can be realized by an electric lead network shown in Fig. 10.10a.

Introduction to Design

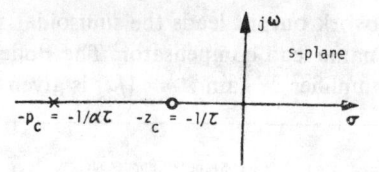

Figure 10.9 The s-plane representation of lead compensator.

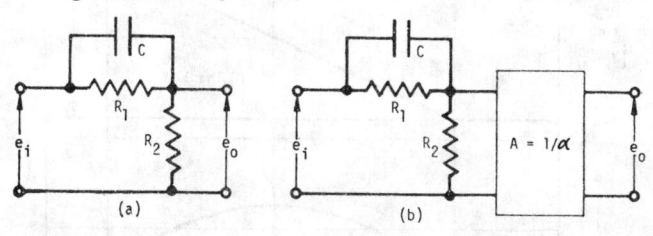

Figure 10.10 (a) Electric lead network; and (b) phase-lead network with amplifier.

Let us derive the transfer function of this lead network, assuming the impedance of the source to be zero and the output load impedance to be infinite. This assumption can be met by suitably designing the impedance levels of the network and the source. From Fig. 10.10a

$$\frac{E_0(s)}{E_i(s)} = \frac{R_2}{R_2 + \dfrac{R_1/sC}{R_1 + 1/sC}} = \frac{(s+1/R_1C)}{s + \dfrac{1}{[R_2/(R_1+R_2)]R_1C}} \tag{10.10}$$

Defining

$$\tau = R_1C \text{ and } \alpha = R_2/(R_1+R_2) < 1 \tag{10.11}$$

we immediately recognize that the network transfer function (10.10) has the same form as that of lead compensator in eqn. (10.8).

It is to be noted that the values of the three network components R_1, R_2 and C are to be determined from the two lead compensator parameters τ and α using eqn. (10.11). Thus there is an additional degree of freedom in the choice of the values of the network components which is used to set the impedance level of the network.

Consider now the sinusoidal transfer function of the lead network. It can be written from eqn. (10.9) as

$$G_c(j\omega) = \alpha \left(\frac{1+j\omega\tau}{1+j\omega\alpha\tau}\right); \quad \alpha < 1$$

At zero frequency the network has a gain of $\alpha < 1$ or an attenuation of $1/\alpha$. In frequency domain compensation technique, it is convenient to cancel the d.c. attenuation of the network with an amplification $A = 1/\alpha$. The lead compensator is then visualized as a combination of a network and an amplifier as shown in Fig. 10.10b. The sinusoidal transfer function of the lead compensator is then given by

$$G_c(j\omega) = \frac{1+j\omega\tau}{1+j\alpha\omega\tau}; \quad \alpha < 1 \tag{10.12}$$

Since $\alpha < 1$, the network output leads the sinusoidal input under steady-state and hence the name lead compensator. The Bode diagram of the lead compensator (with amplifier of gain $A = 1/\alpha$) is given in Fig. 10.11.

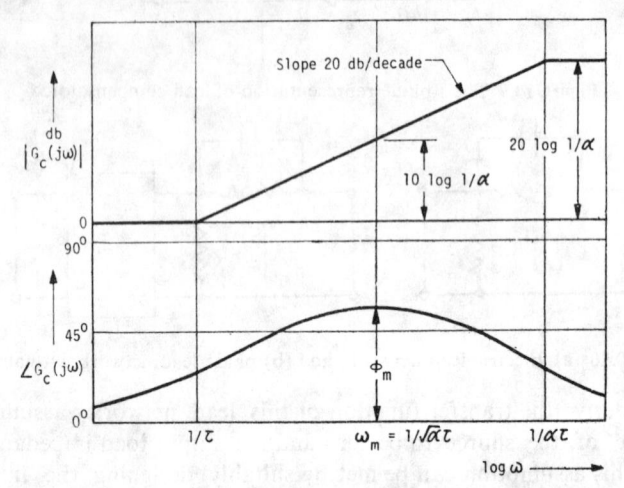

Figure 10.11 Bode plot of phase-lead network with amplifier of gain $A = 1/\alpha$.

From eqn. (10.12), the phase-lead of the compensator at any frequency ω is given by

$$\phi = \tan^{-1}\omega\tau - \tan^{-1}\alpha\omega\tau$$

or

$$\tan \phi = \frac{\omega\tau(1-\alpha)}{1+\alpha\omega^2\tau^2}$$

Using the condition $d\phi/d\omega = 0$, we find that the maximum phase-lead occurs at the frequency

$$\omega_m = 1/(\tau\sqrt{\alpha}) = \sqrt{[(1/\tau)(1/\alpha\tau)]} \qquad (10.13)$$

As seen from eqn. (10.13), ω_m is the geometric mean of the two corner frequencies of the compensator.

At $\omega = \omega_m$, the maximum phase-lead ϕ_m is given by

$$\tan \phi_m = (1-\alpha)/(2\sqrt{\alpha})$$

or

$$\sin \phi_m = (1-\alpha)/(1+\alpha)$$

This gives α in terms of ϕ_m as

$$\alpha = \frac{1-\sin \phi_m}{1+\sin \phi_m} \qquad (10.14)$$

The magnitude of $G_c(j\omega)$ at $\omega = \omega_m$, the frequency of maximum phase-lead, is

$$|G_c(j\omega_m)| = \left|\frac{1+j\omega_m\tau}{1+j\alpha\omega_m\tau}\right| = 1/\sqrt{\alpha} \qquad (10.15)$$

Equation (10.14) is useful in computing the α parameter of the network from the required maximum phase-lead. A plot of ϕ_m versus $1/\alpha$ is shown Fig. 10.12, from which it is observed that to obtain phase-leads more

Introduction to Design

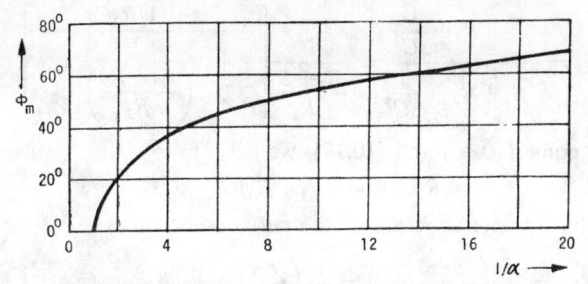

Figure 10.12 ϕ_m vs $1/\alpha$ for a lead network.

than 60° ($\alpha \approx 0.08$), the network attenuation increases rather sharply, quite out of proportion to the increase in phase-lead. Therefore for phase leads greater than 60°, it is advisable to use two cascaded lead networks with moderate values of α rather than a single lead network with too small a value of α.

An important consideration governing the choice of α is the inherent noise in control systems. The nature of this noise is such that the noise signal frequencies are higher than the control signal frequencies. We easily see from Fig. 10.11 that in a lead network the high frequency noise signals are amplified by a factor $1/\alpha > 1$, while the (low frequency) control signals undergo unit amplification (0 db gain). Thus the signal/noise ratio at the output of the lead compensator is poorer than at its input. To prevent the signal/noise ratio at the output from deteriorating excessively, it is recommended that the value of α should not be less than 0.07. A common choice is $\alpha = 0.1$.

Lag Compensator

The s-plane representation of lag compensator is shown in Fig. 10.13a. It has a pole at $-1/\beta\tau$ and a zero at $-1/\tau$ with the zero located to the left of the pole on the negative real axis. The general form of the transfer function of the lag compensator is

$$G_c(s) = \frac{(s+z_c)}{(s+p_c)} = \frac{(s+1/\tau)}{(s+1/\beta\tau)}; \quad \beta = \frac{z_c}{p_c} > 1, \tau > 0 \quad (10.16)$$

Figure 10.13 (a) The s-plane representation of lag compensator; and (b) electric lag network.

The realization of the transfer function of eqn. (10.16) is achieved with an electric lag network shown in Fig. 10.13b from which we can write

$$\frac{E_0(s)}{E_i(s)} = \frac{R_2 + \frac{1}{sC}}{R_1 + R_2 + \frac{1}{sC}} = \frac{1}{\frac{R_1+R_2}{R_2}} \left[s + \frac{s+(1/R_2C)}{\left(\frac{R_1+R_2}{R_2}\right)R_2C} \right] \quad (10.17)$$

Comparing eqns. (10.16) and (10.17), we get

$$\tau = R_2 C, \, \beta = (R_1+R_2)/R_2 > 1$$

Therefore the transfer function of the network becomes

$$G_c(s) = \frac{1}{\beta}\left(\frac{s+1/\tau}{s+1/\beta\tau}\right) = \frac{1}{\beta}\left(\frac{s+z_c}{s+p_c}\right); \, \beta = z_c/p_c > 1 \quad (10.18)$$

$$= \frac{\tau s + 1}{\beta \tau s + 1} \quad (10.19)$$

It is to be noted that compared to the form of transfer function (10.16), $G_c(s)$ realized by the network has a multiplicative factor of $1/\beta$. As in the case of lead network realization, we have an additional degree of freedom in the lag network realization also, which is used for impedance matching.

The sinusoidal transfer function of the lag network is given by

$$G_c(j\omega) = \frac{(1+j\omega\tau)}{(1+j\beta\omega\tau)} \quad (10.20)$$

Since $\beta > 1$, the steady-state output has a lagging phase angle with respect to the sinusoidal input and hence the name lag network. The Bode diagram of the lag network is drawn in Fig. 10.14. The maximum phase lag ϕ_m and the corresponding frequency ω_m are obtained from eqns. (10.14) and (10.13) respectively by replacing α with β.

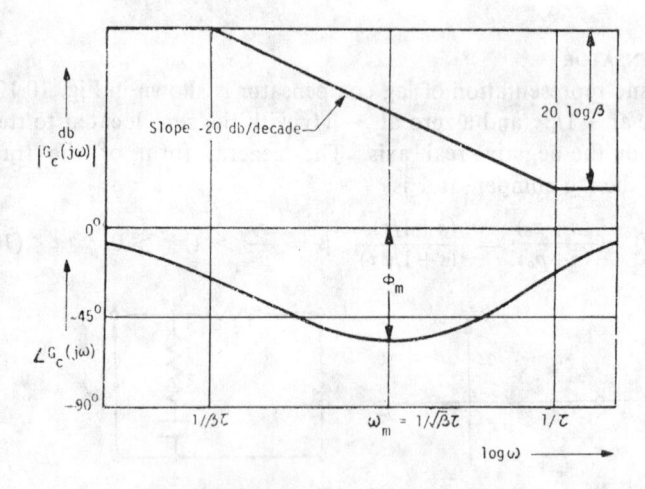

Figure 10.14 Bode plot of phase-lag network.

From Fig. 10.14 it is observed that the lag network has a d.c. gain of unity while it offers a high frequency gain of $1/\beta$. Since $\beta > 1$, it means that the high frequency noise is attenuated in passing through the network whereby

Introduction to Design

the signal to noise ratio is improved, in contrast to the lead network. A typical choice of β is 10.

LAG-LEAD COMPENSATOR*

As discussed earlier, the lag-lead compensator is a combination of a lag compensator and a lead compensator. The lag-section has one real pole and one real zero with the pole to the right of zero. The lead-section also has one real pole and one real zero but the zero is to the right of the pole. The general form of this compensator is

$$G_c(s) = \underbrace{\left(\frac{s+1/\tau_1}{s+1/\beta\tau_1}\right)}_{\text{Lag section}} \underbrace{\left(\frac{s+1/\tau_2}{s+1/\alpha\tau_2}\right)}_{\text{Lead section}}; \quad \beta > 1, \alpha < 1 \qquad (10.21)$$

The eqn. (10.21) can be realized by a single electric lag-lead network shown in Fig. 10.15. From this figure, the transfer function of the network is given by

$$\frac{E_0(s)}{E_i(s)} = \left[\frac{R_2 + 1/sC_2}{R_2 + \dfrac{1}{sC_2} + \dfrac{R_1/sC_1}{R_1 + 1/sC_1}}\right]$$

$$= \left[\frac{\left(s+\dfrac{1}{R_1C_1}\right)\left(s+\dfrac{1}{R_2C_2}\right)}{s^2 + \left(\dfrac{1}{R_1C_1} + \dfrac{1}{R_2C_1} + \dfrac{1}{R_2C_2}\right)s + \dfrac{1}{R_1R_2C_1C_2}}\right] \qquad (10.22)$$

Comparing eqns. (10.21) and (10.22), we have

$$R_1C_1 = \tau_1 \qquad (10.23)$$

$$R_2C_2 = \tau_2 \qquad (10.24)$$

$$R_1R_2C_1C_2 = \alpha\beta\tau_1\tau_2 \qquad (10.25)$$

$$\frac{1}{R_1C_1} + \frac{1}{R_2C_1} + \frac{1}{R_2C_2} = \frac{1}{\beta\tau_1} + \frac{1}{\alpha\tau_2} \qquad (10.26)$$

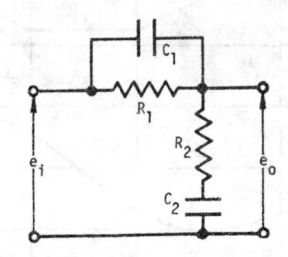

Figure 10.15 Electric lag-lead network.

*When the forward path transfer function has complex poles close to the $j\omega$-axis, phase-lead or phase-lag networks are not effective. In such cases compensation may be achieved by Bridged-T networks.

From eqns. (10.23), (10.24) and (10.25), it is found that

$$\alpha\beta = 1$$

It means that a single lag-lead network does not permit us an independent choice of α and β. Keeping this in view, the transfer function of lag-lead compensator may be written as

$$G_c(s) = \frac{(s+1/\tau_1)(s+1/\tau_2)}{(s+1/\beta\tau_1)(s+\beta/\tau_2)}; \quad \beta > 1 \quad (10.27)$$

$$= \left(\frac{s+z_{c1}}{s+p_{c1}}\right)\left(\frac{s+z_{c2}}{s+p_{c2}}\right); \quad \beta = z_{c1}/p_{c1} = p_{c2}/z_{c2} > 1 \quad (10.28)$$

where $\tau_1 = R_1 C_1$, $\tau_2 = R_2 C_2$ and $\beta > 1$ such that

$$\frac{1}{R_1 C_1} + \frac{1}{R_2 C_1} + \frac{1}{R_2 C_2} = \frac{1}{\beta\tau_1} + \frac{\beta}{\tau_2} \quad (10.29)$$

The s-plane representation of a lag-lead compensator is shown in Fig. 10.16. The sinusoidal transfer function of lag-lead compensator is given by

$$G_c(j\omega) = \frac{(1+j\omega\tau_1)(1+j\omega\tau_2)}{(1+j\omega\beta\tau_1)(1+j\omega\tau_2/\beta)} \quad (10.30)$$

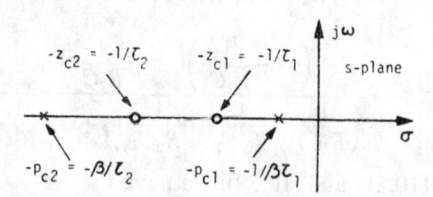

Figure 10.16 The s-plane representation of lag-lead compensator.

The corresponding Bode plot is given in Fig. 10.17.

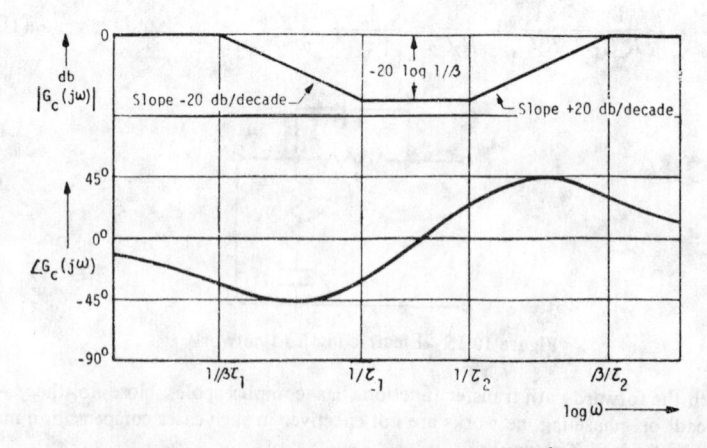

Figure 10.17 Bode plot of lag-lead network.

10.4 Cascade Compensation in Time Domain

Cascade compensation in time domain is conveniently carried out by the root locus technique. In this method of compensation, the original design specifications on dynamic response are converted into ζ and ω_n of a pair of desired complex conjugate closed-loop poles based on the assumption that the system will be dominated by these two complex poles and therefore its dynamic behaviour can be approximated by that of a second-order system. The desired complex pole pair can be located in the s-plane as shown in Fig. 10.18.

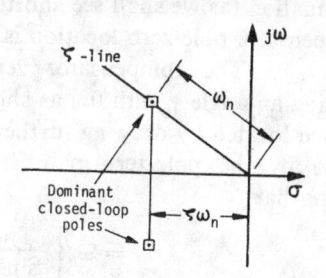

Figure 10.18 Locating the desired dominant closed-loop poles.

A compensator is now designed so that the least-damped complex poles of the resulting transfer function correspond to the desired dominant poles and all other closed-loop poles are either located very close to the open-loop zeros or are relatively far away from the $j\omega$-axis. This ensures that the poles other than the dominant poles make a negligible contribution to the system dynamics.

LEAD COMPENSATION

Consider a unity feedback system with a forward path unalterable transfer function $G_f(s)$. Let the system dynamic response specifications be translated into the desired location s_d for the dominant complex closed-loop poles as shown in Fig. 10.19.

If the angle criterion at s_d is not met, i.e., $\angle G_f(s_d) \neq \pm 180°$, the uncompensated root locus with variable open-loop gain will not pass through the desired root location, indicating the need for compensation. The lead compensator $G_c(s)$ has to be so designed that the compensated root locus passes through s_d. In terms of

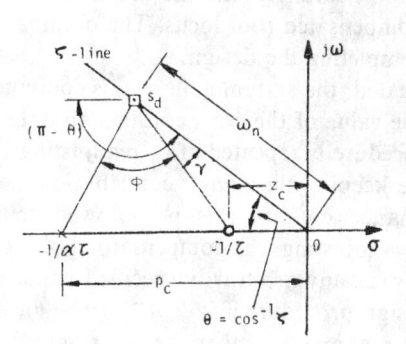

Figure 10.19 Angle contribution of lead compensator.

the angle criterion this requires that

$$\angle G_c(s_d)G_f(s_d) = \angle G_c(s_d) + \angle G_f(s_d) = \pm 180°$$

or
$$\angle G_c(s_d) = \phi = \pm 180° - \angle G_f(s_d) \qquad (10.31)$$

Thus for the root locus of the compensated system to pass through the desired root location, the lead compensator pole-zero pair must contribute an angle ϕ given by eqn. (10.31) and shown in Fig. 10.19.

For a given angle ϕ required for lead compensation, there is no unique location for the pole-zero pair. From the point of view of lead network attenuation (as we shall see shortly, this is a narrow point of view), the best compensator pole-zero location is the one which gives the largest value of $\alpha = z_c/p_c$. The compensator zero is located by drawing a line from s_d making an angle γ with $0s_d$ as shown in Fig. 10.19. The compensator pole is then located by drawing further the requisite angle ϕ to be contributed at s_d by the pole-zero pair. It readily follows from the geometry of this figure that

$$z_c = \omega_n \left[\frac{\sin \gamma}{\sin (\pi - \theta - \gamma)} \right]$$

and

$$p_c = \omega_n \left[\frac{\sin(\gamma+\phi)}{\sin (\pi - \theta - \gamma - \phi)} \right]$$

$$\therefore \quad \alpha = \frac{\sin \gamma \sin (\pi - \theta - \gamma - \phi)}{\sin (\pi - \theta - \gamma) \sin (\gamma+\phi)}$$

The angle γ for largest α is obtained from the condition

$$\frac{d\alpha}{d\gamma} = 0$$

which gives

$$\gamma = \tfrac{1}{2}(\pi - \theta - \phi) \tag{10.32}$$

Though the above method of locating the lead compensator pole-zero yields the largest value of α, it does not guarantee the dominance of the desired closed-loop poles in the compensated root locus. The dominance condition must be checked before completing the design.

With compensator pole-zero so located, the system gain at s_d is computed to determine the error constant. If the value of the error constant so obtained is unsatisfactory, the above procedure is repeated after readjusting the compensator pole-zero location while keeping their angle contribution fixed at ϕ. In this readjustment the dominance condition is not allowed to be violated. It may be noted that by adjusting the compensator pole-zero location, it is not always possible to meet any arbitrarily specified value of the error constant. In the above design procedure if $(\gamma+\phi) > (\pi - \theta)$, the compensator pole lies on the positive real axis and therefore a single lead network becomes unrealizable. In such cases, realization can be achieved by using two or more identical networks in tandem.

Example 10.1: Let us first consider the example of a unity feedback type-2 system with

$$G_f(s) = K/s^2$$

This system has zero steady-state error for both the step and ramp inputs. It can be seen from its root locus plot that the closed-loop poles always lie on the $j\omega$-axis. As explained earlier, the lead compensator is the only choice for this system.

It is desired to compensate the system so as to meet the following transient response specifications.

Settling time ≤ 4 sec
Peak overshoot for step input $\leq 20\%$

These specifications imply that

$$t_s = \frac{4}{\zeta\omega_n} \leq 4 \text{ sec} \quad \text{or} \quad \zeta\omega_n \geq 1$$

and
$$\zeta \geq 0.45$$

The desired location ($\zeta = 0.45$ and $\zeta\omega_n = 1$) for the dominant closed-loop poles (s_d) is shown in Fig. 10.20a. The angle at s_d because of the two poles at origin of the uncompensated system is

$$\angle G_f(s) = -2 \times 117° = -234°$$

Therefore the angle contribution at s_d required of the lead compensator pole-zero pair is

$$\phi = -180° - (-234°) = 54°$$

For the largest value of α, we have from eqn. (10.32)

$$\gamma = \tfrac{1}{2}(\pi - \theta - \phi) = \tfrac{1}{2}(180° - \cos^{-1} 0.45 - 54°) = 31.5°$$

Using angles γ and ϕ as shown in Fig. 10.20a, we obtain

$$z_c = 1/R_1 C = 1.15$$

$$p_c = \frac{1}{\left(\dfrac{R_1}{R_1 + R_2}\right) R_1 C} = 4.3$$

$$\alpha = z_c/p_c = 0.268 \text{ (satisfactory)}$$

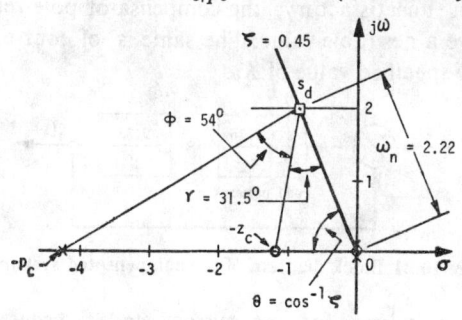

Figure 10.20 (a) Determination of locations for the compensator pole and zero.

Choosing $C = 1\mu F$, we obtain

$$R_1 = 870 \; K\Omega; \; R_2 = 635 \; K\Omega$$

With the addition of the tandem lead compensator, the open-loop transfer function of the system becomes

$$G(s) = G_f(s) G_c(s) = \frac{K(s+1.15)}{s^2(s+4.3)}$$

The root locus for the compensated system is shown in Fig. 10.20b from which the gain at the desired closed-loop pole location is found to be $K = 9.33$. Therefore the open-loop transfer function of the compensated system is

$$G(s) = \frac{9.33(s+1.15)}{s^2(s+4.3)}$$

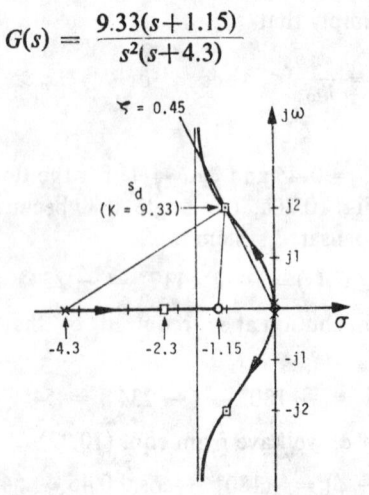

Figure 10 20 (b) Root locus plot of system with $G(s) = K(s+1.15)/s^2(s+4.3)$.

The compensated system block diagram is given in Fig. 10.21.

The acceleration error constant is given by

$$K_a = \lim_{s \to 0} s^2 G(s) = \frac{9.33 \times 1.15}{4.3} = 2.5$$

The steady-state performance of this system is quite satisfactory. In case K_a obtained above is unsatisfactory, the compensator pole-zero pair may be adjusted to achieve a desirable value. The same is, of course, not possible for any arbitrarily specified value of K_a.

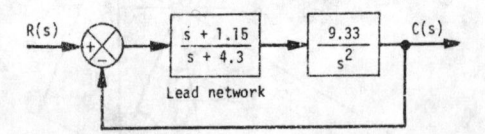

Figure 10.21 Block diagram of the compensated system.

The compensator designed for the system under discussion results in maximum value of α and hence minimum value of additional required amplifier gain $A = 1/\alpha (= 3.75)$. It is of course essential to check the dominance condition before declaring the compensator design to be complete.

To accomplish this, the third closed-loop pole introduced because of compensator zero is determined. From Fig. 10.20b, it is found to be at $s = -2.3$. Because of this closed-loop pole, there will be an exponential term in the transient response in addition to the one due to the complex pair of closed-loop poles.

Introduction to Design

The closed-loop transfer function of the compensated system is obtained as

$$\frac{C(s)}{R(s)} = \frac{9.33(s+1.15)}{(s+1+j2)(s+1-j2)(s+2.3)}$$

For an impulse input,

$$c(t) = \mathcal{L}^{-1}\left[\frac{9.33(s+1.15)}{(s+1+j2)(s+1-j2)(s+2.3)}\right]$$
$$= 3.94\, e^{-t} \sin(2t - \tan^{-1}1.83) - 1.84\, e^{-2.3t}$$

We thus observe that the second term in $c(t)$ contributes appreciably to system dynamics. If this contribution is not tolerable, the compensator zero is shifted further to left and the design procedure repeated. This, of course, means higher network attenuation but as stated earlier, α up to 0.07 is an acceptable value. The above design also illustrates that the best value of α may not result in satisfactory design.

Though the method presented above yields the largest value of α which is a desirable result, it does not guarantee the dominance of desired closed-loop poles in the compensated root locus. No general rule can be laid down which will satisfy this condition for various open-loop pole-zero configurations. However, the following *guidelines* can be of considerable help.

Place the compensating zero on the real axis in the region below the desired closed-loop pole location such that it lies close to the left of any open-loop pole in this region. While giving generally a large value of α, this technique ensures that the closed-loop pole on the real axis caused by the introduction of the compensating zero will be located close to it and would therefore make a negligible contribution to system dynamics. This technique is suggested by *cancellation compensation* wherein the compensating zero is placed exactly in the open-loop pole location completely cancelling the effect of this pole. In case the uncompensated system does not have any open-loop pole in the region below the desired closed-loop poles, the dominance condition must be checked by locating the real axis closed-loop pole created by the compensating zero.

Example 10.2: A type-2 system with an open-loop transfer function

$$G_f(s) = \frac{K}{s^2(s+1.5)}$$

is to be compensated to meet the specifications given in Example 10.1.

Following the procedure of the previous example. the desired dominant roots are found to lie at $s_d = -1 \pm j2$. From Fig. 10.22 the angle contribution required from a lead compensator is

$$\phi = \pm 180° - \angle G_f(s_d)$$
$$= \pm 180° - (-2 \times 117° - 75°) = 129°$$

The large value of ϕ here is an indication that a double lead network is appropriate. Each section of a double lead network has then to contribute an angle of 64.5° at s_d.

Let us locate the compensator zero at $s = -1.7$, i.e., in the region below the desired dominant closed-loop pole location and just to the left of the open-loop pole at $s = -1.5$. Join the compensator zero to s_d and locate the compensator pole by making an angle $\phi = 64.5°$ as shown in Fig. 10.22. The location of the pole is found to be at -19.8.

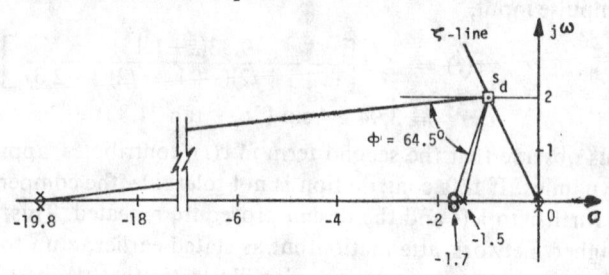

Figure 10.22 Design of double lead compensator.

The open-loop transfer function of the compensated system becomes

$$G(s) = \frac{830(s+1.7)^2}{s^2(s+1.5)(s+19.8)^2}$$

By locating the closed-loop poles of the compensated system, it can be easily verified that one closed-loop pole is located very close to the open-loop zero at -1.7 (contributed by the compensating zero) and therefore makes negligible contribution to system dynamics, while the other closed-loop poles are located far to the left of -1 and hence the dominance of the desired closed-loop poles ($-1 \pm j2$) is preserved.

Example 10.3: Consider a type-1 system with an open-loop transfer function of

$$G_f(s) = \frac{K}{s(s+1)(s+4)}$$

The system is to be compensated to meet the following specifications:

Damping ratio $\zeta = 0.5$

Undamped natural frequency $\omega_n = 2$

Using the transient response specifications the desired dominant closed-loop poles are found to lie at

$$s_d = -1 \pm j1.73$$

From Fig. 10.23, the angle contribution required from the lead compensator pole-zero pair is

$$\phi = -180° - \angle G_f(s_d) = -180° - (-120° - 90° - 30°) = 60°$$

Further it is observed that the open-loop pole at $s = -1$ lies directly below the desired closed-loop pole location. Place the compensator zero close to this pole to its left, say at $s = -1.2$. Such a choice of compensator zero generally ensures the dominance condition.

Join the compensator zero to s_d and locate the compensator pole by making an angle of $\phi = 60°$ as shown in Fig. 10.23. The location of the pole is found to be at -4.95.

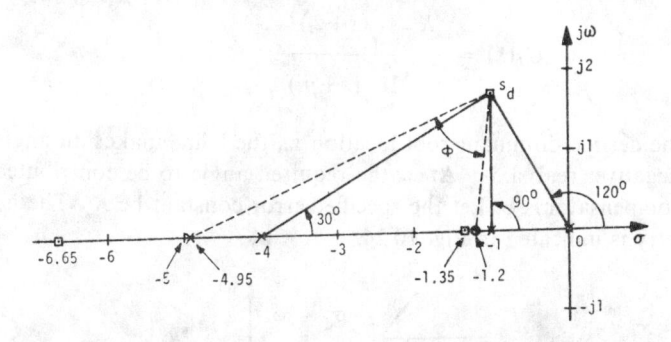

Figure 10.23 Design of lead compensator for Example 10.3.

The open-loop transfer function of the compensated system becomes

$$G(s) = \frac{30.4(s+1.2)}{s(s+1)(s+4)(s+4.95)}$$

The velocity error constant is

$$K_v = \lim_{s \to 0} sG(s) = \frac{30.4 \times 1.2}{1 \times 4 \times 4.95} = 1.84$$

It must be understood here that only a marginal increase in K_v above this value can be achieved by a slight readjustment of the compensating zero. Any large shift in the compensating zero would result in violation of the dominance condition.

Let us now check the dominance condition. From the root locus of the compensated system (Fig. 10.23) the closed-loop transfer function in factored form is obtained as

$$\frac{C(s)}{R(s)} = \frac{30.4(s+1.2)}{(s+1+j1.73)(s+1-j1.73)(s+1.35)(s+6.65)}$$

For an impulse input

$$c(t) = 2.94e^{-t} \sin(1.73t - \tan^{-1} 0.216) - 0.27e^{-1.35t} + 0.89e^{-6.65t}$$

In the above equation, the third term is contributed by the closed-loop pole which is created by the introduction of the compensator zero and lies close to this zero. It is observed that the coefficient of this term is small and its contribution to system dynamics is therefore negligible. Equally the contribution of the pole at $s = -6.65$ is negligible as it is located far to the left of the dominant poles. It is thus established that this compensator design guarantees the dominance of the closed-loop poles at s_d.

Warren-Ross Method of Lead Compensator Design

This method of lead compensator design avoids the trial and error procedure for making the compensated root locus pass through the desired root location s_d while simultaneously satisfying the specified error constant. We take the usual unity feedback type-r system with a forward transfer function of

$$G_f(s) = \frac{K \prod_{i=1}^{m}(s+z_i)}{s^r \prod_{j=r+1}^{n}(s+p_j)} \quad (10.33)$$

For the desired dominant root location s_d, the ζ-line makes an angle θ with the negative real axis. At s_d the required angle to be contributed by the lead compensator is ϕ. Let the specified error constant be K_e^c. The necessary geometry is indicated in Fig. 10.24.

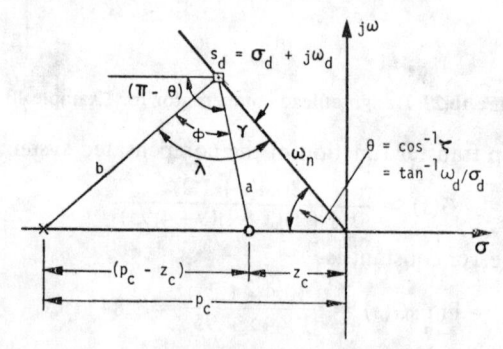

Figure 10.24 Geometry of Warren-Ross method of lead compensator design.

In order to meet the specifications, let us introduce the lead compensator with transfer function

$$G_c(s) = \frac{s+z_c}{s+p_c}$$

The point s_d is now made to lie on the root locus of the compensated system and the gain K^c of the compensated system is given by

$$K^c(s_d) = \frac{|s_d|^r \prod_{j=r+1}^{n}|s_d+p_j|}{\prod_{i=1}^{m}|s_d+z_i|} \cdot \frac{|s_d+p_c|}{|s_d+z_c|}$$

Let
$$K^{uc}(s_d) = |s_d|^r \frac{\prod_{j=r+1}^{n}|s_d+p_j|}{\prod_{i=1}^{m}|s_d+z_i|}$$

$$\therefore \quad K^c(s_d) = K^{uc}(s_d) \frac{|s_d+p_c|}{|s_d+z_c|} = K^{uc}(s_d) \frac{b}{a} \quad (10.34)$$

where $|s_d+p_c| = b$, is the distance of the compensator pole from s_d and $|s_d+z_c| = a$, is the distance of the compensator zero from s_d as shown in Fig. 10.24. It may be noted that $K^{uc}(s_d)$ is the uncompensated system gain at s_d. Since with compensation, the system must meet specified error constant K_e^c, we have*

$$K_e^c = K^c(s_d) \frac{\prod_{i=1}^{m} z_i}{\prod_{j=r+1}^{n} p_j} \frac{z_c}{p_c}$$

or $\quad K_e^c = K^{uc}(s_d) \dfrac{\prod_{i=1}^{m} z_i}{\prod_{j=r+1}^{n} p_j} \dfrac{z_c}{p_c} \dfrac{b}{a}\quad$ (10.35)

It follows from eqn. (10.35) that

$$\frac{z_c}{p_c} = \left[\frac{K_e^c}{K^{uc}(s_d)} \frac{\prod_{j=r+1}^{n} p_j}{\prod_{i=1}^{m} z_i} \right] \frac{a}{b} = K_1(s_d)\frac{a}{b} \quad (10.36)$$

where $\quad K_1(s_d) = \dfrac{K_e^c}{K^{uc}(s_d)} \dfrac{\prod_{j=r+1}^{n} p_j}{\prod_{i=1}^{m} z_i} \quad$ (10.37)

From the geometry of Fig. 10.24 we can write

$$\frac{p_c}{b} = \frac{\sin \lambda}{\sin \theta} \quad \text{and} \quad \frac{a}{z_c} = \frac{\sin \theta}{\sin \gamma}$$

Multiplying the above two equations, we have

$$\frac{a}{b} = \frac{z_c}{p_c} \frac{\sin \lambda}{\sin \gamma} \quad (10.38)$$

Substituting in eqn. (10.36) we get

$$\frac{z_c}{p_c} = K_1(s_d) \frac{z_c}{p_c} \frac{\sin \lambda}{\sin \gamma}$$

or $\quad \sin \gamma = K_1(s_d) \sin(\gamma + \phi)$

from which we obtain

$$\tan \gamma = \frac{K_1(s_d) \sin \phi}{1 - K_1(s_d) \cos \phi} \quad (10.39)$$

With $K_1(s_d)$ and ϕ known from the uncompensated system and the specified K_e^c, we can determine γ from eqn. (10.39). If γ turns out to be negative, a lead network cannot be realized as the network zero lies on the positive real axis. Also if $(\gamma + \phi) > (\pi - \theta)$ a single network is unrealizable as the network pole lies on the positive real axis. In this case the requisite

*It may be noted that K^c and K_e^c are the system gains in pole-zero and time-constant forms respectively.

ϕ can be provided by a number (say t) of identical networks in tandem (connected of course by buffer stages) each providing an angle ϕ/t at s_d. Equation (10.39) is then modified by replacing $K_1(s_d)$ with $[K_1(s_d)]^{1/t}$ and ϕ with ϕ/t.

By obtaining the angle γ from eqn. (10.39) or its modified form the compensator zero is located and then from ϕ (or ϕ/t in multiple network case) the compensator pole can be determined. Since in this design procedure the dominance condition does not appear in the picture, it must be checked by drawing the root locus of the compensated system.

Let us design a lead compensator using Warren-Ross method for the system of Example 10.3 with the additional specification of $K_v = 1.5$.

Using transient response specifications, the desired dominant closed-loop poles were found to lie at $s_d = -1 \pm j1.73$ and the angle contribution ϕ required from the lead compensator pole-zero pair was found to be 60°.

The gain at s_d obtained from Fig. 10.25 is

$$K^{uc}(s_d) = 12.13$$

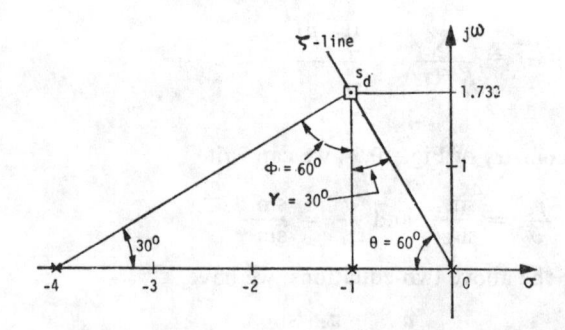

Figure 10.25 Lead compensation by Warren-Ross method.

Since $K_e^c = K_v = 1.5$, we have from eqn. (10.37)

$$K_1(s_d) = \left(\frac{.5}{12.13}\right) \times 1 \times 4 = 0.495$$

Then eqn. (10.39) yields

$$\tan \gamma = \frac{0.495 \times 0.866}{1 - 0.495 \times 0.5} = 0.572$$

or $\gamma = 30°$

Draw a line making an angle of 30° with Os_d (i.e., with the ζ-line) intersecting the real axis at $s = -1$. This determines the location of compensator zero. Then $\phi = 60°$ determines compensator pole at $s = -4$ as shown in Fig. 10.25.

The open-loop transfer function of the compensated system becomes

$$G(s) = \frac{K(s+1)}{s(s+1)(s+4)(s+4)} = \frac{K}{s(s+4)^2}$$

The compensation cancels out the open-loop pole at $s = -1$.

For the compensated system, the value of K at s_d is found to be 24.5. Therefore

$$G(s) = \frac{24.5}{s(s+4)^2}$$

From the compensated system it can be checked that the dominance condition is preserved.

Lag Compensation

As pointed out earlier in Section 10.2, a lag compensator is used to improve the steady-state behaviour of a system while preserving a satisfactory transient response. This compensation scheme therefore is found useful in systems having satisfactory transient response but unsatisfactory steady-state response.

Consider a unity feedback system with a forward-path transfer function of

$$G_f(s) = \frac{K \prod_{i=1}^{m} (s+z_i)}{s^r \prod_{j=r+1}^{n} (s+p_j)}$$

At a certain value of K, this system has a satisfactory transient response, i.e., its root locus plot passes through (or close to) the desired closed-loop pole location s_d indicated in Fig. 10.26. It is required to improve the system error constant to a specified value K_e^c without impairing its transient response. This requires that after compensation the root locus should continue to pass through s_d, while the error constant at s_d is raised to K_e^c. To accomplish this, consider adding a lag compensator pole-zero pair with

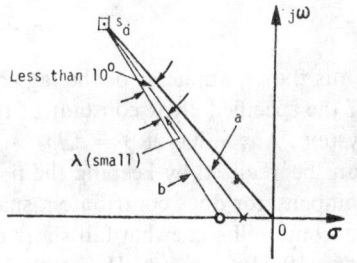

Figure 10.26 Locating the lag compensator pole-zero.

zero to the left of the pole. If this pole-zero pair is located close to each other, it will contribute a negligible angle at s_d such that s_d continues to lie on the root locus of the compensated system. Figure 10.26 shows the location of such a lag compensator pole-zero pair. It should be noticed from this figure that apart from being close to each other, the pole-zero pair is also located close to origin, the reason for which will become obvious from the discussion given below.

The gain of the uncompensated system at s_d is given by

$$K^{uc}(s_d) = \frac{|s_d|^r \prod_{j=r+1}^{n} |s_d+p_j|}{\prod_{i=1}^{m} |s_d+z_i|} \qquad (10.40)$$

For the compensated system, the system gain at s_d is

$$K^c(s_d) = \frac{\mid s_d \mid^r \prod_{j=r+1}^{n} \mid s_d + p_j \mid}{\prod_{i=1}^{m} \mid s_d + z_i \mid} \frac{a}{b} \qquad (10.41)$$

Since the pole and zero are located very close to each other, they are nearly equidistant from s_d, i.e., $a \approx b$. It therefore follows from eqns. (10.40) and (10.41), that

$$K^c(s_d) \approx K^{uc}(s_d)$$

The error constant of the compensated system is given by

$$K_e^c = K^c(s_d) \frac{\prod_{i=1}^{m} z_i}{\prod_{j=r+1}^{n} p_j} \frac{z_c}{p_c} \approx K^{uc}(s_d) \frac{\prod_{i=1}^{m} z_i}{\prod_{j=r+1}^{n} p_j} \frac{z_c}{p_c}$$

$$\approx K_e^{uc} \times \frac{z_c}{p_c}$$

where K_e^{uc} = error constant at s_d of the uncompensated system. Since the error constant K_e^c of the compensated system must equal the specified value we have

$$K_e^c \approx K_e^{uc} \frac{z_c}{p_c}$$

or

$$\beta = \frac{z_c}{p_c} \approx \frac{K_e^c}{K_e^{uc}} \qquad (10.42)$$

Thus the β-parameter of the lag compensator is nearly equal to the ratio of the specified error constant to the error constant of the uncompensated system. Any value of $\beta = z_c/p_c > 1$ with $-p_c$ and $-z_c$ close to each other can be realized by keeping the pole-zero pair close to origin. Since the lag compensator does contribute a small negative angle λ at s_d, the actual error constant will somewhat fall short of the specified value if β obtained from eqn. (10.42) is used. Hence for design purpose, we choose a β somewhat larger than that given by this equation.

Further, the effect of the small lag angle λ is to give the closed-loop pole with specified ζ but slightly lower ω_n. This can be anticipated and counteracted by taking the ω_n of s_d to be somewhat larger than the specified value.

In the light of the above discussion, the steps for lag compensator design are summarized below.

1. Draw the root locus plot for the uncompensated system.
2. Translate the transient response specifications into a pair of complex dominant roots and locate these roots on the uncompensated root locus plot. Since the transient performance of the uncompensated system is satisfactory, the desired dominant roots will lie on (or close to) the uncompensated root locus.

Introduction to Design

3. Calculate the gain of the uncompensated system at the dominant root s_d and evaluate the corresponding error constant.

4. Determine the factor by which the error constant of the uncompensated system should be increased to meet the specified value. Choose the β-parameter of lag compensator to be somewhat greater than this factor.

5. Select zero of the compensator sufficiently close to the origin. As a guide rule, we may construct a line making an angle of 10° (or less) with the desired ζ-line from s_d (see Fig. 10.26). The intersection of this line with the real axis gives location of the compensator zero.

6. The compensator pole can then be located at $-p_c = -z_c/\beta$. It is important to note that the pole-zero pair should contribute an angle λ less than 5° at s_d so that the root locus plot in the region of s_d is not appreciably changed and hence satisfactory transient behaviour of the system is preserved.

Let us illustrate the above steps with the help of an example.

Example 10.4: Let us reconsider the system discussed in Example 10.3 with an open-loop transfer function of

$$G_f(s) = \frac{K}{s(s+1)(s+4)}$$

The system is to be compensated to meet the following specifications:

Damping ratio $\zeta = 0.5$
Settling time $t_s = 10$ sec
Velocity error constant $K_v \geqslant 5$ sec^{-1}

From the transient response specifications, it follows that

$$\text{undamped natural frequency } \omega_n = \frac{4}{10 \times 0.5} = 0.8 \text{ rad/sec}$$

The desired dominant closed-loop poles are then required to be located at

$$s_d = -\zeta\omega_n \pm j\sqrt{(1-\zeta^2)}\,\omega_n = -0.4 \pm j0.7$$

as shown in Fig. 10.27. These transient response specifications are such that s_d lies on (it may also lie close to) the root locus of the uncompensated system.

At s_d, the gain K^{uc} of the uncompensated system is given by

$$K^{uc} = 0.8 \times 0.9 \times 3.7 = 2.66$$

The velocity error constant of the uncompensated system is given by

$$K_v^{uc} = \lim_{s \to 0} sG_f(s) = 2.66/4 = 0.666$$

The system has to be compensated to obtain a velocity error constant of $K_v^c \geqslant 5$. Therefore the β-parameter of the lag network is given by

$$\beta = K_v^c/K_v^{uc} = 5/0.666 = 7.5$$

To counter the effect of a small negative angle contributed by the lag network at s_d, choose a slightly higher value of β, say $\beta = 10$.

From the desired location s_d, draw a line making an angle of say 6° with the desired ζ-line. Its intersection with the real axis determines the compensator zero at $-z_c = -0.1$ as shown in Fig. 10.27. The compensator pole is then located at $-p_c = -z_c/\beta = -0.01$.

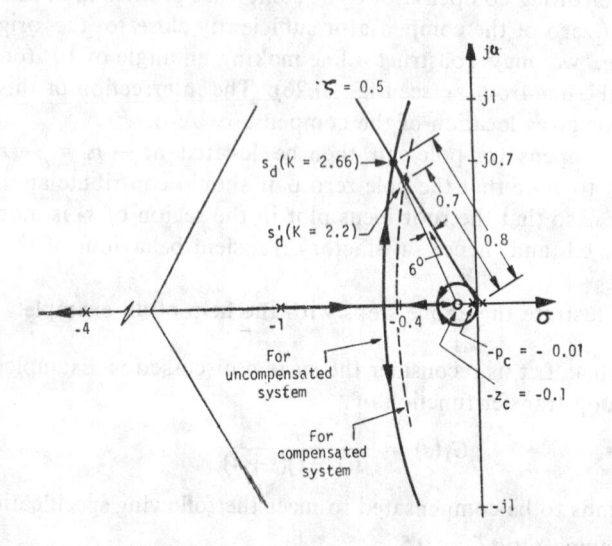

Figure 10.27 Root locus plots for uncompensated and lag compensated systems.

Now for the compensated system locate the point s'_d which lies on the $\zeta = 0.5$ line and the compensated root locus using the angle criterion. For the compensated system at s'_d

$$K^c(s'_d) = 2.2$$

Thus the open-loop transfer function of the compensated system is

$$G(s) = \frac{2.2(s+0.1)}{s(s+1)(s+4)(s+0.01)}$$

For which we obtain

$$K_v = \frac{2.2 \times 0.1}{4 \times 0.01} = 5.5 \text{ (satisfactory)}$$

Comparison of the root locus for the uncompensated and lag compensated systems shows that ω_n has decreased from 0.8 to 0.6 rad/sec. This means a slight increase in the settling time. If this increase is unacceptable, the system must be redesigned by choosing ω_n at s_d to be slightly higher than the specified value 0.8.

LAG-LEAD COMPENSATION

In the preceding section, we have seen that the lead compensator is suitable for systems having unsatisfactory transient response but it provides only a limited improvement in steady-state response. If the steady-state behaviour is highly unsatisfactory, the lead compensator may not be the answer. On the other hand for systems with satisfactory transient response but unsatisfactory steady-state response, the lag compensator is found to be a good choice.

When both transient and steady-state responses are quite unsatisfactory, we must draw upon the combined powers of lag and lead compensators in order to meet the specifications. A more convenient choice is the combined lag-lead compensator introduced earlier in Section 10.3.

The general form of the transfer function of the lag-lead compensator in pole-zero form is

$$G_c(s) = \underbrace{\frac{(s+z_{c1})}{(s+p_{c1})}}_{\text{Lag-section}} \underbrace{\frac{(s+z_{c2})}{(s+p_{c2})}}_{\text{Lead-section}}; \quad \beta = \frac{z_{c1}}{p_{c1}} = \frac{p_{c2}}{z_{c2}} > 1$$

$$= G_{c1}(s)G_{c2}(s)$$

The usual design procedure is to first design the lead-section to meet the specifications on transient response. This design yields a suitable value of β. The error constant is then determined for the lead compensated system. If this is satisfactory the design is complete. If on the other hand, the specified error constant is much higher than that obtained by lead compensation, we then proceed to design the lag-section such that the overall system meets the specifications on steady-state performance. Since β is already determined, the lag-section can increase the error constant by about a factor β. If this increase is sufficient, we proceed to complete the lag-section design as usual. Otherwise, we must redesign the lead-section for a larger value of β. The following example illustrates the procedure.

Example 10.5: Let us again consider the system discussed in Example 10.3. The open-loop transfer function of the uncompensated system is

$$G_f(s) = \frac{K}{s(s+1)(s+4)}$$

This system is now required to be compensated to meet the following specifications:

   Damping ratio $\zeta = 0.5$
   Undamped natural frequency $\omega_n = 2$
   Error constant $K_v \geqslant 5$

Let us design a lead compensator to meet the transient response specifications (refer Example 10.3). The desired location of the dominant closed-loop poles is

$$s_d = -1 \pm j1.73$$

The angle contribution required of the lead compensator pole-zero is $\phi = 60°$. Let the compensator zero be placed at $s = -1$, so as to cancel one of the open-loop poles. The compensating pole is then located by drawing a line at $\phi = 60°$ from the line joining -1 and s_d as shown in Fig. 10.28. The compensating pole is found to be at $s = -4$, giving $\beta = -4/-1 = 4$.

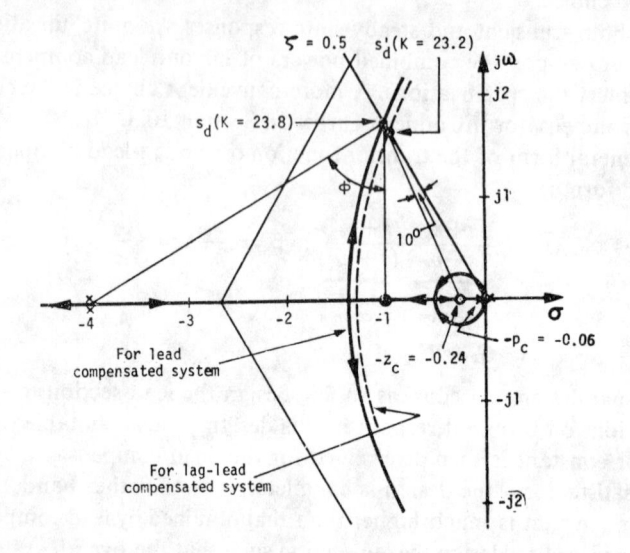

Figure 10.28 Design of lag-lead compensator.

The transfer function of the lead compensated system is

$$G_f(s)G_{c2}(s) = \frac{K^{c2}}{s(s+4)^2} \qquad (10.43)$$

From Fig. 10.28, the gain K^{c2} at s_d is found to be 23.8. Therefore, from eqn. (10.43)

$$G_f(s)G_{c2}(s) = \frac{23.8}{s(s+4)^2}$$

$$K_v^{c2} = \lim_{s \to 0} sG_f(s)G_{c2}(s) = 1.49$$

This does not meet the specified $K_v \geqslant 5$. A lag-section using $\beta = 4$ will increase K_v by about four times, which then satisfies the specification on steady-state performance.

The line from s_d making an angle of $10°$ with the desired ζ-line intersects real axis at -0.24 which gives the location of the zero of the lag-section. The pole of lag-section is then found to be at $-p_{c1} = -0.24/4 = -0.06$.

The open-loop transfer function of the lag-lead compensated system then becomes

$$G(s) = \frac{K^c(s+0.24)}{s(s+0.06)(s+4)^2} \qquad (10.44)$$

The root-locus plot for lag-lead compensated system is shown in Fig. 10.28. From this figure, the gain at s_d' (which is slightly shifted from s_d due to introduction of the lag-section) is given by

$$K^c = 23.2$$

Therefore from eqn. (10.44) the open-loop transfer function of the lag-lead compensated system is

$$G(s) = \frac{23.2(s+0.24)}{s(s+0.06)(s+4)^2}$$

10.5 Cascade Compensation in Frequency Domain

Compensation design can also be conveniently carried out using frequency domain method. Frequency domain specifications are generally given in the following form:

1. Phase margin ϕ_{pm} or resonant peak M_r—indicative of relative stability.
2. Bandwidth ω_b or resonant frequency ω_r—indicative of rise time and settling time.
3. Error constant—indicative of steady-state error.

In case the specifications are given in time domain, we must first translate these into frequency domain to carry out frequency domain compensation. This translation is carried out by using the explicit correlations between the two domains for second-order system. As pointed out earlier in the text, these correlations are valid approximations for higher-order systems dominated by a pair of complex conjugate poles. The correlations given earlier in Chapters 8 and 9 are reproduced below for convenience of use.

$$M_r = \frac{1}{2\zeta\sqrt{(1-\zeta^2)}} \tag{10.45}$$

$$\omega_r = \omega_n\sqrt{(1-2\zeta^2)} \tag{10.46}$$

$$\phi_{pm} = \tan^{-1}\{2\zeta/[\sqrt{(1+4\zeta^4)} - 2\zeta^2]^{1/2}\} \tag{10.47}$$

$$\omega_b = \omega_n[1 - 2\zeta^2 + \sqrt{(2-4\zeta^2+4\zeta^4)}]^{1/2} \tag{10.48}$$

The specifications in terms of M_r and ω_r are found convenient when compensation is carried out by Nyquist plots, while compensation using Bode plots is usually easier to handle for specified phase margin. Gain cross-over frequency could be used as a rough measure of ω_b. Of course, when Nichols charts are used, any type of specification can be handled.

After completing the compensation design in frequency domain, one must recheck the time response specifications by computing the exact time response of the compensated system. This is necessary because the time response specifications are converted into frequency response proceeding on the assumption that the compensated system will have a dominant pair of closed-loop poles. Based on the results of this check, it may sometimes be necessary to repeat the complete design process.

From the above discussion it is realized that in frequency domain compensation, direct control on system time performance is lost. This disadvantage is compensated by the advantages of frequency domain methods such as simplicity in analysis and design, ease in experimental determination of frequency response for real systems.

As said earlier, the frequency domain compensation may be carried out using Nyquist plots, Bode plots or Nichols chart. The advantages of the Bode plots are that they are easier to draw and modify. Further the gain adjustments are conveniently carried out and the error constants are always clearly in evidence. We shall discuss below mainly the design' procedures using Bode plots, while Nichols charts are used to check the values of M_r, ω_r and ω_b wherever necessary.

LEAD COMPENSATION

The lead compensation on Bode plots proceeds by adjusting the system error constant to the desired value. The phase margin of the uncompensated system is then checked. If found unsatisfactory, the lead compensation is designed to meet the specified phase margin. Figure 10.29 shows the Bode plots of a unity feedback system with an open-loop transfer function $G_f(s) = K_v/s(\tau s+1)$. With the K_v adjusted to the specified value, let the gain cross-over frequency and the phase margin of the uncompensated system be ω_{c1} and ϕ_1 respectively. Let it be assumed that ϕ_1 falls short of the specified phase margin ϕ_s. Additional phase margin can be provided by a lead network so placed that its two corner frequencies be on either side of ω_{c1}. This location of the network's Bode plot ensures that the phase-lead is provided over the desired region. As is seen from Fig. 10.29, the addition of the phase-lead network (along with necessary amplification to cancel the attenuation of the network) in this region causes the new cross-over frequency to shift to the right to some unknown value ω_{c2}. This fact

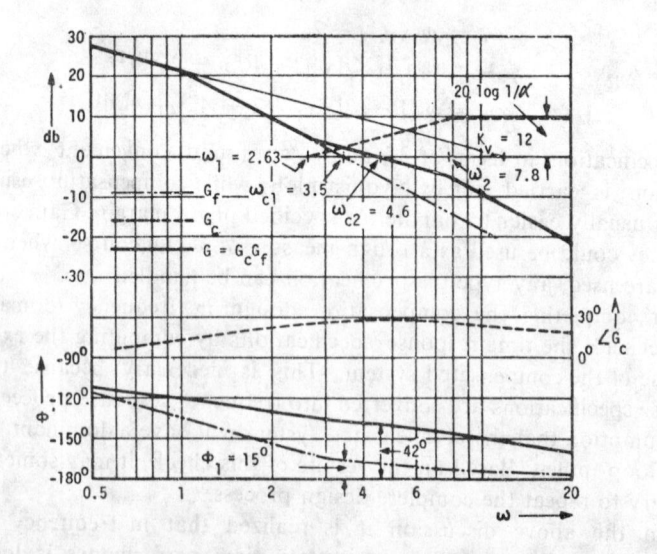

Figure 10.29 Design of lead compensation for type-1 system.

reduces the contribution to phase margin of the fixed part of the system $G_f(s)$ to some value $\phi_2 < \phi_1$. The phase-lead ϕ_l required at ω_{c2} to bring the phase margin to the specified value is given by

$$\phi_l = \phi_s - \phi_2$$

Since ϕ_2 is unknown, we make a guess on the same as

$$\phi_2 = \phi_1 - \epsilon$$

Thus the required phase-lead at the new cross-over frequency ω_{c2} is obtained as

$$\phi_l = \phi_s - \phi_1 + \epsilon$$

where ϵ is the unknown reduction in the phase angle $\angle G_f(s)$ on account of the increase in the cross-over frequency. A guess is made on the value of ϵ depending on the slope in this region of the db-log ω plot of the uncompensated system. For a slope of -40 db/decade, as is in the present example, $\epsilon = 5°$ is a good guess. The guessed value may have to be as high as 15-20° for a slope of -60 db/decade.

In order to provide a phase-lead ϕ_l at ω_{c2} with the largest value of the network parameter α (this is desirable from signal/noise considerations), the frequency of maximum phase-lead ω_m of the network must be made to coincide with ω_{c2}. Thus we set

$$\omega_{c2} = \omega_m$$

$$\therefore \quad \phi_m = \phi_l$$

The α-parameter of the network can then be computed from

$$\alpha = \frac{1 - \sin \phi_l}{1 + \sin \phi_l}$$

Since at ω_m the network provides a db-gain of $10 \log (1/\alpha)$ (see Fig. 10.11) the new cross-over frequency $\omega_{c2} = \omega_m$ can be determined as that frequency at which the uncompensated system has a db-gain of $-10 \log(1/\alpha)$.

With $\omega_{c2} = \omega_m$ thus established, the two corner frequencies of the lead network can be computed as

$$\omega_1 = 1/\tau = \omega_m \sqrt{\alpha}; \; \omega_2 = 1/\alpha\tau = \omega_m/\sqrt{\alpha}$$

The log-magnitude and phase plots of the compensated system can now be completely drawn for checking the actual phase margin achieved. How far this falls short of the specified value depends on how good was the guess of ϵ. If the phase margin achieved is still unsatisfactory, the process is repeated with a higher value of ϵ. The process usually converges to a satisfactory solution in one or two trials within an error of about 2-3° which is acceptable for all practical purposes.

Certain overall observations can be made from the Bode plots of the lead compensated system:

1. The cross-over frequency is increased.

2. The high frequency end of the log-magnitude plot has been raised up by a db-gain of 20 log $(1/\alpha)$.

Since the cross-over frequency is a rough measure of the bandwidth of a closed-loop system, we can say that the lead compensation increases the system bandwidth and hence an improvement in the speed of response of the system results. The actual value of the bandwidth can be determined by transferring the data from Bode plots to Nichols chart. It must of course be noted that too large an increase in system bandwidth so as to include some of the noise frequencies is undesirable and must be guarded against in lead compensation design.

The design procedure for a lead compensator outlined above is quite general and applies to any type and order of system though it has been illustrated through an example of a type-1, second-order system. The complete design procedure is summarized below.

1. Determine the loop gain K to satisfy the specified error constant.
2. Using this value of K, determine the phase margin of the uncompensated system.
3. Determine the phase-lead required using the relation

$$\phi_l = \phi_s - \phi_1 + \epsilon$$

where ϕ_s = specified phase margin; ϕ_1 = phase margin of the fixed part of the system (i.e., the uncompensated system); and ϵ = a margin of safety required by the fact that the cross-over frequency will increase due to compensation.

4. Let $\phi_m = \phi_l$ and determine the α-parameter of the network from

$$\alpha = \frac{1 - \sin \phi_m}{1 + \sin \phi_m}$$

If the required ϕ_m is more than 60°, it is recommended to use two identical networks each contributing a maximum lead of $\phi_l/2$.

5. Calculate the db-gain 10 log $(1/\alpha)$ provided by the network at ω_m. Locate the frequency at which the uncompensated system has a gain of $-10 \log (1/\alpha)$. This is the cross-over frequency $\omega_{c2} = \omega_m$ of the compensated system.

6. Compute the two corner frequencies of the network as

$$\omega_1 = 1/\tau = \omega_m \sqrt{\alpha}; \ \omega_2 = 1/\alpha\tau = \omega_m/\sqrt{\alpha}$$

7. Draw the magnitude and phase plots of the compensated system and check the resulting phase margin. If the phase margin is still low, raise the value of ϵ and repeat from step 3 above.

8. Check any additional specifications on system performance, e.g., bandwidth. Redesign for another choice of cross-over frequency ω_{c2}, till this specification is met. It may be noted that the additional specification can only be met if it is consistent.

Example 10.6: Consider a type-1 unity feedback system with an open-loop transfer function of

$$G_f(s) = \frac{K_v}{s(s+1)}$$

It is specified that $K_v = 12$ sec^{-1} and $\phi_{pm} = 40°$.

The Bode plots of the system with $K_v = 12$ are drawn in Fig. 10.29. The phase margin of the uncompensated system is found to be $\phi_1 = 15°$. If a lead compensator is used, the phase-lead required at the new cross-over frequency is given by

$$\phi_l = 40° - 15° + 5° = 30° = \phi_m$$

$$\therefore \quad \alpha = \frac{1 - \sin 30°}{1 + \sin 30°} = 0.334$$

The magnitude contribution of the compensating network at ω_m is

$$10 \log (1/0.334) = 4.8 \text{ db}$$

Therefore the frequency at which the uncompensated system has a magnitude of -4.8 db becomes the new cross-over frequency $\omega_{c2} = \omega_m$ when the lead network is added. From the Bode plot of Fig. 10.29, we find

$$\omega_{c2} = 4.6 \text{ rad/sec}$$

Lower corner frequency of the network, $\omega_1 = 1/\tau = \omega_m \sqrt{\alpha} = 2.63$ rad/sec

Upper corner frequency of the network, $\omega_2 = 1/\alpha\tau = \omega_m/\sqrt{\alpha} = 7.8$ rad/sec

The transfer function of the lead network (with amplifier), therefore becomes

$$G_c(s) = \frac{0.376\,s+1}{128\,s+1}$$

The amplification necessary to cancel the lead network attenuation is

$$A = 1/\alpha = 3$$

The open-loop transfer function of the compensated system is

$$G(s) = G_f(s)G_c(s) = \frac{12(0.376s+1)}{s(s+1)(0.128s+1)}$$

The log-magnitude vs phase angle for both $G_f(s)$ and $G(s)$ are plotted on Nichols chart in Fig. 10.30, from which we observe the following:

1. Phase margin is increased from 15° to 42°.
2. Bandwidth is increased from 5.5 rad/sec to 9 rad/sec.
3. M_r is reduced from $+12$ db to $+3.0$ db.
4. ω_r is increased from 3.5 rad/sec to 4.6 rad/sec.

Thus, in general, the effect of the lead compensator is to increase the margin of stability and speed of response.

Example 10.7: Let us design lead compensation for a type-2 system with an open-loop transfer function

$$G_f(j\omega) = \frac{K}{(j\omega)^2(j0.2\omega+1)}$$

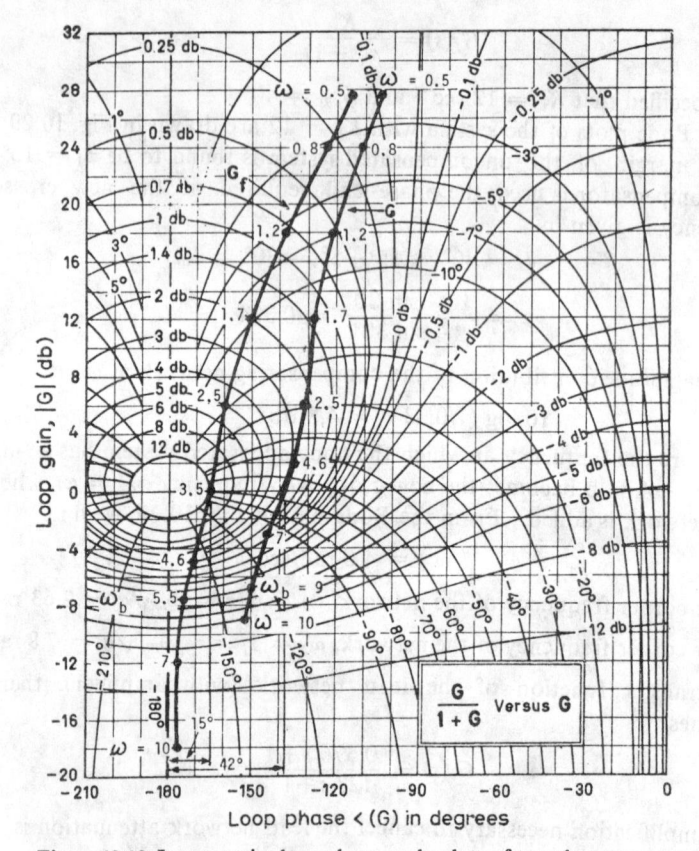

Figure 10.30 Log-magnitude vs phase angle plots of type-1 system.

Assume that the system is required to be compensated to meet the following specifications:
1. Acceleration error constant $K_a = 10$
2. Phase margin $= 35°$

The specification on K_a is met by choosing $K = 10$, such that the open-loop transfer function of the system becomes

$$G_f(j\omega) = \frac{10}{(j\omega)^2(j0.2\omega+1)}$$

The Bode plots of $G_f(j\omega)$ are drawn in Fig. 10.31, from which it is found that the cross-over frequency is $\omega_{c1} = 3.16$ rad/sec and the phase margin is $\phi_1 = -33°$. Further, the uncompensated system is absolutely unstable.

Since the required phase margin is $\phi_s = 35°$, the phase-lead needed at the cross-over frequency ω_{c2} of the compensated system is obtained as

$$\phi_l = 35° - (-33°) + 14° = 82°$$

where $\epsilon = 14°$ is the estimated reduction in the $\angle G_f(j\omega)$ since $\omega_{c2} > \omega_{c1}$. A large value of ϵ has been selected in this example as $\angle G_f(j\omega)$ is

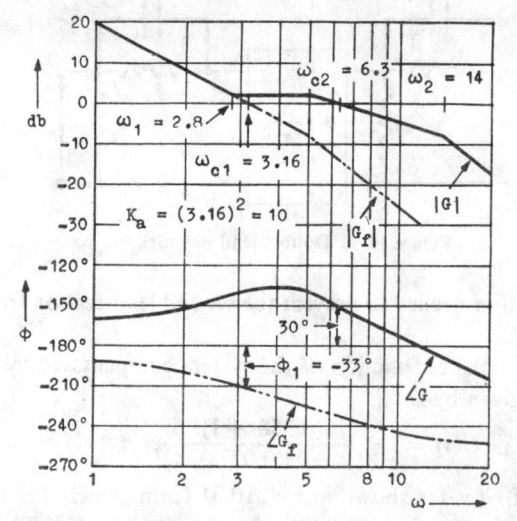

Figure 10.31 Design of lead compensation for type-2 system.

decreasing at a faster rate since the final slope of the log-magnitude curve is -60 db/decade.

As discussed in Section 10.4, using a single lead network to give such a large phase-lead is not advisable. We shall design a double lead network so that each section has to provide a maximum phase lead of $82/2 = 41°$. The attenuation factor of each lead-section is

$$\alpha = \frac{1 - \sin 41°}{1 + \sin 41°} \approx 0.2$$

From the plot of the uncompensated system, it is seen that the gain $2[-10 \log (1/0.2)] = -13.2$ db occurs at $\omega_{c2} = 6.3$ rad/sec. This should be the cross-over frequency of the compensated system. Choosing $\omega_m = \omega_{c2} = 6.3$ rad/sec, we obtain the network corner frequencies as

$$\omega_1 = (\sqrt{\alpha})\omega_m = 6.3 \times 0.446 = 2.8 \text{ rad/sec}$$
$$\omega_2 = (1/\sqrt{\alpha})\omega_m = 2.8/0.2 = 14.0 \text{ rad/sec}$$

Thus the transfer function of each section of the double lead network (with the attenuation cancelled by an amplification $A = 1/\alpha = 5$) is

$$G_c^1(s) = \left(\frac{0.358s + 1}{0.077s + 1}\right)$$

The transfer function of the double lead network becomes

$$G_c(s) = \frac{(0.358s + 1)^2}{(0.077s + 1)^2}$$

The net additional amplifier gain required is $A^2 = 25$ as shown in Fig. 10.32. Cascading this amplifier between the two sections of the lead network

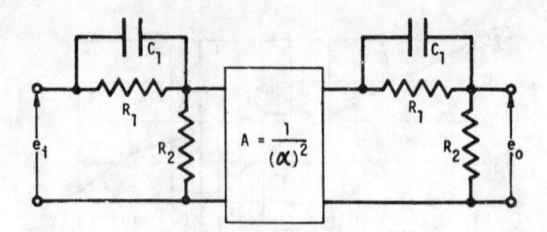

Figure 10.32 Double lead network.

provides the isolation needed to prevent the second lead-section from loading the first one.

The open-loop transfer function of the system compensated by the double lead network is given by

$$G(s) = \frac{10(0.358s+1)^2}{s^2(0.2s+1)(0.077s+1)^2}$$

The Bode plot of $G(j\omega)$ is shown in Fig. 10.31 from which it is found that the phase margin at the new cross-over frequency is $+30°$. This indicates that the system has become stable but the desired phase margin is not yet fully achieved. This is because of the excessive lag of the fixed part of system at the new cross-over frequency. If this phase margin is not acceptable then the compensator should be redesigned with a higher value of ϵ, say $\epsilon = 25°$.

LAG COMPENSATION

The frequency response of a phase-lag network has been presented in Fig. 10.14 from which it is seen that the network acts like a low-pass filter attenuating the high frequencies by a db of $(-20 \log \beta)$. The phase-lag mainly occurs within and around the two corner frequencies of the lag network. It must be recognized here that any phase-lag is undesirable at the cross-over frequency of the compensated system. Therefore, it is the attenuation characteristic of the network which is exploited for compensation purposes.

Consider for illustration a unity feedback system with an open-loop transfer function of the form $G_f(s) = K_v/s(\tau_1 s+1)(\tau_2 s+1)$ whose Bode plots are given in Fig. 10.33 with K_v adjusted to the specified value. The uncompensated system has a cross-over frequency of ω_{c1} and a phase margin of ϕ_1 which happens to be negative. It is desired to raise the phase margin to a specified value ϕ_s without altering K_v, the error constant. It must be noted here that *lag compensation is only possible if there exists a range of frequencies in which the uncompensated system has a phase angle less negative than* $(-180° + \phi_s)$. Such is not the case with systems of type-2 and higher where lead compensation must therefore be used.

It is immediately noticed from the Bode plots of the uncompensated system that the desired phase margin can be attained by modifying the

Introduction to Design

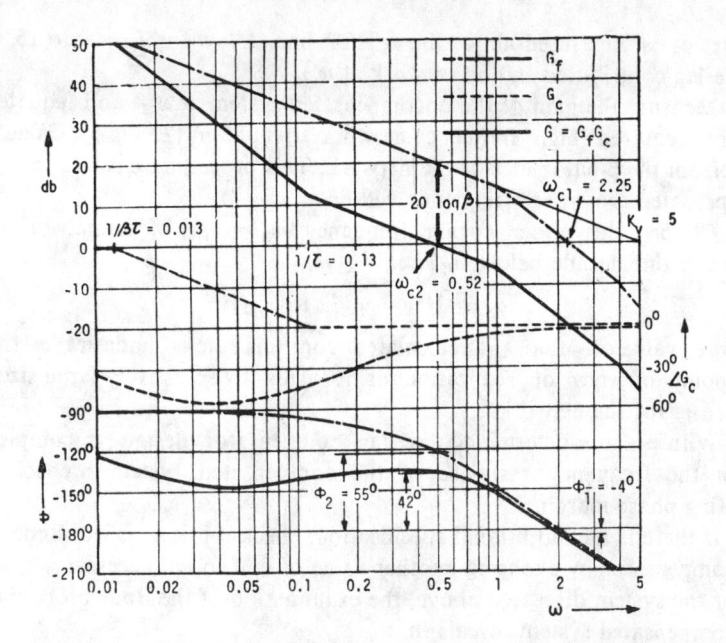

Figure 10.33 Design of lag compensation for type-1 system.

db-log ω plot so as to lower the cross-over frequency, while the phase plot is not allowed to alter significantly in the region of the new cross-over frequency. This can be easily achieved by a lag network wherein the high frequency attenuation ($- 20 \log \beta$) is utilized to lower the cross-over frequency, while the two corner frequencies of the network are placed sufficiently lower than the desired cross-over frequency so that the phase-lag contribution of the network at this cross-over frequency is made sufficiently small. Usually the upper corner frequency of the network is placed one octave to one decade lower than the cross-over frequency ω_{c2} of the compensated system. To nullify the effect of the small phase-lag contribution of the network which will still be present at ω_{c2}, the uncompensated system must contribute an angle ϕ_2 at ω_{c2} towards phase margin where,

$$\phi_2 = \phi_s + \epsilon$$

where ϵ is allowed a value of 5-15°.

It shall be noted that the network contributes 0 db in the low frequency region so that the K_v remains unaltered by introducing the lag network.

With the above explanation, the lag compensator design procedure may be stated as below.

1. Determine the open-loop gain necessary to satisfy the specified error constant. If the phase margin of the uncompensated system with this gain is unsatisfactory, then design a lag network as per succeeding steps.

2. Find the frequency ω_{c2} where the uncompensated system makes a phase margin contribution of

$$\phi_2 = \phi_s + \epsilon$$

where ϕ_2 is measured above the $-180°$ line. Allow for $\epsilon = 5°$ to $15°$ for phase-lag contributed by the network at ω_{c2}.

3. Measure the gain of the uncompensated system at ω_{c2} and equate it to the required high frequency network attenuation (20 log β). Calculate therefrom the β-parameter of the network. This procedure ensures that the compensated cross-over frequency will lie at ω_{c2}.

4. Choose the upper corner frequency ($\omega_2 = 1/\tau$) of the network one octave to one decade below ω_{c2}, i.e.,

$$\omega_2 = 1/\tau = \omega_{c2}/2 \text{ to } \omega_{c2}/10$$

A larger value of τ than that calculated from this rule is undesirable from the point of view of realization as it leads to excessively large size of condenser for the network.

5. With β and τ determined, the lag compensator design is complete. Draw the frequency response of the compensated system to check the resulting phase margin.

6. If there is any additional specification, check if it is satisfied. Redesign the compensator by choosing another value of τ, if found necessary.

For the system discussed above, the examination of the Bode plots of the lag compensated system reveals that

(*i*) the cross-over frequency is reduced;

(*ii*) the high-frequency end of the log-magnitude plot is lowered by (20 log β) db.

Thus we find that the lag compensator reduces the system bandwidth (cross-over frequency being a rough measure of bandwidth) and the additional attenuation of high frequencies improves the signal/noise ratio of the system. If for the specified K_v the reduced bandwidth is unacceptable, lead compensator may be tried.

Example 10.8: Let us reconsider the system discussed in Example 10.4, whose open-loop transfer function is

$$G_f(s) = \frac{K}{s(s+1)(s+4)}$$

The system is to be compensated to meet the following specifications:

Damping ratio $\zeta = 0.4$
Settling time $t_s = 10$ sec
Velocity error constant $K_v \geqslant 5$ sec^{-1}

Using the relations given in eqns. (10.47 and 48) and $t_s = 3/\zeta\omega_n$ for 5% tolerance band, we obtain the following equivalent specifications in frequency domain:

$$\phi_s = 43°; \ \omega_b = 1.02 \text{ rad/sec}; \ K_v \geqslant 5 \text{ sec}^{-1}$$

The open-loop transfer function of the uncompensated system may be written as

$$G_f(s) = \frac{K/4}{s(s+1)(0.25s+1)}$$

The specification on K_v is met by choosing $K = 20$. Thus

$$G_f(j\omega) = \frac{5}{j\omega(j\omega+1)(j0.25\omega+1)}$$

The Bode plot of $G_f(j\omega)$ is shown in Fig. 10.33 from which it is found that the cross-over frequency $\omega_{c1} = 2.25$ rad/sec and the phase margin $\phi_1 = -4°$. The uncompensated system is, therefore, unstable.

It is further seen from the Bode plot that neglecting the phase-lag contribution of the network, the specified phase margin of 43° is obtained, if the cross-over frequency is 0.74 rad/sec. Since this is fairly low, the upper network corner frequency cannot be taken far to its left in order to avoid large time constants. This indicates that phase-lag contribution of the network at the new cross-over frequency will be considerable and may be guessed at 12°. The uncompensated system must, therefore, make a phase margin contribution of

$$\phi_2 = \phi_s + \epsilon = 43° + 12° = 55°$$

at ω_{c2} which is found to be 0.52 rad/sec.

Placing the upper corner frequency of the compensator two octaves below ω_{c2}, we have

$$\omega_2 = 1/\tau = \omega_{c2}/(2)^2 = 0.13 \text{ rad/sec}$$

To bring the log-magnitude curve down to 0 db at ω_{c2}, the lag network must provide an attenuation of 20 db. Therefore

$$20 \log \beta = 20$$
or
$$\beta = 10$$

The lower corner frequency of the network is then fixed at

$$\omega_1 = 1/\beta\tau = 0.013 \text{ rad/sec}$$

The transfer function of the lag network is then

$$G_c(s) = \frac{1}{10}\left(\frac{s+0.13}{s+0.013}\right) = \left(\frac{7.7s+1}{77s+1}\right)$$

Phase-lag introduced by the lag network at ω_{c2} is

$$\text{(phase-lag)}\, \omega_{c2} = \tan^{-1}(7.7\omega_{c2}) - \tan^{-1}(77\omega_{c2})$$
$$= 76 - 88 = -12°$$

Therefore, the safety margin of $\epsilon = 12°$ is justified.

The open-loop transfer function of the compensated system becomes

$$G(s) = \frac{5(7.7s+1)}{s(77s+1)(s+1)(0.25s+1)}$$

The Bode plot of $G(j\omega)$ is shown in Fig. 10.33, from where the phase margin of the compensated system is found to be 42°.

The log-magnitude vs. phase angle curves of uncompensated and compensated systems are shown in Fig. 10.34. It is observed that addition of a lag-compensator has reduced the bandwidth from 3.4 rad/sec to 1.1 rad/sec.

However, since the reduced value lies in the acceptable range, our design is complete.

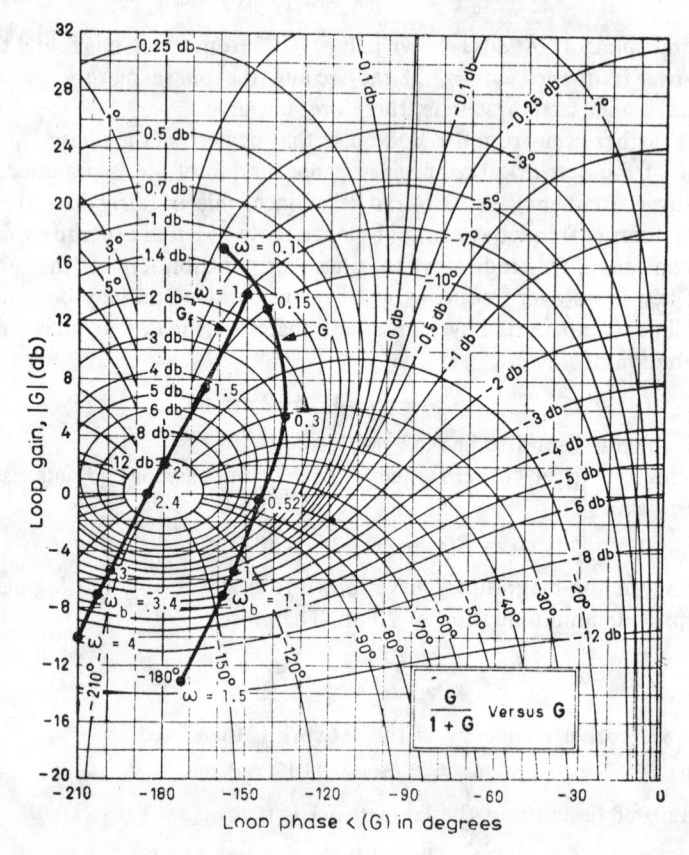

Figure 10.34 Log-magnitude *vs* phase angle curves for Example 10.8.

LAG-LEAD COMPENSATION

As has been discussed earlier in this section, for a specified error constant, the phase margin can be improved to any desired value by employing lead compensation even though the uncompensated system may be absolutely unstable. The lead compensation results in increased bandwidth and faster speed of response. For higher-order systems and for systems with large error constants, large leads are required for compensation, resulting in excessively large bandwidth which is undesirable from noise transmission point of view. For such systems lag compensation is preferred provided the uncompensated system is not absolutely unstable.

The lag compensation, on the other hand, reduces system bandwidth and slows down the speed of response. For large specified error constant and moderately large bandwidth, it may not be possible to meet the specifications through either lead or lag compensation. Under such circumstances we can use a lag-lead compensator wherein for specified error constant,

Introduction to Design

the lag-section is used to provide part of the phase margin and the lead section provides the rest of it as well as gives the desired bandwidth.

To start with this design, we check the phase margin and bandwidth of the uncompensated system for specified error constant. If this bandwidth is smaller than the specified value, lead compensation may be tried. However, if this bandwidth is larger than acceptable, lead compensation would not be desirable, so we try lag compensation provided the uncompensated system is not absolutely unstable. If the lag compensator design results in too low a bandwidth the need for a lag-lead compensation is indicated. A lag-lead compensation design is illustrated through an example below.

From Fig. 10.17, it is seen that a lag-lead network is essentially a band-pass filter having a transfer function

$$G_c(s) = \underbrace{\left[\frac{\tau_1 s+1}{\beta\tau_1 s+1}\right]}_{\text{Lag-section}} \underbrace{\left[\frac{\tau_2 s+1}{(\tau_2/\beta)s+1}\right]}_{\text{Lead-section}}; \quad \beta = \frac{1}{\alpha} > 1$$

We therefore notice that once the lag-section has been designed, τ_1 and β get fixed. Therefore τ_2 is the only variable parameter for lead-section design.

Example 10.9: Let us consider a system with an open-loop transfer function

$$G_f(j\omega) = \frac{K}{j\omega(j0.1\omega+1)(j0.2\omega+1)}$$

The system is to be compensated to meet the following specifications:

Velocity error constant $K_v = 30$ sec^{-1}
Phase margin $\phi_s \geqslant 50°$
Bandwidth $\omega_b = 12$ rad/sec

It easily follows that $K = 30$ satisfies the specification on K_v. The Bode plot of the open-loop transfer function with this value of K is shown in Fig. 10.35 from which it is found that the uncompensated system has a cross-over frequency of 11 rad/sec and a phase margin of $-24°$. The uncompensated system is therefore unstable for the specified K_v. It is also observed that the uncompensated system is conditionally stable, i.e., it is stable for a certain range of values of K_v. From the Nichols chart of Fig. 10.36 it is found that the uncompensated system has a bandwidth of 14 rad/sec. If lead compensation is employed, the system bandwidth will increase still further (this increase will be fairly large as $\angle G_f(j\omega)$ is decreasing rapidly near the cross-over frequency) resulting in an undesirable system which will be sensitive to noise.

If lag compensation is attempted, the bandwidth will decrease sufficiently so as to fall short of the specified value of 12 rad/sec, resulting in a sluggish system. This fact can be verified by designing a lag compensator. We thus find that there is need to go in for lag-lead compensation.

Let us now design a lag-lead compensator to overcome the above menitoned difficulties. Since a full lag compensator will reduce the system

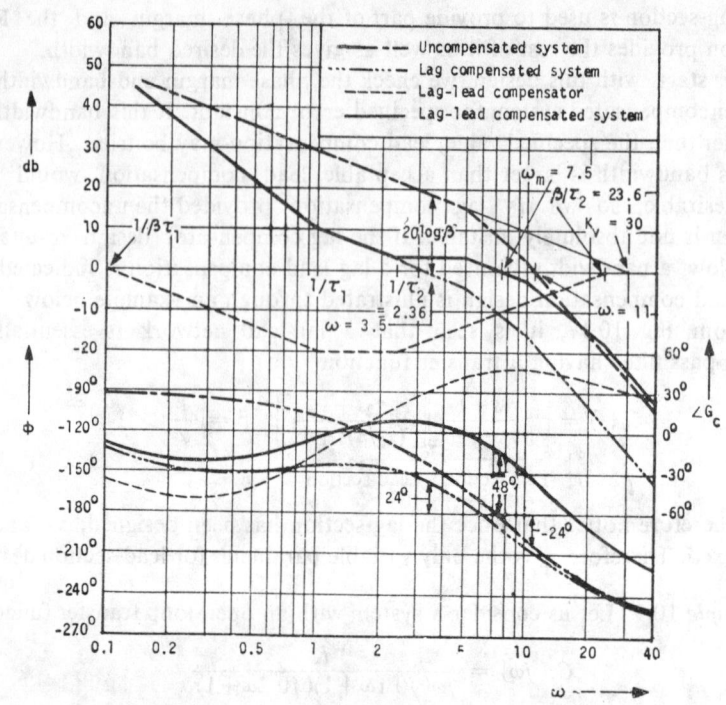

Figure 10.35 Design of lag-lead compensation for type-1 system.

bandwidth excessively, the lag-section of the lag-lead compensator must be designed so as to provide partial compensation only. The lag-section design therefore proceeds by making a choice of the new cross-over frequency due to lag-section compensation only. This choice, of course, must be higher than the cross-over frequency if the system were fully lag compensated. In this example this choice is made as 3.5 rad/sec to start with. It is seen from the Bode plot of the uncompensated system that it must be brought down by 18.5 db for the cross-over frequency to be 3.5 rad/sec. This gives the β-parameter of the lag-section as

$$20 \log \beta = 18.5$$

or
$$\beta = 8.32 \text{ say } 10$$

Let us now choose τ_1 of the lag-section to be 1 rad/sec. The lag-section design is thus complete and its transfer function is given by $G_{c_1}(s) = (s+1)/(10s+1)$. It is found that the lag-section compensated system has a phase margin of 24°.

We now proceed to design the lead-section. This design is constrained by the fact that $\alpha = 1/\beta = 0.1$ is already fixed. The maximum lead provided by the lead-section is therefore

$$\phi_m = \sin^{-1}\left(\frac{1 - 1/\beta}{1 + 1/\beta}\right) = 56°$$

Introduction to Design

To fully utilize the lead effect, we choose the compensated cross-over frequency to coincide with ω_m, which is then obtained as that frequency where the lag-section compensated system has a db of $-10 \log \beta = -10$. This value read off from Bode plot is $\omega_m = 7.5$ rad/sec. Then

$$\omega_m = \sqrt{\beta}/\tau_2 = 7.5$$

or
$$\tau_2 = 0.425 \text{ and } \tau_2/\beta = 0.0425$$

The transfer function of the lead-section, therefore, is

$$G_{c2}(s) = \left(\frac{0.425s + 1}{0.0425s + 1} \right)$$

Combining the transfer functions of the lead- and lag-sections, we obtain

$$G_c(s) = \frac{(s+1)(0.425s+1)}{(10s+1)(0.0425s+1)}$$

The open-loop transfer function of the lag-lead compensated system is

$$G(s) = \frac{30(s+1)(0.425s+1)}{s(0.1s+1)(0.2s+1)(10s+1)(0.0425s+1)}$$

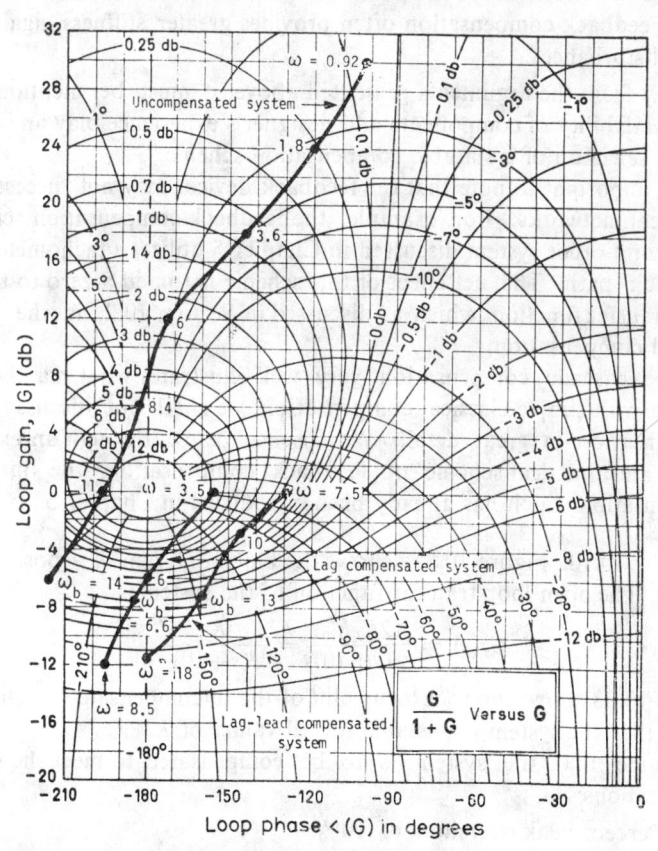

Figure 10.36 Log-magnitude vs phase angle curves for Example 10.9.

The Bode plot for this transfer function is shown in Fig. 10.35. The phase-margin is found to be 48°.

The log-magnitude vs phase angle curve of the lag-lead compensated system is drawn on the Nichols chart in Fig. 10.36, from which the bandwidth of the system is found to be 13 rad/sec.

The design therefore meets the specifications laid down. In case at the first attempt the specifications are not met, we must redesign by adjusting the initial choice of β and τ_1.

10.6 Feedback Compensation

Other than cascade compensation, compensation can also be achieved by placing a compensating device in an internal feedback path around one or more components of the forward path. Though cascade compensation is quite satisfactory and economical in most cases, feedback compensation may be warranted by the following factors:

1. In a feedback compensator energy transfer is from a high energy level to a lower one thereby obviating the need for feedback amplification.
2. In nonelectrical systems, suitable cascade devices may not be available.
3. Feedback compensation often provides greater stiffness against load disturbances.

Apart from the arguments presented above, it must be mentioned here that availability of components and designer's experience play an important role in selection of a suitable compensation scheme.

It is important to note that the feedback devices may not necessarily be electrical networks. For example, the feedback compensation scheme for the second-order system discussed in Chapter 5 utilizes a tachometer in the feedback path. The net effect of this scheme is to add a zero to the open-loop transfer function, which as discussed in Section 10.2, is the principle of lead compensation.

Since cascade compensation very well illustrates what may be accomplished with simple classical compensation, we shall not discuss feedback compensation in great detail. We discuss below, through an example, a simple and practical scheme of feedback compensation, the output rate compensation which has already been introduced in Chapter 5.

Example 10.10: Figure 10.37a shows the block diagram of a position servo system. The open-loop transfer function of the system is

$$G_f(s) = \frac{K_A K_f K_m}{s^2(s+10)} = \frac{K}{s^2(s+10)}$$

Figure 10.38 shows the root locus plot of the uncompensated system. It is found that the system is unstable for all values of K.

Assume that the system is to be compensated to meet the following specifications:

1. Percent peak overshoot $\leqslant 10\%$
2. Settling time $\leqslant 4$ sec

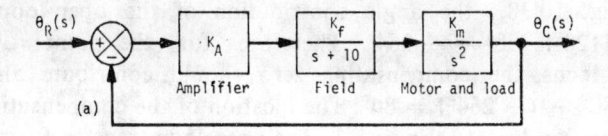

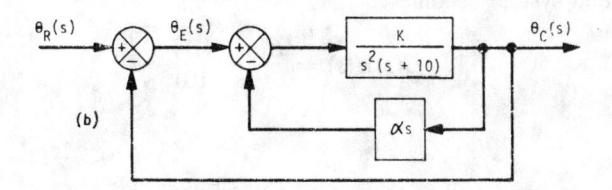

Figure 10.37 (a) Block diagram of a position servo system; and (b) position servo with tachometer feedback.

Let us utilize tachometer feedback to meet the desired specifications. The block diagram of the system with tachometer feedback is shown in Fig. 10.37b. Reducing the minor feedback loop, we have

$$\frac{\theta_C(s)}{\theta_E(s)} = G(s) = \frac{K}{s(s^2+10s+K\alpha)} \tag{10.49}$$

where $G(s)$ is the forward path transfer function of the major feedback loop.

The characteristic equation of the system is

$$1 + \frac{K}{s^3+10s^2+K\alpha s} = 0$$

or
$$s^3+10s^2+K\alpha s+K = 0$$

This characteristic equation can be rewritten in the form

$$1 + \frac{K\alpha(s+1/\alpha)}{s^2(s+10)} = 0 \tag{10.50}$$

From eqn. (10.50) we have

$$G(s)H(s) = \frac{K'(s+1/\alpha)}{s^2(s+10)} \tag{10.51}$$

where $K' = K\alpha$.

Equation (10.51) shows that the net effect of rate feedback is to add a zero at $s = -1\alpha$. Now the design procedure calls for a search for suitable location of this additional zero to satisfy the specifications.

From eqn. (5.17) we find that peak overshoot of 10% corresponds to $\zeta = 0.6$. Then

$$\omega_n = 4/\zeta t_s = (4/0.6 \times 4) = 1.67 \text{ rad/sec}$$

The desired dominant roots are then given as

$$s_d = -1 \pm j1.34$$

From Fig. 10.38, the angle contribution of the open-loop poles at s_d is $-2(128°) - 8° = -264°$. Therefore for the point s_d to be on the root locus, the compensating zero should contribute an angle of $\phi = -180° - (-264°) = 84°$. The location of the compensating zero is thus found to be at $s = -1.1$. The open-loop transfer function of the compensated system becomes

$$G(s)H(s) = \frac{K'(s+1.1)}{s^2(s+10)}$$

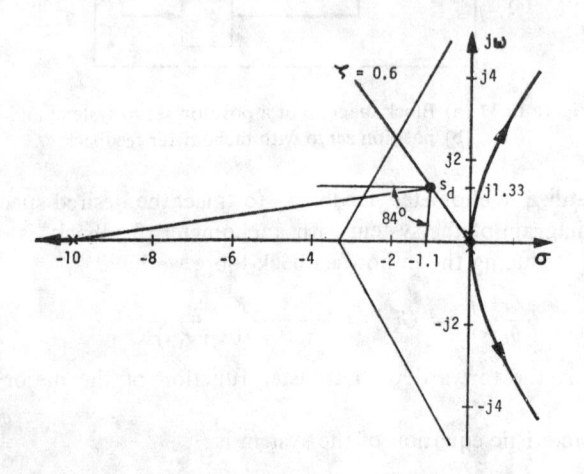

Figure 10.38 Root locus plot of $K/s^2(s+10)$ and location of compensating zero.

The root locus plot of the compensated system is shown in Fig. 10.39. It is now observed that for all values of K', the system is stable. The value of K' at s_d is found to be 17.4.

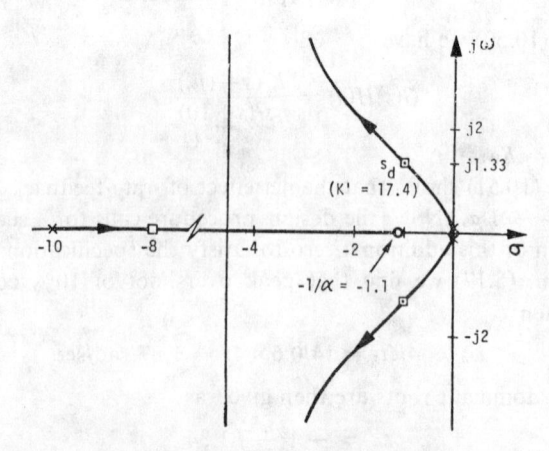

Figure 10.39 Root locus plot of $K'(s+1.1)/s^2(s+10)$.

Introduction to Design

The third closed-loop pole is at $s = -8$ which is far away from the imaginary axis, compared to the dominant closed-loop poles. Hence its effect on the transient behaviour is negligible.

Let us now investigate the steady-state behaviour of the compensated system. From Eqn. (10.49), the velocity error constant K_v is given by

$$K_v = \lim_{s \to 0} \left[s \times \frac{K}{s(s^2 + 10s + K\alpha)} \right] = 1/\alpha$$
$$= 1.1$$

If this value of K_v is acceptable, then the design is complete; otherwise the design procedure given above cannot satisfy the steady-state and transient specifications simultaneously. The control over the velocity error constant can be obtained by introducing an amplifier of adjustable gain in the forward path outside the minor feedback loop as shown in Fig. 10.40.

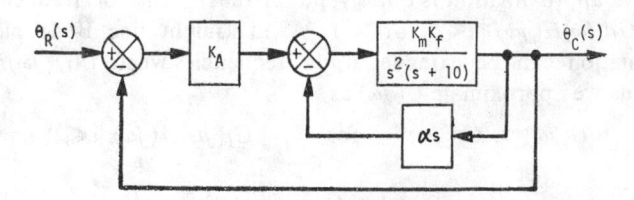

Figure 10.40 Position servo system with tachometer feedback.

The design procedure to satisfy the steady-state and transient specifications simultaneously, is a trial and error procedure which is conveniently carried out in frequency domain, as discussed below.

FEFDBACK COMPENSATION IN FREQUENCY DOMAIN

Compensation by internal feedback loop can be easily carried out in the frequency domain based on an approximation similar to that made for straight line Bode plots. Consider the basic configuration of Fig. 10.41 with an internal compensating loop. The amplifier K_A is included to provide for

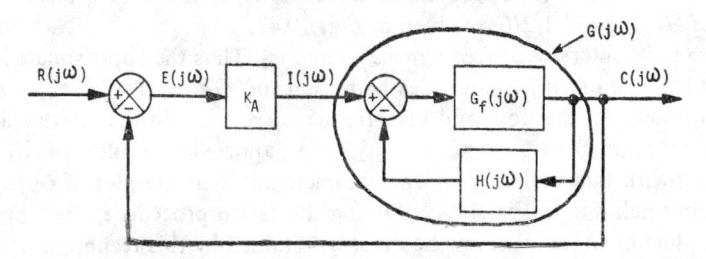

Figure 10.41 Basic feedback compensation scheme.

the reduction in gain caused by the internal feedback loop. The equivalent transfer function of the inner loop is given by

$$\frac{C(j\omega)}{I(j\omega)} = G(j\omega) = \frac{G_f(j\omega)}{1+G_f(j\omega)H(j\omega)}$$

$$= \frac{1}{H(j\omega)}\left[\frac{G_f(j\omega)H(j\omega)}{1+G_f(j\omega)H(j\omega)}\right] \quad (10.52)$$

Knowing $G_f(j\omega)$ and $H(j\omega)$, we can obtain the bracketed term by plotting $G_f(j\omega)H(j\omega)$, on the Nichols chart. This result must then be added to the Bode plots of $1/H(j\omega)$ to obtain $G(j\omega)$. A trial and error procedure for parameter adjustment then becomes highly tedius and time-consuming. An approximate Bode plot of $G(j\omega)$ can however be quickly obtained by the following approximation:

$$G(j\omega) \approx G_f(j\omega) \quad \text{for} \quad |G_f(j\omega)H(j\omega)| \ll 1$$
$$\text{and} \quad G(j\omega) \approx \frac{1}{H(j\omega)} \quad \text{for} \quad |G_f(j\omega)H(j\omega)| \gg 1 \quad (10.53)$$

The above approximation is quite valid for the regions of frequencies for which $|G_f(j\omega)H(j\omega)| \ll 1$ (or $\gg 1$). As in straight line Bode plots, this approximation can be extended up to frequencies where $|G_f(j\omega)H(j\omega)| = 1$. Thus we approximate $G(j\omega)$ as

$$G(j\omega) \approx G_f(j\omega) \quad \text{for} \quad |G_f(j\omega)H(j\omega)| \leqslant 1$$
$$\approx \frac{1}{H(j\omega)} \quad \text{for} \quad |G_f(j\omega)H(j\omega)| \geqslant 1 \quad (10.54)$$

It must be borne in mind that this approximation is not quite valid in the frequency band around the frequencies at which $|G_f(j\omega)H(j\omega)| = 1$. Unlike the straight line Bode plots, there is no direct method of applying corrections at these frequencies.

The condition $|G_f(j\omega)H(j\omega)| = 1$ may be interpreted as

$$20 \log |G_f(j\omega)H(j\omega)| = 0$$
$$\text{or} \quad 20 \log |G_f(j\omega)| = 20 \log \left|\frac{1}{H(j\omega)}\right| \quad (10.55)$$

This condition will be obtained where the log-magnitude plots of $G_f(j\omega)$ and $1/H(j\omega)$ intersect.

For illustration of the approximation technique, consider the typical Bode plots of $G_f(j\omega)$ and $1/H(j\omega)$ given in Fig. 10.42.

These plots intersect at frequencies ω_1 and ω_2. Thus the approximate log-magnitude plot of $G(j\omega)$ coincides with that of $G_f(j\omega)$ for $\omega \leqslant \omega_1$ and $\omega \geqslant \omega_2$, i.e., in the low and high frequency regions. In the intermediate frequency region, i.e., $\omega_1 \leqslant \omega \leqslant \omega_2$, the approximate plot of $G(j\omega)$ coincides with that of $1/H(j\omega)$. The complete approximate plot of $G(j\omega)$ is shown in thick line in Fig. 10.42. During the design procedure, the approximate plots of $G(j\omega)$ can now be quickly obtained by this technique as the

Introduction to Design

gain and time constants of either $G_f(j\omega)$ or $H(j\omega)$ are adjusted to achieve a desirable performance.

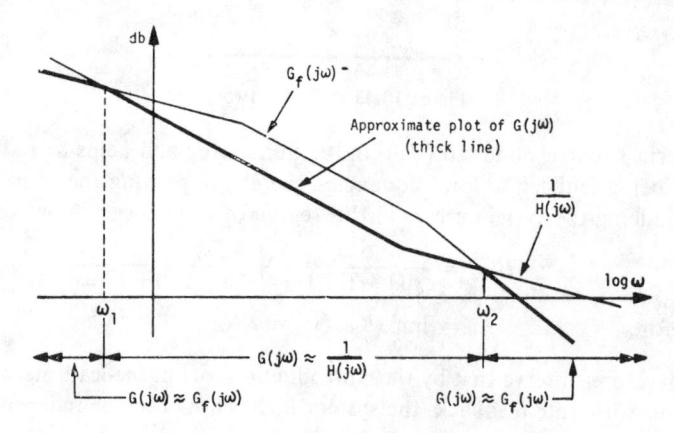

Figure 10.42 Approximate log-magnitude plot of
$$G(j\omega) = \frac{1}{H(j\omega)} \left[\frac{G_f(j\omega)H(j\omega)}{1+G_f(j\omega)H(j\omega)} \right].$$

Before we proceed to outline a design procedure, let us dwell upon the choice of a suitable $H(j\omega)$. A direct feedback will reduce the forward gain considerably. A better choice from gain point of view is rate feedback, i.e., $H(s) = \alpha s$. For a typical case let

$$G_f(s) = \frac{K}{s(\tau_1 s+1)(\tau_2 s+1)}$$

Then, from Fig. 10.41

$$G(s) = \frac{K}{s(\tau_1 s+1)(\tau_2 s+1) + K\alpha s}$$

$$K_v = \lim_{s \to 0} sK_A G(s) = \frac{KK_A}{1+K\alpha}$$

It therefore follows that the effect of rate feedback is to reduce the system K_v because of the term $(1+K\alpha)$ in the denominator. In order to reduce the steady-state errors in the low frequency region, it is desirable to reduce the feedback signal in this region of frequencies. This can be achieved by means of a phase-lead network of the type shown in Fig. 10.43 having a transfer function

$$\frac{E_0(s)}{E_i(s)} = \frac{Ts}{1+Ts}; \quad T = RC$$

With this network connected in tandem with the rate feedback in the feedback path, we have

$$H(s) = \frac{T\alpha s^2}{1+Ts}$$

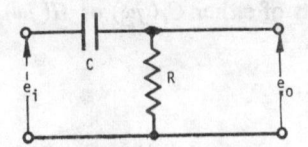

Figure 10.43 A lead network.

The term s^2 corresponds to $(j\omega)^2$ in the numerator and helps to reduce the signal being fedback at low frequencies, thereby improving the system gain and reducing the system errors in this region of frequencies. Now

$$G(s) = \frac{K(1+Ts)}{s[(1+Ts)(1+\tau_1 s)(1+\tau_2 s)+TK\alpha s]}$$

Therefore $\quad K_v = \lim_{s \to 0} sK_A G(s) = KK_A$

We therefore observe that by the introduction of phase-lead network in tandem with rate feedback, the system K_v has now become independent of the feedback.

With the feedback structure so selected we proceed to design the system to meet a set of specifications. This requires the adjustment of two forward path gains K_A and K, and two feedback parameters α and T. The design must necessarily be a trial and error procedure.

Figure 10.44 shows the Bode plots of

$$\frac{1}{H(j\omega)} = \frac{(1+j\omega T)}{j\omega)^2}$$

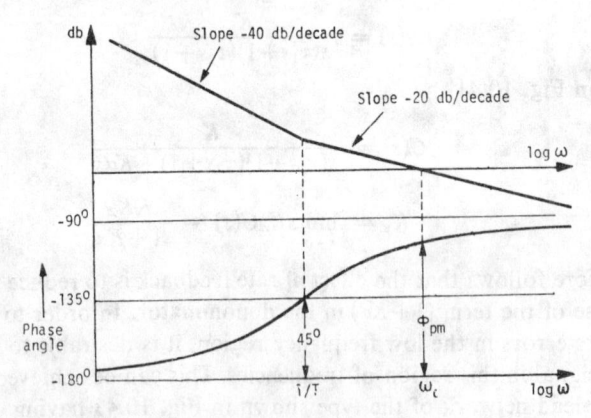

Figure 10.44 Bode plot of $1/H(j\omega) = (1+j\omega T)/T\alpha(j\omega)^2$.

It is seen from this figure that if $1/H(j\omega)$ is made effective around the cross-over frequency of $K_A G(j\omega)$, we can obtain any desired phase margin. Depending upon the specified phase margin we can, in fact, decide the system cross-over frequency.

The design procedure is illustrated through a numerical example given below.

Example 10.11: For the system of Fig. 10.41

$$G_f(s) = \frac{K}{s(2s+1)(0.5s+1)}$$

It is desired to design the system to obtain a phase margin of 35° and K_v of 10.

Figure 10.45 shows the Bode plot of $G_f(j\omega)$ for $K = 1$ (it is an initial trial value). If K is increased to 10 to get the desired K_v, it is found that the uncompensated system has a cross-over frequency of 2.2 rad/sec and a phase margin of $-36°$. This indicates the need for compensation.

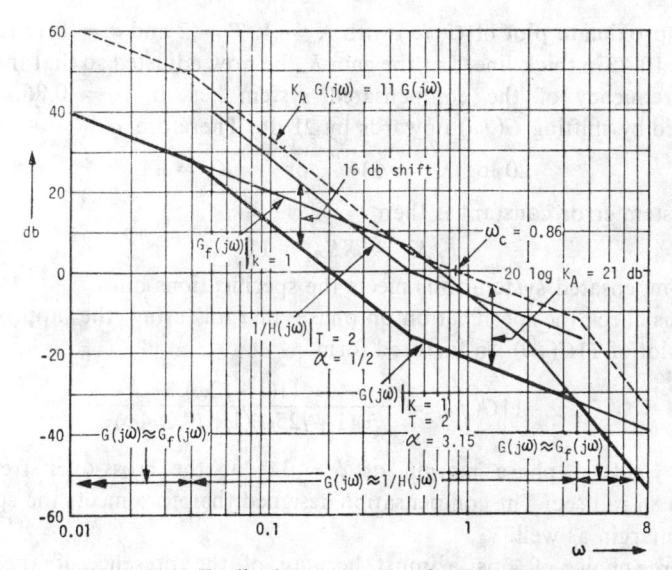

Figure 10.45 Feedback compensation for Example 10.11.

Now to provide compensation, consider the proposition of adding an internal feedback loop (output rate element and phase-lead network in tandem) with

$$H(j\omega) = \frac{\alpha T(j\omega)^2}{(1+j\omega T)}$$

Let us design the system such that for the compensated system, $1/H(j\omega)$ is effective over the region of frequencies around the cross-over frequency ω_c. The cross-over frequency of the compensated system can then be approximately determined from the equation

$$-180° + \tan^{-1}\omega_c T + 180° = 35° + \epsilon$$

where ϵ allows for the reduction of phase margin caused by the presence of the poles of $G(j\omega)$ other than those contributed by $1/H(j\omega)$.

Choosing $\epsilon = 25°$, we have

$$\tan^{-1}\omega_c T = 60°$$

Why it is necessary to choose such a large value of ϵ will become clear towards the end of this problem. Now a choice of $T = 2$ sec, gives $\omega_c = 0.86$ rad/sec.

Draw the Bode plot of $1/H(j\omega)$ (for $T = 2$ and $\alpha T = 1$) and shift it relative to the Bode plot of $G_f(j\omega)|_{K=1}$, such that the frequencies at which the two plots intersect are at least two octaves (i.e., 4 times) away from the expected cross-over frequency $\omega_c = 0.86$. It is found that a downward shift of $1/H(j\omega)$-plot by 16 db satisfies this requirement. Therefore

$$20 \log (1/\alpha T) = -16$$

or

$$\alpha = 3.15$$

The approximate plot of $G(j\omega)$ with $K = 1$, $T = 2$ and $\alpha = 3.15$ is shown in Fig. 10.45 in thick line. Let the gain K_A be now adjusted so that the cross-over frequency of the compensated system lies at $\omega_c = 0.86$. This is achieved by shifting $G(j\omega)$ upwards by 21 db. Therefore

$$20 \log K_A = 21 \quad \text{or} \quad K_A = 11$$

The system error constant is then

$$K_v = KK_A = 11$$

The compensated system thus meets the specifications on K_v.

Let us check the specification on phase margin using the approximate Bode plot of $11G(j\omega)$. In factored form

$$11G(j\omega) = \frac{11(1+j2\omega)}{j\omega(1+j25\omega)(1+j0.294\omega)^2}$$

which yields a phase margin of $\phi = 34°$ at the cross-over frequency $\omega_c = 0.86$ rad/sec. The compensation designed therefore meets the specified phase margin as well.

A large choice of ϵ is a must because of the presence of the factor $(1+j0.294\omega)^2$ in the denominator. Further because of the strong approximations made, the above design must be finally checked by drawing the exact Bode plots.

10.7 Network Compensation of a.c. Systems

The comparison between a.c. and d.c. systems was presented in Section 4.3. Several a.c. components which modulate or demodulate signals (synchros, a.c. servomotor, a.c. tachometer) were described. An a.c. system is generally hybrid in nature with signals in a.c. (suppressed carrier) form in part of the system and in d.c. (demodulated) form in rest of the system— input signal of a synchro pair is d.c. while its output signal is a.c.; input signal of an a.c. servomotor is a.c. while its output signal is d.c. With carrier frequency being much higher than the cut-off signal frequency, the analysis of a.c. system can be carried out like a d.c. system on the basis of envelopes of a.c. signals (i.e., modulating signals). Figure 4.14 depicted an a.c. position control system with tachometer feedback compensation which was analyzed in terms of modulating signals.

Introduction to Design

Network compensation of a.c. system can be carried out in two ways—interposing a d.c. network between a demodulator and modulator as shown in Fig. 10.46a or by using an a.c. network which suitably modifies the envelope of the a.c. signal as in Fig. 10.46b. The former method has the disadvantage of additional cost entailed in demodulator and modulator and further these components introduce lag into the system partly off-setting the compensating effect of d.c. network. In the latter method the a.c. network modifies the signal envelope in accordance with the d.c. network transfer function $G_{dc}(s)$. This method also suffers from a serious drawback discussed later in this article. Therefore tachometric feedback is a preferred method of compensating a.c. systems.

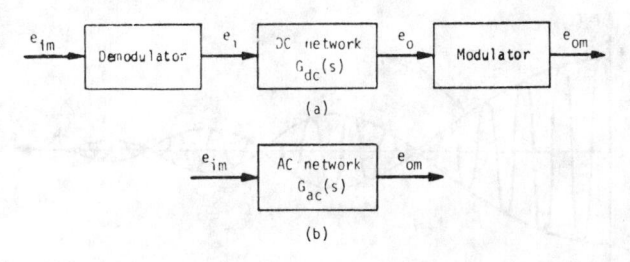

Figure 10.46 Network compensation of a.c. system.

The d.c. compensating network can be designed to compensate the system by any of the various techniques illustrated earlier in this chapter. Knowing $G_{dc}(s)$, the problem is to determine a.c. network transfer function $G_{ac}(s)$ which modifies the signal envelope as $G_{dc}(s)$ modifies the d.c. signal. Before this problem can be tackled, we need to examine the Fourier spectrum of a suppressed carrier modulated signal

$$e_m(t) = e(t) \cos \omega_c t \tag{10.56}$$

where $e(t)$ = modulating signal; and ω_c = carrier frequency.

Signal wave forms are illustrated in Fig. 10.47a. It is noticed that the carrier wave form undergoes a phase reversal whenever the modulating signal changes sign. Further, it is easily noticed that the envelope of the modulated signal has the wave form of $e(t)$, the modulating signal.

Fourier transform of (10.56) gives (see Appendix I),

$$E_m(j\omega) = \mathscr{F}\left[e(t) \frac{e^{j\omega_c t} + e^{-j\omega_c t}}{2}\right]$$
$$= \tfrac{1}{2} E[j(\omega - \omega_c)] + \tfrac{1}{2} E[j(\omega + \omega_c)] \tag{10.57}$$

where

$$E(j\omega) = \mathscr{F}[e(t)]$$
$$E[j(\omega - \omega_c)] = E(j\omega) \text{ centred at } \omega_c$$
$$E[j(\omega + \omega_c)] = E(j\omega) \text{ centred at } -\omega_c$$

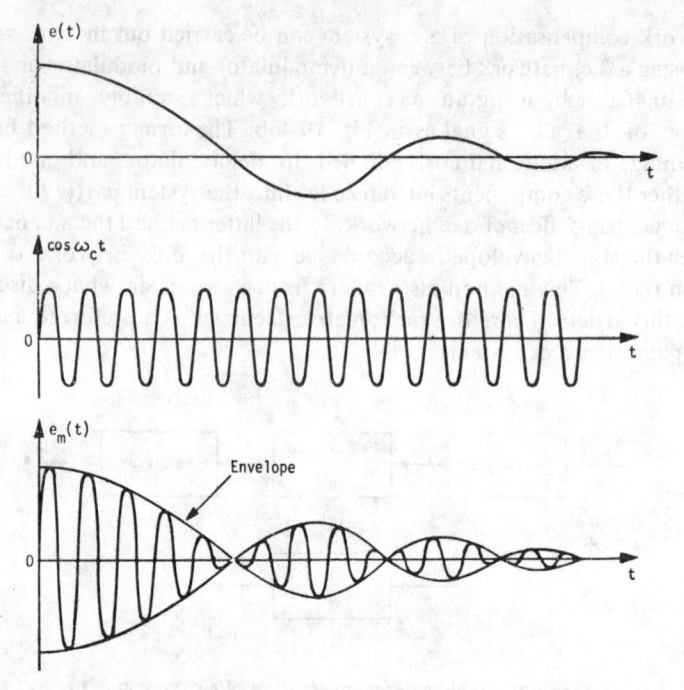

Figure 10.47 (a) Suppressed carrier modulated signal.

The Fourier spectra of modulating (d.c.) and modulated (a.c.) signals are illustrated in Fig. 10.47b for a bandlimited modulating signal with $\omega_c \gg \omega_b$. We immediately observe that

$$\left.\begin{array}{l} E[j(\omega+\omega_c)] \approx 0 \text{ for } \omega \text{ at and around } +\omega_c \\ E[j(\omega-\omega_c)] \approx 0 \text{ for } \omega \text{ at and around } -\omega_c \end{array}\right\} \quad (10.58)$$

This result is an *approximation* as a practical bandlimited signal does contain some energy (however small) in frequencies above ω_b.

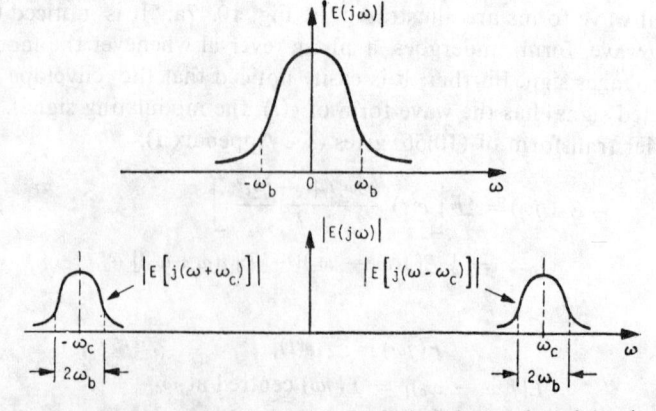

Figure 10.47(b) Fourier spectra of modulating (d.c.) and modulated (a.c.) signals.

Input to a.c. network is suppressed carrier modulated signal

$$e_{im}(t) = e_i(t) \cos \omega_c t$$

The network output should also be a suppressed carrier modulated signal*
with suitably modified envelope, i.e.,

$$e_{om}(t) = e_o(t) \cos \omega_c(t)$$

Now

$$G_{dc}(s) = \frac{\mathcal{L}[e_o(t)]}{\mathcal{L}[e_i(t)]} = \frac{E_o(s)}{E_i(s)} \qquad (10.59)$$

while the a.c. transfer function is

$$G_{ac}(s) = \frac{\mathcal{L}[e_o(t) \cos \omega_c t]}{\mathcal{L}[e_i(t) \cos \omega_c t]} \qquad (10.60)$$

Since

$$\mathcal{L}[e(t) \cos \omega_c t] = \mathcal{L}\left[e(t) \frac{e^{j\omega_c t} + e^{-j\omega_c t}}{2}\right]$$

$$= \tfrac{1}{2} E(s - j\omega_c) + \tfrac{1}{2} E(s + j\omega_c)$$

we have

$$G_{ac}(s) = \frac{E_o(s - j\omega_c) + E_o(s + j\omega_c)}{E_i(s - j\omega_c) + E_i(s + j\omega_c)}$$

As

$$E_o(s) = G_{dc}(s) E_i(s)$$

we can write the a.c. transfer function of the network in terms of the required d.c. transfer function, i.e.,

$$G_{ac}(s) = \frac{G_{dc}(s - j\omega_c) E_i(s - j\omega_c) + G_{dc}(s + j\omega_c) E_i(s + j\omega_c)}{E_i(s - j\omega_c) + E_i(s + j\omega_c)} \qquad (10.61)$$

However, this $G_{ac}(s)$ is not realizable as it is a function of the input signal $E_i(s)$.

Let us now make use of the fact that $e_i(t)$ is a bandlimited signal. To do this, examine the frequency characteristic of a.c. transfer function. Substituting $s = j\omega$ in (10.61),

$$G_{ac}(j\omega) = \frac{G_{dc}[j(\omega - \omega_c)] E_i[j(\omega - \omega_c)] + G_{dc}[j(\omega + \omega_c)] E_i[j(\omega + \omega_c)]}{E_i[j(\omega - \omega_c)] + E_i[j(\omega + \omega_c)]}$$

(10.62)

Utilizing the result (10.58) for a bandlimited signal, we write

$$G_{ac}(j\omega_c) \approx G_{dc}[j(\omega - \omega_c)] \text{ for } \omega \text{ at and around } +\omega_c$$

$$G_{ac}(j\omega_c) \approx G_{dc}[j(\omega + \omega_c)] \text{ for } \omega \text{ at and around } -\omega_c$$

This result can be expressed as

$$G_{ac}(j\omega_c) \approx G_{dc}[j(\omega \mp \omega_c)] \text{ for } \omega \text{ at and around } \pm\omega_c \qquad (10.63)$$

---

*Carrier is assumed to have the same phase at network output as at input. In general, a phase difference could exist. This case is not considered here.

Stated in words, it means that the frequency characteristics of the a.c. network at and around $\pm \omega_c$ are the same as that of the d.c. network at and around $\omega = 0$. Since the network frequency characteristic has become independent of the input due to the modulating input signal having band-limited characteristic, it is realizable.

The method of frequency translation can be used in designing a.c. compensating network from a known d.c. compensating network. Let us translate the frequency ω in the d.c. transfer function to $Z(\omega)$ to obtain the a.c. transfer function, i.e.,

$$G_{ac}(j\omega) = G_{dc}[jZ(\omega)] \tag{10.64}$$

where

$$G_{dc}[jZ] \approx G_{dc}[j(\omega \mp \omega_c)] \text{ for } \omega \text{ at and around } \pm \omega_c.$$

A specific $Z(\omega)$ which has the characteristic (10.64) and results in simple network configurations is

$$Z(\omega) = \frac{\omega}{2}\left(1 - \frac{\omega_c^2}{\omega^2}\right) = \frac{\omega}{2}\left(1 - \frac{\omega_c}{\omega}\right)\left(1 + \frac{\omega_c}{\omega}\right) \tag{10.65}$$

$$\approx \omega \mp \omega_c \text{ for } \omega \text{ at and around } \pm \omega_c$$

If d.c. network has RC configuration, frequency terms in the transfer function are contributed by capacitors. Using the above $Z(\omega)$, each capacitive impedance in d.c. form changes to a.c. form as

$$\frac{1}{j\omega C} \rightarrow \frac{1}{j\frac{\omega}{2}\left(1 - \frac{\omega_c^2}{\omega^2}\right)C}$$

which can be written as

$$\frac{1}{j\omega \frac{C}{2} + \frac{1}{j\omega\left(\frac{2}{C\omega_c^2}\right)}} \tag{10.66}$$

This is the impedance of a parallel LC network with

$$C_1 = \frac{C}{2} \tag{10.67}$$

$$L_1 = \frac{2}{C\omega_c^2}$$

A.C. to d.c. transformation of (10.67) is represented by Fig. 10.48.

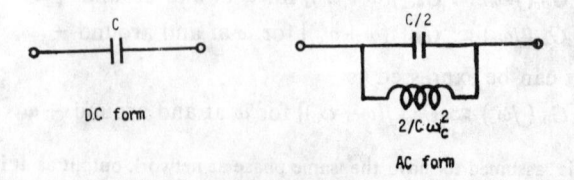

Figure 10.48.

Using the results (10.66) and (10.67), the a.c. lead and lag networks corresponding to d.c. lead and lag networks (of Figs. 10.10a and 10.13) are drawn in Fig. 10.49.

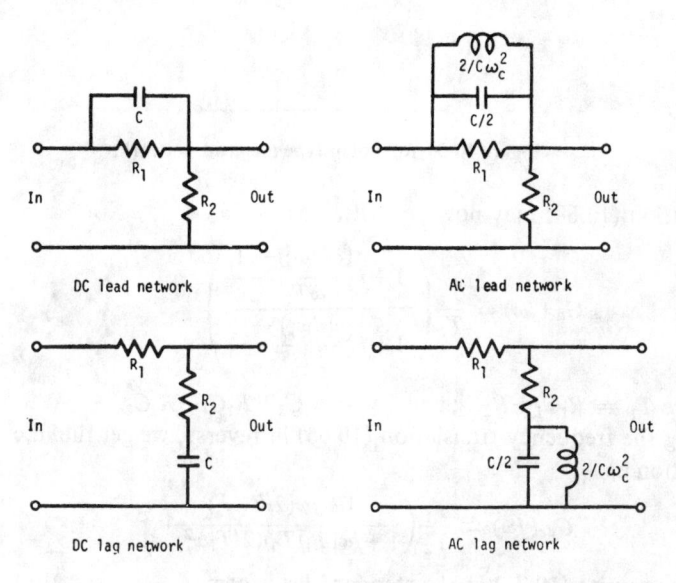

Figure 10.49.

It may be noted that the parallel *LC* combination of the a.c. forms is resonant at the carrier frequency ω_c. If the carrier frequency drifts, it no longer coincides with the network resonant frequency as a consequence of which the network loses its envelope characteristics and therefore its compensating properties. This is a serious drawback of a.c. compensating networks.

A practical inductance coil is never perfectly linear and also has some series resistance. It is therefore preferred to use *RC* compensating networks. A.C. networks of *RC* type can not be designed by the above technique. This technique has to be applied in reverse to a particular *RC* network of resonant (notch) type to obtain its d.c. transfer function.

Consider the *RC* notch type network of Fig. 10.50. For this network

$$G_{ac}(j\omega) = \frac{R_2 + 1/j\omega C_2}{\dfrac{R_1/j\omega C_1}{R_1 + 1/j\omega C_1} + R_2 + 1/j\omega C_2}$$

$$= \frac{(1 - \omega^2 R_1 R_2 C_1 C_2) + j\omega(R_1 C_1 + R_2 C_2)}{(1 - \omega^2 R_1 R_2 C_1 C_2) + j\omega(R_1 C_2 + R_1 C_1 + R_2 C_2)} \quad (10.68)$$

Resonance can be produced by making the real terms of both the numerator and denominator go to zero at $\omega = \omega_c$. This gives

$$\omega_c^2 = \frac{1}{R_1 R_2 C_1 C_2} \quad (10.69)$$

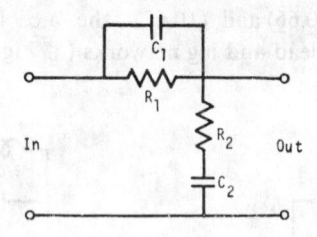

Figure 10.50 *RC* notch type a.c. lead network.

Equation (10.68) may now be written as

$$G_{ac}(j\omega) = \frac{T_1}{T_2}\left(\frac{1+j\dfrac{(\omega^2/\omega_c^2)-1}{\omega T_1}}{1+j\dfrac{(\omega^2/\omega_c^2)-1}{\omega T_2}}\right) \qquad (10.70)$$

where $T_1 = R_1C_1 + R_2C_2$; and $T_2 = R_1C_2 + R_1C_1 + R_2C_2$.
Using the frequency translation (10.65) in reverse, we get the d.c. transfer function as

$$G_{dc}(j\omega) \approx \frac{T_1}{T_2}\left(\frac{1+j\omega(2/T_1\omega_c^2)}{1+j\omega(T_1/T_2)(2/T_1\omega_c^2)}\right) \qquad (10.71)$$

We can write (10.71) in the standard lead form

$$G_{dc}(j\omega) \approx \alpha\left(\frac{1+j\omega T}{1+j\alpha T\omega}\right) \qquad (10.72a)$$

where

$$\alpha = \frac{T_1}{T_2} = \frac{R_1C_1 + R_2C_2}{R_1C_2 + R_1C_1 + R_2C_2} \qquad (10.72b)$$

$$T = \frac{2}{T_1\omega_c^2} = \frac{2R_1R_2C_1C_2}{R_1C_1 + R_2C_2} \qquad (10.72c)$$

R_1, R_2, C_1 and C_2 are the four circuit parameters to be evaluated and eqns. (10.69), (10.72b) and (10.72c) are the three equations to be satisfied. So first we select one of the four parameters arbitrarily and then for given values of ω_c, α and T, the other three parameters are determined.

Problems

10.1 A unity feedback system has an open-loop transfer function of

$$G(s) = \frac{K}{s(s+1)(s+5)}$$

Draw the root locus plot and determine the value of K to give a damping ratio of 0.3. A network having a transfer function of $10(1+10s)/(1+100s)$ is now introduced in tandem. Find the new value of K which gives the same damping ratio for the closed-loop response. Compare the velocity error constant and settling time of the original and the compensated systems.

Introduction to Design

10.2 A servomechanism has an open-loop transfer function of

$$G(s) = \frac{10}{s(1+0.5s)(1+0.1s)}$$

Draw the Bode plot and determine the phase and gain margins. A network having the transfer function $(1+0.23s)/(1+0.023s)$ is now introduced in tandem. Determine the new gain and phase margins. Comment upon the improvement in system response caused by the network.

10.3 A unity feedback control system has an open-loop transfer function of

$$G(s) = 1/s^2$$

Design a suitable compensating network such that a phase margin of 45° is achieved without sacrificing system velocity error constant. Sketch the Bode plot of the uncompensated and compensated systems

10.4 A unity feedback system has an open-loop transfer function of

$$G(s) = \frac{4}{s(2s+1)}$$

It is desired to obtain a phase margin of 40° without sacrificing the K_v of the system. Design a suitable lag-network and compute the value of network components assuming any suitable impedance level.

10.5 A unity feedback system is characterized by the open-loop transfer function

$$G(s) = \frac{K}{s(s+3)(s+9)}$$

(a) Determine the value of K if 20% overshoot to a step input is desired.

(b) For the above value of K determine the settling time and K_v (velocity error constant).

(c) Design a cascade compensator that will give approximately 15% overshoot to a unit step input, while the settling time is decreased by a factor of 2.5 and $K_v \geqslant 20$.

10.6 Consider the system shown in Fig. P-10.6. Design a lead compensator for this system to meet the following specifications:

Damping ratio $\zeta = 0.7$
Settling time $t_s = 1.4$ sec
Velocity error constant $K_v = 2$ sec^{-1}

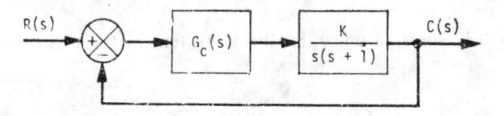

Figure P-10.6.

10.7. Design a phase-lead compensator for the system shown in Fig. P-10.6, to satisfy the following specifications:

(i) The phase margin of the system must be greater than 45°.

(ii) Steady-state error for a unit step input should be less than 1/15 deg per deg/sec of the final output velocity.

(iii) The gain cross-over frequency of the system must be less than 7.5 rad/sec.

10.8 A unity feedback system has an open-loop transfer function

$$G(s) = \frac{K}{s(s+1)(0.2s+1)}$$

Design phase-lag compensation for the system to achieve the following specifications:
Velocity error constant $K_v = 8$
Phase margin $\approx 40°$.

Also compare the cross-over frequency of the uncompensated and compensated systems.

10.9 Figure P-10.9 represents the block diagram of the control system for the pitch rate ($\omega = \dot{\theta}$) control of an aircraft. For improved accuracy, the system is made type-1 by employing an integration gyro which produces an output signal θ_E proportional to the integral of the difference between the command signal $\omega_c = \dot{\theta}_c$ and the pitch rate ω. The command signal is produced by the pilot with the help of a 'control stick'. The error signal adjusts the angle δ_c of the control surfaces of the aircraft by means of a positioning servo. The pitch rate is determined by this angle δ_c and aircraft dynamics.

(i) For the transfer functions represented in the block diagram, design a compensator so that the system has a damping factor of 0.6 and a damped natural frequency of 2.4 rad/sec. What is the system error constant for this setting of the controller?

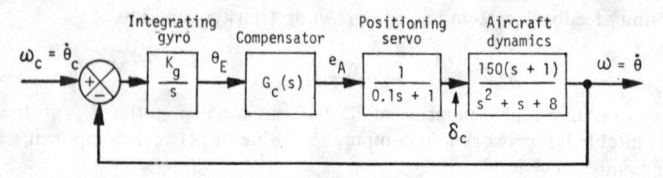

Figure P-10.9.

(ii) Redesign the system so that the system velocity error constant of about 0.5 is achieved without impairing its dynamic response.

(*Hint*: Try lead compensation by root locus method. It may be necessary to use more than two lead networks.)

10.10 A unity feedback type-0 system with transportation lag has a forward path transfer function of

$$G(j\omega) = \frac{10e^{-j0.02\omega}}{(1+j0.5\omega)(1+j0.1\omega)(1+j0.05\omega)}$$

Design a suitable compensation scheme so that the system acquires a damping factor of 0.4 without loss of steady-state accuracy. Estimate the bandwidth and settling time of the compensated system.

(*Hint*: Since design is handled more conveniently in frequency domain in the presence of transportation lag, translate damping factor specification into equivalent phase margin and then proceed.)

10.11 Figure P-10.11 shows a unity feedback system with a forward path transfer function

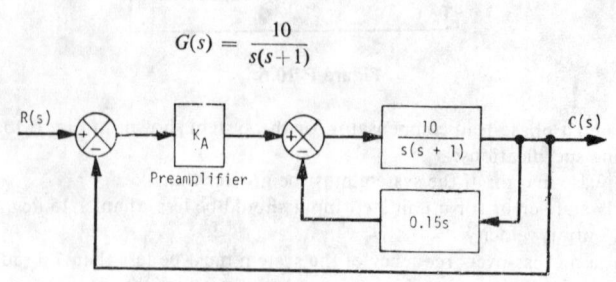

$$G(s) = \frac{10}{s(s+1)}$$

Figure P-10.11.

The system is compensated by means of rate feedback $H(s) = 0.15s$, in the minor feedback loop. Determine the effect of compensation by comparing the phase margins and bandwidths of the compensated and uncompensated schemes. In each case adjust K_A to a value which gives a velocity error constant of $K_v = 10$ for the purpose of comparison.

(*Hint*: Since the system order with internal feedback continues to be second, you can proceed analytically.)

10.12 A position control servomechanism having a forward path transfer function

$$G(s) = \frac{10(1+2s)}{s^2(1+0.2s)}$$

is to be compensated by rate feedback sK_t as indicated in Fig. P-10.12. Estimate a suitable value for K_t so that system has resonant peak $M_r = 1.3$. Estimate also the resonant frequency of the compensated system.

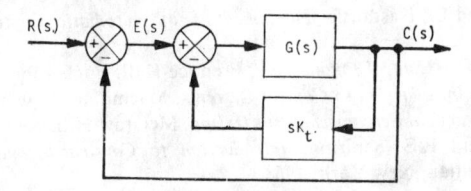

Figure P-10.12.

10.13 A unity feedback system has two transfer functions

$$G_1(s) = \frac{6}{1+0.25s}; \quad G_2(s) = \frac{4}{s(1+s)(1+0.5s)}$$

in tandem in the forward path. With the given values of gain the system is found to be unstable. It is attempted to stabilize the system by introducing a minor feedback having a transfer function $H(s)$ around $G_2(s)$. Show that the system gets stabilized if

$$H(s) = \frac{8s^2}{(1+2s)}$$

Using approximation techniques in frequency domain determine gain and phase margins of the stabilized system.

10.14 The feedback control system shown in Fig. 10.41 has

$$G_f(s) = \frac{3}{s(s+1)}; \quad H(s) = \frac{2s^2}{(1+1.25s)}$$

(a) Using the approximation technique, find

$$G(s) = \frac{G_f(s)}{1+G_f(s)H(s)}$$

in factored form.

(b) With $G(s)$ found in part (a), determine the value of K_A, so that the system has a phase margin of 45°.

10.15 For the bridged-T network of Fig. P-10.15, write the a.c. transfer function $G_{ac}(j\omega)$. Find the expression for the carrier frequency at which $G_{ac}(j\omega)$ has a real value. Write $G_{ac}(j\omega)$ in terms of ω and ω_c so as to use the transformation (10.65) in reverse. What is the corresponding $G_{dc}(j\omega)$? Is it lead type?

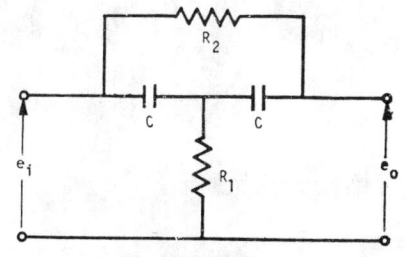

Figure P-10.15.

Further Reading

1. D'Azzo, J.J. and C.H. Houpis, *Linear Control System Analysis and Design*, McGraw-Hill, New York, 1975.
2. Kuo, B.C., *Automatic Control Systems* (3rd Ed.), Prentice-Hall, Englewood Cliffs, N.J., 1975.
3. Thaler, G.J., *Design of Feedback Systems*, Dowden, Hutchinson and Ross, Stroudsburg, Pennsylvania, 1973.
4. Eveleigh, V.W., *Introduction to Control Systems Design*, McGraw-Hill, New York, 1972.
5. Shinners, S.M., *Modern Control System Theory and Application*, Addison-Wesley, Reading, Mass., 1972.
6. Gupta, S.C. and L. Hasdorff, *Fundamentals of Automatic Control*, John Wiley, New York, 1970.
7. Ogata, K., *Modern Control Engineering*, Prentice-Hall, Englewood Cliffs, N.J., 1970.
8. Watkins, B.O., *Introduction to Control Systems*, Macmillan, New York, 1969.
9. Baeck, H.S., *Practical Servomechanism Design*, McGraw-Hill, New York, 1968.
10. Saucedo, R. and E.E. Schiring, *Introduction to Continuous and Digital Control Systems*, Macmillan, New York, 1968.
11. Dorf, R.C., *Modern Control Systems*, Addison-Wesley, Reading, Mass., 1967.
12. Elgerd, O.I., *Control Systems Theory*, McGraw-Hill, New York, 1967.
13. Ivey, K.A., *A.C. Carrier Control Systems*, John Wiley, New York, 1964.
14. Ku, Y.H., *Analysis and Control of Linear Systems*, International Text Book Co., Scranton, Pennsylvania, 1962.
15. Del Toro, V. and S.R. Parker, *Principles of Control Systems Engineering*, McGraw-Hill, New York, 1960.
16. Truxal, J.G., *Automatic Feedback Control System Synthesis*, McGraw-Hill, New York, 1955.

11. Sampled-Data Control Systems

11.1 Introduction

In the control systems dealt so far, the signal at every point in the system is a continuous function of time (the independent variable). In particular the controller elements are such that the controller produces continuous-time control signals from continuous-time input signals. Such a controller is referred to as an *analog* controller*. As the complexity of a control system increases, severe demands of optimal performance (see Chapter 13) and even demands to account for economic factors in control are placed on the controller. The cost of an analog controller rises steeply with increasing control function complexity. In fact constructing a complex control function may even become technically infeasible if one is restricted to use only analog elements. A digital controller, in which either a special purpose computer or a general purpose computer forms the heart, is therefore an ideal choice for complex control systems. A general purpose computer, if used, lends itself to *time-shared* use for other control functions in the plant or process. A digital controller also has the versatility that its control function can be easily modified by changing a few program instructions or even the entire program and a change in instruction can be accomplished either manually or automatically under control of a supervisory function.

Digital controllers have the inherent characteristic that they accept the data in form of short duration pulses (i.e., sampled or discrete data) and produce a similar kind of output as control signal. Figure 11.1 shows a

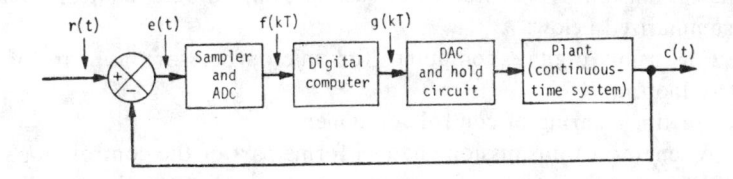

Figure 11.1 Typical system with digital controller (sampled-data control system).

*The term stems from the analog computer whose elements produce continuous-time signals.

simple control scheme employing a digital controller. A sampler and analog-to-digital converter (ADC) is needed at the computer input. The sampler converts the continuous-time error signal into a sequence of pulses which are then expressed in numerical code (such as binary code). Numerically coded output data of digital computer are decoded into continuous-time signal by digital-to-analog converter (DAC) and hold circuit. This continuous-time signal then controls the plant (continuous-time system). The overall system is *hybrid* in which the signal is in sampled form in the digital controller and in continuous form in the rest of the system. A system of this kind is referred to as a *sampled-data control system*.

Even in relatively simple control schemes, sampling may be warranted from other considerations. In fact sampling is a must wherever a high degree of accuracy is a prerequisite. This is the case in most automated machine-tools. For example if it is desired to move the table of a drilling machine within an accuracy of 0.01 mm over a total distance of 1 m, a resolution of 1 in 100,000 is needed which is impossible to measure using an analog type output transducer say a potentiometer. The problem can only be tackled by employing a digitally-coded output sensor.

Sampled-data technique is most appropriate for control systems requiring long distance data transmission. Pulse amplitude modulated* (PAM) data is easily transmitted by means of a carrier over a transmission channel and the data reconstructed at the receiving end. It is well known that pulses may be transmitted with little loss of accuracy; data in analog form will suffer considerable distortion in the trasmission channel. Using sampled-data technique, more than one channel of information may be sequentially sampled and transmitted through a single transmission system (this technique is known as time-multiplexing), decreasing the cost of transmission installation.

Signal sampling reduces the power demand made on the signal and is therefore helpful for signals of weak power origin.

The examples cited above were those of systems in which the sampling operation is purposely introduced. However, there is a class of systems where the signals are available in sampled form only. For example in radar tracking system, the signal sent out and received is in form of pulse trains.

The circumstances that lead to the use of sampled-data control systems are summarized below:

1. For using digital computer (or microprocessor) as part of the control loop.

2. For time-sharing of control components.

3. Whenever a transmission channel forms part of the control loop.

4. Whenever the output of a control component is essentially in discrete form.

Sampling implies that the signal at the output end of the *sampler* is

---

*Pulse width modulation (PWM) is an alternative way of tackling the transmission problem.

available in form of short duration pulses each followed by a skip period when no signal is available so that the control system essentially operates open-loop during the skip period. Uniform periodic sampling is illustrated in Fig. 11.2. It is intuitively obvious that if the sampling rate (or frequency) is too low, significant information contained in the input signal may be missed in the output. We shall show in the next article that the minimum sampling rate has a definite relationship with the highest significant signal frequency (i.e., signal bandwidth).

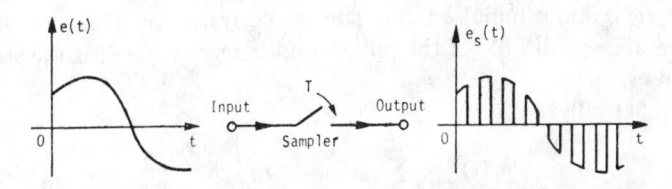

Figure 11.2 Uniform periodic sampling.

Assuming sample width (time) as fixed, other forms of sampling are:
Multiple-order sampling: A particular sampling pattern is repeated periodically.
Multiple-rate sampling: In this case two simultaneous sampling operations with different time periods are carried out on the signal to produce the sampled output.
Random sampling: In this case the sampling instants are random with a particular distribution.

In this book we shall restrict ourselves to uniform periodic sampling.

The mathematical model of a sampled-data control system is essentially in the form of difference equations. The analysis and design of sampled-data systems with linear elements may be effectively carried out by use of the z-transform which was evolved from the Laplace transform as a special form and was later established in its own merit.

11.2 Spectrum Analysis of Sampling Process

The sampling process represented in Fig. 11.2 is equivalent to multiplying the signal $e(t)$ with a periodic pulse train $p_{T,\Delta}(t)$, shown in Fig. 11.3, to produce the sampled signal

$$e_s(t) = e(t)\, p_{T,\Delta}(t)$$

The spectral analysis of such a sampled signal is somewhat involved. The pulse train is therefore approximated by an impulse train with each impulse of strength $1 \times \Delta = \Delta$, the pulse area. The sampled signal is consequently approximated as

$$e_s(t) = e(t)\Delta\delta_T(t) \tag{11.1}$$

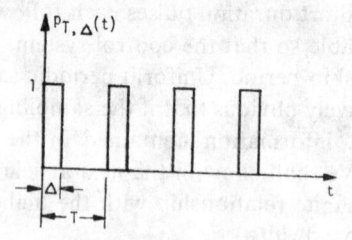

Figure 11.3 Pulse train $p_{T,\Delta}(t)$.

where $\delta_T(t)$ is a unit impulse train. This is illustrated in Fig. 11.4. Notice that the time width (Δ) of the pulse sampler merely modifies the strength of impulses.

Mathematically

$$\delta_T(t) = \sum_{k=-\infty}^{\infty} \delta(t - kT) \qquad (11.2)$$

$$\therefore \quad e_s(t) = \Delta e(t) \sum_{k=-\infty}^{\infty} \delta(t - kT) \qquad (11.3)$$

Taking the Fourier transform (see Appendix I) of eqn. (11.3)

$$\mathcal{F}[e_s(t)] = \frac{\Delta}{2\pi} \mathcal{F}[e(t)] * \mathcal{F}\left[\sum_{k=-\infty}^{\infty} \delta(t - kT)\right] \qquad (11.4)$$

But

$$\mathcal{F}\left[\sum_{k=-\infty}^{\infty} \delta(t - kT)\right] = \omega_s \sum_{k=-\infty}^{\infty} \delta(\omega - k\omega_s)$$

where $\omega_s = \dfrac{2\pi}{T}$ = sampling frequency.

Equation (11.4) can therefore be written as

$$E_s(\omega) = \frac{\Delta}{2\pi} E(\omega) * \omega_s \sum_{k=-\infty}^{\infty} \delta(\omega - k\omega_s) \qquad (11.5)$$

As

$$E(\omega) * \delta(\omega - k\omega_s) = E(\omega - k\omega_s)$$

we have

$$E_s(\omega) = \frac{\Delta}{T} \sum_{k=-\infty}^{\infty} E(\omega - k\omega_s) \qquad (11.6)$$

Equation (11.6) gives the frequency spectrum of the impulse sampled signal in terms of the spectrum of the input signal. Let $E(\omega)$, the spectrum of the input signal, be band-limited† signal with a maximum frequency of ω_m as in Fig. 11.5a. The frequency spectrum of this signal when impulse sampled (see Fig. 11.4c) is plotted in Fig. 11.5b for $\omega_s > 2\omega_m$ and in Fig. 11.5c for $\omega_s < 2\omega_m$.

†All signals of practical significance are band-limited in nature as the energy contained in frequencies above a certain figure is of negligible order.

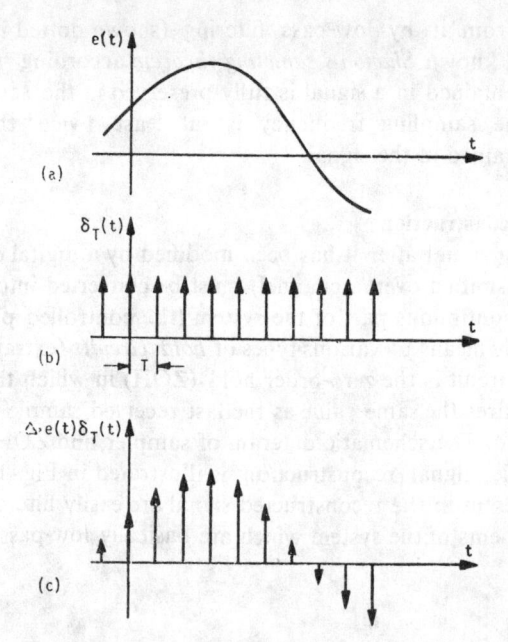

Figure 11.4 Impulse sampling approximation of pulse sampling.

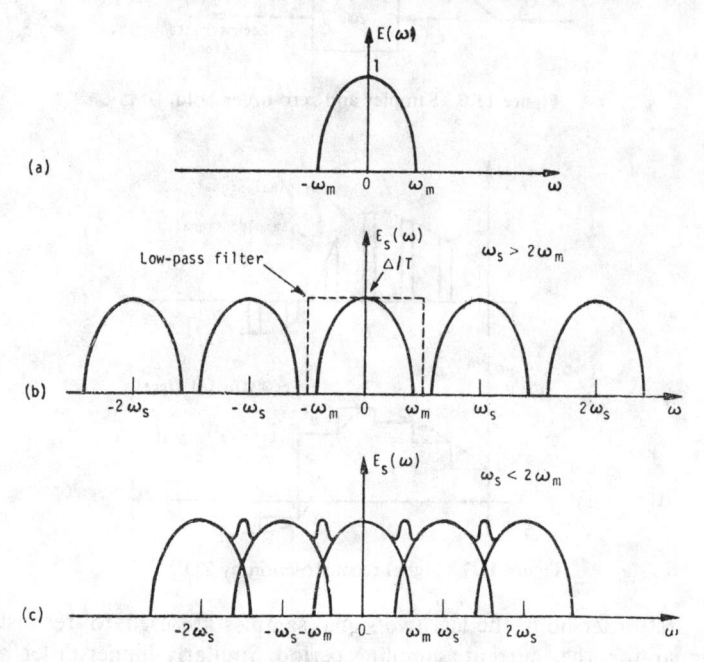

Figure 11.5 Fourier spectra of input signal and its impulse sampled version.

It is immediately observed from Figs. 11.5b and c that so long as $\omega_s \geqslant 2\omega_m$, the original spectrum is preserved in the sampled signal and can be extracted from it by low-pass filtering (shown dotted in Fig. 11.5b). This is the well known *Shanon's sampling theorem* according to which the information contained in a signal is fully preserved in the sampled version so long as the sampling frequency is at least twice the maximum frequency contained in the signal.

11.3 Signal Reconstruction

Sampled-data signal after it has been modified by a digital controller (or has been transmitted over a channel) must be converted into analog form for use in the continuous part of the system (the controlled plant). This is accomplished by means of various types of *hold circuits* (extrapolators). The simplest hold circuit is the zero-order hold (ZOH) in which the reconstructed signal acquires the same value as the last received sample for the entire sampling period. The schematic diagram of sampler and ZOH is shown in Fig. 11.6, while signal reconstruction is illustrated in Fig. 11.7. The high frequencies present in the reconstructed signal are easily filtered out by the controlled elements of the system which are basically low-pass in frequency behaviour.

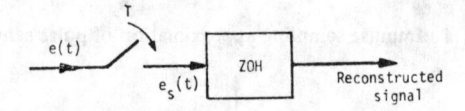

Figure 11.6 Sampler and zero-order hold.

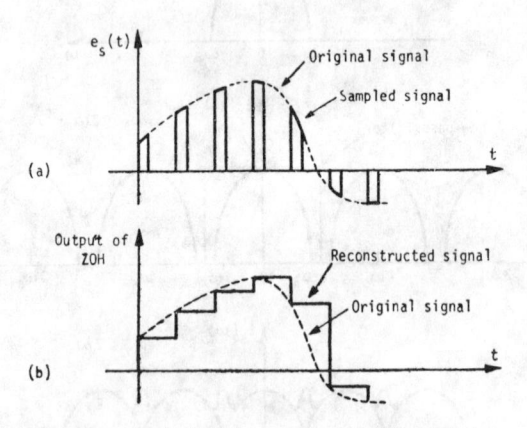

Figure 11.7 Signal reconstruction by ZOH.

In a first-order hold, the last two signal samples are used to reconstruct the signal for the current sampling period. Similarly higher-order holds can be devised. First or higher-order holds offer no particular advantage over the zero-order hold. The simple zero-order hold when used in conjunction with a high samplin rate provides a satisfactory performance.

11.4 Difference Equations

The difference equation model results in sampled-data systems and in numerical analysis of continuous-time systems. Examples of each kind will now be presented here.

Example 11.1: Consider a first-order continuous-time feedback system as in Fig. 11.8. We can immediately write

$$c(t) = A \int_0^t [r(\tau) - c(\tau)]d\tau$$

or
$$c(t) + A \int_0^t c(\tau)d\tau = A \int_0^t r(\tau)d\tau \qquad (11.7)$$

Closed-form solution is not possible if $r(t)$ is an arbitrary function of time. Numerical solution can be obtained by approximating the integral equation (11.7) by a difference equation. Dividing time into discrete intervals of length T, we have

$$c(kT) + A \int_0^{kT} c(\tau)d\tau = A \int_0^{kT} r(\tau)d\tau$$

Approximating integration by discrete summation, we can write

$$c(kT) + A \sum_{m=0}^{k-1} c(mT)T = A \sum_{m=0}^{k-1} r(mT)T \qquad (11.8)$$

Also

$$c(kT+T) + A \sum_{m=0}^{k} c(mT)T = A \sum_{m=0}^{k} r(mT)T \qquad (11.9)$$

Subtracting (11.8) from (11.9),

$$c[(k+1)T] - c(kT) + ATc(kT) = ATr(kT)$$

Writing $c(kT)$ as $c(k)$ and $r(kT)$ as $r(k)$, we have

$$c(k+1) = (1 - AT)c(k) + ATr(k) \qquad (11.10)$$

which is a first-order difference equation with constant coefficients.

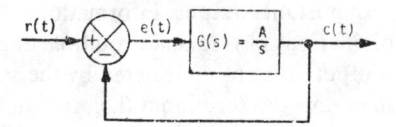

Figure 11.8.

Example 11.2: Introduce now a sampler and ZOH in the forward path of the system of Fig. 11.8 converting it to a sampled-data system of Fig. 11.9.

Because of the action of sampler and ZOH, the output of ZOH during kth time interval is

$$e_k(t) = e(kT); kT \leqslant t < (k+1)T$$

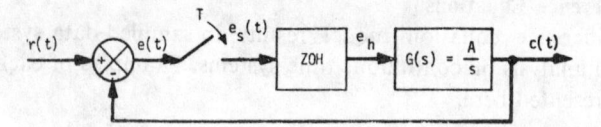

Figure 11.9 First-order sampled-data system.

The output $c(t)$ during this period due to action of the integrator (A/s) is given by

$$c(t) = c(kT) + Ae(kT)(t - kT); \quad kT \leq t < (k+1)T$$

or $\quad c[(k+1)T] = c(kT) + ATe(kT)$

or $\quad c(k+1) = c(k) + ATe(k) \quad (11.11)$

But

$$e(kT) = r(kT) - c(kT)$$

Equation (11.11) can then be written as

$$c(k+1) = (1 - AT)c(k) + ATr(k) \quad (11.12)$$

which indeed is the same difference equation as (11.10).

Let us obtain the solution of the difference equation (11.12). This can be easily carried out recursively

$$c(1) = (1 - AT)c(0) + ATr(0)$$
$$c(2) = (1 - AT)c(1) + ATr(1)$$
$$= (1 - AT)^2 c(0) + (1 - AT)ATr(0) + ATr(1)$$
$$\cdots$$
$$c(k) = \underbrace{(1 - AT)^k c(0)}_{\text{Zero input response}} + \underbrace{AT \sum_{i=0}^{k-1} (1 - AT)^{k-1-i} r(i)}_{\text{Forcing function response}} \quad (11.13)$$

It may be noticed here that the difference equation solution yields $c(k)$, i.e., the values of output $c(t)$ (which is continuous-time function caused by the presence of the continuous part of the system (A/s)) at sampling instants only. No information is available about $c(t)$ during the sampling instants. This is satisfactory so long as the sampling rate is sufficiently fast so as not to miss any important output information. In fact cases can exist where parasitic oscillations of frequency multiple of sampling frequency are present in the output but are undetected by the sampled output (see Problem 11.14). Consider now the zero input (i.e., initial condition) response

$$c(k) = (1 - AT)^k c(0) \quad (11.14)$$

The response decays, i.e., the system is stable if

$$|1 - AT| < 1$$

or $\quad 0 < AT < 2$

or $\quad A < \dfrac{2}{T} \quad (11.15)$

We therefore conclude that the sampled-data system of Fig. 11.9 is conditionally stable. It may be recalled here that the corresponding continuous-time system of Fig. 11.8 is absolutely stable, i.e., stable for all values of A. The instability of the sampled-data system results from the fact that it operates open-loop between sampling instants.

As the difference equation (11.10) approximating the continuous-time system of Fig. 11.8 is identical to eqn. (11.12), it would yield a highly erroneous solution (unstable where the actual system is stable) for $T > 2/A$, which only tells that continuous time must be divided into very much smaller intervals than $T = 2/A$ to be a 'good' approximate solution.

The general form of nth-order linear, constant coefficient difference equation is presented below:

$$c(k+n)+a_1c(k+n-1)+\cdots+a_nc(k)$$
$$= b_0r(k+n)+b_1r(k+n-1)+\cdots+b_nr(k) \qquad (11.16)$$

Often this equation is written in the equivalent form:

$$c(k)+a_1c(k-1)+\cdots+a_nc(k-n)$$
$$= b_0r(k)+b_1r(k-1)+\cdots+b_nr(k-n) \qquad (11.17)$$

The straightforward method of modelling of sampled-data systems gets very messy for higher-order systems. The z-transform technique introduced below is a convenient tool for analysis and design of linear sampled-data systems.

11.5 The z-transform

The one-sided z-transform of a sequence

$$f(k) : \{\ldots 0, 0, f(0), f(1), f(2), \ldots\}$$

which is defined for positive integers k, is defined as the weighted sum

$$\mathcal{Z}[f(k)] = F(z) = \sum_{k=0}^{\infty} f(k)z^{-k} \qquad (11.18)$$

where z is an arbitrary complex number. $F(z)$, a sum of complex numbers, is also a complex number.

Consider a geometric sequence

$$f(k) = \{\ldots 0, 1, a, a^2, \ldots\} = a^k \; ; \; k \geqslant 0$$
$$= 0 \; ; \; k < 0 \qquad (11.19)$$

where a is any real number.

The z-transform of this sequence is

$$\mathcal{Z}[f(k)] = F(z) = \sum_{k=0}^{\infty} a^k z^{-k}$$
$$= \sum_{k=0}^{\infty} (az^{-1})^k \qquad (11.20)$$

The infinite sum (11.20) converges if
$$|az^{-1}| < 1$$
or
$$|z| > |a|$$
and it diverges (is unbounded) if
$$|z| < |a|$$
In the region of convergence,
$$\mathcal{Z}[a^k] = \sum_{k=0}^{\infty} (az^{-1})^k = \frac{1}{1 - az^{-1}} = \frac{z}{z-a}$$

The z-transform of the geometric sequence (11.19) may be compactly expressed as

$$\mathcal{Z}[a^k] = F(z) = \begin{cases} \dfrac{z}{z-a} & ; |z| > |a| \\ \text{unbounded} & ; |z| < |a| \end{cases} \qquad (11.21)$$

The convergence and divergence regions of $F(z)$ in (11.21) are shown in Fig. 11.10. On the circle separating the two regions, the z-transform may or may not converge. We should make separate tests for those z which lie on this boundary.

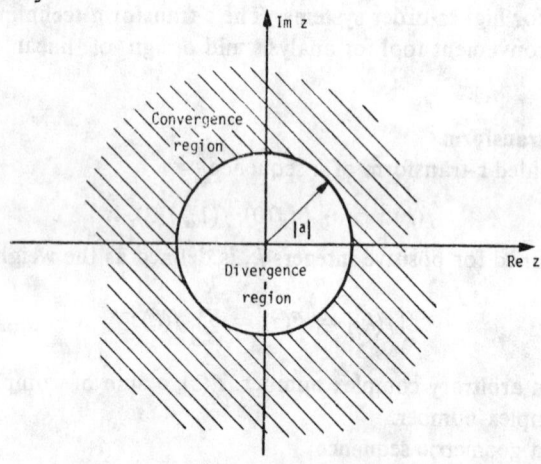

Figure 11.10 Regions of convergence and divergence for the z-transform.

Consider now the sinusoidal sequence defined by
$$f(k) = \sin k\omega T \; ; \; k \geqslant 0$$
$$= 0 \qquad ; k < 0$$

The z-transform is given as
$$\mathcal{Z}[\sin k\omega T] = F(z) = \sum_{k=0}^{\infty} (\sin k\omega T) z^{-k}$$

$$= \sum_{k=0}^{\infty}\left(\frac{e^{jk\omega T} - e^{-jk\omega T}}{2j}\right)z^{-k} \cdot \qquad (11.22)$$

Now using the result of eqn. (11.21),

$$\mathcal{Z}[(e^{j\omega T})^k] = \frac{z}{z - e^{j\omega T}} \; ; \; |z| > 1 \qquad (11.23)$$

From eqns. (11.22) and (11.23), we have the following result:

$$\mathcal{Z}[\sin k\omega T] = \frac{z \sin \omega T}{\{z - e^{j\omega T}\}\{z - e^{-j\omega T}\}}$$

$$= \frac{z \sin \omega T}{z^2 - 2z \cos \omega T + 1} \; ; \; |z| > 1 \qquad (11.24)$$

Convergence region of the z-transform lies outside a circle of unit radius in the complex z-plane.

It can be proved [1] that in general the z-transform of any sequence of numbers $f(k)$ will have a region of convergence specified by

$$|z| > R$$

where the radius of convergence* R depends upon the sequence $f(k)$.

A discrete sequence can be generated from a continuous-time function $f(t)$ by *mathematically sampling* it at time intervals T seconds apart merely by substituting $t = kT$, i.e.,

$$f_s(t) = f(kT)$$

This sampling is different from pulse or impulse sampling of a physical signal by suitable hardware wherein the output is a train of pulses or impulses. Mathematical sampling is merely a mathematical contrivance to generate a sequence of values at discrete instants of time. Diagramatically it is indicated in Fig. 11.11.

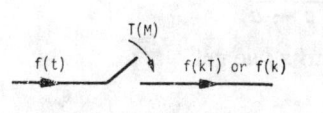

Figure 11.11.

Consider a time function

$$f(t) = e^{j\omega t} u(t)$$

Sampling it mathematically

$$f(k) = e^{j\omega kT} u(k) = \{e^{j\omega T}\}^k \; ; \; k \geqslant 0$$

Recognizing $a = e^{j\omega T}$ in (11.21), we can at once write

$$\mathcal{Z}[e^{j\omega kT}] = \frac{z}{z - e^{j\omega T}} \quad \text{for} \quad |z| > 1$$

We can immediately obtain

$$\mathcal{Z}[\cos \omega kT] = \mathcal{Z}\left[\frac{e^{j\omega kT} + e^{-j\omega kT}}{2}\right]$$

$$= \frac{1}{2}\left[\frac{z}{z - e^{j\omega T}} + \frac{z}{z - e^{-j\omega T}}\right]$$

$$= \frac{z(z - \cos \omega T)}{z^2 - 2z \cos \omega T + 1} \; ; \; |z| > 1$$

---

*Note that we will generally not be concerned with the regions of convergence and divergence. These concepts have been introduced for the sake of completeness.

The Inverse z-Transform

The inverse z-transformation is a process of determining the sequence which generates a given z-transform. It is denoted by

$$f(k) = \mathcal{Z}^{-1}[F(z)]$$

From eqns. (11.21) and (11.24), we have

$$\mathcal{Z}^{-1}\left[\frac{z}{z-a}\right] = a^k$$

$$\mathcal{Z}^{-1}\left[\frac{z \sin \omega T}{z^2 - 2z \cos \omega T + 1}\right] = \sin k\omega T$$

These equations give the following transform pairs:

$$a^k \leftrightarrow \frac{z}{z-a}$$

$$\sin k\omega T \leftrightarrow \frac{z \sin \omega T}{z^2 - 2z \cos \omega T + 1}$$

For illustration purposes, a few transform pairs are generated below:

(i) The process of differentiating a known z-transform pair offers a convenient method for determining a new z-transform pair. For example, consider the known relationship

$$\sum_{k=0}^{\infty} a^k z^{-k} = \frac{z}{z-a}$$

Differentiating both sides with respect to z, we get

$$[-az^{-2} - 2a^2z^{-3} - \ldots] = -\frac{a}{(z-a)^2}$$

Multiplying both sides by $-z^2$,

$$[a + 2a^2z^{-1} + \ldots] = \frac{az^2}{(z-a)^2}$$

or

$$\sum_{k=0}^{\infty} (k+1)a^{k+1}z^{-k} = \frac{az^2}{(z-a)^2}$$

$$(k+1)a^{k+1} \leftrightarrow \frac{az^2}{(z-a)^2} \tag{11.25}$$

(ii) Consider a discrete unit impulse defined as

$$\delta(k) : \{1, 0, 0, \ldots\} = 1 \,;\, k = 0$$
$$= 0 \,;\, k \neq 0$$

$$\mathcal{Z}[\delta(k)] = \sum_{k=0}^{\infty} \delta(k) z^{-k} = 1$$

Thus the z-transform pair is

$$\delta(k) \leftrightarrow 1 \tag{11.26}$$

(iii) A discrete unit step function is defined as

$$u(k) : \{1, 1, 1, ...\} = 1 ; k \geq 0$$
$$= 0 ; k < 0$$
$$\mathcal{Z}[u(k)] = \mathcal{Z}[1^k]$$

Substituting $a = 1$ in eqn. (11.21), we get

$$u(k) \leftrightarrow \frac{z}{z-1} \tag{11.27}$$

PROPERTIES OF THE z-TRANSFORM

In this section we will study some of the important properties of the z-transform which can be used to determine $F(z)$ and its inverse.

Linearity

$$\mathcal{Z}[af(k)+bg(k)] = \sum_{k=0}^{\infty} [af(k)+bg(k)]z^{-k}$$
$$= a\sum_{k=0}^{\infty} f(k)z^{-k} + b\sum_{k=0}^{\infty} g(k)z^{-k}$$
$$= aF(z)+bG(z) \tag{11.28}$$

Shifting

A sequence $f(k)$ when shifted one interval to the left (advanced), can be written as

$$g(k) = f(k+1) ; k \geq -1$$

The shifted sequence is shown in Fig. 11.12b. Taking the z-transform,

$$\mathcal{Z}[f(k+1)] = \sum_{k=0}^{\infty} f(k+1)z^{-k}$$
$$= z\sum_{k=0}^{\infty} f(k+1)z^{-(k+1)}$$

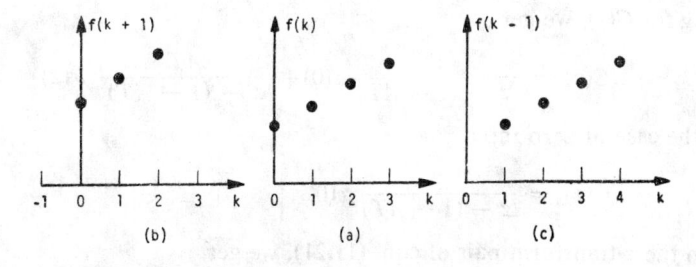

Figure 11.12 Shifted sequences.

Letting $(k+1) = m$,

$$\mathscr{Z}[f(k+1)] = z \sum_{m=1}^{\infty} f(m)z^{-m}$$

$$= z\left[\sum_{m=0}^{\infty} f(m)z^{-m} - f(0)\right]$$

$$= zF(z) - zf(0) \qquad (11.29)$$

In general

$$\mathscr{Z}[f(k+n)] = z^n F(z) - \sum_{i=0}^{n-1} f(i)z^{n-i} \;;\; k \geqslant -n \qquad (11.30)$$

Consider now the sequence

$$g(k) = f(k-n) \;;\; k \geqslant n$$

which is the sequence $f(k)$ shifted n intervals to the right (delayed) as shown in Fig. 11.12c for $n = 1$.

Taking the z-transform,

$$\mathscr{Z}[f(k-n)] = \sum_{k=0}^{\infty} f(k-n)z^{-k} = z^{-n} \sum_{k=0}^{\infty} f(k-n)z^{-(k-n)}$$

Letting $(k-n) = m$, we have

$$\mathscr{Z}[f(k-n)] = z^{-n} \sum_{m=-n}^{\infty} f(m)z^{-m}$$

As $\quad f(m) = 0 \;\text{ for }\; m < 0,$

$$\mathscr{Z}[f(k-n)] = z^{-n} \sum_{m=0}^{\infty} f(m)z^{-m}$$

$$= z^{-n} F(z) \qquad (11.31)$$

The above properties are useful in z-transforming difference equations. Consider the difference equation (11.12) reproduced below:

$$c(k+1) - (1 - AT)c(k) = ATr(k)$$

Taking the z-transform of each term (linearity property)

$$\mathscr{Z}[c(k+1)] - (1 - AT)\,\mathscr{Z}[c(k)] = AT\,\mathscr{Z}[r(k)]$$

$$zC(z) - zc(0) - (1 - AT)C(z) = ATR(z)$$

Solving for $C(z)$, we get

$$C(z) = \frac{z}{z - (1 - AT)} c(0) + \frac{AT}{z - (1 - AT)} R(z) \qquad (11.32)$$

For the case of zero input

$$C(z) = \frac{z}{z - (1 - AT)} c(0)$$

From the z-transform pair of eqn. (11.21), we get

$$c(k) = (1 - AT)^k c(0)$$

which is a result we have earlier obtained by the recursive solution of the difference equation.

Example 11.3: Consider a system described by the difference equation
$$c(k+1)+2c(k) = r(k) \quad ; \quad c(0) = 0$$
Let us obtain the system's impulse response; that is $c(k)$ satisfying
$$c(k+1)+2c(k) = \delta(k)$$

Taking the z-transform
$$(z+2)C(z) = 1$$
$$C(z) = \frac{1}{z+2}$$

From (11.21),
$$\frac{z}{z+2} \leftrightarrow (-2)^k$$

From shifting property (11.30)
$$z^{-1}\left(\frac{z}{z+2}\right) \leftrightarrow (-2)^{k-1} \quad ; \quad k \geqslant 1$$

Hence
$$c(k) = (-2)^{k-i} \quad ; \quad k \geqslant 1$$

Multiplication by k

Let $\mathscr{Z}[f(k)] = F(z)$

Now
$$\mathscr{Z}[kf(k)] = \sum_{k=0}^{\infty} kf(k)z^{-k} = -z\sum_{k=0}^{\infty} -kf(k)z^{-k-1}$$
$$= -z\sum_{k=0}^{\infty} f(k)\frac{d}{dz}z^{-k} = -z\frac{d}{dz}\sum_{k=0}^{\infty} f(k)z^{-k}$$
$$= -z\frac{d}{dz} F(z) \qquad (11.33)$$

Example 11.4: Find the z-transform of the discrete ramp function
$$g(k) = k \quad ; \quad k \geqslant 0$$
$$= 0 \quad ; \quad k < 0.$$

We can write
$$g(k) = ku(k)$$
where $u(k)$ is a discrete step function.

Now
$$\mathscr{Z}[g(k)] = \mathscr{Z}[ku(k)]$$
$$= -z\frac{d}{dz} U(z)$$

Substituting for $U(z)$ from (11.27),

$$\mathcal{Z}[g(k)] = -z\frac{d}{dz}\left(\frac{z}{z-1}\right) = \frac{z}{(z-1)^2}$$

Scale change

Let

$$\mathcal{Z}[f(k)] = F(z)$$

Consider the inverse z-transform of $F(z/a)$, that is,

$$\mathcal{Z}^{-1}\left[F\left(\frac{z}{a}\right)\right] = \mathcal{Z}^{-1}\left[\sum_{k=0}^{\infty} f(k)\left(\frac{z}{a}\right)^{-k}\right]$$

$$= \mathcal{Z}^{-1}\left[\sum_{k=0}^{\infty} a^k f(k) z^{-k}\right] = a^k f(k)$$

Thus, we have the transform pair

$$a^k f(k) \leftrightarrow F(z/a) \tag{11.34}$$

Initial and final values

(a) $$F(z) = \sum_{k=0}^{\infty} f(k)z^{-k} = f(0) + f(1)z^{-1} + f(2)z^{-2} + \cdots$$

Taking the limit as $z \to \infty$, we obtain

$$f(0) = \lim_{z \to \infty} F(z) \tag{11.35}$$

(b) A more useful theorem involves the determination of the final value of the sequence $f(k)$, which is given by

$$\lim_{k \to \infty} f(k) = \lim_{z \to 1} F(z)(z-1) \tag{11.36}$$

provided the limit exists, i.e., the data sequence has a finite value. Consider the z-transform pair

$$F(z) = \frac{z}{z-a} \leftrightarrow f(k) = a^k$$

The pole of $F(z)$ is located at $z = a$. If $|a| > 1$, that is the pole lies outside a unit circle, the sequence $f(k) = a^k$ tends to increase without bound, i.e., the limit does not exist. Therefore for the final value theorem to be applicable, $F(z)$ must not have any poles in the region outside a unit circle, i.e., it should be analytic for $|z| > 1$.

Proof:

$$\mathcal{Z}[f(k+1) - f(k)] = \lim_{m \to \infty} \sum_{k=0}^{m} [f(k+1) - f(k)]z^{-k}$$

By the left shifting property, we get

$$zF(z) - zf(0) - F(z) = \lim_{m \to \infty} \sum_{k=0}^{m} [f(k+1) - f(k)]z^{-k}$$

Letting $z \to 1$ on both sides,

$$\lim_{z \to 1} [(z-1)F(z) - f(0)] = \lim_{z \to 1} \lim_{m \to \infty} \sum_{k=0}^{m} [f(k+1) - f(k)]z^{-k}$$

Interchanging the order of limits on the right-hand side, we have

$$\lim_{z \to 1} (z-1)F(z) = f(0) + \lim_{m \to \infty} \sum_{k=0}^{m} [f(k+1) - f(k)]$$
$$= f(\infty)$$

As an example, consider eqn. (11.32) reproduced below:

$$C(z) = \frac{z}{z - (1 - AT)} c(0) + \frac{AT}{z - (1 - AT)} R(z)$$

Let $\quad r(k) = u(k)$

$$R(z) = \frac{z}{(z-1)}$$

Now

$$c(\infty) = \lim_{z \to 1} (z-1)C(z) = \lim_{z \to 1} \left[\frac{z(z-1)}{z - (1 - AT)} c(0) + \frac{AT(z-1)}{z - (1 - AT)} \times \frac{z}{(z-1)} \right]$$

or $\quad c(\infty) = 1$

The properties of the z-transform are summarized in Table 11.1.

Table 11.1 PROPERTIES OF THE ONE-SIDED z-TRANSFORM

Property	Discrete sequence	z-transform
Linearity	$af(k) + bg(k)$	$aF(z) + bG(z)$
Shifting $n \geqslant 0$	$f(k+n)$	$z^n F(z) - \sum_{i=0}^{n-1} f(i) z^{n-i}$
	$f(k-n)$	$z^{-n} F(z)$
Multiplication by k^n	$k^n f(k)$	$\left(-z \dfrac{d}{dz}\right)^n F(z)$
Scaling or multiplication by a^k	$a^k f(k)$	$F(a^{-1} z)$
Convolution*	$\sum_{m=0}^{k} h(k-m) r(m)$	$H(z) R(z)$
Initial value	$f(0) = \lim_{z \to \infty} F(z)$	
Final value	$f(\infty) = \lim_{z \to 1}(1 - z^{-1}) F(z)$	
	$= \lim_{z \to 1} (z-1) F(z)$	
	if $F(z)$ is analytic for $\|z\| > 1$	

*This property is proved in the next article.

z-TRANSFORM PAIRS

Rather than obtaining the z-transform abinitio every time, it is convenient to look up a table of transform pairs. Some of the commonly used z-transform pairs are listed in Table 11.2.

Table 11.2 COMMON ONE-SIDED z-TRANSFORM PAIRS

$f(t)$ $t \geqslant 0$	$f(k)/f(kT)$ $k \geqslant 0$	$F(z)$
	$\delta(k)$	1
	$u(k)$ or 1	$\dfrac{z}{z-1}$
	a^k	$\dfrac{z}{z-a}$
	ka^k	$\dfrac{az}{(z-a)^2}$
	$k^2 a^k$	$\dfrac{az(z+a)}{(z-a)^3}$
	$(k+1)a^k$	$\dfrac{z^2}{(z-a)^2}$
	$\dfrac{(k+1)(k+2)}{2!} a^k$	$\dfrac{z^3}{(z-a)^3}$
	$\dfrac{(k+1)(k+2)(k+3)}{3!} a^k$	$\dfrac{z^4}{(z-a)^4}$
	$\dfrac{a^k}{k!}$	$e^{az^{-1}}$
	kT	$\dfrac{Tz}{(z-1)^2}$
t^2	$(kT)^2$	$\dfrac{T^2 z(z+1)}{(z-1)^3}$
e^{-at}	e^{-akT}	$\dfrac{z}{z-e^{-aT}}$
te^{-at}	kTe^{-akT}	$\dfrac{zTe^{-aT}}{(z-e^{-aT})^2}$
$\sin \omega t$	$\sin \omega kT$	$\dfrac{z \sin \omega T}{z^2 - 2z \cos \omega T + 1}$
$\cos \omega t$	$\cos \omega kT$	$\dfrac{z(z - \cos \omega T)}{z^2 - 2z \cos \omega T + 1}$

11.6 The z-transfer Function (Pulse Transfer Function)

LINEAR DISCRETE SYSTEMS (LDS)

Consider a linear time-invariant discrete system represented by the block diagram of Fig. 11.13a. The system produces an output sequence $c(k)$ for an input sequence $r(k)$. Such a system can be characterized by its response $h(k)$ (also called the *weighting sequence*) to unit discrete impulse

$$\delta(k) = 1; \quad k = 0$$
$$= 0; \quad k \neq 0 \tag{11.37}$$

Sampled-Data Control Systems

(a) Linear discrete system with arbitrary input

(b) Unit discrete impulse response of LDS

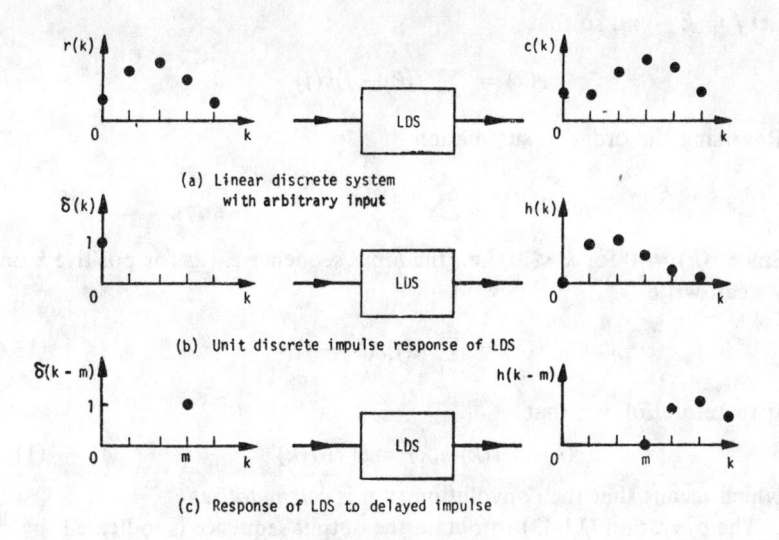

(c) Response of LDS to delayed impulse

Figure 11.13

as shown in Fig. 11.13b. The system is assumed to be *casual* (*nonanticipative*) so that the output appears only after application of the input. Because of this property, if a delayed unit impulse

$$\begin{aligned}\delta(k - m) &= 1; \quad k = m \\ &= 0; \quad k \neq m\end{aligned} \tag{11.38}$$

is applied to the system, the output sequence gets delayed by m intervals; that is it now becomes $h(k - m)$ as shown in Fig. 11.13c.

The input sequence $r(k)$ can be expressed as sum of impulses as

$$r(k) = r(0)\delta(k) + r(1)\delta(k-1) + r(2)\delta(k-2) + \ldots \tag{11.39}$$

The system response to mth impulse $r(m)\delta(k - m)$ is

$$c_m(k) = r(m)h(k - m)$$

Using the principle of superposition, the response to $r(k)$ as expressed in (11.39) can be written as the sum

$$c(k) = \sum_{m=0}^{\infty} c_m(k) = \sum_{m=0}^{\infty} r(m)h(k-m) \tag{11.40}$$

The sum (11.40) is known as the *discrete convolution* and is symbolically indicated as

$$c(k) = r(k) \cdot h(k) \tag{11.41}$$

The sum (11.40) need not be carried beyond $m = k$ as $h(k - m) = 0$ for $m > k$, the system being casual, that is,

$$c(k) = \sum_{m=0}^{k} r(m)h(k-m)$$

Let $j = k - m$, so that

$$c(k) = \sum_{j=k}^{0} r(k-j)h(j)$$

Reversing the order of summation

$$c(k) = \sum_{j=0}^{k} h(j)r(k-j)$$

Since $r(k) = 0$ for $k < 0$, i.e., the input sequence exists for positive k only, we can write

$$c(k) = \sum_{j=0}^{k} h(j)r(k-j) \tag{11.42}$$

It therefore follows that

$$c(k) = r(k) \cdot h(k) = h(k) \cdot r(k) \tag{11.43}$$

which means that the convolution sum is *commutative*.

The operation (11.43) to obtain the output sequence is indicated by the block diagram of Fig. 11.14.

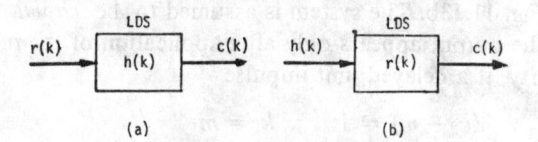

Figure 11.14.

Let us take the z-transform of the convolution sum (11.40)

$$\mathcal{Z}[c(k)] = C(z) = \sum_{k=0}^{\infty} c(k)z^{-k}$$

or

$$C(z) = \sum_{k=0}^{\infty} \left[\sum_{m=0}^{\infty} h(k-m)r(m) \right] z^{-k}$$

Interchanging the order of summation,

$$C(z) = \sum_{m=0}^{\infty} r(m) \sum_{k=0}^{\infty} h(k-m)z^{-k}$$

Let $j = k - m$, then

$$C(z) = \sum_{m=0}^{\infty} r(m) \sum_{j=-m}^{\infty} h(j)z^{-j-m}$$

$$= \sum_{m=0}^{\infty} r(m)z^{-m} \sum_{j=0}^{\infty} h(j)z^{-j}$$

$$= R(z)H(z) \tag{11.44}$$

This is an important result wherein

$$H(z) = \frac{C(z)}{R(z)} \qquad (11.45)$$

is interpreted as the z-transfer function or pulse transfer function of the LDS as shown by the block diagram of Fig. 11.15. The power of the z-transform lies in the fact that the convolution sum becomes a simple multiplication in the z-domain, parallel to the case of the s-domain in linear continuous-time systems.

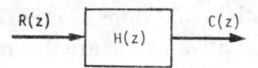

Figure 11.15 The block diagram of LDS in the z-domain.

As for unit impulse input $\delta(k)$,
$$R(z) = \mathcal{Z}[\delta(k)] = 1,$$
the z-transfer function can be interpreted as the z-transform of the unit impulse response, i.e.,
$$C(z) = H(z)$$
which is again parallel to the s-domain case.

When two or more LDS are connected in tandem, their individual z-transfer functions are multiplied to obtain the overall transfer function. This is shown by the block diagram of Fig. 11.16.

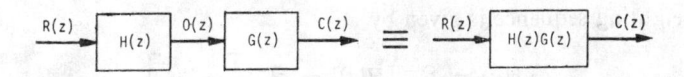

Figure 11.16.

LINEAR DIFFERENCE EQUATIONS

Consider a linear, time-invariant discrete system described by the difference equation (eqn. (11.17))
$$c(k) + a_1 c(k-1) + \ldots + a_n c(k-n)$$
$$= b_0 r(k) + b_1 r(k-1) + \ldots + b_n r(k-n) \qquad (11.46)$$

z-transforming each term of eqn. (11.46) by using the shifting theorem of eqn. (11.31) (it assumes that $c(k) = 0$ and $r(k) = 0$ for $k < 0$, i.e., zero initial conditions and causal input) and employing the linearity property, we have
$$C(z) + a_1 z^{-1} C(z) + \ldots + a_n z^{-n} C(z)$$
$$= b_0 R(z) + b_1 z^{-1} R(z) + \ldots + b_n z^{-n} R(z)$$
which can be written in the transfer function form as
$$\frac{C(z)}{R(z)} = \frac{b_0 + b_1 z^{-1} + \ldots + b_n z^{-n}}{1 + a_1 z^{-1} + \ldots + a_n z^{-n}}$$

Here the z-transfer function of the system described by the difference equations (11.46) is

$$H(z) = \frac{b_0 z^n + b_1 z^{n-1} + \ldots + b_n}{z^n + a_1 z^{n-1} + \ldots + a_n} \tag{11.47}$$

It is obvious that

$$H(z) = \mathscr{Z}[c(k)\,|_{r(k) = \delta(k)}]$$

If instead, the form of difference equation (11.16) is employed, the shifting theorem of eqn. (11.30) will have to be used to obtain the transfer function. The effect of initial conditions is now immediately incorporated. The fact that these alternative forms are dynamically equivalent for zero initial conditions, is illustrated in Example 11.6.

Example 11.5: Determine the impulse response (weighting sequence) of the LDS described by

$$c(k) - \alpha c(k-1) = r(k)$$

Taking the z-transform

$$C(z) - \alpha z^{-1} C(z) = R(z)$$

Note that initial conditions have not appeared by the use of the shifting property (11.31) which already assumes zero initial conditions.

Now

$$H(z) = \frac{C(z)}{R(z)} = \frac{z}{z - \alpha}$$

The weighting sequence is given by

$$h(k) = \mathscr{Z}^{-1}[H(z)] = \mathscr{Z}^{-1}\left[\frac{z}{z-\alpha}\right]$$

Looking up Table 11.2,

$$h(k) = \alpha^k; \quad k \geqslant 0$$
$$= 0; \quad k < 0$$

Example 11.6: Find the z-transform of the output for the LDS described by

$$c(k+2) + a_1 c(k+1) + a_2 c(k)$$
$$= b_0 r(k+2) + b_1 r(k+1) + b_2 r(k) \tag{11.48}$$

Taking the z-transform by use of (11.30), we have

$$(z^2 + a_1 z + a_2) C(z) - c(0) z^2 - c(1) z - a_1 c(0) z$$
$$= (b_0 z^2 + b_1 z + b_2) R(z) - b_0 r(0) z^2 - b_0 r(1) z - b_1 r(0) z$$

or

$$C(z) = \frac{b_0 z^2 + b_1 z + b_2}{z^2 + a_1 z + a_2} R(z)$$
$$+ \frac{z^2[c(0) - b_0 r(0)] + z[c(1) + a_1 c(0) - b_1 r(0) - b_0 r(1)]}{z^2 + a_1 z + a_2} \tag{11.49}$$

$c(k)$ could be obtained by taking the inverse z-transform of eqn. (11.49). Note that the first term of (11.49) incorporates the response due to input $r(k)$ and the second term gives the initial condition response.

If $c(k) = 0$ and $r(k) = 0$ for $k < 0$, we have from eqn. (11.48),

$$k = -2; \quad c(0) = b_0 r(0)$$
$$k = -1; \quad c(1) + a_1 c(0) = b_0 r(1) + b_1 r(0)$$

Hence eqn. (11.49) reduces to

$$C(z) = \frac{b_0 z^2 + b_1 z + b_2}{z^2 + a_1 z + a_2} R(z)$$

or

$$H(z) = \frac{C(z)}{R(z)} = \frac{b_0 z^2 + b_1 z + b_2}{z^2 + a_1 z + a_2}$$

which is of the same form as (11.47).

Note that $H(z)$ could be obtained directly by z-transforming the given difference equation ignoring initial conditions.

11.7 The Inverse z-transform and Response of Linear Discrete Systems

For obtaining the dynamic response of LDS via the z-transform method, the final step is to obtain the inverse transform of the output. We shall study here basically two methods. The third method of inversion integral is beyond the scope of this book [2].

POWER SERIES METHOD

Recalling the definition of the z-transform,

$$\mathscr{Z}[f(k)] = F(z) = \sum_{k=0}^{\infty} f(k) z^{-k}$$

and the fact that $F(z)$ can mostly be expressed in the ratio of polynomial form

$$F(z) = \frac{b_0 z^m + b_1 z^{m-1} + \ldots + b_m}{z^n + a_1 z^{n-1} + \ldots + a_n}; \quad m \leqslant n \text{ (for casual systems)} \quad (11.50)$$

we can immediately recognize the sequence $f(k)$ if $F(z)$ can be written in form of series with increasing powers of z^{-1}. This is easily carried out by dividing the denominator into the numerator.

Example 11.7: Determine the first few terms of the sequence $f(k)$ when

$$F(z) = \frac{z^2 + z}{z^2 - 2z + 1}$$

Dividing out the numerator by the denominator, we obtain

$$F(z) = 1 + 3z^{-1} + 5z^{-2} + 7z^{-3} + 9z^{-4} + \ldots$$

Hence

$$f(0) = 1$$
$$f(1) = 3$$
$$f(2) = 5$$

$$f(3) = 7$$
$$f(4) = 9$$
... ...

It may be recognized that while the power series method gives $f(k)$ for any k, it does not help us to get the general form of $f(k)$. It is therefore useful in numerical studies and not analytical studies.

It is also to be noted that $f(0) = 0$ if in the ratio polynomial $m < n$ (it immediately follows from the initial value theorem).

PARTIAL FRACTION EXPANSION METHOD

The inverse z-transform of $F(z)$; $m \leqslant n$, as in (11.50), can be obtained by partial fractioning it. For $m = n$ case, divide out the numerator polynomial by the denominator polynomial so as to separate the constant term, i.e.,

$$F(z) = d_0 + \frac{N(z)}{M(z)}$$

where the polynomial $N(z)$ is of order less than that of $M(z)$. Assuming distinct poles of $F(z)$, we can expand $F(z)$ into partial fractions as (see Appendix I)

$$F(z) = d_0 + \frac{A_1}{z-a_1} + \frac{A_2}{z-a_2} + \ldots + \frac{A_n}{z-a_n} \qquad (11.51)$$

Now

$$\frac{1}{z-a_i} = z^{-1}\left(\frac{z}{z-a_i}\right) \leftrightarrow (a_i)^{k-1}; k \geqslant 1$$

Hence taking the inverse z-transform of (11.51), we have

$$f(k) = d_0\delta(k) + [A_1(a_1)^{k-1} + A_2(a_2)^{k-1} + \ldots + A_n(a_n)^{k-1}]; \quad k \geqslant 1 \qquad (11.52)$$

In case of repeated roots, we get factors of the type

$$\frac{1}{(z-a)^2} = \frac{1}{a} z^{-1}\left[\frac{az}{(z-a)^2}\right] \leftrightarrow (k-1)a^{k-2}; k \geqslant 1$$

If $\dfrac{N(z)}{M(z)}$ is of the form

$$\frac{zN'(z)}{M(z)}$$

it can be expanded as

$$\frac{zN'(z)}{M(z)} = z\left[\frac{A_1'}{z-a_1} + \frac{A_2'}{z-a_2} + \ldots + \frac{A_n'}{z-a_n}\right]$$

$$= A_1'\frac{z}{z-a_1} + A_2'\frac{z}{z-a_2} + \ldots + A_n'\frac{z}{z-a_n}$$

The inverse z-transform now has the form

$$f(k) = A_1'(a_1)^k + A_2'(a_2)^k + \ldots + A_n'(a_n)^k; \quad k \geqslant 0$$

Both the methods are illustrated in Example 11.8.

Example 11.8: Find the inverse z-transform of
$$\frac{4z^2 - 2z}{z^3 - 5z^2 + 8z - 4}$$

$$\frac{4z^2 - 2z}{z^3 - 5z^2 + 8z - 4} = \frac{4z^2 - 2z}{(z-1)(z-2)^2} = \frac{4z^2 - 2z}{(z-2)^2}\bigg|_{z=1} \frac{1}{(z-1)}$$
$$+ \frac{4z^2 - 2z}{(z-1)}\bigg|_{z=2} \frac{1}{(z-2)^2} + \frac{d}{dz}\left[\frac{4z^2 - 2z}{z-1}\right]\bigg|_{z=2} \frac{1}{(z-2)}$$
$$= \frac{2}{(z-1)} + \frac{12}{(z-2)^2} + \frac{2}{(z-2)}$$

Taking the inverse z-transform, we have
$$f(k) = 2(1)^{k-1} + 2(2)^{k-1} + 6(k-1)(2)^{k-1}; \quad k \geqslant 1$$

Alternatively
$$z\left[\frac{4z-2}{(z-1)(z-2)}\right] = z\left[\frac{4z-2}{(z-2)^2}\bigg|_{z=1} \frac{1}{z-1} + \frac{4z-2}{z-1}\bigg|_{z=2} \frac{1}{(z-2)^2}\right.$$
$$\left. + \frac{d}{dz}\left(\frac{4z-2}{z-1}\right)\bigg|_{z=2} \frac{1}{z-2}\right]$$
$$= 2\frac{z}{z-1} + 6z^{-1}\frac{z^2}{(z-2)^2} - 2\frac{z}{z-2}$$

Taking the inverse z-transform, we get
$$f(k) = 2 + 6k(2)^{k-1}u(k-1) - 2(2)^k; \quad k \geqslant 0$$

It can be easily checked that the two forms of $f(k)$ are equivalent.

Example 11.9: The z-transform for the output in Example 11.2 was given in eqn. (11.32). Assuming zero initial conditions, we have
$$C(z) = \frac{AT}{z - (1 - AT)} R(z)$$

For discrete step input,
$$R(z) = \frac{z}{z-1}$$

$$\therefore \quad C(z) = \frac{ATz}{[z - (1 - AT)](z-1)} = \frac{z}{z-1} - \frac{z}{z - (1 - AT)}$$

Taking the inverse z-transform,
$$c(k) = 1 - (1 - AT)^k$$

Example 11.10: Solve the difference equation
$$x(k+2) - 3x(k+1) + 2x(k) = 4^k; \quad x(0) = 0, x(1) = 1$$

Taking the z-transform of the difference equation,
$$[z^2 X(z) - 3x(1) - z^2 x(0)] - 3[zX(z) - zx(0)] + 2X(z) = \frac{z}{z-4}$$

$$(z^2 - 3z + 2)X(z) = z^2 x(0) + z[x(1) - 3x(0)] + \frac{z}{z-4}$$

or
$$X(z) = \frac{z^2 x(0) + z[x(1) - 3x(0)]}{z^2 - 3z + 2} + \frac{z}{(z-4)(z^2 - 3z + 2)}$$

Substituting the initial conditions

$$X(z) = \frac{z}{(z-1)(z-2)} + \frac{z}{(z-1)(z-2)(z-4)}$$

$$= \left[-\frac{z}{z-1} + \frac{z}{z-2} \right] + \left[\frac{1}{3} \frac{z}{z-1} - \frac{1}{2} \frac{z}{z-2} + \frac{1}{6} \frac{z}{z-4} \right]$$

Taking the inverse z-transform,

$$x(k) = \underbrace{\left[-1 + (2)^k \right]}_{\text{Initial condition response}} + \underbrace{\left[\frac{1}{3} - \frac{1}{2} (2)^k + \frac{1}{6} (4)^k \right]}_{\text{Forcing function response}}$$

11.8 The z-transform Analysis of Sampled-data Control Systems

Sampled-data control systems are in fact hybrid systems as they have both discrete-time and continuous-time signals. Special techniques are therefore required for analysis of these systems.

ANALYSIS OF SAMPLER AND ZERO-ORDER HOLD

Figure 11.17 shows a sampler with zero-order hold (ZOH). As ZOH holds the input signal value for a period T, it means that for a short duration (Δ) input pulse, it produces an output pulse of duration T, the sampling period. This is illustrated in Fig. 11.18.

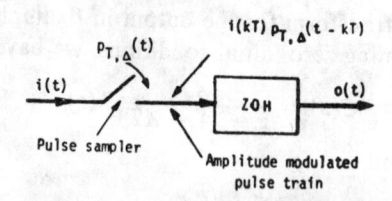

Figure 11.17 Pulse sampler with ZOH.

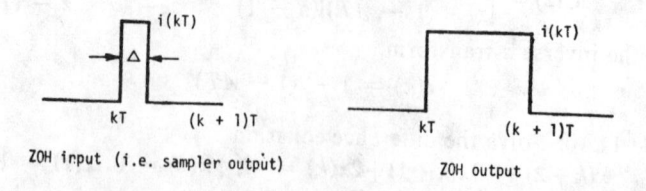

Figure 11.18 ZOH input and output pulses.

The ZOH output pulse appearing at kT instant can be expressed as

$$i(kT)[u(t - kT) - u(t - \overline{k+1}T)]$$

Sampled-Data Control Systems

We can therefore write the ZOH output as

$$o(t) = \sum_{k=0}^{\infty} i(kT)[u(t - kT) - u(t - \overline{k+1}T)] \qquad (11.53)$$

Taking the Laplace transform, we have

$$o(s) = \sum_{k=0}^{\infty} i(kT)\left[\frac{e^{-skT} - e^{-s(k+1)T}}{s}\right]$$

$$= \left(\frac{1 - e^{-sT}}{s}\right)\sum_{k=0}^{\infty} i(kT)e^{-skT} \qquad (11.54)$$

We can give a new input-output interpretation to eqn. (11.54). Taking the inverse Laplace transform of the infinite sum, we get (see Table I.1 in Appendix I)

$$\mathcal{L}^{-1}\left[\sum_{k=0}^{\infty} i(kT)e^{-skT}\right] = \sum_{k=0}^{\infty} i(kT)\delta(t - kT) = i(t)\delta_T(t) \qquad (11.55)$$

Thus the output $o(t)$ of pulse sampler and ZOH can be produced by *impulse sampled* $i(t)$ when passed through a transfer function

$$G_0(s) = \frac{1 - e^{-sT}}{s} \qquad (11.56)$$

This is illustrated in Fig. 11.19. Figures 11.17 and 11.19 represent equivalent operations, but the operation of Fig. 11.19 offers ease of analysis.

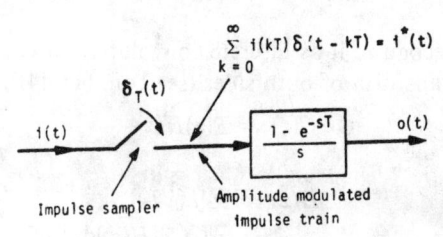

Figure 11.19 Equivalent representation of pulse sampler and ZOH.

It may be noted here that Figs. 11.17 and 11.19 are exactly equivalent and no approximation of any kind has been made here.

Analysis of Systems with Impulse Sampling

In sampled-data systems, our interest generally is to obtain the values of the output at sampling instants. Also we have seen above how pulse sampling and ZOH can be replaced by impulse sampling and $G_0(s)$. Consider now a linear continuous system fed from an impulse sampler as shown in Fig. 11.20a. The output signal $c(t)$ is read off at discrete synchronous sampling instants (kT) by means of a mathematical sampler $T(M)$ or what we can also call a read-out sampler.

The continuous output of the system in the s-domain is given by (Fig. 11.20b)

$$C(s) = H(s)R^*(s)$$

where $R^*(s) = \mathcal{L}[r^*(t)]$ = the Laplace transform of the impulse sampled input signal.

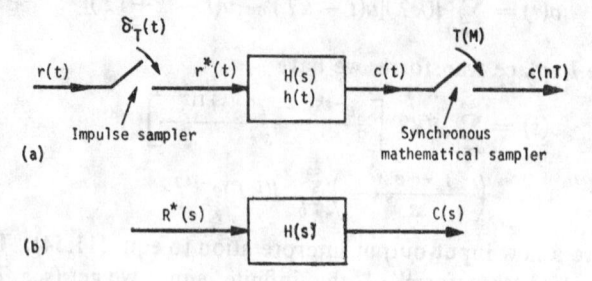

Figure 11.20 Linear continuous system with impulse sampled input.

Let $\mathcal{L}^{-1}[H(s)] = h(t)$ = the impulse response of the linear continuous system.

Now $$r^*(t) = \sum_{k=0}^{\infty} r(kT)\delta(t - kT)$$

Using superposition,

$$c(t) = \sum_{k=0}^{\infty} r(kT)h(t - kT)$$

$\therefore$ $$c(nT) = \sum_{k=0}^{\infty} r(kT)h(nT - kT) \qquad (11.57)$$

We can at once recognize it as discrete convolution of eqn. (11.40).

Taking the z-transform of both sides (see eqn. (11.44)),

$$C(z) = R(z)H(z) \qquad (11.58)$$

where

$$H(z) = \mathcal{Z}[h(kT)]$$
$$= \mathcal{Z}\{\mathcal{L}^{-1}[H(s)]|_{t=kT}\} \qquad (11.59)$$

The z-transforming operation of (11.59) is commonly indicated as

$$H(z) = \mathcal{Z}[H(s)] \qquad (11.60)$$

It follows from eqn. (11.58) that the time-domain block diagram of Fig. 11.20a becomes the z-domain block diagram of Fig. 11.21.

We could of course directly draw the z-domain block diagram of Fig. 11.21 from the s-domain block diagram of Fig. 11.20b which implies that

$$C(s) = H(s)R^*(s) \Leftrightarrow C(z) = H(z)R(z) \qquad (11.61)$$

or $$\mathcal{Z}[H(s)R^*(s)] = H(z)R(z) \qquad (11.62)$$

Figure 11.21 The z-transfrom equivalent of Fig. 11.20a.

To Obtain $H(z)$

The direct approach to obtain
$$H(z) = \mathscr{Z}[H(s)]$$
involves the following steps.
1. Find $h(t) = \mathcal{L}^{-1}[H(s)]$
2. Find $h(kT)$
3. Find $\mathscr{Z}[h(kT)]$

Example 11.11:
$$H(s) = \frac{a}{s(s+a)} = \frac{1}{s} - \frac{1}{s+a}$$
$$h(t) = \mathcal{L}^{-1}\left[\frac{1}{s} - \frac{1}{s+a}\right] = \left[1 - e^{-aT}\right]u(t)$$
$$h(kT) = 1 - e^{-akT}; \quad k \geqslant 0$$
$$= 0 \quad\quad ; k < 0$$

The z-transform of this equation using Table 11.2 is
$$H(z) = \frac{z}{z-1} - \frac{z}{z-e^{-aT}} = \frac{z\{1-e^{-aT}\}}{(z-1)\{z-e^{-aT}\}}$$

Single factor building blocks of the Laplace and z-transform pairs are given in Table 11.3. Expanding any $H(s)$ into partial fraction, $H(z)$ can be found by use of this table.

Table 11.3 Laplace and z-Transform Pairs

$F(s)$	$F(z)$
$1/s$	$\dfrac{z}{z-1}$
$\dfrac{1}{s^2}$	$\dfrac{Tz}{(z-1)^2}$
$\dfrac{1}{s^3}$	$\dfrac{T^2 z(z+1)}{2(z-1)^3}$
$\dfrac{1}{s+a}$	$\dfrac{z}{z-e^{-aT}}$
$\dfrac{1}{(s+a)^2}$	$\dfrac{Tze^{-aT}}{(z-e^{-aT})^2}$
$\dfrac{a}{s(s+a)}$	$\dfrac{z\{1-e^{-aT}\}}{(z-1)\{z-e^{-aT}\}}$
$\dfrac{\omega}{s^2+\omega^2}$	$\dfrac{z \sin \omega T}{z^2 - 2z \cos \omega T + 1}$
$\dfrac{s}{s^2+\omega^2}$	$\dfrac{z(z-\cos \omega T)}{z^2 - 2z \cos \omega T + 1}$
$\dfrac{\omega}{(s+a)^2+\omega^2}$	$\dfrac{ze^{-aT} \sin \omega T}{z^2 - 2ze^{-aT} \cos \omega T + e^{-2aT}}$
$\dfrac{s+a}{(s+a)^2+\omega^2}$	$\dfrac{z^2 - ze^{-aT} \cos \omega T}{z^2 - 2ze^{-aT} \cos \omega T + e^{-2aT}}$

Consider now the case of a continuous system $H(s)$ with continuous input $r(t)$ as in Fig. 11.22a. Imagine that we are interested to read the values of the continuous output at sampling instants, i.e., imagine a mathematical sampler at the output stage. We can equivalently represent it as a block $H(s)R(s)$ with impulse input $\delta(t)$ as in Fig. 11.22b. Now the input and therefore the output does not change by imagining a fictitious impulse sampler through which $\delta(t)$ is applied to $H(s)R(s)$ as in Fig. 11.22c. On application of (11.58), we can at once write

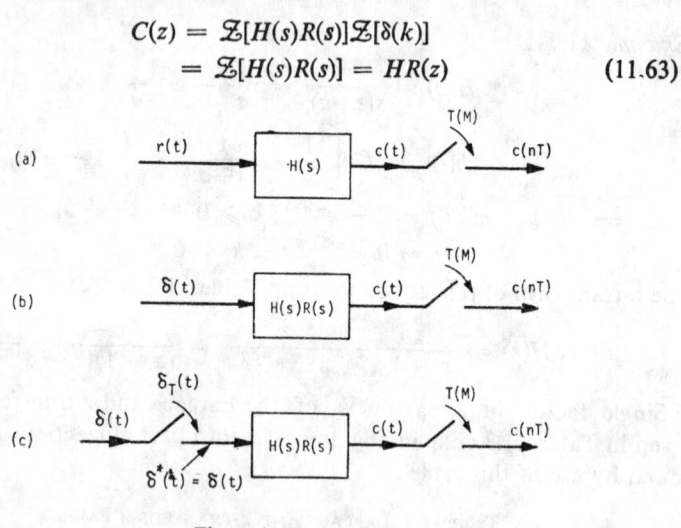

$$C(z) = \mathscr{Z}[H(s)R(s)]\mathscr{Z}[\delta(k)]$$
$$= \mathscr{Z}[H(s)R(s)] = HR(z) \qquad (11.63)$$

Figure 11.22.

The following s-domain and z-domain relationship follows from above:

$$C(s) = H(s)R(s) \Leftrightarrow C(z) = HR(z) \qquad (11.64)$$
or
$$\mathscr{Z}[H(s)R(s)] = HR(z) \qquad (11.65)$$

When impulse sampled input is applied to two (or more) s-domain transfer functions in tandem as in Fig. 11.23a, we get the z-domain transfer function as in Fig. 11.23b where

$$H(z) = H_1H_2(z) \neq H_1(z)H_2(z) \qquad (11.66)$$

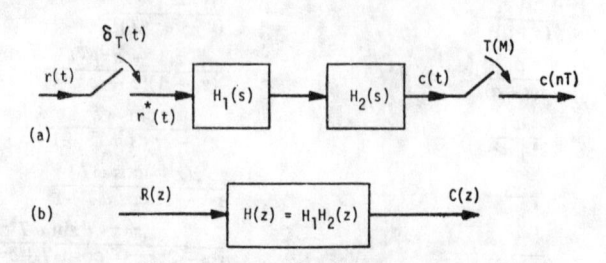

Figure 11.23.

However, if the blocks of Fig. 11.23a are separated by an impulse sampler as in Fig. 11.24, it obviously follows that

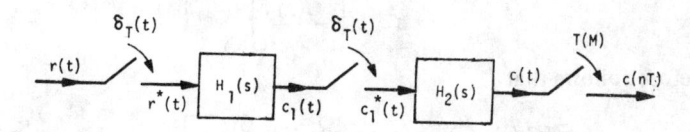

$$H(z) = H_1(z)H_2(z) \tag{11.67}$$

Figure 11.24.

THE z-TRANSFORM ANALYSIS OF SAMPLED-DATA SYSTEMS

Consider an open-loop system wherein a linear continuous part $G(s)$ is fed from a sampler and ZOH as shown in Fig. 11.25a. Its equivalent impulse sampled system is shown in Fig. 11.25b, while the corresponding z-domain block diagram is drawn in Fig. 11.25c. The z-transfer function of the system is given by

$$\mathscr{Z}[G_0(s)G(s)] = \mathscr{Z}\left[\frac{1 - e^{-sT}}{s} G(s)\right] = \mathscr{Z}\left[\frac{G(s)}{s} - \frac{e^{-sT}G(s)}{s}\right]$$

Let

$$\mathscr{L}^{-1}\left[\frac{G(s)}{s}\right] = g_1(t)$$

$$\mathscr{L}^{-1}\left[\frac{e^{-sT}G(s)}{s}\right] = g_1(t - T)$$

Figure 11.25.

Hence

$$\mathcal{Z}\left[\frac{e^{(-sT)}G(s)}{s}\right] = \mathcal{Z}[g_1(kT - T)] = z^{-1}Z[g_1(kT)]$$

$$= z^{-1}\mathcal{Z}\left[\frac{G(s)}{s}\right]$$

It therefore follows that

$$\mathcal{Z}[G_0(s)G(s)] = (1 - z^{-1})\mathcal{Z}\left[\frac{G(s)}{s}\right] \tag{11.68}$$

as depicted in Fig. 11.25d.

For example, if

$$G(s) = \frac{a}{s+a}$$

then

$$\mathcal{Z}\left[\frac{G(s)}{s}\right] = \mathcal{Z}\left[\frac{a}{s(s+a)}\right]$$

This result has already been obtained in Example 11.11. We have therefore

$$\mathcal{Z}[G_0(s)G(s)] = \frac{z\{1 - e^{-aT}\}}{\{z - e^{-aT}\}}$$

In the absence of ZOH in Fig. 11.25a, the linear continuous part of the system is fed from a train of pulses as shown in Fig. 11.26a. The z-transform though no longer valid can still be used if the pulse train is approximated as an impulse train as in Fig. 11.26b. It immediately follows that

$$C(z) = \mathcal{Z}[G(s)]\Delta R(z)$$
$$= \Delta G(z)R(z) = G'(z)R(z) \tag{11.69}$$

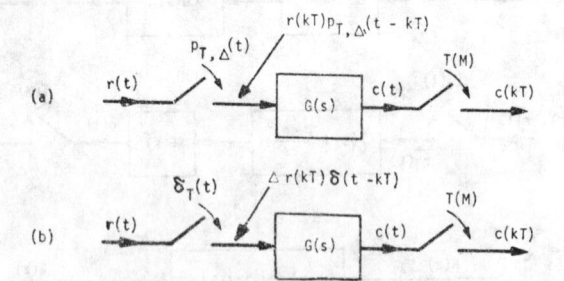

Figure 11.26.

It shall be assumed now onwards that, wherever an impulse approximation of a pulse sampler is employed, the gain Δ is already absorbed in the transfer function.

Consider now the basic sampled-data feedback control system whose block diagram is depicted in Fig. 11.27a. In terms of impulse sampling this

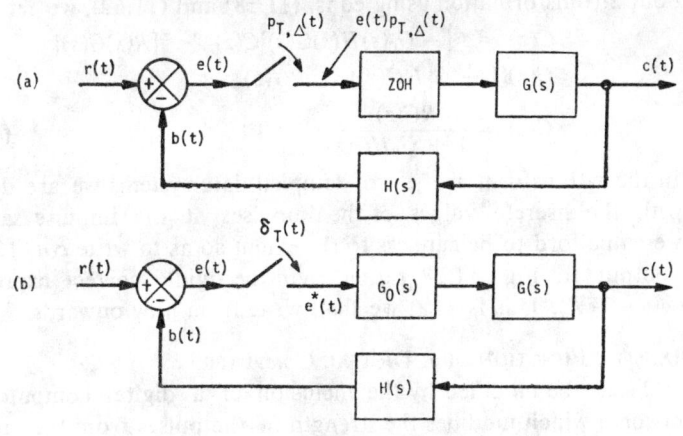

Figure 11.27.

diagram can be redrawn as in Fig. 11.27b. In the z-transform form we can write

$$C(z) = \mathcal{Z}[G_0(s)G(s)]E(z) \qquad (11.70)$$
$$B(z) = \mathcal{Z}[G_0(s)G(s)H(s)]E(z) \qquad (11.71)$$
$$e(t) = r(t) - b(t)$$
or $\qquad e(kT) = r(kT) - b(kT)$
or $\qquad E(z) = R(z) - B(z) \qquad (11.72)$

From eqns. (11.70)-(11.72) it immediately follows that

$$\frac{C(z)}{R(z)} = \frac{\mathcal{Z}[G_0(s)G(s)]}{1+\mathcal{Z}[G_0(s)G(s)H(s)]} = \frac{G_0G(z)}{1+G_0GH(z)} \qquad (11.73)$$

Having become familiar with the technique, from now onwards we may directly draw the block diagrams of sampled-data systems in terms of impulse sampler and $G_0(s)$.

Consider now a sampled-data system with sampler and ZOH in the feedback path as shown in Fig. 11.28. The signal $e(t)$ is contributed by impulse sampled $c(t)$ and continuous input signal $r(t)$. In s-domain, we can write

$$E(s) = -G_0(s)H(s)C^*(s) + R(s)$$
$$C(s) = -G_0(s)G(s)H(s)C^*(s) + G(s)R(s)$$

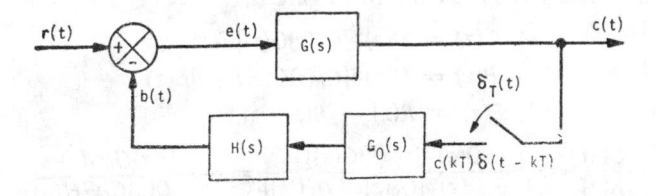

Figure 11.28.

Carrying out z-transformation using eqns. (11.58) and (11.62), we get
$$C(z) = \mathcal{Z}[-G_0(s)H(s)G(s)]C(z) + \mathcal{Z}[R(s)G(s)]$$
or
$$C(z)[1 + G_0HG(z)] = RG(z)$$
$$\therefore \quad C(z) = \frac{RG(z)}{1 + G_0HG(z)} \tag{11.74}$$

Since in the z-transform analysis of sampled-data systems we are dealing merely with the discrete values of the impulses at the impulse sampler output, we can afford to be careless to the extent so as to write $c(nT)$ as the sampler output in Fig. 11.28 rather than the rigidly correct expression of $c(kT) \delta(t - kT)$. This is what we shall write from now onwards.

THE z-TRANSFER FUNCTION OF A DIGITAL COMPUTER

Let Fig. 11.25a be modified by the inclusion of a digital computer (or digital network) which modifies the strength of the pulses from the sampler in accordance with a difference equation of the type (11.46). This is shown by Fig. 11.29a. The action of the digital computer in the z-domain is therefore that of a z-transfer function
$$D(z) = \frac{V(z)}{R(z)}$$
of the type of (11.47). The rest of the system behaves in the manner already illustrated in Fig. 11.25. Thus the system can be represented by the block diagram of Fig. 11.29b. It may be noted that $D(z)$ is an independent block representing the digital computer (or network).

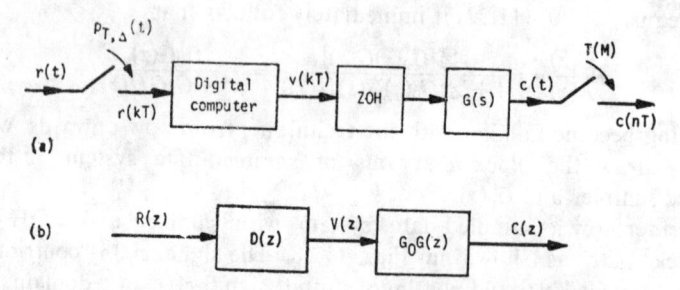

Figure 11.29 Sampled-data system with digital computer.

If a digital computer is included in the forward path of the basic sampled-data control system, we can represent it in form of the block diagram of Fig. 11.30. It easily follows from this figure that
$$C(z) = D(z)\mathcal{Z}[G_0(s)G(s)]E(z)$$
$$B(z) = D(z)\mathcal{Z}[G_0(s)G(s)H(s)]E(z)$$
$$E(z) = R(z) - B(z)$$
$$\frac{C(z)}{R(z)} = \frac{D(z)\mathcal{Z}[G_0(s)G(s)]}{1 + D(z)\mathcal{Z}[G_0(s)G(s)H(s)]} = \frac{D(z)G_0G(z)}{1 + D(z)G_0GH(z)} \tag{11.75}$$

Sampled-Data Control Systems

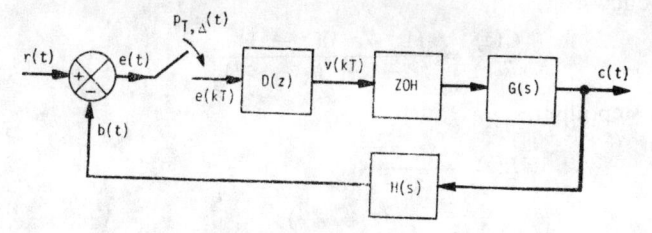

(Figure 11.30 block diagram: r(t) → summing junction → e(t) → sampler $p_{T,\Delta}(t)$ → D(z) → v(kT) → ZOH → G(s) → c(t); feedback through H(s) to b(t))

Figure 11.30.

Inclusion of $D(z)$, the digital computer transfer function, provides us with the technique of compensating a sampled-data control system. This will be discussed later in this chapter.

Example 11.12: For the sampled-data control system shown in Fig. 11.31, find the output $c(k)$ for $r(t) =$ unit step.

$$\frac{C(z)}{R(z)} = \frac{\mathscr{Z}[G_0 G(s)]}{1 + \mathscr{Z}[G_0 GH(s)]}$$

$$\mathscr{Z}[G_0 G(s)] = (1 - z^{-1})\mathscr{Z}\left[\frac{1}{s(s+1)}\right]$$

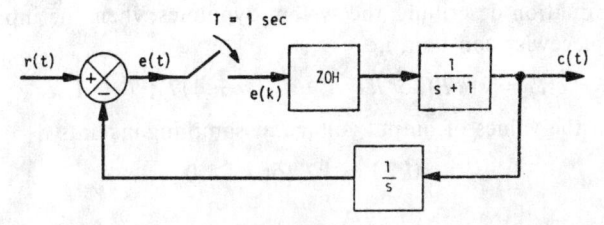

(Figure 11.31 block diagram: T = 1 sec; r(t) → summing → e(t) → sampler → e(k) → ZOH → 1/(s+1) → c(t); feedback through 1/s)

Figure 11.31.

Consulting Table 11.3,

$$\mathscr{Z}\left[\frac{1}{s(s+1)}\right] = \frac{z(1 - e^{-1})}{(z - 1)(z - e^{-1})}$$

Therefore,

$$\mathscr{Z}[G_0 G(s)] = \frac{1 - e^{-1}}{z - e^{-1}}$$

Now

$$\mathscr{Z}[G_0 GH(s)] = (1 - z^{-1})\mathscr{Z}\left[\frac{1}{s^2(s+1)}\right]$$

$$= (1 - z^{-1})\mathscr{Z}\left[\frac{1}{s^2} - \frac{1}{s} + \frac{1}{s+1}\right]$$

$$= (1 - z^{-1})\left[\frac{z}{(z-1)^2} - \frac{z}{z-1} + \frac{z}{z - e^{-1}}\right]$$

$$= \frac{e^{-1}z - 2e^{-1} + 1}{(z - 1)(z - e^{-1})}$$

It then follows:
$$\frac{C(z)}{R(z)} = \frac{(1 - e^{-1})(z - 1)}{z^2 - z + (1 - e^{-1})}$$

For unit step input,
$$R(z) = \frac{z}{z - 1}$$

$\therefore$
$$C(z) = \frac{(1 - e^{-1})z}{z^2 - z + (1 - e^{-1})}$$

$$= \frac{0.632z}{(z - 0.5 - j0.62)(z - 0.5 + j0.62)}$$

$$= \frac{-j0.51z}{(z - 0.5 - j0.62)} + \frac{j0.51z}{(z - 0.5 + j0.62)}$$

Taking the inverse z-transform, we get
$$c(k) = -j0.51(0.5 + j0.62)^k + j0.51(0.5 - j0.62)^k$$

Converting to polar form and combining,
$$c(k) = 1.02(0.785)^k \sin(51.5°k)$$

Example 11.13: Consider the system shown in Fig. 11.32a. Derive the difference equation describing the system dynamics when the input voltage applied is piecewise constant, i.e.,
$$e(t) = e(kT) \text{ for } kT \leqslant t \leqslant (k+1)T\,;\, T = 1 \text{ sec}$$

Also obtain the values of output voltage at sampling instants if
$$e(kT) = kT \text{ for } k \geqslant 0$$

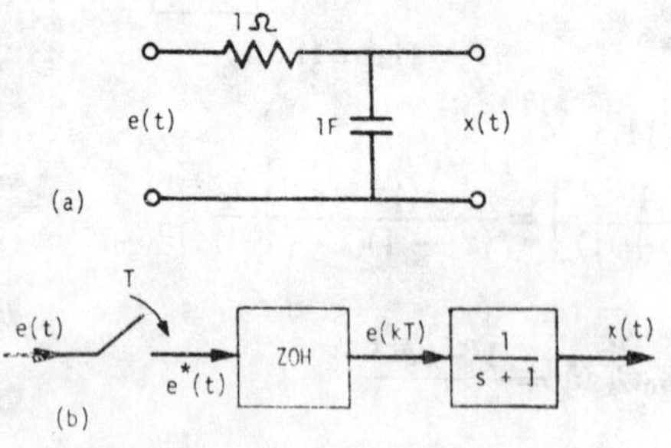

Figure 11.32.

The s-domain transfer function of the system is
$$\frac{1}{s + 1}$$

Equivalently we can draw the sampled-data block diagram for the system as in Fig. 11.32b. Therefore,
$$X(z) = (1 - z^{-1})\mathcal{Z}\left[\frac{1}{s(s+1)}\right]E(z)$$

$$= \frac{(1-z^{-1})z(1-e^{-1})}{(z-1)(z-e^{-1})} E(z)$$

$$= \frac{1-e^{-1}}{(z-e^{-1})} E(z)$$

or $\qquad zX(z) - e^{-1}X(z) = (1-e^{-1}) E(z)$

Taking the inverse z-transform

$$x[(k+1)T] = e^{-1}x(kT) + (1-e^{-1})e(kT)$$

This is the desired difference equation.

Now for $e(kT) = kT = k$

$$E(z) = \frac{Tz}{(z-1)^2} = \frac{z}{(z-1)^2}$$

$$\therefore \quad X(z) = \frac{1-e^{-1}}{(z-e^{-1})} \times \frac{z}{(z-1)^2} = (1-e^{-1})\frac{z}{(z-e^{-1})(z-1)^2}$$

$$= \frac{1}{1-e^{-1}} \frac{z}{z-e^{-1}} - \frac{1}{(1-e^{-1})} \frac{z}{z-1} + \frac{z}{(z-1)^2}$$

Taking the inverse z-transform

$$x(k) = \frac{1}{1-e^{-1}} e^{-k} - \frac{1}{1-e^{-1}} + k$$

11.9 Response between Sampling Instants

The output of sampled-data systems is generally of continuous type while the z-transform technique yields the output values at sampling instants only. It may sometimes be necessary to determine the response between sampling instants. The problem can be tackled through a simple modification of the z-transform technique.

Consider the impulse sampled system of Fig. 11.33 whose output $c(kT)$ at sampling instants can be determined by the z-transform technique. If a fictitious time lead function $e^{\delta s}$ is artificially introduced at the output stage as shown in Fig. 11.34, the output becomes

$$\hat{c}(t) = c(t + \delta)$$

Figure 11.33.

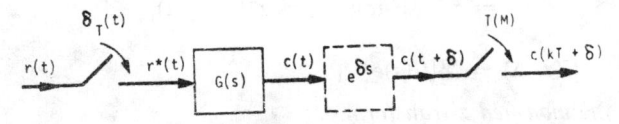

Figure 11.34.

which when sampled gives

$$\hat{c}(kT) = c(kT + \delta); \quad 0 < \delta < T \tag{11.76}$$

It implies that depending upon the value of δ, we can read the values of $c(t)$ anywhere between the sampling instants, i.e., we can read

$$c(\delta), c(T + \delta), c(2T + \delta), \ldots$$

For Fig. 11.34

$$\hat{C}(z) = \mathscr{Z}[c(kT+\delta)] = \mathscr{Z}[G(s)e^{\delta s}]R(z) \tag{11.77}$$

Thus the inverse z-transform of

$$\mathscr{Z}[G(s)e^{\delta s}]R(z)$$

will yield the output values between sampling instants.

Let us take an example wherein

$$G(s) = \frac{1}{s+a}$$

Now

$$g_1(t) = \mathscr{L}^{-1}\left[\frac{1}{s+a}e^{\delta s}\right] = e^{-a(t+\delta)}u(t+\delta)$$

$$g_1(kT) = e^{-a(kT+\delta)}u(kT+\delta)$$

$$\mathscr{Z}[G(s)e^{\delta s}] = \mathscr{Z}[g_1(kT)]$$

$$= e^{-a\delta}\sum_{k=0}^{\infty} e^{-akT}z^{-k} = \frac{e^{-a\delta}z}{z - e^{-aT}}$$

For unit step input

$$\hat{C}(z) = \frac{e^{-a\delta}z^2}{(z-e^{-aT})(z-1)}$$

$$= \frac{e^{-a\delta}e^{-aT}}{e^{-aT} - 1}\left(\frac{z}{z-e^{-aT}}\right) - \frac{e^{-a\delta}}{e^{-aT} - 1}\left(\frac{z}{z-1}\right)$$

Taking the inverse z-transform

$$\hat{c}(kT) = c(kT+\delta) = \frac{e^{-a\delta}e^{-aT}}{e^{-aT} - 1}e^{-akT} - \frac{e^{-a\delta}}{e^{-aT} - 1} 1$$

In case δ is a time delay, we can write

$$\delta = -(1 - \Delta)T; \quad 0 < \Delta < 1$$

so that

$$\mathscr{Z}[G(s)e^{\delta s}] = \mathscr{Z}[G(s)e^{-Ts}e^{\Delta Ts}]$$

$$= z^{-1}\mathscr{Z}[G(s)e^{\Delta Ts}] = z^{-1}G(z, \Delta) \tag{11.78}$$

where

$$G(z, \Delta) = \mathscr{Z}[G(s)e^{\Delta Ts}]$$

is called the *modified z-transform*.

The modified z-transforms for simple factors are listed in Table 11.4.

Table 11.4 MODIFIED z-TRANSFORM PAIRS

$G(s)$	$G(z, \Delta)$
$\dfrac{1}{s}$	$\dfrac{z}{z-1}$
$\dfrac{1}{s^2}$	$\dfrac{\Delta T z^2 + (1-\Delta)Tz}{(z-1)^2}$
$\dfrac{1}{s+a}$	$\dfrac{e^{-a\Delta T}z}{z - e^{-aT}}$
$\dfrac{1}{(s+a)^2}$	$T\left[\dfrac{\Delta e^{-aT\Delta} z}{z - e^{-aT}} + \dfrac{e^{-a(1+\Delta)T} z}{(z - e^{-aT})^2}\right]$

11.10 The z- and s-domain Relationship

Consider a signal $r(t)$ which has discrete values $r(kT)$ at a sampling rate $1/T$. The z-transform of these discrete values is

$$R(z) = \sum_{k=0}^{\infty} r(kT) z^{-k} \qquad (11.79)$$

If the signal $r(t)$ is imagined to be impulse sampled at the same rate, it becomes

$$r^*(t) = \sum_{k=0}^{\infty} r(kT) \delta(t - kT) \qquad (11.80)$$

Taking the Laplace transform, we have

$$R^*(s) = \sum_{k=0}^{\infty} r(kT) e^{-ksT} \qquad (11.81)$$

If we let

$$z = e^{sT} \text{ or } s = \frac{1}{T} \ln z \qquad (11.82)$$

we get

$$R^*(s)\bigg|_{s=\frac{1}{T}\ln z} = \sum_{k=0}^{\infty} r(kT) z^{-k} = R(z) \qquad (11.83)$$

Equation (11.83) exposes an interesting result—the z-transform of a set of discrete values can be obtained by imagining these to be impulses Laplace transforming and then using the transformation (11.82).

The transformation (11.82) maps the s-plane into the z-plane. Consider the mapping of the $j\omega$-axis in the s-plane, i.e.,

$$z = e^{j\omega T}$$
$$= e^{j2\pi\omega/\omega_s} = 1\angle 2\pi (\omega/\omega_s) \qquad (11.84)$$

where $\omega_s = \dfrac{2\pi}{T}$, the sampling frequency.

Thus the section $(-j\omega_s/2) - 0 - (+j\omega_s/2)$ of the $j\omega$-axis maps into the unit circle in the anticlockwise direction $(-\pi, -\pi/2, 0, \pi/2, \pi)$ as shown in Fig. 11.35. In fact every section of the $j\omega$-axis which is integral multiple of ω_s maps into the unit circle. With the direction of mapping indicated, it is then obvious that the left half s-plane can be divided into strips of width ω_s each of which maps into the interior of the unit circle. It suggests that the s-plane stability criterion of the poles of s-transfer function lying in the left half of the s-plane will become the z-plane stability criterion that all the poles of the z-transfer function should lie within the unit circle. This is shown to be so in the next article.

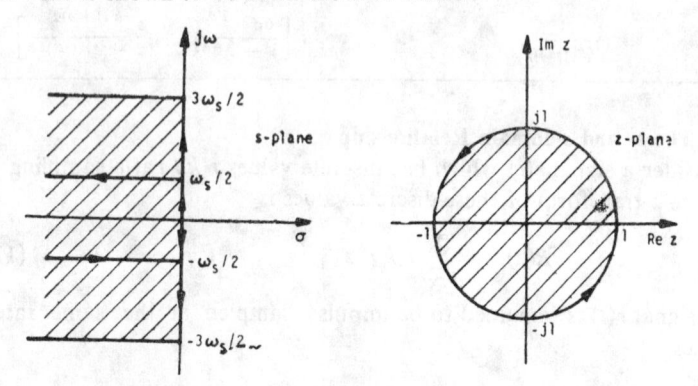

Figure 11.35 Relationship between the s- and z-domains.

11.11 Stability Analysis

STABILITY CRITERION

The input-output relation of a discrete system is given by

$$C(z) = T(z)R(z) \tag{11.85}$$

where $T(z)$ is the z-transfer function of the system, a rational function of two polynomials in z. As in the case of linear continuous systems, the stability is ascertained by examining the system response to impulse input, i.e., $R(z) = 1$. Imagining the poles of $T(z)$ to be simple and distinct, we have

$$C(z) = T(z) = \frac{A_1}{z-a_1} + \frac{A_2}{z-a_2} + \ldots + \frac{A_n}{z-a_n} \tag{11.86}$$

The corresponding discrete time sequence is

$$c(kT) = A_1(a_1)^{k-1} + A_2(a_2)^{k-1} + \ldots + A_n(a_n)^{k-1} \ ; \ k \geqslant 1 \tag{11.87}$$

It is easily seen from eqn. (11.87), that the system response decays to zero if

$$|a_i| < 1 \ ; \ i = 1, 2, \ldots, n$$

i.e., all the poles of the system's z-transfer function must lie within the unit circle in the z-plane (as was rightly guessed in Section 11.10).

The presence of complex conjugate pole pair

$$z = a_i \pm jb_i$$

contributes term in the response of the type

$$\alpha_i^k \cos(k\theta_l + \phi_l)$$

where $\alpha_l = \sqrt{a_i^2 + b_i^2}$; and $\theta_l = \tan^{-1} \dfrac{b_l}{a_l}$ which decays if $\alpha_l < 1$, i.e., the complex pole pair lies within the unit circle.

The presence of real repeated poles contributes additional terms like

$$\frac{1}{(z-a_l)^2} \leftrightarrow (k-1)a^{k-2}\,; \ k \geqslant 2$$

which also decay for $|a_l| < 1$.

The criterion of stability that all the poles of the z-transfer function must lie within the unit circle is therefore of general applicability.

As the poles of the closed-loop transfer function are the same as the roots of the system's characteristic equation, the system stability is determined by the roots of (see eqn. (11.73))

$$1 + \mathscr{Z}[G_0(s)G(s)H(s)] = 0 \qquad (11.88)$$

for systems of Figs. 11.27 and 11.28 and by the roots of (see eqn. (11.75))

$$1 + D(z)\mathscr{Z}[G_0(s)G(s)H(s)] = 0 \qquad (11.89)$$

for system of Fig. 11.30.

Methods of Stability Analysis

1. Jury's Stability Test [3]

It is an algebraic criterion for determining whether or not the roots of the characteristic polynomial lie within a unit circle thereby determining system stability. Consider the characteristic polynomial

$$F_1(z) = a_n z^n + a_{n-1} z^{n-1} + \ldots + a_0 = 0\,; \ a_n > 0 \qquad (11.90)$$

Like the Routh's method, the Jury's test consists of two parts: a simple test for necessity and a second test for sufficiency [4].

The necessary conditions for stability are

$$F_1(1) > 0\,; \ (-1)^n F_1(-1) > 0 \qquad (11.91)$$

The sufficient conditions for stability can be established through two methods.

Method 1: Prepare a table of coefficients* of the characteristic polynomial as below:

---

*The singular case where a row of zeros appears is beyond the scope of this book.

Row	z^0	z^1	z^2	...	z^{n-k}	...	z^{n-1}	z^n
1	a_0	a_1	a_2	...	a_{n-k}	...	a_{n-1}	a_n
2	a_n	a_{n-1}	a_{n-2}	...	a_k	...	a_1	a_0
3	b_0	b_1	b_2	...	...	...	b_{n-1}	
4	b_{n-1}	b_{n-2}	...	...	...	...	b_0	
5	c_0	c_1	c_2	...	...	...	c_{n-2}	
6	c_{n-2}	c_{n-3}	...	...	...	...	c_0	
⋮	⋮	⋮	⋮	⋮	⋮			
$2n-5$	s_0	s_1	s_2	s_3				
$2n-4$	s_3	s_2	s_1	s_0				
$2n-3$	r_0	r_1	r_2					

where
$$b_k = \begin{vmatrix} a_0 & a_{n-k} \\ a_n & a_k \end{vmatrix}$$

$$c_k = \begin{vmatrix} b_0 & b_{n-1-k} \\ b_{n-1} & b_k \end{vmatrix}$$

$$d_k = \begin{vmatrix} c_0 & c_{n-2-k} \\ c_{n-2} & c_k \end{vmatrix}$$

... ...
... ...

The sufficient conditions for stability are

$$\left. \begin{array}{l} |a_0| < |a_n| \\ |b_0| > |b_{n-1}| \\ |c_0| > |c_{n-2}| \\ \quad \vdots \\ |r_0| > |r_2| \end{array} \right\} \quad (n-1) \text{ constraints} \quad (11.92)$$

Example 11.14 : Consider the characteristic equation of the second-order system
$$F_1(z) = a_2 z^2 + a_1 z + a_0 = 0; \quad a_2 > 0$$
The stability constraints are
$$F_1(1) = a_2 + a_1 + a_0 > 0$$
$$(-1)^n F_1(-1) = a_2 - a_1 + a_0 > 0$$
$$|a_0| < a_2$$

Example 11.15 : Consider the characteristic polynomial
$$F_1(z) = 2z^4 + 7z^3 + 10z^2 + 4z + 1$$
$$F_1(1) = 2 + 7 + 10 + 4 + 1 = 24 > 0 \quad \text{satisfied}$$
$$(-1)^4 F_1(-1) = 2 - 7 + 10 - 4 + 1 = 2 > 0 \quad \text{satisfied}$$

Row	z^0	z^1	z^2	z^3	z^4
1	1	4	10	7	2
2	2	7	10	4	1
3	−3	−10	−10	−1	
4	−1	−10	−10	−3	
5	8	20	20		

Employing stability constraints (11.92)

$$|1| < 2 \qquad \text{satisfied}$$
$$|-3| > |-1| \qquad \text{satisfied}$$
$$|8| > |20| \qquad \text{not satisfied}$$

The system is therefore unstable.

Method 2: Rename the coefficients of the characteristic polynomial as

$$F_1(z) = a_0 z^n + a_1 z^{n-1} + \ldots + a_{n-1} z + a_n = 0; \quad a_0 > 0 \quad (11.93)$$

The necessary conditions for stability are the same as in (11.91).

The test for sufficiency is performed as follows:

Two triangularized matrices **X** and **Y** are constructed using the coefficients of the characteristic polynomial as elements, in the following general form,

$$\mathbf{X} = \begin{bmatrix} a_0 & a_1 & a_2 & \ldots & a_{n-2} \\ 0 & a_0 & a_1 & \ldots & a_{n-3} \\ 0 & 0 & a_0 & \ldots & a_{n-4} \\ \vdots & \vdots & \vdots & & \vdots \\ 0 & 0 & 0 & \ldots & a_0 \end{bmatrix}; \quad \mathbf{Y} = \begin{bmatrix} a_2 & a_3 & \ldots & a_{n-1} & a_n \\ a_3 & a_4 & \ldots & a_n & 0 \\ a_4 & \ldots & a_n & 0 & 0 \\ \vdots & & \vdots & \vdots & \vdots \\ a_n & \ldots & 0 & 0 & 0 \end{bmatrix} \quad (11.94)$$

Note that **X** and **Y** are square matrices of order $(n - 1)$ for an *n*th-order characteristic polynomial.

From **X** and **Y**, we construct

$$\mathbf{H}_1 = \mathbf{X} + \mathbf{Y} \quad (11.95)$$

$$\mathbf{H}_2 = \mathbf{X} - \mathbf{Y} \quad (11.96)$$

For the system to be asymptotically stable, $\mathbf{H}_1$ and $\mathbf{H}_2$ must be *positive innerwise*.

A square matrix **H** is said to be *positive innerwise* when all the determinants starting with the centre element(s) and proceeding outwards upto the entire matrix are positive. For example, for

$$\mathbf{H} = \begin{bmatrix} b_1 & b_2 & b_3 & b_4 & b_5 \\ c_1 & c_2 & c_3 & c_4 & c_5 \\ d_1 & d_2 & d_3 & d_4 & d_5 \\ e_1 & e_2 & e_3 & e_4 & e_5 \\ f_1 & f_2 & f_3 & f_4 & f_5 \end{bmatrix}$$

the determinants (indicated by dashed lines) are

$$d_3 \;;\; \begin{vmatrix} c_2 & c_3 & c_4 \\ d_2 & d_3 & d_4 \\ e_2 & e_3 & e_4 \end{vmatrix} \;;\; |\mathbf{H}|$$

The first determinant is a scalar if n is even and is a 2×2 determinant if n is odd.

Example 11.16: Consider the characteristic equation of the second-order system

$$F_1(z) = a_0 z^2 + a_1 z + a_2 = 0;\; a_0 > 0$$

The stability constraints are:

(i) $\quad F_1(1) = a_0 + a_1 + a_2 > 0$
$\quad (-1)^n F_1(-1) = a_0 - a_1 + a_2 > 0$

(ii) $\quad \mathbf{X} = a_0 \quad\quad ;\; \mathbf{Y} = a_2$
$\quad \mathbf{H}_1 = a_0 + a_2 \quad ;\; \mathbf{H}_2 = a_0 - a_2$

For stability,

$$a_0 + a_2 > 0 \quad \text{and} \quad a_0 - a_2 > 0$$

that is $\quad |a_2| < a_0$

Example 11.17: Consider the characteristic polynomial

$$F_1(z) = 2z^4 + 7z^3 + 10z^2 + 4z + 1$$

Necessary conditions for stability are

$$F_1(1) = 2 + 7 + 10 + 4 + 1 = 24 > 0 \text{ satisfied}$$
$$(-1)^4 F_1(-1) = 2 - 7 + 10 - 4 + 1 = 2 > 0 \text{ satisfied}$$

To check the requirement of sufficiency, we construct $\mathbf{X}$ and $\mathbf{Y}$.

$$\mathbf{X} = \begin{bmatrix} 2 & 7 & 10 \\ 0 & 2 & 7 \\ 0 & 0 & 2 \end{bmatrix} ;\; \mathbf{Y} = \begin{bmatrix} 10 & 4 & 1 \\ 4 & 1 & 0 \\ 1 & 0 & 0 \end{bmatrix}$$

Therefore

$$H_1 = X + Y = \begin{bmatrix} 12 & 11 & 11 \\ 4 & 3 & 7 \\ 1 & 0 & 2 \end{bmatrix}; \quad H_2 = X - Y = \begin{bmatrix} -8 & 3 & 9 \\ -4 & -1 & 7 \\ -1 & 0 & 2 \end{bmatrix}$$

H_1 is positive innerwise, if

$\quad\quad 3 > 0 \quad$ satisfied
$\quad\quad |H_1| > 0 \quad$ satisfied

H_2 is positive innerwise, if

$\quad\quad -1 > 0 \quad$ not satisfied
$\quad\quad |H_2| > 0 \quad$ not satisfied

The system is therefore unstable.

Example 11.18: A sampled-data control system of order one with transportation lag is shown in Fig. 11.36. Determine the condition for system stability if $\delta < T$.

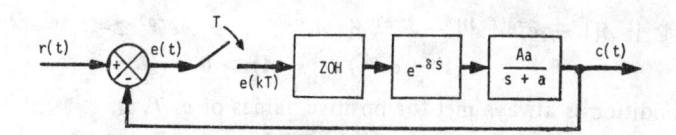

Figure 11.36.

*Solution**

$$G(s) = \left(\frac{1 - e^{-sT}}{s}\right)e^{-\delta s}\left(\frac{Aa}{s+a}\right) = [e^{-\delta s} - e^{-(T+\delta)s}]\frac{Aa}{s(s+a)}$$

$$= A[e^{-\delta s} - e^{-(T+\delta)s}]\left(\frac{1}{s} - \frac{1}{s+a}\right)$$

$$g(t) = A[1 - e^{-a(t-\delta)}]u(t-\delta) - A[1 - e^{-a(t-T-\delta)}]u(t-T-\delta)$$

$$g(kT) = A[1 - e^{-a(kT-\delta)}]u(kT-\delta) - A[1 - e^{-a(kT-T-\delta)}]u(kT-T-\delta)$$

$$= 0 \quad\quad\quad\quad\quad\quad\quad\quad\quad\quad \text{for} \quad k = 0$$

$$= A[1 - e^{-a(T-\delta)}] \quad\quad\quad\quad\quad \text{for} \quad k = 1$$

$$= A[1 - e^{-a(kT-\delta)}] - A[1 - e^{-a(kT-T-\delta)}] \quad \text{for} \quad k \geqslant 2$$

$$= Ae^{a\delta}e^{-akT}[e^{aT} - 1] \quad\quad\quad\quad\quad \text{for} \quad k \geqslant 2$$

*As a time delay which is nonintegral multiple of the sampling period T is encountered, we have to proceed fundamentally to obtain $G(z)$. Obviously

$$G(z) \neq (1 - z^{-1})\mathscr{Z}\left[\frac{Aae^{-\delta s}}{s(s+a)}\right]$$

Similar situation exists for time lead (nonintegral multiple of T) (see Problem 11.19).

$$G(z) = A[1 - e^{-a(T-\delta)}]z^{-1} + Ae^{a\delta}(e^{aT} - 1)\sum_{k=2}^{\infty} e^{-akT} z^{-k}$$

$$= A[1 - e^{-a(T-\delta)}]z^{-1} + Ae^{a\delta}(e^{aT} - 1) z^{-2}\sum_{k=2}^{\infty} e^{-akT} z^{-(k-2)}$$

Letting $k - 2 = r$, we have

$$G(z) = A[1 - e^{-a(T-\delta)}]z^{-1} + Ae^{a\delta}(e^{aT} - 1) z^{-2} \sum_{r=0}^{\infty} e^{-2aT} e^{-arT} z^{-r}$$

$$= A[1 - e^{-a(T-\delta)}]\frac{1}{z} + Ae^{a\delta}(e^{aT} - 1)e^{-2aT}\frac{1}{z(z - e^{-aT})}$$

$$G(z) = A \frac{[1 - e^{-a(T-\delta)}](z - e^{-aT}) + e^{a\delta}e^{-aT}(1 - e^{-aT})}{z(z - e^{-aT})}$$

The characteristic equation is

$$F_1(z) = 1 + G(z)$$
$$= z(z - e^{-aT}) + A[1 - e^{-a(T-\delta)}](z - e^{-aT})$$
$$\quad + Ae^{-a(T-\delta)}(1 - e^{-aT})$$
$$= z^2 + \{A[1 - e^{-a(T-\delta)}] - e^{-aT}\}z + Ae^{-aT}(e^{a\delta} - 1)$$

As per Example 11.14, the conditions of stability for a second-order system are:

(i) $1 + A[1 - e^{-a(T-\delta)}] - e^{-aT} + Ae^{-a(T-\delta)} - Ae^{-aT} > 0$

or $\qquad (1 - e^{-aT})(1 + A) > 0$

This condition is always met for positive values of a, T, A.

(ii) $1 - A + Ae^{-a(T-\delta)} + e^{-aT} + Ae^{-a(T-\delta)} - Ae^{-aT} > 0$

or $\qquad (1 + e^{-aT})(1 - A) + 2Ae^{-a(T-\delta)} > 0$

or $\qquad A < \dfrac{1 + e^{-aT}}{e^{-aT}(2e^{a\delta} - 1) - 1}$

(iii) $A[e^{-a(T-\delta)} - e^{-aT}] < 1$

For positive values of a, T, A, δ and $\delta < T$, this condition can be written as

$$Ae^{-aT}(e^{a\delta} - 1) < 1$$

or $\qquad A < \left(\dfrac{e^{aT}}{e^{a\delta} - 1}\right)$

Conditions (ii) and (iii) when simultaneously met for specified δ, yield the range of A for the system to be stable. The solution has to be obtained graphically or numerically.

2. *Bilinear Transformation*

We have seen above that in determining stability in the z-domain, we need to find if all the roots of the characteristic equation lie within the unit circle. Standard methods of Routh and Nyquist could be applied to this problem if we could find a complex transformation which maps the interior of the unit circle in the z-plane into the left half of the new

Sampled-Data Control Systems 431

plane. The transformation $z = e^{sT}\left[s = \dfrac{1}{T}\ln z\right]$ can not be used for this purpose because of the periodicity of e^{sT} and the multiple strips of the left half s-plane into which the interior of the unit circle maps. A simple transformation which uniquely maps the interior of the unit circle in the z-plane into the left half of the r-plane is the *bilinear transformation*

$$r = \frac{z-1}{z+1}; \quad z = \frac{1+r}{1-r} \tag{11.97}$$

On the unit circle in the z-plane

$z = e^{j\theta}(\theta$ varying anticlockwise from $-\pi$ through 0 to $+\pi)$

$$r = \frac{e^{j\theta}-1}{e^{j\theta}+1} = \frac{e^{j\theta/2}-e^{-j\theta/2}}{e^{j\theta/2}+e^{-j\theta/2}}$$

$$= \tanh j\frac{\theta}{2} = j\tan\frac{\theta}{2} = j\omega_r \tag{11.98}$$

where $\omega_r = \tan\dfrac{\theta}{2}$ varies from $-\infty$ through 0 to $+\infty$.

This mapping is indicated in Fig. 11.37.

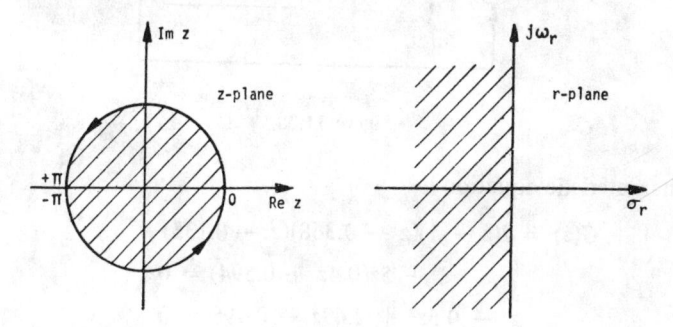

Figure 11.37.

Through use of the bilinear transformation (11.97), the characteristic eqn. (11.90) becomes

$$a_n\left(\frac{1+r}{1-r}\right)^n + a_{n-1}\left(\frac{1+r}{1-r}\right)^{n-1} + \ldots + a_1\left(\frac{1+r}{1-r}\right) + a_0 = 0$$

which can be organized into the form

$$b_n r^n + b_{n-1} r^{n-1} + \ldots + b_1 r + b_0 = 0 \tag{11.99}$$

Now the Routh criterion can be applied to the new characteristic equation (11.99) to determine if all its roots lie in the left half of the r-plane. We could apply the Nyquist criterion as well.

As already shown, the unit circle in the z-plane is the map of sections of width ω_s on the $j\omega$-axis of the s-plane, we can therefore immediately conclude that ω_s-width sections of the $j\omega$-axis in the s-plane map into the whole of the $j\omega_r$-axis in the r-plane.

Heuristically, the Bode plots and gain and phase margin concepts of the frequency domain could be applied once the bilinear transformation has been carried out (see Section 11.12).

Example 11.19: Consider the sampled-data system of Fig. 11.38. Determine its characteristic equation in the z-domain and ascertain its stability via the bilinear transformation.

$$G(z) = \mathscr{Z}\left[\frac{5}{s(s-1)(s+2)}\right] = 5\mathscr{Z}\left[\frac{1}{2s} - \frac{1}{s+1} + \frac{1}{2(s+2)}\right]$$

$$= 5\left[\frac{z}{2(z-1)} - \frac{z}{z-e^{-1}} + \frac{z}{2(z-e^{-2})}\right]$$

$$= \frac{5z(0.4z + 0.594)}{2(z-1)(z-0.368)(z-0.135)}$$

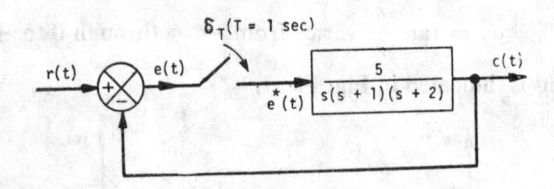

Figure 11.38.

The characteristic equation is

$$1 + G(z) = 2(z-1)(z-0.368)(z-0.135) + 5z(0.4z + 0.594) = 0$$

or
$$z^3 - 0.5z^2 + 2.49z - 0.496 = 0$$

Substituting (11.97), we get

$$\left(\frac{1+r}{1-r}\right)^3 - 0.5\left(\frac{1+r}{1-r}\right)^2 + 2.49\left(\frac{1+r}{1-r}\right) - 0.496 = 0$$

which upon simplification yields

$$3.5r^3 - 2.5r^2 + 0.5r + 2.5 = 0$$

The changes in sign of the characteristic polynomial indicate that the system is unstable. We therefore need not proceed to form the Routh array. Let us check the stability if the system was linear continuous, i.e.,

$$1 + G(s) = 0$$

$$1 + \frac{5}{s(s+1)(s+2)} = 0$$

$$s^3 + 3s^2 + 2s + 5 = 0$$

The Routh array is

$$\begin{array}{c|cc} s^3 & 1 & 2 \\ s^2 & 3 & 5 \\ s^1 & \dfrac{1}{3} & 0 \\ s^0 & 5 & \end{array}$$

It indicates a stable system.

By comparison we find that a stable linear continuous system becomes unstable upon introduction of sampling and ZOH in the forward loop. It can therefore be concluded that *sampling has a destabilizing effect on a system*.

3. The Root Locus Technique

The root locus technique can be easily adopted to stability analysis and design of sampled-data feedback control systems. The characteristic equation of these systems is of the form

$$1 + F(z) = 0$$

where $F(z)$ is a rational function of z. It can be written in the standard pole-zero form

$$1 + \frac{K\Pi(z + z_i)}{\Pi(p + p_j)} = 0$$

for drawing the root locus in the z-plane when the gain K is varied.

Examination of the root locus with respect to the unit circle reveals information on system stability and the range of K for the system to be stable can also be determined. This is illustrated by means of Example 11.20.

Example 11.20: Investigate the stability of the system shown in Fig. 11.39 for sampling period $T = 0.4$ sec, 3 sec.

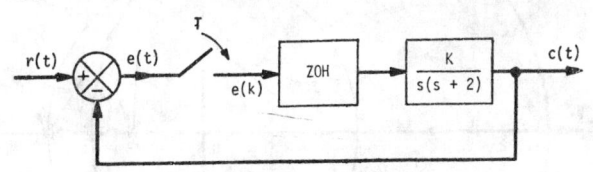

Figure 11.39.

The characteristic equation of the system is

$$1 + G_0G(z) = 0$$

where

$$G_0G(z) = \mathscr{Z}\left[\frac{1 - e^{-sT}}{s} \frac{K}{s(s + 2)}\right]$$

$$= K(1 - z^{-1})\mathscr{Z}\left[\frac{1}{s^2(s + 2)}\right]$$

$$= K(1 - z^{-1})\mathscr{Z}\left[\frac{1}{2s^2} - \frac{1}{4s} + \frac{1}{4(s + 2)}\right]$$

$$= \frac{K}{2}(1-z^{-1})\left[\frac{Tz}{(z-1)^2} - \frac{z}{2(z-1)} + \frac{z}{2(z-e^{-2T})}\right]$$

$$= \frac{K[(2T-1+e^{-2T})z - 2Te^{-2T} - e^{-2T} + 1]}{4(z-1)(z-e^{-2T})} \quad (11.100)$$

Case 1: $\quad T = 0.4$ sec

$$G_0G(z) = \frac{K(z+0.76)}{16(z-1)(z-0.45)} = \frac{K'(z+0.76)}{(z-1)(z-0.45)}$$

Zero : $z = -0.76$

Poles : $z = 1, z = 0.45$

The root locus is plotted in Fig. 11.40. The two breakaway points are calculated below:

$$1 + \frac{K'(z+0.76)}{(z-1)(z-0.45)} = 0$$

or $\quad K' = \dfrac{-(z-1)(z-0.45)}{(z+0.76)}$

$$\frac{dK'}{dz} = \frac{(z^2 - 1.45z + 0.45) + (z+0.76)(-2z+1.45)}{(z+0.76)^2} = 0$$

or $\quad z^2 + 1.52z - 1.55 = 0$

$$z = -2.22, 0.7$$

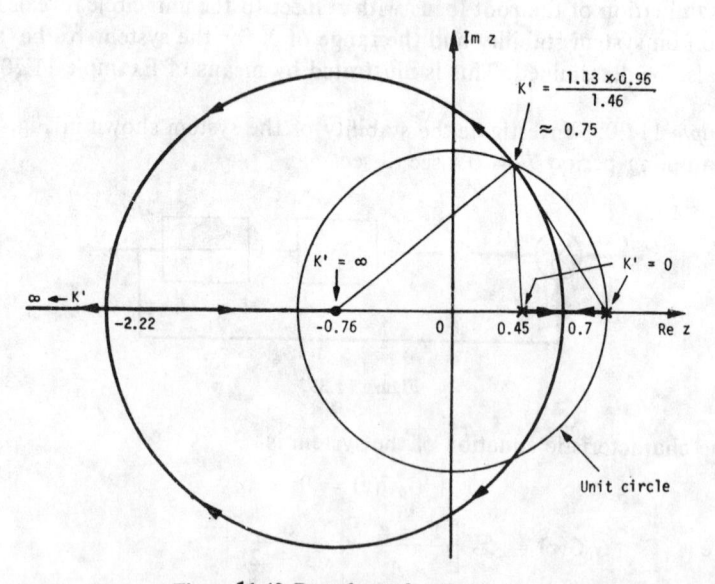

Figure 11.40 Root locus for $T = 0.4$ sec.

The root locus is a circle centred at $z = -0.76$ (zero) and of radius 1.46.

The roots of the characteristic equation lie within the unit circle, i.e., the system is stable for
$$K < 16 \times 0.75 = 12$$

Case 2: $T = 3$ sec

$$G_0G(z) = \frac{1.25K(z + 0.197)}{(z - 1)(z - 0.0025)}$$

Zero : $z = -0.197$
Poles : $z = 1, z = 0.0025$

The root locus is plotted in Fig. 11.41. The two breakaway points calculated as before are $z = -0.684, 0.29$.

The root locus is a circle centred at -0.197 (zero) and of radius 0.487. We immediately observe from the root locus that intersection of loci with unit circle occurs at $z = -1$. The marginal value of K is determined as under.

$$1 + \frac{1.25K(z + 0.197)}{(z - 1)(z - 0.0025)}\bigg|_{z=-1} = 0$$

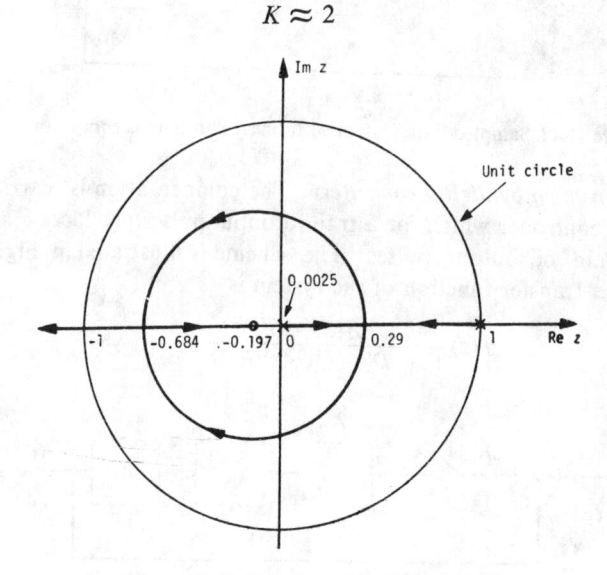

or
$$K \approx 2$$

Figure 11.41 Root locus for $T = 3$ sec.

We observe in the above example that the system which is stable for $K < 12$ when $T = 0.4$ sec becomes unstable for $K > 2$ when $T = 3$ sec. It means that *increasing the sampling period (or decreasing the sampling rate) reduces the margin of stability*.

11.12 Compensation Techniques

Cascade and feedback compensation techniques valid for linear continuous systems equally apply for linear sampled-data control systems. For want of space, we will not discuss here feedback compensation [2]. There are two ways of accomplishing cascade compensation in sampled-data systems.

1. *Compensation by continuous network*: In this scheme the compensating network is introduced in the continuous part of the system as shown in Fig. 11.42. The overall z-transfer function of the system is

$$\frac{C(z)}{R(z)} = T(z) = \frac{\mathscr{Z}[G_0(s)G_c(s)G(s)]}{1 + \mathscr{Z}[G_0(s)G_c(s)G(s)]} \quad (11.101)$$

The pole and zero of $G_c(s)$ are adjusted to achieve a desired closed-loop discrete response. As the open-loop poles and zeros (i.e., the poles and zeros of $\mathscr{Z}[G_0(s)G_c(s)G(s)]$) are implicitly linked to the pole and zero of $G_c(s)$, no direct compensation techniques are possible. Essentially trial and error is the way to proceed. Here this method will not be pursued any further [2].

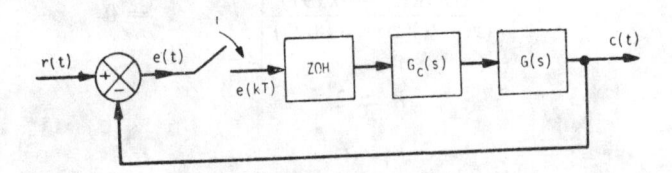

Figure 11.42 Sampled-data system with linear continuous compensator.

2. *Compensation by digital computer*: The compensation is carried out by a digital controller which for a train of input pulses produces a suitably modified train of output pulses. The scheme is illustrated in Fig. 11.43. The overall z-transfer function of the system is

$$\frac{C}{R}(z) = T(z)\frac{D(z)\mathscr{Z}[G_0(s)G(s)]}{1 + D(z)\mathscr{Z}[G_0(s)G(s)]} \quad (11.102)$$

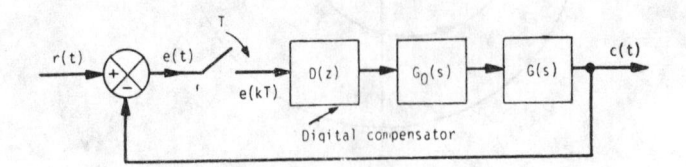

Figure 11.43 Sampled-data control system with digital compensator.

Since the compensator transfer function $D(z)$ enters the characteristic equation independent of $\mathscr{Z}[G_0(s)G(s)]$, it is possible to independently adjust the poles and zeros of $D(z)$ to achieve the desired closed-loop response. The root locus technique can be used in design.

Two methods of designing $D(z)$ will be discussed here.

Time-domain Technique of Designing $D(z)$

Parallel to the case of a continuous system, design specification for a sampled-data system could be laid down as:

1. Zero steady-state error to input $A(kT)^q$ for a given q. For $q = 0, 1, 2$, it corresponds to step, ramp and acceleration inputs in a continuous system.
2. The transient response should settle in finite number of sampling intervals and in as few intervals as possible.

With reference to the compensating scheme of Fig. 11.43, it shall be assumed that the response of all system elements and the overall system is *nonanticipative* (causal). In terms of general transfer function of eqn. (11.50) reproduced below

$$F(z) = \frac{b_0 z^m + b_1 z^{m-1} + \ldots + b_m}{z^n + a_1 z^{n-1} + \ldots + a_n}$$

it implies that $m \leqslant n$ so that when $F(z)$ is expanded in power series of z^{-1}, no positive power terms of z appear (which would mean appearance of output before input is applied—anticipative nature). $F(z)$ can also be written as

$$F(z) = \frac{b_0 z^{-(n-m)} + b_1 z^{-(n-m-1)} + \ldots + b_m z^{-n}}{1 + a_1 z^{-1} + \ldots + a_n z^{-n}} ; n - m \geqslant 0 \quad (11.103)$$

From Fig. 11.43

$$E(z) = R(z) - C(z)$$
$$= [1 - T(z)]R(z) \quad (11.104)$$

The steady-state error is given by

$$e(\infty) = \lim_{z \to 1} (1 - z^{-1})[1 - T(z)]R(z) \quad (11.105)$$

For input of type $A(kT)^q$,

$$R(z) = \frac{B(z)}{(1 - z^{-1})^{q+1}} \quad (11.106)$$

where $B(z)$ is a finite degree polynomial in z^{-1} (see Table 11.2 for $q = 0, 1, 2$).

$$\therefore \quad e(\infty) = \lim_{z \to 1} (1 - z^{-1})[1 - T(z)] \frac{B(z)}{(1 - z^{-1})^{q+1}} \quad (11.107)$$

It immediately follows from (11.107) that $e(\infty) = 0$, if

$$1 - T(z) = (1 - z^{-1})^{q+1}$$

or
$$T(z) = 1 - (1 - z^{-1})^{q+1} \quad (11.108)$$

Now

$$E(z) = B(z) = \text{a finite degree polynomial in } z^{-1}.$$

This further ensures that $e(kT)$ goes to zero in a finite number of sampling intervals which are also the least in number for a specified q.

With $T(z)$ specified as in (11.108), we can obtain $D(z)$ from (11.102) as follows

$$D(z) = \frac{1 - (1 - z^{-1})^{q+1}}{(1 - z^{-1})^{q+1} \mathscr{Z}[G_0(s)G(s)]} \qquad (11.109)$$

Equation (11.109) also ensures that if $\mathscr{Z}[G_0(s)G(s)]$ is nonanticipative, $D(z)$ will also be nonanticipative.

Consider now the system response to three standard inputs wherein the system is compensated with appropriate $D(z)$ (see eqn. (11.104)) in each case.

1. Step input (i.e., $q = 0$)

$$E(z) = (1 - z^{-1}) \frac{A}{(1 - z^{-1})} = A$$

Therefore

$$e(0) = A$$
$$e(kT) = 0 \quad \text{for } k = 1, 2, \ldots$$

2. Ramp input (i.e., $q = 1$)

$$E(z) = (1 - z^{-1})^2 \frac{ATz^{-1}}{(1 - z^{-1})^2} = ATz^{-1}$$

Therefore

$$e(0) = 0$$
$$e(T) = AT$$
$$e(kT) = 0 \quad \text{for } k = 2, 3, 4, \ldots$$

3. Acceleration input (i.e., $q = 2$)

$$E(z) = (1 - z^{-1})^3 \frac{AT^2 z^{-1}(1 + z^{-1})}{(1 - z^{-1})^3}$$
$$= AT^2 z^{-1} + AT^2 z^{-2}$$

Therefore

$$e(0) = 0$$
$$e(T) = AT^2$$
$$e(2T) = AT^2$$
$$e(kT) = 0 \quad \text{for } k = 3, 4, \ldots$$

Example 11.21: In Fig. 11.43, let

$$G(s) = \frac{K}{s(s + 2)}$$

Design an appropriate $D(z)$ when ramp input is applied.

From Example 11.19

$$\mathscr{Z}[G_0(s)G(s)] = \mathscr{Z}\left[\frac{1 - e^{-Ts}}{s} \frac{K}{s(s + 2)}\right]$$

$$= \frac{K[(2T - 1 + e^{-2T})z^{-1} - (2Te^{-2T} + e^{-2T} - 1)z^{-2}]}{4(1 - z^{-1})(1 - e^{-2T}z^{-1})}$$

From eqn. (11.108)

$$T(z) = 1 - (1 - z^{-1})^2 = 2z^{-1} - z^{-2}$$

It immediately follows from eqn. (11.109) that

$$D(z) = \frac{4(2z^{-1} - z^{-2})(1 - z^{-1})(1 - e^{-2T}z^{-1})}{K(1 - z^{-1})^2[(2T - 1 + e^{-2T})z^{-1} - (2Te^{-2T} + e^{-2T} - 1)z^{-2}]}$$

Methods of synthesizing $D(z)$ will be discussed in Chapter 12.

FREQUENCY-DOMAIN TECHNIQUE OF DESIGNING $D(z)$

As already illustrated, use of bilinear transformation permits us to use the Nyquist criterion in the r-plane. Heuristic frequency-domain design criteria of phase and gain margin could be easily extended to the r-plane. This is best illustrated by means of an example.

Example 11.22: Consider the second order sampled-data system of Fig. 11.39. This system is required to meet the following specifications:

$$K_v \geqslant 4 \text{ sec}^{-1}; \text{ phase margin} \geqslant 40°$$

Bandwidth $\leqslant$ 1 Hz (i.e., 6.28 rad/sec)

Design steps are given below:

1. *Selection of sampling frequency*: According to the sampling theorem, the lowest sampling frequency = 2×bandwidth = 2 Hz. However, to be on the safe side, choose sampling frequency = 2.5 Hz so that the sampling period is $T = 0.4$ sec.

2. *z-transfer function for uncompensated system*: From eqn. (11.100), we have for $T = 0.4$ sec,

$$G_0G(z) = \frac{K(z + 0.76)}{16(z - 1)(z - 0.45)} \qquad (11.110)$$

3. *Determination of static gain*: Referring to the unity feedback system of Fig. 11.39, the z-transform of the error signal is

$$E(z) = R(z) - C(z) = \frac{R(z)}{1 + G_0G(z)}$$

Using the final value theorem, we have the steady-state error at the sampling instants

$$e_{ss} = \lim_{k \to \infty} e(kT) = \lim_{z \to 1} (1 - z^{-1}) E(z)$$

The steady-state error with ramp function input, i.e.,

$$R(z) = \frac{Tz}{(z - 1)^2}$$

is given by

$$e_{ss} = \lim_{z \to 1} \frac{T}{(z-1)[1 + G_0G(z)]} = \frac{1}{K_v}$$

where*

$$K_v = \frac{1}{T} \lim_{z \to 1} [(z-1)G_0G(z)] \tag{11.111}$$

For the system under consideration, we have from eqns. (11.110) and (11.111)

$$K_v = \frac{K}{2} \geqslant 4$$

Therefore $K = 8$ may be selected as the static gain for the system.

4. *Bilinear transformation*: The z-transfer function of the uncompensated system is

$$G_0G(z) = \frac{0.5(z + 0.76)}{(z-1)(z-0.45)}$$

Using bilinear transformation

$$z = \frac{1+r}{1-r}$$

we get

$$G_0G(r) = \frac{0.8(-r+1)(r/7.35 + 1)}{r(r/0.38 + 1)}$$

$$G_0G(j\omega_r) = \frac{0.8(-j\omega_r + 1)(j\omega_r/7.35 + 1)}{j\omega_r(j\omega_r/0.38 + 1)}$$

5. *Bode plot for the uncompensated system*: The Bode plot of $G_0G(j\omega_r)$ (i.e., the uncompensated system) is drawn in Fig. 11.44. We immediately find from this figure that

$$\omega_{c1} = 0.5, \phi_1 = 14°$$

These results show that compensation is required to increase the phase margin.

6. *Design of compensation*: We shall design a lag compensator having transfer function

$$D(r) = \frac{1 + \tau r}{1 + \beta \tau r}; \beta > 1$$

*$K_p = \lim_{z \to 1} G_0G(z)$

$K_a = \frac{1}{T^2} \lim_{z \to 1} (z-1)^2 G_0G(z)$

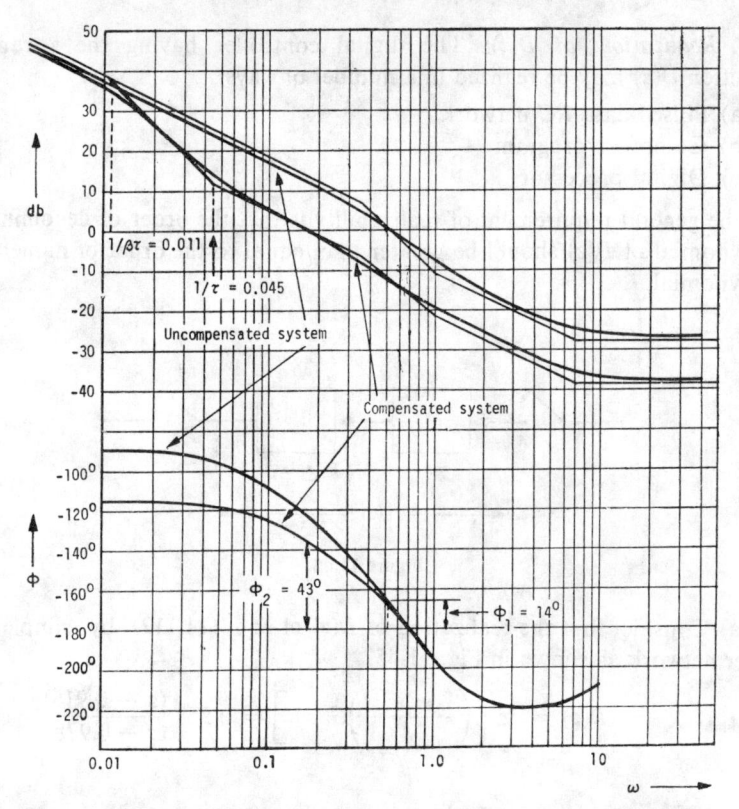

Figure 11.44 Lag compensator design for a sampled-data control system in the r-domain.

in such a way as to realize a phase margin of 40° or more. Following the procedure for designing phase lag compensation for continuous-time systems (Chapter 10) we get

$$D(r) = \frac{1 + 22.2r}{1 + 88.8r}$$

7. *Bode plot for the compensated system*: Bode plot for the compensated system is also shown in Fig. 11.44. We find that the required phase margin has been realized.

We can construct the Nichols plot for the compensated system to verify the satisfaction of bandwidth specification.

8. *Transformation to the z-domain*: Carrying out the reverse transformation

$$r = \frac{z-1}{z+1}$$

we have

$$D(z) = 0.258 \frac{(z - 0.915)}{(z - 0.978)} \tag{11.112}$$

9. *Realization of D(z)*: The digital controller having the z-transfer function $D(z)$ may be realized in a number of ways:
   (a) Pulsed-data *RC* network
   (b) Computer program
   (c) Digital processor

The general requirement of realizability is that the order of denominator polynomial of $D(z)$ should be greater than/equal to the order of numerator polynomial.

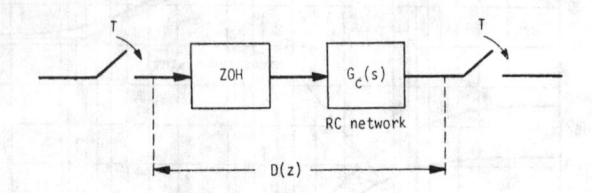

Figure 11.45.

(a) Consider first the realization of $D(z)$ of eqn. (11.112) by simple *RC* filter network as shown in Fig. 11.45.

$$D(z) = \mathscr{Z}\left[\left(\frac{1-e^{-Ts}}{s}\right)G_c(s)\right] = 0.258\frac{(z-0.915)}{(z-0.978)}$$

or

$$\mathscr{Z}\left(\frac{G_c(s)}{s}\right) = \frac{D(z)}{1-z^{-1}} = 0.258\frac{(1-0.915z^{-1})}{(1-z^{-1})(1-0.978z^{-1})}$$

$$= \frac{1}{1-z^{-1}} - \frac{0.775}{1-0.978z^{-1}}$$

Taking the inverse z-transform on both sides, we get

$$\frac{G_c(s)}{s} = \frac{1}{s} - \frac{0.775}{s+a}$$

where

$$0.978 = e^{-aT} = e^{-0.4a}$$

gives

$$a = 0.05$$

The transfer function of the network $G_c(s)$ is

$$G_c(s) = s\left[\frac{1}{s} - \frac{0.775}{s+0.05}\right]$$

$$= \frac{1+4.5s}{1+20s}$$

Evidently, $G_c(s)$ can be synthesized by a simple *RC* network.

Note that realization of $D(z)$ by simple RC series network is possible when $D(z)$ has simple and real poles lying inside or on the unit circle in the z-plane.

(b) The z-transfer function of digital controller is

$$D(z) = \frac{E_2(z)}{E_1(z)} = 0.258 \frac{(z - 0.915)}{(z - 0.978)} = 0.258 \frac{(1 - 0.915z^{-1})}{(1 - 0.978z^{-1})}$$

Therefore

$$(1 - 0.978z^{-1})E_2(z) = 0.258(1 - 0.915z^{-1})E_1(z)$$

or $\quad e_2(k) - 0.978e_2(k - 1) = 0.258e_1(k) - 0.246e_1(k - 1)$

or $\quad e_2(k) = 0.258e_1(k) - 0.246e_1(k - 1) + 0.978e_2(k - 1)$

This difference equation can be solved by a computer program.

(c) Realization of $D(z)$ using a special purpose computer (delay network) will be discussed in Chapter 12.

Problems

11.1 For Fig. 11.9, write the difference equation governing the system response for

(i) $G(s) = \dfrac{1}{s + 1}$
(ii) $G(s) = \dfrac{1}{s^2}$

11.2 Given

$$\mathscr{Z}[x(k)] = X(z)$$

find the z-transform of

(i) $y(k) = \sum_{j=0} x(j)$
(ii) $y(k) = e^{-ak}x(k)$

11.3 Find the z-transform of

(i) k^2
(ii) $ka^{k-1}; k \geqslant 1$

(iii) $k^2 a^{k-1}; k \geqslant 1$
(iv) $\dfrac{a^k}{k!}$

(v) $\sinh \beta k$
(vi) $\cosh \beta k$

(vii) $a^k \cos k\pi = (-a)^k$

11.4 Find the z-transforms of the discrete sequences generated by mathematically sampling (at uniform time interval T) the following continuous-time functions.

(i) t^2
(ii) e^{-at}

(iii) te^{-at}
(iv)* $e^{-at} \sin \omega t$

(v)* $e^{-at} \cos \omega t$

11.5 Find the z-domain transfer function of the following s-domain transfer functions.

(i) $\dfrac{a}{(s + a)^2}$
(ii) $\dfrac{s}{s^2 + \omega^2}$

(iii) $\dfrac{a}{(s + b)^2 + a^2}$
(iv) $\dfrac{a}{s^2 - a^2}$

(v) $\dfrac{s + b}{(s + b)^2 + a^2}$

*You may use the result of Problem 11.2 (ii).

11.6 Find the inverse z-transforms of the following:

(i) $\dfrac{z}{z+a}$

(ii) $\dfrac{1}{z+a}$

(iii) $\dfrac{1}{z-a}$

(iv) $\dfrac{3z^2+2z+1}{z^2-3z+2}$

(v) $\dfrac{3z^2+2z+1}{z^2+3z+2}$

(vi) $\dfrac{2z}{(2z-1)^2}$

(vii) $\dfrac{z-0.4}{z^2+z+2}$

(viii) $\dfrac{z^{-1}}{(1-az^{-1})^2}$

(ix) $\dfrac{z-4}{(z-1)(z-2)^2}$

11.7 A symmetrical ladder network is shown in Fig. P-11.7. Write the difference equation describing the network behaviour. Also specify the boundary conditions. Using the z-transform technique solve for current in any loop. Assume each resistor to be 1 ohm.

(*Hint*: The difference equation is
$$3i_{n+1} - i_n - i_{n+2} = 0; \quad n = 0, 1, \ldots, 10$$
Boundary conditions are
$$i_0 = 4; \quad 3i_{12} = i_{11}$$
The z-transforms of Problem 11.3(v) and (vi) could be used for inverse transforming.)

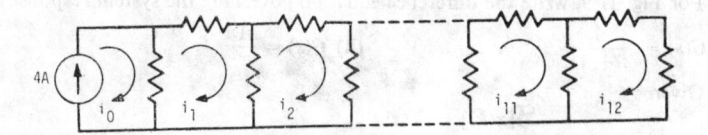

Figure P-11.7.

11.8 The input-output of a sampled-data system is described by the difference equation
$$c(n+2) + 3c(n+1) + 4c(n) = r(n+1) - r(n)$$
Determine the z-transfer function. Also obtain the weighting sequence (discrete impulse response) of the system.

11.9 Solve the difference equation
$$c(k+2) + 3c(k+1) + 2c(k) = u(k); \quad c(0) = 1,$$
$$c(k) = 0 \text{ for } k < 0$$
(*Hint*: $c(1)$ needed in the solution can be obtained by letting $k = -1$
$$c(1) + 3c(0) + 2c(-1) = r(-1)$$
or
$$c(1) = -3$$

11.10 For Fig. 11.9, obtain $C(z)/R(z)$ and write therefrom the difference equation governing the system. Check eqn. (11.12).

Repeat for $G(s)$ given in Problem 11.1 and check the difference equations obtained by the two methods.

11.11 Determine the z-transfer function of two cascaded systems each described by the difference equation
$$c(k) = 0.5c(k-1) + r(k)$$

11.12 The impulse response of a linear continuous system is given by $e^{-at}u(t)$. For impulse sampled unit step input (sampling period = T), find the output at sampled instants.

Repeat for the case where impulse sampler is followed by ZOH. Compare the results.

11.13 Find $C(z)/R(z)$ for the sampled-data closed-loop system of Fig. P-11.13.

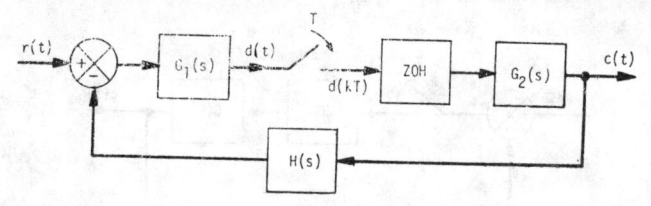

Figure P-11.13.

11.14 A sampled-data control system is shown in Fig. P-11.14. Show that the output of the system at sampling instants is zero.

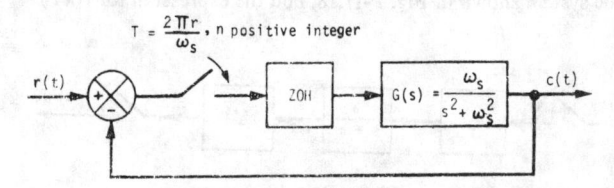

Figure P-11.14.

(*Hint*: $G(z) = 0$ which indicates that the sampled output of the system is zero but the continuous output is not zero.)

11.15 Find $C(z)/R(z)$ for the sampled-data closed-loop system of Fig. P-11.15. Assume both the samplers to be of impulse type.

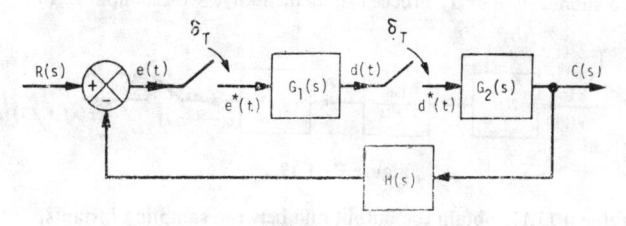

Figure P-11.15.

11.16 For the sampled-data feedback system with a digital network in the feedback path as shown in Fig. P-11.16, find $C(z)/R(z)$.

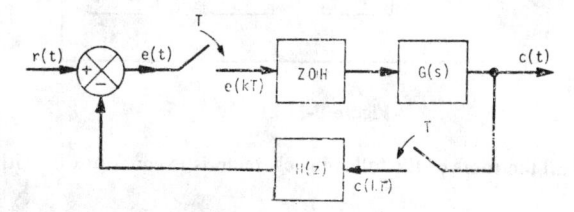

Figure P-11.16

11.17 For the sampled-data system of Fig. P-11.17, find the response to unit step input

Given:
$$G(s) = \frac{1}{s+1}$$

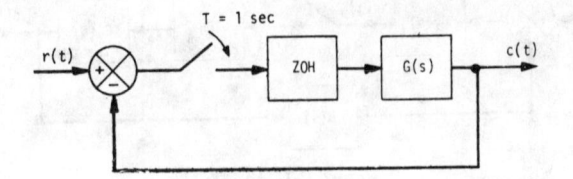

Figure P-11.17.

11.18 For the system shown in Fig. P-11.18, find the expression for $c(kT)$.

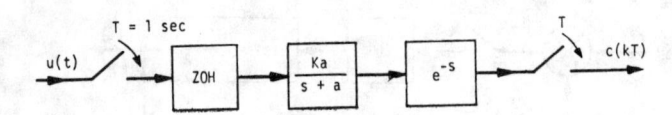

Figure P-11.18.

11.19 For the system of Fig. P-11.19, obtain the expression for $c(kT)$ for $r(t) = $ unit step. Given $0 < \Delta < 1$.

(Hint:
$$\mathcal{Z}\left[\frac{1-e^{-sT}}{s}\frac{Kae^{\Delta T}}{s+a}\right] \neq (1-z^{-1})\mathcal{Z}\left[\frac{Kae^{\Delta Ts}}{s(s+a)}\right]$$

because ΔT is a submultiple of 1. Proceed fundamentally. See Example 11.18.)

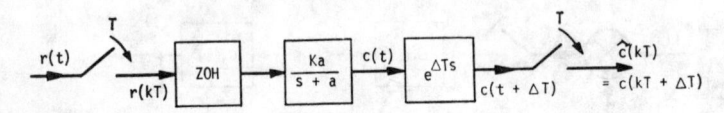

Figure P-11.19.

11.20 In Problem 11.17, obtain the output mid between sampling instants.

11.21 Find the range of K for the system shown in Fig. P-11.21 to be stable.

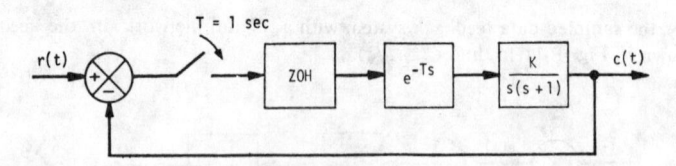

Figure P-11.21.

11.22 Check if all the roots of the following characteristic equations lie within the unit circle.
  (i) $5z^2 - 2z + 2 = 0$
  (ii) $z^3 - 0.2z^2 - 0.25z + 0.05 = 0$
  (iii) $z^4 - 1.7z^3 + 1.04z^2 - 0.268z + 0.024 = 0$

11.23 For the Example 11.20, a compensating digital network with transfer function

$$D(z) = \frac{z + 0.5}{z - 0.6}$$

is introduced in the forward loop. Obtain the root locus plot of the system for $T = 0.4$ sec. Compare the plot with Fig. 11.40 and find the range of gain K for the system to be stable.

11.24 Design a lead compensator in the r-domain for Example 11.22. What is the corresponding z-domain transfer function?

11.25 For the $D(z)$ obtained in Problem 11.24, find $G_c(s)$ as per the compensator given in Fig. 11.45 such that

$$(1 - z^{-1})\mathscr{Z}\left[\frac{G_c(s)}{s}\right] = D(z)$$

Is the s-domain compensator realizable?

(*Hint*: Partial fractionate $\dfrac{1}{1 - z^{-1}} D(z)$ and convert to s-form using Table 11.3.)

References and Further Reading

1. Cadzow, J.A., *Discrete-Time Systems*, Prentice-Hall, Englewood Cliffs, N.J., 1973.
2. Kuo, B.C., *Analysis and Synthesis of Sampled-Data Control Systems*, Prentice-Hall, Englewood Cliffs, N.J., 1963.
3. Jury, E.I., *Inners and Stability of Dynamic Systems*, John Wiley, New York, 1974.
4. Cadzow, J.A. and H.R. Martens, *Discrete-Time and Computer Control Systems*, Prentice-Hall, Englewood Cliffs, N.J., 1970.
5. Gupta, S.C. and L. Hasdorff, *Fundamentals of Automatic Control*, John Wiley, New York, 1970.
6. Kuo, B.C., *Discrete Data Control Systems*, Science-Tech, Champaign, Illinois, 1970.
7. Saucedo, R. and E.E. Schiring, *Introduction to Continuous and Digital Control Systems*, Macmillan, New York, 1968.
8. Freeman, H., *Discrete-Time Systems*, John Wiley, New York, 1965.
9. Lindorff, D.P., *Theory of Sampled-Data Control Systems*, John Wiley, New York, 1965.

12. State Variable Analysis and Design

12.1 Introduction

In the preceding chapters, we studied several methods of analysis and design of feedback systems such as root locus and frequency response methods. These methods require that the physical system be modelled in the form of a transfer function. Though the transfer function model provides us with simple and powerful analysis and design techniques, it suffers from certain drawbacks, e.g., a transfer function is only defined under zero initial conditions. Further, it has certain limitations due to the fact that the transfer function model is only applicable to linear time-invariant systems and there too it is generally restricted to single-input-single-output systems as this approach* becomes highly cumbersome for use in multiple-input-multiple-output systems. Another limitation of the transfer function technique is that it reveals only the system output for a given input and provides no information regarding the internal state of the system. There may be situations where the output of a system is stable and yet some of the system elements may have a tendency to exceed their specified ratings. In addition to this, it may sometimes be necessary and advantageous to provide a feedback proportional to some of the internal variables of a system, rather than the output alone, for the purpose of stabilizing and improving the performance of a system. It is further observed that the classical design methods (root locus and frequency domain methods) based on the transfer function model are essentially trial and error procedures. Such procedures are difficult to visualize and organize even in moderately complex systems and may not lead to a control system which yields an optimum performance in some defined sense.

From the foregoing discussion we feel the need of a more general mathematical representation of a system which, along with the output, yields information about the state of the system variables at some predetermined points along the flow of signals. Such considerations have led to the development of the state variable approach. It is a direct time domain approach which provides a basis for modern control theory and system optimization. It is a very powerful technique for the analysis and design of

*Refer Section 2.4.

linear and nonlinear, time-invariant or time-varying multi-input-multi-output systems. The organization of the state variable approach is such that it is easily amenable to solution through digital computers.

It will be incorrect to conclude from the above introduction that the state variable approach can completely replace the classical approaches. In fact, the classical approaches, provide the control engineer with a deep physical insight into the system and greatly aid the preliminary system design where a complex system is approximated by a more manageable model.

In this chapter our aim is to introduce the state variable approach for linear time-invariant systems in both the continuous-time and discrete-time cases, so as to give a glimpse of its power for the analysis and design of complex systems. It may be pointed out that whereas in transform domain analysis, Laplace transform is needed for continuous-time systems and z-transform is needed for discrete-time systems, the state variable approach offers us a way to look at both the continuous-time and discrete-time systems with the same formulation. In the following, we first discuss the state variable techniques for linear continuous-time systems and then show that the techniques are essentially the same for linear discrete-time systems also.

State variable techniques can be fully appreciated if the readers possess a knowledge of the elementary rules of vector and matrix algebra. Only a most elementary knowledge (Appendix II) will prove sufficient for our specific needs.

12.2 Concepts of State, State Variables and State Model

A mathematical abstraction to represent or model the dynamics of a system utilizes three types of variables called the *input*, the *output* and the *state variables*.

Consider the mechanical system shown in Fig. 12.1 wherein mass M is acted upon by the force $F(t)$. The system is characterized by the relations

$$\frac{d}{dt} v(t) = \frac{1}{M} F(t) \tag{12.1}$$

$$\frac{d}{dt} x(t) = v(t) \tag{12.2}$$

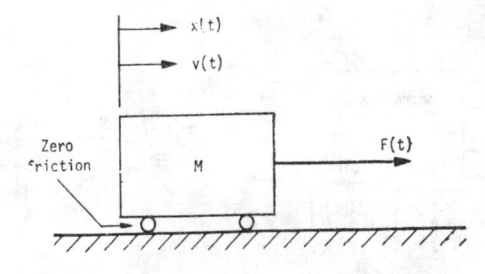

Figure 12.1 A simple mechanical system.

From these relations we get

$$v(t) = \frac{1}{M}\int_{-\infty}^{t} F(t)\,dt = \frac{1}{M}\int_{-\infty}^{t_0} F(t)\,dt + \frac{1}{M}\int_{t_0}^{t} F(t)\,dt$$

$$= v(t_0) + \frac{1}{M}\int_{t_0}^{t} F(t)\,dt \qquad (12.3)$$

$$x(t) = \int_{-\infty}^{t} v(t)\,dt = \int_{-\infty}^{t_0} v(t)\,dt + \int_{t_0}^{t} v(t)\,dt$$

$$= x(t_0) + [t - t_0]v(t_0) + \frac{1}{M}\int_{t_0}^{t} d\tau \int_{t_0}^{\tau} F(t)\,dt \qquad (12.4)$$

We observe that the displacement $x(t)$ (output variable) at any time $t \geqslant t_0$ can be computed if we know the applied force $F(t)$ (input variable) from $t = t_0$ onwards, provided $v(t_0)$ the initial velocity and $x(t_0)$ the initial displacement are known. We may conceive of initial velocity and initial displacement as describing the status or *state* of the system at $t = t_0$. The state of the system of Fig. 12.1 at any time t is given by the variables $x(t)$ and $v(t)$ which are called the *state variables* of the system.

Let us now look into the precise definitions of state and state variables:

The state of a dynamical system is a minimal set of variables (known as state variables) such that the knowledge of these variables at $t = t_0$ *together with the knowledge of the inputs for* $t \geqslant t_0$, *completely determines the behaviour of the system for* $t > t_0$.

In state variable formulation of a system, the state variables are usually represented by $x_1(t), x_2(t), \ldots$; the inputs by $u_1(t), u_2(t)\ldots$; and the outputs by $y_1(t), y_2(t), \ldots$. The state space representation may be visualized in block diagram form as shown in Fig. 12.2. For generality, we have depicted a system which has m inputs, p outputs and n state variables. For notational economy, the different variables may be represented by the input vector $\mathbf{u}(t)$, *output vector* $\mathbf{y}(t)$ and the *state vector* $\mathbf{x}(t)$; where

$$\mathbf{u}(t) = \begin{bmatrix} u_1(t) \\ u_2(t) \\ \vdots \\ u_m(t) \end{bmatrix}, \quad \mathbf{y}(t) = \begin{bmatrix} y_1(t) \\ y_2(t) \\ \vdots \\ y_p(t) \end{bmatrix}, \quad \mathbf{x}(t) = \begin{bmatrix} x_1(t) \\ x_2(t) \\ \vdots \\ x_n(t) \end{bmatrix} \qquad (12.5)$$

In Fig. 12.2, broad arrows have been used to represent vector quantities.

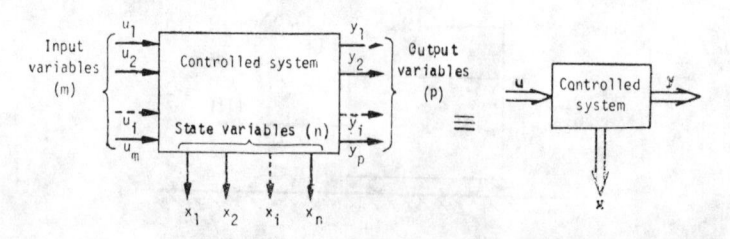

Figure 12.2 Structure of a general control system.

For the system of Fig. 12.1, the state variable representation is given by two first-order differential equations (12.1) and (12.2), the solution of these equations (eqns. (12.3) and (12.4)) gives the two state variables $x(t)$ and $v(t)$ of the system. For a general system of Fig. 12.2, the state variable representation can be arranged in the form of n first-order differential equations

$$\frac{dx_1}{dt} = \dot{x}_1 = f_1(x_1, x_2, \ldots, x_n; u_1, u_2, \ldots, u_m)$$
$$\vdots \qquad \vdots \qquad \qquad \vdots \qquad \qquad \qquad (12.6)$$
$$\frac{dx_n}{dt} = \dot{x}_n = f_n(x_1, x_2, \ldots, x_n; u_1, u_2, \ldots, u_m)$$

Integration of equations (12.6) gives

$$x_i(t) = x_i(t_0) + \int_{t_0}^{t} f_i(x_1, x_2, \ldots, x_n; u_1, u_2, \ldots, u_m)\, dt;$$
$$i = 1, 2, \ldots, n$$

Thus the n state variables and hence the state of the system can be determined uniquely* at any $t > t_0$ if each state variable is known at $t = t_0$ and all the m control forces are known throughout the interval t_0 to t.

The 'n' differential equations (12.6) may be written in vector notation as

$$\dot{\mathbf{x}}(t) = \mathbf{f}(\mathbf{x}(t), \mathbf{u}(t)) \qquad (12.7)$$

where $\mathbf{x}$ is $n \times 1$ state vector, $\mathbf{u}$ is $m \times 1$ input vector, both defined earlier in (12.5), and

$$\mathbf{f}(.) = \begin{bmatrix} f_1(.) \\ f_2(.) \\ \vdots \\ f_n(.) \end{bmatrix} \qquad (12.8)$$

is $n \times 1$ function vector.

For time-varying systems, the function $\mathbf{f}$ is dependent on time as well and the vector equation may be written as

$$\dot{\mathbf{x}}(t) = \mathbf{f}(\mathbf{x}(t), \mathbf{u}(t), t) \qquad (12.9)$$

Equations (12.7) and (12.9) are the *state equations* for time-invariant and time-varying systems respectively. The state vector $\mathbf{x}$ determines a point (called the *state point*) in an n-dimensional space, called the *state space*. The curve traced out by the state point from $t = t_0$ to $t = t_1$ in the direction of increasing time is known as the *state trajectory*. For the two-dimensional cases, the state space reduces to the state plane or phase plane.

The output $\mathbf{y}(t)$ (Fig. 12.2) can in general be expressed in terms of the state $\mathbf{x}(t)$ and input $\mathbf{u}(t)$ as

$$\mathbf{y}(t) = \mathbf{g}(\mathbf{x}(t), \mathbf{u}(t)); \text{ time-invariant systems} \qquad (12.10)$$
$$\mathbf{y}(t) = \mathbf{g}(\mathbf{x}(t), \mathbf{u}(t), t); \text{ time-varying systems} \qquad (12.11)$$

*It is important to note that a unique solution for eqns. (12.6) exists only if f_i and $\partial f_i / \partial x_j$ are defined and are continuous for $i, j = 1, 2, \ldots, n$. It is being assumed here that these conditions are satisfied.

Equations (12.10) and (12.11) are the *output equations* for time-invariant and time-varying systems respectively. It may be noted that the output equation is not a dynamic relation but a static (instantaneous) one. It is called the *read-out* function. We, therefore, need to solve the system state equation and once the system state is known, output can be immediately obtained from the output equation. Solution of the state equations thus provides us information about the system state as well as the system output.

The state and output equations constitute the *state model* of the system.

STATE MODEL OF LINEAR SYSTEMS

State model of a linear time-invariant system is a special case of the general time-invariant model of eqns. (12.7) and (12.10). Derivative of each state variable now becomes a linear combination of system states and inputs, i.e.,

$$\dot{x}_1 = a_{11}x_1 + a_{12}x_2 + \ldots + a_{1n}x_n + b_{11}u_1 + b_{12}u_2 + \ldots + b_{1m}u_m$$
$$\dot{x}_2 = a_{21}x_1 + a_{22}x_2 + \ldots + a_{2n}x_n + b_{21}u_1 + b_{22}u_2 + \ldots + b_{2m}u_m$$
$$\vdots \qquad \vdots \qquad (12.12)$$
$$\dot{x}_n = a_{n1}x_1 + a_{n2}x_2 + \ldots + a_{nn}x_n + b_{n1}u_1 + b_{n2}u_2 + \ldots + b_{nm}u_m$$

where the coefficients a_{ij} and b_{ij} are constants. In the vector-matrix form, eqns. (12.12) may be written as

$$\dot{\mathbf{x}}(t) = \mathbf{A}\mathbf{x}(t) + \mathbf{B}\mathbf{u}(t)$$

where $\mathbf{x}(t)$ is $n \times 1$ state vector, $\mathbf{u}(t)$ is $m \times 1$ input vector, $\mathbf{A}$ is $n \times n$ *system matrix* defined by

$$\mathbf{A} = \begin{bmatrix} a_{11} & a_{12} & \ldots & a_{1n} \\ a_{21} & a_{22} & \ldots & a_{2n} \\ \vdots & \vdots & & \vdots \\ a_{n1} & a_{n2} & \ldots & a_{nn} \end{bmatrix}$$

and $\mathbf{B}$ is $n \times m$ *input matrix* defined by

$$\mathbf{B} = \begin{bmatrix} b_{11} & b_{12} & \ldots & b_{1m} \\ b_{21} & b_{22} & \ldots & b_{2m} \\ \vdots & \vdots & & \vdots \\ b_{n1} & b_{n2} & \ldots & b_{nm} \end{bmatrix}$$

Similarly, the output variables at time t are linear combinations of the values of the input and state variables at time t, i.e.,

$$y_1(t) = c_{11}x_1(t) + \ldots + c_{1n}x_n(t) + d_{11}u_1(t) + \ldots + d_{1m}u_m(t)$$
$$\vdots \qquad \vdots$$
$$y_p(t) = c_{p1}x_1(t) + \ldots + c_{pn}x_n(t) + d_{p1}u_1(t) + \ldots + d_{pm}u_m(t)$$

where the coefficients c_{ij} and d_{ij} are constants. This set of equations may be put in the vector-matrix form

$$\mathbf{y}(t) = \mathbf{C}\mathbf{x}(t) + \mathbf{D}\mathbf{u}(t)$$

where $\mathbf{y}(t)$ is $p \times 1$ output vector, $\mathbf{C}$ is $p \times n$ *output matrix* defined by

$$\mathbf{C} = \begin{bmatrix} c_{11} & c_{12} & \cdots & c_{1n} \\ c_{21} & c_{22} & \cdots & c_{2n} \\ \vdots & \vdots & & \vdots \\ c_{p1} & c_{p2} & \cdots & c_{pn} \end{bmatrix}$$

and $\mathbf{D}$ is $p \times m$ *transmission matrix** defined by

$$\mathbf{D} = \begin{bmatrix} d_{11} & d_{12} & \cdots & d_{1m} \\ d_{21} & d_{22} & \cdots & d_{2m} \\ \vdots & \vdots & & \vdots \\ d_{p1} & d_{p2} & \cdots & d_{pm} \end{bmatrix}$$

The state model of linear time-invariant systems is thus given by the following equations:

$$\dot{\mathbf{x}}(t) = \mathbf{A}\mathbf{x}(t) + \mathbf{B}\mathbf{u}(t); \text{ state equation} \quad (12.13a)$$

$$\mathbf{y}(t) = \mathbf{C}\mathbf{x}(t) + \mathbf{D}\mathbf{u}(t); \text{ output equation} \quad (12.13b)$$

The block diagram representation of the state model is shown in Fig. 12.3.

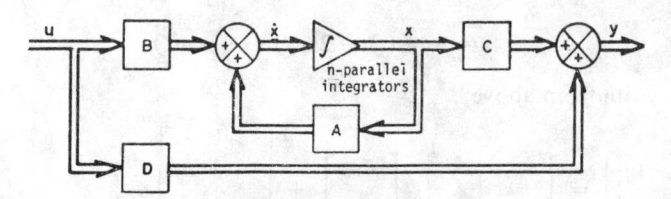

Figure 12.3 Block diagram representation of the state model of a linear multi-input-multi-output system.

For system of Fig. 12.1, let us define

$$x_1(t) = x(t)$$
$$x_2(t) = v(t)$$
$$u(t) = F(t)$$

Equations (12.1) and (12.2) can now be written as

$$\begin{bmatrix} \dot{x}_1 \\ \dot{x}_2 \end{bmatrix} = \begin{bmatrix} 0 & 1 \\ 0 & 0 \end{bmatrix} \begin{bmatrix} x_1 \\ x_2 \end{bmatrix} + \begin{bmatrix} 0 \\ 1/M \end{bmatrix} u \quad (12.14a)$$

The output $y = x(t)$ may be expressed as

$$y = \begin{bmatrix} 1 & 0 \end{bmatrix} \begin{bmatrix} x_1 \\ x_2 \end{bmatrix} \quad (12.14b)$$

Equations (12.14) are of the form of eqns. (12.13).

*Direct coupling of the input to the output is rare in control systems where power amplification is generally desired.

It may be noted here that the state of a system is not uniquely specified. For example, for the system of Fig 12.1, we have taken displacement $x(t)$ and velocity $v(t)$ as the state variables. We may define new variables as

$$z_1(t) = 2x(t) + v(t)$$
$$z_2(t) = x(t) + v(t)$$

or
$$x(t) = z_1(t) - z_2(t)$$
$$v(t) = -z_1(t) + 2z_2(t)$$

Now
$$\frac{dz_1}{dt} = 2\frac{dx}{dt} + \frac{dv}{dt}$$
$$= 2v(t) + \frac{1}{M} F(t)$$
$$= -2z_1(t) + 4z_2(t) + \frac{1}{M} u(t)$$

$$\frac{dz_2}{dt} = \frac{dx}{dt} + \frac{dv}{dt}$$
$$= v(t) + \frac{1}{M} F(t)$$
$$= -z_1(t) + 2z_2(t) + \frac{1}{M} u(t)$$

We can write from above

$$\begin{bmatrix} \dot{z}_1 \\ \dot{z}_2 \end{bmatrix} = \begin{bmatrix} -2 & 4 \\ -1 & 2 \end{bmatrix} \begin{bmatrix} z_1 \\ z_2 \end{bmatrix} + \begin{bmatrix} \frac{1}{M} \\ \frac{1}{M} \end{bmatrix} u \quad (12.14c)$$

The output is given by

$$y = x(t) = z_1(t) - z_2(t)$$

or
$$y = \begin{bmatrix} 1 & -1 \end{bmatrix} \begin{bmatrix} z_1 \\ z_2 \end{bmatrix} \quad (12.14d)$$

We immediately observe that (12.14c) and (12.14d) give an alternative state variable model of the system previously represented by (12.14a) and (12.14b). We, therefore, conclude that the state variables of a system are nonunique and the example further brings out the fact that the state variables need not necessarily be the physical variables of the system.

Though the state model of a system is not unique, however, all such models have one characteristic in common for a given system, namely, the number of elements in the state vector is equal and minimal. This number n is referred to as the *order* of the system.

The state model for a linear time-varying system is of the same form as defined in eqns. (12.13) except for the fact that the coefficients of the matrices **A**, **B**, **C** and **D** are no more constants but are functions of time

In this chapter, we will be mainly concerned with linear time-invariant systems.

STATE MODEL FOR SINGLE-INPUT-SINGLE-OUTPUT LINEAR SYSTEMS

The transfer function analysis deals mainly with single-input-single-output linear time-invariant systems. Here in this chapter we will link the transfer function approach with the state variable approach. If we let $m = 1$ and $p = 1$ in the state model of a multi-input-multi-output linear system, we obtain the following state model for a single-input-single-output linear system:

$$\dot{\mathbf{x}} = \mathbf{A}\mathbf{x} + \mathbf{B}u \quad (12.15a)$$
$$y = \mathbf{C}\mathbf{x} + du \quad (12.15b)$$

where **B** and **C** are now respectively $(n \times 1)$ and $(1 \times n)$ matrices, d is a constant and u is a scalar control variable. The block diagram representation of this state model is shown in Fig. 12.4.

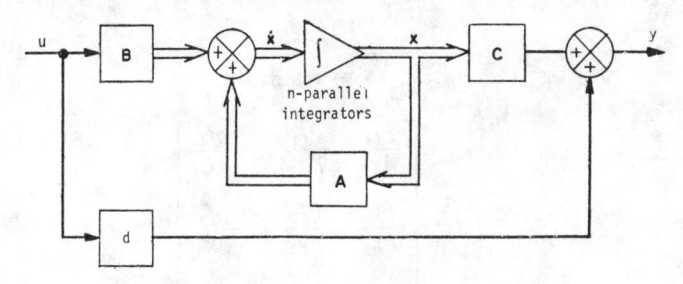

Figure 12.4 Block diagram representation of the state model for a linear single-input-single-output system.

LINEARIZATION OF THE STATE EQUATION

The state equation $\dot{\mathbf{x}} = \mathbf{f}(\mathbf{x}, \mathbf{u})$ of a general time-invariant system can be linearized for small variations about an equilibrium point $(\mathbf{x}_0, \mathbf{u}_0)$. It is assumed that the system is in equilibrium under the conditions $\mathbf{x}_0$ and $\mathbf{u}_0$, i.e.,

$$\dot{\mathbf{x}} = \mathbf{f}(\mathbf{x}_0, \mathbf{u}_0) = 0$$

Since the derivatives of all the state variables are zero at the equilibrium point, the system continues to lie at the equilibrium point unless otherwise disturbed.

The state equation can be linearized about the operating point $(\mathbf{x}_0, \mathbf{u}_0)$ by expanding it into Taylor series and neglecting terms of second- and higher-order. Thus for the ith state equation

$$\dot{x}_i = f_i(\mathbf{x}_0, \mathbf{u}_0) + \sum_{j=1}^{n} \left.\frac{\partial f_i(\mathbf{x}, \mathbf{u})}{\partial x_j}\right|_{\substack{\mathbf{x}=\mathbf{x}_0 \\ \mathbf{u}=\mathbf{u}_0}} (x_j - x_{j0}) + \sum_{k=1}^{m} \left.\frac{\partial f_i(\mathbf{x}, \mathbf{u})}{\partial u_k}\right|_{\substack{\mathbf{x}=\mathbf{x}_0 \\ \mathbf{u}=\mathbf{u}_0}} (u_k - u_{k0})$$

Recognizing that at the operating point, $f_i(\mathbf{x}_0, \mathbf{u}_0) = 0$ and defining the variation about the operating point as

$$\tilde{x}_j = x_j - x_{j0} \qquad \therefore \dot{\tilde{x}}_j = \dot{x}_j$$

$$\tilde{u}_k = u_k - u_{k0}$$

the linearized ith state equation can be written as

$$\dot{\tilde{x}}_i = \sum_{j=1}^{n} \left.\frac{\partial f_i(\mathbf{x}, \mathbf{u})}{\partial x_j}\right|_{\substack{\mathbf{x}=\mathbf{x}_0 \\ \mathbf{u}=\mathbf{u}_0}} \tilde{x}_j + \sum_{k=1}^{m} \left.\frac{\partial f_i(\mathbf{x}, \mathbf{u})}{\partial u_k}\right|_{\substack{\mathbf{x}=\mathbf{x}_0 \\ \mathbf{u}=\mathbf{u}_0}} \tilde{u}_k$$

The above linearized component equation can be written as the vector matrix equation

$$\dot{\tilde{\mathbf{x}}} = \mathbf{A}\tilde{\mathbf{x}} + \mathbf{B}\tilde{\mathbf{u}} \qquad (12.16)$$

where

$$\mathbf{A} = \begin{bmatrix} \frac{\partial f_1}{\partial x_1} & \frac{\partial f_1}{\partial x_2} & \cdots & \frac{\partial f_1}{\partial x_n} \\ \frac{\partial f_2}{\partial x_1} & \frac{\partial f_2}{\partial x_2} & \cdots & \frac{\partial f_2}{\partial x_n} \\ \cdots & & & \\ \frac{\partial f_n}{\partial x_1} & \frac{\partial f_n}{\partial x_2} & \cdots & \frac{\partial f_n}{\partial x_n} \end{bmatrix}$$

$$\mathbf{B} = \begin{bmatrix} \frac{\partial f_1}{\partial u_1} & \frac{\partial f_1}{\partial u_2} & \cdots & \frac{\partial f_1}{\partial u_m} \\ \frac{\partial f_2}{\partial u_1} & \frac{\partial f_2}{\partial u_2} & \cdots & \frac{\partial f_2}{\partial u_m} \\ \cdots & & & \\ \frac{\partial f_n}{\partial u_1} & \frac{\partial f_n}{\partial u_2} & \cdots & \frac{\partial f_n}{\partial u_m} \end{bmatrix}$$

All the partial derivatives in the matrices **A** and **B** defined above (called the *Jacobian matrices*) are evaluated at the equilibrium state $(\mathbf{x}_0, \mathbf{u}_0)$. (Linearization technique will be illustrated through a number of examples in Chapter 14.)

12.3 State Models for Linear Continuous-time Systems

From the discussion of previous section, we know that the state equations of a system are not unique, i.e., there exist more than one set of state variables in terms of which the system behaviour can be completely described. In fact there are infinitely many state models for a given system and any two models are uniquely related. To demonstrate this fact, consider the state model of eqn. (12.13) wherein **x** is an n-dimensional state vector. Consider now another n-dimensional vector **z** such that

$$\dot{\mathbf{x}} = \mathbf{Pz}$$

where **P** is any $n \times n$ nonsingular constant matrix.
Since **P** is a constant matrix, it follows that

$$\dot{\mathbf{x}} = \mathbf{P}\dot{\mathbf{z}}$$

Substituting **x** and $\dot{\mathbf{x}}$ from above in eqn. (12.13a), we get

$$\mathbf{P}\dot{\mathbf{z}} = \mathbf{APz} + \mathbf{Bu}$$

Premultiplying by $\mathbf{P}^{-1}$, we obtain

$$\dot{\mathbf{z}} = \mathbf{P}^{-1}\mathbf{APz} + \mathbf{P}^{-1}\mathbf{Bu}$$

$$= \tilde{\mathbf{A}}\mathbf{z} + \tilde{\mathbf{B}}\mathbf{u} \qquad (12.17a)$$

where $\tilde{\mathbf{A}} = \mathbf{P}^{-1}\mathbf{AP}; \ \tilde{\mathbf{B}} = \mathbf{P}^{-1}\mathbf{B}$

From eqn. (12.13b), we have

$$\mathbf{y} = \mathbf{CPz} + \mathbf{Du}$$

$$= \tilde{\mathbf{C}}\mathbf{z} + \mathbf{Du} \qquad (12.17b)$$

where $\tilde{\mathbf{C}} = \mathbf{CP}$

Equations (12.17a) and (12.17b) give another state model for a given system. Since **P** is assumed to be nonunique nonsingular matrix, the state model is also nonunique. It is important to note that the transformation matrix **P** must be nonsingular, i.e., $|\mathbf{P}| \neq 0$. If this were not the case, the inverse transformation would obviously not exist.

This demonstrates that for a given system, infinitely many state models are possible. However, in a particular case the state model resulting from a specific choice of state variables may prove to be simpler and more useful than other choices of state models. Here, judgement has frequently to be depended upon. Some commonly used state variable formulations are discussed below.

STATE-SPACE REPRESENTATION USING PHYSICAL VARIABLES

The concept of state-space representation is perhaps best introduced by considering an example. We shall consider state variable formulation for a simple electrical system which is an *RLC* network shown in Fig. 12.5. The network has three energy storage elements: a capacitor C and two inductors L_1 and L_2. History of the network is completely specified by the voltage across the capacitor and currents through the inductors at $t = 0$. If we have a knowledge of initial conditions $v(0)$, $i_1(0)$, $i_2(0)$ and the input signal $e(t)$ for $t \geqslant 0$, then the behaviour of the network is completely specified for $t \geqslant 0$. However, if one (or more) of the initial conditions is not known, we are unable to determine the complete response of the network to a given input. Therefore, initial conditions $v(0)$, $i_1(0)$, $i_2(0)$ together with the input

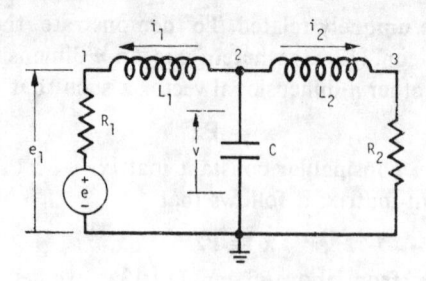

Figure 12.5 An *RLC* network.

signal $e(t)$ for $t \geqslant 0$ constitute the minimal information needed. It then follows that a *natural* selection of the state variables (but by no means the only selection) would be

$$x_1(t) = v(t)$$
$$x_2(t) = i_1(t) \quad (12.18)$$
$$x_3(t) = i_2(t)$$

The differential equations governing the behaviour of the *RLC* network are

$$i_1 + i_2 + C\frac{dv}{dt} = 0$$
$$L_1\frac{di_1}{dt} + R_1 i_1 + e - v = 0 \quad (12.19)$$
$$L_2\frac{di_2}{dt} + R_2 i_2 - v = 0$$

We are interested in expressing the variables $\frac{dv}{dt}$, $\frac{di_1}{dt}$ and $\frac{di_2}{dt}$ as linear combinations of the variables v, i_1, i_2 and e as required in representation (12.13). For this purpose, eqns. (12.19) may be rewritten as follows:

$$\frac{dv}{dt} = -\frac{1}{C}i_1 - \frac{1}{C}i_2$$
$$\frac{di_1}{dt} = \frac{1}{L_1}v - \frac{R_1}{L_1}i_1 - \frac{1}{L_1}e$$
$$\frac{di_2}{dt} = \frac{1}{L_2}v - \frac{R_2}{L_2}i_2$$

In terms of the state variables defined in (12.18) and the input $u(t) = e(t)$, we have the following equations:

$$\begin{bmatrix} \dot{x}_1 \\ \dot{x}_2 \\ \dot{x}_3 \end{bmatrix} = \begin{bmatrix} 0 & -1/C & -1/C \\ 1/L_1 & -R_1/L_1 & 0 \\ 1/L_2 & 0 & -R_2/L_2 \end{bmatrix} \begin{bmatrix} x_1 \\ x_2 \\ x_3 \end{bmatrix} + \begin{bmatrix} 0 \\ -1/L_1 \\ 0 \end{bmatrix} u$$

(12.20a)

State Variable Analysis and Design 459

Assume that voltage across R_2 and current through R_2 are the output variables y_1 and y_2 respectively. The output equations are then given by

$$\begin{bmatrix} y_1 \\ y_2 \end{bmatrix} = \begin{bmatrix} 0 & 0 & R_2 \\ 0 & 0 & 1 \end{bmatrix} \begin{bmatrix} x_1 \\ x_2 \\ x_3 \end{bmatrix} \quad (12.20b)$$

This concludes the state-space representation of the RLC network. Equations (12.20) provide the state model of the system.

Let us consider yet another example in which the electromechanical system of Fig. 12.6a is described in state variable form. Figure 12.6b gives the block diagram of the system which shows that it is a third-order system requiring three state variables to describe its dynamic behaviour. From the block diagram we choose the following set of state variables (in mechanical systems, a *natural* choice of state variables is position and speed),

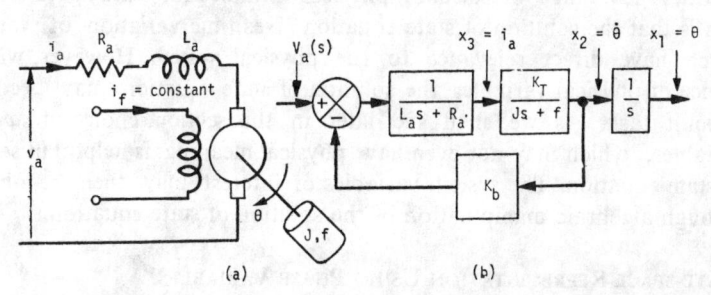

$$x_1 = \theta; \quad x_2 = \dot{\theta}; \quad \text{and} \quad x_3 = i_a$$

Figure 12.6 State variable formulation for an electromechanical system.

We can now write down the following set of three first-order differential equations, relating the inputs and outputs of the first-order factors $\dfrac{1}{s}$, $\dfrac{K_T}{Js+f}$ and $\dfrac{1}{L_a s + R_a}$:

$$\dot{x}_1 = x_2$$
$$J\dot{x}_2 + fx_2 = K_T x_3$$
$$v_a - K_b x_2 = R_a x_3 + L_a \dot{x}_3$$

These three first-order differential equations can be organized into the vector-matrix form given below which provides the state variable description of the system dynamics.

$$\begin{bmatrix} \dot{x}_1 \\ \dot{x}_2 \\ \dot{x}_3 \end{bmatrix} = \begin{bmatrix} 0 & 1 & 0 \\ 0 & -f/J & K_T/J \\ 0 & -K_b/L_a & -R_a/L_a \end{bmatrix} \begin{bmatrix} x_1 \\ x_2 \\ x_3 \end{bmatrix} + \begin{bmatrix} 0 \\ 0 \\ 1/L_a \end{bmatrix} v_a \quad (12.21a)$$

If the motor angle is regarded as the output, it can be written as

$$y = \theta = x_1 = \begin{bmatrix} 1 & 0 & 0 \end{bmatrix} \begin{bmatrix} x_1 \\ x_2 \\ x_3 \end{bmatrix} \qquad (12.21b)$$

Both the examples discussed above have one common feature; the selected state variables are the physical quantities of the systems which can be measured.

In this chapter and in Chapter 13, we will see that in a physical system, in addition to output, other state variables could be utilized for the purpose of feedback. The implementation of design with state variable feedback becomes straightforward if the state variables are available for feedback. The choice of physical variables of a system as state variables therefore helps in implementation of design.

Other advantage of selecting physical variables for state-space formulation is that the solution of state equation gives time variation of variables which have direct relevance to the physical system. However, with the choice of physical variables, the solution of state equation may become a difficult task. As we shall see later in this section, choice of *canonical* variables, which may not even have physical meaning, is helpful in solution of state equation. The system variables of interest may then be obtained through algebraic manipulation of the solution of state equation.

STATE-SPACE REPRESENTATION USING PHASE VARIABLES

In the following, we discuss an alternate state-space representation of control systems using phase variables as state variables. The phase variable state model is easily determined if the system model is already known in the differential equation/transfer function form.*

The general form of an nth-order linear differential equation relating the output $y(t)$ to the input $u(t)$ of a linear continuous-time system is**

$$y^{(n)} + a_1 y^{(n-1)} + \ldots + a_{n-1}\dot{y} + a_n y = b_0 u^{(m)} + \ldots + b_{m-1}\dot{u} + b_m u \qquad (12.22)$$

where for time-invariant systems a_i's and b_j's are constants, m and n are integers with $m \leqslant n$ and $y^{(n)} \triangleq \dfrac{d^n y}{dt^n}$.

The initial conditions are expressed in terms of $y(0), \dot{y}(0), \ldots, y^{(n-1)}(0)$.

---

*Quite often the system dynamics is det rmined experimentally using standard test signals like a step, impulse or sinusoidal signal. A transfer function is conveniently fitted to the experimental data in some best possible manner. The state model of the system is then obtained from the transfer function.

**Note that there is no loss in generality to assume the coefficient of the highest-order derivative of y to be unity.

The transfer function obtained from eqn. (12.22) under the assumption of zero initial conditions is

$$T(s) = \frac{Y(s)}{U(s)} = \frac{b_0 s^m + b_1 s^{m-1} + \ldots + b_{m-1} s + b_m}{s^n + a_1 s^{n-1} + \ldots + a_{n-1} s + a_n} \quad (12.23)$$

When* $m = n$,

$$T(s) = \frac{b_0 s^n + b_1 s^{n-1} + \ldots + b_{n-1} s + b_n}{s^n + a_1 s^{n-1} + \ldots a_{n-1} s + a_n} \quad (12.24)$$

We shall obtain a state model from the transfer function with zero initial conditions and then relax the zero initial conditions to arbitrary initial conditions.

The phase variables are defined as those particular state variables which are obtained from one of the system variables and its derivatives. Often the variable used is the system output and the remaining state variables are then derivatives of the output.

Let us first consider a simple case, where the transfer function does not have zeros. Such a transfer function has the form

$$T(s) = \frac{Y(s)}{U(s)} = \frac{b}{s^n + a_1 s^{n-1} + \ldots + a_{n-1} s + a_n} \quad (12.25)$$

To this transfer function corresponds the differential equation

$$y^{(n)} + a_1 y^{(n-1)} + \ldots + a_{n-1} \dot{y} + a_n y = bu \quad (12.26)$$

By letting

$$\begin{aligned} x_1 &= y \\ x_2 &= \dot{y} \\ &\ldots \ldots \\ x_n &= y^{(n-1)} \end{aligned} \quad (12.27)$$

equation (12.26) is reduced to a set of n first-order differential equations given below:

$$\begin{aligned} \dot{x}_1 &= x_2 \\ \dot{x}_2 &= x_3 \\ &\ldots \ldots \\ \dot{x}_{n-1} &= x_n \\ \dot{x}_n &= -a_n x_1 - a_{n-1} x_2 - \ldots - a_1 x_n + bu \end{aligned}$$

The above equations result in the following state equation:

$$\begin{bmatrix} \dot{x}_1 \\ \dot{x}_2 \\ \vdots \\ \dot{x}_{n-1} \\ \dot{x}_n \end{bmatrix} = \begin{bmatrix} 0 & 1 & 0 & \ldots & 0 \\ 0 & 0 & 1 & \ldots & 0 \\ \vdots & \vdots & \vdots & & \vdots \\ 0 & 0 & 0 & \ldots & 1 \\ -a_n & -a_{n-1} & -a_{n-2} & \ldots & -a_1 \end{bmatrix} \begin{bmatrix} x_1 \\ x_2 \\ \vdots \\ x_{n-1} \\ x_n \end{bmatrix} + \begin{bmatrix} 0 \\ 0 \\ \vdots \\ 0 \\ b \end{bmatrix} u$$

(12.28a)

*Most of the practical control schemes are realized with $m < n$; we will therefore be mainly concerned with this case. However, for the sake of generality, we will obtain a state model for the transfer function (12.24); from the general result, the state model for transfer function (12.23) with $m < n$ may then be obtained by putting appropriate coefficients b_i's equal to zero.

or
$$\dot{\mathbf{x}} = \mathbf{A}\mathbf{x} + \mathbf{B}u$$

It is to be noted that the matrix **A** has a very special form. It has all 1's in the upper off-diagonal, its last row is comprised of the negative of the coefficients of the original differential equation and all other elements are zero. This form of matrix **A** is known as the *Bush form or companion form*.

Also note that **B** has the speciality that all its elements except the last are zero. In fact **A** and **B** and therefore the state equation can be written directly by inspection of the linear differential equation.

The output being $y = x_1$, the output equation is given by
$$y = \mathbf{C}\mathbf{x} \tag{12.28b}$$
where $\mathbf{C} = [1 \ 0 \ ... \ 0]$.

The initial conditions on y give rise to the initial conditions $x_1(0)$, $x_2(0)$, ..., $x_n(0)$ on the state variables as per the definition of the state variables given in eqns. (12.27). Figure 12.7 shows the block diagram representation of the state model derived above.

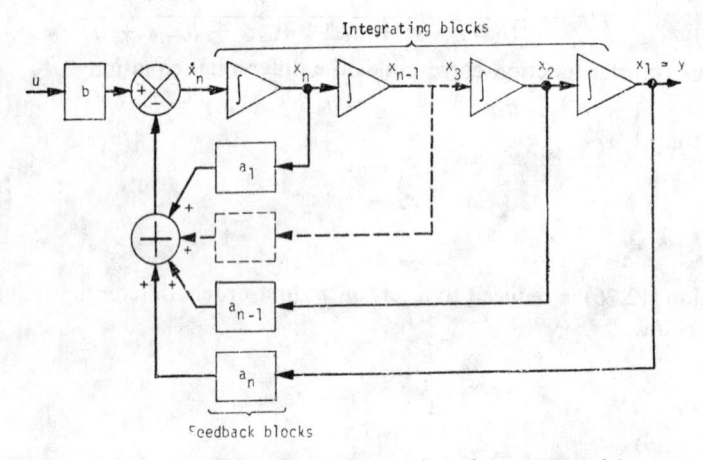

Figure 12.7 Block diagram representation of the state model given by eqns. (12.28).

It is important to note that assuming the initial time equal to zero does not create any loss of generality. Since the system under discussion is time-invariant, the change of state does not depend on initial time but depends only on the length of time during which the control force is applied.

It follows from above that for the transfer functions with poles only, the derivation of the state model through the differential equation is quite straightforward. However, when a transfer function has zeros as well, the resulting differential equation contains terms which are derivatives of the control force u and the method discussed above can no longer be applied as such. An alternate method using signal flow graphs is presented below.

In Chapter 2, it was shown that the transfer function and signal flow graphs are related by Mason's gain formula, reproduced below.

State Variable Analysis and Design

$$T(s) = \frac{1}{\Delta}\sum_k P_k \Delta_k \qquad (12.29)$$

where P_k = path gain of kth forward path; $\Delta = 1 -$ (sum of loop gains of all individual loops) + (sum of gain products of all possible combinations of two non-touching loops) − (sum of gain products of all possible combinations of three non-touching loops) + ...; and $\Delta_k =$ the value of Δ for that part of the graph not touching the kth forward path.

Let us consider the transfer function

$$\frac{Y(s)}{U(s)} = T(s) = \frac{b_0 s^3 + b_1 s^2 + b_2 s + b_3}{s^3 + a_1 s^2 + a_2 s + a_3} \qquad (12.30)$$

This is a third-order transfer function, so we identify three state variables x_1, x_2 and x_3. The signal flow graph must have at least three integrators.

Equation (12.30) may be rearranged as

$$T(s) = \frac{b_0 + b_1/s + b_2/s^2 + b_3/s^3}{1 - (-a_1/s - a_2/s^2 - a_3/s^3)} \qquad (12.31)$$

Comparing eqn. (12.31) with (12.29) we observe that a signal flow graph of eqn. (12.31) may consist of

(i) three feedback loops (touching each other) with gains

$$-a_1/s, -a_2/s^2 \text{ and } -a_3/s^3;$$

(ii) four forward paths which touch the loops and have gains

$$b_0, b_1/s, b_2/s^2 \text{ and } b_3/s^3.$$

A signal flow graph configuration which satisfies the above requirements is shown in Fig. 12.8. From this figure, we have

$$\begin{aligned}
y &= x_1 + b_0 u \\
\dot{x}_1 &= -a_1(x_1 + b_0 u) + x_2 + b_1 u \\
&= -a_1 x_1 + x_2 + (b_1 - a_1 b_0) u \\
\dot{x}_2 &= -a_2 x_1 + x_3 + (b_2 - a_2 b_0) u \\
\dot{x}_3 &= -a_3 x_1 + (b_3 - a_3 b_0) u
\end{aligned} \qquad (12.32)$$

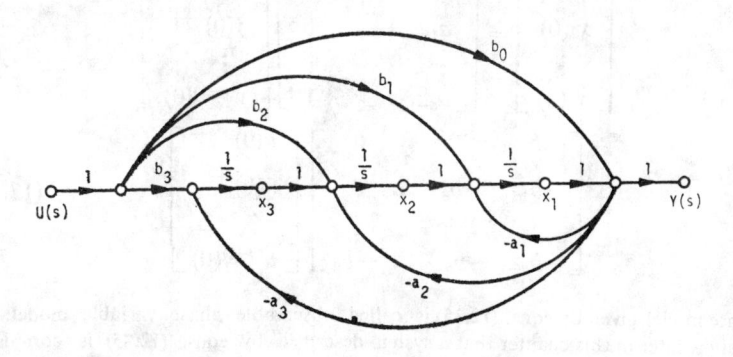

Figure 12.8 Signal flow graph of $T(s)$ of eqn. (12.31).

or
$$\begin{bmatrix} \dot{x}_1 \\ \dot{x}_2 \\ \dot{x}_3 \end{bmatrix} = \begin{bmatrix} -a_1 & 1 & 0 \\ -a_2 & 0 & 1 \\ -a_3 & 0 & 0 \end{bmatrix} \begin{bmatrix} x_1 \\ x_2 \\ x_3 \end{bmatrix} + \begin{bmatrix} b_1 - a_1 b_0 \\ b_2 - a_2 b_0 \\ b_3 - a_3 b_0 \end{bmatrix} u \quad (12.33a)$$

$$y = \begin{bmatrix} 1 & 0 & 0 \end{bmatrix} \begin{bmatrix} x_1 \\ x_2 \\ x_3 \end{bmatrix} + b_0 u \quad (12.33b)$$

The initial conditions $x_1(0)$, $x_2(0)$ and $x_3(0)$ can be obtained in terms of given initial conditions $(y(0), \dot{y}(0), \ddot{y}(0), u(0), \dot{u}(0), \ddot{u}(0))$ from (12.33). The result is given below.

$$\begin{bmatrix} x_1(0) \\ x_2(0) \\ x_3(0) \end{bmatrix} = \begin{bmatrix} 1 & 0 & 0 \\ a_1 & 1 & 0 \\ a_2 & a_1 & 1 \end{bmatrix} \begin{bmatrix} y(0) \\ \dot{y}(0) \\ \ddot{y}(0) \end{bmatrix} + \begin{bmatrix} -b_0 & 0 & 0 \\ -b_1 & -b_0 & 0 \\ -b_2 & -b_1 & -b_0 \end{bmatrix} \begin{bmatrix} u(0) \\ \dot{u}(0) \\ \ddot{u}(0) \end{bmatrix} \quad (12.34)$$

The results given in (12.33) and (12.34) can easily be generalized for an nth-order differential equation (12.22). A general state model for eqn. (12.22) with $m = n$, is given below*:

$$\begin{bmatrix} \dot{x}_1 \\ \dot{x}_2 \\ \vdots \\ \dot{x}_{n-1} \\ \dot{x}_n \end{bmatrix} = \begin{bmatrix} -a_1 & 1 & 0 & \cdots & 0 \\ -a_2 & 0 & 1 & \cdots & 0 \\ \vdots & \vdots & \vdots & & \vdots \\ -a_{n-1} & 0 & 0 & \cdots & 1 \\ -a_n & 0 & 0 & \cdots & 0 \end{bmatrix} \begin{bmatrix} x_1 \\ x_2 \\ \vdots \\ x_{n-1} \\ x_n \end{bmatrix} + \begin{bmatrix} b_1 - a_1 b_0 \\ b_2 - a_2 b_0 \\ \vdots \\ b_{n-1} - a_{n-1} b_0 \\ b_n - a_n b_0 \end{bmatrix} u$$

$$(12.35a)$$

$$y = x_1 + b_0 u \quad (12.35b)$$

$$\begin{bmatrix} x_1(0) \\ x_2(0) \\ \vdots \\ x_n(0) \end{bmatrix} = \begin{bmatrix} 1 & 0 & \cdots & 0 \\ a_1 & 1 & \cdots & 0 \\ \vdots & \vdots & & \vdots \\ a_{n-1} & a_{n-2} & \cdots & 1 \end{bmatrix} \begin{bmatrix} y(0) \\ \dot{y}(0) \\ \vdots \\ y^{(n-1)}(0) \end{bmatrix}$$

$$+ \begin{bmatrix} -b_0 & 0 & \cdots & 0 \\ -b_1 & -b_0 & \cdots & 0 \\ \vdots & \vdots & & \vdots \\ -b_{n-1} & -b_{n-2} & \cdots & -b_0 \end{bmatrix} \begin{bmatrix} u(0) \\ \dot{u}(0) \\ \vdots \\ u^{(n-1)}(0) \end{bmatrix} \quad (12.35c)$$

*State model given by eqns. (12.35) is called observable phase variable model. We shall show later in this chapter that a system described by eqns. (12.35) is completely observable.

State Variable Analysis and Design

An alternate phase variable formulation for the transfer function of eqn. (12.30) may be obtained as follows:

The transfer function $G(s)$ of Fig. 12.9 may be divided into two parts as shown in Fig. 12.10. Thus

$$G(s) = \frac{Y(s)}{U(s)} = \frac{X_1(s)}{U(s)} \cdot \frac{Y(s)}{X_1(s)} \tag{12.36}$$

where

$$\frac{X_1(s)}{U(s)} = \frac{1}{s^3 + a_1 s^2 + a_2 s + a_3} \tag{12.37}$$

and

$$\frac{Y(s)}{X_1(s)} = b_0 s^3 + b_1 s^2 + b_2 s + b_3 \tag{12.38}$$

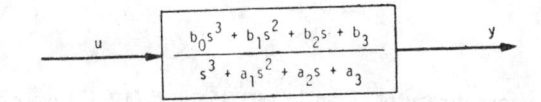

Figure 12.9 A third-order system.

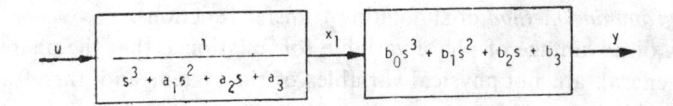

Figure 12.10 An alternate representation of system of Fig. 12.9.

The transfer function of eqn. (12.37) is without zeros and therefore its phase variable representation may be obtained from eqns. (12.28).

$$\begin{bmatrix} \dot{x}_1 \\ \dot{x}_2 \\ \dot{x}_3 \end{bmatrix} = \begin{bmatrix} 0 & 1 & 0 \\ 0 & 0 & 1 \\ -a_3 & -a_2 & -a_1 \end{bmatrix} \begin{bmatrix} x_1 \\ x_2 \\ x_3 \end{bmatrix} + \begin{bmatrix} 0 \\ 0 \\ 1 \end{bmatrix} u \tag{12.39a}$$

From transfer function of eqn. (12.38), we get

$$\begin{aligned} y &= b_0 \dddot{x}_1 + b_1 \ddot{x}_1 + b_2 \dot{x}_1 + b_3 x_1 \\ &= b_0(-a_3 x_1 - a_2 x_2 - a_1 x_3 + u) + b_1 x_3 + b_2 x_2 + b_3 x_1 \\ &= (b_3 - a_3 b_0) x_1 + (b_2 - a_2 b_0) x_2 + (b_1 - a_1 b_0) x_3 + b_0 u \end{aligned} \tag{12.39b}$$

The results given in eqns. (12.39) can easily be generalized for an nth-order differential equation (12.22). A general state model for eqn. (12.22) with $m = n$ is given below*.

---

*State model given by eqns. (12.40) is called controllable phase variable model. We shall show later in this chapter that a system described by eqns. (12.40) is completely controllable.

$$\begin{bmatrix} \dot{x}_1 \\ \dot{x}_2 \\ \vdots \\ \dot{x}_{n-1} \\ \dot{x}_n \end{bmatrix} = \begin{bmatrix} 0 & 1 & 0 & \cdots & 0 \\ 0 & 0 & 1 & \cdots & 0 \\ \vdots & \vdots & \vdots & & \vdots \\ 0 & 0 & 0 & \cdots & 1 \\ -a_n & -a_{n-1} & -a_{n-2} & \cdots & -a_1 \end{bmatrix} \begin{bmatrix} x_1 \\ x_2 \\ \vdots \\ x_{n-1} \\ x_n \end{bmatrix} + \begin{bmatrix} 0 \\ 0 \\ \vdots \\ 0 \\ 1 \end{bmatrix} u$$

(12.40a)

$$y = [(b_n - a_n b_0)\ (b_{n-1} - a_{n-1} b_0)\ \cdots\ (b_2 - a_2 b_0)\ (b_1 - a_1 b_0)] \begin{bmatrix} x_1 \\ x_2 \\ \vdots \\ x_{n-1} \\ x_n \end{bmatrix} + b_0 u$$

(12.40b)

From the general results given in (12.35) and (12.40) we observe that the phase variable formulation can be obtained by inspection from the transfer function and vice-versa. The reader will also note the similarity of phase variable representation as illustrated by Figs. 12.7 and 12.8 with the *direct programming method* of simulating transfer functions.

A disadvantage of phase variable formulation is that the phase variables, in general, are not physical variables of the system and therefore are not available for measurement and control purposes. We have seen that if $G(s)$ has no zeros, the phase variables are given by its output and derivatives of output (eqns. (12.27)). Unfortunately, it becomes difficult to take second or higher derivatives of output. If $G(s)$ has zeros, then the phase variables bear little resemblance to real physical quantities in the system as is seen from eqns. (12.35). Thus, though phase variables are simple to realize mathematically, they are not a practical set of state variables from measurement and control point of view. From the analysis point of view also, phase variables do not offer any advantage. As we shall see shortly, canonical variables are most suitable from analysis point of view.

In spite of these disadvantages, phase variables provide a powerful method of state variable formulation. A link between the transfer function design approach and time-domain design approach is established through phase variables. We shall discuss this link later in this chapter.

STATE-SPACE REPRESENTATION USING CANONICAL VARIABLES

In *canonical-variable* or *normal-form* representation of a system, the matrix A turns out to be a diagonal matrix. This form of state model plays an important role in control theory. Let us derive this model for a general transfer function (12.24), reproduced below:

$$\frac{Y(s)}{U(s)} = T(s) = \frac{b_0 s^n + b_1 s^{n-1} + \cdots + b_{n-1} s + b_n}{s^n + a_1 s^{n-1} + \cdots + a_{n-1} s + a_n}$$

Assume that the denominator is known in factored form and that the poles of the transfer function, located at $\lambda_1, \lambda_2, ..., \lambda_n$, are all distinct. The transfer function can then be expanded into partial fractions as (see Appendix I)

$$\frac{Y(s)}{U(s)} = T(s) = b_0 + \sum_{i=1}^{n} \frac{c_i}{s - \lambda_i} \qquad (12.41)$$

where c_i are the residues of the poles at $s = \lambda_i$.

The block diagram model for the transfer function is shown in Fig. 12.11. Defining the output of each integrator to be a state variable, we can write the state equations as

$$\dot{x}_i = \lambda_i x_i + u; \quad i = 1, 2, ..., n \qquad (12.42a)$$

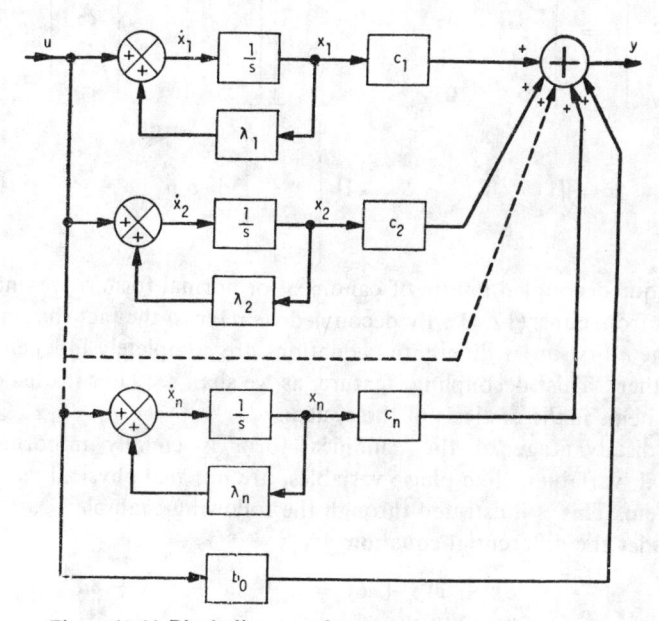

Figure 12.11 Block diagram of a canonical state model.

The output $y(t)$ is given by

$$y = c_1 x_1 + c_2 x_2 + ... + c_n x_n + b_0 u \qquad (12.42b)$$

Equations (12.42) describe the canonical state model of the transfer function of eqn. (12.41). This state model can be expressed in the vector-matrix form as

$$\begin{bmatrix} \dot{x}_1 \\ \dot{x}_2 \\ \vdots \\ \dot{x}_n \end{bmatrix} = \begin{bmatrix} \lambda_1 & 0 & 0 & \cdots & 0 \\ 0 & \lambda_2 & 0 & \cdots & 0 \\ \vdots & \vdots & \vdots & & \vdots \\ 0 & 0 & 0 & \cdots & \lambda_n \end{bmatrix} \begin{bmatrix} x_1 \\ x_2 \\ \vdots \\ x_n \end{bmatrix} + \begin{bmatrix} 1 \\ 1 \\ \vdots \\ 1 \end{bmatrix} u \quad (12.43a)$$

$$y = [c_1 \quad c_2 \quad \cdots \quad c_n] \begin{bmatrix} x_1 \\ x_2 \\ \vdots \\ x_n \end{bmatrix} + b_0 u \qquad (12.43b)$$

It is observed that for the canonical state model described above, *the system matrix* **A** *is a diagonal matrix with the poles of T(s) as its diagonal elements*. It is also observed that elements of column vector **B** of the canonical state model are all unity and the elements of the row vector **C** are the residues of the system poles. An alternate canonical state model is given below.

$$\begin{bmatrix} \dot{x}_1 \\ \dot{x}_2 \\ \vdots \\ \dot{x}_n \end{bmatrix} = \begin{bmatrix} \lambda_1 & 0 & \cdots & 0 \\ 0 & \lambda_2 & \cdots & 0 \\ \vdots & \vdots & & \vdots \\ 0 & 0 & \cdots & \lambda_n \end{bmatrix} \begin{bmatrix} x_1 \\ x_2 \\ \vdots \\ x_n \end{bmatrix} + \begin{bmatrix} c_1 \\ c_2 \\ \vdots \\ c_n \end{bmatrix} u \qquad (12.44a)$$

$$y = [1 \quad 1 \quad \cdots \quad 1] \begin{bmatrix} x_1 \\ x_2 \\ \vdots \\ x_n \end{bmatrix} + b_0 u \qquad (12.44b)$$

The unique decoupled nature of canonical or normal-form representation is obvious from eqns. (12.43). By decoupled we refer to the fact that in normal form, the *n* first-order differential equations are completely independent of each other. This decoupling feature, as we shall see later in this chapter, greatly helps in the analysis of the system.

The disadvantage of the canonical form is equally important. The canonical variables, like phase variables, are not real physical variables of the system. This is illustrated through the following example:

Consider the differential equation

$$\dddot{y} + 6\ddot{y} + 11\dot{y} + 6y = \dddot{u} + 8\ddot{u} + 17\dot{u} + 8u$$

Assuming zero initial conditions, we get

$$\frac{Y(s)}{U(s)} = T(s) = \frac{s^3 + 8s^2 + 17s + 8}{s^3 + 6s^2 + 11s + 6} \qquad (12.45)$$

$$= \frac{s^3 + 8s^2 + 17s + 8}{(s+1)(s+2)(s+3)}$$

$$= 1 - \frac{1}{s+1} + \frac{2}{s+2} + \frac{1}{s+3} \qquad (12.46a)$$

The block diagram simulation for $T(s)$ of eqn. (12.46a) is shown in Fig. 12.12.

From eqn. (12.45), we can write the following:

$$\frac{sY(s)}{U(s)} = s + 2 + \frac{1}{s+1} - \frac{4}{s+2} - \frac{3}{s+3} \qquad (12.46b)$$

$$\frac{s^2 Y(s)}{U(s)} = s^2 + 2s - 6 - \frac{1}{s+1} + \frac{8}{s+2} + \frac{9}{s+3} \qquad (12.46c)$$

State Variable Analysis and Design

Equations (12.46) may be expressed as (refer Fig. 12.12)

$$\begin{bmatrix} \dfrac{Y(s) - U(s)}{U(s)} \\ \dfrac{sY(s) - sU(s) - 2U(s)}{U(s)} \\ \dfrac{s^2Y(s) - s^2U(s) - 2sU(s) + 6U(s)}{U(s)} \end{bmatrix} = \begin{bmatrix} -1 & 2 & 1 \\ 1 & -4 & -3 \\ -1 & 8 & 9 \end{bmatrix} \begin{bmatrix} \dfrac{1}{s+1} \\ \dfrac{1}{s+2} \\ \dfrac{1}{s+3} \end{bmatrix}$$

$$= \begin{bmatrix} -1 & 2 & 1 \\ 1 & -4 & -3 \\ -1 & 8 & 9 \end{bmatrix} \begin{bmatrix} \dfrac{X_1(s)}{U(s)} \\ \dfrac{X_2(s)}{U(s)} \\ \dfrac{X_3(s)}{U(s)} \end{bmatrix}$$

This gives

$$\begin{bmatrix} y - u \\ \dot{y} - \dot{u} - 2u \\ \ddot{y} - \ddot{u} - 2\dot{u} + 6u \end{bmatrix} = \begin{bmatrix} -1 & 2 & 1 \\ 1 & -4 & -3 \\ -1 & 8 & 9 \end{bmatrix} \begin{bmatrix} x_1 \\ x_2 \\ x_3 \end{bmatrix} \quad (12.47)$$

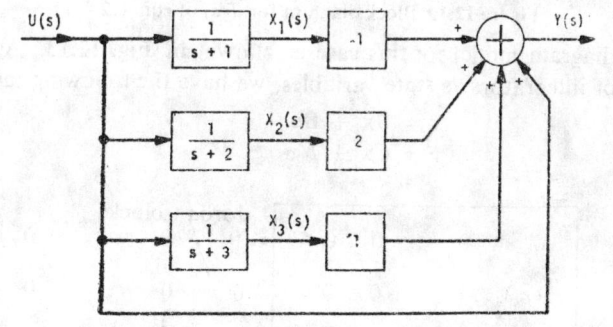

Figure 12.12 Block diagram for $T(s)$ of eqn. (12.46a).

From eqns. (12.47) we find that canonical variables are not physical variables of the system and therefore are not available for measurement and control. (Note that the initial conditions $x_1(0)$, $x_2(0)$, $x_3(0)$ may be obtained in terms of specified initial conditions $y(0)$, $\dot{y}(0)$, $\ddot{y}(0)$, $u(0)$, $\dot{u}(0)$, $\ddot{u}(0)$ from eqns. (12.47).)

The state model (12.44) assumes that the transfer function $T(s)$ has all simple (distinct) poles. The partial fraction technique can still be used if $T(s)$ has some poles that are not simple. To illustrate the method, let us assume that $T(s)$ has a pole of order r at $s = \lambda_1$ and simple poles at $\lambda_{r+1}, ..., \lambda_n$. Then $T(s)$ can be expressed as (see Appendix I),

$$T(s) = b_0 + \sum_{i=1}^{r} \frac{c_i}{(s - \lambda_1)^{r-i+1}} + \sum_{i=r+1}^{n} \frac{c_i}{(s - \lambda_i)} \quad (12.48)$$

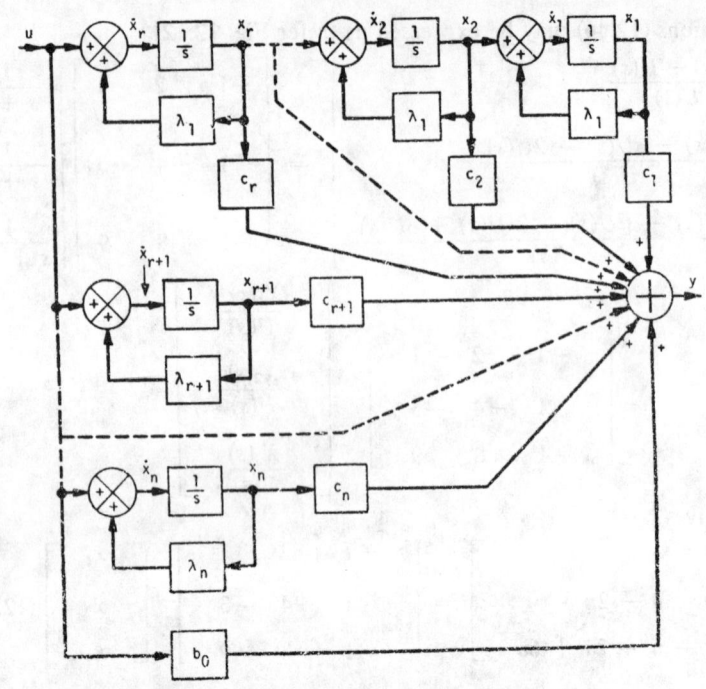

Figure 12.13 Block diagram for $T(s)$ of eqn. (12.48).

A block diagram model for this case is shown in Fig. 12.13. Again with outputs of integrators as state variables, we have the following model:

$$\dot{x} = Jx + Bu \tag{12.49a}$$
$$y = Cx + b_0 u \tag{12.49b}$$

where

$$J = \begin{bmatrix} \lambda_1 & 1 & 0 & \cdots & 0 & 0 & 0 & 0 & \cdots & 0 \\ 0 & \lambda_1 & 1 & \cdots & 0 & 0 & 0 & 0 & \cdots & 0 \\ \vdots & & & & \vdots & \vdots & \vdots & \vdots & & \vdots \\ 0 & 0 & 0 & \cdots & \lambda_1 & 1 & 0 & 0 & \cdots & 0 \\ 0 & 0 & 0 & \cdots & 0 & \lambda_1 & 0 & 0 & \cdots & 0 \\ 0 & 0 & 0 & \cdots & 0 & 0 & \lambda_{r+1} & 0 & \cdots & 0 \\ \vdots & & & & & & & & & \vdots \\ 0 & 0 & 0 & \cdots & 0 & 0 & 0 & 0 & \cdots & \lambda_n \end{bmatrix}$$

$$B = [0 \ 0 \ \ldots \ 0 \ 1 \ 1 \ \ldots \ 1]^T$$
$\qquad\qquad\qquad\quad \uparrow$
$\qquad\qquad\qquad$ rth element

$$C = [c_1 \ c_2 \ \ldots \ c_r \ c_{r+1} \ \ldots \ c_n]$$

Matrix J, called *Jordan canonical matrix*, has the following properties:

(i) The diagonal elements of J are poles of $T(s)$.
(ii) All the elements below the principal diagonal are zero.
(iii) A certain number of unit elements are contained immediately to the right of principal diagonal when the adjacent elements in the principal diagonal are equal.

Equations (12.49) give the Jordan canonical state model for the transfer function of eqn. (12.48). Note that column vector B has unit elements corresponding to last row of the *Jordan blocks* and zero elements corresponding to all other rows of Jordan blocks.

DERIVATION OF TRANSFER FUNCTION FROM STATE MODEL

Having answered the question of obtaining the state model for a given transfer function, we next consider the problem of determining the transfer function from a given state model of single-input-single-output systems.

The link between transfer function and phase variable formulation has already been demonstrated. The transfer function from a state model using phase variables can be obtained by inspection from the general results (12.28), (12.35) or (12.40). An alternate method is to draw signal flow graph for the given phase variable model and then obtain the transfer function using Mason's gain formula.

For a general state model

$$\dot{x} = Ax + Bu \qquad (12.50a)$$
$$y = Cx + du \qquad (12.50b)$$

the transfer function may be obtained as follows:

Taking the Laplace transform of eqns. (12.50), we have

$$sX(s) - x(0) = AX(s) + BU(s)$$
$$Y(s) = CX(s) + dU(s)$$

Solving for $Y(s)$, we obtain

$$Y(s) = C(sI - A)^{-1}x(0) + C(sI - A)^{-1}BU(s) + dU(s)$$

Assuming zero initial conditions, we get the system transfer function as

$$T(s) = \frac{Y(s)}{U(s)} = C(sI - A)^{-1}B + d$$
$$= \frac{C \operatorname{adj}(sI - A)B}{\det(sI - A)} + d \qquad (12.51)$$

An important observation that needs to be made here is that while the state model is nonunique, *the transfer function of the system is unique*, i.e., the transfer function of eqn. (12.51) must work out to be the same irrespective of which particular state model is used to describe the system.

12.4 Diagonalization

We observed in earlier sections that the state model of a system is not unique. Some of the state models presented employed physical variables, phase variables and canonical variables. From application point of view, the physical variables for system representation are most useful as the resulting state variables are real physical variables which can be easily measured and used for control purposes. However, the corresponding state model in this case is generally not convenient for investigation of system properties and evaluation of time response. The canonical state model wherein **A** is in diagonal form is most suitable for this purpose. It is, therefore, useful to study techniques for transforming a general state model into a canonical one. These techniques are often referred to as *diagonalization* techniques.

Consider an nth-order multi-input-multi-output state model

$$\dot{x} = Ax + Bu \quad (12.52a)$$

$$y = Cx + Du \quad (12.52b)$$

Assume that the matrix **A** in this model is nondiagonal. Let us define a new state vector **z** such that

$$x = Mz$$

where **M** is a $(n \times n)$ nonsingular constant matrix. Under this transformation, the original state model modifies to

$$\dot{z} = M^{-1}AMz + M^{-1}Bu \quad (12.53a)$$

$$y = CMz + Du \quad (12.53b)$$

If the matrix **M** can be selected such that $M^{-1}AM$ is a diagonalized matrix, Λ, then the model given by (12.53) is a canonical state model. Under this condition, the matrix **M** is called the *diagonalizing matrix* or the *modal matrix*. Equations (12.53) may be written as

$$\dot{z} = \Lambda z + \tilde{B}u \quad (12.54a)$$

$$y = \tilde{C}z + Du \quad (12.54b)$$

where $\Lambda = M^{-1}AM$ = a diagonal matrix; $\tilde{B} = M^{-1}B$; and $\tilde{C} = CM$.

The determination of the diagonalizing matrix is facilitated by the use of eigenvectors. With this purpose in mind, the concept of eigenvalues and eigenvectors is briefly reviewed below. Refer DeRusso [1] for further details.

Eigenvalues and Eigenvectors

Consider a set of homogeneous equations

$$\begin{aligned} f_{11}x_1 + f_{12}x_2 + \ldots + f_{1n}x_n &= 0 \\ f_{21}x_1 + f_{22}x_2 + \ldots + f_{2n}x_n &= 0 \\ \ldots \quad \ldots \quad \ldots \quad \ldots \quad \ldots & \\ f_{n1}x_1 + f_{n2}x_2 + \ldots + f_{nn}x_n &= 0 \end{aligned} \quad (12.55)$$

or
$$\begin{bmatrix} f_{11} & f_{12} & \cdots & f_{1n} \\ f_{21} & f_{22} & \cdots & f_{2n} \\ \vdots & \vdots & & \vdots \\ f_{n1} & f_{n2} & \cdots & f_{nn} \end{bmatrix} \begin{bmatrix} x_1 \\ x_2 \\ \vdots \\ x_n \end{bmatrix} = \begin{bmatrix} 0 \\ 0 \\ \vdots \\ 0 \end{bmatrix}$$

or
$$\mathbf{Fx} = 0 \qquad (12.56)$$

Equation (12.56) may be expressed as

$$\mathbf{f}_1 x_1 + \mathbf{f}_2 x_2 + \ldots + \mathbf{f}_n x_n = 0 \qquad (12.57)$$

where $\mathbf{f}_i$ is the ith column vector of matrix $\mathbf{F}$.

From eqn. (12.56), we get the solution vector

$$\mathbf{x} = \mathbf{F}^{-1} \cdot 0 = \frac{\mathrm{Adj}(\mathbf{F})}{|\mathbf{F}|} \cdot 0$$

If $|\mathbf{F}| \neq 0$ (i.e., rank r of matrix $\mathbf{F}$ is n), then eqns. (12.55) have only the trivial solution $\mathbf{x} = 0$, i.e., $x_1 = 0$, $x_2 = 0$, ..., $x_n = 0$. Vectors $\mathbf{f}_i$ in (12.57) are then said to be *linearly independent*. If the determinant $|\mathbf{F}|$ does vanish ($r < n$), two or more vectors $\mathbf{f}_i$ are linearly related. In this case, the set of eqns. (12.55) has infinite solutions as demonstrated below:

For a 2×2 matrix

$$\mathbf{F} = \begin{bmatrix} f_{11} & f_{12} \\ f_{21} & f_{22} \end{bmatrix}$$

we have
$$|\mathbf{F}| = \sum_{i=1}^{2} f_{ji} C_{ji}; \quad j = 1 \text{ or } 2 \qquad (12.58a)$$

$$= \sum_{i=1}^{2} f_{ij} C_{ij}; \quad j = 1 \text{ or } 2 \qquad (12.58b)$$

where C_{ij} = co-factor of f_{ij}.

Further,
$$\sum_{i=1}^{2} f_{ji} C_{ki} = \sum_{i=1}^{2} f_{ij} C_{ik} = 0 \text{ for } j \neq k \qquad (12.58c)$$

From eqns. (12.58), we may write for $n \times n$ matrix $\mathbf{F}$,

$$\sum_{i=1}^{n} f_{ji} C_{ki} = |\mathbf{F}| \delta_{jk}; \quad j = 1, 2, \ldots, \text{ or } n$$

$$\sum_{i=1}^{n} f_{ij} C_{ik} = |\mathbf{F}| \delta_{jk}; \quad j = 1, 2, \ldots, \text{ or } n$$

where
$$\delta_{jk} = 1 \quad \text{when } j = k$$
$$= 0 \quad \text{when } j \neq k$$

When $|\mathbf{F}| = 0$, then

$$\sum_{i=1}^{n} f_{ji} C_{ki} = \sum_{i=1}^{n} f_{ij} C_{ik} = 0 \text{ for all values of } j. \qquad (12.59)$$

Consider jth equation from the set of eqns. (12.55)

$$f_{j1} x_1 + f_{j2} x_2 + \ldots + f_{jn} x_n = 0 \qquad (12.60)$$

Let $x_1 = pC_{k1}$, $x_2 = pC_{k2}$, ..., $x_n = pC_{kn}$; where p is a constant. Then from eqn. (12.60), we have

$$p(f_{j1}C_{k1} + f_{j2}C_{k2} + ... + f_{jn}C_{kn}) = 0 \qquad (12.61)$$

The term inside the brackets is equal to zero for all values of j (eqn. (12.59)).

Thus
$$\mathbf{x} = p \begin{bmatrix} C_{k1} \\ C_{k2} \\ \vdots \\ C_{kn} \end{bmatrix} \qquad (12.62)$$

is a solution of eqns. (12.55). We therefore observe that the co-factors C_{k1}, C_{k2}, ..., C_{kn} taken along the kth row yield a solution of eqns. (12.55). Obviously the solution is nonunique because k can have any value and p is an arbitrary constant. If for a particular choice of k(1 to n), all the co-factors turn out to be zero, we then get a trivial solution. Nontrivial solution could then be obtained by taking the co-factors along another row.

Consider for example
$$\mathbf{F} = \begin{bmatrix} 1 & 1 & 0 \\ 0 & 1 & 1 \\ -6 & -11 & -5 \end{bmatrix}$$

Rank r of this matrix is 2. The solution of homogeneous equations

$$\mathbf{Fx} = \mathbf{0}$$

is obtained from

$$\begin{aligned} x_1 + x_2 &= 0 \\ x_2 + x_3 &= 0 \\ -6x_1 - 11x_2 - 5x_3 &= 0 \end{aligned} \qquad (12.63)$$

There are many sets of values of x_i ($i = 1, 2, 3$) which satisfy eqns. (12.63). However, every set of values of x_i yields the unique direction of the solution vector $\mathbf{x}$. There is only one independent solution. If we assume $x_1 = 1$, we get

$$x_2 = -1 \text{ and } x_3 = 1.$$

We can easily verify that in general the number of independent solutions for (12.56) is $(n - r)$.

Let us now turn our attention to the equation

$$\mathbf{Ax} = \mathbf{y}$$

and view it as transformation of $n \times 1$ vector $\mathbf{x}$ to $n \times 1$ vector $\mathbf{y}$ by $n \times n$ matrix operator $\mathbf{A}$. The question which we will like to answer here is: whether there exists a vector $\mathbf{x}$ such that matrix operator $\mathbf{A}$ transforms it to a vector $\lambda \mathbf{x}$ (λ is a constant), i.e., to a vector having the same direction in state space as the vector $\mathbf{x}$. Such a vector $\mathbf{x}$ is a solution of the equation

$$\mathbf{Ax} = \lambda \mathbf{x} \qquad (12.64)$$

Now our question is about the existence of solution of eqn. (12.64), which may be written as

$$(\lambda I - A)x = 0 \qquad (12.65)$$

The set of homogeneous eqns. (12.65) have a nontrivial solution if and only if

$$|\lambda I - A| = 0 \qquad (12.66a)$$

This equation may be expressed in expanded form as

$$q(\lambda) = \lambda^n + a_1\lambda^{n-1} + a_2\lambda^{n-2} + \ldots + a_n = 0 \qquad (12.66b)$$

The values of λ for which eqn. (12.66) is satisfied are called *eigenvalues* of matrix A and eqn. (12.66) is called the *characteristic equation* corresponding to matrix A.

From eqn. (12.51) it is observed that the poles of the transfer function $T(s)$ are given by $q(s) = |sI - A| = 0$. This in fact is the same equation as (12.66) which determines the eigenvalues of A. It is therefore concluded that the eigenvalues of the state model and the poles of the system transfer function are the same. All the conclusions established in Chapter 6 on system stability based upon the location of transfer function poles (i.e., roots of the characteristic equation) are therefore valid for eigenvalues of the state model. Thus a state model is stable if all its eigenvalues have negative real parts. (In Chapter 14, we shall study stability in details.)

Now for $\lambda = \lambda_i$ satisfying eqn. (12.66), we have from eqn. (12.65),

$$(\lambda_i I - A)x = 0 \qquad (12.67)$$

Let $x = m_i$ be the solution of this equation. The solution m_i is called the *eigenvector* of A associated with eigenvalue λ_i.

As said earlier, solution of (12.67) depends on the rank of matrix $(\lambda_i I - A)$. If r is the rank of this matrix, then there are $(n - r)$ independent solutions (eigenvectors).

If the eigenvalues of matrix A are all distinct, then the rank* r of matrix

$$*|\lambda I - A|_{\lambda = \lambda_i} = \begin{vmatrix} \lambda_i - a_{11} & -a_{12} & \ldots & -a_{1n} \\ -a_{31} & \lambda_i - a_{22} & \ldots & -a_{2n} \\ \vdots & \vdots & & \vdots \\ -a_{n1} & -a_{n2} & \ldots & \lambda_i - a_{nn} \end{vmatrix}$$

$$= (\lambda_i - a_{11})C_{11} - a_{12}C_{12} \ldots -a_{1n}C_{1n}$$

$$\frac{\partial}{\partial \lambda_i}\left\{|\lambda I - A|_{\lambda = \lambda_i}\right\} = C_{11}$$

$$\neq 0 \text{ if } r = n - 1 \text{ (Note that } r \neq n)$$
$$= 0 \text{ if } r = n - 2, n - 3, \ldots$$

If $\lambda_1, \lambda_2, \ldots, \lambda_n$ are the eigenvalues of matrix A, then the characteristic polynomial is

$$|\lambda I - A| = (\lambda - \lambda_1)(\lambda - \lambda_2) \ldots (\lambda - \lambda_i) \ldots (\lambda - \lambda_n)$$

Now

$$\frac{\partial}{\partial \lambda_i}\left\{|\lambda I - A|_{\lambda = \lambda_i}\right\} \neq 0 \text{ if } \lambda = \lambda_i \text{ is not repeated}$$

$$= 0 \text{ if } \lambda = \lambda_i \text{ is repeated}$$

Hence if the eigenvalues of A are all distinct, then the rank of $(\lambda I - A)$ is $(n - 1)$.

$(\lambda \mathbf{I} - \mathbf{A})$ is $(n - 1)$ and hence we have only one independent eigenvector corresponding to any particular eigenvalue λ_l. This eigenvector, as observed earlier, may be obtained by taking co-factors of matrix $(\lambda_l \mathbf{I} - \mathbf{A})$ along any row, i.e.,

$$\mathbf{m}_l = \begin{bmatrix} C_{k1} \\ C_{k2} \\ \vdots \\ C_{kn} \end{bmatrix}; k = 1, 2, ..., \text{ or } n$$

where C_{kj} are the co-factors of matrix $(\lambda_l \mathbf{I} - \mathbf{A})$.

Let $\mathbf{m}_1, \mathbf{m}_2, ..., \mathbf{m}_n$ be the eigenvectors corresponding to the eigenvalues $\lambda_1, \lambda_2, ..., \lambda_n$ respectively. Then we have

$$\begin{aligned} \mathbf{AM} &= \mathbf{A}[\mathbf{m}_1 \vdots \mathbf{m}_2 \vdots ... \vdots \mathbf{m}_n] \\ &= [\mathbf{Am}_1 \vdots \mathbf{Am}_2 \vdots ... \vdots \mathbf{Am}_n] \\ &= [\lambda_1 \mathbf{m}_1 \vdots \lambda_2 \mathbf{m}_2 \vdots ... \vdots \lambda_n \mathbf{m}_n] \\ &= \mathbf{M}\Lambda \end{aligned}$$

where

$$\Lambda = \begin{bmatrix} \lambda_1 & 0 & ... & 0 \\ 0 & \lambda_2 & ... & 0 \\ \vdots & \vdots & & \vdots \\ 0 & 0 & ... & \lambda_n \end{bmatrix}$$

$$= \mathbf{M}^{-1}\mathbf{AM}$$

The matrix constructed by placing the eigenvectors (columns) together is therefore a diagonalizing or modal matrix $\mathbf{M}$ of $\mathbf{A}$, i.e.,

$$\mathbf{M} = [\mathbf{m}_1 \vdots \mathbf{m}_2 \vdots ... \vdots \mathbf{m}_n]$$

Note that $\mathbf{A}$ and Λ have the same characteristic equation; therefore the eigenvalues are invariant under this transformation.

When $\mathbf{A}$ is expressed in the form given below,

$$\mathbf{A} = \begin{bmatrix} 0 & 1 & 0 & ... & 0 \\ 0 & 0 & 1 & ... & 0 \\ \vdots & \vdots & \vdots & & \vdots \\ 0 & 0 & 0 & ... & 1 \\ -a_n & -a_{n-1} & -a_{n-2} & ... & -a_1 \end{bmatrix} \quad (12.68a)$$

then the modal matrix can be shown to be a special matrix (called the *Vander Monde matrix*)

$$V = \begin{bmatrix} 1 & 1 & \cdots & 1 \\ \lambda_1 & \lambda_2 & \cdots & \lambda_n \\ \lambda^2 & \lambda_2^2 & \cdots & \lambda_n^2 \\ \vdots & \vdots & & \vdots \\ \lambda_1^{n-1} & \lambda_2^{n-2} & \cdots & \lambda_n^{n-1} \end{bmatrix} \quad (12.68b)$$

Example 12.1: Consider a matrix **A** given below:

$$A = \begin{bmatrix} 0 & 1 & 0 \\ 3 & 0 & 2 \\ -12 & -7 & -6 \end{bmatrix}$$

Corresponding to this matrix, the characteristic equation is

$$|\lambda I - A| = \begin{vmatrix} \lambda & -1 & 0 \\ -3 & \lambda & -2 \\ 12 & 7 & \lambda+6 \end{vmatrix} = 0$$

or

$$(\lambda + 1)(\lambda + 2)(\lambda + 3) = 0$$

Therefore the eigenvalues of matrix **A** are

$$\lambda_1 = -1, \lambda_2 = -2, \lambda_3 = -3$$

The eigenvector m_1 associated with $\lambda_1 = -1$ is obtained from co-factors of the matrix

$$(\lambda_1 I - A) = \begin{bmatrix} -1 & -1 & 0 \\ -3 & -1 & -2 \\ 12 & 7 & 5 \end{bmatrix}$$

The result is (eigenvector has unique direction)

$$m_1 = \begin{bmatrix} C_{11} \\ C_{12} \\ C_{13} \end{bmatrix} = \begin{bmatrix} 9 \\ -9 \\ -9 \end{bmatrix} \text{ or } m_1 = \begin{bmatrix} 1 \\ -1 \\ -1 \end{bmatrix}$$

The eigenvector m_1 could also be obtained from a solution of the homogeneous equations

$$-x_1 - x_2 = 0$$
$$-3x_1 - x_2 - 2x_3 = 0$$
$$12x_1 + 7x_2 + 5x_3 = 0$$

Choosing $x_1 = 1$, we get $x_2 = -1$, $x_3 = -1$, which is the same result as obtained by the method of co-factors. Solution of homogeneous equations is computationally more efficient than obtaining co-factors of a row when the dimension of **A** is large.

Similarly eigenvectors $\mathbf{m}_1$ and $\mathbf{m}_2$ associated with $\lambda_2 = -2$ and $\lambda_3 = -3$ respectively are

$$\mathbf{m}_2 = \begin{bmatrix} 2 \\ -4 \\ 1 \end{bmatrix} ; \quad \mathbf{m}_3 = \begin{bmatrix} 1 \\ -3 \\ 3 \end{bmatrix}$$

The modal matrix **M** obtained by placing the eigenvectors (columns) together is given by

$$\mathbf{M} = [\mathbf{m}_1 \ \vdots \ \mathbf{m}_2 \ \vdots \ \mathbf{m}_3] = \begin{bmatrix} 1 & 2 & 1 \\ -1 & -4 & -3 \\ -1 & 1 & 3 \end{bmatrix}$$

from which we find

$$\mathbf{M}^{-1} = \tfrac{1}{2} \begin{bmatrix} 9 & 5 & 2 \\ -6 & -4 & -2 \\ 5 & 3 & 2 \end{bmatrix}$$

Therefore

$$\mathbf{\Lambda} = \mathbf{M}^{-1}\mathbf{A}\mathbf{M} = \tfrac{1}{2} \begin{bmatrix} 9 & 5 & 2 \\ -6 & -4 & -2 \\ 5 & 3 & 2 \end{bmatrix} \begin{bmatrix} 0 & 1 & 0 \\ 3 & 0 & 2 \\ -12 & -7 & -6 \end{bmatrix} \begin{bmatrix} 1 & 2 & 1 \\ -1 & -4 & -3 \\ -1 & 1 & 3 \end{bmatrix}$$

$$= \begin{bmatrix} -1 & 0 & 0 \\ 0 & -2 & 0 \\ 0 & 0 & -3 \end{bmatrix}$$

which is a diagonal matrix with eigenvalues of **A** as its diagonal elements. In fact $\mathbf{\Lambda}$ could be written directly without the need to compute $\mathbf{M}^{-1}\mathbf{A}\mathbf{M}$.

Generalized Eigenvectors

In our discussion on eigenvectors so far we have assumed that the eigenvalues of matrix **A** are all distinct. Let us now relax this restriction. Consider the matrix

$$\mathbf{A} = \begin{bmatrix} 4 & 1 & -2 \\ 1 & 0 & 2 \\ 1 & -1 & 3 \end{bmatrix}$$

The eigenvalues of this matrix are $\lambda_1 = 1, \lambda_2 = 3, \lambda_3 = 3$. The eigenvector associated with $\lambda = 1$ may be obtained from the row co-factors of the matrix

$$(\lambda_1 I - A) = \begin{bmatrix} -3 & -1 & 2 \\ -1 & 1 & -2 \\ -1 & 1 & -2 \end{bmatrix}$$

$$\mathbf{m}_1 = \begin{bmatrix} C_{11} \\ C_{12} \\ C_{13} \end{bmatrix} = \begin{bmatrix} 0 \\ 0 \\ 0 \end{bmatrix}$$

Co-factors along first row give a null solution. Let us take co-factors along the second row.

$$\mathbf{m}_1 = \begin{bmatrix} C_{21} \\ C_{22} \\ C_{23} \end{bmatrix} = \begin{bmatrix} 0 \\ 8 \\ 4 \end{bmatrix}$$

To obtain eigenvectors associated with the repeated eigenvalue at $\lambda = 3$, we construct the matrix

$$(\lambda_2 I - A) = \begin{bmatrix} \lambda_2 - 4 & -1 & 2 \\ -1 & \lambda_2 & -2 \\ -1 & 1 & \lambda_2 - 3 \end{bmatrix}$$

For $\lambda_2 = 3$, the rank r of 3×3 matrix $(\lambda_2 I - A)$ is 2. Therefore one independent eigenvector associated with $\lambda = 3$ can be obtained. It is given by

$$\mathbf{m}_2 = \begin{bmatrix} C_{11} \\ C_{12} \\ C_{13} \end{bmatrix} = \begin{bmatrix} \lambda_2(\lambda_2 - 3) + 2 \\ (\lambda_2 - 3) + 2 \\ -1 + \lambda_2 \end{bmatrix} = \begin{bmatrix} 2 \\ 2 \\ 2 \end{bmatrix}$$

The vector $\mathbf{m}_3$ for the modal matrix

$$M = [\mathbf{m}_1 \vdots \mathbf{m}_2 \vdots \mathbf{m}_3]$$

may be generated from the independent eigenvector $\mathbf{m}_2$ as follows:

$$\mathbf{m}_3 = \begin{bmatrix} \dfrac{d}{d\lambda_2} C_{11} \\ \dfrac{d}{d\lambda_2} C_{12} \\ \dfrac{d}{d\lambda_2} C_{13} \end{bmatrix}_{\lambda_2 = 3} = \begin{bmatrix} 2\lambda_2 - 3 \\ 1 \\ 1 \end{bmatrix}_{\lambda_2 = 3} = \begin{bmatrix} 3 \\ 1 \\ 1 \end{bmatrix}$$

The vector $\mathbf{m}_3$ is *generalized eigenvector*. The modal matrix $\mathbf{M}$ is then given by

$$\mathbf{M} = \begin{bmatrix} 0 & 2 & 3 \\ 8 & 2 & 1 \\ 4 & 2 & 1 \end{bmatrix}$$

The modal matrix $\mathbf{M}$ now transforms $\mathbf{A}$ to the Jordan matrix, i.e..

$$\mathbf{M}^{-1}\mathbf{AM} = \begin{bmatrix} 1 & 0 & 0 \\ 0 & 3 & 1 \\ 0 & 0 & 3 \end{bmatrix} = \mathbf{J}$$

↑
Jordan blocks

In general, if an eigenvalue λ_1 has multiplicity q and the rank of $n \times n$ matrix $(\lambda_1 \mathbf{I} - \mathbf{A})$ is $(n - 1)$, i.e., there is only one independent eigenvector

$$\mathbf{m}_1 = \begin{bmatrix} C_{k1} \\ C_{k2} \\ C_{k3} \\ \vdots \\ C_{kn} \end{bmatrix}$$

associated with the eigenvalue λ_1, then the remaining $(q - 1)$ vectors for the modal matrix are the generalized eigenvectors given below (for proof, refer Ogata [2])

$$[\mathbf{m}_1 \mathbf{m}_2 \cdots \mathbf{m}_q] = \begin{bmatrix} C_{k1} & \dfrac{1}{1!}\dfrac{d}{d\lambda_1}C_{k1} & \dfrac{1}{2!}\dfrac{d^2}{d\lambda_1^2}C_{k1} & \cdots & \dfrac{1}{(q-1)!}\dfrac{d^{q-1}}{d\lambda_1^{q-1}}C_{k1} \\ C_{k2} & \dfrac{1}{1!}\dfrac{d}{d\lambda_1}C_{k2} & \dfrac{1}{2!}\dfrac{d^2}{d\lambda_1^2}C_{k2} & \cdots & \dfrac{1}{(q-1)!}\dfrac{d^{q-1}}{d\lambda_1^{q-1}}C_{k2} \\ \vdots & \vdots & \vdots & & \vdots \\ C_{kn} & \dfrac{1}{1!}\dfrac{d}{d\lambda_1}C_{kn} & \dfrac{1}{2!}\dfrac{d^2}{d\lambda_1^2}C_{kn} & \cdots & \dfrac{1}{(q-1)!}\dfrac{d^{q-1}}{d\lambda_1^{q-1}}C_{kn} \end{bmatrix}$$

(12.68c)

The matrix $\mathbf{M}^{-1}\mathbf{AM}$ will have a $q \times q$ Jordan block corresponding to the eigenvalue λ_1.

When $\mathbf{A}$ is known in the companion form (eqn. (12.68a)) and has eigenvalues $\lambda_1, \lambda_1, \ldots, \lambda_1, \lambda_{m+1}, \ldots, \lambda_n$ (eigenvalue λ_1 of multiplicity m), then the transformation matrix is the modified Vander Monde matrix (eqn. (12.68b))

$$\mathbf{V} = \begin{bmatrix} 1 & 0 & 0 & 0 & \cdots & 1 & \cdots & 1 \\ \lambda_1 & 1 & 0 & 0 & \cdots & \lambda_{m+1} & \cdots & \lambda_n \\ \lambda_1^2 & 2\lambda_1 & 1 & 0 & \cdots & \lambda_{m+1}^2 & \cdots & \lambda_n^2 \\ \lambda_1^3 & 3\lambda_1^2 & 3\lambda_1 & 1 & \cdots & \lambda_{m+1}^3 & \cdots & \lambda_n^3 \\ \vdots & \vdots & \vdots & \vdots & & \vdots & & \vdots \\ \lambda_1^{n-1} & \dfrac{d}{d\lambda_1}(\lambda_1^{n-1}) & \dfrac{1}{2!}\dfrac{d^2}{d\lambda_1^2}(\lambda_1^{n-1}) & \dfrac{1}{3!}\dfrac{d^3}{d\lambda_1^3}(\lambda_1^{n-1}) & \cdots & \lambda_{m+1}^{n-1} & \cdots & \lambda_n^{n-1} \end{bmatrix}$$
(12.68d)

It is possible in some rare cases, to have a state model whose **A** matrix has repeated eigenvalue λ_l such that rank of $n \times n$ matrix $(\lambda_l \mathbf{I} - \mathbf{A})$ is less than $(n-1)$, i.e., there are more than one independent eigenvectors corresponding to λ_l. However, we shall assume in our discussion that for state models we deal with, for each λ_l of multiplicity q, there exists a modal matrix **M** such that the Jordan matrix $\mathbf{J} = \mathbf{M}^{-1}\mathbf{A}\mathbf{M}$ will have a $q \times q$ Jordan block corresponding to eigenvalue λ_l.

12.5 Solution of State Equations

In the preceding two sections, we discussed various state models used to represent a system. In the present section, we shall develop methods for solution of the state equations from which the system transient response can then be obtained. We shall restrict our discussion to linear time-invariant systems only.

Let us first review the classical method of solution by considering a first-order scalar differential equation

$$\frac{dx}{dt} = ax \, ; \, x(0) = x_0 \tag{12.69}$$

This equation has the solution

$$x(t) = e^{at}x_0$$
$$= \left(1 + at + \frac{1}{2!}a^2t^2 + \cdots + \frac{1}{i!}a^i t^i + \cdots\right)x_0 \tag{12.70}$$

Let us now consider the state equation

$$\dot{\mathbf{x}}(t) = \mathbf{A}\mathbf{x}(t); \, \mathbf{x}(0) = \mathbf{x}_0 \tag{12.71}$$

which represents a homogeneous* linear system (unforced system) with constant coefficients.

*The state equation of a linear system is

$$\dot{\mathbf{x}} = \mathbf{A}\mathbf{x} + \mathbf{B}\mathbf{u}$$

If **A** is a constant matrix and **u** is a zero vector (i.e., no control forces are applied to the system), the equation represents a homogeneous linear system. On the other hand, if **A** is a constant matrix and **u** is nonzero vector, the equation represents a nonhomogeneous linear system (i.e., control forces are applied to the system).

By analogy with the scalar case, we assume a solution of the form
$$\mathbf{x}(t) = \mathbf{a}_0 + \mathbf{a}_1 t + \mathbf{a}_2 t^2 + \ldots + \mathbf{a}_i t^i + \ldots$$
where $\mathbf{a}_i$ are vector coefficients.

By substituting the assumed solution into eqn. (12.71), we get
$$\mathbf{a}_1 + 2\mathbf{a}_2 t + 3\mathbf{a}_3 t^2 + \ldots = \mathbf{A}(\mathbf{a}_0 + \mathbf{a}_1 t + \mathbf{a}_2 t^2 + \ldots)$$
The comparison of vector coefficients of equal powers of t, yields
$$\mathbf{a}_1 = \mathbf{A}\mathbf{a}_0$$
$$\mathbf{a}_2 = \frac{1}{2}\mathbf{A}\mathbf{a}_1 = \frac{1}{2!}\mathbf{A}^2\mathbf{a}_0$$
$$\vdots$$
$$\mathbf{a}_i = \frac{1}{i!}\mathbf{A}^i\mathbf{a}_0$$

In the assumed solution, equating $\mathbf{x}(t = 0) = \mathbf{x}_0$, we find that
$$\mathbf{a}_0 = \mathbf{x}_0$$
The solution $\mathbf{x}(t)$ is thus found to be
$$\mathbf{x}(t) = \left(\mathbf{I} + \mathbf{A}t + \frac{1}{2!}\mathbf{A}^2 t^2 + \ldots + \frac{1}{i!}\mathbf{A}^i t^i + \ldots\right)\mathbf{x}_0$$

Each of the terms inside the brackets is an $n \times n$ matrix. Because of the similarity of the entity inside the bracket with a scalar exponential of eqn. (12.70), we call it a *matrix exponential*, which may be written as

$$e^{\mathbf{A}t} = \mathbf{I} + \mathbf{A}t + \frac{1}{2!}\mathbf{A}^2 t^2 + \ldots + \frac{1}{i!}\mathbf{A}^i t^i + \ldots \qquad (12.72)$$

The solution $\mathbf{x}(t)$ can now be written as
$$\mathbf{x}(t) = e^{\mathbf{A}t}\mathbf{x}_0 \qquad (12.73)$$

From eqn. (12.73) it is observed that the initial state $\mathbf{x}_0$ at $t = 0$, is driven to a state $\mathbf{x}(t)$ at time t. This transition in state is carried out by the matrix exponential $e^{\mathbf{A}t}$. Because of this property, $e^{\mathbf{A}t}$ is known as *state transition matrix* and is denoted by $\Phi(t)$.

Let us now determine the solution of the nonhomogeneous state equation (forced system)
$$\dot{\mathbf{x}}(t) = \mathbf{A}\mathbf{x}(t) + \mathbf{B}\mathbf{u}(t); \quad \mathbf{x}(0) = \mathbf{x}_0 \qquad (12.74)$$
Rewrite this equation in the form
$$\dot{\mathbf{x}}(t) - \mathbf{A}\mathbf{x}(t) = \mathbf{B}\mathbf{u}(t)$$
Multiplying both sides by $e^{-\mathbf{A}t}$, we can write
$$e^{-\mathbf{A}t}[\dot{\mathbf{x}}(t) - \mathbf{A}\mathbf{x}(t)] = \frac{d}{dt}[e^{-\mathbf{A}t}\mathbf{x}(t)] = e^{-\mathbf{A}t}\mathbf{B}\mathbf{u}(t)$$
Integrating both sides with respect to t between the limits 0 and t, we get
$$e^{-\mathbf{A}t}\mathbf{x}(t)\Big|_0^t = \int_0^t e^{-\mathbf{A}\tau}\mathbf{B}\mathbf{u}(\tau)\,d\tau$$

$$e^{-At}x(t) - x(0) = \int_0^t e^{-A\tau}Bu(\tau)\,d\tau$$

Now pre-multiplying both sides by e^{At}, we have

$$x(t) = \underbrace{e^{At}x(0)}_{\text{Homogeneous solution}} + \underbrace{\int_0^t e^{A(t-\tau)}Bu(\tau)\,d\tau}_{\text{Forced solution}} \quad (12.75)$$

If the initial state is known at $t = t_0$ rather than $t = 0$, eqn. (12.75) becomes

$$x(t) = e^{A(t-t_0)}x(t_0) + \int_{t_0}^t e^{A(t-\tau)}Bu(\tau)\,d\tau \quad (12.76)$$

PROPERTIES OF THE STATE TRANSITION MATRIX

In the above discussion, the state transition matrix has been defined as

$$\Phi(t) = e^{At}$$

wherein the initial time has been taken as $t = 0$. In general for a linear time-invariant system, if the initial time is $t = t_0$, state transition matrix becomes

$$\Phi(t - t_0) = e^{A(t-t_0)}$$

Since the state transition matrix depends upon $(t - t_0)$, t_0 is conveniently regarded as zero.

Certain useful properties of the state transition matrix $\Phi(t)$ are given below:

1. $\Phi(0) = e^{A0} = I$
2. $\Phi(t) = e^{At} = (e^{-At})^{-1} = [\Phi(-t)]^{-1}$
   or $\Phi^{-1}(t) = \Phi(-t)$
3. $\Phi(t_1 + t_2) = e^{A(t_1+t_2)} = e^{At_1}e^{At_2}$
   $= \Phi(t_1)\Phi(t_2) = \Phi(t_2)\Phi(t_1)$

In terms of the state transition matrix $\Phi(t)$, the solution of the forced system given by eqn. (12.76) can be written as

$$x(t) = \Phi(t)x(0) + \int_0^t \Phi(t - \tau)Bu(\tau)\,d\tau \quad (12.77a)$$

If instead the initial state is known at t_0:

$$x(t) = \Phi(t - t_0)x(t_0) + \int_{t_0}^t \Phi(t - \tau)Bu(\tau)\,d\tau \quad (12.77b)$$

where

$$\Phi(t - t_0) = e^{A(t-t_0)}$$
$$\Phi(t - \tau) = e^{A(t-\tau)}$$

Example 12.2: Obtain the time response of the following system

$$\begin{bmatrix} \dot{x}_1 \\ \dot{x}_2 \end{bmatrix} = \begin{bmatrix} 1 & 0 \\ 1 & 1 \end{bmatrix} \begin{bmatrix} x_1 \\ x_2 \end{bmatrix} + \begin{bmatrix} 1 \\ 1 \end{bmatrix} u$$

where $u(t)$ is a unit step occuring at $t = 0$ and $\mathbf{x}^T(0) = [1\ 0]$.

Solution:
We have in this case
$$\mathbf{A} = \begin{bmatrix} 1 & 0 \\ 1 & 1 \end{bmatrix};\quad \mathbf{B} = \begin{bmatrix} 1 \\ 1 \end{bmatrix}$$

The state transition matrix $\Phi(t)$ is given by
$$e^{\mathbf{A}t} = \mathbf{I} + \mathbf{A}t + \frac{1}{2!}\mathbf{A}^2 t^2 + \frac{1}{3!}\mathbf{A}^3 t^3 + \cdots$$

$$\mathbf{A}^2 = \begin{bmatrix} 1 & 0 \\ 1 & 1 \end{bmatrix}\begin{bmatrix} 1 & 0 \\ 1 & 1 \end{bmatrix} = \begin{bmatrix} 1 & 0 \\ 2 & 1 \end{bmatrix}$$

Then
$$e^{\mathbf{A}t} = \begin{bmatrix} 1 & 0 \\ 0 & 1 \end{bmatrix} + \begin{bmatrix} t & 0 \\ t & t \end{bmatrix} + \tfrac{1}{2}\begin{bmatrix} t^2 & 0 \\ 2t^2 & t^2 \end{bmatrix} + \cdots$$
$$= \begin{bmatrix} 1 + t + 0.5t^2 + \cdots & 0 \\ t + t^2 + \cdots & 1 + t + 0.5t^2 + \cdots \end{bmatrix}$$
$$= \begin{bmatrix} \phi_{11} & \phi_{12} \\ \phi_{21} & \phi_{22} \end{bmatrix}$$

The terms ϕ_{11} and ϕ_{22} are easily recognized as series expansion of e^t. To recognize ϕ_{21}, more terms of the infinite series should be evaluated. In fact ϕ_{21} is te^t.

The computation of $e^{\mathbf{A}t}$ by expanding it into a power series in t and then adding the corresponding elements in the matrix terms of the infinite series, is practical only for simple cases. We will shortly discuss alternate methods of computation of state transition matrix.

Returning to the example under consideration, we have the state transition matrix as
$$\Phi(t) = \begin{bmatrix} e^t & 0 \\ te^t & e^t \end{bmatrix}$$

The time response of the system is given by
$$\mathbf{x}(t) = \Phi(t)\left[\mathbf{x}_0 + \int_0^t \Phi(-\tau)\mathbf{B}u\, d\tau\right]$$

Now with $u = 1$,
$$\Phi(-\tau)\mathbf{B}u = \begin{bmatrix} e^{-\tau} & 0 \\ -\tau e^{-\tau} & e^{-\tau} \end{bmatrix}\begin{bmatrix} 1 \\ 1 \end{bmatrix} = \begin{bmatrix} e^{-\tau} \\ e^{-\tau}(1-\tau) \end{bmatrix}$$

Therefore
$$\int_0^t \Phi(-\tau)\mathbf{B}u\, d\tau = \begin{bmatrix} \int_0^t e^{-\tau}\, d\tau \\ \int_0^t e^{-\tau}(1-\tau)\, d\tau \end{bmatrix} = \begin{bmatrix} 1 - e^{-t} \\ te^{-t} \end{bmatrix}$$

Then the solution $x(t)$ is given by

$$x(t) = \begin{bmatrix} e^t & 0 \\ te^t & e^t \end{bmatrix} \left\{ \begin{bmatrix} 1 \\ 0 \end{bmatrix} + \begin{bmatrix} 1 - e^{-t} \\ te^{-t} \end{bmatrix} \right\}$$

$$= \begin{bmatrix} 2e^t - 1 \\ 2te^t \end{bmatrix}$$

COMPUTATION OF STATE TRANSITION MATRIX

Computation by Laplace transformation

Let us consider an unforced system whose state equation is

$\dot{x} = Ax$, where A is a constant matrix

Taking the Laplace transform of this equation, we obtain

$$sX(s) - x(0) = AX(s) \tag{12.78}$$

where $X(s)$ is the Laplace transform of the unforced response and $x(0)$ is the initial condition vector. Equation (12.78) may be rearranged as

$$[sI - A]X(s) = x(0)$$

or $\qquad X(s) = [sI - A]^{-1} x(0)$

Taking the inverse Laplace transform, we get

$$x(t) = \mathcal{L}^{-1}[(sI - A)^{-1}] x(0)$$

where $x(t)$ is the unforced response of the system. This solution obviously must be identical with the one obtained earlier in eqn. (12.73) The comparison yields a different approach to determine the state transition matrix which is given below:

$$\Phi(t) = e^{At} = \mathcal{L}^{-1}[(sI - A)^{-1}] = \mathcal{L}^{-1} \Phi(s) \tag{12.79}$$

where $\qquad \Phi(s) = (sI - A)^{-1}$ is called the *resolvant matrix*.

Let us consider now the response when the control force vector u is applied. The state equation for this case is

$$\dot{x} = Ax + Bu$$

Performing the Laplace transformation gives

$$sX(s) - x_0 = AX(s) + BU(s)$$

or $\qquad (sI - A)X(s) = x_0 + BU(s)$

Therefore

$$X(s) = [(sI - A)^{-1}] x_0 + [(sI - A)^{-1}] BU(s)$$

By inverse Laplace transformation

$$x(t) = \mathcal{L}^{-1}[(sI - A)^{-1}] x_0 + \mathcal{L}^{-1}[(sI - A)^{-1} BU(s)]$$
$$= \Phi(t) x_0 + \mathcal{L}^{-1}[\Phi(s) BU(s)] \tag{12.80}$$

Example 12.3: Let us reconsider the system discussed in Example 12.2 in which

$$A = \begin{bmatrix} 1 & 0 \\ 1 & 1 \end{bmatrix}$$

Then

$$(s\mathbf{I} - \mathbf{A}) = \begin{bmatrix} s & 0 \\ 0 & s \end{bmatrix} - \begin{bmatrix} 1 & 0 \\ 1 & 1 \end{bmatrix} = \begin{bmatrix} s-1 & 0 \\ -1 & s-1 \end{bmatrix}$$

The determinant of $(s\mathbf{I} - \mathbf{A})$ is

$$|sI - A| = (s-1)^2$$

Therefore, the resolvant matrix is given by

$$\Phi(s) = (s\mathbf{I} - \mathbf{A})^{-1} = \frac{1}{(s-1)^2} \begin{bmatrix} (s-1) & 0 \\ 1 & (s-1) \end{bmatrix}$$

$$= \begin{bmatrix} \dfrac{1}{(s-1)} & 0 \\ \dfrac{1}{(s-1)^2} & \dfrac{1}{(s-1)} \end{bmatrix}$$

Hence

$$\Phi(t) = \mathcal{L}^{-1}[(s\mathbf{I} - \mathbf{A})^{-1}] = \begin{bmatrix} \mathcal{L}^{-1}\left[\dfrac{1}{(s-1)}\right] & 0 \\ \mathcal{L}^{-1}\left[\dfrac{1}{(s-1)^2}\right] & \mathcal{L}^{-1}\left[\dfrac{1}{(s-1)}\right] \end{bmatrix}$$

$$= \begin{bmatrix} e^t & 0 \\ te^t & e^t \end{bmatrix} = e^{\mathbf{A}t}$$

Computation by canonical transformation

It has been shown in Section 12.4 that a state model with a nondiagonal matrix **A** having distinct eigenvalues can be reduced to canonical form. The canonical form is most convenient for time domain analysis as each component state variable equation is a first order equation and is decoupled from all other component state variable equations.

Consider the state equation

$$\dot{\mathbf{x}} = \mathbf{A}\mathbf{x}; \quad \mathbf{x}(0) = \mathbf{x}_0 \quad (12.81a)$$

The solution vector

$$\mathbf{x}(t) = e^{\mathbf{A}t}\mathbf{x}_0 \quad (12.81b)$$

Assume that matrix **A** is nondiagonal and has distinct eigenvalues. Let us define a new state vector **z** such that

$$\mathbf{x} = \mathbf{M}\mathbf{z} \quad (12.82)$$

where **M** is diagonalizing or modal matrix. Under this transformation, the original state model modifies to

$$\mathbf{z} = \mathbf{M}^{-1}\mathbf{A}\mathbf{M}\mathbf{z}$$
$$= \Lambda \mathbf{z} \quad (12.83a)$$

where
$$\Lambda = \begin{bmatrix} \lambda_1 & 0 & 0 & \cdots & 0 \\ 0 & \lambda_2 & 0 & \cdots & 0 \\ \vdots & \vdots & \vdots & & \vdots \\ 0 & 0 & 0 & \cdots & \lambda_n \end{bmatrix}$$

The solution vector
$$\mathbf{z}(t) = e^{\Lambda t}\, \mathbf{z}(0) \tag{12.83b}$$
Now
$$e^{\Lambda t} = \mathbf{I} + \Lambda t + \frac{1}{2!}\Lambda^2 t^2 + \frac{1}{3!}\Lambda^3 t^3 + \cdots$$

$$= \begin{bmatrix} 1 & & & 0 \\ & 1 & & \\ & & \ddots & \\ 0 & & & 1 \end{bmatrix} + \begin{bmatrix} \lambda_1 & & & 0 \\ & \lambda_2 & & \\ & & \ddots & \\ 0 & & & \lambda_n \end{bmatrix} t + \frac{1}{2!}\begin{bmatrix} \lambda_1^2 & & & 0 \\ & \lambda_2^2 & & \\ & & \ddots & \\ 0 & & & \lambda_n^2 \end{bmatrix} t^2$$

$$+ \frac{1}{3!}\begin{bmatrix} \lambda_1^3 & & & 0 \\ & \lambda_2^3 & & \\ & & \ddots & \\ 0 & & & \lambda_n^3 \end{bmatrix} t^3 + \cdots$$

$$= \begin{bmatrix} 1 + \lambda_1 t + \frac{1}{2!}\lambda_1^2 t^2 + \frac{1}{3!}\lambda_1^3 t^3 + \cdots & & & 0 \\ & 1 + \lambda_2 t + \frac{1}{2!}\lambda_2^2 t^2 + \frac{1}{3!}\lambda_2^3 t^3 + \cdots & & \\ & & \ddots & \\ 0 & & & 1 + \lambda_n t + \frac{1}{2!}\lambda_n^2 t^2 + \frac{1}{3!}\lambda_n^3 t^3 + \cdots \end{bmatrix}$$

or
$$e^{\Lambda t} = \begin{bmatrix} e^{\lambda_1 t} & & & 0 \\ & e^{\lambda_2 t} & & \\ & & \ddots & \\ 0 & & & e^{\lambda_n t} \end{bmatrix} \tag{12.84}$$

From eqns. (12.83b) and (12.82), we have
$$\mathbf{x}(t) = \mathbf{M} e^{\Lambda t}\, \mathbf{M}^{-1} \mathbf{x}_0$$
Comparison of this equation with eqn. (12.81b) yields
$$e^{\mathbf{A}t} = \mathbf{M} e^{\Lambda t}\, \mathbf{M}^{-1}$$
$$= \mathbf{M} \begin{bmatrix} e^{\lambda_1 t} & & & 0 \\ & e^{\lambda_2 t} & & \\ & & \ddots & \\ 0 & & & e^{\lambda_n t} \end{bmatrix} \mathbf{M}^{-1} \tag{12.85}$$

If **A** involves multiple eigenvalues, then e^{At} can be written in a slightly complicated form developed below.

Assume that matrix **A** in eqn. (12.81a) has eigenvalues $\lambda_1, \lambda_1, \lambda_1, \lambda_4, ..., \lambda_n$. For transformation to Jordan canonical form, the transformation matrix is say **M**. The resulting state equation is

$$\mathbf{z} = \mathbf{M}^{-1}\mathbf{AMz}$$
$$= \mathbf{Jz} \qquad (12.86)$$

$$\mathbf{J} = \begin{bmatrix} \lambda_1 & 1 & 0 & 0 & \cdots & 0 \\ 0 & \lambda_1 & 1 & 0 & \cdots & 0 \\ 0 & 0 & \lambda_1 & 0 & \cdots & 0 \\ 0 & 0 & 0 & \lambda_4 & \cdots & 0 \\ \vdots & \vdots & \vdots & \vdots & & \vdots \\ 0 & 0 & 0 & 0 & \cdots & \lambda_n \end{bmatrix}$$

This is equivalent to following set of first-order differential equations:

$$\begin{aligned} \dot{z}_1 &= \lambda_1 z_1 + z_2 \\ \dot{z}_2 &= \lambda_1 z_2 + z_3 \\ \dot{z}_3 &= \lambda_1 z_3 \\ \dot{z}_i &= \lambda_i z_i; \quad i = 4, 5, ..., n \end{aligned} \qquad (12.87)$$

Solving for $z_3, z_4, ..., z_n$ we get

$$z_3(t) = e^{\lambda_1 t} z_3(0)$$
$$z_i(t) = e^{\lambda_i t} z_i(0); \quad i = 4, 5, ..., n$$

Substituting $z_3(t)$ into second equation in (12.87) we get

$$\dot{z}_2 = \lambda_1 z_2 + e^{\lambda_1 t} z_3(0)$$

which yields the solution

$$z_2(t) = e^{\lambda_1 t} z_2(0) + t e^{\lambda_1 t} z_3(0)$$

Similarly we get

$$z_1(t) = e^{\lambda_1 t} z_1(0) + t e^{\lambda_1 t} z_2(0) + \tfrac{1}{2} t^2 e^{\lambda_1 t} z_3(0)$$

The solution vector is therefore

$$\mathbf{z}(t) = \begin{bmatrix} e^{\lambda_1 t} & t e^{\lambda_1 t} & \tfrac{1}{2} t^2 e^{\lambda_1 t} & 0 & \cdots & 0 \\ 0 & e^{\lambda_1 t} & t e^{\lambda_1 t} & 0 & \cdots & 0 \\ 0 & 0 & e^{\lambda_1 t} & 0 & \cdots & 0 \\ 0 & 0 & 0 & e^{\lambda_4 t} & \cdots & 0 \\ \vdots & \vdots & \vdots & \vdots & & \vdots \\ 0 & 0 & 0 & 0 & \cdots & e^{\lambda_n t} \end{bmatrix} \mathbf{z}(0)$$

$$= \begin{bmatrix} 1 & t & \tfrac{1}{2} t^2 & 0 & \cdots & 0 \\ 0 & 1 & t & 0 & \cdots & 0 \\ 0 & 0 & 1 & 0 & \cdots & 0 \\ 0 & 0 & 0 & 1 & \cdots & 0 \\ \vdots & \vdots & \vdots & \vdots & & \vdots \\ 0 & 0 & 0 & 0 & \cdots & 1 \end{bmatrix} e^{\Lambda t} \mathbf{z}(0)$$

where
$$e^{\Lambda t} = \begin{bmatrix} e^{\lambda_1 t} & & & 0 \\ & e^{\lambda_2 t} & & \\ & & \ddots & \\ 0 & & & e^{\lambda_n t} \end{bmatrix}$$

If eigenvalue λ_1 is repeated q times, then we will get
$$z(t) = Q(t)e^{\Lambda t} z(0)$$
where

$$Q(t) = \begin{bmatrix} 1 & t & \frac{1}{2!}t^2 & \cdots & \frac{1}{(q-1)!}t^{q-1} & 0 & \cdots & 0 \\ 0 & 1 & t & \cdots & \frac{1}{(q-2)!}t^{q-2} & 0 & \cdots & 0 \\ \vdots & \vdots & \vdots & & \vdots & \vdots & & \vdots \\ 0 & 0 & 0 & \cdots & t & 0 & \cdots & 0 \\ 0 & 0 & 0 & \cdots & 1 & 0 & \cdots & 0 \\ 0 & 0 & 0 & \cdots & 0 & 1 & \cdots & 0 \\ \vdots & \vdots & \vdots & & \vdots & \vdots & & \vdots \\ 0 & 0 & 0 & \cdots & 0 & 0 & \cdots & 1 \end{bmatrix} \quad (12.88)$$

The solution $x(t)$ in this case is given by
$$x(t) = MQ(t)e^{\Lambda t}M^{-1}x(0)$$
Comparison with eq. (12.81b) yields
$$e^{\Lambda t} = MQ(t)e^{\Lambda t}M^{-1} \quad (12.89)$$

Example 12.4: Consider a system with the state model
$$\dot{x} = \begin{bmatrix} 0 & 1 \\ -2 & -3 \end{bmatrix} x + \begin{bmatrix} 0 \\ 2 \end{bmatrix} u; \, x(0) = \begin{bmatrix} 0 \\ 1 \end{bmatrix}$$
Compute the state transition matrix.

Solution:

The characteristics equation for this system is
$$|\lambda I - A| = \begin{vmatrix} \lambda & -1 \\ 2 & (\lambda + 3) \end{vmatrix} = 0$$
or
$$(\lambda + 1)(\lambda + 2) = 0$$
Therefore, the eigenvalues of matrix A are
$$\lambda_1 = -1, \lambda_2 = -2$$
Since the system matrix A is in the companion form, we can choose

the Vander Monde matrix,
$$M = \begin{bmatrix} 1 & 1 \\ \lambda_1 & \lambda_2 \end{bmatrix} = \begin{bmatrix} 1 & 1 \\ -1 & -2 \end{bmatrix}$$
as the modal matrix. Then
$$\Lambda = M^{-1}AM = \begin{bmatrix} -1 & 0 \\ 0 & -2 \end{bmatrix}$$
$$e^{\Lambda t} = \begin{bmatrix} e^{-t} & 0 \\ 0 & e^{-2t} \end{bmatrix}$$

From eqn. (12.85),
$$e^{At} = Me^{\Lambda t}M^{-1}$$
$$= \begin{bmatrix} 1 & 1 \\ -1 & -2 \end{bmatrix} \begin{bmatrix} e^{-t} & 0 \\ 0 & e^{-2t} \end{bmatrix} \begin{bmatrix} 2 & 1 \\ -1 & -1 \end{bmatrix}$$
$$= \begin{bmatrix} 2e^{-t} - e^{-2t} & e^{-t} - e^{-2t} \\ -2e^{-t} + 2e^{-2t} & -e^{-t} + 2e^{-2t} \end{bmatrix}$$

Computation by techniques based on the Cayley-Hamilton theorem

The state transition matrix may be computed using the technique based on the *Cayley-Hamilton Theorem*. For large systems, this method is far more convenient computationally as compared to the other two methods advanced earlier. To begin with, let us state the Cayley-Hamilton Theorem.

Every square matrix A satisfies its own characteristic equation. In other words, if

$$q(\lambda) = |\lambda I - A| = \lambda^n + a_1\lambda^{n-1} + \ldots + a_{n-1}\lambda + a_n = 0 \quad (12.90)$$

is the characteristic equation of A, then

$$q(A) = A^n + a_1 A^{n-1} + \ldots + a_{n-1} A + a_n I = 0 \quad (12.91)$$

This theorem provides a simple procedure for evaluating the function of a matrix. In the study of linear systems, we are mostly concerned with functions which can be represented as a series of the powers of a matrix. Given an $n \times n$ matrix A with the characteristic equation as in eqn. (12.90) above, let $\lambda_1, \lambda_2, \ldots, \lambda_n$ be the eigenvalues of A. The matrix polynomial

$$f(A) = k_0 I + k_1 A + k_2 A^2 + \ldots + k_n A^n + k_{n+1} A^{n+1} + \ldots \quad (12.92)$$

can be computed by consideration of the scalar polynomial

$$f(\lambda) = k_0 + k_1\lambda + k_2\lambda^2 + \ldots + k_n\lambda^n + k_{n+1}\lambda^{n+1} + \ldots \quad (12.93)$$

If $f(\lambda)$ is divided by the characteristic polynomial $q(\lambda)$, then we have

$$\frac{f(\lambda)}{q(\lambda)} = Q(\lambda) + \frac{R(\lambda)}{q(\lambda)}$$

or
$$f(\lambda) = Q(\lambda)q(\lambda) + R(\lambda) \quad (12.94)$$

where $R(\lambda)$ is the remainder polynomial of the following form:
$$R(\lambda) = \alpha_0 + \alpha_1\lambda + \alpha_2\lambda^2 + \ldots + \alpha_{n-1}\lambda^{n-1} \qquad (12.95)$$
If we evaluate $f(\lambda)$ at the eigenvalues $\lambda_1, \lambda_2, \ldots, \lambda_n$; then $q(\lambda) = 0$ and we have
$$f(\lambda_i) = R(\lambda_i); \quad i = 1, 2, \ldots, n \qquad (12.96)$$
The coefficients $\alpha_0, \alpha_1, \ldots, \alpha_{n-1}$, can be obtained by successively substituting $\lambda_1, \lambda_2, \ldots, \lambda_n$ into eqn. (12.96).

Substituting **A** for the variable λ in eqn. (12.94), we get
$$f(\mathbf{A}) = Q(\mathbf{A})q(\mathbf{A}) + R(\mathbf{A})$$
Since $q(\mathbf{A})$ is identically zero, it follows that
$$f(\mathbf{A}) = R(\mathbf{A})$$
$$= \alpha_0\mathbf{I} + \alpha_1\mathbf{A} + \ldots + \alpha_{n-1}\mathbf{A}^{n-1} \qquad (12.97)$$
which is the desired result.

The formal procedure of evaluation of the matrix polynomial $f(\mathbf{A})$ is given below:
1. Find the eigenvalues of matrix **A**.
2. If all the eigenvalues are distinct, solve n simultaneous equations given by (12.96) for the coefficients $\alpha_0, \alpha_1, \ldots, \alpha_{n-1}$.

If **A** possesses an eigenvalue λ_k of order m, then only one independent equation can be obtained by substituting λ_k into eqn. (12.96). The remaining $(m - 1)$ linear equations, which must be obtained in order to solve for α_i's, can be found by differentiating both sides of eqn. (12.94). Since
$$\left.\frac{d^j q(\lambda)}{d\lambda^j}\right|_{\lambda = \lambda_k} = 0; \; j = 0, 1, \ldots, m - 1,$$
it follows that
$$\left.\frac{d^j f(\lambda)}{d\lambda^j}\right|_{\lambda = \lambda_k} = \left.\frac{d^j R(\lambda)}{d\lambda^j}\right|_{\lambda = \lambda_k}; \; j = 0, 1, \ldots, m - 1 \qquad (12.98)$$

3. The coefficients α_i obtained in step 2 and eqn. (12.97) yield the required result.

Example 12.5: Find $f(\mathbf{A}) = \mathbf{A}^{10}$ for
$$\mathbf{A} = \begin{bmatrix} 0 & 1 \\ -2 & -3 \end{bmatrix}$$

Solution:
The characteristic equation is
$$q(\lambda) = |\lambda\mathbf{I} - \mathbf{A}| = \begin{vmatrix} \lambda & -1 \\ 2 & \lambda + 3 \end{vmatrix} = (\lambda + 1)(\lambda + 2) = 0$$
Matrix **A** has distinct eigenvalues $\lambda_1 = -1, \lambda_2 = -2$.

Since A is of second-order, the polynomial $R(\lambda)$ will be of the following form:
$$R(\lambda) = \alpha_0 + \alpha_1 \lambda$$
The coefficients α_0 and α_1 are evaluated from equations
$$f(\lambda_1) = \lambda_1^{10} = \alpha_0 + \alpha_1 \lambda_1$$
$$f(\lambda_2) = \lambda_2^{10} = \alpha_0 + \alpha_1 \lambda_2$$
The result is $\quad \alpha_0 = -1022, \alpha_1 = -1023$
From eqn. (12.97), we get
$$f(\mathbf{A}) = \mathbf{A}^{10} = \alpha_0 \mathbf{I} + \alpha_1 \mathbf{A}$$
$$= \begin{bmatrix} -1022 & -1023 \\ 2046 & 2047 \end{bmatrix}$$

The Cayley-Hamilton technique allows us to attack the problem of computation of $e^{\mathbf{A}t}$, where A is a constant $n \times n$ matrix.

The power series for the scalar $e^{\lambda t}$,
$$e^{\lambda t} = 1 + \lambda t + \frac{\lambda^2 t^2}{2!} + \ldots + \frac{\lambda^n t^n}{n!} + \ldots$$
converges for all finite λ and t. It follows from this result [2] that the matrix power series
$$e^{\mathbf{A}t} = \mathbf{I} + \mathbf{A}t + \frac{\mathbf{A}^2 t^2}{2!} + \ldots + \frac{\mathbf{A}^n t^n}{n!} + \ldots$$
converges for all A and for all finite t. Therefore the matrix polynomial $f(\mathbf{A}) = e^{\mathbf{A}t}$ can be expressed as a polynomial in A of degree $(n-1)$ using the technique presented earlier. This is illustrated below with the help of an example.

Example 12.6: Find $f(\mathbf{A}) = e^{\mathbf{A}t}$ for
$$\mathbf{A} = \begin{bmatrix} 0 & 1 \\ -1 & -2 \end{bmatrix}$$

Solution:
The characteristic equation is
$$q(\lambda) = |\lambda \mathbf{I} - \mathbf{A}| = \begin{vmatrix} \lambda & -1 \\ 1 & \lambda + 2 \end{vmatrix} = (\lambda + 1)^2 = 0$$
Matrix A has eigenvalues $\lambda_1, \lambda_2 = -1$.

Since A is of second-order, the polynomial $R(\lambda)$ will be of the following form:
$$R(\lambda) = \alpha_0 + \alpha_1 \lambda$$
The coefficients α_0 and α_1 are evaluated from equations (eqns. (12.96) and (12.98))
$$f(-1) = e^{-t} = \alpha_0 - \alpha_1$$

$$-\frac{d}{d\lambda}f(\lambda)\bigg|_{\lambda=-1} = te^{-t} = \frac{d}{d\lambda}R(\lambda)\bigg|_{\lambda=-1} = \alpha_1$$

The result is $\alpha_0 = (1+t)e^{-t}, \ \alpha_1 = te^{-t}$

From eqn. (12.97), we get

$$f(\mathbf{A}) = e^{\mathbf{A}t} = \alpha_0 \mathbf{I} + \alpha_1 \mathbf{A}$$

$$= \begin{bmatrix} (1+t)e^{-t} & te^{-t} \\ -te^{-t} & (1-t)e^{-t} \end{bmatrix}$$

12.6 Concepts of Controllability and Observability

In control problems, two basic questions need to be answered in deciding whether or not a control solution exists. These questions may be posed as:

(i) Can we transfer the system from any initial state to any other desired state in finite time by application of a suitable control force?

(ii) Knowing the output vector for a finite length of time, can we determine the initial state of the system?

The answers to these basic questions were conceptualized by Kalman into what is known as controllability and observability of a system. More precise definitions of controllability and observability are advanced below:

A system is said to be completely state controllable if it is possible to transfer the system state from any initial state $\mathbf{x}(t_0)$ to any other desired state $\mathbf{x}(t_f)$ in specified finite time by a control vector $\mathbf{u}(t)$.

Sometimes it is desired to transfer the system output from an initial value to any other desired value. For a system whose output does not depend upon control vector but depends upon state vector only, the dimensionality of output vector is usually less than that of state vector. Therefore, the output control is comparatively easier. In this section, we shall discuss state controllability only. The conditions for output controllability, if desired, may be derived on the similar lines.

A system is said to be completely observable, if every state $\mathbf{x}(t_0)$ can be completely identified by measurements of the output $\mathbf{y}(t)$ over a finite time interval.

A system which is not completely observable, implies that some of its state variables are shielded from observation.

As pointed out above, the concepts of controllability and observability play an important role in control engineering. These concepts are relatively recent and were originally introduced by Kalman. The mathematical tests of controllability and observability developed by Kalman, though elegant, do not give a physical feel of the problems involved. The alternative test introduced by Gilbert, uses canonical state model and provides better physical insight into the problem. In this section, we shall first discuss the Gilbert's method and then state Kalman's tests without proof. The proof involves concepts which have not been introduced in this book. The interested leader is directed to Ogata [2].

CONTROLLABILITY

As mentioned earlier, the concept of controllability involves the dependence of state variables of the system on the inputs. Consider a single-input linear time-invariant system

$$\dot{x} = Ax + Bu \qquad (12.99)$$

where $x = n$-dimensional state vector; $u =$ control signal (or control force); $A = n \times n$ matrix; and $B = n \times 1$ matrix.

Let the initial system state be $x(0)$ and the final desired state be $x(t_f)$. The system described by eqn. (12.99) is controllable if it is possible to construct a control signal which, in finite time interval $0 < t \leqslant t_f$, will transfer the system state from $x(0)$ to $x(t_f)$.

Let us first assume that the eigenvalues of the matrix A are all distinct so that it can be transformed into the canonical state variable form

$$\dot{z} = \Lambda z + \widetilde{B}u$$

or

$$\begin{bmatrix} \dot{z}_1 \\ \vdots \\ \dot{z}_n \end{bmatrix} = \begin{bmatrix} \lambda_1 & \cdots & 0 \\ & \ddots & \\ 0 & \cdots & \lambda_n \end{bmatrix} \begin{bmatrix} z_1 \\ \vdots \\ z_n \end{bmatrix} + \begin{bmatrix} \widetilde{b}_1 \\ \vdots \\ \widetilde{b}_n \end{bmatrix} u \qquad (12.100)$$

This equation can be written in the component form as

$$\dot{z}_i = \lambda_i z_i + \widetilde{b}_i u, \quad i = 1, 2, \ldots, n$$

which has the solution

$$z_i(t) = e^{\lambda_i t} z_i(0) + e^{\lambda_i t} \int_0^t e^{-\lambda_i \tau} \widetilde{b}_i u(\tau)\, d\tau$$

The system described by eqn. (12.99) is completely controllable if state variable z_i can be transferred from initial state $z_i(0)$ to a final state $z_i(t_f)$ in a finite time t_f. In other words, the system is controllable if it is possible to construct a control signal $u(t)$ such that the following equation is satisfied:

$$\frac{z_i(t_f) - e^{\lambda_i t_f} z_i(0)}{e^{\lambda_i t_f}} = \int_0^{t_f} e^{-\lambda_i \tau} \widetilde{b}_i u(\tau)\, d\tau$$

There are actually, numerous values of $u(\tau)$ which satisfy this equation provided $\widetilde{b}_i \neq 0$, because otherwise the link between input and the corresponding state variable gets broken and hence it is no longer possible to control that particular state variable.

It therefore follows that the necessary condition of complete controllability is simply that the vector $\widetilde{B}$ should not have any zero elements. If any element of this vector is zero, then the corresponding state variable is not controllable. It can be further shown that the condition stated here is in fact both necessary and sufficient.

The result just obtained, can be extended to the case where the control force u is an m-dimensional vector. For the system described by

$$\dot{z} = \Lambda z + \widetilde{B}u$$

where
$$\tilde{\mathbf{B}} = \begin{bmatrix} \tilde{b}_{11} & \tilde{b}_{12} & \cdots & \tilde{b}_{1m} \\ \tilde{b}_{21} & \tilde{b}_{22} & \cdots & \tilde{b}_{2m} \\ \vdots & \vdots & & \vdots \\ \tilde{b}_{n1} & \tilde{b}_{n2} & \cdots & \tilde{b}_{nm} \end{bmatrix}$$

the necessary and sufficient condition for controllability is that the matrix $\tilde{\mathbf{B}}$ must have no rows with all zeros. It is observed from the above equation that if any row of the matrix $\mathbf{B}$ is zero, it is not possible to influence the corresponding state variable by the control forces and hence the particular state variable is uncontrollable.

If the matrix $\mathbf{A}$ in eqn. (12.99) does not possess distinct eigenvalues, diagonalization is no longer possible. In such cases, we may transform $\mathbf{A}$ into Jordan canonical form

$$\dot{\mathbf{z}} = \mathbf{J}\mathbf{z} + \tilde{\mathbf{B}}\mathbf{u}$$

For a system with eigenvalues $\lambda_1, \lambda_1, \lambda_1, \lambda_4, \lambda_4, \lambda_6, \ldots, \lambda_n$, the Jordan matrix $\mathbf{J}$ has the form

$$\mathbf{J} = \begin{bmatrix} \begin{array}{|ccc|} \hline \lambda_1 & 1 & 0 \\ 0 & \lambda_1 & 1 \\ 0 & 0 & \lambda_1 \\ \hline \end{array} & \begin{matrix} 0 & 0 \\ 0 & 0 \\ 0 & 0 \end{matrix} & \begin{matrix} 0 \\ 0 \\ 0 \end{matrix} & \cdots & \begin{matrix} 0 \\ 0 \\ 0 \end{matrix} \\ \begin{matrix} 0 & 0 & 0 \\ 0 & 0 & 0 \end{matrix} & \begin{array}{|cc|} \hline \lambda_4 & 1 \\ 0 & \lambda_4 \\ \hline \end{array} & \begin{matrix} 0 \\ 0 \end{matrix} & \cdots & \begin{matrix} 0 \\ 0 \end{matrix} \\ \begin{matrix} 0 & 0 & 0 \end{matrix} & \begin{matrix} 0 & 0 \end{matrix} & \boxed{\lambda_6} & \cdots & 0 \\ \vdots & \vdots & \vdots & & \vdots \\ \begin{matrix} 0 & 0 & 0 \end{matrix} & \begin{matrix} 0 & 0 \end{matrix} & 0 & \cdots & \boxed{\lambda_n} \end{bmatrix}$$

(Jordan Blocks)

The conditions of complete controllability can now be stated as:

The elements of any row of $\tilde{\mathbf{B}}$ that correspond to the last row of each Jordan block are not all zero.

The Gilbert's method of testing controllability discussed above requires that the system be transformed into canonical state or Jordan canonical form. The test of controllability due to Kalman which can be applied to any state model, (canonical or otherwise), is stated below:

A general nth order multi-input linear time-invariant system (with an m-dimensional control vector),

$$\dot{\mathbf{x}} = \mathbf{A}\mathbf{x} + \mathbf{B}\mathbf{u}$$

is completely controllable if and only if the rank of the composite matrix

$$Q_c = [B \vdots AB \vdots \ldots \vdots A^{n-1}B] \qquad (12.101)$$

is n. Since matrices A and B are involved in (12.101), we may say that the pair (AB) is controllable if rank of Q_c is n.

Example 12.7: Consider the system with state equation

$$\begin{bmatrix} \dot{x}_1 \\ \dot{x}_2 \\ \dot{x}_3 \end{bmatrix} = \begin{bmatrix} 0 & 1 & 0 \\ 0 & 0 & 1 \\ -6 & -11 & -6 \end{bmatrix} \begin{bmatrix} x_1 \\ x_2 \\ x_3 \end{bmatrix} + \begin{bmatrix} 0 \\ 0 \\ 1 \end{bmatrix} u$$

Let us first transform this state equation into canonical form.

The characteristic equation is given by

$$|A - \lambda I| = \begin{vmatrix} -\lambda & 1 & 0 \\ 0 & -\lambda & 1 \\ -6 & -11 & -6-\lambda \end{vmatrix} = 0$$

which gives the eigenvalues

$$\lambda_1 = -1, \quad \lambda_2 = -2, \quad \lambda_3 = -3$$

Choosing Vander Monde matrix as modal matrix, we have

$$M = \begin{bmatrix} 1 & 1 & 1 \\ \lambda_1 & \lambda_2 & \lambda_3 \\ \lambda_1^2 & \lambda_2^2 & \lambda_3^2 \end{bmatrix} = \begin{bmatrix} 1 & 1 & 1 \\ -1 & -2 & -3 \\ 1 & 4 & 9 \end{bmatrix}$$

$$\widetilde{B} = M^{-1}B = \begin{bmatrix} 3 & 5/2 & 1/2 \\ -3 & -4 & -1 \\ 1 & 3/2 & 1/2 \end{bmatrix} \begin{bmatrix} 0 \\ 0 \\ 1 \end{bmatrix} = \begin{bmatrix} 1/2 \\ -1 \\ 1/2 \end{bmatrix}$$

Therefore, the state equation in canonical form is given by

$$\begin{bmatrix} \dot{z}_1 \\ \dot{z}_2 \\ \dot{z}_3 \end{bmatrix} = \begin{bmatrix} -1 & 0 & 0 \\ 0 & -2 & 0 \\ 0 & 0 & -3 \end{bmatrix} \begin{bmatrix} z_1 \\ z_2 \\ z_3 \end{bmatrix} + \begin{bmatrix} 1/2 \\ -1 \\ 1/2 \end{bmatrix} u$$

Since no element of $\widetilde{B}$ is zero, the system is completely controllable.

Let us now test controllability of this system by the Kalman's test. We have

$$B = \begin{bmatrix} 0 \\ 0 \\ 1 \end{bmatrix}; \quad A = \begin{bmatrix} 0 & 1 & 0 \\ 0 & 0 & 1 \\ -6 & -11 & -6 \end{bmatrix}$$

Then
$$AB = \begin{bmatrix} 0 & 1 & 0 \\ 0 & 0 & 1 \\ -6 & -11 & -6 \end{bmatrix} \begin{bmatrix} 0 \\ 0 \\ 1 \end{bmatrix} = \begin{bmatrix} 0 \\ 1 \\ -6 \end{bmatrix}$$

$$A^2B = \begin{bmatrix} 0 & 1 & 0 \\ 0 & 0 & 1 \\ -6 & -11 & -6 \end{bmatrix} \begin{bmatrix} 0 \\ 1 \\ -6 \end{bmatrix} = \begin{bmatrix} 1 \\ -6 \\ 25 \end{bmatrix}$$

The composite matrix defined in eqn. (12.101) is given by
$$Q_c = [B \vdots AB \vdots A^2B]$$
$$= \begin{bmatrix} 0 & 0 & 1 \\ 0 & 1 & -6 \\ 1 & -6 & 25 \end{bmatrix}$$

It is easily seen that det $Q_c \neq 0$, i.e., its rank is $r = n = 3$. The system is therefore completely controllable.

Controllable phase variable form

Notice that if the system equation is given by
$$z = A_c z + B_c u \qquad (12.102)$$
where z is $n \times 1$ vector and u is a scalar, and
$$A_c = \begin{bmatrix} 0 & 1 & 0 & \cdots & 0 \\ 0 & 0 & 1 & \cdots & 0 \\ \vdots & \vdots & \vdots & & \vdots \\ 0 & 0 & 0 & \cdots & 1 \\ -a_n & -a_{n-1} & -a_{n-2} & \cdots & -a_n \end{bmatrix}; \quad B_c = \begin{bmatrix} 0 \\ 0 \\ \vdots \\ 0 \\ 1 \end{bmatrix}$$

then it is easily verified that the rank of the matrix $[B_c \vdots A_c B_c \vdots \ldots \vdots A_c^{n-1} B_c]$ is n. Therefore a system described by state equations of the form (12.102) is always controllable.

The converse is also true, i.e., if a linear time-invariant system given by
$$\dot{x} = Ax + Bu \qquad (12.103)$$
where x is $n \times 1$ state vector, A is $n \times n$ system matrix, B is $n \times 1$ control vector, u is scalar control input, is controllable, then it can be transformed into the form (12.102).

To prove this, let us assume that such a transformation exists and is given by
$$z = Px \qquad (12.104)$$

where $\mathbf{P} = \begin{bmatrix} p_{11} & p_{12} & \cdots & p_{1n} \\ p_{21} & p_{22} & \cdots & p_{2n} \\ \vdots & \vdots & & \vdots \\ p_{n1} & p_{n2} & & p_{nn} \end{bmatrix} = \begin{bmatrix} \mathbf{P}_1 \\ \mathbf{P}_2 \\ \vdots \\ \mathbf{P}_n \end{bmatrix}$

$\mathbf{P}_i = [p_{i1} \quad p_{i2} \quad \cdots \quad p_{in}]; \quad i = 1, 2, \ldots, n$

In the component form we have from eqn. (12.104)

$$z_1(t) = p_{11}x_1 + p_{12}x_2 + \ldots + p_{1n}x_n$$
$$= \mathbf{P}_1 \mathbf{x}(t)$$

Taking derivative on both sides of this equation, we have

$$\dot{z}_1(t) = \mathbf{P}_1\dot{\mathbf{x}} = \mathbf{P}_1\mathbf{A}\mathbf{x} + \mathbf{P}_1\mathbf{B}u$$

But $\dot{z}_1 = z_2$ is a function of $\mathbf{x}$ only as per transformation (12.104). Therefore $\mathbf{P}_1\mathbf{B} = 0$ and

$$z_2 = \mathbf{P}_1\mathbf{A}\mathbf{x}$$

Taking derivative on both sides once again, we have

$$\dot{z}_2 = z_3 = \mathbf{P}_1\mathbf{A}^2\mathbf{x} \text{ and } \mathbf{P}_1\mathbf{A}\mathbf{B} = 0$$

This process gives

$$\dot{z}_{n-1} = z_n = \mathbf{P}_1\mathbf{A}^{n-1}\mathbf{x} \text{ with } \mathbf{P}_1\mathbf{A}^{n-2}\mathbf{B} = 0$$

Thus

$$\mathbf{z}(t) = \mathbf{P}\mathbf{x} = \begin{bmatrix} \mathbf{P}_1 \\ \mathbf{P}_1\mathbf{A} \\ \vdots \\ \mathbf{P}_1\mathbf{A}^{n-1} \end{bmatrix} \mathbf{x}$$

and $\mathbf{P}_1$ should satisfy the conditions

$$\mathbf{P}_1\mathbf{B} = \mathbf{P}_1\mathbf{A}\mathbf{B} = \ldots = \mathbf{P}_1\mathbf{A}^{n-2}\mathbf{B} = 0$$

The transformation (12.104) transforms (12.103) to

$$\dot{\mathbf{z}} = \mathbf{P}\mathbf{A}\mathbf{P}^{-1}\mathbf{z} + \mathbf{P}\mathbf{B}u$$

Comparing this with (12.102), we get

$$\mathbf{B}_c = \mathbf{P}\mathbf{B}$$

or

$$\begin{bmatrix} 0 \\ 0 \\ \vdots \\ 1 \end{bmatrix} = \begin{bmatrix} \mathbf{P}_1\mathbf{B} \\ \mathbf{P}_1\mathbf{A}\mathbf{B} \\ \vdots \\ \mathbf{P}_1\mathbf{A}^{n-1}\mathbf{B} \end{bmatrix}$$

or

$$\mathbf{P}_1[\mathbf{B} \vdots \mathbf{A}\mathbf{B} \vdots \ldots \vdots \mathbf{A}^{n-1}\mathbf{B}] = [0 \quad 0 \quad \ldots \quad 1]$$

This gives
$$P_1 = [0 \ 0 \ ... \ 1][B \vdots AB \vdots ... \vdots A^{n-1}B]^{-1}$$

The matrix $Q_c = [B \vdots AB \vdots ... \vdots A^{n-1}B]$ is nonsingular since the state model (12.103) is controllable. Therefore, the controllable state model (12.103) can be transformed to the form given by eqn. (12.102) by the transformation

$$z = Px$$

where
$$P = \begin{bmatrix} P_1 \\ P_1A \\ \vdots \\ P_1A^{n-1} \end{bmatrix}; \ P_1 = [0 \ 0 \ ... \ 1]Q_c^{-1} \quad (12.105)$$

The equation (12.102) is said to be in the *controllable phase variable form*.

Example 12.8: Consider a linear system described by the differential equation

$$\ddot{y} + 2\dot{y} + y = \dot{u} + u$$

With
$$x_1 = y, \quad x_2 = \dot{y} - u$$

as the state variables, we get the state model

$$\begin{bmatrix} \dot{x}_1 \\ \dot{x}_2 \end{bmatrix} = \begin{bmatrix} 0 & 1 \\ -1 & -2 \end{bmatrix} \begin{bmatrix} x_1 \\ x_2 \end{bmatrix} + \begin{bmatrix} 1 \\ -1 \end{bmatrix} u \quad (12.106a)$$

$$y = x_1 \quad (12.106b)$$

Let us test controllability of this system by Kalman's test.
From eqn. (12.106a), we have

$$B = \begin{bmatrix} 1 \\ -1 \end{bmatrix}; \ A = \begin{bmatrix} 0 & 1 \\ -1 & -2 \end{bmatrix}$$

Then
$$AB = \begin{bmatrix} 0 & 1 \\ -1 & -2 \end{bmatrix} \begin{bmatrix} 1 \\ -1 \end{bmatrix} = \begin{bmatrix} -1 \\ 1 \end{bmatrix}$$

The composite matrix defined in eqn. (12.101) is given by

$$Q_c = [B \vdots AB] = \begin{bmatrix} 1 & -1 \\ -1 & 1 \end{bmatrix}$$

The rank r of this matrix is 1. The system is therefore not completely controllable. One state of the system is uncontrollable (r out of n states are controllable).

By the method discussed in Section 12.3, the given differential equation can be transformed to the following controllable phase variable model:

$$\begin{bmatrix} \dot{x}_1 \\ \dot{x}_2 \end{bmatrix} = \begin{bmatrix} 0 & 1 \\ -1 & -2 \end{bmatrix} \begin{bmatrix} x_1 \\ x_2 \end{bmatrix} + \begin{bmatrix} 0 \\ 1 \end{bmatrix} u$$

$$y = \begin{bmatrix} 1 & 1 \end{bmatrix} \begin{bmatrix} x_1 \\ x_2 \end{bmatrix}$$

Thus state controllability depends on how the state variables are defined for a given system.

OBSERVABILITY

Consider the state model of an nth order single-output linear time-invariant system,

$$\dot{x} = Ax + Bu$$
$$y = Cx$$

The state equation may be transformed to the canonical form by the linear transformation $x = Mz$. The resulting state and output equations are

$$\dot{z} = \Lambda z + \tilde{B}u \qquad (12.107a)$$

$$y = \tilde{C}z$$
$$= \tilde{c}_1 z_1 + \tilde{c}_2 z_2 + \ldots + \tilde{c}_n z_n \qquad (12.107b)$$

Since diagonalization decouples the states, no state now contains any information regarding any other state, i.e., each state must be independently observable. It therefore follows that for a state to be observed through the output y, its corresponding coefficient in eqn. (12.107b) should be nonzero.

If any particular c_i is zero, the corresponding z_i can have any value without its effect showing up in the output y. Thus the necessary (it is also sufficient) condition for complete state observability is that none of the $\tilde{c}_i$'s (i.e., none of the elements of $\tilde{C} = CM$) should be zero.

The result may be extended to the case of multi-input-multi-output systems where the output vector, after canonical transformation is given by

$$\begin{bmatrix} y_1 \\ y_2 \\ \vdots \\ y_p \end{bmatrix} = \begin{bmatrix} \tilde{c}_{11} & \tilde{c}_{12} & \ldots & \tilde{c}_{1n} \\ \tilde{c}_{21} & \tilde{c}_{22} & \ldots & \tilde{c}_{2n} \\ \vdots & \vdots & & \vdots \\ \tilde{c}_{p1} & \tilde{c}_{p2} & \ldots & \tilde{c}_{pn} \end{bmatrix} \begin{bmatrix} z_1 \\ z_2 \\ \vdots \\ z_n \end{bmatrix}$$

or
$$y = \tilde{C}z$$

The necessary condition for complete observability is that none of the columns of the matrix $\tilde{C}$ be zero.

The Kalman's test of observability is as follows:

A general nth order multi-input-multi-output linear time-invariant system
$$\dot{x} = Ax + Bu$$
$$y = Cx$$
is completely observable if and only if the rank of composite matrix
$$Q_0 = [C^T \;\vdots\; A^T C^T \;\vdots\; \ldots \;(A^T)^{n-1} C^T] \quad (12.108)$$
is n. This condition is also referred to as the pair (AC) being observable.

Duality property

Comparing eqn. (12.108) with (12.101), the following observations may be made:

1. The pair (AB) is controllable implies that the pair $(A^T B^T)$ is observable.
2. The pair (AC) is observable implies that the pair $(A^T C^T)$ is controllable.

Thus the concepts of controllability and observability are dual concepts.

Observable phase variable form

If the system equations are given by*
$$\dot{z} = A_0 z + B_0 u \quad (12.109a)$$
$$y = C_0 z + du \quad (12.109b)$$

where z is $n \times 1$ vector, y and u are scalars, and

$$A_0 = \begin{bmatrix} 0 & 0 & \ldots & 0 & -a_n \\ 1 & 0 & \ldots & 0 & -a_{n-1} \\ 0 & 1 & \ldots & 0 & -a_{n-2} \\ \vdots & \vdots & & \vdots & \vdots \\ 0 & 0 & \ldots & 1 & -a_1 \end{bmatrix} ; \; B_0 = \begin{bmatrix} \beta_n \\ \beta_{n-1} \\ \beta_{n-2} \\ \vdots \\ \beta_1 \end{bmatrix}$$

$$C_0 = [\,0 \; 0 \; \ldots \; 0 \; 1\,]$$

then it is easily verified using (12.108) that the system is always observable. The converse is also true, i.e., a linear time-invariant observable system can be transformed into the form (12.109). This result can easily be proved using the duality property. Equations (12.109) are said to be in the *observable phase variable form*.

Example 12.9: Let us examine the observability of the system given below.

$$\begin{bmatrix} \dot{x}_1 \\ \dot{x}_2 \\ \dot{x}_3 \end{bmatrix} = \begin{bmatrix} 0 & 1 & 0 \\ 0 & 0 & 1 \\ 0 & -2 & -3 \end{bmatrix} \begin{bmatrix} x_1 \\ x_2 \\ x_3 \end{bmatrix} + \begin{bmatrix} 0 \\ 0 \\ 1 \end{bmatrix} u = Ax + Bu \quad (i)$$

*Note that (12.35) can be transformed to (12.109) with the transformation
$$z_i = x_{n+1-i}; \quad i = 1, 2, \ldots, n$$

$$y = \begin{bmatrix} 3 & 4 & 1 \end{bmatrix} \begin{bmatrix} x_1 \\ x_2 \\ x_3 \end{bmatrix} = \mathbf{Cx} \qquad (ii)$$

The characteristic equation is

$$|\mathbf{A} - \lambda \mathbf{I}| = \begin{vmatrix} -\lambda & 1 & 0 \\ 0 & -\lambda & 1 \\ 0 & -2 & -3-\lambda \end{vmatrix} = 0$$

or $\qquad \lambda(\lambda + 1)(\lambda + 2) = 0$

Therefore the eigenvalues of matrix $\mathbf{A}$ are

$$\lambda_1 = 0, \lambda_2 = -1, \lambda_3 = -2$$

The Vander Monde matrix is then

$$\mathbf{M} = \begin{bmatrix} 1 & 1 & 1 \\ 0 & -1 & -2 \\ 0 & 1 & 4 \end{bmatrix}$$

Under the linear transformation $\mathbf{x} = \mathbf{Mz}$, the output is given by

$$y = \mathbf{CMz} = \begin{bmatrix} 3 & 0 & -1 \end{bmatrix} \begin{bmatrix} z_1 \\ z_2 \\ z_3 \end{bmatrix}$$

It is found that the system is not completely observable, since the stat variable z_2 is hidden from observation.

Let us apply the Kalman's test to the same system. From eqns. (i) and (ii)

$$\mathbf{A}^T\mathbf{C}^T = \begin{bmatrix} 0 & 0 & 0 \\ 1 & 0 & -2 \\ 0 & 1 & -3 \end{bmatrix} \begin{bmatrix} 3 \\ 4 \\ 1 \end{bmatrix} = \begin{bmatrix} 0 \\ 1 \\ 1 \end{bmatrix};$$

$$(\mathbf{A}^T)^2\mathbf{C}^T = \begin{bmatrix} 0 & 0 & 0 \\ 1 & 0 & -2 \\ 0 & 1 & -3 \end{bmatrix} \begin{bmatrix} 0 \\ 1 \\ 1 \end{bmatrix} = \begin{bmatrix} 0 \\ -2 \\ -2 \end{bmatrix}$$

Therefore the composite matrix defined in eqn. (12.108) is given by

$$\mathbf{Q}_0 = [\mathbf{C}^T \vdots \mathbf{A}^T\mathbf{C}^T \vdots (\mathbf{A}^T)^2\mathbf{C}^T] = \begin{bmatrix} 3 & 0 & 0 \\ 4 & 1 & -2 \\ 1 & 1 & -2 \end{bmatrix}$$

Since $\begin{vmatrix} 3 & 0 \\ 4 & 1 \end{vmatrix} \neq 0$ and $\begin{vmatrix} 3 & 0 & 0 \\ 4 & 1 & -2 \\ 1 & 1 & -2 \end{vmatrix} = 0$

the rank of the matrix $\mathbf{Q}_0$ is $r = 2$, while $n = 3$. Hence one of the state variables is unobservable.

EFFECT OF POLE-ZERO CANCELLATION IN TRANSFER FUNCTION

In the analysis and design of linear time-invariant control systems, we have used transfer functions extensively. Although controllability and observability are concepts of modern control theory, they are closely related to the properties of the transfer function.

For an nth-order system with distinct eigenvalues, let us assume that the input-output transfer function is of the form

$$T(s) = \frac{Y(s)}{U(s)} = \frac{b_0 s^m + b_1 s^{m-1} + \ldots + b_{m-1} s + b_m}{s^n + a_1 s^{n-1} + \ldots + a_{n-1} s + a_n}; \quad m < n$$

$$= \frac{K(s - \beta_1)(s - \beta_2) \cdots (s - \beta_m)}{(s - \lambda_1)(s - \lambda_2) \ldots (s - \lambda_n)}$$

$$= \sum_{k=1}^{n} \frac{c_k}{(s - \lambda_k)}; \quad c_k \text{ are residues of poles at } s = \lambda_k$$

Assume that the transfer function has identical pair of pole and zero at $\beta_i = \lambda_i$; thus $c_i = 0$. The effect of this cancellation on controllability and observability properties depends on how the state variables are defined. If state variables are selected so as to get state model of the form (12.44), then $c_i = 0$ will appear in control vector $\mathbf{B}$ and the state x_i is uncontrollable. On the other hand if state variables are selected so as to get state model of the form (12.43), then $c_i = 0$ will appear in output vector $\mathbf{C}$ and the state x_i is shielded from observation.

Thus if the input-output transfer function of a linear time-invariant system has pole-zero cancellation, the system will be either not state controllable or unobservable, depending on how the state variables are defined. If the transfer function does not have pole-zero cancellation, the system can always be represented by completely controllable and observable state model.

The importance of controllability and observability concepts will be obvious from the discussion in next section and also in Chapter 13. The reader may note here that practically all physical systems are controllable and observable. However, mathematical models of these systems may not be so. Mathematical models may lack these properties, particularly if linearization has been necessary. If this happens, the model is not an accurate representation of the system. Such a model may not give an optimal control solution and if one exists it may not be implementable. One therefore should seek another state model which is controllable and observable. The reader will surely get a better feel of these concepts after going through Chapter 13 on Optimal Control Systems.

12.7 Pole Placement by State Feedback

Chapter 10 was an exposition of the design of linear control systems using classical control theory techniques, viz., the Bode diagrams, root locus plots, Nichols chart, etc. We observed that in classical design approach, the form of the compensator is preselected based on the performance of the uncompensated system and the design specifications. The parameters of the compensator are then adjusted to bring the system performance within acceptable bounds. We also observed that compensated system required feedback of only one variable, the output. Of course, in systems compensated through inner feedback loop, more than one dynamic variables are fedback.

The modern approach to control system design takes a different view of the problem. It demands not only acceptable performance but optimal (best) performance in some defined sense. The quality or goodness of a system is represented by selecting a suitable performance index J (see Section 5.8). An optimal control is then obtained by minimizing* the selected performance index. The core of this design approach is the selection of a suitable performance index. The performance indices most frequently employed are based on error or functions of error, control energy or functions of control energy, duration of control effort, etc. The final choice of a performance index is the result of physical understanding of the system and sound engineering judgement and skill. Because of the importance of this design approach, a full chapter is devoted to it in this book (Chapter 13). We shall see that in a physical system, in addition to output, other state variables could be utilized for purpose of feedback to get optimum results.

While the classical approach is restrictive in terms of the dynamic variables used for feedback, the optimal control theory's dilemma is the choice of a suitable performance index which is not so clear cut. It is, therefore, logical to extend the power of classical design approach by providing full state feedback. This is the subject of discussion in this article. Of late, interest in literature in this method of design has indeed picked up. We shall restrict our discussion to single-input-single-output systems. For further details the reader is directed to Chen [3].

Consider the single-input-single-output system with nth-order state model

$$\dot{x} = Ax + Bu \qquad (12.110a)$$

$$y = Cx \qquad (12.110b)$$

The state variable feedback for this system is essentially a scalar function which takes the form

---

*The optimal control problem is standardized by minimization of the performance index. Maximization, if required, is given by

$$J_{max} = (-J)_{min}$$

$$\sigma = \mathbf{Kx} = [k_1 \ k_2 \ \ldots \ k_n] \begin{bmatrix} x_1 \\ x_2 \\ \vdots \\ x_n \end{bmatrix} \qquad (12.111a)$$

The block diagram of the system with state variable feedback is shown in Fig. 12.14. From this figure it is observed that

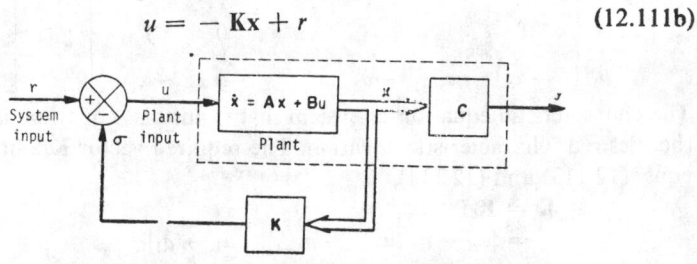

$$u = -\mathbf{Kx} + r \qquad (12.111b)$$

Figure 12.14 State variable feedback system.

where u is plant input and r is the system input.

From eqns. (12.110) and (12.111), we get state equation for the feedback system as

$$\dot{\mathbf{x}} = (\mathbf{A} - \mathbf{BK})\mathbf{x} + \mathbf{B}r \qquad (12.112)$$

It can be shown that if the pair (AB) is controllable, then by the state feedback given by (12.111), the eigenvalues of $(\mathbf{A} - \mathbf{BK})$ in (12.112) can be arbitrarily assigned.

To prove this, let us first transform (12.110a) into controllable phase variable form using the transformation

$$\mathbf{z} = \mathbf{Px}$$

given by (12.105). The resulting phase variable form is

$$\dot{\mathbf{z}} = \begin{bmatrix} 0 & 1 & 0 & \ldots & 0 & 0 \\ 0 & 0 & 1 & \ldots & 0 & 0 \\ \vdots & \vdots & \vdots & & \vdots & \vdots \\ 0 & 0 & 0 & \ldots & 0 & 1 \\ -a_n & -a_{n-1} & -a_{n-2} & \ldots & -a_2 & -a_1 \end{bmatrix} \mathbf{z} + \begin{bmatrix} 0 \\ 0 \\ \vdots \\ 0 \\ 1 \end{bmatrix} u$$

$$= \mathbf{A}_c \mathbf{z} + \mathbf{B}_c u$$

Under this transformation, eqn. (12.111b) modifies to

$$u = -\mathbf{KP}^{-1}\mathbf{z} + r$$
$$= -\mathbf{K}_c \mathbf{z} + r$$

where $\qquad \mathbf{K}_c = \mathbf{KP}^{-1} \qquad (12.113)$

Since the eigenvalues are invariant under transformation of state model, the set of eigenvalues of $(\mathbf{A} - \mathbf{BK})$ is the same as that of $(\mathbf{A}_c - \mathbf{B}_c \mathbf{K}_c)$.

Let the desired eigenvalues be $\lambda_1, \lambda_2, \ldots, \lambda_n$ and

$$\lambda^n + \alpha_1 \lambda^{n-1} + \alpha_2 \lambda^{n-2} + \ldots + \alpha_n = 0$$

be the corresponding characteristic equation.

If we choose
$$\mathbf{K}_c = [\alpha_n - a_n, \alpha_{n-1} - a_{n-1}, \ldots, \alpha_1 - a_1] \tag{12.114}$$
then the controllable phase variable form of the feedback system is

$$\dot{\mathbf{z}} = \begin{bmatrix} 0 & 1 & 0 & \cdots & 0 & 0 \\ 0 & 0 & 1 & \cdots & 0 & 0 \\ \vdots & \vdots & \vdots & & \vdots & \vdots \\ 0 & 0 & 0 & \cdots & 0 & 1 \\ -\alpha_n & -\alpha_{n-1} & -\alpha_{n-2} & & -\alpha_2 & -\alpha_1 \end{bmatrix} \mathbf{z} + \begin{bmatrix} 0 \\ 0 \\ \vdots \\ 0 \\ 1 \end{bmatrix} r$$

The characteristic equation of system matrix in this equation is same as the desired characteristic equation. The required vector $\mathbf{K}$, obtained from eqns. (12.113) and (12.114), is
$$\mathbf{K} = \mathbf{K}_c \mathbf{P}$$
$$= [\alpha_n - a_n \;\; \alpha_{n-1} - a_{n-1} \ldots \alpha_1 - a_1]\mathbf{P} \tag{12.115}$$

We have seen that if a system is completely controllable, its eigenvalues can be arbitrarily located by complete state feedback. Thus any unstable system can be stabilized by complete state feedback if all the states are controllable. More restrictively, even if the system is not completely controllable, it is still *stabilizable* as long as the uncontrollable states are stable.

Example 12.10: Consider the position control system of Fig. 12.15 using a d.c. motor in armature control mode. The signal flow graph of the system with usual symbols is drawn in Fig. 12.16a. Being a third-order system three physical variable θ_c, $\dot{\theta}_c$ and i_a are selected as state variables, i.e.,
$$\mathbf{x}^T = [\theta_c \;\; \dot{\theta}_c \;\; i_a]$$

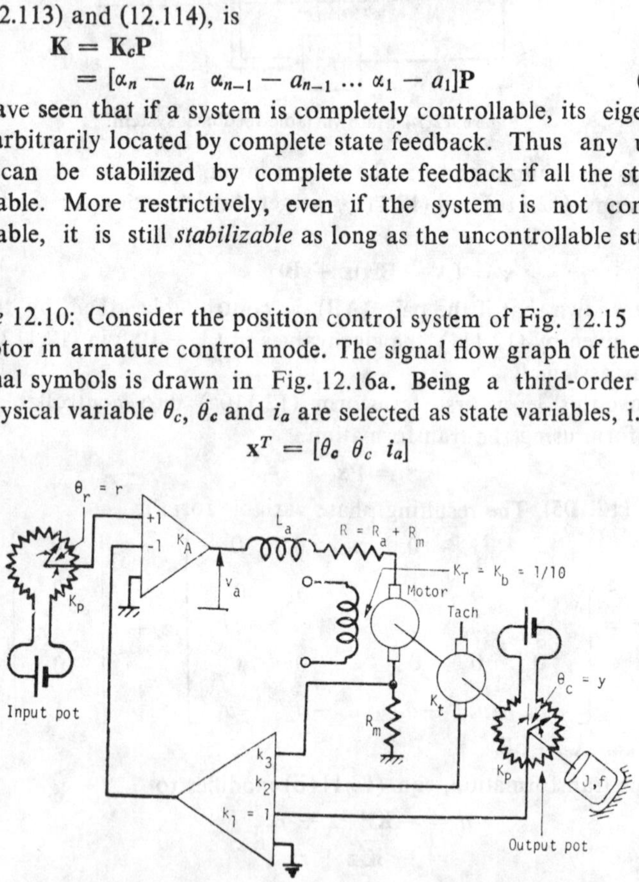

Figure 12.15 A position control system with state variable feedback.

Full state feedback is employed with position feedback, being obtained from a potentiometer, rate feedback from a tachometer and current feedback from a current sample across a resistance (small) in series with armature circuit. In the modified signal flow graph of Fig. 12.16b the outer

State Variable Analysis and Design   507

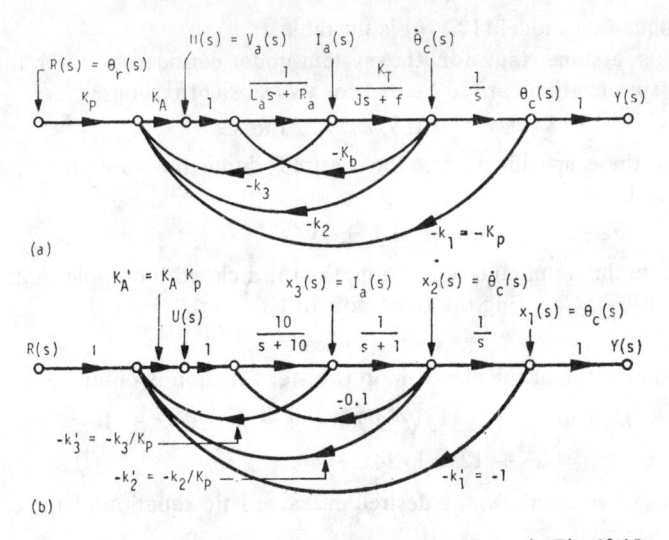

Figure 12.16 Signal flow graphs of the system shown in Fig. 12.15.

feedback (position feedback) gain $-k_1 = -K_P$ is reduced to unity by combining potentiometer constant K_P with the gain K_A in the forward path. Three adjustable gains (k'_2, k'_3 and K'_A), one in forward path and two in feedback paths, are now available for achieving design specifications. It must be noted that the inner feedback loop is inherent in the motor operation. Suitable transfer functions for fixed part of the system are shown in Fig. 12.16b. The armature circuit time constant has been assumed to be larger than normal for purpose of illustrating the design technique.

With all the three state feedback loops open, the system is described by

$$\begin{bmatrix} \dot{x}_1 \\ \dot{x}_2 \\ \dot{x}_3 \end{bmatrix} = \begin{bmatrix} 0 & 1 & 0 \\ 0 & -1 & 1 \\ 0 & -1 & -10 \end{bmatrix} \begin{bmatrix} x_1 \\ x_2 \\ x_3 \end{bmatrix} + \begin{bmatrix} 0 \\ 0 \\ 10 \end{bmatrix} u$$

$$= \mathbf{A}\mathbf{x} + \mathbf{B}u \qquad (12.116)$$

Let us first test controllability of pair (**AB**). The composite matrix

$$\mathbf{Q}_c = [\mathbf{B} \ \vdots \ \mathbf{AB} \ \vdots \ \mathbf{A}^2\mathbf{B}]$$

$$= \begin{bmatrix} 0 & 0 & 10 \\ 0 & 10 & -110 \\ 10 & -100 & 990 \end{bmatrix}$$

has rank $r = 3$. Therefore, the pair (**AB**) is controllable. The eigenvalues of **A** in (12.116) are given by the characteristic equation

$$|\lambda \mathbf{I} - \mathbf{A}| = 0$$

or

$$\lambda^3 + 11\lambda^2 + 11\lambda = 0 \qquad (12.117a)$$

Note that state model (12.116) is unstable.

Let us assume that for the system under consideration, the following design specifications are to be met for its transient response

$$\zeta = 0.5; \omega_n = 2 \text{ rad/sec}$$

From these specifications, the desired dominant closed-loop poles are found to be at

$$s_d = -1 \pm j\sqrt{3}$$

To ensure the dominance condition, the third closed-loop pole should lie at $s \leqslant -10\zeta\omega_n$. Choosing the third pole to be

$$s_3 = -10\zeta\omega_n = -10$$

the denominator of the closed-loop transfer function is obtained as

$$D(s) = (s + 1 + j\sqrt{3})(s + 1 - j\sqrt{3})(s + 10)$$
$$= s^3 + 12s^2 + 24s + 40$$

In terms of eigenvalues, the desired characteristic equation of the compensated system is

$$\lambda^3 + 12\lambda^2 + 24\lambda + 40 = 0 \quad (12.117b)$$

Following the procedure given earlier, we have from (12.117)

$$\mathbf{K}_c = [K_{c1} \quad K_{c2} \quad K_{c3}] = [(40 - 0) \quad (24 - 11) \quad (12 - 11)]$$
$$= [40 \quad 13 \quad 1]$$

From eqn. (12.115), we have

$$\mathbf{K} = \mathbf{K}_c \mathbf{P}$$

where (eqn. (12.105))

$$\mathbf{P} = \begin{bmatrix} \mathbf{P}_1 \\ \mathbf{P}_1 \mathbf{A} \\ \mathbf{P}_1 \mathbf{A}^2 \end{bmatrix}$$

$$\mathbf{P}_1 = [0 \quad 0 \quad 1]\mathbf{Q}_c^{-1}$$

$$= [0 \quad 0 \quad 1]\begin{bmatrix} 1.1 & 1 & 0.1 \\ 1.1 & 0.1 & 0 \\ 0.1 & 0 & 0 \end{bmatrix}$$

$$= [0.1 \quad 0 \quad 0]$$
$$\mathbf{P}_1 \mathbf{A} = [0 \quad 0.1 \quad 0]$$
$$\mathbf{P}_1 \mathbf{A}^2 = [0 \quad -0.1 \quad 0.1]$$

Thus

$$\mathbf{P} = \begin{bmatrix} 0.1 & 0 & 0 \\ 0 & 0.1 & 0 \\ 0 & -0.1 & 0.1 \end{bmatrix}$$

and
$$\mathbf{K} = [40 \quad 13 \quad 1]\,\mathbf{P}$$
$$= [4 \quad 1.2 \quad 0.1]$$
$$= [K_1 \quad K_2 \quad K_3]$$

From Fig. 12.16b, we find that
$$K_1 = k'_1 K'_A = 4$$
$$K_2 = k'_2 K'_A = 1.2$$
$$K_3 = k'_3 K'_A = 0.1$$

With $k'_1 = 1$, we have
$$K'_A = 4,\, k'_2 = 0.3,\, k'_3 = 0.025$$

The above choice of gains meets the design specifications.

OBSERVER SYSTEMS

In the problem of pole placement by state feedback we found that the control law is (eqn. (12.111b))
$$u = -\mathbf{K}\mathbf{x} + r$$

As we have seen earlier, to implement this control law, we need to feedback all the state variables. In Example 12.10, fortunately the state variables we selected were accessible for feedback. However, in many practical situations, all the state variables are not accessible for measurement and control purposes; only inputs and outputs are measurable. In the following we show that the available inputs and outputs can be used to drive a device whose outputs will approximate the state vector. This device is called a *state observer*. The output of observer can then be used to implement the feedback control law. Figure 12.17 shows the overall system structure including the observer. Intuitively, the observer should have the same state equations as the original system and design criterion should be to minimize the difference between the system output $y = \mathbf{C}\mathbf{x}$ and the output $\hat{y} = \mathbf{C}\hat{\mathbf{x}}$ as constructed by the observed vector $\hat{\mathbf{x}}$. Note that this is equivalent to minimization of $\mathbf{x} - \hat{\mathbf{x}}$. Since $\mathbf{x}$ is inaccessible, we attempt to minimize $y - \hat{y}$. The difference $(y - \hat{y})$ is multiplied by an $n \times 1$ vector $\mathbf{G}$ and fed

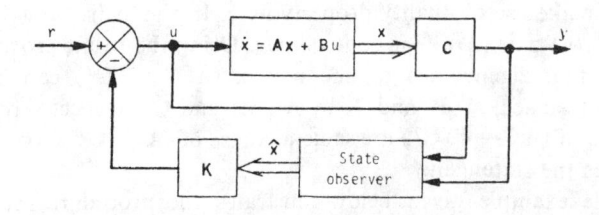

Figure 12.17 A linear system with state observer.

into the input of the integrators of the observer (Fig. 12.18). The problem is to design **G** so that $(y - \hat{y})$ is minimized.

From Fig. 12.18, we have

$$\dot{\hat{x}} = A\hat{x} + G(Cx - C\hat{x}) + Bu$$
$$= (A - GC)\hat{x} + GCx + Bu$$

or $\dot{x} - \dot{\hat{x}} = (A - GC)(x - \hat{x})$

Define

$$\tilde{x} = x - \hat{x}$$

This gives

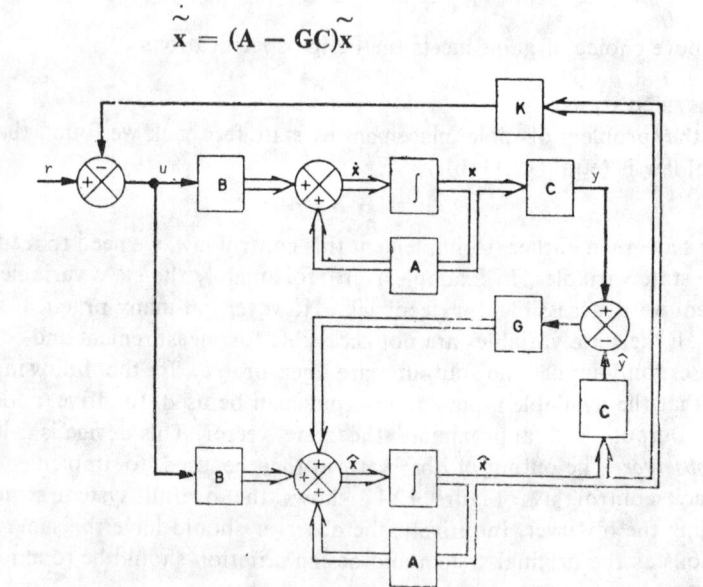

$$\dot{\tilde{x}} = (A - GC)\tilde{x}$$

Figure 12.18 A linear system with state observer.

$\tilde{x}$ can be made to decay fast if eigenvalues of the matrix $(A - GC)$ are properly selected. In the following we show that if the pair (AC) is observable, then the eigenvalues of the matrix $(A - GC)$ can be chosen arbitrarily.

We will make use of duality property here. If the pair (AC) is observable, then the pair $(A^T C^T)$ is controllable. Using the result proved earlier, we can say that eigenvalues of the matrix $(A^T - C^T G^T)$ can be chosen arbitrarily (replace **A**, **B** and **K** by A^T, C^T and G^T respectively). But the eigenvalues of $(A^T - C^T G^T)$ are same as those of $(A - GC)^T$ or $(A - GC)$. This proves the statement.

A simple example given below illustrates the procedure for design of observers.

Example 12.11: Consider a linear system described by the equations

$$\dot{x} = \begin{bmatrix} 1 & 2 & 0 \\ 3 & -1 & 1 \\ 0 & 2 & 0 \end{bmatrix} x + \begin{bmatrix} 2 \\ 1 \\ 1 \end{bmatrix} u \qquad (12.118a)$$

$$= Ax + Bu$$
$$y = [0 \ 0 \ 1]x \qquad (12.118b)$$
$$= Cx$$

It can easily be verified that the state model (12.118) is observable.

It is desired to design a state observer so that eigenvalues of the matrix $(A - GC)$ are at $-4, -3 \pm j1$.

Let

$$G = \begin{bmatrix} g_1 \\ g_2 \\ g_3 \end{bmatrix}$$

Then

$$(A - GC) = \begin{bmatrix} 1 & 2 & -g_1 \\ 3 & -1 & 1-g_2 \\ 0 & 2 & -g_3 \end{bmatrix}$$

The characteristic equation is

$$|\lambda I - (A - GC)| = 0$$

or

$$\begin{vmatrix} \lambda-1 & -2 & g_1 \\ -3 & \lambda+1 & g_2-1 \\ 0 & -2 & \lambda+g_3 \end{vmatrix} = 0$$

This gives

$$\lambda^3 + g_3\lambda^2 + (2g_2 - 9)\lambda + 2 + 6g_1 - 2g_2 - 7g_3 = 0 \qquad (12.119a)$$

The desired characteristic equation is

$$(\lambda + 3 + j1)(\lambda + 3 - j1)(\lambda + 4) = 0$$

or

$$\lambda^3 + 10\lambda^2 + 34\lambda + 40 = 0 \qquad (12.119b)$$

Equating the coefficients in eqns. (12.119a) and (12.119b) we get

$$g_1 = 25.2, \ g_2 = 21.5, \ g_3 = 10$$

12.8 State Variables and Linear Discrete-time Systems

State variable methods for the analysis and design of continuous-time systems developed in earlier sections of this chapter can be extended for the analysis and design of discrete-time systems.

In Chapter 11, we discussed analysis and design of discrete-time systems using z-transfer function. There we observed that while dealing with discrete-time systems, we often encounter two different situations. The first situation involves systems that are completely discrete with respect to time in the sense that they receive and send out discrete signals only. It is hard to find examples of physical systems of this nature, other than digital computers. Most of the physical systems are of continuous-time type. In socio-economic and some other nonphysical systems, a natural discrete unit of time exists and such systems can effectively be modelled using difference equations.

The second situation is that the components of the system are continuous-time elements but the signals at certain points of the system are discrete with respect to time because of the sample and hold operations. Both the cases will be studied in this section.

The general form of state model for a multivariable discrete-time system is

$$\mathbf{x}[(k+1)T] = \mathbf{f}[\mathbf{x}(kT), \mathbf{u}(kT)] \qquad (12.120a)$$

$$\mathbf{y}(kT) = \mathbf{g}[\mathbf{x}(kT), \mathbf{u}(kT)] \qquad (12.120b)$$

where $\mathbf{x}(kT)$ = state vector, $\mathbf{u}(kT)$ = input vector and $\mathbf{y}(kT)$ = output vector.

From these equations we see that given the initial state $\mathbf{x}(0)$ and values of inputs $\mathbf{u}(0), \mathbf{u}(T), \mathbf{u}(2T), ..., \mathbf{u}(kT)$; we can uniquely deduce the evolution of state $\mathbf{x}(T), \mathbf{x}(2T), ..., \mathbf{x}[(k+1)T]$ and the evolution of the output $\mathbf{y}(0), \mathbf{y}(T), ..., \mathbf{y}(kT)$. Also note that eqn. (12.120a) is a set of first-order difference equations which represents dynamics of discrete-time process and eqn. (12.120b) is an algebraic equation.

For an nth-order linear time-invariant system, eqns. (12.120) reduce to the following form:

$$\mathbf{x}(k+1) = \mathbf{A}\mathbf{x}(k) + \mathbf{B}\mathbf{u}(k); \quad \mathbf{x}(kT) \triangleq \mathbf{x}(k) \qquad (12.121a)$$

$$\mathbf{y}(k) = \mathbf{C}\mathbf{x}(k) + \mathbf{D}\mathbf{u}(k) \qquad (12.121b)$$

$\mathbf{x}(k) = n \times 1$ state vector

$\mathbf{u}(k) = m \times 1$ input vector

$\mathbf{y}(k) = p \times 1$ output vector

$\mathbf{A} = n \times n$ system matrix

$\mathbf{B} = n \times m$ input matrix

$\mathbf{C} = p \times n$ output matrix

$\mathbf{D} = p \times m$ transmission matrix

STATE MODELS FROM LINEAR DIFFERENCE EQUATIONS/TRANSFER FUNCTIONS

The general form of nth-order linear difference equation relating output $y(k)$ to the input $u(k)$ of a discrete-time system is

$$y(k+n) + a_1 y(k+n-1) + ... + a_{n-1} y(k+1) + a_n y(k)$$
$$= b_0 u(k+m) + ... + b_{m-1} u(k+1) + b_m u(k)$$

where, for time-invariant systems a_i's and b_j's are constants; k, m, n are integers with $m \leqslant n$. The initial conditions are expressed in terms of $y(0)$,

$y(1), ..., y(n - 1)$. The transfer function obtained from this equation, under the assumption of zero initial conditions, is

$$T(z) = \frac{Y(z)}{U(z)} = \frac{b_0 z^m + ... + b_{m-1} z + b_m}{z^n + a_1 z^{n-1} + ... + a_{n-1} z + a_n}$$

When $m = n$,

$$T(z) = \frac{Y(z)}{U(z)} = \frac{b_0 + b_1 z^{-1} + ... + b_n z^{-n}}{1 + a_1 z^{-1} + ... + a_n z^{-n}} \quad (12.122)$$

The implementation of this transfer function by a digital computer can be done in many ways. Two methods are illustrated in terms of state diagrams in the following:

Method 1: The signal flow graph for $T(z)$ of (12.122) is shown in Fig. 12.19 (z^{-1} represents a delay element having a delay of T sec). Taking outputs of delay elements as state variables, we can write

$x_1(k + 1) = x_2(k)$
$x_2(k + 1) = x_3(k)$
$\vdots$
$x_n(k + 1) = -a_n x_1(k) - a_{n-1} x_2(k) ... - a_1 x_n(k) + u$
$y(k) = b_n x_1(k) + b_{n-1} x_2(k) ... + b_1 x_n(k)$
$\quad + b_0(-a_n x_1(k) - ... - a_1 x_n(k) + u(k))$
$\quad = (b_n - a_n b_0) x_1(k) + (b_{n-1} - a_{n-1} b_0) x_2(k)$
$\quad + ... + (b_1 - a_1 b_0) x_n(k) + b_0 u(k)$

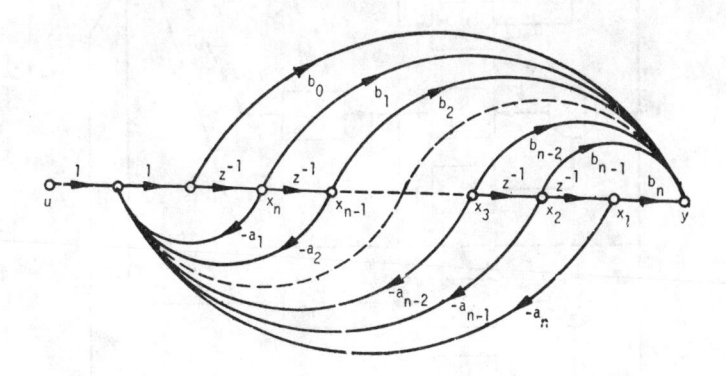

Figure 12.19 Signal flow graph for transfer function of eqn. (12.122).

These equations can be written as

$$\mathbf{x}(k + 1) = \mathbf{A}\mathbf{x}(k) + \mathbf{B}u(k) \quad (12.123a)$$
$$y(k) = \mathbf{C}\mathbf{x}(k) + b_0 u(k) \quad (12.123b)$$

where

$$A = \begin{bmatrix} 0 & 1 & 0 & \cdots & 0 \\ 0 & 0 & 1 & \cdots & 0 \\ \vdots & \vdots & \vdots & & \vdots \\ 0 & 0 & 0 & \cdots & 1 \\ -a_n & -a_{n-1} & -a_{n-2} & \cdots & -a_1 \end{bmatrix}; \quad B = \begin{bmatrix} 0 \\ 0 \\ \vdots \\ 0 \\ 1 \end{bmatrix}$$

$$C = [(b_n - a_n b_0) \ (b_{n-1} - a_{n-1} b_0) \ \ldots \ (b_1 - a_1 b_0)]$$

This is phase variable form of state model (Section 12.3).

Method 2: Another state space representation can be obtained by the partial fraction technique. First let us assume that $T(z)$ of (12.122) has n distinct poles located at $\lambda_1, \lambda_2, \ldots, \lambda_n$. Then $T(z)$ can be expressed as

$$T(z) = b_0 + \sum_{i=1}^{n} \frac{c_i}{z - \lambda_i}; \quad c_i \text{ is the residue of pole at } \lambda_i$$

The block diagram for this $T(z)$ is shown in Fig. 12.20. Taking outputs of delay elements as state variables, we get the following state model:

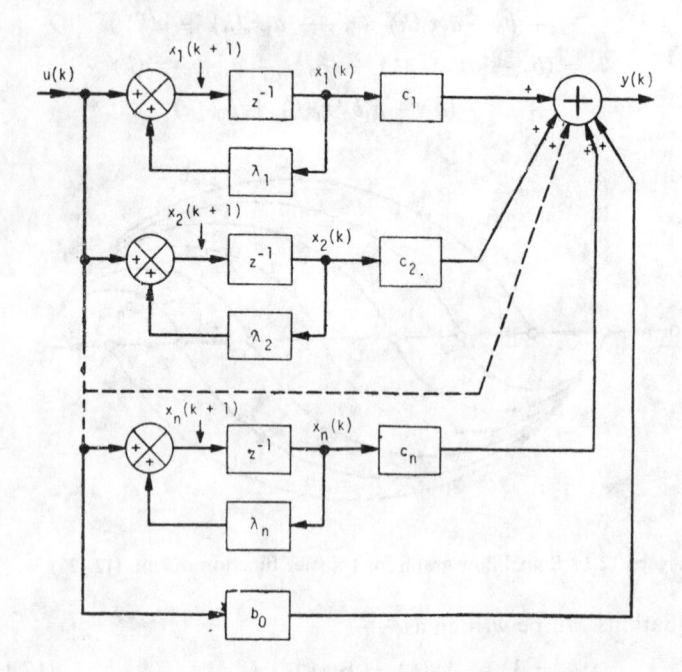

Figure 12.20 Block diagram of canonical state model.

$$\begin{bmatrix} x_1(k+1) \\ x_2(k+1) \\ \vdots \\ x_n(k+1) \end{bmatrix} = \begin{bmatrix} \lambda_1 & 0 & 0 & \cdots & 0 \\ 0 & \lambda_2 & 0 & \cdots & 0 \\ \vdots & \vdots & \vdots & & \vdots \\ 0 & 0 & 0 & \cdots & \lambda_n \end{bmatrix} \begin{bmatrix} x_1(k) \\ x_2(k) \\ \vdots \\ x_n(k) \end{bmatrix} + \begin{bmatrix} 1 \\ 1 \\ \vdots \\ 1 \end{bmatrix} u(k)$$

(12.124a)

$$y(k) = [c_1 \quad c_2 \quad \cdots \quad c_n] \begin{bmatrix} x_1(k) \\ x_2(k) \\ \vdots \\ x_n(k) \end{bmatrix} + b_0 u(k) \quad (12.124b)$$

This is the canonical state model (Section 12.3).

When some of the poles of $T(z)$ are repeated, we get Jordan canonical form. The method of obtaining this form is identical to that discussed earlier in Section 12.3.

DERIVATION OF TRANSFER FUNCTION FROM STATE MODEL

Taking the z-transform on both sides of eqn. (12.121a) (assuming scalar input),

$$zX(z) - zx(0) = AX(z) + BU(z)$$

Solving for $X(z)$, we get

$$X(z) = (zI - A)^{-1} zx(0) + (zI - A)^{-1} BU(z) \quad (12.125)$$

Taking the z-transform of eqn. (12.121b) we get (for scalar output, $D = d$),

$$Y(z) = CX(z) + dU(z)$$

Solving for $Y(z)$, we obtain

$$Y(z) = C(zI - A)^{-1} zx(0) + C(zI - A)^{-1} BU(z) + dU(z)$$

Assuming zero initial conditions, we get the system transfer function as

$$T(z) = \frac{Y(z)}{U(z)} = C(zI - A)^{-1} B + d$$

$$= \frac{C \; dj(zI - A) B}{|zI - A|} + d \quad (12.126)$$

Setting the denominator equal to zero, we have the characteristic equation

$$|zI - A| = 0 \quad (12.127)$$

The roots of the characteristic equation are the eigenvalues of matrix A.

LINEAR TRANSFORMATION OF STATE VECTOR

We know that canonical state model is very useful in system analysis (Section 12.5).

The transformation

$$x(k) = Mz(k); \quad M \quad \text{modal matrix}$$

transforms (12.121) to the following form

$$z(k+1) = \Lambda z(k) + \tilde{B}u(k); \Lambda = M^{-1}AM; \tilde{B} = M^{-1}B \quad (12.128a)$$

$$y(k) = \tilde{C}z(k) + Du(k); \tilde{C} = CM \quad (12.128b)$$

The techniques of obtaining modal matrix have already been discussed in Section 12.4. If matrix A has repeated eigenvalues, it can be transformed to Jordan canonical form.

SOLUTION OF STATE EQUATIONS

From the state model (12.121) we can write

$$x(1) = Ax(0) + Bu(0)$$
$$x(2) = Ax(1) + Bu(1)$$
$$= A^2x(0) + ABu(0) + Bu(1)$$
$$\cdots \quad \cdots \quad \cdots \quad \cdots \quad \cdots$$
$$x(k) = A^k x(0) + A^{k-1}Bu(0) + A^{k-2}Bu(1) + \ldots + Bu(k-1)$$

Let us define for $k > 0$

$$\Phi(k) \stackrel{\Delta}{=} A^k$$

and

$$\Phi(0) \stackrel{\Delta}{=} I \text{ (identity matrix)}$$

Then we have

$$x(k) = \Phi(k)x(0) + \sum_{i=0}^{k-1} \Phi(k-i-1)Bu(i) \quad (12.129)$$

The matrix $\Phi(k)$ is referred to as the state transition matrix for discrete-time system (12.121). Properties of the state transition matrix are summarized below:

1. $\Phi(0) = I$
2. $\Phi^{-1}(k) = \Phi(-k)$
3. $\Phi(k, k_0) = \Phi(k - k_0) = A^{(k-k_0)}; k > k_0$

Computation of Φ

Some methods of computing Φ are given below:

1. Comparing eqns. (12.125) and (12.129), we observe that

$$\Phi(k) = \mathscr{Z}^{-1}[(zI - A)^{-1}z]$$

2. If M is the diagonalizing matrix of A, then

$$M^{-1}AM = \Lambda = \begin{bmatrix} \lambda_1 & 0 & \cdots & 0 \\ 0 & \lambda_2 & \cdots & 0 \\ \vdots & \vdots & & \vdots \\ 0 & 0 & \cdots & \lambda_n \end{bmatrix}$$

$$\Phi(k) = M\Lambda^k M^{-1}$$

$$= M \begin{bmatrix} \lambda_1^k & 0 & \cdots & 0 \\ 0 & \lambda_2^k & \cdots & 0 \\ \vdots & \vdots & & \vdots \\ 0 & 0 & \cdots & \lambda_n^k \end{bmatrix} M^{-1}$$

If **A** has repeated eigenvalues, it may first be converted to Jordan matrix **J** and then state transition matrix can be easily computed. The technique is similar to the one discussed for continuous-time systems in Section 12.5.

3. Use of technique based on Caley-Hamilton theorem: This technique has already been given in Section 12.5. It is illustrated for a discrete-time system in Example 12.12.

Example 12.12: A discrete-time system has state and output equation given by

$$x_1(k+1) = 1/4 x_1(k) + u(k)$$
$$x_2(k+1) = 1/8 x_1(k) + 1/8 x_2(k) + u(k)$$

$$y(k) = [1/2 \quad 0] \begin{bmatrix} x_1(k) \\ x_2(k) \end{bmatrix}$$

Solve for the output $y(k)$ when $u(k)$ = unit impulse and $\mathbf{x}(0) = \mathbf{0}$.

Solution:

It is easily seen that

$$\mathbf{A} = \begin{bmatrix} 1/4 & 0 \\ 1/8 & 1/8 \end{bmatrix}; \quad \mathbf{B} = \begin{bmatrix} 1 \\ 1 \end{bmatrix}; \quad \mathbf{C} = [1/2 \quad 0]$$

The characteristic equation is

$$q(\lambda) = |\lambda \mathbf{I} - \mathbf{A}| = \begin{vmatrix} \lambda - 1/4 & 0 \\ -1/8 & \lambda - 1/8 \end{vmatrix} = 0$$

which gives the eigenvalues as

$$\lambda = 1/4, \ 1/8 \text{ or } 1/2^2, \ 1/2^3$$

Now

$$f(\lambda) = \lambda^k = \alpha_0 + \lambda \alpha_1$$
$$(\tfrac{1}{2})^{2k} = \alpha_0 + 1/4 \alpha_1$$
$$(\tfrac{1}{2})^{3k} = \alpha_0 + 1/8 \alpha_1$$

Solving we get

$$\alpha_0 = -(\tfrac{1}{2})^{2k} + (\tfrac{1}{2})^{3k-1}$$
$$\alpha_1 = (\tfrac{1}{2})^{2k-3} - (\tfrac{1}{2})^{3k-3}$$
$$\Phi(k) = \mathbf{A}^k = \alpha_0 \mathbf{I} + \alpha_1 \mathbf{A}$$

$$= \begin{bmatrix} -(\tfrac{1}{2})^{2k} + (\tfrac{1}{2})^{3k-1} & 0 \\ 0 & -(\tfrac{1}{2})^{2k} + (\tfrac{1}{2})^{3k-1} \end{bmatrix} + [(\tfrac{1}{2})^{2k-3} - (\tfrac{1}{2})^{3k-3}] \begin{bmatrix} (\tfrac{1}{2})^2 & 0 \\ (\tfrac{1}{2})^3 & (\tfrac{1}{2})^3 \end{bmatrix}$$

$$\Phi(k) = \begin{bmatrix} -(\tfrac{1}{2})^{2k} + (\tfrac{1}{2})^{2k-1} & 0 \\ (\tfrac{1}{2})^{2k} - (\tfrac{1}{2})^{3k} & -(\tfrac{1}{2})^{3k} + (\tfrac{1}{2})^{3k-1} \end{bmatrix}$$

From eqn. (12.129),

$$\begin{aligned}
x(k) &= \Phi(k-1)B \cdot 1 \\
&= \begin{bmatrix} -(\tfrac{1}{2})^{2(k-1)} + (\tfrac{1}{2})^{2k-3} & 0 \\ (\tfrac{1}{2})^{2(k-1)} - (\tfrac{1}{2})^{3k-3} & -(\tfrac{1}{2})^{3(k-1)} + (\tfrac{1}{2})^{3k-4} \end{bmatrix} \begin{bmatrix} 1 \\ 1 \end{bmatrix} \\
&= \begin{bmatrix} -(\tfrac{1}{2})^{2(k-1)} + (\tfrac{1}{2})^{2k-3} \\ (\tfrac{1}{2})^{2(k-1)} \end{bmatrix}
\end{aligned}$$

$$\begin{aligned}
y(k) &= [\tfrac{1}{2} \quad 0]\, x(k) \\
&= -(\tfrac{1}{2})^{2k-1} + (\tfrac{1}{2})^{2(k-1)} = (\tfrac{1}{2})^{2(k-1)}[-\tfrac{1}{2} + 1] = (\tfrac{1}{2})^{2k-1}
\end{aligned}$$

LINEAR CONTINUOUS-TIME SYSTEMS WITH SAMPLED INPUTS

As has been pointed out earlier, discrete-time systems may arise in various ways. A specific example is that of calculating the response of a continuous-time system by means of a digital computer. Under this situation, a continuous-time system must first be converted into an equivalent discrete-time system.

Consider the continuous-time state equation

$$\dot{x} = Ax + Bu \tag{12.130}$$

We shall impose the following restrictions on the system.

(i) The input vector u can change only at sampling instants.
(ii) The sampling period T is constant at all times.

The continuous-time system with these restrictions is shown in Fig. 12.21. Because of sample and hold operations,

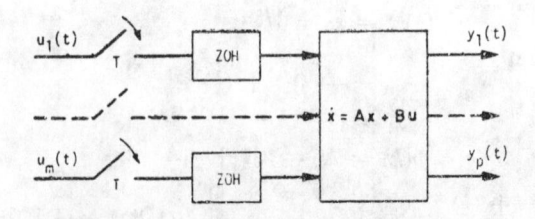

$$u_i(t) = u_i(kT); \quad kT \leqslant t < (k+1)T$$
$$k = 0, 1, 2, \ldots$$
$$i = 1, 2, \ldots, m$$

Figure 12.21 Open-loop sampled-data system.

Solution of (12.130) obtained from (12.76) with t_0 as the initial time, is

$$\mathbf{x}(t) = e^{\mathbf{A}(t-t_0)}\mathbf{x}(t_0) + \int_{t_0}^{t} e^{\mathbf{A}(t-\tau)}\mathbf{B}\mathbf{u}(\tau)\, d\tau$$

In our case, the input is sampled, so we will develop the solution going from one sampling instant $t_0 = kT$ to the next sampling instant $t = (k+1)T$. Thus

$$\mathbf{x}(t) = e^{(\mathbf{A}t-kT)}\mathbf{x}(kT) + \int_{kT}^{t} e^{\mathbf{A}(t-\tau)}\mathbf{B}\mathbf{u}(kT)\, d\tau; \quad (12.131)$$
$$kT \leq t < (k+1)T$$

If we are interested only in response at the sampling instants, we set $t = (k+1)T$. This gives

$$\mathbf{x}(k+1) = \Phi(T)\mathbf{x}(k) + \theta(T)\mathbf{u}(k) \quad (12.132a)$$

where

$$\Phi(T) = e^{\mathbf{A}T} \quad (12.132b)$$

and

$$\theta(T) = \int_{kT}^{(k+1)T} e^{\mathbf{A}((k+1)T-\tau)}\mathbf{B}\, d\tau$$

Letting $\sigma = \tau - kT$, we have

$$\theta(T) = \int_{0}^{T} e^{\mathbf{A}(T-\sigma)}\mathbf{B}\, d\sigma$$

With $\lambda = T - \sigma$, we get

$$\theta(T) = \int_{0}^{T} e^{\mathbf{A}\lambda}\mathbf{B}\, d\lambda \quad (12.132c)$$

If we are interested in value of $\mathbf{x}(t)$ between sampling instants, we first solve for $\mathbf{x}(kT)$ for any k using (12.132) and then use (12.131) to determine $\mathbf{x}(t)$ for $kT < t < (k+1)T$.

Example 12.13: Consider the sampled-data control system shown in Fig. 12.22.

For sample and hold element,

$$u(t) = e(t_0); \quad t_0 \leq t < t_0 + T$$
$$e(t_0) = r(t_0) - y(t_0)$$

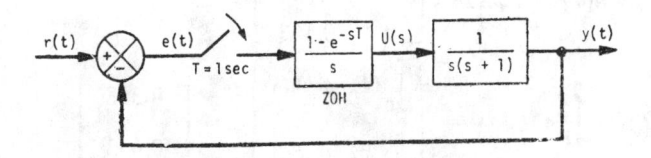

Figure 12.22 A sampled-data feedback system.

The selection of state variables for the continuous-time plant is shown in Fig. 12.23. The state model for the continuous-time plant is

$$\begin{bmatrix} \dot{x}_1 \\ \dot{x}_2 \end{bmatrix} = \begin{bmatrix} 0 & 1 \\ 0 & -1 \end{bmatrix} \begin{bmatrix} x_1 \\ x_2 \end{bmatrix} + \begin{bmatrix} 0 \\ 1 \end{bmatrix} e(t_0); \; t_0 \leq t < t_0 + T$$

$$= \mathbf{A}\mathbf{x} + \mathbf{B}e(t_0) \quad (12.133a)$$

$$y = \mathbf{C}\mathbf{x} \quad (12.133b)$$

$$= \begin{bmatrix} 1 & 0 \end{bmatrix} \begin{bmatrix} x_1 \\ x_2 \end{bmatrix}$$

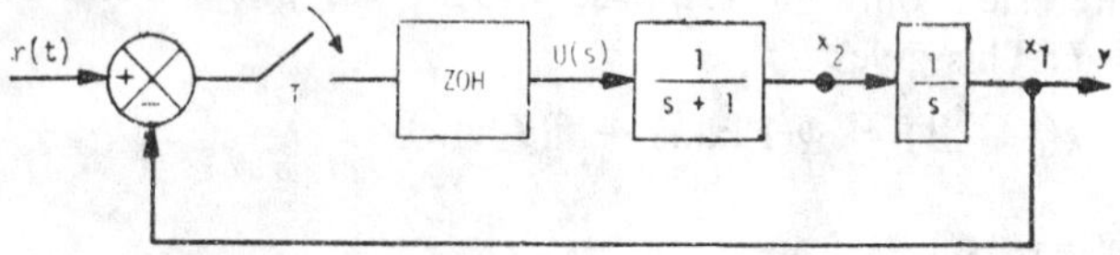

Figure 12.23 Rearrangement of block diagram of Fig. 12.22.

Therefore
$$e(t_0) = r(t_0) - x_1(t_0)$$

The solution of eqn. (12.133a) is given by

$$\mathbf{x}(t) = e^{\mathbf{A}(t-t_0)}\mathbf{x}(t_0) + \int_{t_0}^{t} e^{\mathbf{A}(t-\tau)} \mathbf{B} e(t_0) \, d\tau; \; t_0 \leq t < t_0 + T$$

Letting $t_0 = kT$ and $t = (k+1)T$, we get

$$\mathbf{x}(k+1) = e^{\mathbf{A}T}\mathbf{x}(kT) + \int_{kT}^{(k+1)T} e^{\mathbf{A}[(k+1)T-\tau]} \mathbf{B} e(kT) \, d\tau$$

$$= \Phi(T)\mathbf{x}(kT) + \theta(T)[r(kT) - x_1(kT)] \quad (12.134)$$

For $\mathbf{A} = \begin{bmatrix} 0 & 1 \\ 0 & -1 \end{bmatrix}$, $e^{\mathbf{A}t}$ can be computed by any of the two techniques presented in Section 12.5. We get the following result:

$$e^{\mathbf{A}t} = \Phi(t) = \begin{bmatrix} 1 & 1-e^{-t} \\ 0 & e^{-t} \end{bmatrix}$$

This gives

$$\Phi(T) = e^{\mathbf{A}T} = \begin{bmatrix} 1 & 0.632 \\ 0 & 0.368 \end{bmatrix}$$

$$\theta(T) = \int_0^T e^{\mathbf{A}\lambda} \mathbf{B} d\lambda = \begin{Bmatrix} \int_0^1 (1-e^{-\lambda}) d\lambda \\ \int_0^1 e^{-\lambda} d\lambda \end{Bmatrix} = \begin{bmatrix} 0.368 \\ 0.632 \end{bmatrix}$$

Therefore from eqn. (12.134) we have

$$\mathbf{x}(k+1) = \begin{bmatrix} 1 & 0.632 \\ 0 & 0.368 \end{bmatrix} \begin{bmatrix} x_1(k) \\ x_2(k) \end{bmatrix} + \begin{bmatrix} 0.368 \\ 0.632 \end{bmatrix} [r(k) - x_1(k)]$$

$$= \begin{bmatrix} 0.632 & 0.632 \\ -0.632 & 0.368 \end{bmatrix} \begin{bmatrix} x_1(k) \\ x_2(k) \end{bmatrix} + \begin{bmatrix} 0.368 \\ 0.632 \end{bmatrix} r(k) \quad (12.135)$$

Another approach for obtaining the state model of system of Fig. 12.22 is through z-transformation.

Using results presented in Chapter 11, we get

$$G(z) = \mathscr{Z}\left\{\frac{1 - e^{-sT}}{s^2(s+1)}\right\} = \frac{0.368(z + 0.717)}{(z-1)(z-0.368)}$$

$$= \frac{0.368z^{-1} + 0.264z^{-2}}{1 - (1.368z^{-1} - 0.368z^{-2})} \quad (12.136)$$

The signal flow graph of this transfer function is shown in Fig. 12.24. Taking outputs of delay elements as state variables, we get the following formulation.

$$x_1(k + 1) = x_2(k)$$
$$x_2(k + 1) = -0.368x_1(k) + 1.368x_2(k) + e(k)$$
$$y(k) = 0.264x_1(k) + 0.368x_2(k)$$

Now

$$e(k) = r(k) - y(k)$$
$$= r(k) - 0.264x_1(k) - 0.368x_2(k)$$

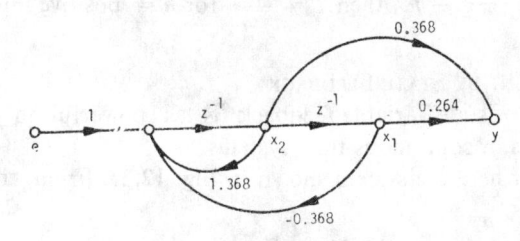

Figure 12.24 A state variable feedback sampled-data system.

Therefore
$$x_2(k + 1) = -0.632x_1(k) + x_2(k) + r(k)$$

The state model of the system of Fig. 12.22 is

$$\begin{bmatrix} x_1(k+1) \\ x_2(k+1) \end{bmatrix} = \begin{bmatrix} 0 & 1 \\ -0.632 & 1 \end{bmatrix} \begin{bmatrix} x_1(k) \\ x_2(k) \end{bmatrix} + \begin{bmatrix} 0 \\ 1 \end{bmatrix} r(k) \quad (12.137a)$$

$$y(k) = [0.264 \quad 0.368] \begin{bmatrix} x_1(k) \\ x_2(k) \end{bmatrix} \quad (12.137b)$$

It may be noted that choice of states for the system in Fig. 12.21 is different in state models given by (12.135) and (12.137). However, the output given by eqns. (12.133b) and (12.137b) will be the same.

CONTROLLABILITY AND OBSERVABILITY

The controllability and observability concepts presented in Section 12.6 carry over directly to the linear discrete-time systems. Kalman's and Gilbert's tests can be used for discrete-time systems as well.

In sampled-data systems, there is an additional requirement. Assume that transfer function of continuous-time plant has a partial fraction expansion which contains the term $G(s) = \dfrac{\omega}{(s + \sigma)^2 + \omega^2}$. The z-transform of this term is

$$G(z) = \mathscr{Z}\left[\dfrac{\omega}{(s+\sigma)^2 + \omega^2}\right] = \dfrac{z^{-1}e^{-\sigma T}\sin\beta T}{1 - 2z^{-1}e^{-\sigma T}\cos\beta T + z^{-2}e^{-2\sigma T}}$$

Now if $T = \dfrac{2\pi n}{\omega}$ where n = positive integer, then

$$G(z) = 0$$

The system may even be unstable for $\sigma < 0$, but this fact could not be inferred from the observations of the output. In this situation the system is neither completely controllable nor completely observable. Therefore an additional requirement for complete controllability and observability of sampled-data systems is that if the characteristic roots of the continuous-time plant are $-\sigma \pm j\omega$, then $T \neq \dfrac{2n\pi}{\omega}$ for n = positive integer.

POLE PLACEMENT BY STATE FEEDBACK

The notion of state variable feedback is as powerful in sampled-data systems as it is in continuous-time systems.

We consider here the system shown in Fig. 12.25. From this figure, we have

$$\mathbf{\dot{x}} = \mathbf{A}\mathbf{x} + \mathbf{B}u \quad (12.138a)$$

$$u = -\mathbf{K}\mathbf{x} + r \quad (12.138b)$$

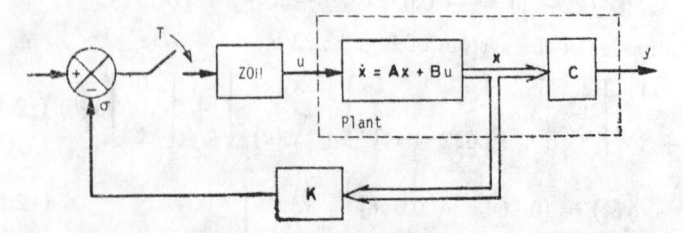

$$y = \mathbf{C}\mathbf{x} \quad (12.138c)$$

Figure 12.25 A state variable feedback sampled-data system.

The discrete form of eqn. (12.138a) as given by (12.132) is
$$x(k+1) = \Phi(T)x(k) + \theta(T)u(k)$$
Now
$$u(k) = -Kx(k) + r(k)$$
Therefore
$$x(k+1) = (\Phi(T) - \theta(T)K)x(k) + \theta(T)r(k) \qquad (12.139)$$
We note that this result is similar to the equivalent result for continuous-time systems as seen in (12.112). The method of pole placement discussed in Section 12.7 is applicable to the system of Fig. 12.25.

Problems

12.1 For the system shown in Fig. P-12.1, choose $v_1(t)$ and $v_2(t)$ as state variables and write down the state equations satisfied by them. Bring these equations in the vector-matrix form.

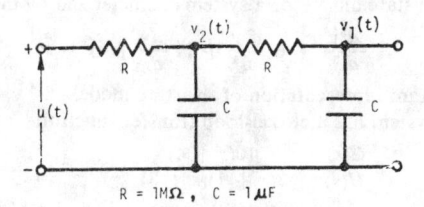

Figure P-12.1.

12.2 A schematic diagram representing a d.c. motor and load is given in Fig. P-12.2. The field is maintained constant during operation. Assume the motor is operating in the linear region. Determine the state equations in the vector-matrix form for the state variables

(i) $x^T = [\theta \;\; \dot\theta \;\; i_a]$ (ii) $x^T = [\theta \;\; \dot\theta \;\; \ddot\theta]$

Moment of inertia of motor and load = J
Back emf constant = K_b
Coefficient of friction of motor and load =
Motor torque constant = K_T

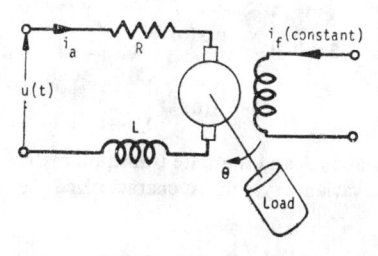

Figure P-12.2.

12.3 Consider the mechanical system depicted in Fig. P-12.3, consisting of two platforms coupled to each other and to ground via springs and dashpot dampers. Choosing suitable state variables, construct a state model of the system.

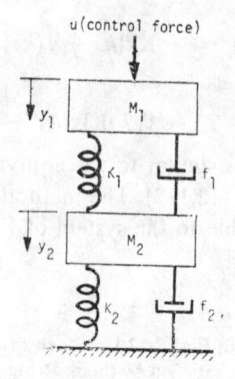

Figure P-12.3.

12.4 Construct the state model for a system characterized by the differential equation

$$\frac{d^3y}{dt^3} + 6\frac{d^2y}{dt^2} + 11\frac{dy}{dt} + 6y = u$$

Give the block diagram representation of the state model.

12.5 A feedback system has a closed-loop transfer function

$$\frac{C(s)}{U(s)} = \frac{10(s+4)}{s(s+1)(s+3)}$$

Construct three different state models for this system and give block diagram representation for each state model.

12.6 A feedback system is characterized by the closed-loop transfer function

$$T(s) = \frac{s^2 + 3s + 3}{s^3 + 2s^2 + 3s + 1}$$

Draw a suitable signal flow graph and therefrom construct a state model of the system.

12.7 Given

$$A_1 = \begin{bmatrix} \sigma & 0 \\ 0 & \sigma \end{bmatrix}; \quad A_2 = \begin{bmatrix} 0 & \omega \\ -\omega & 0 \end{bmatrix}; \quad A = \begin{bmatrix} \sigma & \omega \\ -\omega & \sigma \end{bmatrix}$$

Compute e^{At}.

(*Hint*: $e^{At} = e^{(A_1 + A_2)t} = e^{A_1 t} \cdot e^{A_2 t}$; this holds in general if $A_1 A_2 = A_2 A_1$.)

12.8 For a system represented by the state equation

$$\dot{x}(t) = Ax(t)$$

the response is $x(t) = \begin{bmatrix} e^{-2t} \\ -2e^{-2t} \end{bmatrix}$ when $x(0) = \begin{bmatrix} 1 \\ -2 \end{bmatrix}$

and $x(t) = \begin{bmatrix} e^{-t} \\ -e^{-t} \end{bmatrix}$ when $x(0) = \begin{bmatrix} 1 \\ -1 \end{bmatrix}$

Determine the system matrix A and the state transition matrix.

12.9 A linear time-invariant system is characterized by the homogeneous state equation

$$\begin{bmatrix} \dot{x}_1 \\ \dot{x}_2 \end{bmatrix} = \begin{bmatrix} 1 & 0 \\ 1 & 1 \end{bmatrix} \begin{bmatrix} x_1 \\ x_2 \end{bmatrix}$$

(a) Compute the solution of the homogeneous equation, assuming the initial state vector

$$\mathbf{x}_0 = \begin{bmatrix} 1 \\ 0 \end{bmatrix}$$

(Employ both the Laplace transform method and canonical transformation method.)

(b) Consider now that the system has a forcing function and is represented by the following nonhomogeneous state equation

$$\begin{bmatrix} \dot{x}_1 \\ \dot{x}_2 \end{bmatrix} = \begin{bmatrix} 1 & 0 \\ 1 & 1 \end{bmatrix} \begin{bmatrix} x_1 \\ x_2 \end{bmatrix} + \begin{bmatrix} 0 \\ 1 \end{bmatrix} u$$

where u is a unit step function.

Compute the solution of this equation assuming initial conditions of part (a).

12.10 A linear time-invariant system is described by the following state model.

$$\begin{bmatrix} \dot{x}_1 \\ \dot{x}_2 \\ \dot{x}_3 \end{bmatrix} = \begin{bmatrix} 0 & 1 & 0 \\ 0 & 0 & 1 \\ -6 & -11 & -6 \end{bmatrix} \begin{bmatrix} x_1 \\ x_2 \\ x_3 \end{bmatrix} + \begin{bmatrix} 0 \\ 0 \\ 2 \end{bmatrix} u$$

$$y = \begin{bmatrix} 1 & 0 & 0 \end{bmatrix} \begin{bmatrix} x_1 \\ x_2 \\ x_3 \end{bmatrix}$$

Transform this state model into a canonical state model and therefrom obtain the explicit solutions for the state vector and output when the control force u is a unit step function and initial state vector is

$$\mathbf{x}_0^T = \begin{bmatrix} 0 & 0 & 2 \end{bmatrix}$$

12.11 A feedback system is represented by a signal flow graph shown in Fig. P-12.11.
(a) Construct a state model of the system.
(b) Diagonalize the coefficient matrix A of the state model of part (a).
(c) Determine the stability of the system.
(d) Determine the transfer function $C(s)/U(s)$ from the state model of part (a).

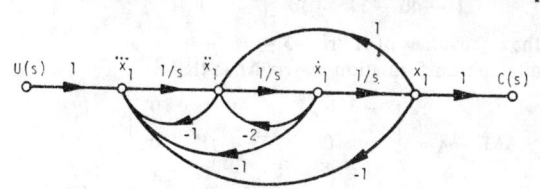

Figure P-12.11

12.12 For the state equation

$$\dot{\mathbf{x}} = \mathbf{A}\mathbf{x}$$

$$\mathbf{A} = \begin{bmatrix} 0 & 1 & 0 \\ 3 & 0 & 2 \\ -12 & -7 & -6 \end{bmatrix}$$

find the initial condition vector $\mathbf{x}(0)$ which will excite only the mode corresponding to the eigenvalue with the most negative real part.

(*Hint*: The problem is conveniently handled in the canonical form.)

12.13 Find the transformation matrices **P** which will convert the following equations to diagonal form. Given that the eigenvalues are λ_1, λ_2, and λ_3.

(i)
$$\begin{bmatrix} \dot{x}_1 \\ \dot{x}_2 \\ \dot{x}_3 \end{bmatrix} = \begin{bmatrix} 0 & 1 & 0 \\ 0 & 0 & 1 \\ -a_1 & -a_2 & -a_3 \end{bmatrix} \begin{bmatrix} x_1 \\ x_2 \\ x_3 \end{bmatrix}$$

(ii)
$$\begin{bmatrix} \dot{x}_1 \\ \dot{x}_2 \\ \dot{x}_3 \end{bmatrix} = \begin{bmatrix} -a_1 & 1 & 0 \\ -a_2 & 0 & 1 \\ -a_3 & 0 & 0 \end{bmatrix} \begin{bmatrix} x_1 \\ x_2 \\ x_3 \end{bmatrix}$$

12.14 Prove that for

$$\mathbf{A} = \begin{bmatrix} -a_1 & 1 & 0 & \ldots & 0 & 0 \\ -a_2 & 0 & 1 & \ldots & 0 & 0 \\ \vdots & \vdots & \vdots & & \vdots & \vdots \\ -a_{n-1} & 0 & 0 & \ldots & 0 & 1 \\ -a_n & 0 & 0 & \ldots & 0 & 0 \end{bmatrix}$$

with distinct eigenvalues $\lambda_1, \lambda_2, \ldots, \lambda_n$, the diagonalizing matrix is given by

$$\mathbf{M} = \begin{bmatrix} 1 & \ldots & 1 \\ \lambda_1 + a_1 & \ldots & \lambda_n + a_1 \\ \lambda_1^2 + a_1\lambda_1 + a_2 & \ldots & \lambda_n^2 + a_1\lambda_n + a_2 \\ \vdots & & \vdots \\ \lambda_1^{n-1} + a_1\lambda_1^{n-2} + \ldots + a_{n-2}\lambda_1 + a_{n-1} & \ldots & \lambda_n^{n-1} + a_1\lambda_n^{n-2} + \ldots + a_{n-2}\lambda_n + a_{n-1} \end{bmatrix}$$

12.15 Consider a state model

$$\dot{\mathbf{x}} = \mathbf{A}\mathbf{x} + \mathbf{B}u$$

where
$$\mathbf{A} = \begin{bmatrix} 0 & 1 & 0 \\ 0 & 0 & 1 \\ -40 & -34 & -10 \end{bmatrix}; \mathbf{B} = \begin{bmatrix} 0 \\ 0 \\ 1 \end{bmatrix}$$

(i) Show that the eigenvalues of **A** are $-3 \pm j1$, -4.
(ii) Suggest a suitable transformation matrix **M** so that

$$\mathbf{M}^{-1}\mathbf{A}\mathbf{M} = \mathbf{\Lambda} = \begin{bmatrix} -3+j1 & 0 & 0 \\ 0 & -3-j1 & 0 \\ 0 & 0 & -4 \end{bmatrix}$$

(iii) Suggest a suitable transformation matrix **Q** so that

$$\mathbf{Q}^{-1}\mathbf{\Lambda}\mathbf{Q} = \begin{bmatrix} -3 & 1 & 0 \\ -1 & -3 & 0 \\ 0 & 0 & -4 \end{bmatrix}$$

(*Hint*:
$$\mathbf{Q} = \begin{bmatrix} \tfrac{1}{2} & -j/2 & 0 \\ \tfrac{1}{2} & j/2 & 0 \\ 0 & 0 & 1 \end{bmatrix})$$

12.16 Write the state equations of the system shown in Fig. P-12.16, in which x_1, x_2 and x_3 constitute the state vector. Determine whether the system is completely controllable and observable.

State Variable Analysis and Design 527

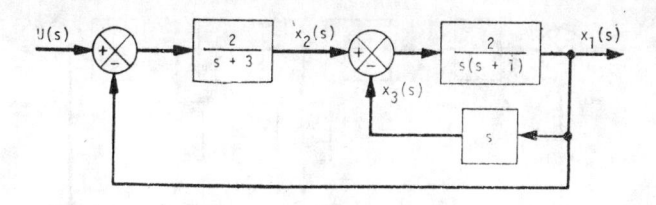

Figure P-12-16.

12.17 Block diagram representation of a linear time-invariant system is given in Fig. P-12.17. Check whether the system is completely observable.

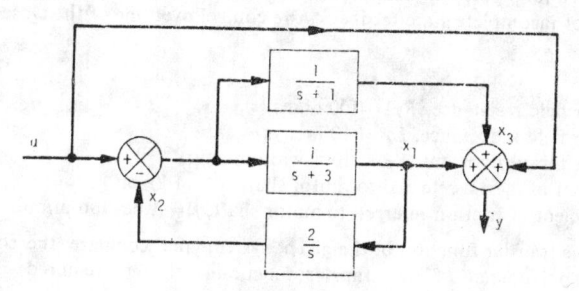

Figure P-12.17.

12.18 Consider the system with state variable feedback shown in Fig. P-12.18. Determine the values of K and k^T so that the system meets the following specifications:

Peak overshoot $\leqslant 15\%$
Settling time $\leqslant 4$ sec
Velocity error constant $\geqslant 1.5$

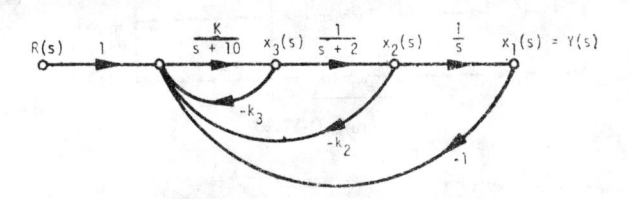

Figure P-12.18.

12.19 Find the forward path gain K_A and the k^T for the system shown in Fig. P-12.19 such that the closed-loop transfer function is

$$\frac{24}{[(s+1)^2 + (\sqrt{3})^2](s+6)}$$

Draw the root locus plot and comment upon the stability of the compensated system.

12.20 Consider the position control system of Fig. P-12.20 employing a d.c. motor in field control mode with state variables defined on the diagram. Incomplete state feedback ($x_3 = i_f$ is not fedback) is used to achieve the location of the dominant closed-loop poles at $-1 \pm j\sqrt{3}$. Obtain the values of the gains K_A and k_2. Find the location of

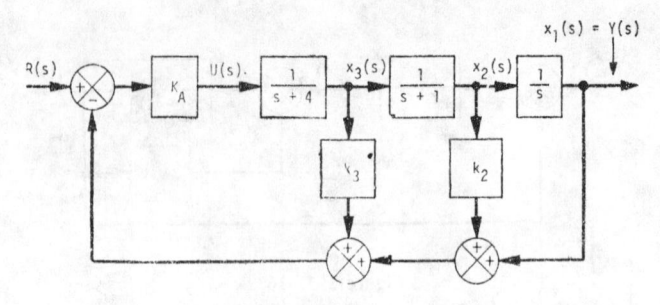

Figure P-12.19.

the third closed-loop pole and check if the dominance condition is met. It may be noted that because of incomplete state feedback, the control over one of the closed-loop poles is lost.

Given:

Motor field resistance, $R_f = 100$ ohms
Motor field inductance, $L_f = 10$ henrys
Motor torque constant, $K_T = 10$ newton-m/amp
Moment of inertia referred to motor shaft, $J = 1$ kg-m^2
Coefficient of friction referred to motor shaft, $f = 1$ newton-m/rad/sec

(*Hint*: Find the transfer function of the given system and compare the coefficients of denominator polynomial of the transfer function with the required characteristic equation.)

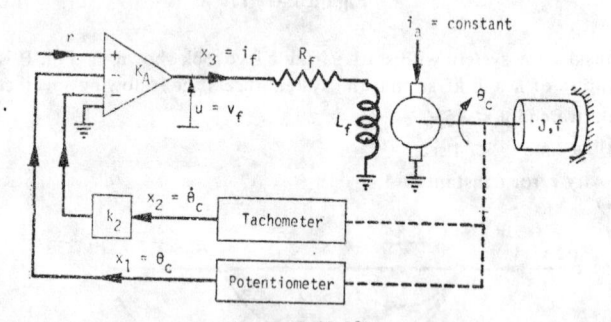

Figure P-12.20.

12.21 Consider a linear system

$$\mathbf{x} = \begin{bmatrix} 0 & 1 \\ 0 & -5 \end{bmatrix} \mathbf{x} + \begin{bmatrix} 0 \\ 100 \end{bmatrix} u$$

$$y = \begin{bmatrix} 1 & 0 \end{bmatrix} \mathbf{x}$$

The feedback controller for the system is given by

$$u = \begin{bmatrix} -k_1 & -k_2 \end{bmatrix} \begin{bmatrix} x_1 \\ x_2 \end{bmatrix} + r$$

Assume that the states x_1 and x_2 are not accessible for feedback. An observer system is to be designed to reconstruct $\mathbf{x}$. The structure of the overall system is as shown in Fig. 12.18.

Design the feedback matrix $\mathbf{G}$ so that $\mathbf{x} - \hat{\mathbf{x}}$ will decay as fast as e^{-10t}

12.22 A discrete-time system has the transfer function
$$T(z) = \frac{4z^3 - 12z^2 + 13z - 7}{(z-1)^2(z-2)}$$
Determine the state model of the system in

(i) phase variable form
(ii) Jordan canonical form

12.23 A discrete-time system is described by the difference equation
$$y(k+2) + 5y(k+1) + 6y(k) = u(k)$$
$$y(0) = y(1) = 0;\ T = 1\ \text{sec}$$

(i) Determine a state model in canonical form.
(ii) Find the state transition matrix.
(iii) For input $u(k) = 1$ for $k \geqslant 0$, find the output $y(k)$.

12.24 A continuous-time plant is described by the state equation
$$\dot{x} = \begin{bmatrix} 0 & 1 \\ -2 & -3 \end{bmatrix} x + \begin{bmatrix} 0 \\ 1 \end{bmatrix} u;\ x(0) = \begin{bmatrix} 1 \\ 0 \end{bmatrix}$$

The state equation is to be solved for $x(t)$ using digital computer. Obtain suitable recursive relations. Take sampling interval $T = 1$ sec.

12.25 Show that if there is pole-zero cancellation in the input-output transfer function of a linear time-invariant discrete-time system, the system will be either uncontrollable or unobservable.

12.26 The linear continuous-time plant of a sampled-data system is described by the state equation
$$\dot{x} = \begin{bmatrix} 0 & 1 \\ -4 & 0 \end{bmatrix} x + \begin{bmatrix} 0 \\ 2 \end{bmatrix} u$$
Determine the values of sampling period T which make the system uncontrollable.

References and Further Reading

1. DeRusso, M.P., R.J. Roy and C.M. Close, *State Variables for Engineers*, John Wiley, New York, 1965.
2. Ogata, K., *State Space Analysis of Control Systems*, Prentice-Hall, Englewood Cliffs, N.J., 1967.
3. Chen, C.T., *Introduction to Linear System Theory*, Holt, Rinehart and Winston, New York, 1970.
4. Kuo, B.C., *Discrete-Data Control Systems*, Science Tech., Champaign, Illinois, 1974.
5. Wiberg, D.M., *State Space and Linear Systems*, Schaum's Outline Series, McGraw-Hill, New York, 1971.
6. Gupta, S.C. and L. Hasdorff, *Fundamentals of Automatic Control*, John Wiley, New York, 1970.
7. Schultz, D.G. and J.L. Melsa, *State Functions and Linear Control Systems*, McGraw-Hill, New York, 1967.
8. Athans, M. and P.L. Falb, *Optimal Control*, McGraw-Hill, New York, 1966.
9. Dorf, R.C., *Time-Domain Analysis and Design of Control Systems*, Addison-Wesley, Reading, Mass., 1965.
10. Pipes, L., *Matrix Methods for Engineering*, Prentice-Hall, Englewood Cliffs, N.J., 1963.

13. Optimal Control Systems

13.1 Introduction

There are, as discussed in Chapter 10, basically two approaches to the design of control systems. In one approach we select the configuration of the overall system by introducing compensators and then choose the parameters of the compensators to meet the given specifications on performance. In the other approach, for a given plant we find an overall system that meets the given specifications and then compute the necessary compensators.

The classical design method based on first approach mentioned above, has already been discussed at length in earlier chapters. This method relies heavily on the Laplace transform and z-transform. The designer is given a set of specifications in time domain or in frequency domain and system configuration. Peak overshoot, settling time, gain margin, phase margin, steady-state error, etc., are among most commonly used specifications. These design specifications are selected because of convenience in graphical interpretation with respect to root locus or frequency plots. Compensators are selected that give as closely as possible, the desired system performance. In general, it may not be possible to satisfy all the desired specifications. Then, through a trial and error procedure, an acceptable system performance (within certain tolerable error) is achieved. There are generally many designs that can yield this acceptable performance, i.e., the solution is not unique. This trial and error design procedure works satisfactorily for single-input-single-output systems. The gap between the classical design procedure and its application to multi-input-multi-output systems has been bridged recently.

The trial and error uncertainties are eliminated in the parameter optimization method. The point of departure in the parameter optimization procedure is that the performance specifications consist of a single performance index. Integral square error performance index is very common but other performance indices can be used as well (refer Chapter 5). For a fixed system configuration, parameters that minimize the performance index are selected.

The ready availability of the digital computer has led to the development of modern *optimal control theory* which has become the major technique for the design of automatic control systems. This theory is based on the second design approach mentioned above. In this chapter, we attempt to introduce optimal control theory. For further reading, references are given at the end of the chapter.

13.2 Parameter Optimization: Servomechanisms

The parameter optimization method has already been used in Chapter 5 (Examples 5.5-5.7). There we considered a relatively simple case: for fixed configuration of overall system, we determined suitable value of system gain K so as to minimize the given index of performance. The minimization was carried out analytically. The analytical approach of parameter optimization consists of the following steps:

(i) Compute the performance index J as a function of the free parameters $K_1, K_2, ..., K_n$ of the system with fixed configuration;

$$J = J(K_1, K_2, ..., K_n) \tag{13.1}$$

(ii) Determine the solution set K_i of the equations

$$\frac{\partial J}{\partial K_i} = 0; \quad i = 1, 2, ..., n \tag{13.2}$$

Equations (13.2) give the necessary conditions for J to be minimum. From the solution set of eqns. (13.2), find the subset that satisfies the sufficient conditions which require that the *Hessian matrix* given below is positive definite.

$$\mathbf{H} = \begin{bmatrix} \dfrac{\partial^2 J}{\partial K_1^2} & \dfrac{\partial^2 J}{\partial K_1 \partial K_2} & \cdots & \dfrac{\partial^2 J}{\partial K_1 \partial K_n} \\ \dfrac{\partial^2 J}{\partial K_2 \partial K_1} & \dfrac{\partial^2 J}{\partial K_2^2} & \cdots & \dfrac{\partial^2 J}{\partial K_2 \partial K_n} \\ \cdots & \cdots & \cdots & \cdots \\ \dfrac{\partial^2 J}{\partial K_n \partial K_1} & \dfrac{\partial^2 J}{\partial K_n \partial K_2} & \cdots & \dfrac{\partial^2 J}{\partial K_n^2} \end{bmatrix} \tag{13.3}$$

Since $\dfrac{\partial^2 J}{\partial K_i \partial K_j} = \dfrac{\partial^2 J}{\partial K_j \partial K_i}$, the matrix $\mathbf{H}$ is always symmetric.

(iii) If there are two or more sets of K_i satisfying the necessary as well as sufficient conditions of minimization given by eqns. (13.2) and (13.3) respectively, then compute the corresponding J for each set. The set that has the smallest J gives the optimal parameters.

The minimization problem will be more easily solved if we can express performance index in terms of transform domain quantities [1, 2]. For quadratic performance index (eqn. (5.47)), this can be done by using the Parseval's theorem (see Appendix I) which allows us to write

$$\int_0^\infty x^2(t)\, dt = \frac{1}{2\pi j} \int_{-j\infty}^{j\infty} X(s) X(-s)\, ds \tag{13.4}$$

in which $X(s)$ = Laplace transform of $x(t)$, where $x(t)$ is defined for $t \geqslant 0$ and $x(t)$ is zero for $t < 0$.

The value of right hand integral in eqn. (13.4) can easily be found from published tables [3], provided that $X(s)$ can be written in the form $B(s)/A(s)$;

$$B(s) \triangleq b_0 + b_1 s + \ldots + b_{n-1} s^{n-1}$$

$$A(s) \triangleq a_0 + a_1 s + \ldots + a_n s^n \tag{13.5}$$

where $A(s)$ has zeros only in the left half of the complex plane.
Results upto fourth-order are given in Table 13.1.

Table 13.1 TABULATION OF DEFINITE INTEGRAL FOR CONTINUOUS-TIME SYSTEMS

$$J_n = \frac{1}{2\pi j} \int_{-j\infty}^{j\infty} \frac{B(s)B(-s)}{A(s)A(-s)} ds$$

$$B(s) = \sum_{k=0}^{n-1} b_k s^k$$

$$A(s) = \sum_{k=0}^{n} a_k s^k; \quad A(s) \text{ has zeros in left half plane only.}$$

$$J_1 = \frac{b_0^2}{2a_0 a_1}$$

$$J_2 = \frac{b_1^2 a_0 + b_0^2 a_2}{2a_0 a_1 a_2}$$

$$J_3 = \frac{b_2^2 a_0 a_1 + (b_1^2 - 2b_0 b_2) a_0 a_3 + b_0^2 a_2 a_3}{2a_0 a_3 (-a_0 a_3 + a_1 a_2)}$$

$$J_4 = \frac{b_3^2(-a_0^2 a_3 + a_0 a_1 a_2) + (b_2^2 - 2b_1 b_3) a_0 a_1 a_4 + (b_1^2 - 2b_0 b_2) a_0 a_3 a_1 + b_0^2(-a_1 a_4^2 + a_2 a_3 a_4)}{2a_0 a_4 (-a_0 a_3^2 - a_1^2 a_4 + a_1 a_2 a_3)}$$

For sampled-data systems, the following relation is very useful in analytical design.

$$\sum_{k=0}^{\infty} x^2(kT) = \frac{1}{2\pi j} \oint_{\substack{\text{unit} \\ \text{circle}}} X(z) X(z^{-1}) \frac{dz}{z} \tag{13.6}$$

in which $X(z) = z$-transform of $x(kT)$; where $x(kT)$ is defined for $k \geqslant 0$ and is 0 for $k < 0$.

The value of right hand side of eqn. (13.6) can be found from published tables [4], provided that $X(z)$ can be written in the form

$$X(z) = \frac{b_0 z^n + b_1 z^{n-1} + \ldots + b_n}{a_0 z^n + a_1 z^{n-1} + \ldots + a_n} \tag{13.7}$$

Results upto third-order are given in Table 13.2.

Table 13.2 TABULATION OF DEFINITE INTEGRAL FOR SAMPLED-DATA SYSTEMS

$$J_n = \frac{1}{2\pi j} \oint_{\text{unit circle}} X(z)X(z^{-1})\frac{dz}{z}$$

$$X(z) = \frac{b_0 z^n + b_1 z^{n-1} + \cdots + b_n}{a_0 z^n + a_1 z^{n-1} + \cdots + a_n}$$

$$J_1 = \frac{(b_0^2 + b_1^2)a_0 - 2b_0 b_1 a_1}{a_0(a_0^2 - a_1^2)}$$

$$J_2 = \frac{B_0 a_0 e_1 - B_1 a_0 a_1 + B_2(a_1^2 - a_2 e_1)}{a_0[(a_0^2 - a_2^2)e_1 - (a_0 a_1 - a_1 a_2)a_1]}$$

where

$$B_0 = b_0^2 + b_1^2 + b_2^2$$
$$B_1 = 2(b_0 b_1 + b_1 b_2)$$
$$B_2 = 2b_0 b_2$$
$$e_1 = a_0 + a_2$$

$$J_3 = \frac{a_0 B_0 Q_0 - a_0 B_1 Q_1 + a_0 B_2 Q_2 - B_3 Q_3}{[(a_0^2 - a_3^2)Q_0 - (a_0 a_1 - a_2 a_3)Q_1 + (a_0 a_2 - a_1 a_3)Q_2]a_0}$$

where

$$B_0^2 = b_0^2 + b_1^2 + b_2^2 + b_3^2$$
$$B_1 = 2(b_0 b_1 + b_1 b_2 + b_2 b_3)$$
$$B_2 = 2(b_0 b_2 + b_1 b_3)$$
$$B_3 = 2b_0 b_3$$
$$Q_0 = (a_0 e_1 - a_3 a_2)$$
$$Q_1 = (a_0 a_1 - a_1 a_3)$$
$$Q_2 = (a_1 e_2 - a_2 e_1)$$
$$Q_3 = (a_1 - a_3)(e_2^2 - e_1^2) + a_0(a_0 e_2 - a_3 e_1)$$
$$e_1 = a_0 + a_2$$
$$e_2 = a_1 + a_3$$

SERVOMECHANISM OR TRACKING PROBLEM

In this section we shall be concerned with the parameter optimization in *servomechanisms* or *tracking systems* wherein the objective of design is to maintain the actual output ($c(t)$) of the system as close as possible to the desired output which is usually the reference input ($r(t)$) to the system. We may define an error $e(t)$ by setting $e(t) = c(t) - r(t)$. Loosely speaking, the design objective in a servomechanism or tracking problem is to keep error $e(t)$ 'small'. This objective, as we observed in Chapter 5, can be realized by minimizing the performance index

$$J = \int_0^\infty e^2(t)\,dt$$

if control $u(t)$ is not constrained in magnitude.

Example 13.1: Referring to the block diagram of Fig. 13.1, consider that $G(s) = 100/s^2$ and $R(s) = 1/s$. Determine the optimal value of parameter K such that

$$J = \int_0^\infty e^2(t)\, dt \text{ is minimum}$$

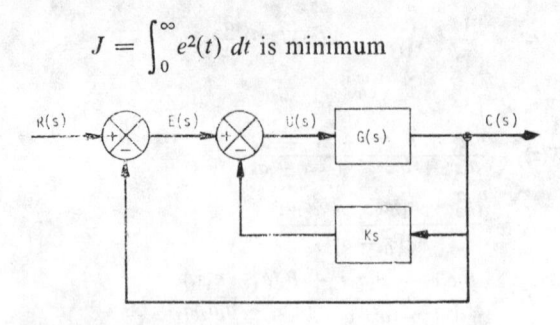

Figure 13.1 Control system for parameter optimization.

Solution:
From Fig. 13.1,

$$E(s) = \frac{s + 100K}{s^2 + 100Ks + 100}$$

From Table 13.1,

$$J = \int_0^\infty e^2(t)\, dt = \frac{1}{2\pi j}\int_{-j\infty}^{j\infty} E(s)E(-s)\, ds$$

$$= \frac{1 + 100K^2}{200K} \qquad (13.8)$$

$$\frac{\partial J}{\partial K} = 0 \text{ gives } K = 0.1$$

It can be checked that $\frac{\partial^2 J}{\partial K^2} > 0$. Therefore $K = 0.1$ is the optimal value of the free parameter of system of Fig. 13.1. The minimum value of J obtained from eqn. (13.8) is $J_{min} = 0.1$.

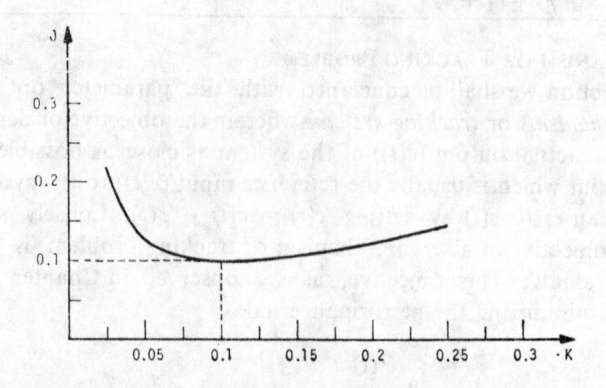

Figure 13.2 The performance index versus the parameter K.

A curve of performance index as a function of K is shown in Fig. 13.2. It is clear that this system is not very sensitive to changes in K. The sensitivity of an optimum system is given by

$$S_K^{opt} = \frac{\Delta J/J}{\Delta K/K}$$

where K is the design parameter.

For the system under consideration

$$S_K^{opt} \approx \frac{0.00045/0.1}{0.01/0.1} = 0.045$$

Example 13.2: Figure 13.3 illustrates a typical sampled-data system. The transfer functions $G_p(s)$ and $G_0(s)$ of the controlled plant and hold circuit respectively are known. The data-processing unit $D(z)$ which operates on the sampled error signal $e_s(t)$ is to be designed.

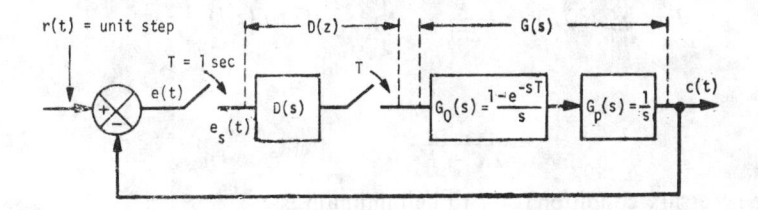

Figure 13.3 Control system for parameter optimization.

For the design of continuous-time servo systems, we have used the integral square error criterion. This criterion can be effectively utilized for the design of sampled-data servo systems as well*. However, for sampled-data systems, the *sum square error criterion* which requires the minimization of sum square error

$$J = \sum_{k=0}^{\infty} e^2(kT) \tag{13.9}$$

is more convenient. We need to concern ourselves only with information at sampling instants; therefore we need only the z-transform while the integral square error criterion needs modified z-transform.

Assuming the processing unit $D(z)$ in Fig. 13.3, to be simply an amplifier of gain K, let us find K so that the sum square error given by eqn. (13.9) is minimized (see Problem 13.4).

From Fig. 13.3, we have

$$G(s) = G_0(s)G_p(s) = \frac{1 - e^{-sT}}{s^2}$$

$$r(t) = \text{unit step}$$

---

*Refer Gupta [5] for the design of compensators for sampled-data systems using integral square error criterion.

Therefore
$$G(z) = (1 - z^{-1})\mathscr{Z}\left[\frac{1}{s^2}\right] = \frac{1}{z-1}$$
$$R(z) = \frac{z}{z-1}$$
Now
$$E(z) = R(z) - \frac{G(z)D(z)}{1 + G(z)D(z)} R(z)$$
$$= \frac{z}{z-1}\left(1 - \frac{K}{z+K-1}\right)$$
$$= \frac{z}{z+K-1}$$

From Table 13.2, we obtain
$$J = \sum_{i=0}^{\infty} e^2(iT) = \frac{1}{2\pi j} \oint_{\substack{\text{unit}\\\text{circle}}} E(z)E(z^{-1}) \frac{dz}{z}$$

$$= \frac{1}{2K - K^2}$$

The necessary condition for J to be minimum is
$$\frac{\partial J}{\partial K} = \frac{-(2-2K)}{(2K-K^2)^2} = 0 \quad \text{or} \quad K = 1$$

The sufficient condition $\frac{\partial^2 J}{\partial K^2} > 0$ is satisfied for $K = 1$. Therefore $K = 1$ is the optimum value of the parameter and corresponding value of performance index is
$$J_{\min} = 1$$

Compensator design subject to constraints

In Chapter 5, we observed that the optimal design of servo systems obtained by minimizing the performance index
$$J = \int_0^\infty e^2(t)\, dt \tag{13.10}$$
may be unsatisfactory because it may lead to excessively large magnitudes of some control signals. A more realistic solution to the problem is reached if the performance index is modified to account for the physical constraints like saturation in physical devices. Therefore, a more realistic performance index should be to minimize
$$J = \int_0^\infty e^2(t)\, dt \tag{13.11}$$

subject to the constraint

$$\max |u(t)| \leqslant M \tag{13.11a}$$

for some constant M. The constant M is determined by the linear range of the system plant. Similar constraints could be imposed on other components of the system, e.g., compensators.

However this will make the design extremely complicated. We assume here that these components are of such quality that they are not saturated in the control process. The constraint is therefore imposed only on the plant.

Although the criterion given by (13.11) can be used in the design, it is not convenient to work with. This has already been demonstrated in Example 13.1 wherein evaluation of J became easier because of its quadratic nature. Further, if the criterion given by eqn. (13.11) is used, the resulting optimal system is not necessarily a linear system; i.e., in order to implement the optimal design, nonlinear and/or time-varying devices are erquired [6]. Therefore, we would very much like to replace the performance criterion given by (13.11) by the following *quadratic performance index*:

$$J = \int_0^\infty [e^2(t) + \lambda u^2(t)] \, dt \tag{13.12}$$

where λ, a positive constant, is called the *weighting factor*. If λ is a small positive number, more weight is imposed on the error. As $\lambda \to 0$, the contribution of $u(t)$ becomes less significant and the performance index reduces to (13.10). In this case, the magnitude of $u(t)$ will be very large and the constraint given by (13.11a) may be violated. If $\lambda \to \infty$, performance criterion given by (13.12) reduces to

$$J = \int_0^\infty u^2(t) \, dt \tag{13.13}$$

and the optimal system that minimizes this J is one with $u = 0$. From these two extreme cases, we conclude that if λ is properly chosen, then the constraint of (13.11a) will be satisfied.

The integral square control index J given by (13.13) may be regarded as a measure of energy required by the control. Since $u^2(t)$ is proportional to power consumed for control, the time integral of $u^2(t)$ is a measure of energy consumed by the system.

For sampled-data servo systems, the performance index J is considered to include the following two summations.

$$J_e = \sum_{k=0}^\infty e^2(kT)$$

$$J_u = \sum_{k=0}^\infty u^2(kT)$$

The second summation puts a constraint on the control signal. The optimization problem is to find free parameters of the system so that

$$J = J_e + \lambda J_u = \sum_{k=0}^{\infty} [e^2(kT) + \lambda u^2(kT)] \tag{13.14}$$

is minimized.

Example 13.3: Referring to the block diagram of Fig. 13.4, consider that $G(s) = 100/s^2$ and $R(s) = 1/s$. Determine the optimal values of the parameters K_1 and K_2 such that (i) $J_e = \int_0^{\infty} e^2(t)\,dt$ is minimized (ii) $J_u = \int_0^{\infty} u^2(t)\,dt = 0.1$.

Solution:

From Fig. 13.4,

$$E(s) = \frac{s + 100K_1K_2}{s^2 + 100K_1K_2 s + 100K_1}$$

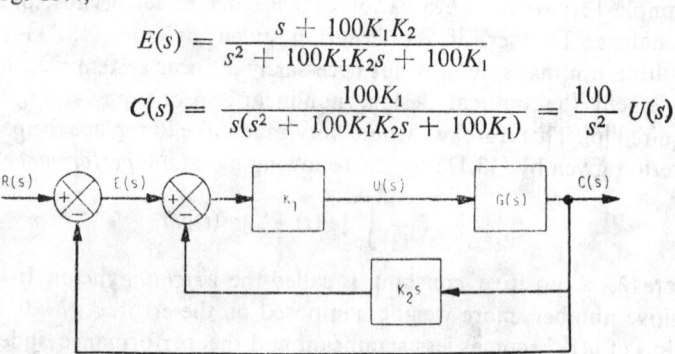

$$C(s) = \frac{100K_1}{s(s^2 + 100K_1K_2 s + 100K_1)} = \frac{100}{s^2} U(s)$$

Figure 13.4 Control system for parameter optimization.

Therefore

$$U(s) = \frac{sK_1}{s^2 + 100K_1K_2 s + 100K_1}$$

From Table 13.1,

$$J_e = \int_0^{\infty} e^2(t)\,dt = \frac{1}{2\pi j}\int_{-j\infty}^{j\infty} E(s)E(-s)\,ds$$

$$= \frac{1 + 100K_1K_2^2}{200K_1K_2}$$

$$J_u = \int_0^{\infty} u^2(t)\,dt = \frac{1}{2\pi j}\int_{-j\infty}^{j\infty} U(s)U(-s)\,ds$$

$$= \frac{K_1}{200K_2}$$

The energy constraint on the system is thus expressed by the equation

$$J_u = \frac{K_1}{200K_2} = 0.1 \tag{13.15}$$

The performance index for the system is (eqn. (13.12))

$$J = J_e + \lambda J_u$$

$$= \frac{1 + 100K_1K_2^2}{200K\,K_2} + \lambda\,\frac{K_1}{200K_2}$$

$$\frac{\partial J}{\partial K_i} = 0 \text{ for } i = 1, 2, \text{ gives}$$

$$\lambda K_1^2 = 1 \tag{13.16}$$

$$100 K_1 K_2^2 - 1 - \lambda K_1^2 = 0 \tag{13.17}$$

From eqns. (13.15) – (13.17), we get

$$\lambda = 0.25, K_1 = 2, K_2 = 0.1$$

The Hessian matrix

$$\mathbf{H} = \begin{bmatrix} \dfrac{\partial^2 J}{\partial K_1^2} & \dfrac{\partial^2 J}{\partial K_1 \partial K_2} \\ \dfrac{\partial^2 J}{\partial K_2 \partial K_1} & \dfrac{\partial^2 J}{\partial K_2^2} \end{bmatrix} = \begin{bmatrix} \dfrac{1}{100 K_1^3 K_2} & \dfrac{1 - \lambda K_1^2}{200 K_1^2 K_2^2} \\ \dfrac{1 - \lambda K_1^2}{200 K_1^2 K_2^2} & \dfrac{1}{K_2} - \dfrac{(100 K_1 K_2^2 - 1)}{100 K_1 K_2^3} \end{bmatrix}$$

For $K_1 = 2, K_2 = 0.1$;

$$\mathbf{H} = \begin{bmatrix} \dfrac{1}{80} & 0 \\ 0 & 5 \end{bmatrix} \text{ is positive definite}$$

Therefore $K_1 = 2$, $K_2 = 0.1$ satisfy the necessary as well as sufficient conditions for J to be minimum.

13.3 Optimal Control Problems: Transfer Function Approach

Following steps are involved in the solution of an optimal control problem uning transfer function approach:

(i) Given a plant with transfer function $G(s)$, find the transfer function of the overall system which is optimal with respect to given performance criterion.

(ii) Compute the compensators for the system obtained from step (i).

SERVOMECHANISM PROBLEM: CONTINUOUS-TIME SYSTEMS

The optimal servomechanism or tracking problem for single-input-single-output linear time-invariant systems can be formulated in the following way:

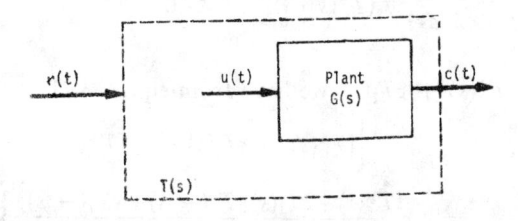

Figure 13.5 An optimal control problem.

For a given plant (Fig. 13.5) with irreducible proper transfer function $G(s)$, find the transfer function $T^*(s)$ of the overall system that minimizes the quadratic performance index

$$J = \int_0^\infty [(r(t) - c(t))^2 + \lambda u^2(t)]\, dt$$

where λ is a positive constant, $u(t)$ is the actuating signal, $c(t)$ is the actual output of the system and $r(t)$ is the desired output (reference input).

In the frequency domain, the performance index is given as

$$J = \frac{1}{2\pi j} \int_{-j\infty}^{j\infty} \left\{ [R(s) - C(s)][R(-s) - C(-s)] + \lambda U(s)U(-s) \right\} ds \quad (13.18)$$

For the system under consideration

$$\frac{C(s)}{U(s)} = G(s); \quad \frac{C(s)}{R(s)} = T(s) \quad (13.19)$$

Therefore

$$U(s) = \frac{T(s)R(s)}{G(s)} \quad (13.20)$$

Substituting eqns. (13.19) and (13.20) in eqn. (13.18), we get

$$J(T(s)) = \frac{1}{2\pi j} \int_{-j\infty}^{j\infty} \left\{ [T(s) - 1][T(-s) - 1] + \lambda \frac{T(s)T(-s)}{G(s)G(-s)} \right\} R(s)R(-s)\, ds \quad (13.21)$$

Now the problem is to find a stable $T^*(s)$ that minimises $J(T(s))$.

Observe that if $T^*(s)$ were the optimum transfer function, the performance index corresponding to $T^*(s)$ must, by definition, have a smaller value than that corresponding to any other transfer function, i.e., for some constant ϵ, the following holds:

$$J(T^*(s)) \leqslant J[T^*(s) + \epsilon T_1(s)] = J_\epsilon(T^*(s))$$

where $T_1(s)$ is some arbitrary stable transfer function. The necessary and sufficient conditions for $T^*(s)$ to be the minimizing transfer function are

$$\left. \frac{d}{d\epsilon} J_\epsilon(T^*(s)) \right|_{\epsilon=0} = 0$$

$$\left. \frac{d^2}{d\epsilon^2} J_\epsilon(T^*(s)) \right|_{\epsilon=0} > 0 \quad (13.22)$$

With $T(s) = T^*(s) + \epsilon T_1(s)$, we have from eqn. (13.21),

$$J_\epsilon(T^*(s)) = \frac{1}{2\pi j} \int_{-j\infty}^{j\infty} \left\{ [T^*(s) + \epsilon T_1(s) - 1][T^*(-s) + \epsilon T_1(-s) - 1] \right.$$
$$\left. + \lambda \frac{[T^*(s) + \epsilon T_1(s)][T^*(-s) + \epsilon T_1(-s)]}{G(s)G(-s)} \right\} R(s)R(-s)\, ds$$
$$(13.23)$$

Writing the right hand side of this equation in terms of increasing order of ϵ, we get

$$J_\epsilon = J_1 + \epsilon(J_2 + J_3) + \epsilon^2 J_4 \tag{13.24}$$

where

$$J_1 = \frac{1}{2\pi j} \int_{-j\infty}^{j\infty} \left\{ (T^*(s) - 1)(T^*(-s) - 1) + \lambda \frac{T^*(s)T^*(-s)}{G(s)G(-s)} \right\} R(s)R(-s)\,ds$$

$$J_2 = \frac{1}{2\pi j} \int_{-j\infty}^{j\infty} \left\{ (T^*(-s) - 1) + \lambda \frac{T^*(-s)}{G(s)G(-s)} \right\} R(s)R(-s)T_1(s)\,ds$$

$$J_3 = \frac{1}{2\pi j} \int_{-j\infty}^{j\infty} \left\{ (T^*(s) - 1) + \lambda \frac{T^*(s)}{G(s)G(-s)} \right\} R(s)R(-s)T_1(-s)\,ds$$

$$J_4 = \frac{1}{2\pi j} \int_{-j\infty}^{j\infty} \left\{ 1 + \frac{\lambda}{G(s)G(-s)} \right\} R(s)R(-s)T_1(s)T_1(-s)\,ds$$

Assuming

$$G(s) = \frac{N(s)}{D(s)} \tag{13.25}$$

we can write the integrand of J_4 as

$$\left[\frac{N(s)N(-s) + \lambda D(s)D(-s)}{N(s)N(-s)} \right] R(s)R(-s)T_1(s)T_1(-s) \tag{13.26}$$

Consider the polynomial

$$P(s) = N(s)N(-s) + \lambda D(s)D(-s)$$

It is clear that $P(s) = P(-s)$. Hence if s_1 is a root of $P(s) = 0$, so is $-s_1$. We next show that $P(s)$ has no roots on the imaginary axis:

$$P(j\omega) = N(j\omega)N(-j\omega) + \lambda D(j\omega)D(-j\omega)$$
$$= |N(j\omega)|^2 + \lambda |D(j\omega)|^2$$

Since $\lambda > 0$ and $N(s)$ and $D(s)$ have no common factor, $P(j\omega) \neq 0$ for all ω. Hence $P(s)$ has no roots on the $j\omega$-axis and all the roots are symmetrical with respect to this axis. Consequently, we may write

$$P(s) = D^*(s)D^*(-s) = N(s)N(-s) + \lambda D(s)D(-s) \tag{13.27}$$

where $D^*(s)$ consists of all left half s-plane roots of $P(s)$ and $D^*(-s)$ consists of all right half s-plane roots of $P(s)$. The factorization in eqn. (13.27) is called *spectral factorization*.

Let us now look at the integrand (13.26) of J_4. We find that the integrand is symmetrical with respect to the imaginary axis and therefore from eqn. (13.4) we conclude that J_4 is always positive.

One more observation from eqn. (13.24): By substituting $-s$ for s in J_2, it becomes J_3. Thus $J_2 = J_3$.

From (13.22) and (13.24), following necessary and sufficient conditions are obtained for the minimization of performance index:

$$J_2 + J_3 = 0$$

$$J_4 > 0$$

These conditions are satisfied if

$$\frac{1}{2\pi j} \int_{-j\infty}^{j\infty} \left[(T^*(s) - 1) + \lambda \frac{T^*(s)}{G(s)G(-s)} \right] R(s)R(-s)T_1(-s)\, ds = 0 \quad (13.28)$$

Due to the requirement of stability, all the poles of $T(s)$, $T^*(s)$ and $T_1(s)$ should be located in the left half of s-plane and correspondingly the poles of $T(-s)$, $T^*(-s)$ and $T_1(-s)$ would then be located in the right half of s-plane. One solution of eqn. (13.28) is that all the poles of the integrand are in the right half of s-plane and the line integral is evaluated as a contour integral around the left half of s-plane. Since the contour does not enclose any singularity of integrand, the contour integration is zero

Thus the condition given by eqn. (13.28) is satisfied if all the poles of the function

$$X = \left[(T^*(s) - 1) + \lambda \frac{T^*(s)}{G(s)G(-s)} \right] R(s)R(-s)$$

lie in the right half of s-plane.

The function X may be written as

$$X = \left[1 + \frac{\lambda}{G(s)G(-s)} \right] T^*(s)R(s)R(-s) - R(s)R(-s)$$

$$= A(s)A(-s)B(s)B(-s)T^*(s) - B(s)B(-s) \quad (13.29)$$

where $A(s)$ and $B(s)$ are ratio polynomials with no poles or zeros in the right half s-plane and

$$A(s)A(-s) = \frac{\lambda}{G(s)G(-s)} + 1 \quad (13.30)$$

$$B(s)B(-s) = R(s)R(-s) \quad (13.31)$$

Dividing eqn. (13.29) by $A(-s)B(-s)$ we get

$$A(s)B(s)T^*(s) - \frac{B(s)}{A(-s)} = \frac{X}{A(-s)B(-s)} \quad (13.32)$$

The function $B(s)/A(-s)$ can be expressed in the partial fraction expansion given below:

$$\frac{B(s)}{A(-s)} = C_+(s) + C_-(s) \quad (13.33)$$

where $C_+(s)$ is composed of the part involving all the poles in the left half s-plane†.

Equation (13.32) may be written as

$$A(s)B(s)T^*(s) - C_+(s) = \frac{X}{A(-s)B(-s)} + C_-(s) \qquad (13.34)$$

Since all the poles of the expression on left hand side of eqn. (13.34) are in left half s-plane and all the poles of the expression in the right hand side are in right half s-plane, the two sides of eqn. (13.34) are therefore independent and must each be equal to zero. We obtain as its solution as

$$T^*(s) = \frac{C_+(s)}{A(s)B(s)}$$

Now

$$A(s) = \frac{N(s)}{D^*(s)}$$

where $N(s)$ is given by eqn. (13.25) and $D^*(s)$ is given by eqn. (13.27).

$$T^*(s) = \frac{N(s)C_+(s)}{D^*(s)B(s)} \qquad (13.35)$$

Once the transfer function $T^*(s)$ of the overall system is obtained, the designer is free to choose a compensation scheme for the system.

Example 13.4. Consider the linear plant of a system characterized by the transfer function

$$G(s) = 100/s^2$$

The objective of the system is to make the output $c(t)$ follow a unit step input $r(t)$ minimizing

$$J = \int_0^\infty \{(r(t) - c(t))^2 + 0.25u^2(t)\}\, dt$$

where $u(t)$ is the actuating signal of the plant††.

From eqn. (13.25),

$$G(s) = \frac{N(s)}{D(s)} = \frac{100}{s^2}$$

Therefore,

$$N(s) = N(-s) = 100; \qquad D(s) = D(-s) = s^2$$

---

†In $B(s)/A(-s)$, poles in the left half of s-plane are contributed by $B(s)$ only which is given by eqn. (13.31). When $R(s)$ has poles on the $j\omega$-axis, they may be considered to lie on axis parallel to the $j\omega$-axis, shifted by $\pm\delta$. The final result is obtained by letting $\delta \to 0$. Exactly same result is obtained if without using δ, we split the $j\omega$-axis poles of $R(s)R(-s)$ equally between $B(s)$ and $B(-s)$ and treat the ones belonging to $B(s)$ as if they were in the left half of s-plane and the ones belonging to $B(-s)$ as if they were in the right half s-plane.

††It may be pointed out here that the choice of the weighting factor λ is quite a problem in design. Its value depends on the linear range of the plant. Here we have arbitrarily taken $\lambda = 0.25$.

From eqn. (13.27),
$$D^*(s)D^*(-s) = N(s)N(-s) + 0.25D(s)D(-s)$$
$$= 100 + 0.25s^4$$

Let $\quad D^*(s) = a_0 + a_1 s + a_2 s^2$

Therefore
$$(a_0 + a_1 s + a_2 s^2)(a_0 - a_1 s + a_2 s^2) = 100 + 0.25s^4$$

Comparing the coefficients, we obtain
$$a_0 = 100, \quad a_2 = 0.5 \text{ and } \quad a_1 = 10$$

This gives
$$D^*(s) = 0.5s^2 + 10s + 100$$
$$A(-s) = \frac{D^*(-s)}{N(-s)} = \frac{0.5s^2 - 10s + 100}{100}$$

Given $\quad R(s) = 1/s$

From eqn. (13.31),
$$B(s) = \frac{1}{s} \text{ (read footnote on page 543)}$$

We next obtain the partial fraction expansion of $\frac{B(s)}{A(-s)}$:

$$\frac{B(s)}{A(-s)} = \frac{100}{s(0.5s^2 - 10s + 100)} = C_+(s) + C_-(s)$$

This gives
$$C_+(s) = \frac{1}{s}$$

The optimal transfer function is (eqn. (13.35))
$$T^*(s) = \frac{100}{0.5s^2 + 10s + 100}$$
$$= \frac{200}{s^2 + 20s + 200} \qquad (13.36)$$

Design of compensators

There is a great deal of flexibility in the selection of compensators and hence the final system configuration. Some schemes for the problem under consideration are given below:

(i) If the configuration of the compensator is chosen as shown in Fig. 13.6, then the compensator $G_c(s)$ can be obtained from[†]

$$T^*(s) = \frac{G_c(s)G(s)}{1 + G_c(s)G(s)}$$

†It may be noted that compensator $G_c(s)$ should not cancel an unstable pole of $G(s)$. If the cancelled or missing pole is a stable one, it will not cause a serious effect on the response of the system.

Optimal Control Systems

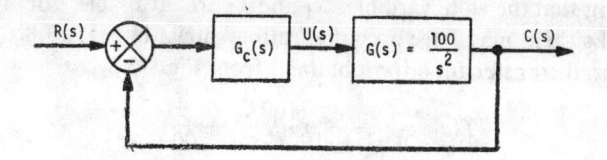

Figure 13.6 An optimal control system for Example 13.4.

as

$$G_c(s) = \frac{T^*(s)}{G(s)[1 - T^*(s)]} \qquad (13.37)$$

$$= \frac{20s}{s + 20}$$

(ii) Scheme of Fig. 13.7 employing both cascade and feedback compensation realizes the transfer function $T^*(s)$ given by eqn. (13.36).

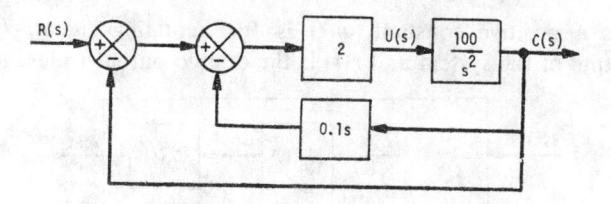

Figure 13.7 An optimal control system for Example 13.4.

(iii) Given

$$G(s) = \frac{100}{s^2} = \frac{C(s)}{U(s)}$$

A state model is given by equations

$$\dot{x}_1 = x_2$$
$$\dot{x}_2 = 100\,u$$
$$c = x_1$$

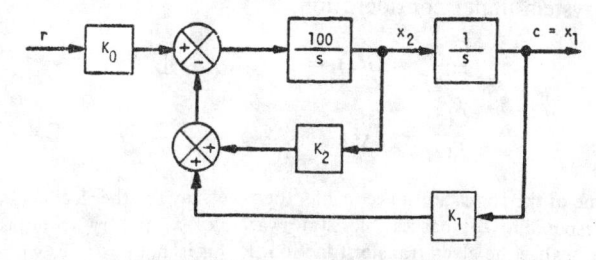

Figure 13.8 An optimal control system for Example 13.4.

Assuming that the state variables x_1 and x_2 are available for feedback† the overall system may have a configuration shown in Fig. 13.8.

The overall transfer function obtained from Fig. 13.8 is

$$T(s) = \frac{100K_0}{s^2 + 100K_2 s + 100K_1}$$

Comparing the coefficients of $T(s)$ with $T^*(s)$, we get

$$K_0 = 2, \; K_1 = 2 \text{ and } K_2 = 0.2$$

SERVOMECHANISM PROBLEM: DISCRETE-TIME SYSTEMS

For discrete-time systems, the optimal tracking problem can be formulated in the following way:

For a given plant with irreducible proper transfer function $G_p(s)$ and hold circuit with transfer function $G_0(s)$ (Fig. 13.9), find the transfer function $T^*(z)$ of the overall system that minimizes the quadratic performance index

$$J = \sum_{k=0}^{\infty} \{[r(kT) - c(kT)]^2 + \lambda u^2(kT)\}$$

where λ is a positive constant, $u(t)$ is the actuating signal, $c(t)$ is the actual output of the system and $r(t)$ is the desired output (reference input).

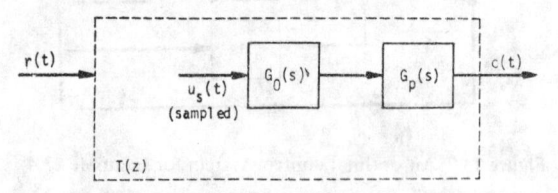

Figure 13.9 An optimal control problem.

In the z-domain, the performance index is given as

$$J = \frac{1}{2\pi j} \oint_{\substack{\text{unit} \\ \text{circle}}} \left\{ [R(z) - C(z)][R(z^{-1}) - C(z^{-1})] + \lambda U(z)U(z^{-1}) \right\} z^{-1} \, dz \quad (13.38)$$

For the system under consideration,

$$\frac{C(z)}{U(z)} = G(z) = \mathscr{Z}[G_0(s)G(s)]$$

$$\frac{C(z)}{R(z)} = T(z)$$

---

†When some of the states are inaccessible, then we may set the feedback coefficients $K_i = 0$ corresponding to inaccessible state variables x_i and try to adjust remaining coefficients to realize the given transfer function. If this is not possible, we may go for reconstruction of state variables. This has been discussed in Chapter 12.

Therefore
$$U(z) = \frac{T(z)R(z)}{G(z)} \tag{13.39}$$

Equation (13.38) may be expressed as

$$J(T(z)) = \frac{1}{2\pi j} \oint_{\text{unit circle}} \left[(T(z) - 1)(T(z^{-1}) - 1) + \lambda \frac{T(z)T(z^{-1})}{G(z)G(z^{-1})} \right] R(z)R(z^{-1})z^{-1} \, dz \tag{13.40}$$

The procedure for derivation of the optimum form of $T(z)$ is similar to the procedure used for derivation of $T^*(s)$ earlier in this section. Instead of s and $-s$, we have z and z^{-1}. Instead of the imaginary axis, the left half s-plane and the right half s-plane, we have the unit circle, inside and outside of the unit circle respectively.

Following the steps of eqns. (13.21) to (13.28), we have the following result:

A necessary as well as sufficient condition for $T^*(z)$ to minimize J given by eqn. (13.40) is that the function

$$X = \left\{ (T^*(z) - 1) + \lambda \frac{T^*(z)}{G(z)G(z^{-1})} \right\} R(z)R(z^{-1})z^{-1} \tag{13.41}$$

not have any pole inside the unit circle.

The function X may be written as

$$X = \left[1 + \frac{\lambda}{G(z)G(z^{-1})} \right] T^*(z)R(z)R(z^{-1})z^{-1} - R(z)R(z^{-1})z^{-1}$$
$$= A(z)A(z^{-1})B(z)B(z^{-1})z^{-1}T^*(z) - B(z)B(z^{-1})z^{-1} \tag{13.42}$$

where $A(z)$ and $B(z)$ are polynomials with no poles or zeros outside the unit circle ($A(z^{-1})$ and $B(z^{-1})$ have no poles and zeros inside the unit circle) and

$$A(z)A(z^{-1}) = 1 + \frac{\lambda}{G(z)G(z^{-1})} \tag{13.43}$$

$$B(z)B(z^{-1}) = R(z)R(z^{-1}) \tag{13.44}$$

Dividing eqn. (13.42) by $A(z^{-1})B(z^{-1})$, we get

$$A(z)B(z)T^*(z)z^{-1} - \frac{B(z)}{A(z^{-1})} = \frac{X}{A(z^{-1})B(z^{-1})} \tag{13.45}$$

The function $\dfrac{B(z)z^{-1}}{A(z^{-1})}$ can be expressed in the partial fraction form as follows:

$$\frac{B(z)z^{-1}}{A(z^{-1})} = C_+(z) + C_-(z) \tag{13.46}$$

where $C_+(z)$ is composed of the part with all the poles inside the unit circle.

Equation (13.45) may now be written as

$$A(z)B(z)z^{-1}T^*(z) - C_+(z) = \frac{X}{A(z^{-1})B(z^{-1})} + C_-(z)$$

The expression on the left hand side does not have any poles outside the unit circle, while the expression on the right hand side does not have any poles inside the unit circle; the two sides are thus independent and must therefore equal zero. We obtain the solution as

$$T^*(z) = \frac{zC_+(z)}{A(z)B(z)} \qquad (13.47)$$

Example 13.5: For the discrete-time system shown in Fig. 13.9, given

$$G_p(s) = \frac{1}{s}, \quad G_0(s) = \frac{1-e^{-sT}}{s}, \quad r(t) = \text{unit step}, \quad T = 1 \text{ sec}$$

find optimal transfer function $T^*(z)$ so that output $c(t)$ follows input $r(t)$ minimizing

$$J_e = \sum_{k=0}^{\infty} [r(kT) - c(kT)]^2 \qquad (13.48)$$

with

$$J_u = \sum_{k=0}^{\infty} u^2(kT) = 0.5 \qquad (13.49)$$

Solution:
For the given system

$$G(z) = \mathscr{Z}\left[\frac{1-e^{-sT}}{s} \cdot \frac{1}{s}\right] = (1-z^{-1})\mathscr{Z}\left[\frac{1}{s^2}\right] = \frac{1}{z-1}$$

$$R(z) = \frac{z}{z-1}$$

The performance index for the system is (eqn. (13.14))

$$J = J_e + \lambda J_u$$

where positive constant λ will be determined from *control effort constraint* given by eqn. (13.49).

From eqn. (13.43),

$$A(z)A(z^{-1}) = 1 + \lambda(z-1)(z^{-1}-1)$$
$$= (1 + 2\lambda) - \lambda z - \lambda z^{-1}$$

Let $\quad A(z) = a_0 + a_1 z$
Therefore
$$(a_0 + a_1 z)(a_0 + a_1 z^{-1}) = (1 + 2\lambda) - \lambda z - \lambda z^{-1}$$

Comparing the coefficients we obtain

$$a_0^2 + a_1^2 = 1 + 2\lambda$$
$$a_0 a_1 = -\lambda$$

Solving for a_0 and a_1, we get

$$a_0 = \frac{1}{2}(1 - \sqrt{1 + 4\lambda})$$

$$a_1 = \frac{1}{2}(1 + \sqrt{1 + 4\lambda})$$

Therefore

$$A(z) = \frac{1}{2}(1 - \sqrt{1 + 4\lambda}) + \frac{1}{2}(1 + \sqrt{1 + 4\lambda})z$$

From eqn. (13.44),

$$B(z)B(z^{-1}) = \left(\frac{z}{z-1}\right)\left(\frac{z^{-1}}{z^{-1}-1}\right) \qquad (13.50)$$

In order to avoid factoring difficulties, we can either introduce the factors $[z - (1 \pm \delta)]$ and let $\delta \to 0$ at the end, or split the poles on unit circle in two parts, half of them belonging to $B(z)$ will be considered to lie inside the unit circle and the remaining half belonging to $B(z^{-1})$ will be considered to lie outside the unit circle.

We can therefore write from eqn. (13.50),

$$B(z) = \frac{z}{z-1}, \quad B(z^{-1}) = \frac{1}{1-z}$$

We next obtain partial fraction expansion of $\dfrac{B(z)z^{-1}}{A(z^{-1})}$ (eqn. (13.46)).

$$\frac{B(z)z^{-1}}{A(z^{-1})} = \frac{2}{(z-1)[(1 - \sqrt{1+4\lambda}) + (1 + \sqrt{1+4\lambda})z^{-1}]}$$

$$= C_+(z) + C_-(z)$$

This gives

$$C_+(z) = \frac{1}{z-1}$$

The optimal transfer function is (eqn. (13.47))

$$T^*(z) = \frac{2}{(1 - \sqrt{1+4\lambda}) + (1 + \sqrt{1+4\lambda})z} \qquad (13.51)$$

From eqn. (13.39), we obtain the optimum actuating signal

$$U^*(z) = \frac{T^*(z)R(z)}{G(z)} = \frac{2z}{(1 - \sqrt{1+4\lambda}) + (1 + \sqrt{1+4\lambda})z}$$

The control effort constraint (13.49) may now be expressed in the z-domain.

$$\frac{1}{2\pi j}\oint_{\substack{\text{unit}\\\text{circle}}} U^*(z)U^*(z^{-1})z^{-1}\,dz = 0.5$$

Using Table 13.2, we get the following equation:

$$\frac{1}{\sqrt{1 + 4\lambda}} = 0.5$$

This gives

$$\lambda = 0.75$$

From eqn. (13.51), the optimal transfer function is

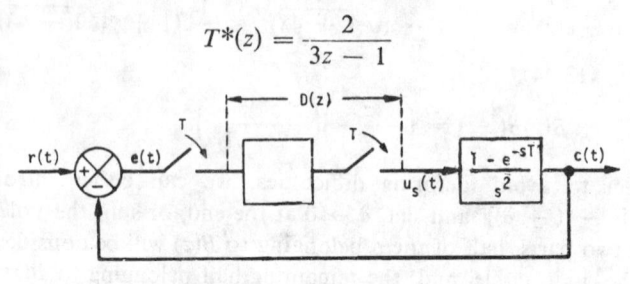

$$T^*(z) = \frac{2}{3z - 1}$$

Figure 13.10 An optimal control system for Example 13.5.

If the configuration of the compensator is chosen as shown in Fig. 13.10, then $D(z)$, the transfer function of data processing unit is obtained from

$$T^*(z) = \frac{D(z)G(z)}{1 + D(z)G(z)}$$

as

$$D(z) = \frac{T^*(z)}{G(z)[1 - T^*(z)]} = \frac{2}{3} \qquad (13.52)$$

Thus the optimum controller in this case is just a gain of 2/3.

OUTPUT REGULATOR PROBLEM

We now introduce another important class of optimal control problems: *the output regulator problem*. It is a special case of the tracking problem in which $r(t) = 0$. For zero input, the output is zero if all the initial conditions are zero. The response $c(t)$ is due to nonzero initial conditions that, in turn, are caused by disturbances. The primary objective of the design is to damp out the response due to initial conditions quickly without excessive overshoot and oscillations.

Consider for example the system of Problem 3.8 (Fig. P-3.8). The disturbance torque of the sea causes the ship to roll. The response (roll angle $\ell(t)$) to this disturbance is highly oscillatory. The oscillations in the rolling motion are to be damped out quickly without excessive overshoot. If there is no constraint on 'control effort', the controller which minimizes the performance index

$$J = \int_0^\infty (\theta(t) - \theta_d(t))^2 \, dt$$

will be the optimum one; θ_d is the desired roll which is clearly zero. Therefore the problem of stabilization of ship against rolling motion is a regulator problem. If, for disturbance torque applied at $t = t_0$, the controller is required to regulate the roll motion within finite time $(t_f - t_0)$, a suitable performance criterion for design of optimum controller is to minimize

$$J = \int_{t_0}^{t_f} \theta^2(t)\, dt \qquad (13.53)$$

LIMITATIONS OF TRANSFER FUNCTION APPROACH

Since the responses of regulating systems are due to nonzero initial conditions, the transfer functions do not give useful formulation of such systems.

For the servomechanism or tracking problem in single-input-single-output linear time-invariant systems, the transfer function approach is straightforward. However, the limitation of this approach for multi-input-multi-output linear time-invariant systems is obvious. The spectral factorization becomes quite complex because of involvement of matrix operations. In addition, the transfer function approach is restricted to systems with quadratic performance index with no constraint on time.

The transfer function design technique becomes ineffective for time-varying and nonlinear systems.

All these limitations of transfer function approach are absent in the *state variable approach*. With the availability of digital computer, the state variable approach has become the most powerful method for solving modern complex control problems. However, the transfer function approach is quite simple for the class of systems for which it is applicable.

13.4 Optimal Control Problems: State Variable Approach

Following steps are involved in the solution of an optimal control problem using state variable approach:

(i) Given a plant in the form of state equations

$$\dot{\mathbf{x}}(t) = \mathbf{A}\mathbf{x}(t) + \mathbf{B}\mathbf{u}(t): \quad \text{Continuous-time systems} \qquad (13.54a)$$

$$\mathbf{x}(k+1) = \mathbf{A}\mathbf{x}(k) + \mathbf{B}\mathbf{u}(k): \quad \text{Discrete-time systems} \qquad (13.54b)$$

where $\mathbf{x} = n \times 1$ state vector; $\mathbf{u} = m \times 1$ control vector; $\mathbf{A} = n \times n$ constant matrix; and $\mathbf{B} = n \times m$ constant matrix, find the control function $\mathbf{u}^*$ which is optimal with respect to given performance criterion.

(ii) Realize the control function obtained from step (i).

THE STATE REGULATOR PROBLEM

We recall from earlier section that when a system variable $x_1(t)$ (the output) is required to be 'near' zero, the performance measure is (eqn. (13.53)),

$$J = \int_{t_0}^{t_f} x_1^2(t)\, dt$$

A performance index written in terms of two state variables of a system would then be

$$J = \int_{t_0}^{t_f} (x_1^2(t) + x_2^2(t))\,dt$$

Therefore if the state $\mathbf{x}(t)$ of a system described by (13.54a) is required to be close to $\mathbf{x}_d = \mathbf{0}$, a design criterion would be to determine a control function that minimizes

$$J = \int_{t_0}^{t_f} (\mathbf{x}^T \mathbf{x})\,dt$$

In practical systems, the control of all the states of the system is not equally important. For example, if in addition to roll angle $\theta(t)$ of a ship (Fig. P-3.8), the pitch angle $\phi(t)$ is also required to be zero, the performance index of eqn. (13.53) gets modified to

$$J = \int_{t_0}^{t_f} (\theta^2(t) + \lambda\phi^2(t))\,dt$$

where λ, a positive constant, is the weighting factor. The roll motion contributes much discomfort to the passengers; in the design of passenger ships, the value of λ will be less than one.

A weighted quadratic performance index is

$$J = \int_{t_0}^{t_f} (\mathbf{x}^T \mathbf{Q} \mathbf{x})\,dt$$

where $\mathbf{Q}$ is positive definite, real, symmetric, constant matrix. The simplest form of $\mathbf{Q}$ one can use is a diagonal matrix:

$$\mathbf{Q} = \begin{bmatrix} q_1 & 0 & \cdots & 0 \\ 0 & q_2 & \cdots & 0 \\ \vdots & \vdots & \cdots & \vdots \\ 0 & 0 & \cdots & q_n \end{bmatrix}$$

The ith entry of $\mathbf{Q}$ represents the amount of weight the designer places on the constraint on state variable $x_i(t)$. The larger the value of q_i relative to other values of q, the more control effort is spent to regulate $x_i(t)$.

To minimize the deviation of the final state $\mathbf{x}(t_f)$ of the system from the desired state $\mathbf{x}_d = \mathbf{0}$, a possible performance measure is

$$J = \mathbf{x}^T(t_f)\mathbf{H}\mathbf{x}(t_f)$$

where $\mathbf{H}$ is positive definite, real, symmetric, constant matrix. In the infinite time state regulator problem ($t_f \to \infty$), the final state should approach the equilibrium state $\mathbf{x} = \mathbf{0}$; so the terminal constraint is no longer necessary.

The optimal design obtained by minimizing

$$J = \mathbf{x}^T(t_f)\mathbf{H}\mathbf{x}(t_f) + \int_{t_0}^{t_f} \mathbf{x}^T(t)\mathbf{Q}\mathbf{x}(t)\,dt$$

may be unsatisfactory in practice. A more realistic solution to the problem is reached if the performance index is modified by adding a penalty term for physical constraints. One of the ways of accomplishing this is to introduce the following quadratic control term† in the performance index:

$$J = \int_{t_0}^{t_f} \mathbf{u}^T(t)\mathbf{R}\mathbf{u}(t)\, dt$$

where **R** is a positive definite, real, symmetric, constant matrix. By giving sufficient weight to control terms, the amplitude of controls which minimize the overall performance index may be kept within practical bounds, although at the expense of increased error in $\mathbf{x}(t)$.

Continuous-time systems

We now formulate the *state regulator problem* in continuous-time systems:

Consider the controlled process described by the state equations (13.54a). Find optimal control law $\mathbf{u}^*(t)$, $t \in [t_0, t_f]$, where t_0 and t_f are specified initial and final times respectively, so that the quadratic performance index††

$$J = \frac{1}{2}\mathbf{x}^T(t_f)\mathbf{H}\mathbf{x}(t_f) + \frac{1}{2}\int_{t_0}^{t_f}(\mathbf{x}^T(t)\mathbf{Q}\mathbf{x}(t) + \mathbf{u}^T(t)\mathbf{R}\mathbf{u}(t))\, dt \quad (13.55)$$

is minimized, subject to an initial state

$$\mathbf{x}(t_0) = \mathbf{x}_0$$

If the terminal time t_f is not constrained, the performance index is†††

$$J = \frac{1}{2}\int_{t_0}^{\infty}(\mathbf{x}^T(t)\mathbf{Q}\mathbf{x}(t) + \mathbf{u}^T(t)\mathbf{R}\mathbf{u}(t))\, dt \quad (13.56)$$

For this infinite-time state regulator problem, solution exists only if (13.54a) represents a completely controllable system (proved in next section).

In the performance measures given by (13.55) and (13.56), **R** is restricted to be positive definite. This requirement is, as we shall see, a condition for existence of a finite control.

The matrices **Q** and **H** may be positive definite or semidefinite. However, if both of them are simultaneously zero matrices, i.e., $\mathbf{Q} = \mathbf{H} = \mathbf{0}$, then the performance index becomes

$$J = \frac{1}{2}\int_{t_0}^{t_f} \mathbf{u}^T(t)\mathbf{R}\mathbf{u}(t)\, dt$$

which is minimum when $\mathbf{u}(t) = \mathbf{0}$. To exclude this trivial case, we shall assume that **Q** and **H** are not both zero matrices, although we shall allow either **Q** or **H** to be the zero matrix individually.

---

†Recall the statement made earlier in this chapter—if the performance index is quadratic, the resulting optimal system is a linear system.

††Note that multiplication by a constant 1/2, does not affect the minimization problem. The constant helps in mathematical manipulations as we shall see later.

†††To may be taken as zero in linear time-invariant systems.

Discrete-time systems

The state regulator problem in discrete-time systems is formulated as follows:

Consider the controlled process described by state equations (13.56b). Find the optimal control sequence $\mathbf{u}^*(k)$, $k = 0, 1, \ldots, N-1$, where N is a specified fixed positive integer number, so that the quadratic performance index

$$J = \frac{1}{2} \mathbf{x}^T(N)\mathbf{H}\mathbf{x}(N) + \frac{1}{2} \sum_{k=0}^{N-1} [\mathbf{x}^T(k)\mathbf{Q}\mathbf{x}(k) + \mathbf{u}^T(k)\mathbf{R}\mathbf{u}(k)] \quad (13.57)$$

is minimized, subject to an initial state

$$\mathbf{x}(0) = \mathbf{x}_0$$

Note that eqns. (13.54b) and (13.57) may be the result of a discrete approximation to a continuous-time state regulator problem or the formulation for the state regulator problem in a sampled-data system.

If a process is to be controlled for a large number of stages ($N \to \infty$), then a performance index is

$$J = \frac{1}{2} \sum_{k=0}^{\infty} [\mathbf{x}^T(k)\mathbf{Q}\mathbf{x}(k) + \mathbf{u}^T(k)\mathbf{R}\mathbf{u}(k)] \quad (13.58)$$

For this infinite-time problem, solution exists only if (13.54b) represents a completely controllable system (proved in next section).

In eqns. (13.57) and (13.58), **R** is positive definite, **Q** and **H** are positive definite (or semidefinite). However both **Q** and **H** can not be zero matrices simultaneously.

THE OUTPUT REGULATOR PROBLEM

In the state regulator problem, we are concerned with making all the components of the state vector $\mathbf{x}(t)$ small. In the output regulator problem on the other hand, we are concerned with making the components of the output vector small. If the controlled process is observable then, as we shall see, we can reduce the output regulator problem to the state regulator problem.

The output regulator problem is formulated as follows:

Consider an observable controlled process described by the equations

$$\dot{\mathbf{x}}(t) = \mathbf{A}\mathbf{x}(t) + \mathbf{B}\mathbf{u}(t)$$
$$\mathbf{y}(t) = \mathbf{C}\mathbf{x}(t) \quad (13.59)$$

where $\mathbf{x}(t) = n \times 1$ state vector; $\mathbf{u}(t) = m \times 1$ control vector; $\mathbf{y}(t) = p \times 1$ output vector; $\mathbf{A} = n \times n$ constant matrix; $\mathbf{B} = n \times m$ constant matrix; and $\mathbf{C} = p \times n$ constant matrix.

Find optimal control law $\mathbf{u}^*(t)$, $t \in [t_0, t_f]$, where t_0 and t_f are specified initial and final times respectively, so that the quadratic performance index

$$J = \frac{1}{2} \mathbf{y}^T(t_f)\mathbf{H}\mathbf{y}(t_f) + \frac{1}{2} \int_{t_0}^{t_f} (\mathbf{y}^T(t)\mathbf{Q}\mathbf{y}(t) + \mathbf{u}^T(t)\mathbf{R}\mathbf{u}(t)) \, dt$$

$$(13.60)$$

is minimized, subject to an initial state

$$\mathbf{x}(t_0) = \mathbf{x}_0$$

The output regulator problem in discrete-time systems can be formulated on similar lines.

THE TRACKING PROBLEM

Recall that we discussed in detail the physical aspects of the tracking problem in the earlier section and there we presented its formulation. In the state variable approach, this problem is formulated as follows:

Consider an observable controlled process described by eqn. (13.59). Suppose that the vector $\mathbf{z}(t)$ is the desired output; we assume that the dimension of $\mathbf{z}(t)$ is equal to that of $\mathbf{y}(t)$. Define the error vector $\mathbf{e}(t)$ by setting

$$\mathbf{e}(t) = \mathbf{z}(t) - \mathbf{y}(t)$$

Find optimal control law $\mathbf{u}^*(t)$, $t\epsilon[t_0, t_f]$ where t_0 and t_f are specified initial and final times respectively, so that the quadratic performance index

$$J = \frac{1}{2} \mathbf{e}^T(t_f)\mathbf{H}\mathbf{e}(t_f) + \frac{1}{2} \int_{t_0}^{t_f} (\mathbf{e}^T(t)\mathbf{Q}\mathbf{e}(t) + \mathbf{u}^T(t)\mathbf{R}\mathbf{u}(t))\, dt$$

(13.61)

is minimized.

The tracking problem in discrete-time systems can be formulated on the similar lines.

APPROACHES FOR MINIMIZATION OF PERFORMANCE INDEX

Once the performance index for a system has been chosen, the next task is to determine a control function that minimizes this index. Two approaches for accomplishing this minimization are available:

(i) *Minimum Principle of Pontryagin*: This is based on the concepts of calculus of variations.

(ii) *Dynamic Programming Developed by Bellman*: This is based on the *principle of invariant imbedding* and leads to the *principle of optimality* that follows the basic laws of nature and does not need complex mathematical development to explain its validity.

For the linear regulator problem, minimization using ordinary differential calculus has been achieved [7, 8]. However, we shall use dynamic programming approach in our discussion. This will help the reader to directly study advanced literature on optimal control theory.

In the following sections, we shall solve various optimal control problems. We shall use the results of Appendix II very often. Many details of the proofs will be omitted. For these details, the reader may refer advanced literature on the subject.

Optimal control problems can be solved with the help of digital computer. However, we shall illustrate the optimal control theory by simple examples which can be solved analytically.

13.5 The State Regulator Problem

DISCRETE-TIME SYSTEMS

The optimization method based on dynamic programming views the control problem as a multistage decision problem and the control input as a time sequence of decisions [6]. A sampled-data system gives rise to a sequence of transformations of the original state vector; the decision process therefore consists of selecting the transformations at each discrete interval of time.

Consider a process described by state equations (13.54b) with performance index (13.57). This process is called a multistage decision process of N stages where the choice of $u(k)$ at each sampling instant is considered the decision of interest. If we select from the myriad of possible choices, the choice of decisions $u(0)$, $u(1)$, ..., $u(N-1)$ which minimizes the given performance index J, we have selected the *optimal policy or sequence*.

For a given sampled-data system described by (13.54b), the optimal control sequence is a function of the initial state $x(0)$ and the number of stages N. We take the decision at the stage corresponding to $k = 0$, to determine optimal path not merely from the given state $x(0)$ but from all possible states which the system can possess. In general, at the ith stage, we determine optimal path not merely from the state $x(i)$ which is a function of the initial state $x(0)$ and the controls $u(0)$, $u(1)$, ..., $u(i-1)$, but from all possible states of the ith stage. Thus we do not regard the control problem as an isolated problem with fixed value of $x(0)$ and N but rather imbed it within a family of problems. This is the *principle of invariant imbedding* introduced by Bellman.

Another important feature of dynamic programming formulation is that we do not attempt to find all the values of optimal control sequence simultaneously; we find one control value at a time until the entire optimal policy is determined. This sequential decision process is based on the *principle of optimality* which states that an optimal control policy has the property that whatever the initial state and initial decisions are, the remaining decisions must form an optimal policy with respect to the state resulting from the initial decisions. The principle of optimality reduces the N-stage decision process into N single-stage decision processes.

It is permissible to find the last decision as the initial calculation step in multistage decision process, i.e., the calculations of optimal decisions may proceed from the last decision back to the first decision. The system state at the Nth stage is a function of the state at $(N-1)$th stage and the control $u(N-1)$:

$$x(N) = f(x(N-1), u(N-1))$$

Optimal Control Systems

As per the principle of optimality, regardless of how the state reaches $x(N-1)$, once $x(N-1)$ is known then, using $x(N-1)$ as the initial state for the last stage of state transition, the control $u(N-1)$ must be so chosen that (eqn. (13.57))

$$J_{N-1,N}(x(N-1), u(N-1)) = \frac{1}{2}x^T(N)Hx(N) + \frac{1}{2}[x^T(N-1)Qx(N-1) + u^T(N-1)Ru(N-1)] \quad (13.62)$$

is a minimum.

Note that $J_{N-1,N}$ is the performance index of operation during the interval $(N-1)\Delta t \leq t \leq N\Delta t$ where Δt is the sampling interval. $J_{N-1,N}$ is also performance index of the one-stage process with initial state $x(N-1)$. Thus optimal policy and minimum performance index of one-stage process are imbedded in the results for the N-stage process.

Equation (13.62) may be written as

$$J_{N-1,N}(x(N-1), u(N-1)) = \frac{1}{2}[Ax(N-1) + Bu(N-1)]^T P(0)[Ax(N-1) + Bu(N-1)] + \frac{1}{2}[x^T(N-1)Qx(N-1) + u^T(N-1)Ru(N-1)] \quad (13.63)$$

where $P(0) = H$.

The necessary condition for minimum $J_{N-1,N}$ is

$$\frac{\partial J_{N-1,N}}{\partial u(N-1)} = 0$$

This gives (Appendix II, eqns. (II.21), (II.24))

$$Ru^*(N-1) + B^T P(0)[Ax(N-1) + Bu^*(N-1)] = 0$$

or

$$u^*(N-1) = -[R + B^T P(0)B]^{-1}B^T P(0)Ax(N-1)$$
$$= K(N-1)x(N-1) \quad (13.64)$$

where

$$K(N-1) = -[R + B^T P(0)B]^{-1}B^T P(0)A$$

This control satisfies the sufficient condition for minimum $J_{N-1,N}$ since (eqn. (II.19) in Appendix II)

$$\frac{\partial^2 J_{N-1,N}}{\partial u^2(N-1)} = R + B^T P(0)B$$

is positive definite. This is because R is a positive definite matrix and $B^T P(0) B$ is positive definite (or semidefinite as $P(0) = H$ is positive definite or semidefinite).

The minimum value of performance index obtained from eqns. (13.63) and (13.64) is

$$J^*_{N-1,N}(\mathbf{x}(N-1)) = \frac{1}{2}\mathbf{x}^T(N-1)\left\{[\mathbf{A}+\mathbf{BK}(N-1)]^T\mathbf{P}(0)[\mathbf{A}\right.$$
$$\left.+\mathbf{BK}(N-1)] + \mathbf{K}^T(N-1)\mathbf{RK}(N-1) + \mathbf{Q}\right\}\mathbf{x}(N-1)$$
$$= \frac{1}{2}\mathbf{x}^T(N-1)\mathbf{P}(1)\mathbf{x}(N-1) \tag{13.65}$$

where
$$\mathbf{P}(1) = [\mathbf{A}+\mathbf{BK}(N-1)]^T\mathbf{P}(0)[\mathbf{A}+\mathbf{BK}(N-1)]$$
$$+ \mathbf{K}^T(N-1)\mathbf{RK}(N-1) + \mathbf{Q}$$

Imbedding a two-stage process in the N-stage process and using the principle of optimality, we get the following result:

Regardless of how the state reaches $\mathbf{x}(N-2)$, once $\mathbf{x}(N-2)$ is known then, using $\mathbf{x}(N-2)$ as the initial state for the operation in last two stages, the control sequence $\mathbf{u}(N-2)$, $\mathbf{u}(N-1)$ should be so chosen that

$$J_{N-2,N}(\mathbf{x}(N-2), \mathbf{u}(N-2), \mathbf{u}(N-1)) = \frac{1}{2}\mathbf{x}^T(N)\mathbf{P}(0)\mathbf{x}(N)$$
$$+ \frac{1}{2}\sum_{k=N-2}^{N-1}[\mathbf{x}^T(k)\mathbf{Q}\mathbf{x}(k) + \mathbf{u}^T(k)\mathbf{R}\mathbf{u}(k)]$$
$$= \frac{1}{2}\mathbf{x}^T(N-2)\mathbf{Q}\mathbf{x}(N-2) + \frac{1}{2}\mathbf{u}^T(N-2)\mathbf{R}\mathbf{u}(N-2) + J_{N-1,N}$$

is a minimum.

For a two-stage process, whatever the initial state $\mathbf{x}(N-2)$ and initial control $\mathbf{u}(N-2)$, the remaining control $\mathbf{u}(N-1)$ must be optimal with respect to the value of $\mathbf{x}(N-1)$ that results from application of $\mathbf{u}(N-2)$. Therefore the performance index for the two-stage process is (using eqns. (13.54b), (13.65))

$$J_{N-2,N}(\mathbf{x}(N-2), \mathbf{u}(N-2))$$
$$= \frac{1}{2}\mathbf{x}^T(N-2)\mathbf{Q}\mathbf{x}(N-2) + \frac{1}{2}\mathbf{u}^T(N-2)\mathbf{R}\mathbf{u}(N-2) + J^*_{N-1,N}$$
$$= \frac{1}{2}\mathbf{x}^T(N-2)\mathbf{Q}\mathbf{x}(N-2) + \frac{1}{2}\mathbf{u}^T(N-2)\mathbf{R}\mathbf{u}(N-2)$$
$$+ \frac{1}{2}[\mathbf{A}\mathbf{x}(N-2)+\mathbf{B}\mathbf{u}(N-2)]^T\mathbf{P}(1)[\mathbf{A}\mathbf{x}(N-2)+\mathbf{B}\mathbf{u}(N-2)]$$

The minimization procedure (as applied earlier to one-stage process) gives the following result:

$$\mathbf{u}^*(N-2) = -[\mathbf{R}+\mathbf{B}^T\mathbf{P}(1)\mathbf{B}]^{-1}\mathbf{B}^T\mathbf{P}(1)\mathbf{A}\mathbf{x}(N-2)$$
$$= \mathbf{K}(N-2)\mathbf{x}(N-2)$$
$$J^*(\mathbf{x}(N-2)) = \frac{1}{2}\mathbf{x}^T(N-2)\left\{[\mathbf{A}+\mathbf{BK}(N-2)]^T\mathbf{P}(1)[\mathbf{A}+\mathbf{BK}(N-2)]\right.$$
$$\left.+\mathbf{K}^T(N-2)\mathbf{RK}(N-2)+\mathbf{Q}\right\}\mathbf{x}(N-2)$$
$$= \frac{1}{2}\mathbf{x}^T(N-2)\mathbf{P}(2)\mathbf{x}(N-2)$$

Optimal Control Systems

By induction, we can write the complete solution for the N-stage process as

$$\mathbf{u}^*(N-j) = \mathbf{K}(N-j)\mathbf{x}(N-j); j = 1, 2, ..., N \quad (13.66\text{a})$$

$$\mathbf{K}(N-j) = -[\mathbf{R} + \mathbf{B}^T\mathbf{P}(j-1)\mathbf{B}]^{-1}\mathbf{B}^T\mathbf{P}(j-1)\mathbf{A} \quad (13.66\text{b})$$

$$\mathbf{P}(0) = \mathbf{H} \quad (13.66\text{c})$$

$$\mathbf{P}(j) = [\mathbf{A} + \mathbf{BK}(N-j)]^T\mathbf{P}(j-1)[\mathbf{A} + \mathbf{BK}(N-j)] \quad (13.66\text{d})$$
$$+ \mathbf{K}^T(N-j)\mathbf{RK}(N-j) + \mathbf{Q}$$

$$J^*_{N-j,N} = \frac{1}{2}\mathbf{x}^T(N-j)\mathbf{P}(j)\mathbf{x}(N-j) \quad (13.66\text{e})$$

Equations (13.66) form the recursive relations for determination of $\mathbf{K}$ and $\mathbf{P}$ matrices. Starting with $j = 1$, $\mathbf{K}(N-1)$ is evaluated from eqn. (13.66b) with $\mathbf{P}(0) = \mathbf{H}$, a given matrix. Equation (13.66d) is then solved for $\mathbf{P}(1)$. This constitutes one cycle of procedure, which we then continue by calculating $\mathbf{K}(N-2)$, $\mathbf{P}(2)$ and so on. The matrices $[\mathbf{K}(N-1), \mathbf{K}(N-2), ..., \mathbf{K}(0)]$ are stored. The optimal controller is then realized by determining the appropriate gain settings (eqn. (13.66a)) as the system transits from stage to stage.

For the N-stage process with specified initial state $\mathbf{x}(0) = \mathbf{x}_0$, the minimum value of performance index is (eqn. (13.66e))

$$J^*_{0,N}(\mathbf{x}_0) = \frac{1}{2}\mathbf{x}_0^T\mathbf{P}(N)\mathbf{x}_0 \quad (13.67)$$

The following important points are observed in this result:

(i) The optimal control at each stage is a linear combination of the states; thus giving linear state variable feedback control policy.
(ii) Feedback is time-varying, although $\mathbf{A}$, $\mathbf{B}$, $\mathbf{R}$ and $\mathbf{Q}$ are all constant matrices, i.e., optimal control policy converts a linear time-invariant plant with time-invariant quadratic performance index into a linear time-varying feedback system.
(iii) Following the procedure given above, results for optimal control of linear time-varying plant can easily be obtained [6].

Realization of optimal control policy

Once the optimal control policy has been determined, its realization is the second phase of the optimal control problem.

Since the optimal control is

$$\mathbf{u}^*(k) = \mathbf{K}(k)\mathbf{x}(k)$$

its realization seeks feedback of the state variables. If the plant states are available for measurement, $\mathbf{u}^*(k)$ is implemented by feedback of the state variables through time-varying gain elements; the engineering construction of time-varying functions is done by means of a digital computer.

If the plant states are not available for measurement, then it is possible to construct a physical device—*state observer*, which produces at its output the plant states, when driven by both the plant input and output (discussed in Chapter 12). This is however possible only if the plant equations satisfy the conditions of observability.

It may be noted that it is not necessary that the system (13.54b) be controllable. Assume that some states of the system are uncontrollable. We expect that this will impose problems in the regulator system if the uncontrollable states are unstable because these unstable states will be reflected in the performance index. However the contribution of the unstable uncontrollable states (unstable controllable states are stabilizable) to the performance index is always finite provided that the control interval $[t_0, t_f]$ is finite. We shall see in the next section that we shall require controllability as $t_f \to \infty$ to ensure that the value of performance index is finite.

Example 13.6: The second-order linear system

$$\dot{x}_1(t) = -0.2\, x_1(t) + x_2(t) + u(t)$$
$$\dot{x}_2(t) = -0.5\, x_2(t) + 0.5\, u(t) \quad (13.68)$$

is to be controlled to minimize the performance index

$$J = \frac{1}{2} \int_0^5 (x_1^2(t) + x_2^2(t) + u^2(t))\, dt \quad (13.69)$$

Let us determine the optimal control policy for this system using the results given by (13.66).

For determination of the optimal control policy using the results (13.66), the system differential equations must be approximated by difference equations and integral in performance index must be approximated by a summation. To do this, divide the time interval $0 \leqslant t \leqslant t_f$ into N equal increments of time duration T. Then from eqn. (13.68),

$$\frac{x_1(t+T) - x_1(t)}{T} \simeq -0.2 x_1(t) + x_2(t) + u(t)$$

or

$$x_1(k+1) = (1 - 0.2T)x_1(k) + Tx_2(k) + Tu(k)$$

Assuming $T = 1$ sec (this gives $N = 5$), we get

$$x_1(k+1) = 0.8 x_1(k) + x_2(k) + u(k) \quad (13.70a)$$

Similarly we obtain

$$x_2(k+1) = 0.5 x_2(k) + 0.5 u(k) \quad (13.70b)$$

The performance index given by (13.69) may be approximated to a summation as follows:

$$J = \frac{1}{2}\left\{\int_0^T (x_1^2(t) + x_2^2(t) + u^2(t))\, dt + \int_T^{2T} (x_1^2(t) + x_2^2(t) + u^2(t))\, dt + \ldots + \int_{(N-1)T}^{NT} (x_1^2(t) + x_2^2(t) + u^2(t))\, dt\right\}$$

Optimal Control Systems

$$\simeq \frac{1}{2}\Big\{[x_1^2(0) + x_2^2(0) + u^2(0)]T + [x_1^2(1) + x_2^2(1) + u^2(1)]T$$

$$+ \ldots + [x_1^2(N-1) + x_2^2(N-1) + u^2(N-1)]T\Big\}$$

$$= \frac{1}{2} \sum_{k=0}^{4} (x_1^2(k) + x_2^2(k) + u^2(k)) \tag{13.71}$$

Comparing eqns. (13.70) and (13.71) with (13.54b) and (13.57) respectively, we obtain

$$\mathbf{A} = \begin{bmatrix} 0.8 & 1 \\ 0 & 0.5 \end{bmatrix}, \mathbf{B} = \begin{bmatrix} 1 \\ 0.5 \end{bmatrix}$$

$$\mathbf{H} = \mathbf{O}, \mathbf{Q} = \begin{bmatrix} 1 & 0 \\ 0 & 1 \end{bmatrix}, \mathbf{R} = 1$$

The following results[†] are obtained from (13.66):

$$\mathbf{K}(4) = [0 \quad 0]$$
$$\mathbf{K}(3) = -[0.355 \quad 0.555]$$
$$\mathbf{K}(2) = -[0.395 \quad 0.677]$$
$$\mathbf{K}(1) = -[0.395 \quad 0.687]$$
$$\mathbf{K}(0) = -[0.395 \quad 0.687]$$

The block diagram of the optimal system is shown in Fig. 13.11.

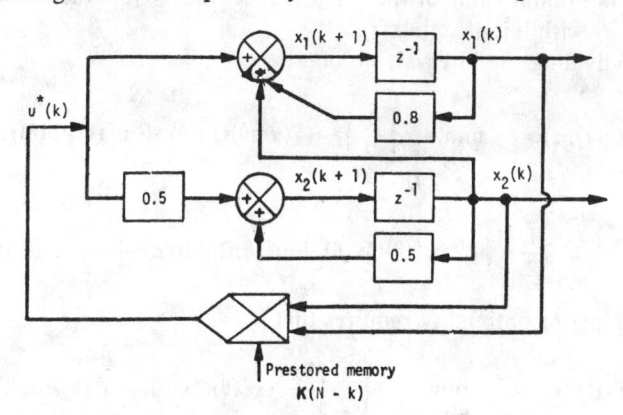

Figure 13.11 Block diagram of optimal control system for Example 13.6.

CONTINUOUS-TIME SYSTEMS

In Example 13.6, we approximated a continuous-time system by a discrete-time system. This approach leads to recursive relations (13.66), ideally suited for digital computer solution. In this section, we consider an alternative approach that leads to nonlinear differential equations.

[†]The required calculations may best be carried out using a digital computer.

Consider a process described by the state equations (13.54a). We are required to find the optimal control law $u^*(t)$ so that quadratic performance index given by (13.55) is minimized.

Let us use the principle of imbedding to include this problem in a larger class of problems by considering the performance measure

$$J(x(t)\ t, \underset{t \leqslant \tau \leqslant t_f}{u\ (\tau)}) = \frac{1}{2} x^T(t_f) H x(t_f) + \frac{1}{2} \int_t^{t_f} (x^T(\theta) Q x(\theta) + u^T(\theta) R u(\theta))\, d\theta;\ t \leqslant t_f$$

The optimal control $u^*(t)$ is to be found by minimizing J with respect to admissible controls while treating the other variables as constants. Then the optimal value of the performance index is

$$J^*(x(t), t, \underset{t \leqslant \tau \leqslant t_f}{u^*(\tau)}) = J^*(x(t), t)$$

$$= \underset{\substack{u(\tau) \\ t \leqslant \tau \leqslant t_f}}{\min} \left\{ \frac{1}{2} \int_t^{t_f} (x^T(\theta) Q x(\theta) + u^T(\theta) R u(\theta))\, d\theta \right.$$

$$\left. + \frac{1}{2} x^T(t_f) H x(t_f) \right\} \qquad (13.72)$$

J^* is the minimum value of the performance index for the time interval $t \leqslant \tau \leqslant t_f$ with initial state $x(t)$.

By subdividing the interval, we obtain

$$J^*(x(t), t) = \underset{\substack{u(\tau) \\ t \leqslant \tau \leqslant t_f}}{\min} \left\{ \frac{1}{2} \int_t^{t+\Delta t} (x^T(\theta) Q x(\theta) + u^T(\theta) R u(\theta))\, d\theta \right.$$

$$\left. + \frac{1}{2} \int_{t+\Delta t}^{t_f} (x^T(\theta) Q x(\theta) + u^T(\theta) R u(\theta))\, d\theta + \frac{1}{2} x^T(t_f) H x(t_f) \right\}$$

The principle of optimality requires that

$$J^*(x(t), t) = \underset{\substack{u(\tau) \\ t \leqslant \tau \leqslant t + \Delta t}}{\min} \left\{ \frac{1}{2} \int_t^{t+\Delta t} (x^T(\theta) Q x(\theta) + u^T(\theta) R u(\theta))\, d\theta \right.$$

$$\left. + J^*(x(t + \Delta t), t + \Delta t) \right\}$$

where $J^*(x(t + \Delta t), t + \Delta t)$ is the minimum value of the performance index for the process with initial state $x(t + \Delta t)$.

Expanding $J^*(x(t + \Delta t), t + \Delta t)$ in Taylor's series about $(x(t), t)$ and neglecting terms of order higher than first, we get (see **Appendix II**)

$$J^*(\mathbf{x}(t), t) = \min_{\substack{\mathbf{u}(\tau) \\ t \leqslant \tau \leqslant t + \Delta t}} \left\{ \frac{1}{2} \int_t^{t+\Delta t} (\mathbf{x}^T(\theta)\mathbf{Q}\mathbf{x}(\theta) + \mathbf{u}^T(\theta)\mathbf{R}\mathbf{u}(\theta))\, d\theta \right.$$

$$\left. + J^*(\mathbf{x}(t), t) + \left[\frac{\partial J^*(\mathbf{x}(t), t)}{\partial t}\right]\Delta t + \left[\frac{\partial J^*(\mathbf{x}(t), t)}{\partial \mathbf{x}}\right]^T [\mathbf{x}(t + \Delta t) - \mathbf{x}(t)] \right\}$$

$$= J^*(\mathbf{x}(t), t) + \left[\frac{\partial J^*(\mathbf{x}(t), t)}{\partial t}\right]\Delta t + \min_{\mathbf{u}(t)} \left\{ \frac{1}{2} (\mathbf{x}^T(t)\mathbf{Q}\mathbf{x}(t) + \mathbf{u}^T(t)\mathbf{R}\mathbf{u}(t))\Delta t \right.$$

$$\left. + \left[\frac{\partial J^*(\mathbf{x}(t), t)}{\partial \mathbf{x}}\right]^T [\mathbf{A}\mathbf{x}(t) + \mathbf{B}\mathbf{u}(t)]\Delta t \right\}$$

(Note that $\dot{\mathbf{x}}(t) = \dfrac{\mathbf{x}(t + \Delta t) - \mathbf{x}(t)}{\Delta t} \cdot$)

or

$$-\frac{\partial J^*(\mathbf{x}(t), t)}{\partial t} = \min_{\mathbf{u}(t)} \left\{ \frac{1}{2} (\mathbf{x}^T(t)\mathbf{Q}\mathbf{x}(t) + \mathbf{u}^T(t)\mathbf{R}\mathbf{u}(t)) \right.$$

$$\left. + \left[\frac{\partial J^*(\mathbf{x}(t), t)}{\partial \mathbf{x}}\right]^T [\mathbf{A}\mathbf{x}(t) + \mathbf{B}\mathbf{u}(t)] \right\} \tag{13.73}$$

To carry out the minimization process, we differentiate both sides of above equation with respect to $\mathbf{u}(t)$. This gives (Appendix II, eqns. (II.21) and (II.24))

$$0 = \mathbf{R}\mathbf{u}^*(t) + \mathbf{B}^T \frac{\partial J^*}{\partial \mathbf{x}} (\mathbf{x}(t), t)$$

Solving for $\mathbf{u}^*(t)$, we get

$$\mathbf{u}^*(t) = -\mathbf{R}^{-1}\mathbf{B}^T \frac{\partial J^*(\mathbf{x}(t), t)}{\partial \mathbf{x}} \tag{13.74}$$

The second derivative of $-\dfrac{\partial J^*(\mathbf{x}(t), t)}{\partial t}$ with respect to $\mathbf{u}(t)$ gives a positive definite matrix $\mathbf{R}$. Therefore $\mathbf{u}^*(t)$ given by eqn. (13.74) satisfies the necessary and sufficient conditions for minimum.

Substituting eqn. (13.74) into eqn. (13.73), we get

$$-\frac{\partial J^*(\mathbf{x}(t), t)}{\partial t} = \frac{1}{2}\mathbf{x}^T(t)\mathbf{Q}\mathbf{x}(t) + \frac{1}{2}\left(\frac{\partial J^*(\mathbf{x}(t), t)}{\partial \mathbf{x}}\right)^T \mathbf{B}\mathbf{R}^{-1}\mathbf{B}^T \frac{\partial J^*(\mathbf{x}(t), t)}{\partial \mathbf{x}}$$

$$+ \left(\frac{\partial J^*(\mathbf{x}(t), t)}{\partial \mathbf{x}}\right)^T \mathbf{A}\mathbf{x} - \left(\frac{\partial J^*(\mathbf{x}(t), t)}{\partial \mathbf{x}}\right)^T \mathbf{B}\mathbf{R}^{-1}\mathbf{B}^T \frac{\partial J^*(\mathbf{x}(t), t)}{\partial \mathbf{x}}$$

$$= \frac{1}{2}\mathbf{x}^T(t)\mathbf{Q}\mathbf{x}(t) - \frac{1}{2}\left(\frac{\partial J^*(\mathbf{x}(t), t)}{\partial \mathbf{x}}\right)^T \mathbf{B}\mathbf{R}^{-1}\mathbf{B}^T \frac{\partial J^*(\mathbf{x}(t), t)}{\partial \mathbf{x}}$$

$$+ \left(\frac{\partial J^*(\mathbf{x}(t), t)}{\partial \mathbf{x}}\right)^T \mathbf{A}\mathbf{x} \tag{13.75}$$

From eqn. (13.72), we get the boundary condition:

$$J^*(\mathbf{x}(t_f), t_f) = \frac{1}{2}\mathbf{x}^T(t_f)\mathbf{H}\mathbf{x}(t_f) \tag{13.76}$$

Earlier in this section we found that in the linear regulator problem for discrete-time systems, the minimum value of the performance index is a time-varying quadratic function of state (eqn. (13.66f)). Therefore, it is reasonable to assume

$$J^*(\mathbf{x}(t), t) = \frac{1}{2} \mathbf{x}^T(t)\mathbf{P}(t)\mathbf{x}(t) \tag{13.77}$$

where $\mathbf{P}(t)$ is a real symmetric positive definite matrix for $t < t_f$ and

$$\mathbf{P}(t_f) = \mathbf{H} \tag{13.78}$$

Substituting the assumed solution in eqn. (13.75), we get

$$-\frac{1}{2}\mathbf{x}^T(t)\frac{d\mathbf{P}(t)}{dt}\mathbf{x}(t) = \frac{1}{2}\mathbf{x}^T(t)\mathbf{Q}\mathbf{x}(t) - \frac{1}{2}\mathbf{x}^T(t)\mathbf{P}(t)\mathbf{B}\mathbf{R}^{-1}\mathbf{B}^T\mathbf{P}(t)\mathbf{x}(t)$$
$$+ \mathbf{x}^T(t)\mathbf{P}(t)\mathbf{A}\mathbf{x}(t)$$

or

$$2\mathbf{x}^T(t)\mathbf{P}(t)\mathbf{A}\mathbf{x}(t) - \mathbf{x}^T(t)\mathbf{P}(t)\mathbf{B}\mathbf{R}^{-1}\mathbf{B}^T\mathbf{P}(t)\mathbf{x}(t) + \mathbf{x}^T(t)\mathbf{Q}\mathbf{x}(t)$$
$$+ \mathbf{x}^T(t)\dot{\mathbf{P}}(t)\mathbf{x}(t) = 0$$

or

$$\mathbf{x}^T(t)[2\mathbf{P}(t)\mathbf{A} - \mathbf{P}(t)\mathbf{B}\mathbf{R}^{-1}\mathbf{B}^T\mathbf{P}(t) + \mathbf{Q} + \dot{\mathbf{P}}(t)]\mathbf{x}(t) = 0 \tag{13.79}$$

The only way this equation can be satisfied for an arbitrary $\mathbf{x}(t)$ is that the quantity inside the brackets be equal to zero. However, we know (Appendix II) that in the scalar function $\mathbf{z}^T\mathbf{W}\mathbf{z}$, only the symmetric part of the matrix $\mathbf{W}$ (eqn (II.5) in Appendix II),

$$\mathbf{W}_S = \frac{\mathbf{W} + \mathbf{W}^T}{2}$$

is of importance.

Examination of eqn. (13.79) reveals that all the terms within brackets are already symmetric except the first term.

$$\text{Symmetric part of } [2\mathbf{P}(t)\mathbf{A}] = 2\frac{\mathbf{P}(t)\mathbf{A} + \mathbf{A}^T\mathbf{P}(t)}{2}$$
$$= \mathbf{P}(t)\mathbf{A} + \mathbf{A}^T\mathbf{P}(t)$$

Therefore in order for eqn. (13.79) to be satisfied, it is necessary that the differential equation

$$\dot{\mathbf{P}}(t) + \mathbf{Q} - \mathbf{P}(t)\mathbf{B}\mathbf{R}^{-1}\mathbf{B}^T\mathbf{P}(t) + \mathbf{P}(t)\mathbf{A} + \mathbf{A}^T\mathbf{P}(t) = 0 \tag{13.80}$$

is satisfied subject to the boundary condition (13.78).

Equation (13.80) is a matrix differential equation of the Riccati type and is thus referred to as *the Riccati equation*. The Riccati equation is nonlinear and for this reason, we usually cannot obtain closed-form solutions; therefore we must compute $\mathbf{P}(t)$ using a digital computer.

Equation (13.80) is a set of n^2 nonlinear equations. Numerical integration is carried out backwards; from $t = t_f$ to $t = t_0$ with the boundary condition $\mathbf{P}(t_f) = \mathbf{H}$. Actually since the $n \times n$ matrix $\mathbf{P}(t)$ is symmetric, we need to integrate only $\dfrac{n(n+1)}{2}$ differential equations.

Once $\mathbf{P}(t)$ is known for $t_0 \leqslant t \leqslant t_f$, the *optimal control law* is obtained from eqns. (13.74) and (13.77):

$$\mathbf{u}^*(t) = -\mathbf{R}^{-1}\mathbf{B}^T\mathbf{P}(t)\mathbf{x}(t)$$
$$= \mathbf{K}(t)\mathbf{x}(t) \qquad (13.81)$$

where

$$\mathbf{K} = -\mathbf{R}^{-1}\mathbf{B}^T\mathbf{P}(t)$$

Example 13.7: A first-order system is described by the differential equation

$$\dot{x}(t) = 2x(t) + u(t)$$

It is desired to find the control law that minimizes the performance index

$$J = \frac{1}{2}\int_0^{t_f}\left(3x^2 + \frac{1}{4}u^2\right)dt$$

$$t_f = 1 \text{ sec}$$

For this problem, the matrices $\mathbf{A}, \mathbf{B}, \mathbf{H}, \mathbf{Q}$ and $\mathbf{R}$ reduce to scalars and are given by

$$\mathbf{A} = 2, \mathbf{B} = 1, \mathbf{H} = 0, \mathbf{Q} = 3, \mathbf{R} = \tfrac{1}{4}$$

In addition, the matrix $\mathbf{P}(t)$ also reduces to a scalar function of time $p(t)$. The matrix Riccati equation then becomes the scalar differential equation

$$\dot{p}(t) + 3 - 4p^2(t) + 4p(t) = 0$$

with the boundary condition $p(t_f) = 0$.

The solution $p(t)$ can be obtained by separation of variables.

$$\int_{t_f}^{t}\frac{dp}{4\left(p - \frac{3}{2}\right)\left(p + \frac{1}{2}\right)} = \int_{t_f}^{t}dt$$

or

$$\frac{1}{8}\left\{\ln\left[\frac{p(t) - \frac{3}{2}}{p(t_f) - \frac{3}{2}}\right] - \ln\left[\frac{p(t) + \frac{1}{2}}{p(t_f) + \frac{1}{2}}\right]\right\} = t - t_f$$

This gives

$$p(t) = \frac{-\dfrac{3}{2}(1 - e^{8(t - t_f)})}{1 + 3e^{8(t - t_f)}} \qquad (13.82)$$

The optimal control law is (eqn. (13.81)),

$$u^*(t) = -4p(t)x(t) \tag{13.83}$$

The block diagram of the optimal system is shown in Fig. 13.12.

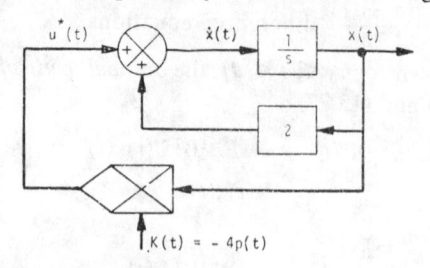

Figure 13.12 Block diagram of optimal control system for Example 13.7.

13.6 The Infinite-time Regulator Problem

CONTINUOUS-TIME SYSTEMS

The infinite-time state regulator problem may be stated as follows: Consider the controlled process (eqn. (13.54a))

$$\dot{x}(t) = Ax(t) + Bu(t) \tag{13.84}$$

Find the control law $u^*(t)$, so that quadratic performance index (eqn. (13.56))

$$J = \frac{1}{2} \int_0^\infty (x^T(t)Qx(t) + u^T(t)Ru(t)) \, dt; \; Q \text{ and } R \text{ are positive definite matrices} \tag{13.85}$$

is minimized, subject to the initial condition

$$x(0) = x_0$$

Let us discuss the salient points of the infinite-time state regulator problem.

(i) For many applications, the final time t_f has no special significance and there exist no obvious values of t_f which should be specified in the performance index. In such cases, we are satisfied to let $t_f \to \infty$, as is done in eqn. (13.85).

(ii) When $t_f \to \infty$, $x(\infty) \to 0$ for the optimal system to be stable. Therefore the terminal penalty term (in eqn. (13.55)) has no significance; consequently it does not appear in eqn. (13.85) i.e., we have set $H = 0$ in the general quadratic performance index (13.55).

(iii) In the finite-time regulator problem, there is no restriction on the controllability of the plant. This is because J is always finite and instability does not impose any problems in finite-interval control. This may not be so in the infinite-interval case. J can become infinite if

(a) one or more states of the plant are uncontrollable,
(b) the uncontrollable states are unstable, and
(c) the unstable states are reflected in system performance index.

Optimal Control Systems

The performance index J would be infinite for all controls; we can not, therefore, distinguish the optimal control from other controls. J will be finite (i.e., the solution to the infinite-time regulator problem will exist) if the states that are not asymptotically stable, are controllable (note that controllable unstable states of the plant can be stabilized by the feedback controller) or broadly, we may say that the controlled process given by eqn. (13.84) should satisfy the conditions of controllability.

(*iv*) Consider the matrix Riccati equation (eqn. (13.80)),

$$\dot{\mathbf{P}}(t) + \mathbf{Q} - \mathbf{P}(t)\,\mathbf{B}\mathbf{R}^{-1}\mathbf{B}^T\mathbf{P}(t) + \mathbf{P}(t)\mathbf{A} + \mathbf{A}^T\mathbf{P}(t) = 0$$

with the boundary condition $\mathbf{P}(t_f) = \mathbf{0}$.

We solve the Riccati equation backward in time with t_f as the starting time and $\mathbf{P}(t_f)$ as the initial condition. At time $t = t_f - \epsilon$, where ϵ is a small positive number, the transient due to initial condition will dominate the solution $\mathbf{P}(t)$; the transient will die out for large values of ϵ resulting in steady-state solution of Riccati equation for some interval of time. This time interval increases as t_f becomes large. As $t_f \to \infty$, the time interval of steady-state solution of Riccati equation grows without bound, i.e., $\mathbf{P}(t) \to$ a constant $\mathbf{P}^0$ for all finite time t.

Consider the system of Example 13.7. For terminal time t_f; the solution of the Riccati equation is (eqn (13.82))

$$p(t) = \frac{\dfrac{3}{2}(1 - e^{8(t-t_f)})}{1 + 3\,e^{8(t-t_f)}}$$

$p(t)$ is plotted in Fig. 13.13 for $t_f = 1$ sec and 10 sec.

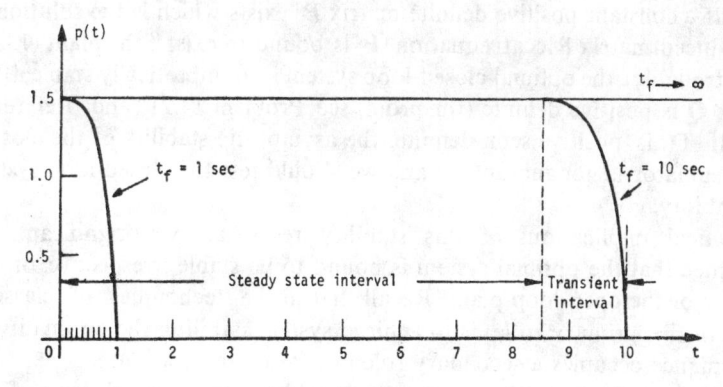

Figure 13.13.

The solution of Riccati equation with infinite terminal time is

$$p^0 = \lim_{t_f \to \infty} p(t) = 1.5$$

The optimal control law is then (eqn. (13.83))

$$u^*(t) = -6x(t)$$

The practical implication of this result is obvious.

The optimal law is implemented using time-invariant gain elements in contrast to the finite-time case wherein gain elements must be time-varying to implement the optimal control given by eqn. (13.83).

In general, if the states of the system (13.84) that are not asymptotically stable are controllable, then $\lim_{t_f \to \infty} \mathbf{P}(t)$ tends to a unique positive definite constant matrix $\mathbf{P}^0$ which is the solution of the equation (eqn. (13.80) with $\dot{\mathbf{P}}(t) = 0$)

$$\mathbf{A}^T\mathbf{P}^0 + \mathbf{P}^0\mathbf{A} - \mathbf{P}^0\mathbf{B}\mathbf{R}^{-1}\mathbf{B}^T\mathbf{P}^0 + \mathbf{Q} = 0$$

called the *reduced matrix Riccati equation* (for proof of this result refer [7]).

The solution of the reduced matrix Riccati equation may not be unique. The desired unique answer is obtained by enforcing the requirement that $\mathbf{P}^0$ be positive definite.

Another point to be noted is that computation of $\mathbf{P}^0$ is often convenient by numerical integration of Riccati equation (13.80) for arbitrary large values of t_f till $\mathbf{P}(t)$ converges to a constant matrix $\mathbf{P}^0$.

The optimal control law in the infinite-time regulator problem is (eqn. (13.81))

$$\mathbf{u}^*(t) = -\mathbf{R}^{-1}\mathbf{B}^T\mathbf{P}^0\mathbf{x}(t)$$
$$= \mathbf{K}\mathbf{x}(t)$$

where
$$\mathbf{K} = -\mathbf{R}^{-1}\mathbf{B}^T\mathbf{P}^0$$

The resulting optimal system is thus a linear time-invariant system.

(v) If a constant positive definite matrix $\mathbf{P}^0$ exists which is the solution of the reduced matrix Riccati equation ($\mathbf{P}^0$ is bound to exist if the plant (13.84) is controllable), the optimal closed-loop system is asymptotically stable if the matrix $\mathbf{Q}$ is positive definite (for proof, see Problem 14.21 and Reference [9]). If $\mathbf{Q}$ is positive semi-definite, the asymptotic stability of the closed-loop regulator is not guaranteed and we should test the closed-loop system for stability.

Practical implications of this stability result are very significant. We guarantee that the optimal system is bound to be stable irrespective of the stability of the open-loop plant. Recall that in the techniques of classical control, the primary role is to achieve system stability; the optimality of performance occupies a secondary role in the design procedure.

(vi) For the infinite-time regulator problem, the optimal value of the performance index is obtained as follows:

From eqn. (13.77), we have

$$J^*(\mathbf{x}(t), t) = \frac{1}{2}\mathbf{x}^T(t)\mathbf{P}(t)\mathbf{x}(t)$$

For the infinite-time case, we may write

$$J^* = \int_0^\infty -\frac{1}{2}\frac{d}{dt}(\mathbf{x}^T(t)\mathbf{P}^0\mathbf{x}(t))\,dt$$

$$= -\frac{1}{2} \mathbf{x}^T(t)\mathbf{P}^0\mathbf{x}(t) \Big|_0^\infty$$

$$= \frac{1}{2} \mathbf{x}^T(0)\mathbf{P}^0\mathbf{x}(0)$$

(Note that $\mathbf{x}(t) \underset{t \to \infty}{=} \mathbf{0}$ as per stability result obtained earlier.)

Results of this section are summarised below:

Given the controllable linear time-invariant plant (13.84) and the performance index (13.85), then a unique optimal control exists and is given by

$$\mathbf{u}^*(t) = \mathbf{Kx}; \quad \mathbf{K} = -\mathbf{R}^{-1}\mathbf{B}^T\mathbf{P}^0 \qquad (13.86a)$$

where $\mathbf{P}^0$ is a constant positive definite matrix which is the solution of the reduced matrix Riccati equation

$$\mathbf{A}^T\mathbf{P}^0 + \mathbf{P}^0\mathbf{A} - \mathbf{P}^0\mathbf{B}\mathbf{R}^{-1}\mathbf{B}^T\mathbf{P}^0 + \mathbf{Q} = \mathbf{0} \qquad (13.86b)$$

The minimum value of performance index is

$$J^* = \frac{1}{2} \mathbf{x}_0^T \mathbf{P}^0 \mathbf{x}_0 \qquad (13.87)$$

Example 13.8: Let us obtain the control law which minimizes the performance index

$$J = \int_0^\infty (x_1^2 + u^2)\, dt$$

for the system

$$\begin{bmatrix} \dot{x}_1 \\ \dot{x}_2 \end{bmatrix} = \begin{bmatrix} 0 & 1 \\ 0 & 0 \end{bmatrix} \begin{bmatrix} x_1 \\ x_2 \end{bmatrix} + \begin{bmatrix} 0 \\ 1 \end{bmatrix} u$$

We have for this problem

$$\mathbf{A} = \begin{bmatrix} 0 & 1 \\ 0 & 0 \end{bmatrix},\ \mathbf{B} = \begin{bmatrix} 0 \\ 1 \end{bmatrix},\ \mathbf{Q} = \begin{bmatrix} 2 & 0 \\ 0 & 0 \end{bmatrix},\ \mathbf{R} = [2]$$

The reduced matrix Riccati equation is given as

$$\begin{bmatrix} 0 & 0 \\ 1 & 0 \end{bmatrix}\begin{bmatrix} p_{11} & p_{12} \\ p_{12} & p_{22} \end{bmatrix} + \begin{bmatrix} p_{11} & p_{12} \\ p_{12} & p_{22} \end{bmatrix}\begin{bmatrix} 0 & 1 \\ 0 & 0 \end{bmatrix}$$

$$- \begin{bmatrix} p_{11} & p_{12} \\ p_{12} & p_{22} \end{bmatrix}\begin{bmatrix} 0 \\ 1 \end{bmatrix}\left[\frac{1}{2}\right]\begin{bmatrix} 0 & 1 \end{bmatrix}\begin{bmatrix} p_{11} & p_{12} \\ p_{12} & p_{22} \end{bmatrix} + \begin{bmatrix} 2 & 0 \\ 0 & 0 \end{bmatrix}$$

$$= \begin{bmatrix} 0 & 0 \\ 0 & 0 \end{bmatrix}$$

(Note that we have utilized the fact that the $\mathbf{P}^0$ is symmetric.)

Upon simplification we get,

$$\frac{-p_{12}^2}{2} + 2 = 0$$

$$p_{11} - \frac{p_{12}p_{22}}{2} = 0$$

$$\frac{-p_{22}^2}{2} + 2p_{12} = 0$$

The solution of these equations yields the positive definite matrix

$$\mathbf{P}^0 = \begin{bmatrix} 2\sqrt{2} & 2 \\ 2 & 2\sqrt{2} \end{bmatrix}$$

From eqn. (13.86a), the optimal control law is given by

$$u^*(t) = -\mathbf{R}^{-1}\mathbf{B}^T\mathbf{P}^0\mathbf{x}(t)$$

$$= -\left[\frac{1}{2}\right][0 \quad 1]\begin{bmatrix} 2\sqrt{2} & 2 \\ 2 & 2\sqrt{2} \end{bmatrix}\begin{bmatrix} x_1(t) \\ x_2(t) \end{bmatrix}$$

$$= -x_1(t) - \sqrt{2}\, x_2(t)$$

It can easily be verified that the closed-loop system is asymptotically stable.

Discrete-time Systems

Given the controllable sampled-data system

$$\mathbf{x}(k+1) = \mathbf{A}\mathbf{x}(k) + \mathbf{B}\mathbf{u}(k)$$

Find the control sequence $\mathbf{u}^*(k)$ so that quadratic performance index

$$J = \frac{1}{2}\sum_{k=0}^{\infty}[\mathbf{x}^T(k)\mathbf{Q}\mathbf{x}(k) + \mathbf{u}^T(k)\mathbf{R}\mathbf{u}(k)]$$

is minimized, subject to the initial condition

$$\mathbf{x}(0) = \mathbf{x}_0$$

Results for this infinite-stage discrete-time regulator problem directly follow those of the continuous-time problem. The matrices **P** and **K** in the recursive relations (13.66) are time-invariant when $N \to \infty$. Therefore

$$\mathbf{u}^*(k) = \mathbf{K}\mathbf{x}(k) \tag{13.88a}$$

$$\mathbf{K} = -[\mathbf{R} + \mathbf{B}^T\mathbf{P}\mathbf{B}]^{-1}\mathbf{B}^T\mathbf{P}\mathbf{A} \tag{13.88b}$$

$$\mathbf{P} = [\mathbf{A} + \mathbf{B}\mathbf{K}]^T\mathbf{P}[\mathbf{A} + \mathbf{B}\mathbf{K}] + \mathbf{K}^T\mathbf{R}\mathbf{K} + \mathbf{Q} \tag{13.88c}$$

The minimum value of the performance index is

$$J^* = \frac{1}{2}\mathbf{x}_0^T\mathbf{P}\mathbf{x}_0 \tag{13.88d}$$

The optimal control sequence for the infinite-stage process may be obtained from the algebraic equations given above. An alternative way is to solve the recursive relations (13.66) using digital computer for as many stages as required for $P(N - k)$ to coverage to a constant matrix.

13.7 The Output Regulator and the Tracking Problems

In this section we shall discuss briefly the output regulator problem and the tracking problem (tracking problem has been discussed in Section 13.3 using the transfer function approach). Under certain restrictions, both these problems are reducible to the form of the state regulator problem.

THE OUTPUT REGULATOR PROBLEM

The output regulator problem may be stated as follows:

Consider the controlled process

$$\dot{x}(t) = Ax(t) + Bu(t); \quad x(0) = x_0 \qquad (13.89)$$
$$y(t) = Cx(t)$$

where $x(t) = n \times 1$ state vector; $u(t) = m \times 1$ control vector; $y(t) = p \times 1$ output vector; $A = n \times n$ constant matrix; $B = n \times m$ constant matrix; and $C = p \times n$ constant matrix.

Find the control law $u^*(t)$ so that the following quadratic performance index is minimized,

$$J = \frac{1}{2} y^T(t_f) H y(t_f) + \frac{1}{2} \int_0^{t_f} (y^T(t) Q y(t) + u^T(t) R u(t)) \, dt \qquad (13.90)$$

where R is a positive definite constant symmetric matrix, H and Q are positive definite (or semidefinite) constant, symmetric matrices with the restriction that both of them are not zero matrices at a time.

From the performance index given by (13.90) it is obvious that in the output regulator problem we desire to bring and keep output $y(t)$ near zero without using an excessive amount of control energy.

Substituting $y(t) = Cx(t)$ in eqn. (13.90), we get

$$J = \frac{1}{2} x^T(t_f) C^T H C x(t_f) + \frac{1}{2} \int_0^{t_f} (x^T(t) C^T Q C x(t) + u^T(t) R u(t)) \, dt \qquad (13.91)$$

Comparing eqn. (13.91) with eqn. (13.55), we observe that the two indices are identical in form; H and Q in eqn. (13.55) are replaced by $C^T H C$ and $C^T Q C$ respectively to get eqn. (13.91). If we assume that the system (13.89) is observable, then C can not be zero. Therefore $C^T H C$ and $C^T Q C$ will be positive definite (or semidefinite) matrices whenever H and Q are positive definite (or semidefinite).

We have the following solution for the output regulator problem which directly follows from the results of Section 13.5.

For the observable system (13.89) with the performance index (13.90), a unique optimal control exists and is given by

$$u^*(t) = - R^{-1} B^T P(t) x(t)$$
$$= K(t) x(t) \qquad (13.92)$$

where the symmetric positive definite matrix $P(t)$ is the solution of the matrix Riccati equation

$$\dot{P}(t) + C^TQC - P(t)BR^{-1}B^TP(t) + P(t)A + A^TP(t) = 0 \qquad (13.93)$$

with the boundary condition

$$P(t_f) = C^THC$$

THE TRACKING PROBLEM

Here we shall study a class of tracking problems which are reducible to the form of the output regulator problem.

Consider an observable process described by eqns. (13.89). It is desired to bring and keep output $y(t)$ close to the desired output $r(t)$. We define an error vector

$$e(t) = y(t) - r(t)$$

The design objective is to find the control law $u^*(t)$ so that the performance index

$$J = \frac{1}{2} e^T(t_f)He(t_f) + \frac{1}{2}\int_0^{t_f} (e^T(t)Qe(t) + u^T(t)Ru(t))\, dt \qquad (13.94)$$

is minimized.

To reduce this problem to the form of the output regulator problem, we consider only those $r(t)$ that can be generated by arbitrary initial conditions $z(0)$ in the system

$$\begin{aligned}\dot{z}(t) &= Az(t) \\ r(t) &= Cz(t)\end{aligned} \qquad (13.95)$$

The matrices A and C are same as those of the plant (13.89).

We now define a new variable

$$w = x - z$$

Then

$$\begin{aligned}\dot{w} &= Aw + Bu \\ e &= Cw\end{aligned} \qquad (13.96)$$

Applying results of the output regulator problem gives immediately that the optimal control for the tracking problem under consideration is

$$\begin{aligned}u^*(t) &= -R^{-1}B^TP(t)w \\ &= K(t)[x - z]\end{aligned} \qquad (13.97)$$

where $P(t)$ is the solution of the Riccati equation (eqn. (13.93))

$$\dot{P}(t) + C^TQC - P(t)BR^{-1}B^TP(t) + P(t)A + A^TP(t) = 0$$

with the boundary condition

$$P(t_f) = C^THC$$

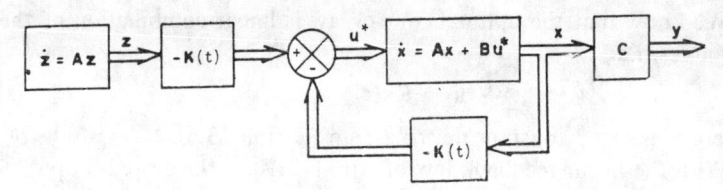

Figure 13.14 The augmented system for the tracking problem.

Figure 13.14 shows the augmented system (13.96) separated into its component systems (13.89) and (13.95) and controlled by the law† (13.97).

13.8 Parameter Optimization: Regulators

We have studied so far the optimal regulator problem wherein (i) for a given plant, we find a control function u* which is optimal with respect to given performance criterion, and then (ii) realize the control function $u^*(t) = Kx(t)$.

An optimal solution obtained through steps (i) and (ii) may not be the best solution in all circumstances. For example, all the elements of matrix K may not be free; some gains are fixed by the physical constraints of the system and are therefore relatively inflexible. Similarly if all the states $x(t)$ are not accessible for feedback, one has to go for a state observer whose complexity is comparable to that of the system itself. It is natural to seek a procedure that relies on the use of feedback from only the accessible state variables, constraining the gain elements of matrix K corresponding to the inaccessible state variables to have zero value [10]. Thus, whether one chooses an optimal or suboptimal solution depends on many factors such as cost of equipment, availability of equipment, total space requirement etc., in addition to the performance required out of the system.

In this section, we present a simple method of obtaining the solution of a control problem when some elements of feedback matrix K are constrained.

The process considered here may be represented by

$$\dot{x} = Ax + Bu \qquad (13.98)$$

where $x = n \times 1$ state vector; $u = m \times 1$ control vector; $A = n \times n$ constant matrix; and $B = n \times m$ constant matrix.

The process is assumed to be completely controllable. The performance index is

$$J = \frac{1}{2} \int_0^\infty (x^T Q x + u^T R u) \, dt \qquad (13.99)$$

where Q and R are, respectively, positive semidefinite and positive definite, real, symmetric, constant matrices.

---

†Results for the output regulator problem and the tracking problem for the case when final time is not constrained, can easily be obtained from the results of Section 13.6. Identical results immediately follow for the discrete-time systems also.

We know that the optimal control law is linear combination of the state variables, i.e.,

$$u = Kx(t) \tag{13.100}$$

where K is $m \times n$ constant matrix (from Section 13.6).

With the linear feedback law of eqn. (13.100), the closed-loop system is described by

$$\dot{x} = Ax + BKx$$
$$= (A + BK)x \tag{13.101}$$

Substituting for the control vector u from eqn. (13.100) in the performance index J of eqn. (13.99), we have

$$J = \frac{1}{2}\int_0^\infty (x^TQx + x^TK^TRKx)\,dt$$

$$= \frac{1}{2}\int_0^\infty x^T(Q + K^TRK)x\,dt \tag{13.102}$$

This integral is guaranteed to converge because of the fact that a controllable system is stabilizable; the eigenvalues of the matrix $A + BK$ (eqn. (13.101)) may be assumed to have negative real parts and $\underset{t\to\infty}{x(t)} = 0$.

Let us postulate the existence of a real symmetric positive definite matrix P such that

$$x^T(Q + K^TRK)x = -\frac{d}{dt}(x^TPx) \tag{13.103}$$

or

$$x^T(Q + K^TRK)x = -\dot{x}^TPx - x^TP\dot{x}$$

Substituting for $\dot{x}$ and $\dot{x}^T$ from eqn. (13.101), we have

$$x^T(Q + K^TRK)x = -x^T[(A + BK)^TP + P(A + BK)]x$$

Since the above equality holds for arbitrary $x(t)$, we have

$$(A + BK)^TP + P(A + BK) + K^TRK + Q = 0 \tag{13.104}$$

From this equation we can determine elements of P as functions of the elements of the feedback matrix K.

By virtue of eqn. (13.103), the performance index is given by

$$J = -\frac{1}{2}\int_0^\infty \frac{d}{dt}(x^TPx)\,dt = -\frac{1}{2}x^TPx\Big|_0^\infty$$

$$= \frac{1}{2}x^T(0)\,Px(0) \tag{13.105}$$

If $K_1, K_2, ..., K_n$ are the free elements of matrix P, we have

$$J = J(K_1, K_2, ..., K_n) \tag{13.106}$$

The necessary and sufficient conditions for J to be minimum are given by eqns. (13.2) and (13.3) respectively. Solution set K_i of eqn. (13.106) satisfying

Optimal Control Systems

these conditions is obtained which gives the suboptimal solution to the control problem. Of course, K_l must satisfy the further constraint that the closed-loop system be asymptotically stable.

If all the parameters of **P** are free, the procedure given above will yield an optimal solution. Procedure of Section 13.6 will also yield the same solution (optimal solution to the regulator problem under consideration is unique). It is more convenient to find **P** from the matrix Riccati equation for the optimal solution.

In the special case where the performance index is independent of the control **u**, we have

$$J = \frac{1}{2} \int_0^\infty \mathbf{x}^T \mathbf{Q} \mathbf{x} \, dt$$

In this case the matrix **P** is obtained from eqn. (13.104) by putting $\mathbf{R} = 0$ resulting in the modified matrix equation

$$(\mathbf{A} + \mathbf{BK})^T \mathbf{P} + \mathbf{P}(\mathbf{A} + \mathbf{BK}) + \mathbf{Q} = 0 \qquad (13.107)$$

Even though **R** is originally assumed to be positive definite, substituting $\mathbf{R} = 0$ is a valid operation here as the positive definiteness of **R** has not been used in this derivation.

Example 13.9: Consider the second-order system of Fig. 13.15a wherein it is desired to find optimum ζ which minimizes the integral square error, i.e.,

$$J = \int_0^\infty e^2(t) \, dt$$

for the initial conditions $c(0) = 1$, $\dot{c}(0) = 0$.

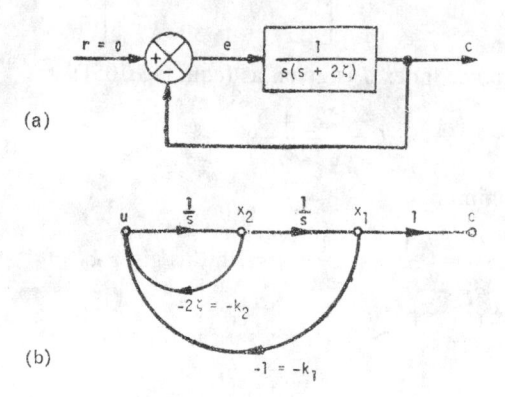

Figure 13.15 (a) A second-order control problem;
(b) State variable formulation

The problem is reframed in the state form of Fig. 13.15b as one of obtaining feedback control law with the constraint $k_1 = 1$. For this form we have

$$\begin{bmatrix} \dot{x}_1 \\ \dot{x}_2 \end{bmatrix} = \begin{bmatrix} 0 & 1 \\ 0 & 0 \end{bmatrix} \begin{bmatrix} x_1 \\ x_2 \end{bmatrix} + \begin{bmatrix} 0 \\ 1 \end{bmatrix} u; \; x_1(0) = 1, \; x_2(0) = 0$$

$$u = -\begin{bmatrix} k_1 & k_2 \end{bmatrix} \begin{bmatrix} x_1 \\ x_2 \end{bmatrix}$$

Now

$$J = \int_0^\infty e^2(t) \, dt = \int_0^\infty x_1^2 \, dt \quad \text{since } e = -c = -x_1$$

Therefore

$$\mathbf{Q} = \begin{bmatrix} 2 & 0 \\ 0 & 0 \end{bmatrix}$$

Substituting various values in eqn. (13.107) we have

$$\begin{bmatrix} 0 & -1 \\ 1 & -k_2 \end{bmatrix} \begin{bmatrix} p_{11} & p_{12} \\ p_{12} & p_{22} \end{bmatrix} + \begin{bmatrix} p_{11} & p_{12} \\ p_{12} & p_{22} \end{bmatrix} \begin{bmatrix} 0 & 1 \\ -1 & -k_2 \end{bmatrix} + \begin{bmatrix} 2 & 0 \\ 0 & 0 \end{bmatrix}$$
$$= \begin{bmatrix} 0 & 0 \\ 0 & 0 \end{bmatrix}$$

Solving we get

$$\mathbf{P} = \begin{bmatrix} \dfrac{1 + k_2^2}{k_2} & 1 \\ 1 & \dfrac{1}{k_2} \end{bmatrix}$$

The performance index J is given as (eqn. (13.105))

$$J = \frac{k_2^2 + 1}{2k_2}$$

For J to be minimum,

$$\frac{\partial J}{\partial k_2} = \left(\frac{1}{2} - \frac{1}{2k_2^2} \right) = 0; \text{ this gives } k_2 = 1$$

$$\frac{\partial^2 J}{\partial k_2^2} = \frac{1}{k_2^3} > 0; \text{ this is satisfied for } k_2 = 1$$

Therefore, optimal value of parameter $k_2 = 1$. Since $k_2 = 2\zeta$, we find $\zeta = 0.5$ minimizes integral square error for the given initial conditions.

It can easily be verified that the suboptimal control derived above results in a closed-loop system which is asymptotically stable.

The control obtained by minimization of J given by (13.105) will vary from one $\mathbf{x}(0)$ to another. However, for practical reasons, it is desirable to have just one control irrespective of what $\mathbf{x}(0)$ is. One way to solve this problem is to assume that $x(0)$ is a random variable, uniformly distributed on the surface of the n-dimensional unit sphere [13]:

$$E[x(0)x^T(0)] = \frac{1}{n} I \qquad (13.108)$$

where E denotes expected value.

We now define a new performance index

$$\begin{aligned}\bar{J} &= E[J] \\ &= E\left[\frac{1}{2} x^T(0)Px(0)\right] \\ &= \frac{1}{2n} \text{tr } [P] \qquad (13.109)\end{aligned}$$

where $\text{tr}[P]$ = trace of P = sum of all diagonal elements of matrix P.

For the system of Example 13.9, the parameter k_2 that optimizes $\bar{J}$ of (13.109) is $\sqrt{2}$. Therefore

$$u = -\begin{bmatrix} 1 & \sqrt{2} \end{bmatrix} \begin{bmatrix} x_1 \\ x_2 \end{bmatrix}$$

is the suboptimal control law that is independent of initial conditions.

Problems

13.1 Consider the feedback system shown in Fig. P-13.1. The output is required to track unit step. Find the value of α that minimizes integral square error.

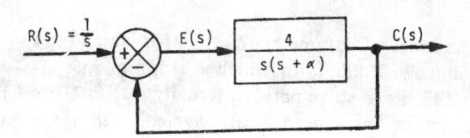

Figure P-13.1.

13.2 Referring to the block diagram of Fig. P-13.2, consider that $G(s) = 100/s^2$ and $R(s) = 1/s$. Determine the optimal values of parameters K_1 and K_2 such that

$$J = \int_0^\infty (e^2(t) + 0.25 \, u^2(t)) \, dt$$

is minimum.

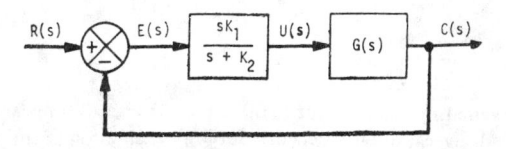

Figure P-13.2.

578   Control Systems Engineering

13.3 Figure P-13.3 illustrates a typical sampled-data system. Find optimal value of K so that

$$J = \sum_{k=0}^{\infty} e^2(kT)$$

is minimized.

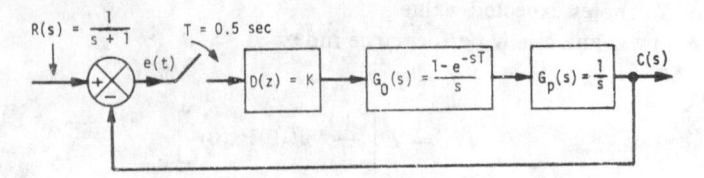

[Block diagram: $R(s) = \frac{1}{s+1}$, $T = 0.5$ sec, $e(t)$, $D(z) = K$, $G_0(s) = \frac{1-e^{-sT}}{s}$, $G_p(s) = \frac{1}{s}$, $C(s)$]

Figure P-13.3.

13.4 Given the system shown in Fig. P-13.4.
(i) Find the sum square error

$$J = \sum_{k=0}^{\infty} e^2(kT)$$

as a function of the free parameters K and a.

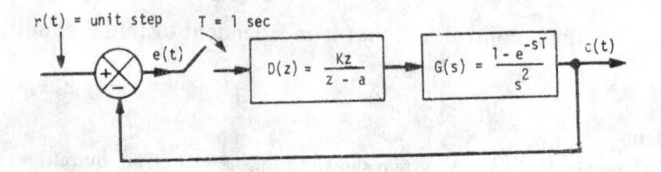

[Block diagram: $r(t)$ = unit step, $T = 1$ sec, $e(t)$, $D(z) = \frac{Kz}{z-a}$, $G(s) = \frac{1-e^{-sT}}{s^2}$, $c(t)$]

Figure P-13.4.

(ii) Determine the range of parameters K and a for which the system is stable.
(iii) Determine values of K and a for which J is minimum.

[*Hint*: It may be difficult to solve part (iii) analytically. Using digital computer, we can search for a minimum J as K and a are varied within the range determined in part (ii).]

13.5 Consider the linear plant of a system characterized by the transfer function

$$G(s) = \frac{10}{s^2}$$

The objective of the system is to make the output $c(t)$ follow a step input $r(t)$ of amplitude 0.5 minimizing

$$J_e = \int_0^{\infty} (r(t) - c(t))^2 \, dt$$

with

$$J_u = \int_0^{\infty} u^2(t) \, dt = 2.5$$

where $u(t)$ is the actuating signal of the plant. Find the overall transfer function of the optimal system and suggest suitable schemes for implementation of this transfer function.

13.6 For the discrete-data system shown in Fig. P-13.6, given

$$G_p(s) = \frac{1}{s+1}, \quad G_0(s) = \frac{1-e^{-sT}}{s}$$

Find $D(z)$ so that
$$J = \sum_{k=0}^{\infty} [e^2(kT) + u^2(kT)]$$
is minimized.

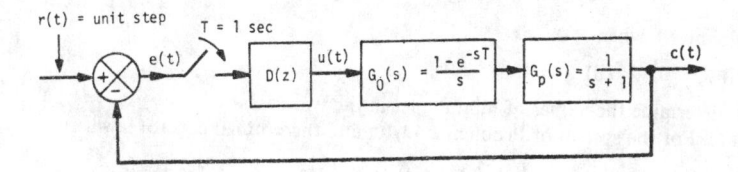

Figure P-13.6.

13.7 Determine the optimal control law for the system described by
$$\dot{x} = \begin{bmatrix} 0 & 1 \\ -2 & -3 \end{bmatrix} x + \begin{bmatrix} 0 \\ 1 \end{bmatrix} u$$

such that the following performance index is minimized
$$J = \int_0^{\infty} (x^T x + u^2) \, dt$$

13.8 Consider a system described by the equations
$$\begin{bmatrix} \dot{x}_1 \\ \dot{x}_2 \end{bmatrix} = \begin{bmatrix} 0 & 1 \\ 0 & 0 \end{bmatrix} \begin{bmatrix} x_1 \\ x_2 \end{bmatrix} + \begin{bmatrix} 0 \\ 1 \end{bmatrix} u; \quad x_1(0) = x_2(0) = 1$$

Choose the feedback law
$$u = -x_1 - Kx_2$$

(i) Find the value of K so that
$$J = \frac{1}{2} \int_0^{\infty} (x_1^2 + x_2^2) \, dt$$

is minimized.
(ii) Find minimum value of J.
(iii) Find sensitivity of J with respect to K.

13.9 Consider the plant
$$\begin{bmatrix} \dot{x}_1 \\ \dot{x}_2 \end{bmatrix} = \begin{bmatrix} 1 & 0 \\ -1 & 2 \end{bmatrix} \begin{bmatrix} x_1 \\ x_2 \end{bmatrix} + \begin{bmatrix} 1 \\ 0 \end{bmatrix} u$$

(i) Prove that the system is unstable.
(ii) Prove that the system is controllable.
(iii) Select any values for matrices Q and R with the constraint that they are positive definite and design a controller for the plant so as to minimize
$$J = \frac{1}{2} \int_0^{\infty} (x^T Q x + u^T R u) \, dt$$

Check that the resulting overall system is stable.

13.10 A plant is described by the equations
$$\begin{bmatrix} \dot{x}_1 \\ \dot{x}_2 \end{bmatrix} = \begin{bmatrix} 0 & 1 \\ 0 & 0 \end{bmatrix} \begin{bmatrix} x_1 \\ x_2 \end{bmatrix} + \begin{bmatrix} 0 \\ 1 \end{bmatrix} u; \quad x_1(0) = 1, x_2(0) = 0$$

Choose the feedback law
$$u = -K[x_1 + x_2]$$
Find the value of K so that
$$J = \frac{1}{2}\int_0^\infty (x_1^2 + x_2^2 + \lambda u^2)\, dt$$
is minimized, when

(i) $\lambda = 0$; (ii) $\lambda = 1$.

Also determine the values of minimum J in the two cases.

13.11 For the system of Problem P-13.10, find the optimal control law
$$u = -K_1 x_1 - K_2 x_2$$
that minimizes the given performance index for

(i) $\lambda = 0$; (ii) $\lambda = 1$.

13.12 For the plant described by the equations
$$x_1(k+1) = 0.8x_1(k) + x_2(k) + u(k)$$
$$x_2(k+1) = 0.5x_2(k) + 0.5u(k)$$
find the optimal control law that minimizes the performance index
$$J = \frac{1}{2}\sum_{k=0}^\infty [x_1^2(k) + x_2^2(k) + u^2(k)]$$

13.13 Consider a system
$$x(k+1) = 0.368x(k) + 0.632u(k)$$
Find the control sequence so that the following performance index is minimized:
$$J = x^2(N) + \sum_{k=1}^3 [x^2(k) + u^2(k)]$$
Also find the control sequence when $N \to \infty$.

13.14 The linear discrete system
$$x_1(k+1) = x_1(k) + x_2(k)$$
$$x_2(k+1) = x_2(k) + u(k)$$
is to be controlled to minimize the performance measure
$$J = \sum_{k=0}^2 [4x_1^2(k) + u^2(k)]$$
Obtain the optimal control sequence $[u(0), u(1), u(2)]$; the initial state is
$$x_1(0) = 1,\, x_2(0) = 0$$

13.15 It is desired to determine the control law that causes the plant
$$\dot{x}_1 = x_2$$
$$\dot{x}_2 = -x_1 - 2x_2 + u$$
to minimize the performance measure
$$J = 10x_1^2(t_f) + \frac{1}{2}\int_0^{t_f}(x_1^2 + 2x_2^2 + u^2)\, dt$$

The final time t_f is 1 sec.

(i) Determine the discrete approximation for the system. Use sampling interval $T = 0.02$ sec.

(ii) Determine the optimal control law from the discrete formulation.
(*Hint*: Part (ii) may best be solved using digital computer.)

13.16 A first-order system
$$\dot{x} = -x + u$$
is to be controlled to minimize
$$J = \frac{1}{2}\int_0^1 (x^2 + u^2)\, dt$$
Find the optimal control law.

13.17 Consider the second-order system
$$\dot{x}_1 = x_2$$
$$\dot{x}_2 = u$$
and the performance index
$$J = \frac{1}{2}\left[\, x_1^2(t_f) + 2x_2^2(t_f)\,\right] + \int_0^5 \left[\, x_1^2(t) + 2x_2^2(t) + x_1(t)\,x_2(t) + \frac{1}{4}u^2(t)\,\right] dt$$
The optimal control is
$$u^* = -\mathbf{R}^{-1}\mathbf{B}^T\mathbf{P}(t)\mathbf{x}$$
Obtain **R** and **B** and set up differential equations whose solution will yield the matrix $\mathbf{P}(t)$.

13.18 Consider the system
$$\dot{x}_1 = x_2$$
$$\dot{x}_2 = u$$
$$y = x_1$$
Find the control law which minimizes
$$J = \frac{1}{2}\int_0^\infty (y^2 + u^2)\, dt$$

13.19 Consider the system shown in Fig. P-13.19. The performance index to be minimized is
$$J = \frac{1}{2}\int_0^\infty ((y - r)^2 + u^2)\, dt$$
Convert this problem to the form of the output regulator problem and find the optimum value of K.

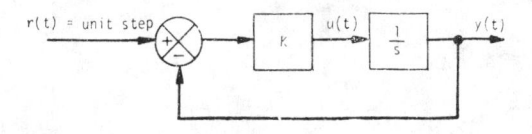

Figure P-13.19.

References and Further Reading

1. Chang, S.S.L., *Synthesis of Optimal Control Systems*, McGraw-Hill, New York, 1961.
2. Newton, G.C., Jr., L.A. Gould and J.F. Kaiser, *Analytical Design of Linear Feedback Controls*, John Wiley, New York, 1957.
3. Jury, E.I., and A.G. Dewey, "A General Formulation of the Total Square Integrals for Continuous Systems", *IEEE Trans. on Aut. Cont.*, Jan. 1965, AC-10, pp. 119-120.
4. Jury, E.I., "A Note on the Evaluation of the Total Square Integral", *IEEE Trans. on Aut. Cont.*, Jan. 1965, AC-10, pp. 110-111.
5. Gupta, S.C. and L. Hasdorff, *Fundamentals of Automatic Control*, John Wiley, New York, 1970.
6. Kirk, D.E., *Optimal Control Theory : An Introduction*, Prentice-Hall, Englewood Cliffs, N.J., 1970.
7. Wiberg, D.M., *State Space and Linear Systems*, Schaum's Outline Series, McGraw-Hill, New York, 1971.
8. Cadzow, J.A. and H.R. Martens, *Discrete-time and Computer Control Systems*, Prentice-Hall, Englewood Cliffs, N.J., 1970.
9. Schultz, D.G. and J.L. Melsa, *State Functions and Linear Control Systems*, McGraw-Hill, New York, 1967.
10. Kuo, B.C., *Automatic Control Systems*, 3rd ed., Prentice-Hall, Englewood Cliffs, N.J., 1975.
11. Athans, M. and P.L. Falb, *Optimal Control: An Introduction to the Theory and its Applications*, McGraw-Hill, New York, 1966.
12. Hsu, J.C. and A.U. Meyer, *Modern Control Principles and Applications*, McGraw-Hill, New York, 1968.
13. Anderson, B.D.O. and J.B. Moore, *Linear Optimal Control*, Prentice-Hall, Englewood Cliffs, N.J., 1971.
14. Kuo, B.C., *Discrete-Data Control Systems*, Science Tech., Champaign, Illinois, 1974.
15. Chen, C.T., *Analysis and Synthesis of Linear Control Systems*, Holt, Rinehart and Winston, New York, 1975.
16. Sage, A.P. and C.C. White, III, *Optimum Systems Control*, Prentice-Hall, Englewood Cliffs, N.J., 1977.

14. Nonlinear Systems

14.1 Introduction

In the preceding chapters we concerned ourselves exclusively with the linear, time-invariant (continuous and discrete) models of systems which, if examined in detail, are always found to be nonlinear to some extent. This does not mean that the powerful techniques of linear theory are not useful as these are restricted in use to the linearized versions of real systems. In fact, for many control applications the linear theory has produced excellent results which are supported experimentally. This again should not mean that we restrict ourselves to linear theory only, which would impose severe burden on system design in two ways: firstly, the constraint of linear operation over wide range demands unnecessarily high quality and therefore high cost components; secondly, the restriction to linear theory may inhibit the designer's curiosity to deliberately introduce nonlinear components or operate the otherwise linear components in nonlinear region with a view to improve system response.

Many practical systems are sufficiently nonlinear so that the important features of their performance may be completely overlooked if they are analyzed and designed through linear techniques as such. For such systems we must necessarily employ special analytical, graphical and numerical techniques which take account of system nonlinearities.

BEHAVIOUR OF NONLINEAR SYSTEMS

The most fundamental property of a linear system is the validity of the principle of superposition. It is on account of this property that it can perhaps be guaranteed that a linear system designed to perform satisfactorily when excited by a standard test signal, will exhibit satisfactory behaviour under any circumstances. Furthermore, the amplitude of the test signal is unimportant since any change in input signal amplitude results simply in change of response scale with no change in the basic response characteristics.

In contrast to the linear case, the response of nonlinear systems to a particular test signal is no guide to their behaviour to other inputs, since *the principle of superposition no longer holds*. In fact, the nonlinear system response may be highly sensitive to input amplitude, e.g., a nonlinear system giving its best response for a certain step input may exhibit highly unsatisfactory behaviour when the input amplitude is changed. Further, the Laplace- and z-transforms which are simple and powerful tools of linear theory become inapplicable and hence the analysis of nonlinear systems for different time-varying inputs is rendered quite difficult.

The stability of linear systems is determined solely by the location of the system poles (or eigenvalues) and is independent entirely of whether or not the system is driven. Furthermore, the stability of undriven linear systems is independent of the magnitude of the finite initial state.

The situation is not so clear-cut in nonlinear systems. Here the stability is very much dependent on the input and also the initial state. Further, the nonlinear systems may exhibit *limit cycles* which are self-sustained oscillations of fixed frequency and amplitude. Determination of existence of limit cycles is not an easy task as these may depend upon both the type and amplitude of the excitation signal. The stability study of nonlinear systems in fact requires the information about the type and amplitude of anticipated inputs, initial conditions, etc., in addition to the usual requirement of physical and mathematical models of systems.

Even for a stable nonlinear system, the transient and frequency response may exhibit certain peculiar features which are not found in linear systems. To illustrate some of the commonly encountered phenomena, let us consider the spring-mass-damper system of Fig. 14.1a. If the components are assumed to be linear, the system equation with a sinusoidal forcing function is given by

$$M\ddot{x} + f\dot{x} + Kx = F \cos \omega t \qquad (14.1)$$

The well known frequency response curve of this system is shown in Fig. 14.2. This response curve is obtained by measuring the amplitude x of the response, as the frequency of the forcing function is gradually varied keeping its amplitude constant.

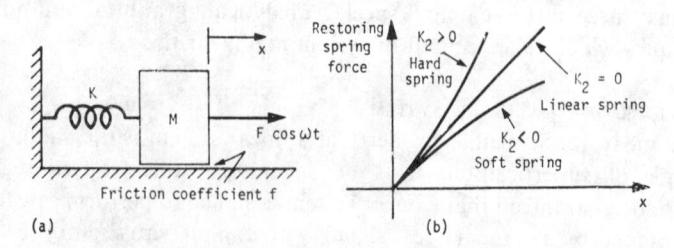

Figure 14.1 (a) A spring-mass-damper system; (b) Spring characteristics.

Let us now assume that the restoring force of the spring is nonlinear, given by $K_1 x + K_2 x^3$. The nonlinear spring characteristic is shown in Fig. 14.1b. The spring is linear if $K_2 = 0$, while it is called a *hard spring* if $K_2 > 0$ and a *soft spring* if $K_2 < 0$. For the case of the nonlinear spring, eqn. (14.1) modifies to

$$M\ddot{x} + f\dot{x} + K_1 x + K_2 x^3 = F \cos \omega t \qquad (14.2)$$

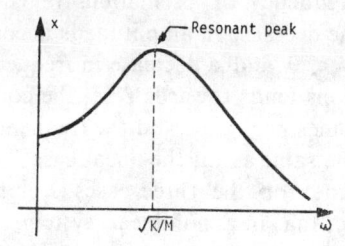

Figure 14.2 Frequency response curve of spring-mass-damper system.

If an experiment similar to that for the linear case is now performed, the frequency response curve of the form shown in Fig. 14.3a may be obtained for the hard spring case, i.e., $K_2 > 0$. The comparison of this figure with Fig. 14.2 reveals that the presence of the nonlinear term $K_2 x^3$ ($K_2 > 0$) in eqn. (14.2) has caused the resonance peak to bend towards higher frequencies. As the input frequency is gradually increased from zero, holding the input amplitude fixed, the measured response follows the curve through the points A, B and C, but at C an increment in frequency results in a discontinuous jump down to the point D, after which with further increase in frequency, the response curve follows through DE. If the frequency is now decreased, the response follows the curve EDF with a jump up to B occurring at F and then the response curve moves towards A. It is observed that in certain range of frequencies, the response function is double-valued as seen from Fig. 14.3a. This phenomenon which is peculiar to nonlinear systems is known as *jump resonance*.

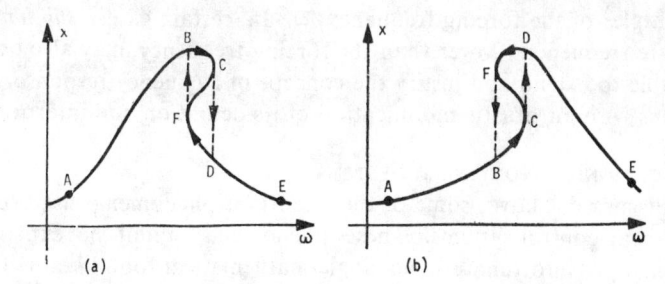

Figure 14.3 (a) Jump resonance in nonlinear systems (hard spring case); (b) Jump resonance in nonlinear systems (soft spring case).

For the soft spring case ($K_2 < 0$), the resonant peak bends towards lower frequencies and a similar jump response phenomenon takes place with the difference that there is a jump upward when the frequency is increased and a jump downward when the frequency is decreased as shown in Fig. 14.3b.

If for the spring-mass-damper system represented by eqn. (14.2), the forcing function is removed and the unforced system is activated by initial conditions only, the response x results in damped oscillations. For the linear case ($K_2 = 0$), the amplitude of successive peaks decreases and the undamped natural frequency of oscillations remains unchanged, while in the nonlinear case, the decrease in amplitude is accompanied by an increase in frequency for $K_2 < 0$ and a decrease in frequency for $K_2 > 0$. As the amplitude of oscillations tends towards zero, the contribution of the nonlinear term $K_2 x^3$ becomes negligible and the frequency of oscillations tends towards $\sqrt{(K_1/M)}$, the same as for the linear case. The frequency of oscillations versus amplitude for the three cases is plotted in Fig. 14.4. These plots illustrate the fact that in a nonlinear system, the response (the frequency of oscillations in the present case) is sensitive to amplitude.

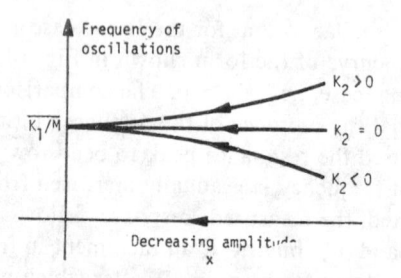

Figure 14.4 Frequency *vs* amplitude for the oscillations of spring-mass-damper system with nonlinear spring.

When a linear system is excited by a sinusoidal input of frequency ω, the steady-state output is always sinusoidal of the same frequency. This is not the case in nonlinear systems, where if the input is a sine wave, the output in general is nonsinusoidal containing frequencies (harmonics) which are multiples of the forcing frequency ω. In certain cases, *sub-harmonics*, which are frequencies lower than the forcing frequency, may also exist. It is impossible to extend and justify the concept of frequency response to such situations without drastic modification of its definition and interpretation.

INVESTIGATION OF NONLINEAR SYSTEMS

We discussed above, some of the important phenomena that occur only in nonlinear control systems. These phenomena cannot be explained by linear theory. Unfortunately, no single mathematical tool, like the Laplace- and z-transforms for linear systems, exists which could analyze the vast variety of nonlinear systems and nonlinear phenomena. Results derived for a class of nonlinear systems cannot be extended to others and no general performance and design criteria exist. Because of this difficulty, an analyst first considers the possibility of approximating a nonlinear system by a linear model. This type of approximation is valid only if the operation is in a restricted range about the operating point. For example, a nonlinear spring having a restoring force of $K_1 x + K_2 x^3$ may be approximated by a linear one provided the condition $K_2 x^3 \ll K_1 x$ is satisfied over the entire operating range of displacement x. This type of approximation gives satisfactory results for systems having small nonlinearities.

Nonlinear Systems

Many physical systems are decidedly nonlinear, even within the restricted region about the operating point. For example, a dry friction element produces a damping force of the nature shown in Fig. 14.5a. From this figure, it is seen that no single straight line can represent such a curve throughout the range of speeds usually encountered. However, the advantages of linear theory can be extended to this case by piecewise linearization as shown in Fig. 14.5b. In such cases, the response obeys a certain linear differential equation in one region of operation and a different one in another region. Depending upon the value of the region parameter (speed in dry friction case), we switch over from one differential equation to the other such that the end conditions of the first are carried over as the initial conditions of the other. Such systems are known as *piecewise linear systems*.

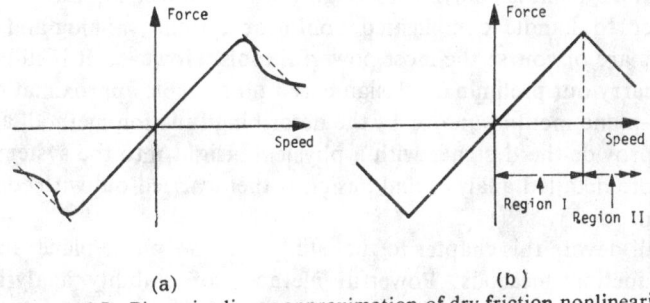

Figure 14.5 Piecewise linear approximation of dry friction nonlinearity.

At the cost of considerably increased computational work, piecewise linearization could be applied to any nonlinear system by dividing the whole region of operation into small pieces. In fact, this is how we numerically solve a nonlinear differential equation.

In more complicated situations, where the system is distinctly nonlinear, piecewise linearization is cumbersome and time consuming. For such cases, two methods of analysis are available, which have an appreciable degree of generality. One of these is the *phase-plane method*, which is one of the most powerful tools available today. This is basically a graphical method from which information about transient behaviour and stability is easily obtained by constructing phase trajectories. From the point of view of practical utility, the method is restricted to second-order systems, excited by step or ramp inputs. Higher-order systems may first be approximated by their second-order equivalents for investigation by the phase-plane method. Because of this approximation, the results obtained by this analysis should be checked by analog or digital simulation.

The other method, known as the *describing function method*, is based on harmonic linearization. We assume here that the input to the nonlinear component is sinusoidal and depending upon the filtering properties of the linear part of the overall system, the output is adequately represented by the fundamental frequency term in the Fourier series. This approximation is not restricted to small signal dynamics. Unfortunately, the approximation is intuitive and has no direct mathematical justification. The degree of accuracy achieved, is heavily dependent upon the filtering property of the linear part of the system. It is no wonder that this linearization may not always work. It usually gives sufficiently accurate information about system stability and limit cycles but does not give a trustworthy information about transient response.

In fact, the phase-plane and describing function methods use complimentary approximations. The phase-plane method retains the nonlinearity as such and uses the second-order approximation of a higher-order linear part, while on the other hand, the describing function method retains the linear part and harmonically linearizes the nonlinearity. The two techniques thus compliment each other. Greater confidence can therefore be placed in results if the system is analyzed through both these techniques.

In order to handle complicated nonlinear systems, analog and digital computers are of course the most powerful tools. However, it is still worthwhile to carry out preliminary design based on suitable approximations by the phase-plane method and/or by the describing function method, as these methods provide the designer with a physical insight into the system behaviour. More detailed analysis and design is then carried out with computer simulation.

We shall devote this chapter to the study of the phase-plane and describing function methods. Powerful methods of stability analysis—the second method of Liapunov and the method of Popov will also be briefly presented.

14.2 Common Physical Nonlinearities

In control systems, nonlinearities can be classified as *incidental and intentional*. Incidential nonlinearities are those which are inherently present in the system. The designer strives to design the system so as to limit the adverse effects of these nonlinearities. Common examples of incidental nonlinearities are saturation, dead-zone, coulomb friction, stiction, backlash, etc. The intentional nonlinearities, on the other hand, are those which are deliberately inserted in the system to modify system characteristics. The most common example of this type of nonlinearity is a relay. In the following paragraphs we shall discuss in brief the basic features of the commonly encountered nonlinearities.

SATURATION

This is perhaps the most common of all nonlinearities. All practical systems, when driven by sufficiently large signals, exhibit the phenomenon of saturation due to limitations of physical capabilities of their components. Many components such as amplifiers* have output proportional to input in a limited range of input signals. When the input exceeds this range, the output

---

*Other examples of saturation are torque and speed saturation in electric motors.

Nonlinear Systems

tends to become nearly constant as shown in Fig. 14.6. Though the change over from one range to another is gradual, it is sufficiently accurate in most cases to approximate the saturation phenomenon by straight line segments as shown.

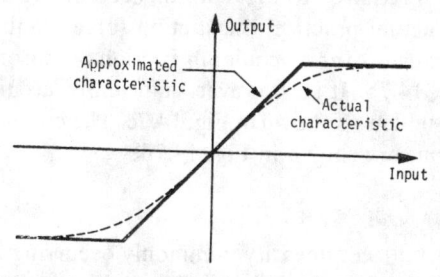

Figure 14.6 Piecewise linear approximation of saturation nonlinearity.

FRICTION

Retarding frictional forces exist whenever mechanical surfaces come in sliding contact. The predominant frictional force called the *viscous friction* is proportional to the relative velocity of sliding surfaces, i.e.,

$$\text{viscous friction force} = f\dot{x}$$

where f is a constant and $\dot{x}$ is the relative velocity. Viscous friction is thus linear in nature. Figure 14.7a is the graphical representation of such friction. In addition to the viscous friction, there exist two nonlinear

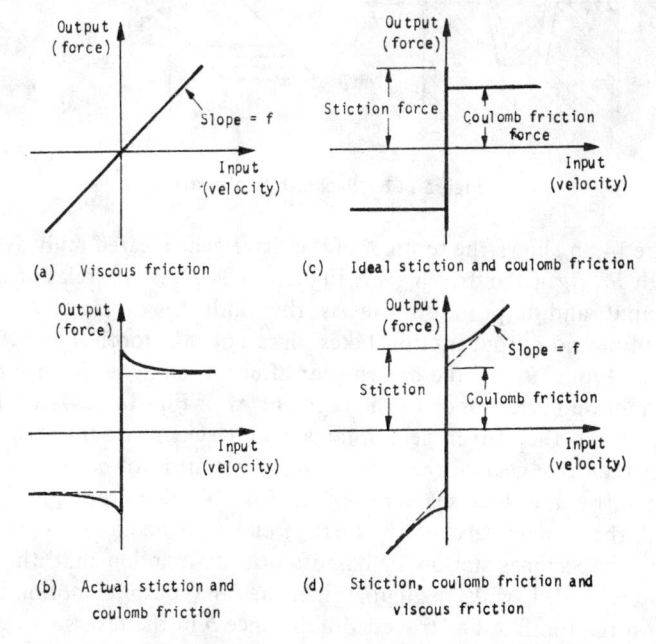

Figure 14.7 Characteristics of various types of frictions.

frictions. One is the *coulomb friction* which is a constant retarding force (always opposing the relative motion) and the other is the *stiction* which is the force required to initiate motion. The force of stiction is always greater than that of coulomb friction since due to interlocking of surface irregularities, more force is required to move an object from rest than to maintain it in motion. In actual practice, the stiction force gradually decreases with velocity and changes over to coulomb friction at reasonably low velocities as shown in Fig. 14.7b. It is however sufficiently accurate to regard this change over as sudden as shown in Fig. 14.7c. The composite characteristics of various frictions are shown in Fig. 14.7d.

BACKLASH

Another important nonlinearity commonly occurring in physical systems is hysteresis in mechanical transmission such as gear trains and linkages. This nonlinearity is somewhat different from magnetic hysteresis and is commonly referred to as backlash. Backlash in fact is the play between the teeth of the drive gear and those of the driven gear. Consider a gear box as shown in Fig. 14.8a having backlash as illustrated in Fig. 14.8b.

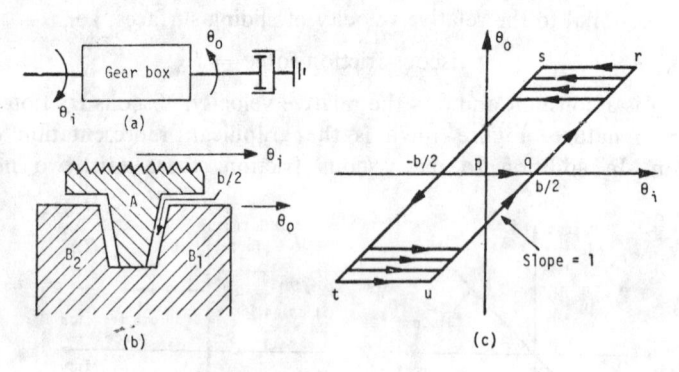

Figure 14.8 Backlash nonlinearity.

Figure 14.8b shows the tooth A of the drive gear located midway between the teeth B_1, B_2 of the driven gear. Figure 14.8c gives the relationship between input and output motions. As the tooth A is driven clockwise from this position, no output motion takes place until the tooth A makes contact with the tooth B_1 of the driven gear after travelling a distance $b/2$. This output motion corresponds to the segment pq of Fig. 14.8c. After the contact is made the driven gear rotates counter-clockwise through the same angle as the drive gear, if the gear ratio is assumed to be unity. This is illustrated by the line segment qr of Fig. 14.8c. As the input motion is reversed, the contact between the teeth A and B_1 is lost and the driven gear immediately becomes stationary based on the assumption that the load is friction-controlled with negligible inertia. The output motion therefore ceases till the tooth A has travelled a distance b in the reverse direction as illustrated by the segment rs. After the tooth A establishes contact with the tooth B_2, the driven gear now moves in clockwise direction as shown by the segment st. As the input motion is reversed the driven gear is again at standstill for the segment tu and then follows the drive gear along uq. This completes one cycle of the output motion.

It is easily seen from Fig. 14.8c that for a given input, the output is multivalued. Which particular output will result for a given input depends upon the history of the input. This type of nonlinearity has thus inherent memory and is referred to as *memory type* nonlinearity. The width of the input-output curve equals the total backlash b while its height corresponds to the limits of the input angle θ_i.

In a servo system, the gear backlash may cause sustained oscillations or chattering phenomenon and the system may even turn unstable for large backlash. Backlash can be reduced by using high quality gears. It can be almost eliminated by using spring loaded split gear as the driven gear. Such gears are employed in instrumentation systems.

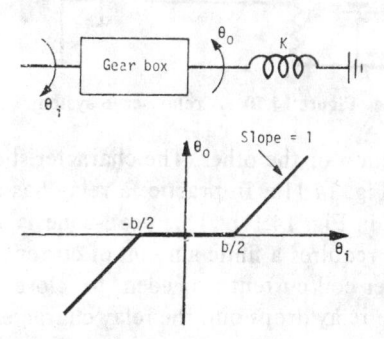

Figure 14.9 Dead-zone nonlinearity.

DEAD-ZONE

Figure 14.9 shows the case when gears with backlash drive a torsional spring load. The drive and the driven gears must now move together except in the region $\pm b/2$ where the spring remains untwisted and there is no contact between the teeth of the drive and driven gears. Figure 14.9 also illustrates the relationship between input and output motions which exhibit the phenomenon of dead-zone.

In addition to the example cited above dead-zone nonlinearity occurs in many other devices which are insensitive to small signals.

RELAY

A relay is a nonlinear power amplifier which can provide large power amplification inexpensively and is therefore deliberately introduced in control systems. A relay-controlled system can be switched abruptly between several discrete states which are usually off, full forward and full reverse. Relay-controlled systems find wide applications in the control field.

A simple relay servo system is shown in Fig. 14.10. The error signal controls the relay current which moves the solenoid so as to make the

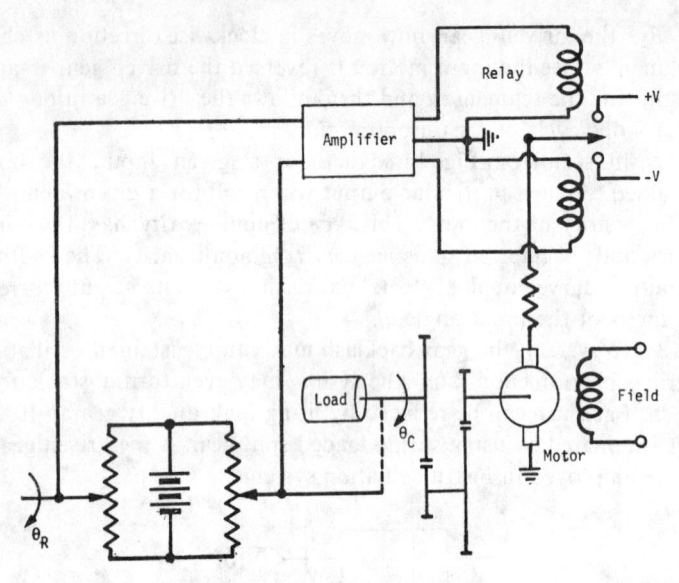

Figure 14.10 A relay servo system.

contact in one direction or the other. The characteristic of such an ideal relay is shown in Fig. 14.11a. In practice a relay has a definite amount of dead-zone as shown in Fig. 14.11b. This dead-zone is caused by the fact that the relay coil requires a finite amount of current to actuate the relay. Further since a larger coil current is needed to close the relay than the current at which the relay drops out, the relay characteristic always exhibits hysteresis. Relay characteristic with hysteresis alone is illustrated in Fig. 14.11c, while the more practical characteristic with both dead-zone and hysteresis is illustrated in Fig. 14.11d.

(a) Ideal relay

(b) Relay with dead-zone

(c) Relay with pure hysteresis

(d) Relay with dead-zone and hysteresis

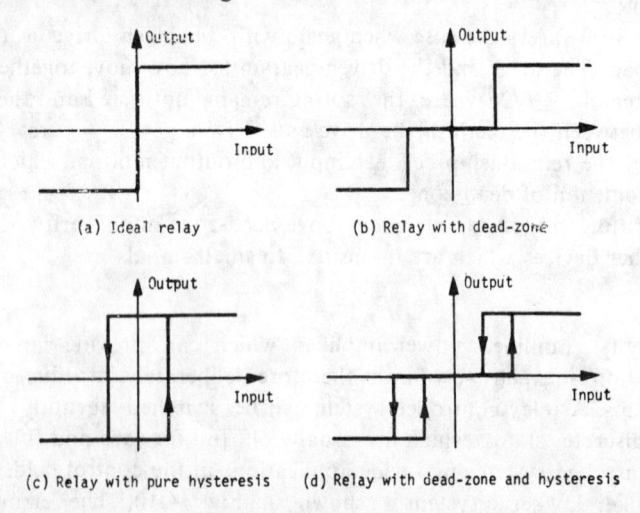

Figure 14.11 Relay nonlinearity.

MULTIVARIABLE NONLINEARITY

Some nonlinearities such as the torque-speed characteristics of a servomotor, transistor characteristics, etc., are functions of more than one variable. Such nonlinearities are called multivariable nonlinearities.

14.3 The Phase-plane Method: Basic Concepts

Consider an unforced linear spring-mass-damper system whose dynamics is described by

$$M \frac{d^2x}{dt^2} + f \frac{dx}{dt} + Kx = 0 \tag{14.3}$$

Let this system be activated by initial conditions only.

Equation (14.3) may be written in the standard form

$$\frac{d^2x}{dt^2} + 2\zeta\omega_n \frac{dx}{dt} + \omega_n^2 x = 0 \tag{14.4}$$

where ζ and ω_n are familiar quantities; the damping factor and undamped natural frequency of the system.

Let the state of this system be described by two variables, the displacement x and the velocity dx/dt, i.e., in state variable notation

$$x_1 = x; \quad x_2 = dx/dt$$

The state variables defined in this fashion (refer Chapter 12) are called *phase variables*. With this definition, the scalar differential equation (14.4) can be written in the state variable form,

$$\frac{dx_1}{dt} = x_2 \tag{14.5}$$

$$\frac{dx_2}{dt} = -\omega_n^2 x_1 - 2\zeta\omega_n x_2 \tag{14.6}$$

These equations may then be solved for phase variables x_1 and x_2. The time response plots of x_1, x_2 for various values of damping with initial conditions $x_1(0) = x_1^0$ and $x_2(0) = 0$ are shown in Fig. 14.12a. When the differential equations describing the dynamics of the system are nonlinear, it is in general not possible to obtain a closed form solution of x_1, x_2. For example, if the spring force is nonlinear ($K_1 x + K_2 x^3$), eqns. (14.5) and (14.6) take the form

$$\frac{dx_1}{dt} = x_2$$

$$\frac{dx_2}{dt} = -\frac{f}{M} x_2 - \frac{K_1}{M} x_1 - \frac{K_2}{M} x_1^3$$

Solving these equations by integration is no more an easy task. In such situations, a graphical method known as the *phase-plane method* is found to be very helpful. It overcomes the above said difficulty and provides essentially the same information as obtained from time response curves.

594   Control Systems Engineering

The coordinate plane with axes that correspond to the dependent variable $x_1 = x$ and its first derivative $x_2 = \dot{x}$, is called the *phase-plane*. The curve described by the state point (x_1, x_2) in the phase-plane with time as running parameter is called a *phase-trajectory*. A phase-trajectory can be easily constructed by graphical or analytical techniques to be discussed later. For the linear mass-spring-damper system under discussion, the phase-trajectories for different values of ζ corresponding to the time response curves of Fig. 14.12a are shown in Fig. 14.12b.

(a) Time response curves   (b) Phase-trajectories

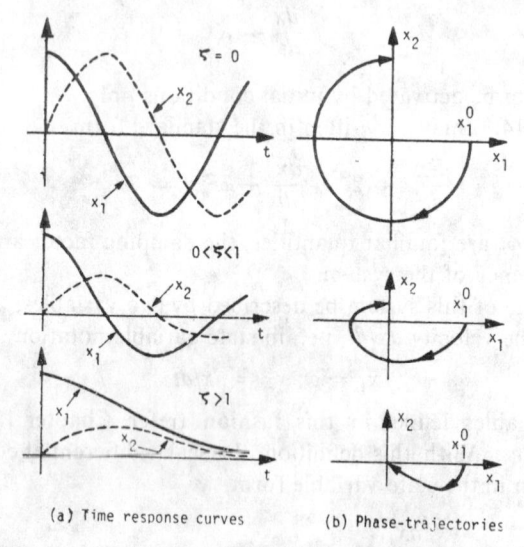

Figure 14.12 Time response curves and phase-trajectories for linear spring-mass-damper system.

In Fig. 14.13, trajectories of the critically-damped linear system for a given initial condition of displacement (curve A), an initial condition of velocity (curve B) and initial condition of both displacement and velocity (curve C) are shown in thick lines. Trajectories shown in dotted lines correspond to other initial conditions. Such a family of trajectories is called a *phase-portrait*. It is observed that in a phase-portrait, larger or smaller initial conditions produce geometrically similar trajectories.

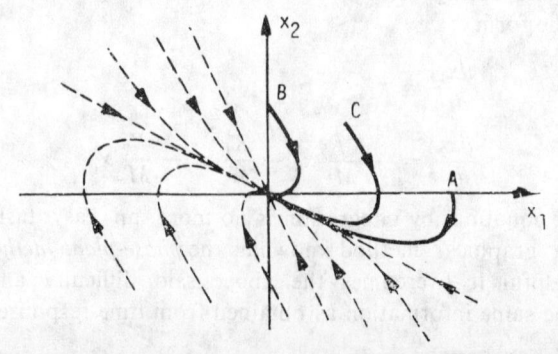

Figure 14.13 Phase-portrait for a critically-damped system.

The power of the phase-plane method lies in the fact that an easily constructable family of trajectories provides a graphical picture of the total system response such that a glance at the phase-portrait is sufficient to answer many important questions regarding system behaviour. More complete information about system behaviour, if required, can then be obtained by performing simple graphical constructions on the phase-portrait.

The phase-plane for second-order systems defined above, is indeed a special case of phase-space or state-space defined for nth order systems. Much work has been done to extend this approach of analysis for third-order systems [1]. Though a phase-trajectory for a third-order system can be graphically visualized through its projections on two planes, say (x_1, x_2)- and (x_2, x_3)-planes, this complicacy causes the technique to lose its major power of quick graphical visualization of the total system response. The phase-trajectories are therefore generally restricted to second-order systems only.

For linear time-invariant systems the entire phase-plane is covered with trajectories with one and only one curve passing through each point of the phase-plane except for certain critical points through which either infinite number or none of the trajectories pass. Such points (called singular points) are discussed in the next section.

If the parameters of a system vary with time or if a time-varying driving function is imposed, two or more trajectories may pass through a single point in phase-plane. In such cases, the phase-portrait becomes complex and more difficult to work with and interpret. Therefore, the use of phase-plane method is restricted to second-order systems with constant parameters and constant or zero input. However, it may be mentioned that investigators have made fruitful use of the phase-plane method in investigating second-order time-invariant systems under simple time-varying inputs such as ramp. Some simple time-varying systems have also been analyzed by this method [2].

Our discussion will be limited to autonomous systems (defined in the next section) only.

14.4 Singular Points

Consider a general time-invariant system described by the state equation
$$\dot{x} = f(x, u) \tag{14.7}$$
If the input vector u is constant, it is possible to write the above equation in the form
$$\dot{x} = F(x) \tag{14.8}$$
A system represented by an equation of this form is called an *autonomous system*. For such a system, consider the points in the phase-space at which the derivatives of all the state variables are zero. Such points are called *singular points*. These are in fact *equilibrium points* already defined in Chapter 12. If the system is placed at such a point, it will continue to lie

there if left undisturbed (the derivatives of all the phase variables being zero, the system state remains unchanged).

For studying the system dynamic response at an equilibrium (singular) point to small perturbation, the system is linearized (using linearization techniques presented in Chapter 12) at that point. The linearized model of system of eqn. (14.8) may be written as

$$\dot{x} = Ax \tag{14.9}$$

For this linear autonomous system, the equilibrium states are given by those states x_e satisfying $Ax_e = 0$. From this we see that $x_e = 0$ will be the only solution of the above equation provided that the determinant of A is nonzero. Equivalently, if all the eigenvalues of the system are different from 0, then the origin is the only singular point.

In general, if x_e is a singular point, it is convenient to shift the origin of coordinates to x_e. To achieve this, we define new phase variables as

$$\tilde{x} = x - x_e \tag{14.10}$$

The system of eqn. (14.8) in terms of new phase variables is represented as

$$\dot{\tilde{x}} = F(\tilde{x}) \tag{14.11}$$

with the equilibrium point lying at $\tilde{x} = 0$.

Consider now, a linearized autonomous second-order system described by equation of the form (14.9). The examination of phase-plane trajectory can best be done in the canonical form of representation. Using the linear transformation

$$x = Mz; \quad M \text{ is a modal matrix} \tag{14.12}$$

equation (14.9), being of second-order, may be converted into the canonical form

$$\begin{bmatrix} \dot{z}_1 \\ \dot{z}_2 \end{bmatrix} = \begin{bmatrix} \lambda_1 & 0 \\ 0 & \lambda_2 \end{bmatrix} \begin{bmatrix} z_1 \\ z_2 \end{bmatrix} \tag{14.13}$$

where λ_1 and λ_2 are eigenvalues of matrix A, which are assumed to be distinct.

It should be noticed that the transformation given by eqn. (14.12) merely changes the coordinate system from (x_1, x_2) to (z_1, z_2) having the same origin, as shown in Fig. 14.14. The new coordinate system (z_1, z_2) in general may not be rectangular.

Equation (14.13) may be written in the component form as

$$\dot{z}_1 = \lambda_1 z_1 \tag{14.14}$$

$$\dot{z}_2 = \lambda_2 z_2 \tag{14.15}$$

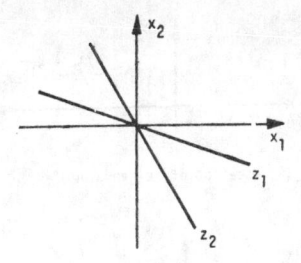

Figure 14.14 Change of coordinate axes by linear transformation.

The trajectory traced out by the representative point P in the (z_1, z_2)-plane has at any instant a resultant velocity vector associated with the motion as shown in Fig. 14.15. The slope of the velocity vector is given by

$$\tan \theta = \frac{dz_2/dt}{dz_1/dt} = \frac{dz_2}{dz_1} = \frac{\lambda_2 z_2}{\lambda_1 z_1}$$

which upon integration gives

$$z_2 = c(z_1)^{\lambda_2/\lambda_1} \tag{14.16}$$

where c is the constant of integration.

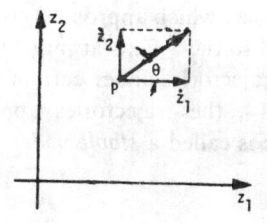

Figure 14.15 Velocity vector interpretation.

A plot of eqn. (14.16) in the (z_1, z_2)-plane gives the phase-trajectory. Since the derivatives of the phase-variables are zero at the equilibrium points, the slope of the trajectory at all such points is indeterminate (0/0).

We shall classify below singular points on the basis of location of the eigenvalues in the s-plane of the linearized version of the system.

NODAL POINT

Consider a linear second-order system whose eigenvalues are real, distinct and negative as shown in Fig. 14.16a. It is obvious from eqns. (14.14) and (14.15) that the singular point is located at the origin. For this case the equation of the phase-trajectory follows directly from eqn. (14.16) as

$$z_2 = c(z_1)^{k_1}$$

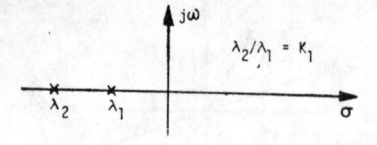

(a) Location of eigenvalues

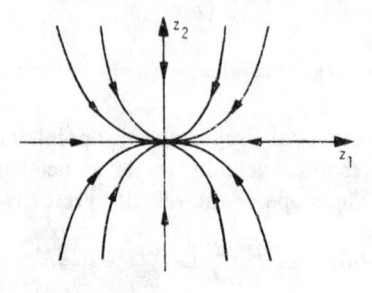

(b) Stable node in (z_1, z_2)-plane

Figure 14.16.

where $k_1 = \lambda_2/\lambda_1 \geqslant 0$.

For various initial conditions, the phase-portrait is drawn in Fig. 14.16b. The trajectories are parabolas which approach† the origin tangentially to the z_1-axis and are parallel to the z_2-axis at large distances from the origin. The axes themselves are trajectories under certain initial conditions. On the (x_1, x_2)-plane of Fig. 14.14, the trajectories appear as in Fig. 14.17. A singular point of this type is called a *stable node*.

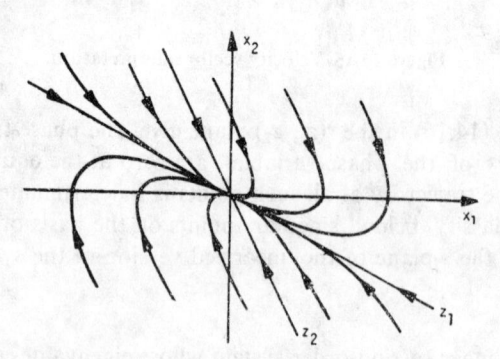

Figure 14.17 Stable node in (x_1, x_2)-plane.

---

†From eqns. (14.14) and (14.15), $z_1 = z_1(0)e^{\lambda_1 t}$ and $z_2 = z_2(0)e^{\lambda_3 t}$. Since $\lambda_1, \lambda_3 < 0$, both z_1 and z_2 decay with time and the state point travels towards origin for any initial state.

When λ_1 and λ_2 are real and distinct but lie in the right half of the s-plane, the phase portrait in the (x_1, x_2)-plane is as shown in Fig. 14.18. Such a singular point is known as an *unstable node*.*

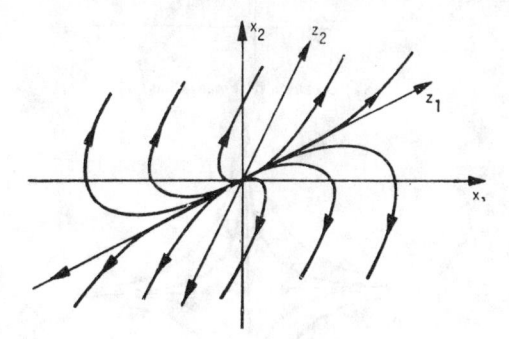

Figure 14.18 Unstable node in (x_1, x_2)-plane.

SADDLE POINT

Consider now a system with eigenvalue locations as shown in Fig. 14.19a. For this case, the equation of the phase-trajectory becomes

$$z_2 = c(z_1)^{-k_2}$$

or
$$(z_1)^{k_2} z_2 = c$$

where
$$k_2 = -\lambda_2/\lambda_1 > 0$$

For various initial conditions, the phase-portrait in the (z_1, z_2)-plane is shown in Fig. 14.19b. The trajectories are equilateral hyperbolas with directions as indicated.†† The origin which is the singular point is called a *saddle point* in this case. On the (x_1, x_2)-plane, the saddle point phase-portrait appears as shown in Fig. 14.19c. It is important to note that the *saddle point is always unstable*.

FOCUS POINT

Consider a system with complex conjugate eigenvalues

$$\lambda_1, \lambda_2 = \sigma \pm j\omega$$

---

*For systems with nonzero repeated eigenvalues λ, the transformation (14.12) gives

$$\begin{bmatrix} \dot{z}_1 \\ \dot{z}_2 \end{bmatrix} = \begin{bmatrix} \lambda & 1 \\ 0 & \lambda \end{bmatrix} \begin{bmatrix} z_1 \\ z_2 \end{bmatrix}$$

The trajectory equation is

$$\frac{dz_2}{dz_1} = \frac{\lambda z_2}{\lambda z_1 + z_2} = \frac{1}{1/\lambda + z_1/z_2}$$

This equation can be solved graphically (see Section 14.6). The resulting singular point is a node.

††These directions are established by examining $z_1 = z_1(0)e^{\lambda_1 t}$ and $z_2 = z_2(0)e^{\lambda_2 t}$. As time increases, z_1 reaches $\infty (\lambda_1 > 0)$ and z_2 reaches $0 (\lambda_2 < 0)$ and hence the directions are as shown.

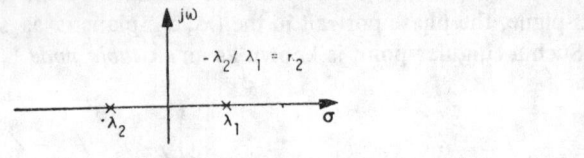

(a) Location of eigenvalues, $-\lambda_2/\lambda_1 = r_2$

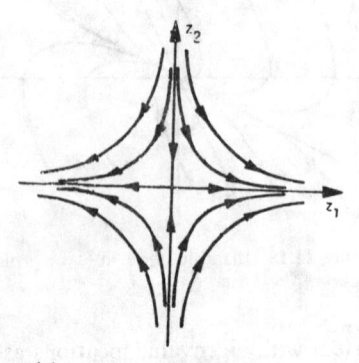

(b) Saddle point in (z_1, z_2)-plane

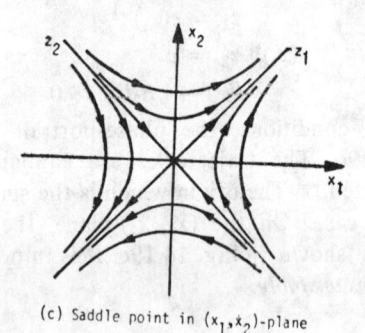

(c) Saddle point in (x_1, x_2)-plane

Figure 14.19.

The canonical state model of such a system is given by

$$\begin{bmatrix} \dot{z}_1 \\ \dot{z}_2 \end{bmatrix} = \begin{bmatrix} \sigma + j\omega & 0 \\ 0 & \sigma - j\omega \end{bmatrix} \begin{bmatrix} z_1 \\ z_2 \end{bmatrix}$$

If the following transformation is made

$$\begin{bmatrix} z_1 \\ z_2 \end{bmatrix} = \begin{bmatrix} 1/2 & -j/2 \\ 1/2 & j/2 \end{bmatrix} \begin{bmatrix} y_1 \\ y_2 \end{bmatrix}$$

then we obtain

$$\begin{bmatrix} \dot{y}_1 \\ \dot{y}_2 \end{bmatrix} = \begin{bmatrix} \sigma & \omega \\ -\omega & \sigma \end{bmatrix} \begin{bmatrix} y_1 \\ y_2 \end{bmatrix}$$

Therefore
$$\dot{y}_1 = \sigma y_1 + \omega y_2$$
$$\dot{y}_2 = -\omega y_1 + \sigma y_2$$

Dividing, we get

$$\frac{dy_2}{dy_1} = \frac{y_2 - ky_1}{y_1 + ky_2}; \quad k = \omega/\sigma \tag{14.17}$$

Let us define

$$\left.\begin{array}{c}\dfrac{dy_2}{dy_1} = \tan \psi \\ y_2/y_1 = \tan \theta\end{array}\right\} \tag{14.18}$$

and

Then from eqn. (14.17)

$$\tan \psi = \frac{\tan \theta - k}{1 + k \tan \theta}$$

or
$$\tan(\theta - \psi) = k \tag{14.19}$$

This is the equation of a spiral. A plot of this equation for negative values of σ (i.e., for complex conjugate eigenvalues with negative real parts) shown in Fig. 14.20a, is a family of equiangular spirals. The origin which is a singular point in this case is called a *stable focus*.

When the eigenvalues are complex conjugate with positive real parts, the phase-portrait consists of expanding spirals as shown in Fig. 14.20b, and the singular point is an *unstable focus*.

(a) Stable focus in (y_1, y_2)-plane (b) Unstable focus in (y_1, y_2)-plane

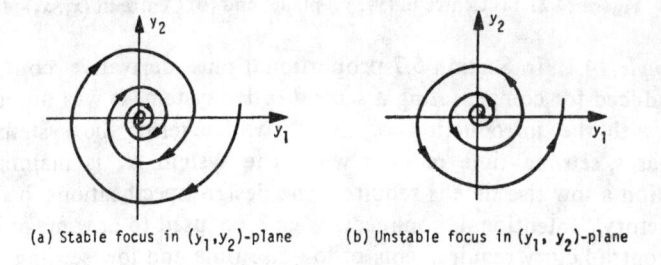

Figure 14.20.

When transformed to the (x_1, x_2)-plane, the phase-portrait in the above two cases is essentially spiralling in nature, except that the spirals are now somewhat twisted in shape.

CENTRE OR VORTEX POINT

Consider now the case of complex conjugate eigenvalues with zero real parts, i.e.,

$$\lambda_1, \lambda_2 = \pm j\omega$$

From eqn. (14.17) we obtain
$$\frac{dy_2}{dy_1} = -\frac{y_1}{y_2}$$
or
$$y_1\, dy_1 + y_2\, dy_2 = 0$$
which has the solution
$$y_1^2 + y_2^2 = c^2$$

Thus the equation of a spiral given by (14.19) becomes that of a circle when $\sigma = 0$. For various values of integration constant c, which depends upon the initial conditions, the phase-portrait is a family of circles as in Fig. 14.21a. This type of singular point is known as a *centre or vortex point*†.

When transformed to the (x_1, x_2)-plane, the phase-portrait is a family of ellipses as shown in Fig. 14.21b.

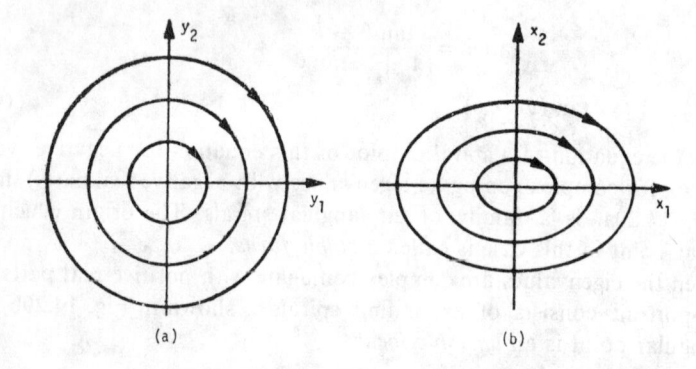

Figure 14.21 (a) Centre in (y_1, y_2)-plane; and (b) Centre in (x_1, x_2)-plane.

Example 14.1: In Section 5.7 proportional plus derivative controller was introduced for compensating a second-order system. It was observed there that with the introduction of the derivative term, the system damping increases, settling time reduces while the system K_v is maintained. If in addition a low rise time is required, the design specifications become contradictory. Intentional nonlinearity can be used to advantage in meeting the contradictory requirements of low rise time and low settling time with fixed K_v. This is easily achieved by making the derivative term become dependent upon the magnitude of error as shown in Fig. 14.22. By suitably shaping the nonlinear function $f(e)$, the derivative term $K_D f(e)\, \dot{e}$ is kept low (or even made negative) under large error conditions giving a small rise time. This term is made to become large for small errors so that the system settles in short time. The differential equation describing the nonlinear system is

$$[e + K_D f(e) \dot{e}] K_v = \tau \ddot{c} + \dot{c}$$
$$e = r - c$$

†When one or both eigenvalues are zero, there will be many singular points. For example, with $\lambda_1 = 0$, we have from eqn. (14.13) $\dot{z}_1 = 0;\ \dot{z}_2 = \lambda_2 z_2$. Thus every state $(z_1, z_2) = (c, 0)$ is an equilibrium state where c is an arbitrary constant.

Nonlinear Systems

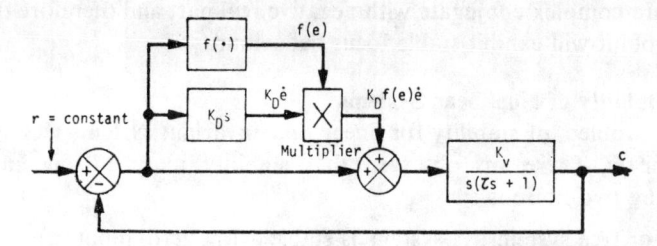

Figure 14.22 Nonlinear proportional plus derivative controller for second-order system.

If r = constant
$$\dot{e} = -\dot{c}; \quad \ddot{e} = -\ddot{c}$$
$$\therefore \quad \tau\ddot{e} + [1 + Kf(e)]\dot{e} + K_v e = 0 \tag{14.20a}$$
where
$$K = K_D K_v$$

Defining the state variables as $x_1 = e$, $x_2 = \dot{e}$, the state equation corresponding to eqn. (14.20a) is

$$\dot{x}_1 = x_2$$
$$\dot{x}_2 = -\frac{K_v}{\tau} x_1 - \frac{1 + Kf(x_1)}{\tau} x_2 \tag{14.20b}$$

Letting $\dot{x}_1 = \dot{x}_2 = 0$, the only equilibrium point is found to be at $(x_1 = 0, x_2 = 0)$

Linearizing eqn. (14.20b) about the equilibrium point by using eqn. (12.16) of Section 12.2, we have

$$\tilde{\mathbf{x}} = \mathbf{A}\tilde{\mathbf{x}}$$

where

$$\mathbf{A} = \begin{bmatrix} 0 & 1 \\ -\dfrac{K_v}{\tau} & -\dfrac{1 + Kf(0)}{\tau} \end{bmatrix} \tag{14.21}$$

The eigenvalues of $\mathbf{A}$ are given by

$$\lambda_1, \lambda_2 = -\left(\frac{1 + Kf(0)}{2\tau}\right) \pm \sqrt{\left\{\frac{1}{4}\left[\frac{1 + Kf(0)}{\tau}\right]^2 - \frac{K_v}{\tau}\right\}}$$

A desirable design feature could be that the system rises and settles fast without oscillations, for which the equilibrium point must be a stable node. For $f(0) > 0$, stable node, i.e., $\lambda_1, \lambda_2 < 0$, is achieved by the condition

$$f(0) > \left[\frac{2\sqrt{(K_v\tau)} - 1}{K}\right] \tag{14.22a}$$

The stable node requirement fixes the point $f(0)$ on the $f(e)$ function.

In case the above inequality is reversed, i.e.,

$$f(0) < \left[\frac{2\sqrt{(K_v\tau)} - 1}{K}\right] \tag{14.22b}$$

λ_1, λ_2 are complex conjugate with negative real part and therefore the equilibrium point will exhibit stable focus behaviour.

14.5 Stability of Nonlinear Systems

The problem of stability for linear time-invariant systems was studied in Chapter 6. There we observed that stability may be defined as per the following two notions :

(i) For free system : A system is stable with zero input and arbitrary initial conditions if the resulting trajectory tends towards the equilibrium state.

(ii) For forced system : A system is stable if with bounded input, the system output is bounded.

These two notions were found to be essentially equivalent for linear time-invariant systems. In nonlinear systems, unfortunately, there is no definite correspondence between the two notions. Many important results have been obtained for free systems. (Our discussion will be limited to this class of nonlinear systems.*) Even for this class of systems, the concept of stability is not clear-cut. The linear autonomous systems (with nonzero eigenvalues) have only one equilibrium state and their behaviour about the equilibrium state completely determines the qualitative behaviour in the entire state-plane. In nonlinear systems, on the other hand, system behaviour for small deviations about the equilibrium point may be different from that for large deviations. Therefore, *local stability* does not imply stability in the overall state-plane and the two concepts should be considered separately. Secondly, in a nonlinear system with multiple equilibrium states, the system trajectories may move away from one equilibrium point and tend to other as time progresses. Thus it appears that in case of nonlinear systems, it is simpler to speak of system stability relative to the equilibrium state rather than using a general term 'stability of a system'.

There are many types of stability definitions in the literature. We shall concentrate on three of these : stability, asymptotic stability and asymptotic stability in-the-large.

Consider an autonomous system described by the state equation (14.8). Assume that the system has only one equilibrium point. This is the case with all properly designed systems. Furthermore, without loss of generality, let the origin of state space be taken as the equilibrium point.

The system of eqn. (14.8) is *stable* at the origin if, for every initial state $x(t_0)$ which is sufficiently close to origin, $x(t)$ remains near the origin for all t. It is *asymptotically stable* if $x(t)$ in fact approaches the origin as $t \to \infty$. It is *asymptotically stable in-the-large* if it is asymptotically stable for every initial state regardless of how near or far it is from the origin.

Following are the mathematically precise definitions of the different types of stability.

*For stability study of forced nonlinear systems, reference [3] may be consulted.

Nonlinear Systems

The system of eqn. (14.8) is *stable* at the origin if, for every real number $\epsilon > 0$, there exists a real number $\delta(\epsilon) > 0$ such that $\| x(t_0) \| \leqslant \delta$ results in $\| x(t) \| \leqslant \epsilon$ for all $t \geqslant t_0$.

This definition (from Russian mathematician A.M. Liapunov) of stability uses the concept of vector norm. The euclidean norm for a vector **x** with n components $x_1, x_2, ..., x_n$ is (eqn. (II.12), Appendix II),

$$\| x \| = (x_1^2 + x_2^2 + ... + x_n^2)^{1/2}$$

$\| x \| \leqslant R$ defines a hyper-spherical region $S(R)$ of radius R surrounding the equilibrium point $x = 0$. In terms of the euclidean norm, the above definition of stability implies that for any $S(\epsilon)$ that we may designate, the designer must produce $S(\delta)$ so that system state initially in $S(\delta)$ will never leave $S(\epsilon)$. This is illustrated in Fig. 14.23a.

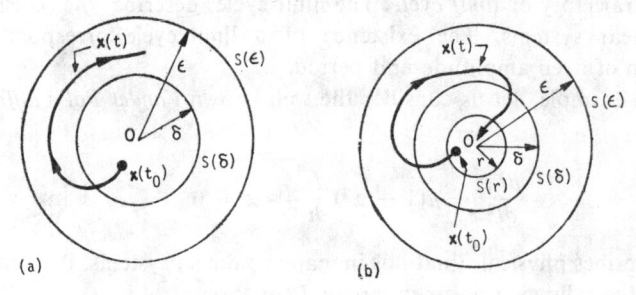

Figure 14.23 Stability definitions.

Consider for example, a second-order linear autonomous system whose phase-trajectories for various initial conditions are shown in Fig. 14.21b. For a specified value of ϵ, we can find a closed phase-trajectory whose maximum distance from the origin is ϵ. We then select a value of δ which is less than the minimum distance from that curve to the origin. The $\delta(\epsilon)$, so chosen will satisfy the conditions that guarantee stability. On the other hand, the equilibrium state (saddle point) shown in Fig. 14.19c is not stable since for a specified ϵ, a value of δ can not be produced to satisfy the condition of the definition.

A system is said to be *locally stable* (or *stable in-the-small*) if the region $S(\epsilon)$ is small.

Let us now turn to the other two definitions of stability.

The system of eqn. (14.8) is *asymptotically stable* at the origin if
(a) it is stable, and
(b) there exists a real number $r > 0$ such that $\| x(t_0) \| \leqslant r$ results in $x(t) \to 0$ as $t \to \infty$.

Property (b) implies that every motion starting in $S(r)$ converges to origin as $t \to \infty$. This is illustrated in Fig. 14.23b. Note that the singular points in the phase portraits of Figs. 14.17 and 14.20a are asymptotically stable.

The system of eqn. (14.8) is *asymptotically stable in-the-large* (*globally asymptotically stable*) at the origin if

(a) it is stable, and
(b) every initial state $x(t_0)$ results in $x(t) \to 0$ as $t \to \infty$.

Thus asymptotic stability in-the-large guarantees that every motion will approach the origin. In fact the singular points in the phase-portraits of Figs. 14.17 and 14.20a are asymptotically stable in-the-large.

Limit Cycles

In the definitions of stability of nonlinear systems given above, the system stability has been defined in terms of disturbed steady-state coming back to its equilibrium position or at least staying within tolerable limits from it. This does include the possibility that a disturbed nonlinear system even while staying within tolerance limits, may exhibit a special behaviour of following a closed trajectory or *limit cycle*. The limit cycles describe the oscillations of nonlinear systems. The existence of a limit cycle corresponds to an oscillation of fixed amplitude and period.

As an example, let us consider the well known *Vander Pol's differential equation*

$$\frac{d^2x}{dt^2} - \mu(1 - x^2)\frac{dx}{dt} + x = 0$$

which describes physical situations in many nonlinear systems. By comparing this with the following linear differential equation

$$\frac{d^2x}{dt^2} + 2\zeta\frac{dx}{dt} + x = 0$$

we observe that Vander Pol's equation has damping factor $-(\mu/2)(1 - x^2)$ which depends upon x. If it is assumed that initially $|x| \gg 1$, then the damping factor has large positive value. The system therefore behaves like an overdamped system with consequent decrease of the amplitude of x. In this process, the damping factor also decreases and the system state finally enters a limit cycle as shown by the outer trajectory of Fig. 14.24a.

On the other hand, if initially $|x| \ll 1$, the damping is negative hence the amplitude of x increases till the system state again enters the limit cycle as shown by the inner trajectory of Fig. 14.24a. The limit cycle shown in Fig. 14.24a is a stable one, since the paths in its neighbourhood converge toward the limit cycle.

On the other hand, if the paths in the neigbourhood of a limit cycle (closed trajectory) diverge away from it, it indicates that the limit cycle is unstable. Consider, for example, the Vander Pol's equation with the sign of its damping term reversed, i.e.,

$$\frac{d^2x}{dt^2} + \mu(1 - x^2)\frac{dx}{dt} + x = 0$$

Nonlinear Systems

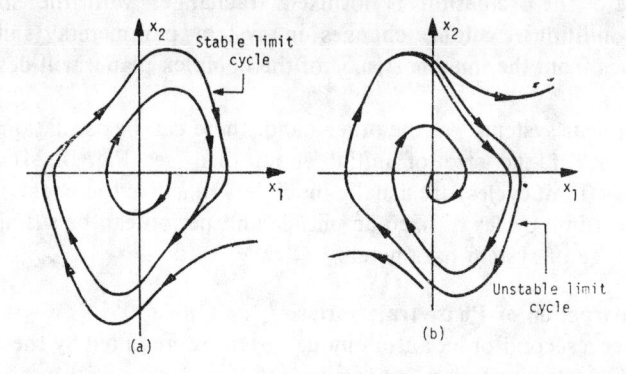

Figure 14.24 Limit cycle behaviour of nonlinear systems.

The phase-trajectories for this equation are shown in Fig. 14.24b, from which it is observed that an unstable limit cycle occurs.

It is important to note that a limit cycle, in general is an undesirable characteristic of a control system. It may be tolerated only if its amplitude is within specified limits.

It may be noted that the Vander Pol's equation with damping term $+ \mu(1 - e^2)$ in fact describes the behaviour of the nonlinear system presented in Example 14.1 if $K_v = \tau = 1$. The identity between the two is established by choosing

$$1 + Kf(e) = \mu(1 - e^2)$$

or
$$f(e) = \frac{\mu(1 - e^2) - 1}{K}$$

For stable node behaviour around origin (equilibrium point)

$$f(0) = \frac{\mu - 1}{K} > \frac{2\sqrt{(K_v \tau)} - 1}{K}\bigg|_{K_v = \tau = 1} = \frac{1}{K}$$

or
$$\mu > 2$$

and for stable focus behaviour

$$\mu < 2$$

As in the case of Vander Pol's equation, the nonlinear system of Example 14.1 with $f(e)$ defined above will exhibit an unstable limit cycle behaviour as shown in Fig. 14.24b. Though the nonlinearity makes the transient response more desirable than in a linear case (except for small inputs to which the response is sluggish), the price is paid in terms of unstable behaviour if the system perturbation is large enough to make the initial state point lie outside the limit cycle.

It may be noted that in linear autonomous systems, when oscillations occur, the resulting trajectories will be closed curves (Fig. 14.21 gives trajectories for a second-order system in which oscillations occur). The amplitude of the oscillations is not fixed. It changes with the size of the initial conditions. Slight changes in system parameters (shifting the eigenvalues from the imaginary axis of the complex plane) will destroy the oscillations.

In nonlinear systems, on the other hand, there can be oscillations that are independent of the size of initial conditions (see Fig. 14.24) and these oscillations (limit cycles) are usually much less sensitive to system parameter variations. Limit cycles of fixed amplitude and period can be sustained over a finite range of system parameters.

14.6 Construction of Phase-trajectories

Consider a second-order autonomous system represented by the following vector differential equation

$$\dot{\mathbf{x}} = \mathbf{F}(\mathbf{x})$$

In the component form, the above equation may be written as two scalar first-order differential equations

$$\dot{x}_1 = f_1(x_1, x_2)$$
$$\dot{x}_2 = f_2(x_1, x_2)$$

Dividing, we get

$$\frac{dx_2}{dx_1} = \frac{f_2(x_1, x_2)}{f_1(x_1, x_2)} \qquad (14.23)$$

Equation (14.23) defines the slope of the phase-trajectory. Every point (x_1, x_2) of the phase-plane, has associated with it the slope of the trajectory which passes through that point. The only exception are the singular points at which the trajectory slope is indeterminate. To construct the phase-trajectory, the slope equation must be integrated. Except for simple cases, it is not possible to perform this integration analytically. Therefore the integration must generally be carried out graphically or numerically (may be by use of analog or digital computers).

The direction of phase-trajectory at any point (x_1, x_2) can be ascertained by examining the signs of

$$\Delta x_1 = f_1(x_1, x_2)\,\Delta t$$
$$\Delta x_2 = f_2(x_1, x_2)\,\Delta t$$

in eqn. (14.23) for a small positive increment Δt in time. In the commonly met special case, where

$$x_2 = \frac{dx_1}{dt}$$

x_1 increases with time in the upper half of the phase-plane since $x_2 = dx_1/dt > 0$ and the state point therefore moves from left to right ($\rightarrow$) in this region. In the lower half of the phase-plane $x_2 = dx_1/dt < 0$, x_1 decreases with time and the state point must therefore move from right to left ($\leftarrow$).

Further with this definition of x_2, the trajectories must cross the x_1-axis at 90° since $\dfrac{dx_2}{dx_1}\bigg|_{x_2=0} = \dfrac{f_2(x_1, x_2)}{x_2}\bigg|_{x_2=0} = \infty$ except at a singular point.

In this section we shall present analytical and graphical methods of integrating the slope equation.

CONSTRUCTION BY ANALYTICAL METHOD

This is a useful approach for systems described by simple and piecewise linear differential equations. One may obviously raise the question that when time solutions are available by direct integration, where is the necessity of drawing phase-plane plots for such systems? In fact even for such systems, the phase-plane plot provides a powerful qualitative aid for investigating system behaviour and the design of system parameters to achieve a desired response. Furthermore, the existence of limit cycles is sharply brought into focus by the phase-portrait.

We will illustrate the analytical method with the help of the following example.

Example 14.2: Consider the system shown in Fig. 14.25. The nonlinear element is an ideal relay whose characteristics are shown in the figure.

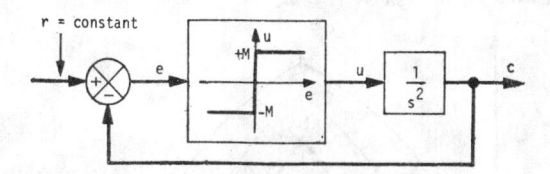

Figure 14.25 An ON-OFF control system.

From Fig. 14.25, the differential equation describing the dynamics of the system is given by

$$\ddot{c} = u \tag{14.24}$$

Choosing the state vector, $[x_1\ x_2]^T = [c\ \dot{c}]^T$, we have

$$\left.\begin{aligned}\dot{x}_1 &= x_2 \\ \dot{x}_2 &= u\end{aligned}\right\} \tag{14.25}$$

The output of the ON-OFF controller is given by

$$\begin{aligned}u &= M\ \mathrm{sgn}\ e \\ &= M\ \mathrm{sgn}\ (r - x_1)\end{aligned} \tag{14.26}$$

where r is the constant reference input and 'sgn' stands for 'sign of'. From eqns. (14.25) and (14.26) we have the trajectory equation as

$$\frac{dx_2}{dx_1} = \frac{M\ \mathrm{sgn}\ (r - x_1)}{x_2}$$

Separating the variables and integrating we get

$$\int_{x_2(0)}^{x_2} x_2\, dx_2 = M \int_{x_1(0)}^{x_1} \operatorname{sgn}(r - x_1)\, dx_1$$

Simplifying the above equation, we get

$$x_2^2 = 2Mx_1 - 2Mx_1(0) + x_2^2(0);\ x_1 < r \qquad (14.27)$$

Region I (*e* positive, relay output $+M$)

$$x_2^2 = -2Mx_1 + 2Mx_1(0) + x_2^2(0);\ x_1 > r \qquad (14.28)$$

Region II (*e* negative, relay output $-M$)

The phase-plane can be divided into two regions—Region I ($x_1 < r$) where the trajectory is defined by eqn. (14.27) and Region II ($x_1 > r$) where the trajectory is defined by eqn. (14.28). The trajectories in each of these regions are parabolas, symmetrical about the x_1-axis, suitably oriented with vertices placed in accordance with initial conditions [vertex of the parabolic trajectory of Region I is at $x_1 = x_1(0) - (1/2M)x_2^2(0)$ and that of Region II is at $x_1 = x_1(0) + (1/2M)x_2^2(0)$]. Typical trajectories are drawn in Fig 14.26a and we observe that each complete trajectory closes on to itself resulting in a limit cycle. Figure 14.26a shows a family of such trajectories.

A typical trajectory traced by the representative point *P* starting from **x(0)** is shown in thick line. We find that the relay switches from $-M$ to $+M$ at $t = t_1$ and vice versa at $t = t_2$.

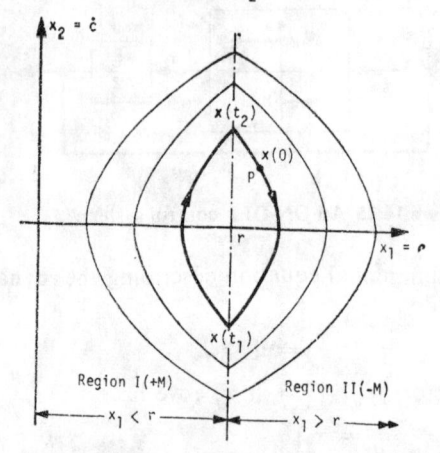

Figure 14.26 (a) Phase-trajectories for the system shown in Fig. 14.25.

As mentioned in the previous chapter, a singular point may be shifted to origin. For the example under consideration, eqn. (14.24) may be written as

$$\ddot{e} = -u$$

Choosing the state vector $[x_1\ x_2]^T = [e\ \dot{e}]^T$, we have

$$\dot{x}_1 = x_2$$
$$\dot{x}_2 = -M \operatorname{sgn} x_1$$

The slope equation now becomes

$$\frac{dx_2}{dx_1} = -\frac{M \operatorname{sgn} x_1}{x_2}$$

The phase-portrait corresponding to this slope equation is shown in Fig. 14.26b wherein the centre of the portrait now lies at the origin.

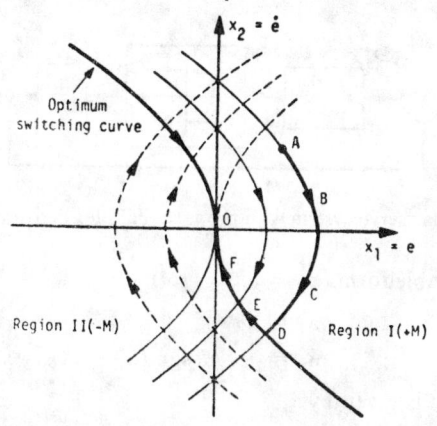

Figure 14.26 (b) Minimum time trajectory.

MINIMUM TIME TRAJECTORY

Assume now that the system under consideration in Example 14.2 starts with initial condition corresponding to the point A as shown in Fig. 14.26b. The representative point follows the trajectory through B and C. Further assume that the relay switches when the representative point reaches D. The representative point will then follow the parabolic trajectory to origin.

This is the shortest time (*minimum settling time*) path to origin. It obviously follows from Fig. 14.26b that irrespective of the starting point, the system takes minimum time to reach origin if the relay switches as soon as the state point reaches the optimum switching curve shown in thick line in Fig. 14.26b. In fact the *optimum switching curve* is part of the parabolic trajectory passing through origin. To implement the optimum settling time, we must incorporate an additional element in the system which generates the necessary switching curve and thereby determines the required switching instants of the relay.

It should be carefully noted that the example discussed above is perhaps oversimplified. In practical systems, the elements used for switching may not switch at prescribed instants due to hysteresis and dead-zone of the relay. The above method, of course, does give some feel of the power of ON-OFF controllers in optimizing the performance of control systems. Such systems are also called *bang-bang* control systems.

Example 14.3: On lines similar to those of linear systems, consider now the proposition of compensating the ideal relay system of Example 14.2 by means of proportional plus derivative controller ahead of the relay as shown in Fig. 14.27. The modified system is described by

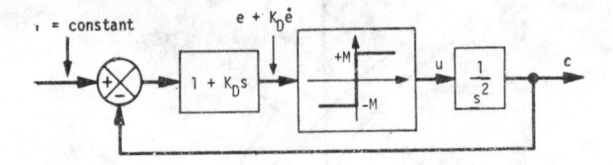

$$\ddot{e} = -u$$
$$u = M \text{ sgn}(e + K_D \dot{e})$$

Figure 14.27 Relay servo system with proportional plus derivative controller.

In the state variable form ($x_1 = e$, $x_2 = \dot{e}$)

$$\dot{x}_1 = x_2$$
$$\dot{x}_2 = -M \text{ sgn}(x_1 + K_D x_2)$$

The slope equation is given by

$$\frac{dx_2}{dx_1} = -\frac{M \text{ sgn}(x_1 + K_D x_2)}{x_2}$$

which upon integration gives

$$x_2^2 = -2Mx_1 + 2Mx_1(0) + x^2(0)$$
$$\text{Region I } [(x_1 + K_D x_2) > 0; u = +M]$$
$$x_2^2 = 2Mx_1 - 2Mx_1(0) + x_2^2(0)$$
$$\text{Region II } [(x_1 + K_D x_2) < 0; u = -M]$$

The state-space is divided into two regions by the line

$$x_1 + K_D x_2 = 0$$

In each half of the phase-plane separated by the switching line, the system trajectories would be parabolas. Assume that the system under consideration starts with initial conditions corresponding to the point A as shown in Fig. 14.28. The relay switches when the representative point reaches B. By geometry of the situation we see that the trajectory resulting from the reversal of the drive at point B will bring the representative point on a parabola passing much closer to the origin. This will continue until the trajectory intersects the switching line at a point closer to origin than points A_1 and A_2 which are points of intersection of the switching line with parabolas passing through the origin. In Fig. 14.28, point C corresponds to this situation. Here, an instant after the relay is switched, the system trajectory will recross the switching line and the relay must switch back. The relay will thus chatter while the system stays on the switching line. In a true second-order system, the chattering frequency will be infinite and amplitude will be zero; the representative point slides along the switching

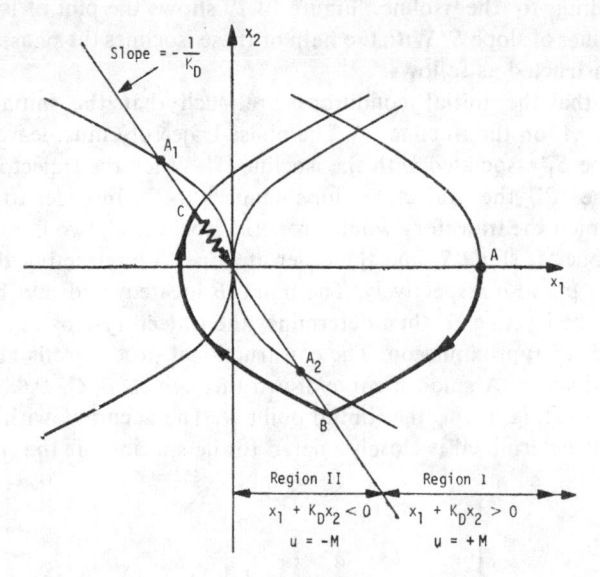

Figure 14.28.

line. For a second-order system which is an approximation of a higher-order system, the amplitude and frequency of chatter depend on approximations involved.

Compensation has thus converted the oscillating system into an asymptotically stable one; though the goal has been achieved in an inefficient way. It can easily be verified that if K_D is small (slope of the switching line nearly vertical) the chattering of the relay can be reduced though at the expense of high settling time.

CONSTRUCTION BY GRAPHICAL METHODS

A number of methods for graphical construction of phase-trajectories have been developed. We shall discuss here two of the important graphical methods—the isocline method and the delta method.

The isocline method

As mentioned earlier, the trajectory slope S at any point in the phase-plane is given by [see eqn. (14.23)]

$$\frac{dx_2}{dx_1} = S = \frac{f_2(x_1, x_2)}{f_1(x_1, x_2)} \quad (14.29)$$

For a specific trajectory slope S_1, we have

$$f_2(x_1, x_2) = S_1 f_1(x_1, x_2)$$

This equation defines the locus of all such points in the phase-plane at which the slope of the phase-trajectory is S_1. Such a locus is called an *isocline*. Every trajectory crossing an isocline must do so at a fixed slope corresponding to the isoline. Figure 14.29 shows the plot of isoclines for various values of slope S. With the help of these isoclines the phase-trajectory can be constructed as follows.

Assume that the initial conditions are such that the initial point is located at A on the isocline '1'. The phase-trajectory must leave the point A at a slope S_1 associated with the isocline '1'. When the trajectory reaches the isocline '2', the trajectory slope changes to S_2. In order to locate the point at which the trajectory would cross the isocline '2', two lines are drawn from A, one at slope S_1 and the other at slope S_2 intersecting the isocline '2' at points p and q respectively. The point B located midway between p and q on the isocline '2' then determines the trajectory crossing point to a good degree of approximation. The constructional procedure is now repeated at B and so on. A smooth curve drawn through $A, B, C, D, \ldots$ gives the trajectory starting from the initial point A. The accuracy with which the trajectory is determined is closely related to the spacing of the isoclines—

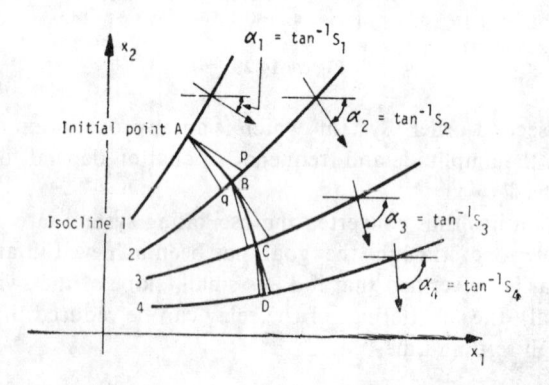

Figure 14.29 Construction of trajectory by the isocline method.

the closer are the isoclines, the more accurate is the phase-trajectory. It should be noted that once the isoclines are drawn, any number of trajectories could be easily determined.

Example 14.4: Consider the system shown in Fig. 14.30. The nonlinear element is a relay with dead-zone, whose characteristics are shown in the figure.

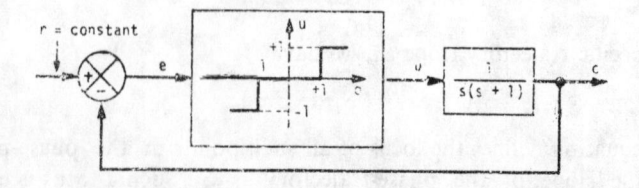

Figure 14.30 'ON-OFF' control system with dead-zone.

The differential equation describing the dynamics of the system is given by
$$\ddot{c} + \dot{c} = u$$
$$e = r - c$$
Since r is constant, the first equation can be written as
$$\ddot{e} + \dot{e} = -u$$
The relay output equation is given by
$$u = \phi(e)$$
where
$$\phi(e) = \begin{cases} +1 & \text{for } e > 1 \\ 0 & \text{for } -1 < e < 1 \\ -1 & \text{for } e < -1 \end{cases}$$

Choosing the phase variables as $x_1 = e$; $x_2 = \dot{e}$, we obtain the following first-order equations:
$$\dot{x}_1 = x_2$$
$$\dot{x}_2 = -x_2 - \phi(x_1)$$
The slope of the phase-trajectory is then given by
$$\frac{dx_2}{dx_1} = -\frac{x_2 + \phi(x_1)}{x_2} = S$$
The phase-plane may be divided into three regions:

Region I (defined by $x_1 > 1$): In this region, the slope equation is given by
$$S = -\frac{x_2 + 1}{x_2}$$
or
$$x_2 = \frac{-1}{S+1}$$
which defines the isoclines to be a family of straight lines parallel to the horizontal axis (i.e., x_1-axis).

Region II (defined by $-1 < x_1 < 1$): In this region, the slope equation is given by
$$S = -1$$

Region III (defined by $x_1 < -1$): For this region, the slope equation is
$$S = -\frac{(x_2 - 1)}{x_2}$$
or
$$x_2 = \frac{1}{S+1}$$

Therefore, the isoclines are parallel to the horizontal axis, as in the case of Region I.

Some representative isoclines are shown in Fig. 14.31. For a step input $r = 3$ and zero initial conditions, the initial point is located at A. Using the method of construction discussed above, the phase-trajectory is easily drawn and is indicated by thick line in Fig. 14.31.

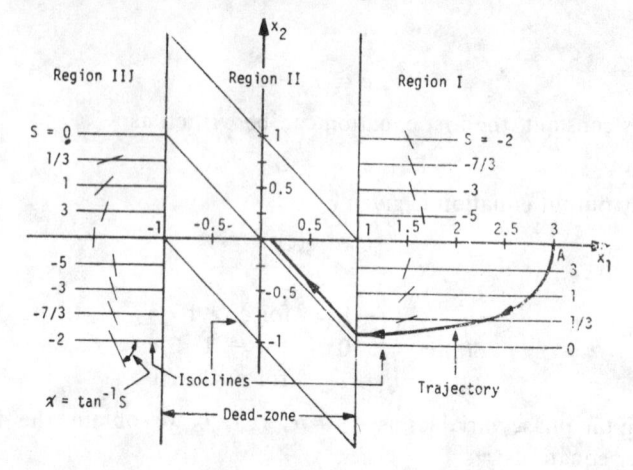

Figure 14.31 Isoclines and a typical trajectory for the system of Fig. 14.30.

It is important to note that a small dead-zone region is not always undesirable in relay-controllers. Let us investigate the behaviour of the system shown in Fig. 14.30 using ideal relay (no dead-zone) as a controller. For such a relay, the width of Region II (corresponding to dead-zone) in the phase-plane reduces to zero. The phase-trajectory of such a system with $r = 3$ is shown in Fig. 14.32. Comparison of Figs. 14.31 and 14.32 reveals

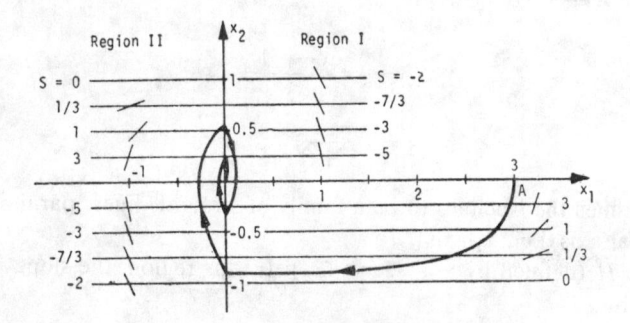

Figure 14.32 Phase-trajectory for the system of Fig. 14.30 when dead-zone is absent.

that dead-zone in relay characteristic helps to reduce system oscillations, thereby reducing settling time. However, the ideal relay controller drives the system to origin (i.e., zero steady-state error), while the relay controller with dead-zone drives it to a point within the dead-zone width. The maximum possible error equals half the dead-zone width. A large dead-zone would of course cause the steady-state performance of the system to deteriorate.

Nonlinear Systems

The delta method

The delta method of constructing phase-trajectories is applied to systems of the form

$$\ddot{x} + f(x, \dot{x}, t) = 0 \tag{14.30}$$

where $f(x, \dot{x}, t)$ may be linear or nonlinear and may even be time-varying but must be continuous and single-valued. With the help of this method, phase-trajectory for any system with step or ramp (or any time-varying) input can be conveniently drawn. The method results in considerable time saving when a single or a few phase-trajectories are required rather than a complete phase-portrait.

In applying the delta method, eqn. (14.30) is first converted to the form

$$\ddot{x} + K^2[x + \delta(x, \dot{x}, t)] = 0 \tag{14.31}$$

where K is a constant. When $\delta(x, \dot{x}, t) = 0$, K is the undamped natural frequency of the system and may be indicated as such by ω_n. Equation (14.31) is then written as

$$\ddot{x} + \omega_n^2[x + \delta(x, \dot{x}, t)] = 0 \tag{14.32}$$

In general $\delta(x, \dot{x}, t)$ depends upon the variables x, $\dot{x}$ and t, but for short intervals the changes in these variables are negligible. Thus over a short interval, we have

$$\ddot{x} + \omega_n(x + \delta) = 0 \tag{14.33}$$

where δ is regarded constant.

Let us choose the state variables as

$$x_1 = x; \quad x_2 = \dot{x}/\omega_n$$

Equation (14.33) then reduces to the following first-order equations

$$\dot{x}_1 = \omega_n x_2$$
$$\dot{x}_2 = -\omega_n(x_1 + \delta)$$

Therefore, the slope equation over a short interval is given by

$$\frac{dx_2}{dx_1} = -\frac{x_1 + \delta}{x_2} \tag{14.34}$$

With δ known at any point P on the trajectory and assumed constant for a short interval, we can draw a short segment of the trajectory by using the trajectory slope dx_2/dx_1 indicated in eqn. (14.34). A simple geometrical construction given below can be used for this purpose.

At any time t, the point $P(x_1, x_2)$ is known in the phase-plane as shown in Fig. 14.33. Cut-off $OC = \delta$ on the negative* side of the x_1-axis such that $CB = x_1 + \delta$. The line CP has then a slope $x_2/(x_1 + \delta)$. The line PQ perpendicular to CP at P will have a slope $-(x_1 + \delta)/x_2$, which is also the slope of the trajectory at the point P. If a circular arc, with C as centre and CP as radius is drawn, the tangent to this arc at the point P will have the same slope as that of the line PQ.

*If δ is negative, OC is cut-off on the positive side of the x_1-axis.

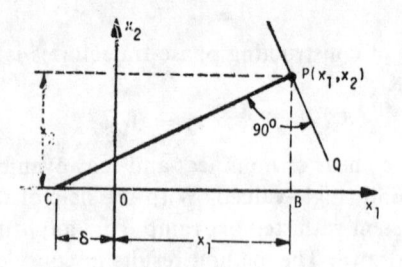

Figure 14.33 Geometry of the δ-method.

This suggests the following method of constructing a phase-trajectory.
1. From the initial point (x_1^0, x_2^0) and initial time t_0 (in case δ is time-dependent) calculate the value of δ. Auxiliary plots may be used to read off δ.
2. Draw a short arc segment through the initial point with $(-δ, 0)$ as centre, thereby determining a new point (x_1^1, x_2^1) on the trajectory.
3. Repeat the process at (x_1^1, x_2^1) and continue.

The above constructional procedure is illustrated with the help of the following example.

Example 14.5: Consider a system with nonlinear damping, described by the differential equation

$$\ddot{x} + 4 |\dot{x}| \dot{x} + 4x = 0 \tag{14.35}$$

Equation (14.35) may be converted to the form

$$\ddot{x} + \omega_n^2(x + δ) = 0$$

where
$$\omega_n = 2$$
$$δ = |\dot{x}| \dot{x} \tag{14.36}$$

Let us choose the state variables as

$$x_1 = x; \quad x_2 = \dot{x}/2$$

The slope equation is then given by (14.34). From eqn. (14.36), δ can be expressed in terms of state variables as

$$δ = 4 |x_2| x_2 \tag{14.37}$$

A plot of δ versus x_2 is shown in Fig. 14.34 [the plotting of δ-curve is not necessary as δ may be computed directly from eqn. (14.37)]. Let the initial condition be given as $x_1(0) = x(0) = 1$ and $x_2(0) = \dot{x}(0)/2 = 0$. This locates *A* as the initial point of the trajectory in Fig. 14.34. At this point, $δ = 0$. Therefore the first short arc *AB* is centred at (0, 0) with a radius of unity. The value of x_2 at the point *B* is used to compute the new value of δ. With $(-δ, 0)$ as centre, another short arc *BC* is drawn. This process is continued till the complete phase-trajectory as shown in Fig. 14.34 has been drawn. Here C_1, C_2, C_3, C_4 are the centres of the arcs *AB*, *BC*, *CD* and *DE* respectively.

Nonlinear Systems

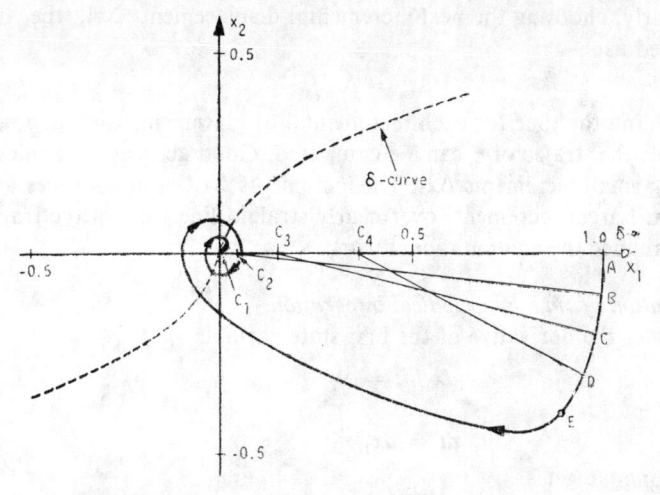

Figure 14.34 Phase-trajectory for Example 14.5 constructed by δ-method.

Computation of Time

The variation of time along the path of a phase-trajectory can be found by various approximation techniques, some of which are presented below.

Computation of time by average speeds

Any trajectory can be approximated by short line segments by taking small increments along the x_1-axis as shown in Fig. 14.35. For a small incremental displacement Δx_1^0, the average system speed

$$\dot{x}_{av} = x_{2av}^0$$

can be used for computing the time required for this incremental displacement since the trajectory segment during this increment is nearly a straight line. Thus

$$\Delta t^\circ = \Delta x_1^0 / x_{2av}^0$$

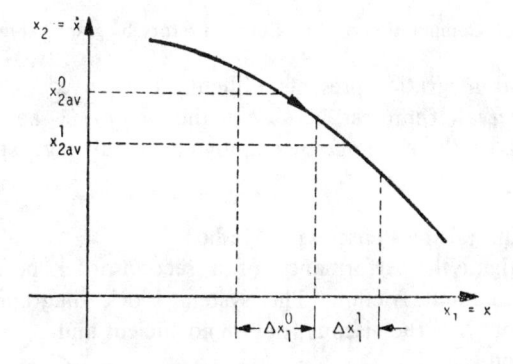

Figure 14.35 Computation of time along trajectory by average speeds.

Similarly, choosing the next incremental displacement Δx_1^1, the time is computed as

$$\Delta t^1 = \Delta x^1 / x_{2av}^1$$

By adding the time for each incremental displacement, the time at any point of the trajectory can be computed. Good accuracy is achieved by choosing small increments Δx_1. The increments Δx_1, need not necessarily be constant. Larger increments over nearly straight line parts of the trajectory help to reduce the computational work.

Computation of time by graphical integration

Consider the derivative of the first state variable

$$dx_1/dt = x_2$$

or

$$dt = dx_1/x_2$$

Integrating we get

$$t = \int_{x_1(0)}^{x_1(t)} \frac{1}{x_2} dx_1$$

From the above expression for time, it follows that the time required for the system state to change from $x_1(0)$ to $x_1(t)$ is represented by the area under $1/x_2$ versus x_1 plot as shown in Fig. 14.36. The area could be evaluated by a planimeter or by rectangular or trapazoidal rules.

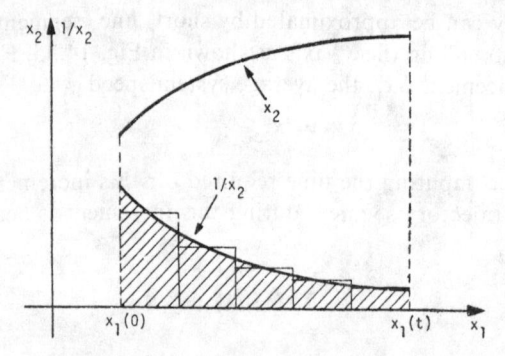

Figure 14.36 Computation of time along trajectory by graphical integration.

The graphical integration presents difficulties when x_2 passes through zero in the interval of interest, because at this point, $1/x_2$ becomes infinity. This difficulty is easily overcome by employing the average speed method in this region.

14.7 System Analysis by Phase-plane Method

Let us investigate the performance of a second-order position control system with coulomb friction. The system block diagram is shown in Fig. 14.37, where f_v is the viscous friction coefficient and f_c sgn($\dot{c}$) is the coulomb friction.

Nonlinear Systems

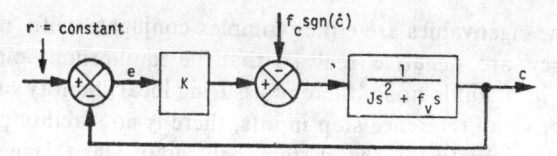

Figure 14.37 Position control system with viscous and coulomb frictions.

The dynamics of the system is described by the following differential equation:

$$Ke - f_c \operatorname{sgn}(\dot{c}) = J\ddot{c} + f_v \dot{c} \tag{14.38}$$

For constant input, $\dot{c} = -\dot{e}$ and $\ddot{c} = -\ddot{e}$. Therefore

$$J\ddot{e} + f_v \dot{e} + f_c \operatorname{sgn}(\dot{e}) + Ke = 0$$

or

$$\frac{J}{f_v}\ddot{e} + \dot{e} + \frac{f_c}{f_v} \operatorname{sgn}(\dot{e}) + \frac{K}{f_v} e = 0$$

Letting $J/f_v = \tau$, we get

$$\tau\ddot{e} + \dot{e} + \frac{f_c}{f_v} \operatorname{sgn}(\dot{e}) + \frac{K}{f_v} e = 0 \tag{14.39}$$

In terms of the state variables

$$x_1 = e;\ x_2 = \dot{e}$$

the above equation is converted to the following two first-order differential equations:

$$\dot{x}_1 = x_2 \tag{14.40}$$

$$\tau \dot{x}_2 = -\frac{K}{f_v} x_1 - x_2 - \frac{f_c}{f_v} \operatorname{sgn}(x_2) \tag{14.41}$$

Let us first determine the singular point of the system. At a singular point (x_1^0, x_2^0), we have from eqns. (14.40) and (14.41)

$$0 = x_2^0$$

$$0 = -\frac{K}{f_v} x_1^0 - x_2^0 - \frac{f_c}{f_v} \operatorname{sgn}(x_2)$$

It is observed from these relations that the singular points of the system are given by

$$x_1^0 = -f_c/K \operatorname{sgn}(x_2)$$
$$x_2^0 = 0$$

Let us now investigate the stability of the singular points. From eqns. (14.40) and (14.41), the matrix **A** of the system linearized around the singular point is

$$\mathbf{A} = \begin{bmatrix} 0 & 1 \\ -K/\tau f_v & -1/\tau \end{bmatrix}$$

The characteristic equation is given by

$$\lambda^2 + \frac{1}{\tau}\lambda + \frac{K}{\tau f_v} = 0$$

Therefore the eigenvalues are either complex conjugate with negative real parts or they are negative real so that the equilibrium point is either a stable focus or a stable node, thereby assuring local stability of the system. Therefore for small reference step inputs, there is no stability problem.

Let us now investigate the system behaviour when large inputs are applied. For this investigation we need the phase-portrait.

From eqns. (14.40) and (14.41), the trajectory slope equation is obtained as

$$\frac{dx_2}{dx_1} = S = \frac{-\frac{K}{f_v} x_1 - x_2 - \frac{f_c}{f_v} \text{sgn}(x_2)}{x_2 \tau}$$

or

$$x_2 = \frac{-\frac{K}{f_v}\left[x_1 + \frac{f_c}{K} \text{sgn}(x_2)\right]}{1 + S\tau} \qquad (14.42)$$

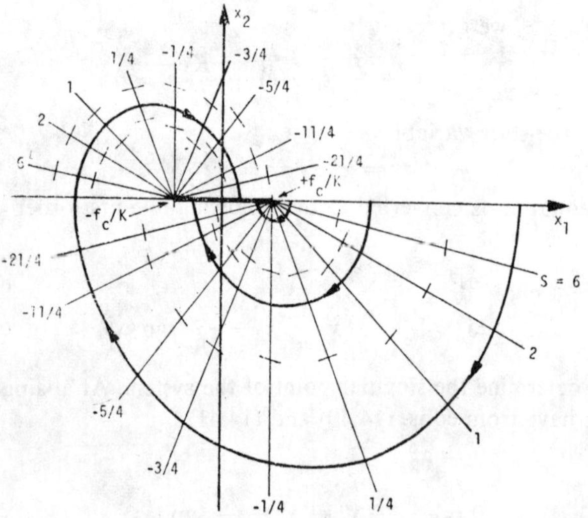

Figure 14.38 Phase-portrait of the system shown in Fig. 14.37, with $K/f_v = 5$, $\tau = 4$.

This equation defines the isoclines in the phase-plane. When $f_c = 0$, the isoclines are the same as for a linear system. The effect of coulomb friction is merely to shift the focal point by a distance f_c/K to the left for $x_2 > 0$ and by the same distance to the right for $x_2 < 0$. The x_1-axis divides the phase-plane into two regions since $\text{sgn}(x_2)$ reverses at $x_2 = 0$.

Figure 14.38 shows the isoclines and the phase-portrait for the system under discussion. It is observed that for small as well as large inputs, the resulting trajectories terminate on a line along the x_1-axis from $-f_c/K$ to $+f_c/K$, i.e., the line joining the singular points. Therefore, we conclude that the system under discussion is locally as well as globally stable.

Let us now investigate the effect of stiction on the system behaviour. Assume a discontinuous stiction, i.e., it appears only when velocity is zero as shown in Fig. 14.7c.

When stiction f_s is present, the singular points of the system shift to $x_1 = \pm f_s/K$ as shown in Fig. 14.39. No motion will result for inputs less than f_s/K since the driving torque will be less than the opposing torque due to stiction. For inputs greater than f_s/K, the system is pulled free and as soon as it happens, the stiction f_s changes over to coulomb friction f_c. The singular points instantaneously shift to $\pm f_c/K$ and the phase trajectories are controlled by the isoclines with coulomb friction only as shown in Fig. 14.39. It is found from this figure that system stability is guaranteed in the presence of stiction as well.

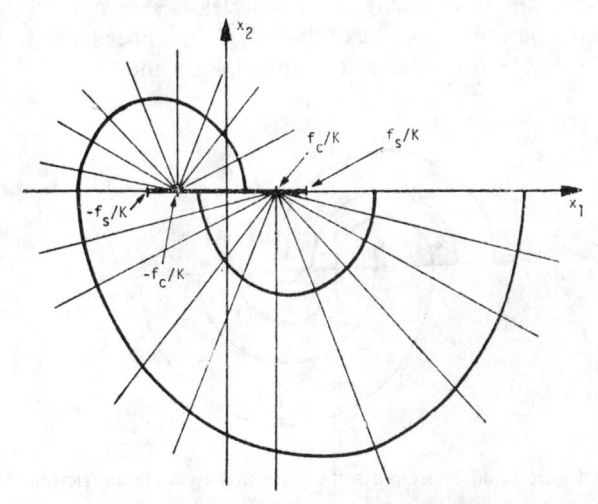

Figure 14.39 Phase-portrait of the system with viscous and coulomb friction and stiction.

If in the system under discussion, the viscous friction is assumed to be zero, the differential equation governing the system behaviour becomes

$$J\ddot{e} + Ke + f_c \operatorname{sgn}(\dot{e}) = 0$$

or
$$\ddot{e} + \omega_n^2 \left[e + \frac{f_c}{K} \operatorname{sgn}(\dot{e}) \right] = 0 \qquad (14.43)$$

where
$$\omega_n = \sqrt{(K/J)}$$

Define
$$x_1 = e; \quad x_2 = \dot{e}/\omega_n$$

giving the state variable form of eqn. (14.43) as

$$\dot{x}_1 = \omega_n x_2$$
$$\dot{x}_2 = -\omega_n \left[x_1 + \frac{f_c}{K} \operatorname{sgn}(\omega_n x_2) \right]$$

Let
$$x_1' = x_1 + \frac{f_c}{K} \operatorname{sgn}(\omega_n x_2), \text{ therefore } \dot{x}_1' = \dot{x}_1$$

We have now the slope equation as

$$dx_2/dx_1' = -x_1'/x_2$$

which upon integration gives a circular trajectory
$$x_1'^2 + x_2^2 = C$$
or
$$\left[x_1 + \frac{f_c}{K} \operatorname{sgn}(\omega_n x_2)\right]^2 + x_2^2 = C \qquad (14.44)$$
with trajectory centres located at

$(-f_c/K, 0)$ for $x_2 > 0$ (upper half of phase-plane)
$(+f_c/K, 0)$ for $x_2 < 0$ (lower half of phase-plane)

The phase-portrait is a family of semicircles as shown in Fig. 14.40 with centres located on the $(x_1 = e)$-axis at $-f_c/K$ for upper half of the phase-plane and at $+f_c/K$ for lower half of the phase-plane.

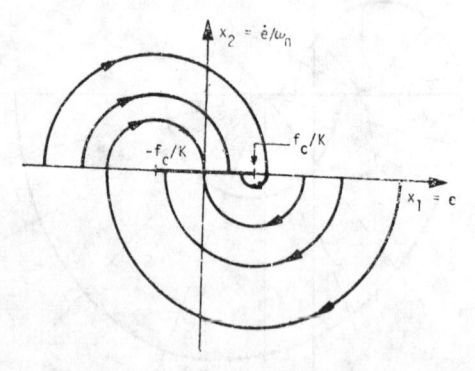

Figure 14.40 Phase-portrait of system with coulomb friction.

From Fig. 14.40, it is observed that though the system with coulomb friction is stable, there is a possibility of large steady-state error. Such large errors can be eliminated by the *dither* phenomenon explained below.

Consider the situation shown in Fig. 14.41 which corresponds to the case of large steady-state error with the final state of the system lying at A. Assume that a negative velocity, low amplitude disturbance signal AB is

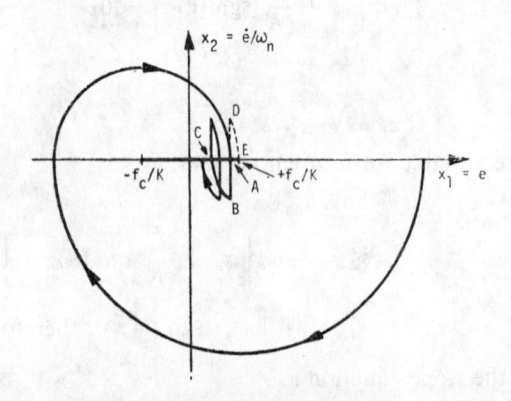

Figure 14.41 Dither phenomenon.

Nonlinear Systems 625

applied to the system. The phase-trajectory will be governed by the focal point at $e = + f_c/K$ and the system goes to the state C. On the other hand, if a disturbance signal AD of positive velocity is applied to the system, the motion is governed by the focal point at $e = - f_c/K$ and the system goes to the state E. Since AC is greater than AE, alternate positive and negative disturbances will drive the system closer towards the origin (as shown in Fig. 14.41) thereby eliminating large steady-state error. This phenomenon is commonly observed in indicating instruments where coulomb friction error can be reduced by knocking the instrument repeatedly.

14.8 The Describing Function Method: Basic Concepts

The phase-plane method discussed in the preceding sections uses a state model of the system which is obtained from the exact differential equations governing the system dynamics. This method, though very powerful, is generally restricted to systems described by second-order differential equations. For higher-order systems, therefore the only alternative is to approximate the system by a second-order one, in order to be able to use the phase-plane method for analysis and design.

Certain approximate techniques exist which are capable of determining the behaviour of a wider class of systems than is possible by the phase-plane method. These techniques are known as the *describing function techniques*. To discuss the basic concept underlying these techniques, let us consider the block diagram of a nonlinear system shown in Fig. 14.42 wherein the blocks $G_1(s)$ and $G_2(s)$ represent the linear elements, while the block N represents the nonlinear element.

Let us assume that input x to the nonlinearity is sinusoidal, i.e.,
$$x = X \sin \omega t$$
With such an input, the output of the nonlinear element will in general be nonsinusoidal periodic function which may be expressed in terms of Fourier series as follows:
$$y = A_0 + A_1 \sin \omega t + B_1 \cos \omega t + A_2 \sin 2\omega t + B_2 \cos 2\omega t + \ldots$$
It is important to note that in writing the above expression, is has been assumed that the nonlinearity N does not generate subharmonics. Furthermore, if the nonlinearity is assumed to be symmetrical, the average value of y is zero, so that the output y is then given by
$$y = A_1 \sin \omega t + B_1 \cos \omega t + A_2 \sin 2\omega t + B_2 \cos 2\omega t + \ldots$$

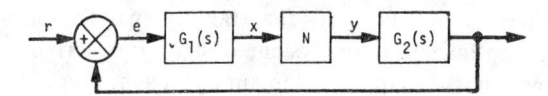

Figure 14.42 A nonlinear system.

In the absence of an external input (i.e., $r = 0$), the output y of N is fedback to its input through the linear elements $G_2(s)$ and $G_1(s)$ in tandem.

If $G_2(s)G_1(s)$ has low-pass characteristics* (this is usually the case in control systems), it can be assumed to a good degree of approximation that all the harmonics of y are filtered out in the process such that the input $x(t)$ to the nonlinear element N is mainly contributed by the fundamental component of y, i.e., $x(t)$ remains sinusoidal. Under such conditions the harmonic content of y can be thrown away for the purpose of analysis and the fundamental component of y, i.e.,

$$y_1 = A_1 \sin \omega t + B_1 \cos \omega t$$
$$= Y_1 \sin(\omega t + \phi_1)$$

need only be considered.

The above procedure heuristically linearizes the nonlinearity, since for a sinusoidal input, only a sinusoidal output of the same frequency is now assumed to be produced. This type of linearization is valid for large signals as well so long as the harmonic filtering condition is satisfied.

Under the above assumption, the nonlinearity can be replaced by a *describing function* $K_N(X, \omega)$ which is defined to be the complex function embodying amplification and phase shift of the fundamental frequency component of y relative to x, i.e.,

$$K_N(X, \omega) = (Y_1/X) \angle \phi_1$$

when the input to the nonlinearity is

$$x = X \sin \omega t$$

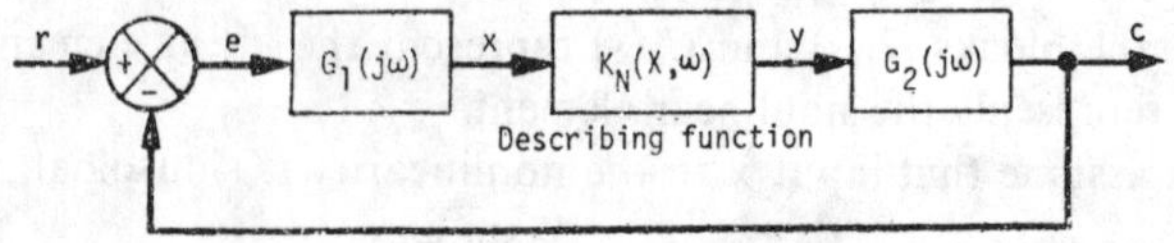

Figure 14.43 Nonlinear system with nonlinearity replaced by describing function.

As indicated above, the describing function is in general dependent upon the amplitude and frequency of the input.

From Fig. 14.43, it is observed that it is much easier to handle a nonlinear system when the nonlinearity is replaced by its describing function. All the linear theory frequency domain techniques now become applicable. It is important to remind ourselves here that this simplicity has been achieved at the cost of certain limitations, the foremost being the assumption that the linear elements in the feedback path around the nonlinearity have low-pass characteristics. Because of such limitations, the describing function method is mainly used for stability analysis and is not directly applied to the optimization of system design. Furthermore, though the describing function is a frequency domain approach, no general correlation is possible between time and frequency responses.

Before coming to the stability study by the describing function method, it is worthwhile to derive the describing functions of some common nonlinearities.

---

*Filtering characteristics of the linear part of a nonlinear system improve as the order of the system goes up.

14.9 Derivation of Describing Functions

As discussed in the previous section, the describing function of a nonlinear element is given by

$$K_N(X, \omega) = (Y_1/X) \angle \phi_1 \qquad (14.45)$$

where X = amplitude of the input sinusoid; Y_1 = amplitude of the fundamental harmonic component of the output; and ϕ_1 = phase shift of the fundamental harmonic component of the output with respect to the input.

Therefore for computing the describing function of a nonlinear element, we are simply required to find the fundamental harmonic component of its output for an input $x = X \sin \omega t$. The fundamental component of the output can be written as

$$y_1 = A_1 \sin \omega t + B_1 \cos \omega t$$

where the coefficients A_1 and B_1 of the Fourier series are

$$B_1 = \frac{1}{\pi} \int_0^{2\pi} y \cos \omega t \, d(\omega t) \qquad (14.46)$$

$$A_1 = \frac{1}{\pi} \int_0^{2\pi} y \sin \omega t \, d(\omega t) \qquad (14.47)$$

The amplitude and phase angle of the fundamental component of the output are given by

$$Y_1 = \sqrt{(A_1^2 + B_1^2)} \qquad (14.48)$$

$$\phi_1 = \tan^{-1}(B_1/A_1) \qquad (14.49)$$

Illustrative derivations of the describing functions of commonly encountered nonlinearities are given below.

Dead-zone and Saturation

Idealized characteristic of a nonlinearity having dead-zone and saturation and its response to sinusoidal input are shown in Fig 14.44. The output wave may be described as follows.

$$y = \begin{cases} 0 & ; 0 \leqslant \omega t \leqslant \alpha \\ K(x - D/2) & ; \alpha \leqslant \omega t \leqslant \beta \\ K(S - D/2) & ; \beta \leqslant \omega t \leqslant (\pi - \beta) \\ K(x - D/2) & ; (\pi - \beta) \leqslant \omega t \leqslant (\pi - \alpha) \\ 0 & ; (\pi - \alpha) \leqslant \omega t \leqslant \pi \end{cases}$$

where $\alpha = \sin^{-1} D/2X$ and $\beta = \sin^{-1} S/X$

Using eqns. (14.46) and (14.47) and recognizing that the output has half-wave and quarter-wave symmetries, we have

$$B_1 = 0$$

$$A_1 = \frac{4}{\pi} \int_0^{\pi/2} y \sin \omega t \, d(\omega t)$$

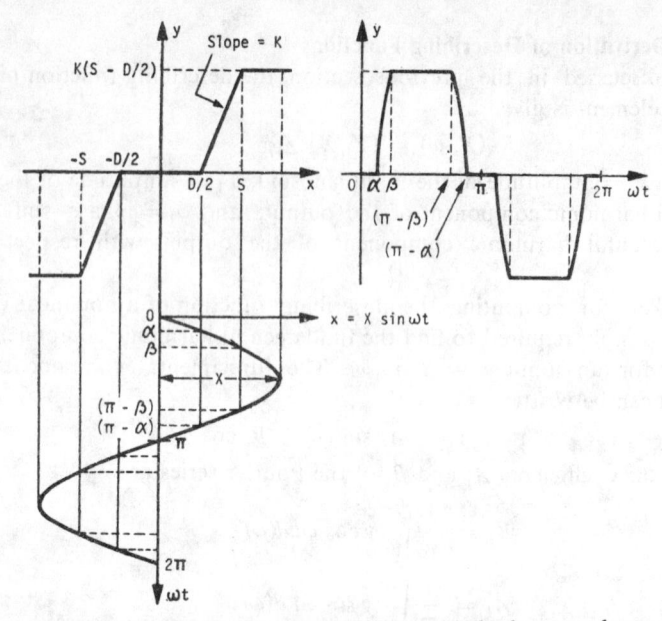

Figure 14.44 Sinusoidal response of nonlinearity with dead-zone and saturation.

$$= \frac{4}{\pi}\left[\int_{\alpha}^{\beta} K(X \sin \omega t - D/2) \sin \omega t \, d(\omega t)\right.$$
$$\left. + \int_{\beta}^{\pi/2} K(S - D/2) \sin \omega t \, d(\omega t)\right]$$
$$= \frac{K}{\pi}\left\{2X(\beta - \alpha) - X(\sin 2\beta - \sin 2\alpha)\right.$$
$$\left. + 4\left[\frac{D}{2}(\cos \beta - \cos \alpha) + \left(S - \frac{D}{2}\right)\cos \beta\right]\right\}$$
$$= \frac{KX}{\pi}[2(\beta - \alpha) + (\sin 2\beta - \sin 2\alpha)]$$

Therefore the describing function is given by

$$\frac{K_N(X)}{K} = \begin{cases} 0 & ; X < D/2; \alpha = \beta = \pi/2 \\ 1 - \frac{2}{\pi}(\alpha + \sin \alpha \cos \alpha) & ; D/2 < X < S; \beta = \pi/2 \\ \frac{1}{\pi}[2(\beta - \alpha) + (\sin 2\beta - \sin 2\alpha)] & ; X > S \end{cases} \quad (14.50)$$

Two special cases immediately follow from eqn. (14.50).

CASE I. *Saturation nonlinearity* ($D/2 = 0, \alpha = 0$)

$$\frac{K_N(X)}{K} = \begin{cases} 1 & ; X < S \\ \frac{2}{\pi}(\beta + \sin \beta \cos \beta) \\ = \frac{2}{\pi}\{\sin^{-1}(S/X) + (S/X)\sqrt{[1 - (S/X)^2]}\}; X > S \end{cases} \quad (14.51)$$

Nonlinear Systems

This function is sketched in Fig. 14.45.

CASE II. *Dead-zone nonlinearity* $(S \to \infty; \beta = \pi/2)$

$$\frac{K_N(X)}{K} = \begin{cases} 0 & ; X < D/2 \\ 1 - \dfrac{2}{\pi}(\alpha + \sin \alpha \cos \alpha) & \\ = 1 - \dfrac{2}{\pi}\{\sin^{-1}(D/2X) + (D/2X)\sqrt{[1-(D/2X)^2]}\} & \\ & ; X > D/2 \end{cases} \quad (14.52)$$

This function is sketched in Fig. 14.46.

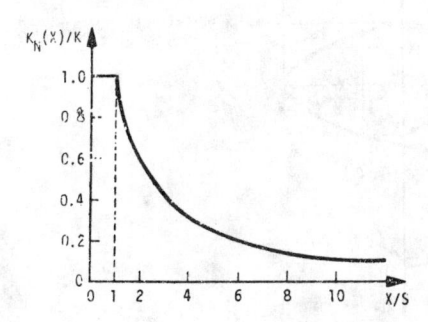

Figure 14.45 Describing function of saturation nonlinearity.

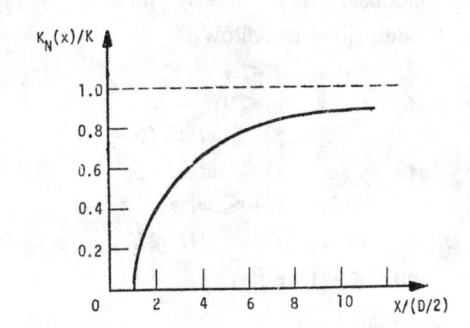

Figure 14.46 Describing function of dead-zone nonlinearity.

It is found that the describing functions of saturation and dead-zone non-linearities are frequency-invariant having zero phase shift. In fact all nonlinearities, whose input-output characteristics are represented by a planer graph, would result in describing functions independent of frequency but amplitude dependent. An element described by a nonlinear differential equation on the other hand has both frequency and amplitude dependent describing function. A frequency-invariant describing function having zero phase shift would be produced by a *memoryless* nonlinearity whose output is independent of the history of input as is the case with saturation and dead-zone nonlinearities.

Relay with Dead-zone and Hysteresis

The characteristics of a relay with dead-zone and hysteresis and its response to sinusoidal input are shown in Fig. 14.47a.

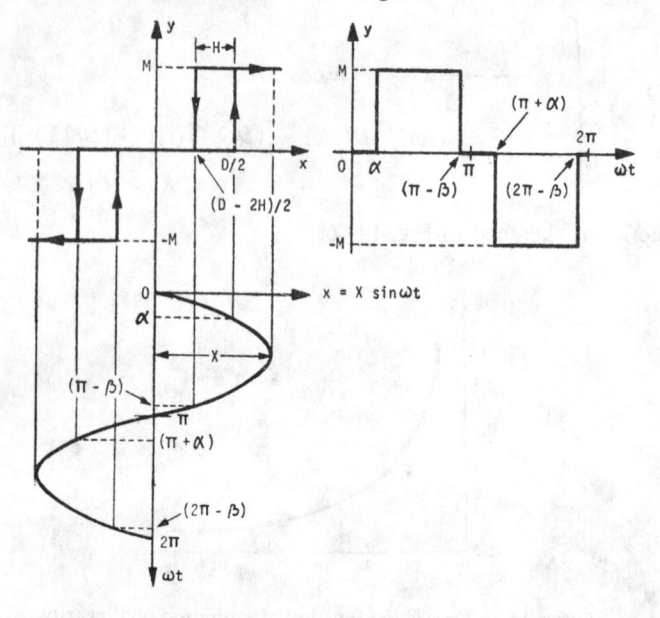

Figure 14.47 (a) Sinusoidal response of relay with dead-zone and hysteresis.

The output y may be described as follows:

$$y = \begin{cases} 0 & ; \; 0 \leqslant \omega t \leqslant \alpha \\ +M & ; \; \alpha \leqslant \omega t \leqslant (\pi - \beta) \\ 0 & ; \; (\pi - \beta) \leqslant \omega t \leqslant (\pi + \alpha) \\ -M & ; \; (\pi + \alpha) \leqslant \omega t \leqslant (2\pi - \beta) \\ 0 & ; \; (2\pi - \beta) \leqslant \omega t \leqslant 2\pi \end{cases}$$

where $\alpha = \sin^{-1} D/2X$; $\beta = \sin^{-1} (D - 2H)/2X$.

Using eqns. (14.46) and (14.47) we have

$$B_1 = \frac{2}{\pi} \int_0^\pi y \cos \omega t \, d(\omega t)$$

$$= \frac{2}{\pi} \int_\alpha^{\pi - \beta} M \cos \omega t \, d(\omega t) = \frac{2M}{\pi} (\sin \beta - \sin \alpha)$$

$$= \frac{2M}{\pi}\left(-\frac{H}{X}\right)$$

$$A_1 = \frac{2}{\pi} \int_0^\pi y \sin \omega t \, d(\omega t)$$

$$= \frac{2}{\pi} \int_\alpha^{\pi - \beta} M \sin \omega t \, d(\omega t) = \frac{2M}{\pi} (\cos \alpha + \cos \beta)$$

$$= \frac{2M}{\pi}\left\{\sqrt{\left[1 - \left(\frac{D}{2X}\right)^2\right]} + \sqrt{\left[1 - \left(\frac{D - 2H}{2X}\right)^2\right]}\right\}$$

Therefore

$$B_1/X = \frac{2M}{\pi X}\left(-\frac{H}{X}\right) \quad (14.53)$$

$$A_1/X = \frac{2M}{\pi X}\left\{\sqrt{\left[1 - \left(\frac{D}{2X}\right)^2\right]} + \sqrt{\left[1 - \left(\frac{D-2H}{2X}\right)^2\right]}\right\} \quad (14.54)$$

$$K_N(X) = \begin{cases} \sqrt{[(A_1/X)^2 + (B_1/X)^2]} \angle \tan^{-1} B_1/A_1; & X > D/2 \\ 0 & ; X < D/2 \end{cases} \quad (14.55)$$

It is seen that K_N is a functiox of X, the input amplitude and is independent of frequency. Further, it being a *memory type* nonlinearity (i.e., output is dependent upon the history of input), K_N has both magnitude and angle (a lagging angle which produces the effect of the pole of a linear system). $|(DK_N/M)|$ and $\angle K_N$ for various values of H/D are plotted against X/D in Fig. 14.47b. Equation (14.55) is the general expression for the describing function of a relay element, from which the describing functions of various simplified relay characteristics given in Fig. 14.48 are directly obtainable and are given below.

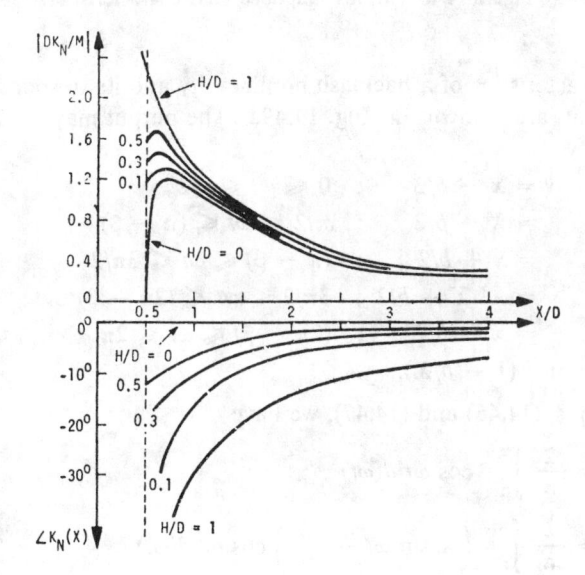

Figure 14.47 (b) Describing function of relay with dead-zone and hysteresis.

I. *Ideal relay* (Fig. 14.48a). Letting $D = H = 0$ in eqn. (14.55)

$$K_N(X) = 4M/\pi X \quad (14.56)$$

II. *Relay with dead-zone* (Fig. 14.48b). Letting $H = 0$ in eqn. (14.55)

$$\frac{D}{M} K_N(X) = \begin{cases} 0 & ; X < D/2 \\ \frac{4D}{\pi X}\sqrt{[1 - (D/2X)^2]} & ; X > D/2 \end{cases} \quad (14.57)$$

The plot of this function is the curve corresponding to $H/D = 0$ in Fig. 14.47b.

III. *Relay with hysteresis* (Fig. 14.48c). Letting $H = D$ in eqn. (14.55)

$$\frac{H}{M} K_N(X) = \begin{cases} 0 & ; X < H/2 \\ \frac{4H}{\pi X} \angle -\sin^{-1} H/2X & ; X > H/2 \end{cases} \quad (14.58)$$

The magnitude and phase characteristics are the curves corresponding to $H/D = 1$ in Fig. 14.47b.

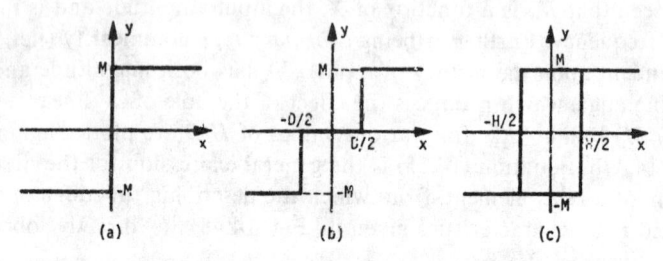

Figure 14.48 Various simplified relay characteristics.

BACKLASH

The characteristics of a backlash nonlinearity and its response to sinusoidal input are shown in Fig. 14.49a. The output may be described as follows:

$$\begin{aligned}
y &= x - b/2 & ; & \quad 0 \leqslant \omega t \leqslant \pi/2 \\
&= X - b/2 & ; & \quad \pi/2 \leqslant \omega t \leqslant (\pi - \beta) \\
&= x + b/2 & ; & \quad (\pi - \beta) \leqslant \omega t \leqslant 3\pi/2 \\
&= -X + b/2 & ; & \quad 3\pi/2 \leqslant \omega t \leqslant (2\pi - \beta) \\
&= x - b/2 & ; & \quad (2\pi - \beta) \leqslant \omega t \leqslant 2\pi
\end{aligned}$$

where $\beta = \sin^{-1}(1 - b/X)$.

Using eqns. (14.46) and (14.47), we have

$$B_1 = \frac{2}{\pi} \int_0^\pi y \cos \omega t \, d(\omega t)$$

$$= \frac{2}{\pi} \int_0^{\pi/2} \left(X \sin \omega t - \frac{b}{2} \right) \cos \omega t \, d(\omega t)$$

$$+ \frac{2}{\pi} \int_{\pi/2}^{\pi-\beta} \left(X - \frac{b}{2} \right) \cos \omega t \, d(\omega t)$$

$$+ \frac{2}{\pi} \int_{\pi-\beta}^\pi \left(X \sin \omega t + \frac{b}{2} \right) \cos \omega t \, d(\omega t)$$

$$= -\frac{X}{\pi} \cos^2 \beta$$

$$A_1 = \frac{2}{\pi} \int_0^\pi y \sin \omega t \, d(\omega t)$$

Nonlinear Systems

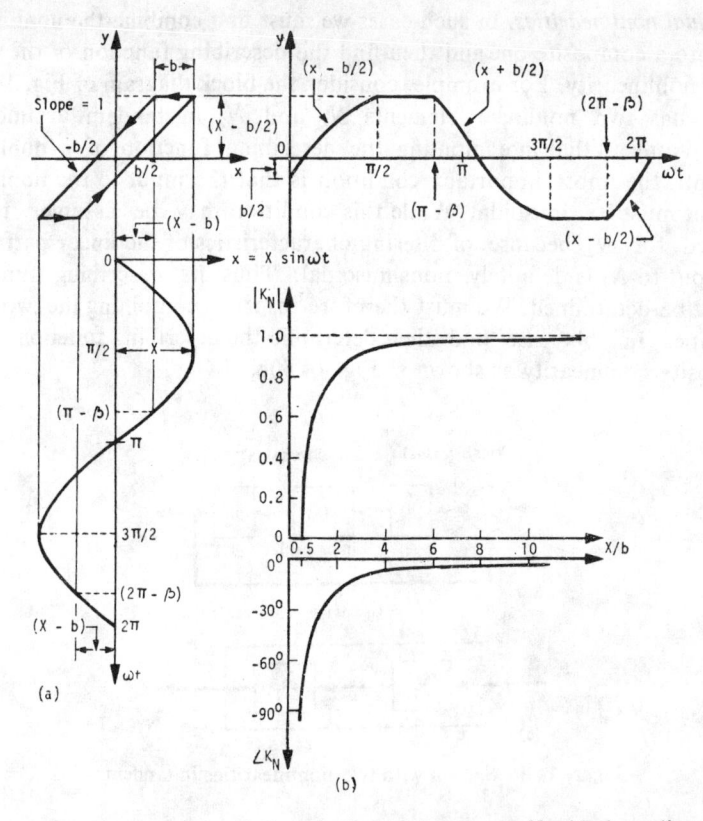

Figure 14.49. (a) Characteristic and sinusoidal response of backlash nonlinearity; (b) Describing function of backlash nonlinearity.

$$= \frac{2}{\pi}\int_0^{\pi/2}\left(X\sin\omega t - \frac{b}{2}\right)\sin\omega t\, d(\omega t)$$

$$+ \frac{2}{\pi}\int_{\pi/2}^{\pi-\beta}\left(X - \frac{b}{2}\right)\sin\omega t\, d(\omega t)$$

$$+ \frac{2}{\pi}\int_{\pi-\beta}^{\pi}\left(X\sin\omega t + \frac{b}{2}\right)\sin\omega t\, d(\omega t)$$

$$= \frac{X}{\pi}\left[\left(\frac{\pi}{2} + \beta\right) + \frac{1}{2}\sin 2\beta\right]$$

Therefore,

$$K_N(X) = \begin{cases} 0 & ; X < b/2 \\ \sqrt{[(A_1/X)^2 + (B_1/X)^2]}\, \angle \tan^{-1} B_1/A_1 & ; X > b/2 \end{cases} \quad (14.59)$$

The magnitude and phase characteristics of this describing function are shown in Fig. 14.49b.

When two nonlinearities are placed in tandem, the resultant describing function cannot be obtained by multiplying the describing functions of the individual nonlinearities. In such cases we must first combine the nonlinearities into a composite one and then find the describing function of the composite nonlinearity. For example, consider the block diagram of Fig. 14.50a which has two nonlinear elements N_1 and N_2 in tandem around the loop. We recall that in computing the describing function of a nonlinear element, the most important condition is that the input to the nonlinear element must be sinusoidal. While this condition may be assumed to be satisfied for N_2 because of filtering characteristics of the linear part $G(s)$, the input to N_1 is definitely nonsinusoidal. Thus its describing function cannot be determined. We must therefore resort to combining the two nonlinearities in this case and then determine the describing function of the composite nonlinearity as shown in Fig. 14.50b.

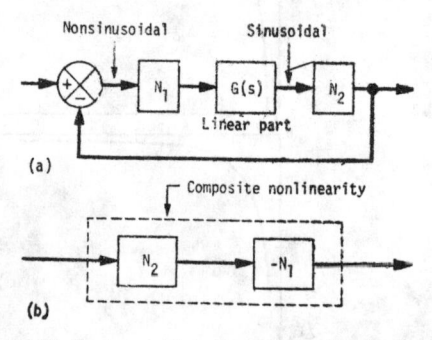

Figure 14.50 System with two nonlinearities in tandem.

In the above analysis, the input to nonlinearity is restricted to be sinusoidal. This condition may be relaxed for inputs consisting of a low frequency signal superimposed by high frequency (dither) signal. For such inputs, the describing functions are called the *dual-input describing functions*. This topic, however, is beyond the scope of this book [3].

14.10 Stability Analysis by Describing Function Method

As mentioned in Section 14.8, the widest use of describing functions is in stability investigations and prediction of limit cycles in feedback systems.

Consider the system shown in Fig. 14.43 wherein the nonlinearity has been replaced by its describing function. The system having thus been linearized, its characteristic equation can be written as

$$1 + G_1(j\omega)G_2(j\omega)K_N(X, \omega) = 0$$

Notice that the characteristic equation is independent of the physical location of the nonlinearity in the system. This has happened due to the assumption made in replacing the nonlinearity by its describing function. According to the Nyquist stability criterion, the system will exhibit sustained oscillations or limit cycle when

$$G_1(j\omega) \, G_2(j\omega) \, K_N(X, \omega) = -1 \qquad (14.60)$$

This condition implies that the plot of $G_1(j\omega)G_2(j\omega)K_N(X, \omega)$ passes through the critical point -1. The describing function technique has thus helped us to extend the Nyquist criterion to nonlinear systems. This extension of the Nyquist criterion is completely heuristic and has no mathematical basis. However, it is a powerful tool because it works.

The condition of eqn. (14.60) can be written in the modified form

$$G_1(j\omega)\,G_2(j\omega) = -1/K_N(X, \omega) \qquad (14.61)$$

This modified condition differs from the condition (14.60) in the fact that now we plot the frequency response of the linear part $G_1(j\omega)G_2(j\omega)$, while the critical point -1 becomes the critical locus $-1/K_N(X, \omega)$. The intersections of the plot $G_1(j\omega)G_2(j\omega)$ with the critical locus determine the amplitude and frequency of the limit cycles.

Consider a simple frequency-invariant describing function wherein the locus of $-1/K_N(X)$ is a single curve in the complex plane as shown in Fig. 14.51. Let the $G(j\omega)$-plot be superimposed on $-1/K_N(X)$-locus. It is assumed here that the system under investigation is open-loop stable. The closed-loop system stability is investigated below.

The values of X for which the $-1/K_N(X)$-locus is enclosed by the $G(j\omega)$-plot [i.e., the $-1/K_N(X)$-locus lies in the region to the right of an observer traversing the $G(j\omega)$-plot for positive frequencies in the direction of increasing ω] correspond to unstable conditions. Similarly, the values of X for which the $-1/K_N(X)$-locus is not enclosed by the $G(j\omega)$-plot [i.e., the $-1/K_N(X)$-locus lies in the region to the left of an observer traversing the $G(j\omega)$-plot for positive frequencies in the direction of increasing ω] correspond to stable conditions.

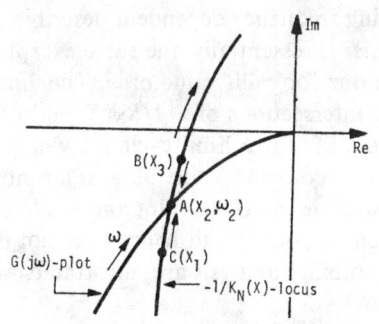

Figure 14.51 Prediction and stability of limit cycle.

The $-1/K_N(X)$-locus and $G(j\omega)$-plot intersect at the point ($\omega = \omega_2$, $X = X_2$) which corresponds to the condition of limit cycle (self-sustained oscillation). The system is unstable for $X < X_2$ and is stable for $X > X_2$. The stability of the limit cycle can be judged by the perturbation technique described below.

Suppose the system is originally operating at A under the state of a limit cycle. Assume that a slight perturbation is given to the system so that the input to the nonlinear element increases to X_3, i.e., the operating point is shifted to B. Since B is in the range of stable operation, the amplitude of input to the nonlinear element progressively decreases and hence the operating point moves back towards A. Similarly, a perturbation which decreases the amplitude of input to the nonlinearity shifts the operating point to C which lies in the range of unstable operation. The input amplitude now progressively increases and the operating point again returns to A. Therefore the system has a stable limit cycle at A. Figure 14.52a shows the case of an unstable limit cycle. For systems having $G(j\omega)$-plots and $-1/K_N(X)$-loci as shown in Figs. 14.52b and c, there are two limit cycles, one stable and the other unstable.

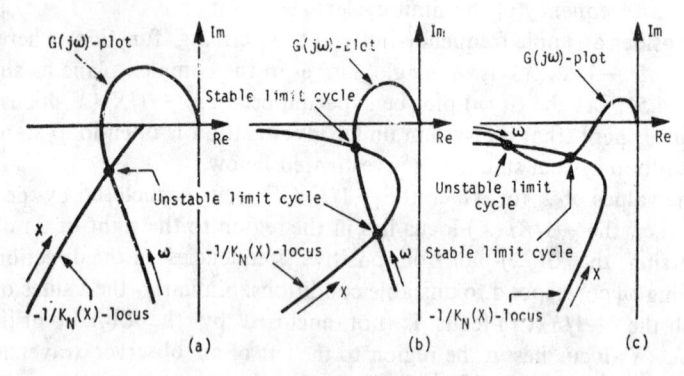

Figure 14.52 Prediction and stability of limit cycles.

For systems having frequency dependent describing functions, the method of stability analysis is essentially the same except that $-1/K_N(X, \omega)$ has infinitely many loci one for each value of ω. The limit cycles, if any, are determined by the intersections of $-1/K_N(X, \omega)$-loci and $G(j\omega)$-plot. Only such intersections qualify for a limit cycle for which the $-1/K_N(X, \omega)$-loci and $G(j\omega)$-plot have a common value of ω. If an intersection point belongs to different values of ω for $G(j\omega)$-plot and $-1/K_N(X, \omega)$-loci, it does not contribute a solution to (14.61) and hence does not determine a limit cycle.

The concepts of stability analysis are illustrated with the help of two examples given below.

Example 14.6: Let us investigate the stability of a relay-controlled system shown in Fig. 14.53.

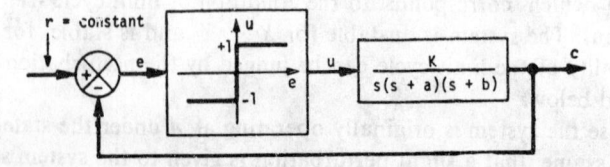

Figure 14.53 A relay-controlled system.

Using the describing function of an ideal relay given by eqn. (14.56), we have

$$-\frac{1}{K_N(E)} = -\frac{\pi}{4} E$$

where E is the maximum amplitude of the sinusoidal signal e. Figure 14.54 shows the locus of $-1/K_N(E)$ as a function of E and the plots of $G(j\omega)$ for two different values of K.

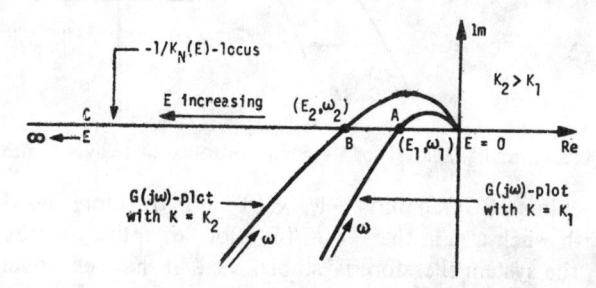

Figure 14.54 Stability analysis of a system controlled by ideal relay.

Consider the $G(j\omega)$-plot for $K = K_1$ which intersects the $-1/K_N(E)$-locus at A resulting in a limit cycle of amplitude E_1 and frequency ω_1. As an observer traverses the $G(j\omega)$-plot in the direction of increasing ω, the portion OA of $-1/K_N(E)$-locus lies to its right and the portion AC lies to its left. Using the arguments presented previously, we can conclude that the limit cycle is a stable one. As the gain is increased to $K_2 > K_1$, the intersection point shifts to B resulting in a limit cycle of amplitude $E_2 > E_1$ and frequency $\omega_2 = \omega_1$. It should be observed that the system has a limit cycle for all positive values of gain.

Since for the ideal relay, $-1/K_N(E)$-locus is the negative real axis, the frequency of limit cycle can also be determined from the condition

$$\angle G(j\omega_l) = -180°$$

where ω_l is the limit cycle frequency. Knowing ω_l, the amplitude of the limit cycle can be found from

$$\frac{1}{K_N(E)} = \frac{\pi}{4} E = |G(j\omega_l)|$$

Let us now investigate system stability when the relay controller shown in Fig. 14.53 has dead-zone. The describing function of this type of relay is given in eqn (14.57) which can be written as

$$-\frac{1}{K_N(E)} = -\frac{\pi D}{4}\left\{\frac{1}{\dfrac{D}{E}\sqrt{\left[1-\left(\dfrac{D}{2E}\right)^2\right]}}\right\}$$

A plot of this function is shown in Fig. 14.55 from which we observe that when the gain K is small enough, the $G(j\omega)$-plot crosses the negative real axis at a point to the right of $-\pi D/4$ such that no intersection takes place

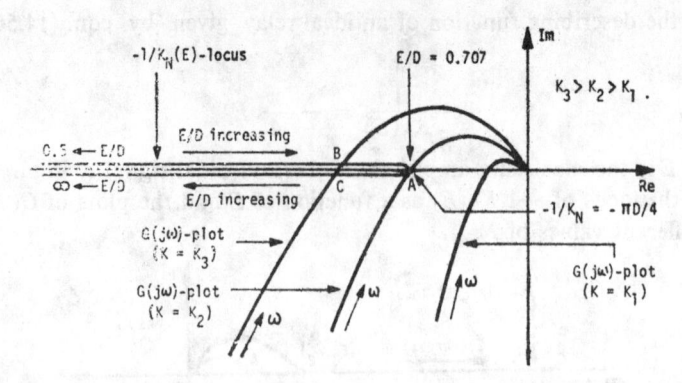

Figure 14.55 Stability analysis of a system controlled by relay with dead-zone.

between the plots of $G(j\omega)$ and $-1/K_N(E)$ and therefore no limit cycle results. With such a gain the $-1/K_N(E)$-plot lies entirely to the left of the $G(j\omega)$-plot, the system therefore is stable, i.e., it has effectively positive damping.

If the gain K is now increased to a value K_2 such that the $G(j\omega)$-plot intersects the $-1/K_N(E)$-plot at the point A (i.e., on the negative real axis at $-\pi D/4$), then there exists a limit cycle of amplitude $-\pi D/4$ and frequency ω_A. Now assume that the system is operating at the point A. Any increase in the amplitude of E takes the operating point to the left so that it is not enclosed by the $G(j\omega)$-plot which means the system has positive damping. This reduces E till the operating point comes back to A. Any decrease in the amplitude of E again takes the operating point to the left of the $G(j\omega)$-plot, i.e., the system has positive damping which further reduces E finally driving the system to rest. Since random disturbances are always present in any system, the system under discussion cannot remain at A. Therefore, the limit cycle represented by A is unstable.

When the gain K is further increased to K_3, the $G(j\omega)$- and $-1/K_N(E)$-plots intersect at two points B and C. By arguments similar to those advanced in the case of ideal relay, it can be shown that the point B represents an unstable limit cycle and C represents a stable limit cycle. It may be noted that though the points B and C lie at the same place on negative real axis, they belong to different values of E/D.

Let us now assume that the relay-controller shown in Fig. 14.53 has hysteresis. The describing function for this type of relay given in eqn. (14.58) can be written as

$$-\frac{1}{K_N(E)} = -\frac{\pi}{4} H\left(\frac{E}{H}\right) \angle \sin^{-1} H/2E;\ E/H > 1/2$$

$$= \frac{\pi H}{4}\left(\frac{E}{H}\right) \angle [-180° + \sin^{-1} H/2E]$$

This function is plotted in Fig. 14.56, and is found to be a straight line parallel to the real axis. The point of intersection of the $G(j\omega)$-plot with

Nonlinear Systems

$-1/K_N(E)$ plot determines the amplitude and frequency of the limit cycle which is found to be stable.

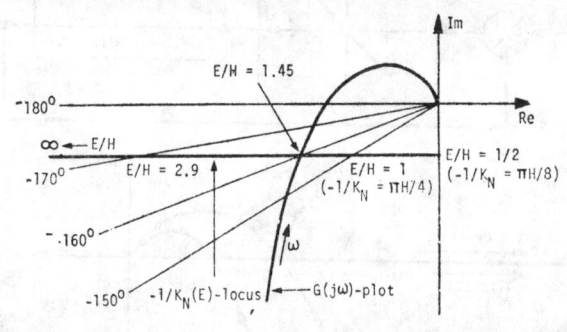

Figure 14.56 Stability analysis of a system controlled by relay with hyteresis.

Lastly, consider the case when the relay-controller shown in Fig. 14.53 has both dead-zone and hysteresis. The describing function for this case is given by eqn. (14.55). The plots of $-1/K_N(E)$ for various values of H are shown in Fig. 14.57. It is seen that for a given system gain, as H is decreased, the plot of $-1/K_N(E)$ is shifted towards left and closer to the $-180°$ axis. Also it is observed that depending upon the values of H and system gain, there may exist no limit cycle, one limit cycle or two limit cycles.

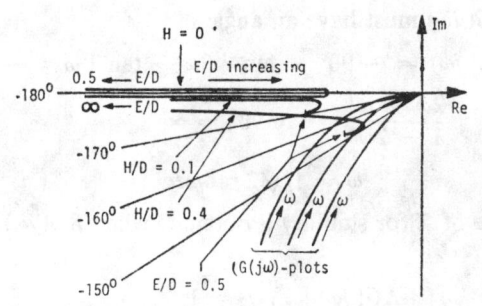

Figure 14.57 Stability analysis of a system controlled by relay with dead-zone and hysteresis.

Example 14.7: Consider a third-order system with a saturating amplifier of Fig. 14.58a having gain K in its linear region. Determine the largest value of gain K for the system to stay stable. What would be the frequency, amplitude and nature of the limit cycle for a gain of $K = 3$?

Solution:
It is convenient to regard the amplifier to have unit gain and the gain K to be attached to the linear part. From eqn. (14.51) for $S = 1$

$$-1/K_N(X) = -1 \qquad ; X < 1$$

$$= \frac{-\pi}{2} \Big/ \{\sin^{-1}(1/X) + (1/X)\sqrt{[1-(1/X)^2]}\}; X > 1$$

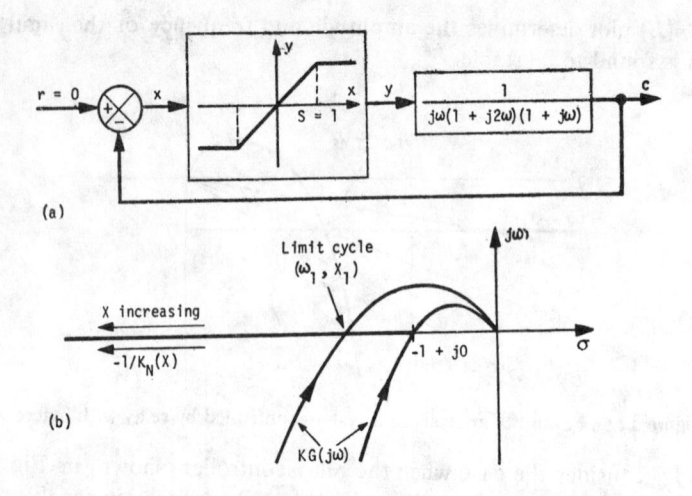

Figure 14.58 Nonlinear system with saturating amplifier.

The plot of $-1/K_N(X)$ thus starts from $(-1+j0)$ and travels along the negative real axis for increasing X as shown in Fig. 14.58b. Now for the equation

$$KG(j\omega) = -1/K_N(X)$$

to be satisfied, $G(j\omega)$ must have an angle of

$$\angle G(j\omega) = -90° - \tan^{-1} 2\omega - \tan^{-1} \omega = -180°$$

or
$$\frac{2\omega + \omega}{1 - 2\omega^2} = \tan 90° = \infty$$

or
$$\omega = 1/\sqrt{2} \text{ rad/sec}$$

The largest value of K for stability is obtained when $KG(j\omega)$ passes through $(-1+j0)$, i.e.,

$$|KG(j\omega)|_{\omega=1/\sqrt{2}} = 1$$

or
$$\frac{K}{(1/\sqrt{2})(\sqrt{3})(\sqrt{3}/\sqrt{2})} = 1$$

i.e., $K = 3/2$

Now for $K = 3$, $KG(j\omega)$ plot intersects $-1/K_N(X)$-plot resulting in a limit cycle at (ω_1, X_1) where $\omega_1 = 1/\sqrt{2}$ while X_1 is obtained from

$$\frac{\pi/2}{\sin^{-1} 1/X_1 + 1/X_1\sqrt{[1-(1/X_1)^2]}} = 3\left(\frac{2}{3}\right) = 2$$

Reading from Fig. 14.45 corresponding to $K_N(X_1) = 1/2$, we get $X_1 \approx 6.5$.

STABILITY ANALYSIS BY GAIN-PHASE PLOTS

The stability analysis using describing functions can be carried out by db gain-phase plots also as illustrated by the following example.

Example 14.8: Consider a second-order system with backlash whose block diagram is shown in Fig. 14.59.

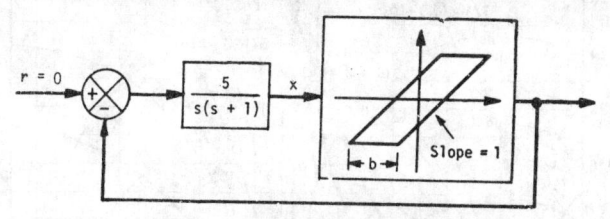

Figure 14.59 Nonlinear system with backlash.

The describing function of backlash nonlinearity is given by eqn. (14.59), from which the following table is prepared.

Table 14.1

b/X	$\lvert K_N(X) \rvert$	$\angle K_N(X)$	20 log [1/ $\lvert K_N(X) \rvert$]
0	1	0	0
0.2	0.954	$-6.7°$	0.4
0.4	0.882	$-13.4°$	1.1
0.5	0.838	$-16.5°$	1.52
0.6	0.794	$-19.7°$	2.0
0.8	0.698	$-26.0°$	3.12
1.0	0.592	$-32.5°$	4.56
1.2	0.482	$-39.5°$	6.36
1.4	0.367	$-46.6°$	8.7
1.6	0.248	$-55.2°$	12.12
1.7	0.186	$-60.4°$	14.62
1.8	0.125	$-66.0°$	18.08
1.9	0.064	$-69.8°$	23.82
2.0	0	$-90.0°$	∞

The plot of the describing function on db gain-phase plane is shown in Fig. 14.60.

The transfer function of the linear part of the system

$$G(j\omega) = \frac{5}{j\omega(j\omega + 1)}$$

is also plotted in Fig. 14.60. It is observed from this figure that the two plots intersect at points A and B. Application of the limit cycle stability test* discussed earlier in this section reveals that the point A corresponds to a stable limit cycle and the point B corresponds to an unstable limit cycle. The stable limit cycle is predicted at $\omega = 1.5$ and $b/X = 1$, while the unstable limit cycle is at $\omega = 0.4$ and $b/X = 1.85$.

*Since the plots in db gain-phase plane are conformal maps of the corresponding plots in the complex plane, the sense of directions in the complex plane is preserved in the db gain-phase plane and hence the same limit cycle stability test applies [i.e., perturbation and enclosure of $-1/K_N(X)$-locus by $G(j\omega)$-plot].

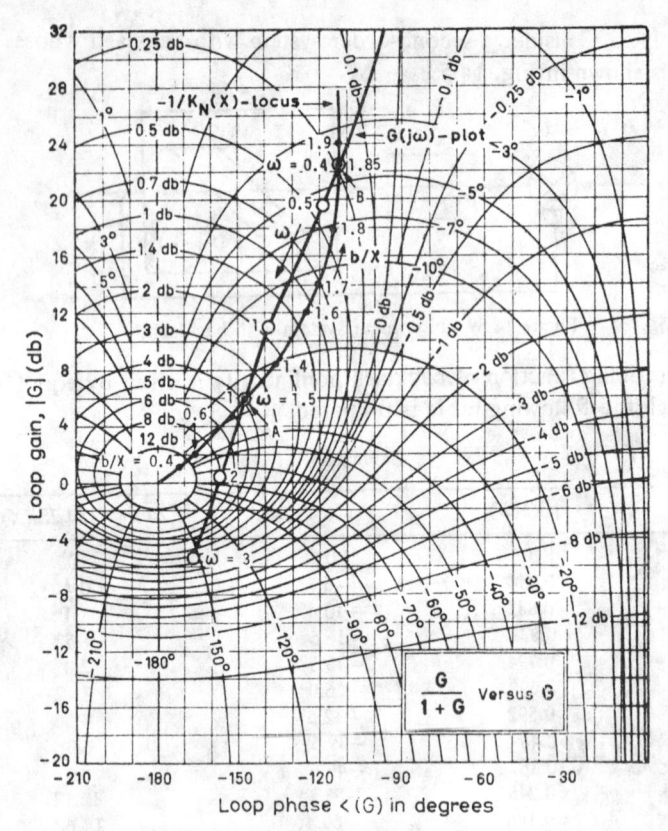

Figure 14.60 db gain-phase plots of $-1/K_N(X)$ and $G(j\omega)$ of the system shown in Fig. 14.59.

14.11 Jump Resonance

In the preceding sections we have seen how a nonlinearity can be heuristically linearized using harmonic analysis. While no simple way exists for estimating the order of error involved in this kind of linearization, it is understood that the concept of describing function is valid whenever strong linear filtering is present in the loop preceding the nonlinearity. It was this fact which helped us to extend the Nyquist stability criterion, which is a technique of linear systems, to the class of nonlinear systems in which the assumption of linear filtering holds. The use of describing function can also be extended to obtain the complete frequency response of a nonlinear system for sinusoidal inputs. Of course in such a response the error will be more in the low frequency region than in the high frequency region. Furthermore, due to heuristic linearization no formal correlation exists between the frequency domain and time domain responses. Certain empirical results do exist in the literature but these are restricted to very special situations. In this section we will illustrate an **analytico-graphical** method of obtaining the frequency response of a nonlinear system and at the same

Nonlinear Systems

time illustrate that under certain conditions such a frequency response exhibits the phenomenon of jump resonance (refer Section 14.1, Fig. 14.3).

Consider a nonlinear system shown in Fig. 14.61 wherein the nonlinear part has been replaced by its describing function.

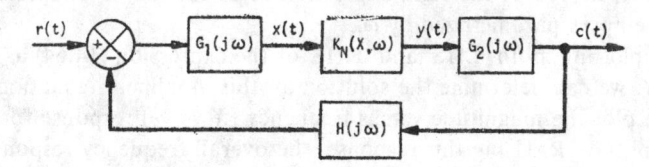

Figure 14.61 Nonlinear system with nonlinearity replaced by its describing function.

Let
$$r(t) = R \sin \omega t$$
$$x(t) = X \sin (\omega t + \phi)$$

It immediately follows that

$$\frac{C(j\omega)}{R(j\omega)} = \frac{G_1(j\omega)K_N(X, \omega)G_2(j\omega)}{1 + G_1(j\omega)K_N(X, \omega)G_2(j\omega)H(j\omega)} \qquad (14.62)$$

We can also write

$$\frac{X(j\omega)}{R(j\omega)} = \frac{G_1(j\omega)}{1 + G_1(j\omega)K_N(X, \omega)G_2(j\omega)H(j\omega)} \qquad (14.63)$$

Further we can write

$$G_1(j\omega) = g_1(\omega) \exp \{j\theta_1(\omega)\}$$
$$G_1(j\omega)G_2(j\omega)H(j\omega) = g_2(\omega) \exp \{j\theta_2(\omega)\}$$
$$K_N(X, \omega) = g_N(X, \omega) \exp \{j\theta_N(X, \omega)\}$$

In order to simplify the analysis we assume

(i) $\theta_N(X, \omega) = 0$
(ii) $g_N(X, \omega) = g_N(X)$, i.e., independent of frequency.

So we can write

$$\frac{X(j\omega)}{R(j\omega)} = \frac{g_1(\omega) \exp \{j\theta_1(\omega)\}}{1 + g_2(\omega)g_N(X) \exp \{j\theta_2(\omega)\}}$$

In terms of magnitudes, we can write the above expression as

$$\frac{X}{R} = \frac{g_1}{\sqrt{1 + 2g_Ng_2 \cos \theta_2 + g_N^2 g_2^2}}$$

Solving for g_N, we get

$$g_N = \frac{-\cos \theta_2 \pm \sqrt{\cos^2 \theta_2 - 1 + (R^2 g_1^2/X^2)}}{g_2}$$

or

$$Xg_N(X) = -\frac{X \cos \theta_2(\omega)}{g_2(\omega)} \pm \frac{1}{g_2(\omega)} \sqrt{R^2 g_1^2(\omega) - X^2 \sin^2 \theta_2(\omega)} \qquad (14.64)$$

The nice thing about the above expression is that its right hand side (RHS) depends only on the linear part of the control system and the amplitude X (i.e., the input to the nonlinear part). Similarly the left hand side (LHS) depends only on the nonlinear part of the control system and the amplitude X. Moreover the RHS expression as a function of X represents a family of ellipses parametrized by ω.

So by plotting both LHS and RHS of the expression (14.64) as functions of X, we can determine the solution of this nonlinear equation and therefrom plot the magnitude versus frequency (X vs. ω) response for fixed input amplitude R. Using this response, the overall frequency response of the nonlinear system for a particular input amplitude is obtained from eqn. (14.62).

Example 14.9: Obtain the frequency response of the system with saturation nonlinearity shown in Fig. 14.62. Given

$$K = 150, S = 2, R = 1.5$$

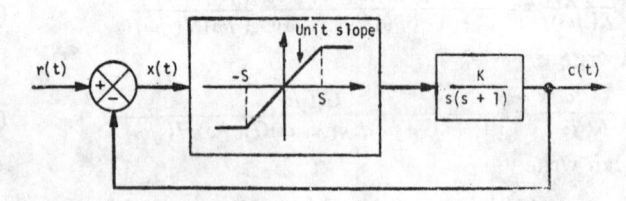

Figure 14.62 Control system with saturation nonlinearity.

Solution

For this particular system we have with reference to Fig. 14.61,

$$g_1 = 1, \theta_1 = 0$$

$$g_2(\omega) = \frac{150}{\omega\sqrt{1+\omega^2}}, \theta_2(\omega) = -90° - \tan^{-1}\omega.$$

In expression (14.64),

$$\text{RHS} = -\frac{X\omega\sqrt{1+\omega^2}\cos(-90° - \tan^{-1}\omega)}{150}$$

$$\pm \frac{\omega\sqrt{1+\omega^2}}{150}\sqrt{2.25 - X^2\sin^2(-90° - \tan^{-1}\omega)}$$

$$\text{LHS} = X\frac{2}{\pi}\left\{\sin^{-1}(2/X) + (2/X)\sqrt{1-(2/X)^2}\right\}; X > S$$

$$= X \qquad\qquad\qquad\qquad\qquad\qquad ; X < S$$

The plots of RHS vs. X with ω as parameter and the plot of LHS vs. X are drawn in Fig. 14.63a. The intersection of these plots yields the solution of X for each ω (for sake of clarity the plots are not drawn to scale). We

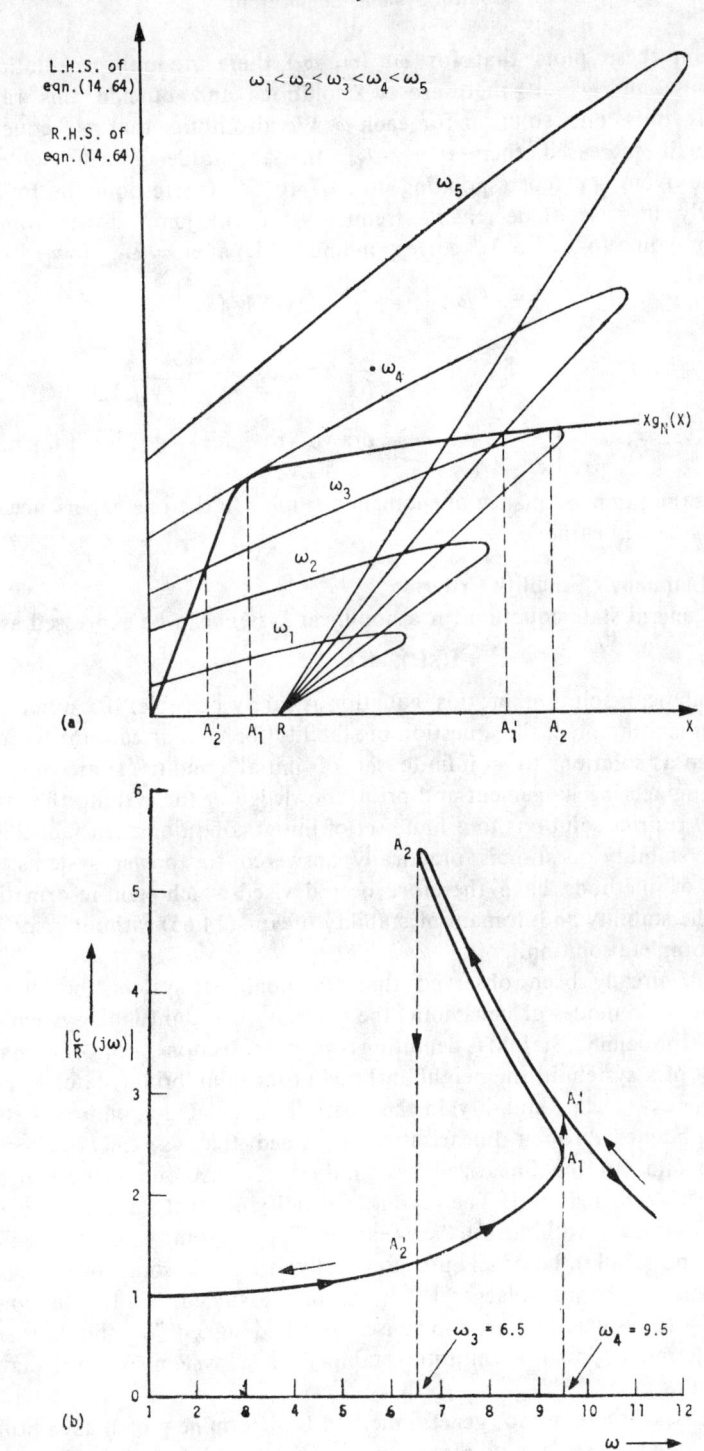

Figure 14.63 Jump resonance

see from these plots that for $\omega \epsilon$ (ω_3, ω_4), there are multiple solutions; for $\omega = \omega_3$ and $\omega = \omega_4$ there are two solutions and outside this range there is only one solution for each ω. We also notice that as frequency is gradually increased, then at $\omega = \omega_4$, the amplitude X will suddenly increase from X_1 (corresponding to A_1) to X_1' (corresponding to A_1'). Similarly in case of decreasing frequency, X will jump down from X_2 (corresponding to A_2) to X_2' (corresponding to A_2') at $\omega = \omega_3$. Now

$$\left| \frac{C}{R}(j\omega) \right| = \frac{1}{R}\left[Xg_N(X)g_2(\omega) \right]$$

$$= \frac{1}{R}\left[Xg_N(X) \frac{150}{\omega\sqrt{1+\omega^2}} \right]$$

The plot $\left| \dfrac{C}{R}(j\omega) \right|_{R=1.5}$ vs. ω is drawn (to scale) in Fig. 14.63b. It exhibits the jump resonance phenomenon similar to the one experienced in X (as explained earlier).

14.12 Liapunov's Stability Criterion

The general state equation for a nonlinear system can be expressed as

$$\dot{\mathbf{x}} = \mathbf{f}(\mathbf{x}(t), \mathbf{u}(t), t) \tag{14.65}$$

The analytical solution of this equation is rarely possible. If a numerical solution is attempted, the question of stability behaviour can not be fully answered as solutions to an infinite set of initial conditions are needed. Using engineering judgement and prior knowledge of the system, the analyst may restrict solutions to a finite set of initial conditions. (In fact this is how the stability question is practically answered for power systems.) A variety of methods have therefore been devised which yield information about the stability and domain of stability of eqn. (14.65) without resorting to its complete solution.

It has already been observed that for nonlinear systems, because of many possible modes of behaviour, the concept of 'stability of system' is difficult to define. Stability definitions given in Section 14.5, concern the stability of a system in the neighbourhood of the equilibrium state. For an autonomous system, stability in-the-small of an isolated equilibrium state can be obtained through linearization provided that (a) linearization is possible and (b) the linearized system does not have one or more eigenvalues with zero real part. The region of validity of local stability is generally not known without further analysis. Thus in some cases, the region may be too small to be of any use practically; while in some other cases the region may be much larger than the one assumed by the designer, giving rise to systems that are too conservatively designed. We should therefore determine global asymptotic stability of a system at an isolated equilibrium point.

At present there is no **general method** to determine global asymptotic stability. We may study a nonlinear system to establish whether a limit

Nonlinear Systems

cycle exists or not because the most common mode of instability in practical systems is self-oscillation. Exact methods for establishing the existence of limit cycles are rare and where they exist, the methods are difficult to apply. Quite frequently the approximate methods like the describing function method are used.

There is always a doubt about the validity of approximate methods. Therefore, exact approaches for stability investigation wherever possible must be used. *Liapunov's direct method* and *Popov's method* are two such approaches which provide us with qualitative aspects of system behaviour. Exact conditions on the optimality of a control can be obtained using dynamic programming and calculus of variations.* These conditions in many cases provide us with quantitative aspects of the system such as optimal control function.

In this section we shall discuss Liapunov's direct method of stability analysis and Popov's method will be presented in Section 14.13.

Basic Stability Theorems

The direct method of Liapunov is based on the concept of energy and the relation of stored energy with system stability. Consider an autonomous physical system described as

$$\dot{\mathbf{x}}(t) = \mathbf{f}(\mathbf{x}(t))$$

and let $\mathbf{x}(\mathbf{x}(t_0), t)$ be a solution. Further, let $V(\mathbf{x})$ be the total energy associated with the system. If the derivative $dV(\mathbf{x})/dt$ is negative for all $\mathbf{x}(\mathbf{x}(t_0), t)$ except the equilibrium point, then it follows that energy of the system decreases as t increases and finally the system will reach the equilibrium point. This holds because energy is non-negative function of system state which reaches a minimum only if the system motion stops. These ideas are well illustrated by the following example.

The spring-mass-damper system shown in Fig. 14.64 is governed by the equation

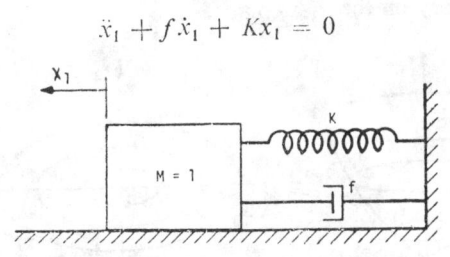

$$\ddot{x}_1 + f\dot{x}_1 + Kx_1 = 0$$

Figure 14.64 A spring-mass-damper system.

A state model of the system is

$$\dot{x}_1 = x_2 \qquad (14.66)$$
$$\dot{x}_2 = -Kx_1 - fx_2$$

---

*Chapter 13 provides an introduction to optimal control theory. Only linear regulator problem has been discussed there. References listed at the end of this chapter give detalied account of optimal control theory.

At any instant, the total energy V in the system consists of the kinetic energy of the moving mass and the potential energy stored in the spring

$$V(x_1, x_2) = \frac{1}{2} x_2^2 + \frac{1}{2} K x_1^2$$

Thus

$$V(\mathbf{x}) > 0 \quad \text{when } \mathbf{x} \neq \mathbf{0}$$
$$V(\mathbf{0}) = 0$$

This means that total energy is positive unless the system is at rest at the equilibrium state $\mathbf{x}_e = \mathbf{0}$, where the energy is zero.

The rate of change of energy is given by

$$\frac{d}{dt} V(x_1, x_2) = \frac{\partial V}{\partial x_1} \frac{dx_1}{dt} + \frac{\partial V}{\partial x_2} \frac{dx_2}{dt}$$
$$= -x_2^2 f$$

Thus dV/dt is negative at all points except where x_2 is zero at which dV/dt is zero. Therefore under positive damping, the system energy can not increase. From eqns. (14.66), we observe that $\dot{x}_2 = -Kx_1$ at the points where $x_2 = 0$; thus the system can not remain in nonequilibrium state for which $x_2 = 0$. Therefore the energy can not remain constant except at the equilibrium point where it will be zero.

A **visual analogy** [4] may be obtained by considering the surface

$$V = \frac{1}{2} x_2^2 + \frac{1}{2} K x_1^2$$

This is a cup-shaped surface as shown in Fig. 14.65a. The constant-V loci are ellipses on the surface of the cup. Let (x_{10}, x_{20}) be the initial condition. If one plots trajectory on the surface shown, the representative point $\mathbf{x}(t)$ crosses the constant-V curves and moves towards the lowest point of the cup which is the equilibrium point. Figure 14.65b shows the projection of a typical trajectory on the x_1-x_2 plane.

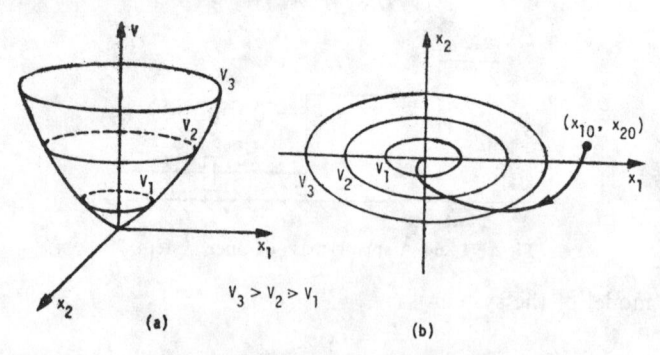

Figure 14.65 Constant-V curves.

In the example given above, it was easy to associate the energy function V with the given system. However, in general, there is no obvious way of associating an energy function with a given set of equations describing a system. In fact there is nothing sacred or unique about the total energy of the system which allows us to determine system stability in the way described above. Other non-negative scalar functions of system state can also answer the question of stability. This idea was introduced and formalized by the Russian mathematician A.M. Liapunov; the scalar function is now known as the *Liapunov function* and the method of investigating stability using Liapunov's function is known as the *Liapunov's direct method*.

The Liapunov's method in its simplest form is given by the following theorem.

Theorem 1: Consider the system

$$\dot{x} = f(x); \quad f(0) = 0$$

Suppose there exists a scalar function $V(x)$ which, for some real number $\epsilon > 0$, satisfies the following properties for all x in the region $\| x \| \leqslant \epsilon$:

(a) $V(x) > 0$; $x \neq 0$
(b) $V(0) = 0$

i.e., $V(x)$ is positive definite scalar function (see Appendix II)

(c) $V(x)$ has continuous partial derivatives with respect to all components of x.

(d) $\dfrac{dV}{dt} \leqslant 0$ (i.e., dV/dt is negative semidefinite scalar function)

Then the system is stable at the origin.

Proof

The vector x satisfying the condition $\| x \| = \epsilon$ is a closed set of vectors. By property (c), $V(x)$ has continuous partial derivatives with respect to all components of x. This implies that $V(x)$ is continuous for $\| x \| \leqslant \epsilon$. Therefore $V(x)$ is continuous on the closed set of vectors given by $\| x \| = \epsilon$; it must therefore assume minimum and maximum values on this set. Let m be the minimum value of $V(x)$ on this set as indicated by contour $C_1(V = m)$ in Fig. 14.66 for the two-dimensional case. According to property (a) above, $m > 0$.

$V(x)$ is continuous at $x = 0$. Therefore as per the mathematical properties of continuous functions, for every real number $r > 0$, there exists a real number $\delta > 0$ such that $| V(x) - V(0) | \leqslant r$ for all x satisfying $\| x - 0 \| \leqslant \delta$. In particular, let us take $r = m/2$. This means there exists a $\delta > 0$ such that $V(x) \leqslant m/2$ for all $\| x \| \leqslant \delta$. This is shown by contour $C_2(V = m/2)$ and δ-circle in Fig. 14.66. Obviously $\delta < \epsilon$ (otherwise the assumption of min $V(x)\Big|_{\| x \| = \epsilon} = m$ will be contradicted).

Consider any $x(t_0)$ such that $\| x(t_0) \| \leqslant \delta$. Since $\dfrac{dV}{dt} \leqslant 0$, we have

$$V(x(t)) \leqslant V(x(t_0)) \leqslant m/2 \quad \text{for all } t > t_0$$

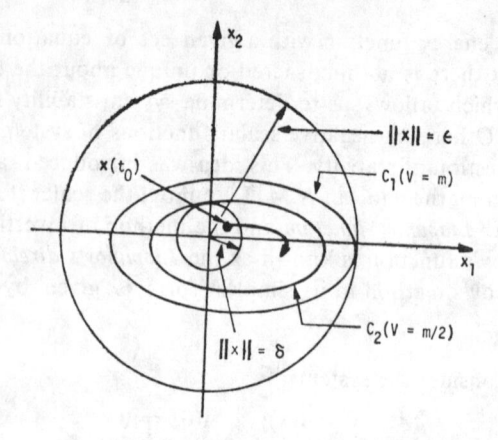

Figure 14.66.

Thus $V(\mathbf{x}(t))$ can never reach the contour C_1 and hence $\mathbf{x}(t)$ can not reach the closed set $\|\mathbf{x}\| = \epsilon$ since $V(\mathbf{x}) \geqslant m$ on this set. Therefore $\|\mathbf{x}(t)\| \leqslant \epsilon$ for all $\|\mathbf{x}(t_0)\| \leqslant \delta$ and $t \geqslant t_0$. This guarantees stability as per the definition given in Section 14.15.

Theorem 2: If the property (d) of Theorem 1 is replaced with (d) $dV/dt < 0$, $\mathbf{x} \neq \mathbf{0}$ (i.e., dV/dt is negative definite scalar function), then the system is asymptotically stable.

It is intuitively obvious since a continuous V function, $V > 0$ except at $\mathbf{x} = \mathbf{0}$, satisfies the condition $dV/dt < 0$ we expect that $\mathbf{x}$ will eventually approach the origin. We shall avoid the rigorous proof of this theorem.

Theorem 3: If all the conditions of Theorem 2 hold and in addition.

$$V(\mathbf{x}) \to \infty \text{ as } \|\mathbf{x}\| \to \infty$$

then the system is asymptotically stable in-the-large at the origin

Note that the additional requirement for this case is that $V(\mathbf{x})$ must approach ∞ as the distance from point $\mathbf{x}$ to the origin approaches ∞ irrespective of the direction. Essentially, this requirement is to assure that points of constant values of $V(\mathbf{x})$ form closed surfaces in state space. If $V(\mathbf{x})$ does not form closed surfaces, then it is possible for system trajectories to go towards infinity.

Consider a positive definite function

$$V(x_1, x_2) = x_2^2 + \frac{x_1^2}{1 + x_1^2}$$

with

$$\frac{dV}{dt} = \frac{\partial V}{\partial x_1}\frac{dx_1}{dt} + \frac{\partial V}{\partial x_2}\frac{dx_2}{dt} < 0$$

for some hypothetical dynamic system. It is possible for a system trajectory to go to ∞ as shown in Fig. 14.67. The reason is that x_1 can become infinity without causing V to become infinite, so Theorem 3 is not satisfied.

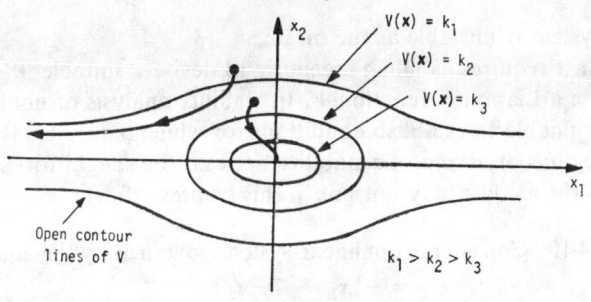

Figure 14.67.

LIAPUNOV FUNCTIONS

The determination of stability via Liapunov's direct method centres around the choice of a positive definite function $V(\mathbf{x})$ called the Liapunov function. Unfortunately, there is no universal method for selecting Liapunov function which is unique for a specific problem. Some Liapunov functions may provide better answer than others. Several techniques have been devised for systematic construction of Liapunov functions (two such techniques are presented in this section), each is applicable to a particular class of systems.

In addition, if a Liapunov function of the required type can not be found, it in no way implies that the system is unstable. (Stability theorems presented in this section merely provide *sufficient conditions* for stability.) It only means that our attempt in trying to establish the stability of the system has failed.

For a given function $V(\mathbf{x})$, there is no general method which will easily allow us to ascertain whether it is positive definite. However, if $V(\mathbf{x})$ is in the quadratic form in x_i's, we can use Sylvester's theorem (see Appendix II) to ascertain definiteness of the function.

Inspite of all these limitations, Liapunov's direct method is a most powerful technique available today for stability analysis of nonlinear systems.

INSTABILITY

It may be noted that instability in a nonlinear system can be established by direct recourse to the instability theorem of the direct method. The basic instability theorem is presented below:

Theorem 4: Consider the system

$$\dot{\mathbf{x}} = \mathbf{f}(\mathbf{x}); \mathbf{f}(0) = 0$$

Suppose there exists a scalar function $W(\mathbf{x})$ which, for some real number $\epsilon > 0$, satisfies the following properties for all $\mathbf{x}$ in the region $\|\mathbf{x}\| \leqslant \epsilon$:

(a) $W(\mathbf{x}) > 0; \mathbf{x} \neq \mathbf{0}$
(b) $W(\mathbf{0}) = 0$
(c) $W(\mathbf{x})$ has continuous partial derivatives with respect to all components of $\mathbf{x}$.
(d) $\dfrac{dW}{dt} \geqslant 0$

Then the system is unstable at the origin.

Note that it requires as much ingenuity to devise a suitable W function as to devise a Liapunov function V. In stability analysis of nonlinear systems, it is valuable to establish conditions for which the system is unstable. Then the regions of asymptotic stability need not be sought for such conditions and the analyst may not put in this fruitless effort.

Example 14.10: Consider a nonlinear system governed by the equations
$$\dot{x}_1 = -x_1 + 2x_1^2 x_2$$
$$\dot{x}_2 = -x_2 \tag{i}$$

We might choose V as
$$V = x_1^2 + x_2^2 \tag{ii}$$

From eqns. (i) and (ii)
$$\frac{dV}{dt} = \frac{\partial V}{\partial x_1}\frac{dx_1}{dt} + \frac{\partial V}{\partial x_2}\frac{dx_2}{dt}$$
$$= -2x_1^2(1 - 2x_1 x_2) - 2x_2^2$$

Thus dV/dt is negative definite if
$$1 - 2x_1 x_2 > 0 \tag{iii}$$

Investigation of the inequality (iii) reveals that there are curves dividing the stable region from the region where $\dot{V} > 0$. The dividing lines lie in the first and third quadrants and are rectangular hyperbolas as shown in Fig. 14.68. In the second and fourth quadrants the inequality is satisfied for all values of x_1 and x_2. Figure 14.68 shows the regions of stability and possible instability. It is easily observed that the origin of the system is asymptotically stable.

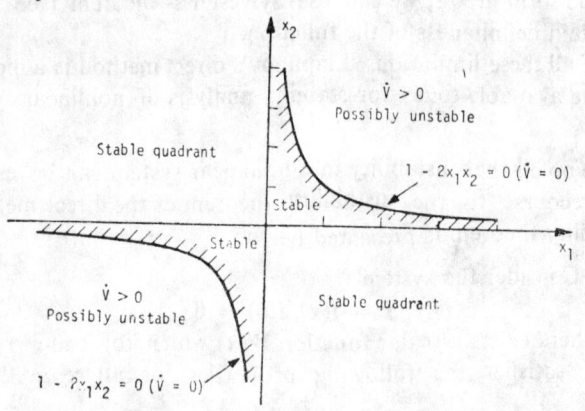

Figure 14.68 Stability regions for a nonlinear system.

Example 14.11: Consider a nonlinear system described by the equations
$$\dot{x}_1 = x_2$$
$$\dot{x}_2 = -x_2 - x_1^3$$

Clearly origin is the equilibrium point.
A choice of possible Liapunov function is
$$V = x_1^4 + x_1^2 + 2x_1x_2 + 2x_2^2$$
$$= x_1^4 + (x_1 + x_2)^2$$
The derivative of the Liapunov function
$$\frac{dV}{dt} = \frac{\partial V}{\partial x_1}\dot{x}_1 + \frac{\partial V}{\partial x_2}\dot{x}_2$$
$$= -2x_1^4 - 2x_2^2$$
is clearly negative definite.
Further $V(\mathbf{x}) \to \infty$ as $\|\mathbf{x}\| \to \infty$.
Therefore the origin of the system is asymptotically stable in-the-large.

THE DIRECT METHOD OF LIAPUNOV AND THE LINEAR SYSTEM

In case of linear systems, the direct method of Liapunov provides a simple approach to stability analysis. It must be emphasized here that compared to the results presented in Chapter 6, no new results are obtained by the use of the direct method for the stability analysis of linear systems. However, the study of linear systems using the direct method is quite useful because it extends our thinking to nonlinear systems.

Consider a linear autonomous system described by the state equation
$$\dot{\mathbf{x}} = \mathbf{A}\mathbf{x} \tag{14.67}$$

The linear system is asymptotically stable in-the-large at the origin if and only if given any symmetric, positive definite matrix $\mathbf{Q}$ (see Appendix II), there exists a symmetric positive definite matrix $\mathbf{P}$ which is the unique solution of
$$\mathbf{A}^T\mathbf{P} + \mathbf{P}\mathbf{A} = -\mathbf{Q} \tag{14.68}$$

Proof

To prove the sufficiency of the result of above theorem, let us assume that a symmetric positive definite matrix $\mathbf{P}$ exists which is the unique solution of eqn. (14.68). Consider the scalar function (eqn. (II.14), Appendix II),
$$V(\mathbf{x}) = \mathbf{x}^T\mathbf{P}\mathbf{x}$$
Note that
$$V(\mathbf{x}) > 0 \quad \text{for } \mathbf{x} \neq \mathbf{0}$$
and
$$V(\mathbf{0}) = 0$$
The time derivative of $V(\mathbf{x})$ is
$$\dot{V}(\mathbf{x}) = \dot{\mathbf{x}}^T\mathbf{P}\mathbf{x} + \mathbf{x}^T\mathbf{P}\dot{\mathbf{x}}$$
Using eqns. (14.67) and (14.68) we get
$$\dot{V}(\mathbf{x}) = \mathbf{x}^T\mathbf{A}^T\mathbf{P}\mathbf{x} + \mathbf{x}^T\mathbf{P}\mathbf{A}\mathbf{x}$$
$$= \mathbf{x}^T(\mathbf{A}^T\mathbf{P} + \mathbf{P}\mathbf{A})\mathbf{x}$$
$$= -\mathbf{x}^T\mathbf{Q}\mathbf{x}$$

Since Q is positive definite, $\dot{V}(x)$ is negative definite. Norm of x may be defined as (eqn. (II.13), Appendix II)
$$\| x \| = (x^T P x)^{1/2}$$
Then
$$V(x) = \| x \|^2$$
$$V(x) \to \infty \text{ as } \| x \| \to \infty$$

The system is therefore asymptotically stable in-the-large at the origin (refer Theorem 3 of this section).

In order to show that the result is also necessary, suppose that the system is asymptotically stable and P is negative definite. Consider the scalar function
$$V(x) = -x^T P x \tag{14.69}$$
Therefore
$$\dot{V}(x) = -[\dot{x}^T P x + x^T P \dot{x}]$$
$$= x^T Q x$$
$$> 0$$

There is a contradiction since $V(x)$ given by eqn. (14.69) satisfies instability theorem (refer Theorem 4 of this section).

Thus the conditions for the positive definiteness of P are necessary and sufficient for asymptotic stability of the system of eqn. (14.67).

HURWITZ CRITERION AND LIAPUNOV'S DIRECT METHOD

In the following we prove that the Liapunov's direct method, as applied to linear time-invariant systems, is the same as the Hurwitz stability criterion described in Chapter 6.

We first consider a third-order system described by the state equation
$$\begin{bmatrix} \dot{x}_1 \\ \dot{x}_2 \\ \dot{x}_3 \end{bmatrix} = \underbrace{\begin{bmatrix} 0 & 1 & 0 \\ -b_3 & 0 & 1 \\ 0 & -b_2 & -b_1 \end{bmatrix}}_{A} \begin{bmatrix} x_1 \\ x_2 \\ x_3 \end{bmatrix}$$

The matrix A in the above equation is known as *Schwarz matrix*. This equation can be written in the component form as
$$\dot{x}_1 = x_2$$
$$\dot{x}_2 = -b_3 x_1 + x_3$$
$$\dot{x}_3 = -b_2 x_2 - b_1 x_3$$

These equations reduce to the following third-order equation
$$\dddot{x}_1 + b_1 \ddot{x}_1 + (b_2 + b_3)\dot{x}_1 + b_1 b_3 x_1 = 0$$
or
$$\dddot{x}_1 + a_1 \ddot{x}_1 + a_2 \dot{x}_1 + a_3 x_1 = 0$$
where b's are related to a's by the following relations:
$$b_1 = a_1, \quad b_2 = \frac{a_1 a_2 - a_3}{a_1}, \quad b_3 = \frac{a_3}{a_1}$$

or
$$b_1 = \Delta_1;\ \Delta_1 = a_1$$
$$b_2 = \frac{\Delta_2}{\Delta_1};\ \Delta_2 = \begin{vmatrix} a_1 & 1 \\ a_3 & a_2 \end{vmatrix}$$
$$b_3 = \frac{\Delta_3}{\Delta_1\Delta_2};\ \Delta_3 = \begin{vmatrix} a_1 & 1 & 0 \\ a_3 & a_2 & a_1 \\ 0 & 0 & a_3 \end{vmatrix}$$

On similar lines, it can be verified that an nth-order linear time-invariant system described by the differential equation
$$\frac{d^n x_1}{dt^n} + a_1 \frac{d^{n-1} x_1}{dt^{n-1}} + \cdots + a_n x_1 = 0$$
can be represented by the following state model:
$$\dot{\mathbf{x}} = \mathbf{A}\mathbf{x}$$
where
$$\mathbf{A} = \begin{bmatrix} 0 & 1 & 0 & 0 & \cdots & 0 & 0 & 0 \\ -b_n & 0 & 1 & 0 & \cdots & 0 & 0 & 0 \\ 0 & -b_{n-1} & 0 & 1 & \cdots & 0 & 0 & 0 \\ \vdots & \vdots & \vdots & \vdots & & \vdots & \vdots & \vdots \\ 0 & 0 & 0 & 0 & \cdots & -b_3 & 0 & 1 \\ 0 & 0 & 0 & 0 & \cdots & 0 & -b_2 & -b_1 \end{bmatrix}$$
(schwarz matrix)
$$b_1 = \Delta_1,\ b_2 = \frac{\Delta_2}{\Delta_1},\ b_3 = \frac{\Delta_3}{\Delta_1\Delta_2},\ \ldots,\ b_i = \frac{\Delta_{i-3}\Delta_i}{\Delta_{i-2}\Delta_{i-1}}\ (i = 4, 5, \ldots, n)$$

Δ_i's are determinants taken on principal minors of the arrangement given by (6.8) in Section 6.3.

Consider the scalar function
$$V(\mathbf{x}) = \mathbf{x}^T \mathbf{P} \mathbf{x}$$
where
$$\mathbf{P} = \begin{bmatrix} b_1 b_2 \cdots b_n & 0 & \cdots & 0 & 0 & 0 \\ 0 & b_1 b_2 \cdots b_{n-1} & \cdots & 0 & 0 & 0 \\ \vdots & \vdots & & \vdots & \vdots & \vdots \\ 0 & 0 & \cdots & b_1 b_2 b_3 & 0 & 0 \\ 0 & 0 & \cdots & 0 & b_1 b_2 & 0 \\ 0 & 0 & \cdots & 0 & 0 & b_1 \end{bmatrix}$$

The derivative of V is obtained as
$$\dot{V}(\mathbf{x}) = \mathbf{x}^T (\mathbf{A}^T \mathbf{P} + \mathbf{P}\mathbf{A})\mathbf{x} = -2 b_1^2 x_n^2$$

It is observed that $V(x)$ is positive definite and $\dot{V}(x)$ is negative definite, if all the elements in the main diagonal of matrix P are positive. Therefore conditions of asymptotic stability of linear system under consideration are $b_i > 0$ $(i = 1, 2, ..., n)$; or equivalently

$$\Delta_i > 0 \qquad (i = 1, 2, ..., n)$$

These conditions are the same as obtained by the Hurwitz stability criterion in Section 6.3 [5].

Example 14.12: Let us determine stability of the system described by the following equation:

$$\dot{x} = Ax$$

$$A = \begin{bmatrix} -1 & -2 \\ 1 & -4 \end{bmatrix}$$

We will first solve eqn. (14.68) for P for an arbitrary choice of positive definite real symmetric matrix Q. We may choose $Q = I$, the identity matrix. Equation (14.68) then becomes

$$A^T P + PA = -I$$

or

$$\begin{bmatrix} -1 & 1 \\ -2 & -4 \end{bmatrix} \begin{bmatrix} p_{11} & p_{12} \\ p_{12} & p_{22} \end{bmatrix} + \begin{bmatrix} p_{11} & p_{12} \\ p_{12} & p_{22} \end{bmatrix} \begin{bmatrix} -1 & -2 \\ 1 & -4 \end{bmatrix} = \begin{bmatrix} -1 & 0 \\ 0 & -1 \end{bmatrix}$$

(14.70)

Note that we have taken $p_{12} = p_{21}$; this is because solution matrix P is known to be a positive definite real symmetric matrix for a stable system.

From eqn. (14.70) we get

$$-2p_{11} + 2p_{12} = -1$$
$$-2p_{11} - 5p_{12} + p_{22} = 0$$
$$-4p_{12} - 8p_{22} = -1$$

Solving for p's, we obtain

$$P = \begin{bmatrix} p_{11} & p_{12} \\ p_{12} & p_{22} \end{bmatrix} = \begin{bmatrix} \dfrac{23}{60} & \dfrac{-7}{60} \\ \dfrac{-7}{60} & \dfrac{11}{60} \end{bmatrix}$$

Using Sylvester's theorem (Appendix II) we find that P is positive definite. Therefore origin of the system under consideration is asymptotically stable in-the-large.

It may be pointed out here that the Routh approach is applicable to the characteristic polynomial of a system. Therefore if the system model is known in the state variable form, we have to first convert the system model into phase variable form and therefrom obtain the characteristic polynomial. This conversion is not so easy for systems of high-order. Liapunov's direct method which is applicable to any state variable formulation of a system, has thus an important role to play in stability analysis of linear systems.

METHODS OF CONSTRUCTING LIAPUNOV FUNCTIONS FOR NONLINEAR SYSTEMS

As has been said earlier in this section the Liapunov theorems give only sufficient conditions on system stability and furthermore there is no unique way of constructing a Liapunov function except in the case of linear systems where a Liapunov function can always be constructed and both necessary and sufficient conditions established. Because of this drawback a host of methods have become available in literature and many refinements have been suggested to enlarge the region in which the system is found to be stable. Since this treatise is meant as a first exposure of the student to the Liapunov's direct method, only two of the relatively simpler techniques of constructing a Liapunov function would be advanced here [5, 6].

Krasovskii's Method

Consider the system

$$\dot{\mathbf{x}} = \mathbf{f}(\mathbf{x}); \mathbf{f}(0) = 0 \text{ (i.e., singular point at origin)}$$

Define a Liapunov function as

$$V = \mathbf{f}^T \mathbf{P} \mathbf{f} \qquad (14.71)$$

where $\mathbf{P}$ = a symmetric positive definite matrix.

Now

$$\dot{V} = \dot{\mathbf{f}}^T \mathbf{P} \mathbf{f} + \mathbf{f}^T \mathbf{P} \dot{\mathbf{f}} \qquad (14.72)$$

$$\dot{\mathbf{f}} = \frac{\partial \mathbf{f}}{\partial \mathbf{x}} \cdot \frac{\partial \mathbf{x}}{\partial t} = \mathbf{J}\mathbf{f}$$

where

$$\mathbf{J} = \begin{bmatrix} \frac{\partial f_1}{\partial x_1} & \frac{\partial f_1}{\partial x_2} & \cdots & \frac{\partial f_1}{\partial x_n} \\ \frac{\partial f_2}{\partial x_1} & \frac{\partial f_2}{\partial x_2} & \cdots & \frac{\partial f_2}{\partial x_n} \\ \cdots & \cdots & \cdots & \cdots \\ \frac{\partial f_n}{\partial x_1} & \frac{\partial f_n}{\partial x_2} & \cdots & \frac{\partial f_n}{\partial x_n} \end{bmatrix} \text{ is Jacobian matrix}$$

Substituting $\dot{\mathbf{f}}$ in (14.72), we have

$$\dot{V} = \mathbf{f}^T \mathbf{J}^T \mathbf{P} \mathbf{f} + \mathbf{f}^T \mathbf{P} \mathbf{J} \mathbf{f}$$
$$= \mathbf{f}^T (\mathbf{J}^T \mathbf{P} + \mathbf{P} \mathbf{J}) \mathbf{f}$$

Let $\mathbf{Q} = \mathbf{J}^T \mathbf{P} + \mathbf{P} \mathbf{J}$

Since V is positive definite, for the system to be asymptotically stable $\mathbf{Q}$ should be negative definite. If in addition, $V(\mathbf{x}) \to \infty$ as $\|\mathbf{x}\| \to \infty$, the system is asymptotically stable in-the-large.

Example 14.13: Consider the nonlinear system shown in Fig. 14.69 where the nonlinear element is described as

$$u = g(e)$$

With
$$r = 0$$
$$c = -e$$
$\therefore$
$$Ku = -\ddot{e} - \dot{e}$$

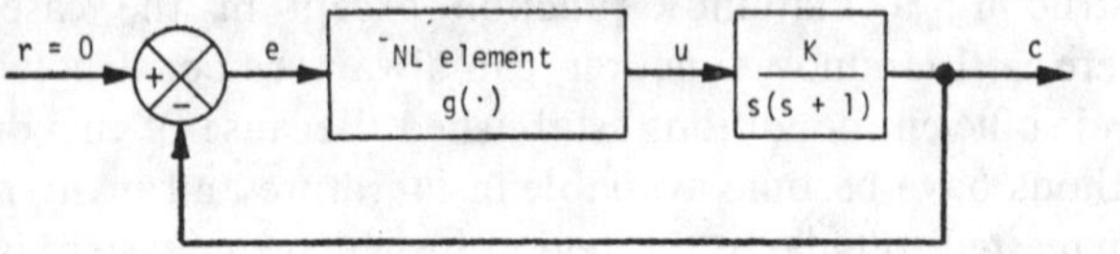

Figure 14.69.

Define
$$x_1 = e$$
$$x_2 = \dot{x}_1$$
$\therefore$
$$\dot{x}_1 = x_2$$
$$\dot{x}_2 = -x_2 - Kg(x_1) \qquad \text{(i)}$$

The equilibrium point lies at origin if $g(0) = 0$.
Now

$$\mathbf{J} = \begin{bmatrix} 0 & 1 \\ -K\dfrac{dg(x_1)}{dx_1} & -1 \end{bmatrix} \qquad \text{(ii)}$$

Let

$$\mathbf{P} = \begin{bmatrix} p_{11} & p_{12} \\ p_{12} & p_{22} \end{bmatrix} \qquad \text{(iii)}$$

For it to be positive definite

$$p_{11} > 0 \qquad \text{(iv)}$$
$$p_{11}p_{22} - p_{12}^2 > 0 \qquad \text{(v)}$$

Now

$$\mathbf{Q} = \mathbf{J}^T\mathbf{P} + \mathbf{PJ}$$

$$= \begin{bmatrix} 0 & -K\dfrac{dg(x_1)}{dx_1} \\ 1 & -1 \end{bmatrix} \begin{bmatrix} p_{11} & p_{12} \\ p_{12} & p_{22} \end{bmatrix}$$

$$+ \begin{bmatrix} p_{11} & p_{12} \\ p_{12} & p_{22} \end{bmatrix} \begin{bmatrix} 0 & 1 \\ -K\dfrac{dg(x_1)}{dx_1} & -1 \end{bmatrix}$$

or $\mathbf{Q} = \begin{bmatrix} -2p_{12}K\dfrac{dg(x_1)}{dx_1} & p_{11} - p_{12} - p_{22}K\dfrac{dg(x_1)}{dx_1} \\ p_{11} - p_{12} - p_{22}K\dfrac{dg(x_1)}{dx_1} & 2(p_{12} - p_{22}) \end{bmatrix}$ (vi)

Nonlinear Systems

For the system to be asymptotically stable, $-Q$ should be positive definite, i.e.,

$$2p_{12}K \frac{dg(x_1)}{dx_1} > 0 \qquad \text{(vii)}$$

$$-4p_{12}K \frac{dg(x_1)}{dx_1}(p_{12} - p_{22}) - \left(p_{11} - p_{12} - p_{22}K\frac{dg(x_1)}{dx_1}\right)^2 > 0$$

or

$$4p_{12}K \frac{dg(x_1)}{dx_1}(p_{22} - p_{12}) > \left(p_{11} - p_{12} - p_{22}K\frac{dg(x_1)}{dx_1}\right)^2 \qquad \text{(viii)}$$

Assume that $K > 0$ and choose $p_{12} > 0$. Inequality (vii) then yields the condition

$$\frac{dg(x_1)}{dx_1} > 0 \qquad \text{(ix)}$$

Choose $p_{11} = p_{12}$ and $p_{22} = \beta p_{12}$, where $\beta > 1$. Inequality (viii) then gives the condition

$$4(\beta - 1) > K\beta^2 \frac{dg(x_1)}{dx_1} \qquad \text{(x)}$$

The inequalities (ix) and (x) together constitute the conditions under which the system is asymptotically stable.

Take for example

$$g(x_1) = x_1^3 \quad \text{(nonlinearity symmetrically lies in first and third quadrants)}$$

$$\therefore \quad \frac{dg(x_1)}{dx_1} = 3x_1^2 > 0 \text{ (which is always met)}$$

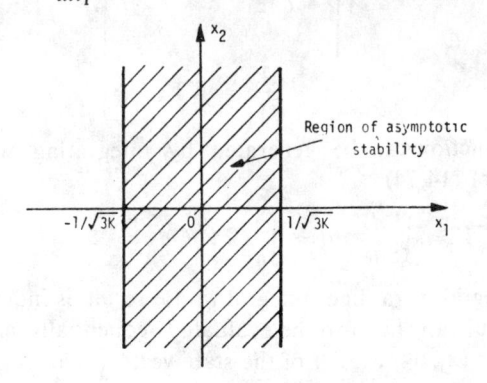

Figure 14.70 Region of asymptotic stability for Example 14.13 with $g(e) = e^3$.

The condition (x) gives

$$x_1^2 < \frac{4}{3K}\left(\frac{1}{\beta} - \frac{1}{\beta^2}\right) \qquad \text{(xi)}$$

It can be easily shown that the largest value of x_1 occurs when $\beta = 2$.

Therefore
$$x_1^2 < \frac{1}{3K}$$

or
$$-\frac{1}{\sqrt{3K}} < x_1 < \frac{1}{\sqrt{3K}} \qquad (\text{xii})$$

This region of asymptotic stability is illustrated in Fig. 14.70.

Variable gradient method

The quadratic form approach used so far to Liapunov function formulation is too restrictive. Here we shall advance the variable gradient method which provides considerable flexibility in selecting a Liapunov function.

For the autonomous system
$$\dot{\mathbf{x}} = \mathbf{f}(\mathbf{x}); \qquad \mathbf{f}(0) = 0$$

let $V(\mathbf{x})$ be a candidate for a Liapunov function.
The time derivative of V can be expressed as

$$\dot{V}(\mathbf{x}) = \frac{\partial V}{\partial x_1} \dot{x}_1 + \frac{\partial V}{\partial x_2} \dot{x}_2 + \ldots + \frac{\partial V}{\partial x_n} \dot{x}_n \qquad (14.73)$$

which can be expressed in terms of the gradient of V as

$$\dot{V} = (\boldsymbol{\nabla} V)^T \dot{\mathbf{x}} \qquad (14.74)$$

where
$$\boldsymbol{\nabla} V = \begin{bmatrix} \dfrac{\partial V}{\partial x_1} = \boldsymbol{\nabla} V_1 \\ \dfrac{\partial V}{\partial x_2} = \boldsymbol{\nabla} V_2 \\ \vdots \\ \dfrac{\partial V}{\partial x_n} = \boldsymbol{\nabla} V_n \end{bmatrix}$$

The Liapunov function can be generated by integrating with respect to time both sides of (14.74)

$$V = \int_0^t \frac{dV}{dt} dt = \int_0^{\mathbf{x}} (\boldsymbol{\nabla} V)^T d\mathbf{x} \qquad (14.75)$$

The above integral is a line integral whose result is independent of the path. The integral can therefore be evaluated sequentially along the component directions $(x_1, x_2, \ldots, x_n)$ of the state vector; that is

$$V = \int_0^{\mathbf{x}} (\boldsymbol{\nabla} V)^T d\mathbf{x} = \int_0^{x_1} \boldsymbol{\nabla} V_1(x_1, 0, \ldots, 0) \, dx_1$$
$$+ \int_0^{x_2} \boldsymbol{\nabla} V_2(x_1, x_2, 0, \ldots, 0) \, dx_2 + \ldots$$
$$+ \int_0^{x_n} \boldsymbol{\nabla} V_n(x_1, x_2, \ldots, x_n) \, dx_n \qquad (14.76)$$

Nonlinear Systems

Let us define

$$e_1 = \begin{bmatrix} 1 \\ 0 \\ 0 \\ \vdots \\ 0 \end{bmatrix}; e_2 = \begin{bmatrix} 0 \\ 1 \\ 0 \\ \vdots \\ 0 \end{bmatrix}; \ldots; e_n = \begin{bmatrix} 0 \\ 0 \\ \vdots \\ \\ 1 \end{bmatrix}$$

The integral given in (14.76) states that the path starts from the origin and moves along the vector e_1 to x_1. From this point, the path moves in the direction of the vector e_2 to x_2. In this way, the path finally reaches the point $(x_1, x_2, \ldots, x_n)$.

For the scalar function V to be unique, the curl of its gradient must be zero [7], i.e.,

$$\nabla \times (\nabla V) = 0 \qquad (14.77)$$

This results in $\dfrac{n}{2}(n-1)$ equations to be satisfied by the components of the gradient vector. These are

$$\frac{\partial \nabla V_i}{\partial x_j} = \frac{\partial \nabla V_j}{\partial x_i} \text{ for all } i, j \qquad (14.78)$$

To begin with, a completely general form given below is assumed for the gradient vector ∇V.

$$\nabla V = \begin{bmatrix} \nabla V_1 \\ \nabla V_2 \\ \vdots \\ \nabla V_n \end{bmatrix} = \begin{bmatrix} a_{11}x_1 + a_{12}x_2 + \cdots + a_{1n}x_n \\ a_{21}x_1 + a_{22}x_2 + \cdots + a_{2n}x_n \\ \cdots \quad \cdots \quad \cdots \quad \cdots \\ \cdots \quad \cdots \quad \cdots \quad \cdots \\ \cdots \quad \cdots \quad \cdots \quad \cdots \\ a_{n1}x_1 + a_{n2}x_2 + \cdots + a_{nn}x_n \end{bmatrix} \qquad (14.79)$$

The a_{ij}'s are completely undetermined quantities and could be constants or functions of both state variables and t. It is convenient to choose a_{nn} as a constant.

The procedure to formulate a Liapunov function from which the stability of the equilibrium state $\mathbf{x} = \mathbf{0}$ may be determined is given in the following steps.

1. Form $\dot{V}$ as per eqn. (14.74) with ∇V as in eqn. (14.79). Choose a_{ij}'s to constrain it to be negative definite or at least negative semidefinite.
2. Determine the remaining unknown a_{ij}'s to satisfy the curl equations (14.78).
3. Recheck $\dot{V}$ in case step (2) has altered it.
4. Determine V by integrating as in eqn. (14.76).
5. Determine the region of stability where V is positive definite.

The above procedure is best illustrated by means of an example.

Example 14.14: Consider the nonlinear system shown in Fig. 14.71. The system is described by the state equations

$$\dot{x}_1 = -3x_2 - f(x_1)$$
$$\dot{x}_2 = -x_2 + f(x_1) \quad \text{(i)}$$

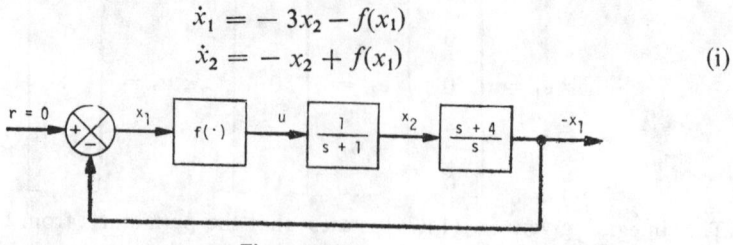

Figure 14.71.

Let us consider the special case wherein the nonlinearity can be expressed as

$$f(x_1) = g(x_1)x_1$$

Therefore the state description of the system becomes

$$\dot{x}_1 = -3x_2 - g(x_1)x_1$$
$$\dot{x}_2 = -x_2 + g(x_1)x_1 \quad \text{(ii)}$$

It is immediately obvious that the equilibrium point lies at origin.
Assume

$$\nabla V = \begin{bmatrix} a_{11}x_1 + a_{12}x_2 \\ a_{21}x_1 + a_{22}x_2 \end{bmatrix} \quad \text{(iii)}$$

Form V as per eqn. (14.74)

$$V = -x_1^2(a_{11} - a_{21})g(x_1)$$
$$+ x_1 x_2[-3a_{11} - a_{12}g(x_1) - a_{21} + a_{22}g(x_1)] \quad \text{(iv)}$$
$$- x_2^2(3a_{12} + a_{22})$$

One of the ways of keeping $\dot{V}$ negative definite is to choose a_{ij}'s such that

$$(a_{11} - a_{21})g(x_1) > 0$$
$$-3a_{11} - a_{12}g(x_1) - a_{21} + a_{22}g(x_1) = 0 \quad \text{(v)}$$
$$3a_{12} + a_{22} > 0$$

These conditions get simplified by making the choice

$$a_{12} = a_{21} = 0 \quad \text{(vi)}$$

so that

$$a_{11}g(x_1) > 0$$
$$a_{11} = \frac{a_{22}}{3} g(x_1) \quad \text{(vii)}$$
$$a_{22} > 0$$

The first of these conditions is satisfied when the other two are met with, because

$$\frac{a_{22}}{3} g^2(x_1) > 0$$

Nonlinear Systems

We can therefore choose a_{22} to be any positive constant. Thus

$$\nabla V = \begin{bmatrix} \frac{1}{3} a_{22} g(x_1) x_1 \\ a_{22} x_2 \end{bmatrix} \quad \text{(viii)}$$

$$\dot{V} = -\frac{1}{3} a_{22} x_1^2 g^2(x_1) - a_{22} x_2^2 \quad \text{(ix)}$$

It may also be noticed that the gradient vector in (viii) meets the curl conditions (14.78).

The Liapunov function can now be obtained by taking the line integral of the gradient vector along the path defined in eqn. (14.76). Thus

$$V = \frac{1}{3} a_{22} \int_0^{x_1} g(x_1) x_1 \, dx_1 + a_{22} \int_0^{x_2} x_2 \, dx$$

$$= \frac{1}{3} a_{22} \int_0^{x_1} g(x_1) x_1 \, dx_1 + \frac{1}{2} a_{22} x_2^2 \quad \text{(x)}$$

If $g(x_1) > 0$, i.e., $f(x_1) = g(x_1) x_1$ lies in first and third quadrants, V is positive definite. Under this condition the system is asymptotically stable. Also if

$$\lim_{x_1 \to \infty} \int_0^{x_1} g(x_1) x_1 \, dx_1 \to \infty$$

the system would be asymptotically stable in-the-large.

14.13 Popov's Stability Criterion

A departure from the use of Liapunov's direct method was introduced by V.M. Popov, who obtained a frequency domain stability criterion as a sufficient condition for asymptotic stability of an important class of nonlinear systems. The basic structure of this class is shown in Fig. 14.72. It consists of a nonlinear element and a linear time-invariant plant. Many systems with a single nonlinearity can be reduced to this basic form.

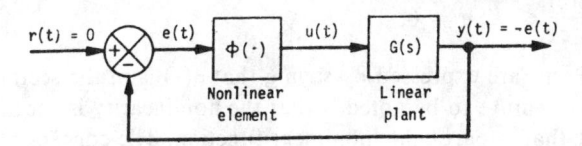

Figure 14.72 Popov's basic feedback control system.

LINEAR PLANT

The linear time-invariant plant is described by the equations

$$\dot{\mathbf{x}} = \mathbf{A}\mathbf{x} + \mathbf{B}u \quad (14.80)$$

$$y = \mathbf{C}\mathbf{x}$$

where $\mathbf{A}$ is $n \times n$ constant matrix, $\mathbf{B}$ is $n \times 1$ constant matrix and $\mathbf{C}$ is $1 \times n$ constant matrix.

The transfer function between the output $y(t)$ and input $u(t)$ is given by (refer Chapter 12)

$$\frac{Y(s)}{U(s)} = G(s) = \mathbf{C}(s\mathbf{I} - \mathbf{A})^{-1}\mathbf{B} \qquad (14.81)$$

In Chapter 6, we discussed the concept of stability as applied to linear systems. We found that a linear system is stable if the system output in response to either an impulse or initial conditions tends to zero as $t \to \infty$. The impulse response of a system with transfer function $G(s)$ is given by

$$g(t) = \mathcal{L}^{-1}[G(s)]$$

The impulse response tends to zero as $t \to \infty$ if the real parts of the poles of $G(s)$ are less than zero.

The linear system is said to be *output stable of degree* α if the poles of $G(s)$ are such that $\mathrm{Re}(s) < -\alpha$ (Note that α represents the degree of damping of the linear system). It means that for $\alpha > 0$, the impulse response tends to zero faster than the function $e^{-\alpha t}$; for $\alpha < 0$, impulse response may diverge but $g(t)\, e^{\alpha t}$ ultimately tends to zero.

The product $g(t)\, e^{\alpha t}$ may tend to zero in following modes*:

(i) $\int_0^\infty [e^{\alpha t} g(t)]^2 \, dt < \infty$

(ii) $\int_0^\infty e^{\alpha t} |g(t)| \, dt < \infty$

The linear system is said to be *output stable* if $\alpha = 0$. Assume that the linear plant in the system of Fig. 14.72 is output stable.

NONLINEAR ELEMENT

The output versus input characteristic of the nonlinear element of Fig. 14.72 is assumed to be restricted to lie within a sector as shown in Fig. 14.73. In other words,

(i) $\phi(0) = 0$ \hfill (14.82)

(ii) $0 \leqslant \dfrac{\phi(e)}{e} \leqslant K;\ e \neq 0,\ 0 < K < \infty$

These restrictions are expressed by saying that $\phi(\cdot)$ is in the sector $[0, K]$.

An essential feature to be noted is that the nonlinearity is specified by a sector rather than a particular nonlinear function. The concept of stability for every characteristic in a given sector is called *absolute stability*.

In the previous section of this chapter, we discussed the concept of stability in the sense of Liapunov. We observed that global asymptotic

---

*(i) $e^{\alpha t} g(t)$ belongs to the class of all square-integrable functions; referred to as class $\mathcal{L}_2$.

(ii) $e^{\alpha t} g(t)$ belongs to the class of all absolute integrable functions; referred to as class $\mathcal{L}_1$.

Nonlinear Systems

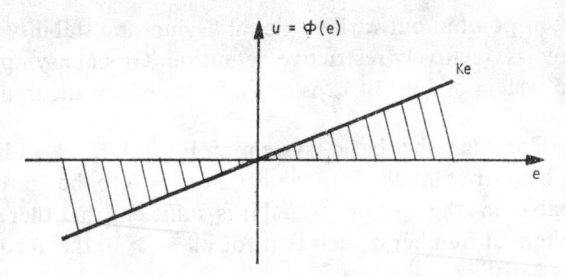

Figure 14.73 Region of absolute stability.

stability requires that all state variables approach zero asymptotically and in addition state variables obey certain restrictions on their dynamic behaviour. In many practical applications, we are not interested in the dynamic behaviour of all the state variables except for the effect of their initial conditions on the behaviour of the output.

To define the concept of stability for basic feedback system of Fig. 14.72, we concentrate on the asymptotic behaviour of the control and output variables of the linear plant. In terms of this goal, stability is defined as follows:

(a) Feedback system of Fig. 14.72 is *control asymptotic* of degree α if there exists a real α such that for every set of initial conditions

$$\int_0^\infty [e^{\alpha t} u(t)]^2 \, dt < \infty \tag{14.83}$$

The system is *absolutely control asymptotic of degree* α in the sector $[0, K]$ if the inequality (14.83) is satisfied for all nonlinear functions in this sector.

(b) Feedback system of Fig. 14.72 is *output asymptotic* of degree α if there exists a real α such that for every set of initial conditions

$$\int_0^\infty [e^{\alpha t} y(t)]^2 \, dt < \infty \tag{14.84}$$

The system is *absolutely output asymptotic of degree* α in the sector $[0, K]$ if the inequality (14.84) is satisfied for all nonlinear functions in this sector.

For $\alpha = 0$, the system of Fig. 14.72 is absolutely control asymptotic (absolutely output asymptotic) if the inequality (14.83) [inequality (14.84)] is satisfied for all nonlinear functions in the sector $[0, K]$.

Theorem 1: (For proof refer [3]). Consider the basic feedback system of Fig. 14.72. The linear plant is assumed to be output stable. For the system to be absolutely control and output asymptotic in the sector $[0, K]$ it is sufficient that there exist a finite real number q such that for all $\omega \geqslant 0$, the following inequality is satisfied*

$$\text{Re}\left[(1 + j\omega q)G(j\omega)\right] + \frac{1}{K} > 0 \tag{14.85}$$

*If $K = \infty$, q is required to be non-negative finite real number.

As has been pointed out earlier, global asymptotic stability in the sense of Liapunov is relatively restrictive definition. Global asymptotic stability of the basic system of Fig. 14.72 is given by following theorem.

Theorem 2: Consider the basic system of Fig. 14.72. The linear plant is assumed to be output stable. For the origin $x = 0$ to be globally asymptotically stable in the sector $[0, K]$ it is sufficient that there exist a non-negative finite real number q such that for all $\omega \geqslant 0$, the inequality given by (14.85) is satisfied.

GEOMETRICAL INTERPRETATION

The Popov's inequality (14.85) has an elegant geometrical interpretation similar to the Nyquist criterion.

Inequality (14.85) may be expressed as

$$\text{Re } G(j\omega) - q\omega \text{ Im } G(j\omega) + \frac{1}{K} > 0$$

or
$$U - qW + \frac{1}{K} > 0 \qquad (14.86)$$

where $U =$ the real part of $G(j\omega)$; and $W = \omega$ times the imaginary part of $G(j\omega)$.

If we replace the inequality in (14.86) by equality sign, we get

$$U - qW + \frac{1}{K} = 0 \qquad (14.87)$$

It is now easy to show that the inequality (14.86) represents those points in the U-W plane which are to the right of the line (given by eqn. (14.87)) passing through the point $(-1/K, 0)$ and having a slope $1/q$. This is illustrated in Fig. 14.74.

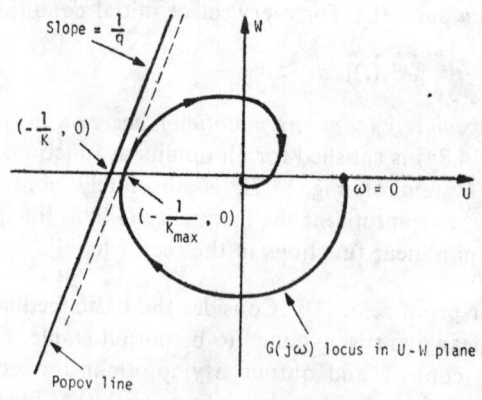

Figure 14.74 Geometrical interpretation of the inequality (14.85).

The procedure thus consists of making a plot of $G(j\omega)$ in the U-W plane for $0 \leqslant \omega \leqslant \infty$. The line given by eqn. (14.87) is then fitted to the curve such that the curve remains wholly to the right of the line while

Nonlinear Systems

$-1/K$ is made as small as possible so that sector $[0, K]$ can be identified. The Popov line corresponding to K (max) is shown dotted in Fig. 14.74.

It is instructive to compare the Popov stability criterion for the class of nonlinear systems corresponding to Fig. 14.72 with the Nyquist stability criterion for a similar class of linear systems. It is immediately observed that what was the Nyquist point $(-1 + j0)$ for linear systems becomes the Popov line for nonlinear systems with respect to which the answer to system stability is sought.

In the theorems stated above, certain restrictions on linear plant and nonlinear element of basic system of Fig. 14.72 were imposed. Improved results with some of these restrictions relaxed are available in the literature. A survey paper on stability theory that also contains an extensive bibliography is [8].

Problems

14.1 A linear second-order servo is described by the equation

$$\ddot{e} + 2\zeta\omega_n \dot{e} + \omega_n^2 e = 0$$

where
$$\zeta = 0.15, \omega_n = 1$$
$$e(0) = 1.5, \dot{e}(0) = 0$$

Determine the singular points. Construct the phase trajectory, using the method of isoclines.

14.2 A simple servo is described by the following equations:

Reaction torque $= \ddot{\theta}_c + 0.5\dot{\theta}_c$
Drive torque $= 2 \, \text{sign} \, (e + 0.5\dot{e})$
$e = \theta_R - \theta_c$
$e(0) = 2$
$\dot{e}(0) = 0$

Construct the phase trajectory using the delta method.

14.3 The position control system shown in Fig. P-14.3 has coulomb friction C at the output shaft. Plot the phase-trajectory for a unit step input and zero initial conditions. Calculate and plot the time response of the error and find the value of the steady-state error.

Given
$$\sqrt{(K/J)} = 1.2 \, \text{rad/sec}$$
$$C/K = 0.3 \, \text{rad}$$
where
$$K = K_A K_1$$

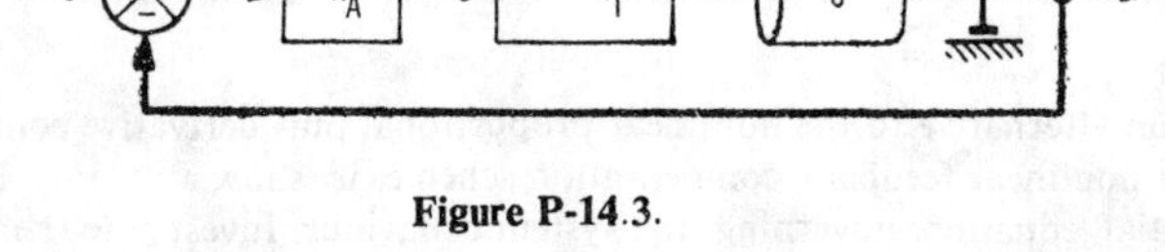

Figure P-14.3.

14.4 A second-order servo containing a relay with dead-zone and hysteresis is shown in the block diagram of Fig. P-14.4. Obtain the phase-trajectory of the system for the initial conditions $e(0) = 0.65$, $\dot{e}(0) = 0$. Does the system have a limit cycle? If so, determine its amplitude and period.

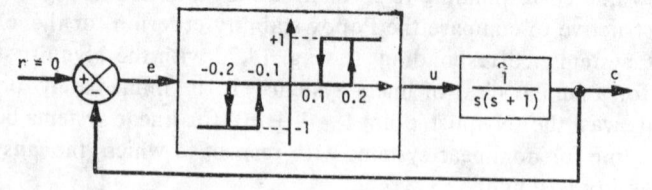

Figure P-14.4.

14.5 For the control system shown in Fig. P-14.5 with tachometer not included in the loop, i.e., with switch S open, plot the phase-trajectory with an initial condition of
$$e(0) = 2, \dot{e}(0) = 0$$
Given: $\tau_m = 0.5$ sec.; $K = 8$; $\delta = 0.5$

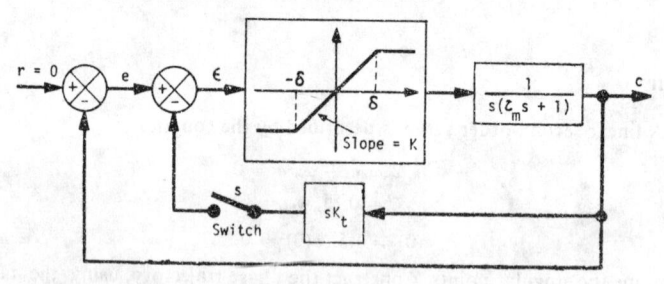

Figure P-14.5.

14.6 Control system shown in Fig. P-14.5, now includes tachometer ($K_t = 0.5$) in the feedback loop, i.e., switch S closed. Plot the phase-trajectory and comment upon the effect of the tachometer feedback.

14.7 For the nonlinear system shown in Fig. P-14.7, draw the phase-trajectory starting from $(e, \dot{e}) = (2, 2)$ in the $(e, \dot{e})$-plane.

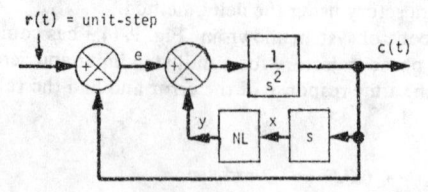

Figure P-14.7.

The characteristics of the nonlinearity (NL) are given below:

$y = 2x$ if $|x| \leqslant 1$
$y = +2$ if $x > 1$
$y = -2$ if $x < -1$

14.8 As an alternative to the nonlinear proportional plus derivative compensation of Fig. 14.22, a nonlinear feedback compensation scheme is shown in Fig. P-14.8. Write the differential equation governing the system behaviour. Investigate the nature of the equilibrium point. Would a limit cycle exist? If so, investigate the stability of the limit cycle.

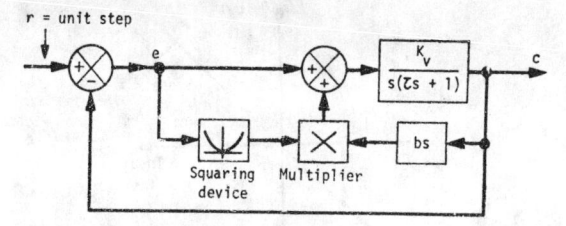

Figure P-14.8.

Draw the phase trajectory for $c = 0$, $\dot{c} = 0$ and $r =$ unit step. The δ-method would be more convenient for use here.

Given: $\qquad K_v = 1, \tau = 1, b = 1$

14.9 The relay control system of Fig. P-14.9 uses a relay with dead-zone and an inner rate feedback loop. Draw a set of isoclines and plot the phase-trajectory for zero input with initial conditions

$$c(0) = 2, \quad \dot{c}(0) = 5.0$$

Also compute and plot time response of the system. Find therefrom the settling time and steady-state error.

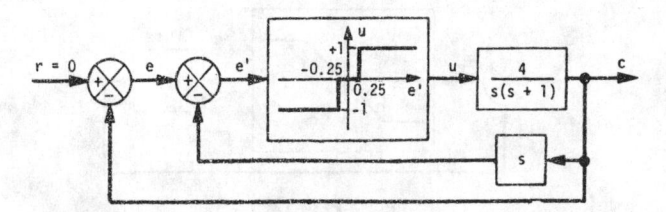

Figure P-14.9.

14.10 Nonlinear spring characteristic can be approximated by two piecewise linear gains as shown in Fig. P-14.10 ($K_2 > K_1$ for hard spring, $K_2 < K_1$ for soft spring). Derive the expression for its describing function.

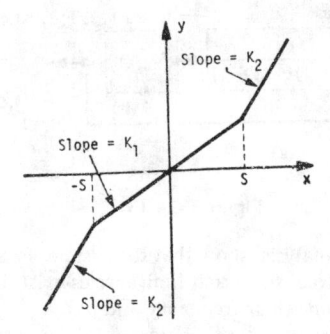

Figure P-14.10.

14.11 Derive the describing function of the element whose input-output characteristic is shown in Fig. P-14.11. Show that the required describing function equals the sum of the describing functions of relay with dead-zone and amplifier with dead-zone (use the results already derived in this chapter).

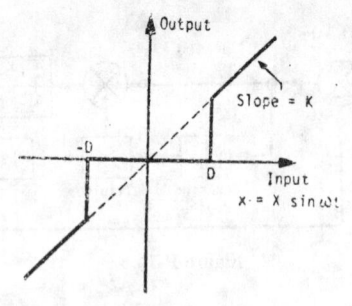

Figure P-14.11.

Can we generalize the above result for any nonlinearity whose characteristic is the sum of the characteristics of two nonlinearities?

14.12 A two-phase servomotor is driven by an amplifier as shown in Fig. P-14.12. The transfer function of the motor is

$$G(s) = \frac{K'e^{-0.1s}}{s(0.1s+1)}$$

Investigate the stability of the system for $K' = 0.1$. What is the largest value of K' for no limit cycle to exist?

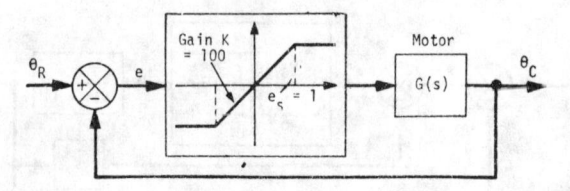

Figure P-14.12.

14.13 An instrument servo system used for positioning a load may be adequately represented by the block diagram shown in Fig. P-14.13a. The backlash characteristics are shown in Fig. P-14.13b.

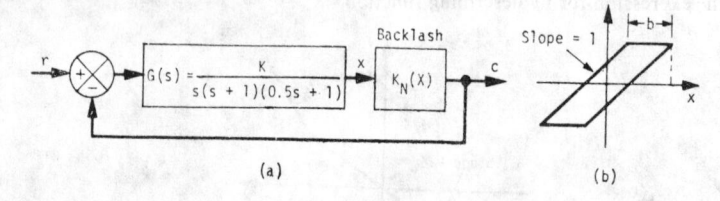

Figure P-14.13.

Using the db gain-phase analysis, show that the system is stable for $K = 1$. If the value of K is now raised to 5, show that limit cycles exist. Investigate the stability of these limit cycles and determine their frequency and b/X.
Given:

b/X	0	0.2	0.4	1	1.4	1.6	1.8	1.9	2.0
$\lvert K_N(X) \rvert$	1	0.954	0.882	0.592	0.367	0.248	0.125	0.064	0
$\angle K_N(X)$	0	$-6.7°$	$-13.4°$	$-32.5°$	$-46.6°$	$-55.2°$	$-66°$	$-69.8°$	$-90°$

How is the system behaviour altered if the backlash nonlinearity is present in the feedback path instead of the forward path in Fig. 14.13a.

14.14 Consider a nonlinear system described by the equations

$$\dot{x}_1 = x_2$$
$$\dot{x}_2 = -(1 - |x_1|)x_2 - x_1$$

Find the region in the state-plane for which the equilibrium state of the system is asymptotically stable.

(*Hint*: A Liapunov function is $V = x_1^2 + x_2^2$)

14.15 Consider the system shown in Fig. P-14.15. Determine the restrictions which must be placed on the nonlinear function $\phi(\cdot)$ in order for the system state with $r(t) = 0$ to be asymptotically stable in-the-large. Use Liapunov's direct method.

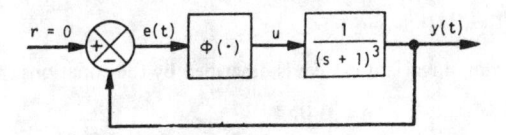

Figure P-14.15.

(*Hint*: Barbashin's result [5] may be used:
The system

$$\dot{x}_1 = x_2$$
$$\dot{x}_2 = x_3$$
$$\dot{x}_3 = -f(x_1) - g(x_2) - ax_3$$

is asymptotically stable in-the-large at the origin provided that

(i) $f(0) = 0 = g(0)$ and f, g are differentiable functions
(ii) $a > 0$
(iii) $\dfrac{f(x_1)}{x_1} \geqslant \epsilon_1 > 0$ if $x_1 \neq 0$
(iv) $\dfrac{ag(x_2)}{x_2} - \dfrac{df(x_1)}{dx_1} \geqslant \epsilon_2 > 0$ if $x_2 \neq 0$

A Liapunov function is

$$V(\mathbf{x}) = aF(x_1) + f(x_1)x_2 + G(x_2) + \frac{(ax_2 + x_3)^2}{2}$$

where

$$F(x_1) \triangleq \int_0^{x_1} f(\tau)\,d\tau; \quad G(x_2) \triangleq \int_0^{x_2} g(\tau)\,d\tau$$

14.16 Consider a nonlinear system described by the equations

$$\dot{x}_1 = -3x_1 + x_2$$
$$\dot{x}_2 = x_1 - x_2 - x_2^3$$

Investigate the stability of equilibrium state.
(*Hint*: Use Krasovskii's method with **P** as identity matrix.)

14.17 A linear autonomous system is described by the state equation

$$\dot{\mathbf{x}} = \mathbf{A}\mathbf{x}$$

$$\mathbf{A} = \begin{bmatrix} -4K & 4K \\ 2K & -6K \end{bmatrix}$$

Find restrictions on the parameter K to guarantee stability of the system.

14.18 Consider the system of Fig. P-14.15. The nonlinear function is replaced by a linear function $\phi(e) = Ke$. Find the restrictions on the parameter K to guarantee system stability. Use the Liapunov's direct method and the Routh criterion. Compare the results.

14.19 Check the stability of the system described by

$$\dot{x}_1 = x_2$$
$$\dot{x}_2 = -x_1 - b_1 x_2 - b_2 x_2^3; \quad b_1, b_2 > 0$$

(*Hint*: Use the variable gradient method with a_{ij}'s as constants. Also try by choosing $a_{12} = x_1/x_2$; $a_{21} = x_2/x_1$.)

14.20 Check the stability of the system described by

$$\dot{x}_1 = x_2$$
$$\dot{x}_2 = -x_1 - x_1^2 x_2$$

14.21 A linear time-invariant system is described by the equations

$$\dot{x} = Ax + Bu$$

The performance index is

$$J = \int_0^\infty (x^T Q x + u^T R u)\, dt$$

where Q and R are constant symmetric, positive definite matrices. The optimal control for the system is

$$u^* = -R^{-1} B^T P x$$

where P is given by the reduced matrix Riccati equation

$$0 = Q + A^T P + PA - PBR^{-1} B^T P$$

Show that if P exists, the closed-loop system is asymptotically stable. (*Hint*: Choose a Liapunov function $V = x^T P x$.)

14.22 Solve Problem 14.15 using the Popov's stability criterion.

14.23 Consider the system shown in Fig. P-14.23. The linear plant is described by the equations

$$\dot{x} = Ax + Bu$$
$$y = Cx$$

$$A = \begin{bmatrix} 0 & 1 \\ -10 & -7 \end{bmatrix}; \quad B = \begin{bmatrix} 1 \\ -4 \end{bmatrix} \quad C = \begin{bmatrix} 1 & 0 \end{bmatrix}$$

Establish an upper limit on K in

$$0 \leqslant \frac{\phi(e)}{e} \leqslant K$$

which ensures global asymptotic stability of the origin.

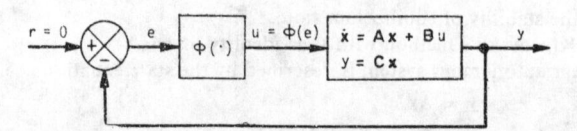

Figure P-14.23.

References and Further Reading

1. Ku, Y.H., *Analysis and Control of Nonlinear Systems*, The Ronald Press, New York, 1958.
2. Graham, D. and D. McRuer, *Analysis of Nonlinear Control Systems*, John Wiley, New York, 1961.
3. Hsu, J.C. and A.U. Meyer, *Modern Control Principles and Applications*, McGraw-Hill, New York, 1968.
4. Schultz, D.G. and J.L. Melsa, *State Functions and Linear Control Systems*, McGraw-Hill, New York, 1967.
5. Ogata, K., *State Space Analysis of Control Systems*, Prentice-Hall, Englewood Cliffs, N.J., 1967.
6. Dorf, R.C., *Time-Domain Analysis and Design of Control Systems*, Addison-Wesley, New York, 1965.
7. Spiegel, M.R., *Vector Analysis*, Schaum, New York, 1959.
8. Brockett, R.W., "The Status of Stability Theory for Deterministic Systems", *IEEE Trans. on Automatic Control*, Vol. AC-11, no. 3, pp. 596-606, July 1966.
9. Narendra, K.S. and J.H. Taylor, *Frequency Domain Criteria for Absolute Stability*, Academic Press, New York, 1973.
10. Holtzman, J.M., *Nonlinear System Theory: A Functional Analysis Approach*, Prentice-Hall, Englewood Cliffs, N.J., 1970.
11. Willems, J.L., *Stability Theory of Dynamical Systems*, John Wiley, New York, 1970.
12. Minorsky, N., *Theory of Nonlinear Control Systems*, McGraw-Hill, New York, 1969.
13. Siljak, D.D., *Nonlinear Systems: The Parameter Analysis and Design*, John Wiley, New York, 1969.
14. Lefschetz, S., *Stability of Nonlinear Control Systems*, Academic Press, New York, 1965.
15. Leondes, C.T., *Advances in Control Systems*, Vol. 2, Academic Press, New York, 1965.
16. Aizerman, M.A. and F.R. Gantmacher, *Absolute Stability of Regulator Systems*, Holden-Day, San Francisco, 1964.
17. Gibson, J.E., *Nonlinear Automatic Control*, McGraw-Hill, New York, 1963.
18. Thaler, G.J. and M.P. Pastel, *Analysis and Design of Nonlinear Feedback Control Systems*, McGraw-Hill, New York, 1962.
19. LaSalle, J.P. and S. Lefschetz, *Stability by Liapunov's Direct Method with Applications*, Academic Press, New York, 1961.
20. Gille, J.C., M.J. Pelegrin and P. Decauline, *Feedback Control Systems: Analysis, Synthesis and Design*, McGraw-Hill, New York, 1959.

APPENDIX I

Fourier and Laplace Transforms

I The Fourier Transform

The Fourier transform of a single-valued function $f(t)$ is defined as the integral

$$F(\omega) = \int_{-\infty}^{\infty} f(t) e^{-j\omega t} \, dt = \mathcal{F}[f(t)] \qquad (I.1)$$

where $\mathcal{F}$ is a symbol indicating the Fourier transform operation. The necessary condition for existence of the Fourier transform is

$$\int_{-\infty}^{\infty} |f(t)| \, dt < \infty \qquad (I.2)$$

There are many useful functions, e.g., sine waves, step functions etc., which are not absolutely integrable, i.e., they do not satisfy condition (I.2). Many such functions do, however, possess Fourier transforms if we allow the transforms to include impulse functions.
The inverse Fourier transform is given by

$$f(t) = \frac{1}{2\pi} \int_{-\infty}^{\infty} F(\omega) e^{j\omega t} \, d\omega = \mathcal{F}^{-1}[F(\omega)] \qquad (I.3)$$

where $\mathcal{F}^{-1}$ is the symbol for the inverse Fourier transform operation. Equation (I.3) may be interpreted as decomposition of $f(t)$ in terms of a continuum of elementary basic functions $\exp(j\omega t)$. $F(\omega)$ is therefore the continuous frequency spectrum of $f(t)$.
Important properties of the Fourier transform are given in Table I.1. Most of these have similar counterparts for the Laplace transform presented later.

Fourier Transform of Periodic Functions

A periodic function $f(t)$ of period T can be expanded in terms of *Fourier series* as

$$f(t) = \sum_{n=-\infty}^{\infty} F_n e^{jn\omega_0 t}; \quad \omega_0 = \frac{2\pi}{T}$$

Appendix I

Table I.1 PROPERTIES OF FOURIER TRANSFORMS

Transform pair	$f(t) \leftrightarrow F(\omega)$
Linearity	$a_1 f_1(t) + a_2 f_2(t) \leftrightarrow a_1 F_1(\omega) + a_2 F_2(\omega)$
Symmetry	$F(t) \leftrightarrow 2\pi f(-\omega)$
Scale change	$f(at) \leftrightarrow \dfrac{1}{|a|} F\left(\dfrac{\omega}{a}\right)$
Real translation	$f(t - t_0) \leftrightarrow e^{-j\omega t_0} F(\omega)$
Complex translation	$e^{j\omega_0 t} f(t) \leftrightarrow F(\omega - \omega_0)$
Real differentiation	$\dfrac{d^n}{dt^n} f(t) \leftrightarrow (j\omega)^n F(\omega)$
Real integration	$\displaystyle\int_{-\infty}^{t} f(\tau)\, d\tau \leftrightarrow \dfrac{F(\omega)}{j\omega} + \pi F(0)\delta(\omega)$
Real multiplication (complex convolution)	$f_1(t) f_2(t) \leftrightarrow \dfrac{1}{2\pi} F_1(\omega) * F_2(\omega)$
Complex multiplication (real convolution)	$f_1(t) * f_2(t) \leftrightarrow F_1(\omega) F_2(\omega)$

where

$$F_n = \frac{1}{T} \int_{-T/2}^{T/2} f(t)\, e^{-jn\omega_0 t}\, dt$$

are the *complex Fourier coefficients* of $f(t)$.
The Fourier transform of $f(t)$ is therefore

$$F(\omega) = \mathcal{F}[f(t)] = \mathcal{F}\left[\sum_{n=-\infty}^{\infty} F_n e^{jn\omega_0 t}\right]$$

$$= \sum_{n=-\infty}^{\infty} F_n \mathcal{F}[e^{jn\omega_0 t}] \quad (I.4)$$

From basic definition (I.1), we have

$$\mathcal{F}[\delta(t)] = \int_{-\infty}^{\infty} \delta(t)\, e^{-j\omega t}\, dt = 1$$

Using the real translation property (Table I.1),

$$\mathcal{F}[\delta(t - t_0)] = e^{-j\omega t_0}$$

Using the symmetry property (Table I.1),

$$\mathcal{F}[e^{j\omega_0 t}] = 2\pi \delta(\omega - \omega_0)$$

Therefore, eqn. (I.4) may be written as

$$F(\omega) = 2\pi \sum_{n=-\infty}^{\infty} F_n \delta(\omega - n\omega_0) \quad (I.5)$$

As an example, consider the unit impulse train

$$f(t) = \sum_{k=-\infty}^{\infty} \delta(t - kT)$$

$$= \sum_{n=-\infty}^{\infty} F_n\, e^{jn\omega_0 t}\,; \quad \omega_0 = \frac{2\pi}{T}$$

where
$$F_n = \frac{1}{T} \int_{-T/2}^{T/2} f(t) e^{-jn\omega_0 t} dt = \frac{1}{T}$$

Therefore
$$F(\omega) = \frac{2\pi}{T} \sum_{n=-\infty}^{\infty} \delta(\omega - n\omega_0) \tag{I.6}$$

I.2 The Laplace Transform

The single-sided Laplace transform (commonly referred to as the Laplace transform) of a *causal function** ($f(t) = 0$ for $t < 0$) is defined as

$$F(s) = \mathcal{L}[f(t)] = \int_0^\infty f(t) e^{-st} dt \tag{I.7}$$

where
$$s = \sigma + j\omega$$

is the complex frequency variable.

In eqn. (I.7), $\mathcal{L}$ is the symbol indicative of the Laplace transforming operation.

The condition for the existence of the Laplace transform of $f(t)$ is

$$\int_0^\infty |f(t)| e^{-\sigma t} dt < \infty \tag{I.8}$$

where σ is any real number.

As an example, consider $f(t) = e^{-at}$, $a > 0$. Then

$$\mathcal{L}[f(t)] = \mathcal{L}(e^{-at}) = \int_0^\infty e^{-at} e^{-st} dt$$
$$= \int_0^\infty e^{-(a+s)t} dt = \frac{1}{s+a}$$

It is easily observed in this case that the condition (I.8) is satisfied if $(a + \sigma) > 0$ or $\sigma > -a$.

The inverse Laplace transform is given by

$$f(t) = \mathcal{L}^{-1}[F(s)] = \frac{1}{2\pi j} \int_{\sigma_1 - j\infty}^{\sigma_1 + j\infty} F(s) e^{st} ds \tag{I.9}$$

where $\sigma_1 > \sigma$, the convergence factor in (I.8).

In eqn. (I.9), $\mathcal{L}^{-1}$ is the symbol indicative of the inverse Laplace transforming operation.

Important properties of the Laplace transform are given in Table I.2.

The inverse Laplace transformation (I.9) involves complex integration which is quite tedious. However, inverse transformation can be easily carried out by expanding $F(s)$ into partial fractions and then using a table

---

*In majority of engineering applications, the time functions of interest are causal. Noncausal functions can be treated through the double-sided Laplace transform which is similar in this respect to the Fourier transform but has a stronger convergence property.

Appendix I

Table 1.2 PROPERTIES OF THE LAPLACE TRANSFORMS

Transform pair	$f(t) \leftrightarrow F(s)$
Linearity	$a_1 f_1(t) + a_2 f_2(t) \leftrightarrow a_1 F_1(s) + a_2 F_2(s)$
Scale change	$f\left(\dfrac{t}{a}\right) \leftrightarrow aF(as); \ a > 0$
Real translation	$f(t - t_0) \leftrightarrow e^{-st_0} F(s)$
Complex translation	$e^{-at} f(t) \leftrightarrow F(s + a)$
Real differentiation	$\dfrac{d^n}{dt^n} f(t) \leftrightarrow s^n F(s) - s^{n-1} f(0) - s^{n-2} \dfrac{df}{dt}(0)$ $- \cdots - \dfrac{d^{n-1} f}{dt^{n-1}}(0)$
Real integration	$\displaystyle\int_{-\infty}^{t} f(t)\, dt \leftrightarrow \dfrac{F(s)}{s} + \dfrac{\left.\int_{-\infty}^{t} f(t)\, dt\right|_{t=0}}{s}$
Multiplication by t	$t^n f(t) \leftrightarrow (-1)^n \dfrac{d^n F(s)}{ds^n}$
Multiplication by $\dfrac{1}{t}$	$\dfrac{1}{t} f(t) \leftrightarrow \displaystyle\int_{s}^{\infty} F(s)\, ds$
Real multiplication (complex convolution)	$f_1(t) f_2(t) \leftrightarrow F_1(s) * F_2(s)$
Complex multiplication (real convolution)	$f_1(t) * f_2(t) \leftrightarrow F_1(s) F_2(s)$
Initial value theorem	$\displaystyle\lim_{t \to 0} f(t) = \lim_{s \to \infty} sF(s)$
Final value theorem	$\displaystyle\lim_{t \to \infty} f(t) = \lim_{s \to 0} sF(s)$

By use of the real integration property and the final value theorem, we obtain

$$\int_0^\infty f(t)\, dt = \lim_{t \to \infty} \int_0^t f(t)\, dt = \lim_{s \to 0} s\, \frac{F(s)}{s} = \lim_{s \to 0} F(s)$$

of transform pairs. This table is based on the fact that the Laplace transform has the uniqueness property, i.e., for any given $f(t)$ there is one and only one $F(s)$ and vice-versa. The transform pairs for some of the commonly encountered functions are given in Table I.3.

PARSEVAL'S THEOREM

Consider a causal signal $f(t)$. Its integral square value (also called energy) is

$$\int_0^\infty f^2(t)\, dt \tag{I.10}$$

Table I.3 LAPLACE TRANSFORM PAIRS

	$F(s)$	$f(t); t \geqslant 0$
1	1	Unit impulse, $\delta(t)$
2	$1/s$	Unit step, $u(t)$
3	$\dfrac{n!}{s^{n+1}}$	t^n (n = integer)
4	$\dfrac{1}{s+a}$	e^{-at}
5	$\dfrac{1}{(s+a)^n}$	$\dfrac{1}{(n-1)!}\, t^{n-1}e^{-at}$ (n = integer)
6	$\dfrac{a}{s(s+a)}$	$1 - e^{-at}$
7	$\dfrac{1}{(s+a)(s+b)}$	$\dfrac{1}{b-a}(e^{-at} - e^{-bt})$
8	$\dfrac{s}{(s+a)(s+b)}$	$\dfrac{1}{b-a}(be^{-bt} - ae^{-at})$
9	$\dfrac{s}{s(s+a)(s+b)}$	$\dfrac{1}{ab}\left[1 + \dfrac{1}{a-b}(be^{-at} - ae^{-bt})\right]$
10	$\dfrac{\omega}{s^2+\omega^2}$	$\sin \omega t$
11	$\dfrac{s}{s^2+\omega^2}$	$\cos \omega t$
12	$\dfrac{\omega}{(s+a)^2+\omega^2}$	$e^{-at}\sin \omega t$
13	$\dfrac{s+a}{(s+a)^2+\omega^2}$	$e^{-at}\cos \omega t$
14	$\dfrac{1}{s^2(s+a)}$	$\dfrac{1}{a^2}(at - 1 + e^{-at})$
15	$\dfrac{\omega_n^2}{s^2 + 2\zeta\omega_n s + \omega_n^2}$	$\dfrac{\omega_n}{\sqrt{(1-\zeta^2)}}\, e^{-\zeta\omega_n t}\sin[\omega_n\sqrt{(1-\zeta^2)}]t\,;\,\zeta<1$
16	$\dfrac{s}{s^2 + 2\zeta\omega_n s + \omega_n^2}$	$\dfrac{-1}{\sqrt{(1-\zeta^2)}}\, e^{-\zeta\omega_n t}\sin[\omega_n\sqrt{(1-\zeta^2)}t - \phi]$
		$\phi = \tan^{-1}\dfrac{\sqrt{(1-\zeta^2)}}{\zeta}\,;\,\zeta<1$
17	$\dfrac{\omega_n}{s(s^2 + 2\zeta\omega_n s + \omega_n^2)}$	$1 - \dfrac{1}{\sqrt{(1-\zeta^2)}}\, e^{-\zeta\omega_n t}\sin[\omega_n\sqrt{(1-\zeta^2)}t + \phi]$
		$\phi = \tan^{-1}\dfrac{\sqrt{(1-\zeta^2)}}{\zeta}\,;\,\zeta<1$

This concept is meaningful only if the signal energy is finite. It immediately follows that the signal is then absolutely integrable, i.e.,

$$\int_0^\infty |f(t)|\, dt < \infty$$

Let us express the signal energy in terms of its Laplace transform. We can write (I.10) as

$$\int_0^\infty f^2(t)\, dt = \int_0^\infty f(t) \left[\frac{1}{2\pi j} \int_{\sigma-j\infty}^{\sigma+j\infty} F(s) e^{st}\, ds \right] dt$$

Interchanging the order of integration

$$\int_0^\infty f^2(t)\, dt = \frac{1}{2\pi j} \int_{\sigma-j\infty}^{\sigma+j\infty} F(s) \left[\int_0^\infty f(t) e^{st}\, dt \right] ds$$

By definition of the Laplace transform we recognise

$$\int_0^\infty f(t) e^{st}\, dt = F(-s)$$

Hence

$$\int_0^\infty f^2(t)\, dt = \frac{1}{2\pi j} \int_{\sigma-j\infty}^{\sigma+j\infty} F(s) F(-s)\, ds$$

Since the signal is absolutely integrable, i.e., $\sigma = 0$, we can write

$$\int_0^\infty f^2(t)\, dt = \frac{1}{2\pi j} \int_{-j\infty}^{j\infty} F(s) F(-s)\, ds \tag{I.11}$$

This result is known as the Parseval's theorem.

SOLUTION OF DIFFERENTIAL EQUATIONS BY THE LAPLACE TRANSFORM

The solution of a linear differential equation with constant coefficients (pertaining to a linear system) can be conveniently obtained through the Laplace transform technique in the following three steps:

(1) Using the real differentiation and the linearity properties of Table I.2 and the transform pairs of Table I.3, transform the differential equation into an algebraic equation in the complex variable s.

(2) Through algebraic manipulation solve for the unknown in terms of the complex variable s.

(3) The time domain solution of the unknown is obtained by inverse Laplace transforming the s-domain solution.

The above procedure is best illustrated by means of an example.

Example: Solve the second-order differential equation

$$\frac{d^2y}{dt^2} + 5\frac{dy}{dt} + 6y = 3t \tag{I.12}$$

for which at $t = 0$, $y = 2$ and $dy/dt = -1$.

Solution:

Laplace transforming both sides of eqn. (I.12) using the real differentiation and the linearity properties and the table of transform pairs, we get

$$[s^2 Y(s) - 2s + 1] + 5[sY(s) - 2] + 6Y(s) = 3/s^2$$

which gives
$$Y(s) = \frac{2s^3 + 9s^2 + 3}{s^2(s^2 + 5s + 6)} \tag{I.13}$$

Equation (I.13) presents the solution in terms of the s-variable. If the solution in time domain is desired, the inverse Laplace transform of $Y(s)$ is to be obtained.

The function given by eqn. (I.13) does not appear in Table I.3. We may, therefore, expand $Y(s)$ into partial fractions and write it in terms of simple functions of s. Then by linearity property, the inverse Laplace transform may be obtained.

The partial function expansion of $Y(s)$ is
$$Y(s) = \frac{-5/12}{s} + \frac{1/2}{s^2} - \frac{10/3}{s+3} + \frac{23/4}{s+2}$$

Using the table of transform pairs, we get
$$y(t) = -\frac{5}{12} + \frac{1}{2} t - \frac{10}{3} e^{-3t} + \frac{23}{4} e^{-2t} \tag{I.14}$$

which is the required solution in time domain.

We observe from the above example that the final step in solving a differential equation by the Laplace transform is to obtain the inverse Laplace transform of a *rational algebraic function* (a ratio of two polynomials; see eqn. (I.13)) by breaking it up into partial fractions. Partial fraction expansion of rational algebraic functions is briefly reviewed in the next section.

I.3 Partial Fraction Expansions

As seen above, in obtaining the solution of linear differential equations with constant coefficients (or analysis of linear time-invariant systems) by the Laplace transform method, we encounter rational algebraic fractions that are ratio of two polynomials in s, such as

$$F(s) = \frac{P(s)}{Q(s)}$$
$$= \frac{b_0 s^m + b_1 s^{m-1} + \ldots + b_m}{a_0 s^n + a_1 s^{n-1} + \ldots + a_n} \tag{I.15}$$

In practical systems, the order of polynomial in numerator is equal to or less than that of denominator. In terms of the orders m and n, rational algebraic fractions are subdivided as follows:

(i) Improper fraction if $m \geqslant n$
(ii) Proper fraction if $m < n$

A proper fraction can further be expanded into partial fractions as will be seen shortly. An improper fraction can be separated into a sum of a polynomial in s and a proper fraction, i.e.,

$$F(s) = \underbrace{\frac{p(s)}{Q(s)}}_{\text{Improper}} = d(s) + \underbrace{\frac{P(s)}{Q(s)}}_{\text{Proper}}$$

This can be achieved by performing a long devision.

To obtain the partial fraction expansion of a proper fraction, first of all we factorize the polynomial $Q(s)$ into n first-order factors (see Appendix III). The roots may be real, complex, distinct or repeated. Various cases are discussed below.

PARTIAL FRACTION EXPANSION WHEN $Q(s)$ HAS DISTINCT ROOTS

In this case, eqn. (I.15) may be written as

$$F(s) = \frac{P(s)}{Q(s)} = \frac{P(s)}{(s+p_1)(s+p_2)\ldots(s+p_k)\ldots(s+p_n)} \quad (I.16)$$

which when expanded, gives

$$F(s) = \frac{A_1}{s+p_1} + \frac{A_2}{s+p_2} + \ldots + \frac{A_k}{s+p_k} + \ldots + \frac{A_n}{s+p_n}$$

(I.17)

The coefficient A_k is called the *residue* at the pole $s = -p_k$.

To evaluate A_k, multiply $F(s)$ in eqn. (I.16) by $(s+p_k)$ and let $s = -p_k$. This gives

$$A_k = (s+p_k)\frac{P(s)}{Q(s)}\bigg|_{s=-p_k} = \frac{P(s)}{\frac{d}{ds}Q(s)}\bigg|_{s=-p_k} \quad (I.18)$$

$$= \frac{P(-p_k)}{(p_1-p_k)(p_2-p_k)\ldots(p_{k-1}-p_k)(p_{k+1}-p_k)\ldots(p_n-p_k)}$$

(I.19)

From eqn. (I.17), we obtain

$$f(t) = \mathcal{L}^{-1}[F(s)] = A_1 e^{-p_1 t} + A_2 e^{-p_2 t} + \ldots + A_k e^{-p_k t} + \ldots + A_n e^{-p_n t}$$

(I.20)

Consider, for example

$$F(s) = \frac{2(s^2+3s+1)}{s(s+1)(s+2)} = \frac{P(s)}{Q(s)}$$

$$= \frac{A_1}{s} + \frac{A_2}{s+1} + \frac{A_3}{s+2}$$

$$A_1 = sF(s)\bigg|_{s=0} = \frac{2(s^2+3s+1)}{(s+1)(s+2)}\bigg|_{s=0} = 1$$

$$A_2 = (s+1)F(s)\bigg|_{s=-1} = \frac{2(s^2+3s+1)}{s(s+2)}\bigg|_{s=-1} = 2$$

$$A_3 = (s+2)F(s)\bigg|_{s=-2} = \frac{2(s^2+3s+1)}{s(s+1)}\bigg|_{s=-2} = -1$$

Partial Fraction Expansion when $Q(s)$ has Complex Conjugate Roots

Suppose that there is a pair of complex conjugate roots in $Q(s)$, given by
$$s = -a - j\omega \quad \text{and} \quad s = -a + j\omega$$
Then $F(s)$ may be written as

$$F(s) = \frac{P(s)}{Q(s)} = \frac{P(s)}{(s + a + j\omega)(s + a - j\omega)(s + p_3)(s + p_4) \ldots (s + p_n)} \quad (I.21)$$

which when expanded gives

$$F(s) = \frac{A_1}{(s + a + j\omega)} + \frac{A_1^*}{(s + a - j\omega)} + \frac{A_3}{s + p_3} + \frac{A_4}{s + p_4} + \ldots + \frac{A_n}{s + p_n} \quad (I.22)$$

where A_1 and A_1^* are residues at the poles $s = -(a + j\omega)$ and $s = -(a - j\omega)$ respectively. These residues form a complex conjugate pair. From eqn. (I.22), the inverse Laplace transform of $F(s)$ may be obtained as follows.

$$f(t) = \mathcal{L}^{-1}[F(s)] = \mathcal{L}^{-1}\left[\frac{A_1}{s + a + j\omega} + \frac{A_1^*}{s + a - j\omega}\right]$$
$$+ A_3 e^{-p_3 t} + \ldots + A_n e^{-p_n t}$$
$$= 2\text{Re}[A_1 e^{-(a+j\omega)t}] + A_3 e^{-p_3 t} + \ldots + A_n e^{-p_n t} \quad (I.23)$$

As per eqn. (I.18), the residue A_1 is given by

$$A_1 = \frac{P(s)}{Q(s)}(s + a + j\omega)\bigg|_{s = -(a + j\omega)}$$

Consider, for example

$$F(s) = \frac{2s + 3}{(s + 1)(s^2 + 4s + 5)} = \frac{P(s)}{Q(s)}$$
$$= \frac{2s + 3}{(s + 2 + j1)(s + 2 - j1)(s + 1)}$$
$$= \frac{A_1}{s + 2 + j1} + \frac{A_1^*}{s + 2 - j1} + \frac{A_3}{s + 1}$$

$$A_1 = (s + 2 + j1)F(s)\bigg|_{s = -(2 + j1)}$$
$$= \frac{2s + 3}{(s + 2 - j1)(s + 1)}\bigg|_{s = -(2 + j1)}$$
$$= -\frac{1}{4} + j\frac{3}{4}$$

$$A_1^* = -\frac{1}{4} - j\frac{3}{4}$$

$$A_3 = (s+1)F(s)\bigg|_{s=-1} = \frac{2s+3}{(s^2+4s+5)}\bigg|_{s=-1} = \frac{1}{2}$$

PARTIAL FRACTION EXPANSION WHEN $Q(s)$ HAS REPEATED ROOTS

Assume that root p_1 of $Q(s)$ is of multiplicity r and all other roots are distinct. The function $F(s)$ may be written as

$$F(s) = \frac{P(s)}{Q(s)} = \frac{P(s)}{(s+p_1)^r (s+p_{r+1})(s+p_{r+2})\ldots(s+p_n)} \quad (\text{I}.24)$$

which when expanded, gives

$$F(s) = \frac{A_{1(r)}}{(s+p_1)^r} + \frac{A_{1(r-1)}}{(s+p_1)^{r-1}} + \frac{A_{1(r-2)}}{(s+p_1)^{r-2}} + \ldots + \frac{A_{12}}{(s+p_1)^2}$$

$$+ \frac{A_{11}}{(s+p_1)} + \frac{A_{r+1}}{(s+p_{r+1})} + \frac{A_{r+2}}{(s+p_{r+2})} + \ldots + \frac{A_n}{(s+p_n)} \quad (\text{I}.25)$$

The coefficients of repeated roots may be obtained using the following relation:

$$A_{1(r-i)} = \frac{1}{i!}\left[\frac{d^i}{ds^i}\left\{(s+p_1)^r \frac{P(s)}{Q(s)}\right\}\right]_{s=-p_1} ; i = 0, 1, 2, \ldots, r-1 \quad (\text{I}.26)$$

Then from eqn. (I.25) we obtain

$$f(t) = \mathcal{L}^{-1}[F(s)] = \left[\frac{A_{1(r)}}{(r-1)!} t^{r-1} + \frac{A_{1(r-1)}}{(r-2)!} t^{r-2} + \ldots \right.$$
$$\left. + A_{12}t + A_{11}\right]e^{-p_1 t} + A_{r+1} e^{-p_{r+1}t} + \ldots + A_n e^{-p_n t} \quad (\text{I}.27)$$

Consider, for example

$$F(s) = \frac{1}{s^2(s+1)} = \frac{P(s)}{Q(s)}$$

$$= \frac{A_{12}}{s^2} + \frac{A_{11}}{s} + \frac{A_3}{s+1}$$

$$A_{12} = s^2 F(s)\bigg|_{s=0} = \frac{1}{s+1}\bigg|_{s=0} = 1$$

$$A_{11} = \frac{d}{ds}(s^2 F(s))\bigg|_{s=0} = \frac{d}{ds}\left(\frac{1}{s+1}\right)\bigg|_{s=0}$$

$$= -\frac{1}{(s+1)^2}\bigg|_{s=0} = -1$$

$$A_3 = 1$$

Further Reading

1. Chirlian, P.M., *Signals, Systems and the Computer*, Intext, New York, 1973.
2. Gabel, R.A. and R.A. Roberts, *Signals and Linear Systems*, John Wiley, New York, 1973.

APPENDIX II

Elements of Matrix Analysis

This appendix is a brief review of some properties of vectors and matrices which have been used in Chapters 12, 13 and 14.

II.1 Basic Definitions

Matrix. A matrix is an ordered rectangular array of elements which may be real numbers, complex numbers, functions or operators. The matrix

$$\mathbf{A} = \begin{bmatrix} a_{11} & a_{12} & \cdots & a_{1n} \\ a_{21} & a_{22} & \cdots & a_{2n} \\ \vdots & \vdots & & \vdots \\ a_{m1} & a_{m2} & \cdots & a_{mn} \end{bmatrix} = [a_{ij}] \qquad (\text{II}.1)$$

is a rectangular array of mn elements.

a_{ij} denotes (i, j)th element, i.e., the element located in ith row and jth column. The matrix defined above has m rows and n columns. This matrix is said to be a *rectangular matrix* of order $m \times n$.

When $m = n$, i.e., the number of rows is equal to that of columns, the matrix is said to be a *square matrix* of order n.

An $m \times 1$ matrix, i.e., a matrix having only one column is called a *column vector*.

A $1 \times n$ matrix, i.e., a matrix having only one row is called a *row vector*.

Diagonal matrix. A diagonal matrix is a square matrix whose elements off the main diagonal are all zeros ($a_{ij} = 0$ for $i \neq j$). The following matrix is a diagonal matrix

$$\mathbf{A} = \begin{bmatrix} a_{11} & 0 & \cdots & 0 \\ 0 & a_{22} & \cdots & 0 \\ \vdots & \vdots & & \vdots \\ 0 & 0 & \cdots & a_{nn} \end{bmatrix}$$

Elements of Matrix Analysis

Unit (identity) matrix. A unit matrix **I** is a diagonal matrix whose diagonal elements are all equal to unity ($a_{ij} = 1$ for $i = j$; $a_{ij} = 0$ for $i \neq j$)

$$\mathbf{I} = \begin{bmatrix} 1 & 0 & \cdots & 0 \\ 0 & 1 & \cdots & 0 \\ \vdots & \vdots & & \vdots \\ 0 & 0 & \cdots & 1 \end{bmatrix}$$

Null matrix. A null matrix **0** is a matrix whose elements are all equal to zero.

$$\mathbf{0} = \begin{bmatrix} 0 & 0 & \cdots & 0 \\ 0 & 0 & \cdots & 0 \\ \vdots & \vdots & & \vdots \\ 0 & 0 & \cdots & 0 \end{bmatrix}$$

Determinant of a matrix. For each square matrix, there exists a determinant which is formed by taking the determinant of the elements of the matrix. For example, if

$$\mathbf{A} = \begin{bmatrix} 3 & -2 & 1 \\ -2 & 6 & 4 \\ 1 & 4 & 8 \end{bmatrix} \quad \text{(II.2)}$$

then, $\det(\mathbf{A}) = |\mathbf{A}| = 3\begin{vmatrix} 6 & 4 \\ 4 & 8 \end{vmatrix} - (-2)\begin{vmatrix} -2 & 4 \\ 1 & 8 \end{vmatrix} + 1\begin{vmatrix} -2 & 6 \\ 1 & 4 \end{vmatrix}$

$$= 3(32) + 2(-20) + (-14) = 42 \quad \text{(II.3)}$$

Transpose of a matrix. The transpose of matrix **A** denoted by $\mathbf{A}^T$ is the matrix formed by interchanging the rows and columns of **A**. Namely, if **A** is given by (II.1) then $\mathbf{A}^T$ is given by

$$\mathbf{A}^T = \begin{bmatrix} a_{11} & a_{21} & \cdots & a_{m1} \\ a_{12} & a_{22} & \cdots & a_{m2} \\ \vdots & \vdots & & \vdots \\ a_{1n} & a_{2n} & \cdots & a_{mn} \end{bmatrix} = [a_{ji}]$$

Note that

$$(\mathbf{A}^T)^T = \mathbf{A}$$

Symmetric matrix. A square matrix is symmetric if it is equal to its transpose, i.e.,

$$\mathbf{A}^T = \mathbf{A}$$

For example, matrix **A** given by (II.2) is symmetric.

Skew-symmetric matrix. A square matrix is skew-symmetric if it is equal to its negative transpose, i.e.,

$$\mathbf{A}^T = -\mathbf{A}$$

The matrix

$$A = \begin{bmatrix} 0 & 1 & -3 \\ -1 & 0 & -2 \\ 3 & 2 & 0 \end{bmatrix}$$

is skew-symmetric.

Minor. If the ith row and jth column of determinant **A** are deleted, the remaining $(n-1)$ rows and $(n-1)$ columns form a determinant M_{ij}. This determinant is called the minor of the element a_{ij}.

Principal minor. A minor of $|\mathbf{A}|$ whose diagonal elements are also diagonal elements of $|\mathbf{A}|$, is called a principal minor of $|\mathbf{A}|$.

Laplace expansion formula. The value of determinant of a matrix **A** can be obtained by the so-called Laplace expansion formula.

$$|\mathbf{A}| = \sum_{j=1}^{n} (-1)^{i+j} a_{ij} M_{ij} \quad \text{for any integer } i; \; 1 \leqslant i \leqslant n$$

$$= \sum_{i=1}^{n} (-1)^{i+j} a_{ij} M_{ij} \quad \text{for any integer } j; \; 1 \leqslant j \leqslant n$$

where M_{ij} is the minor of element a_{ij}.

Cofactor. The cofactor C_{ij} of element a_{ij} of the matrix **A** is defined as

$$C_{ij} = (-1)^{i+j} M_{ij}$$

Adjoint matrix. The adjoint matrix of a square matrix **A** is found by replacing each element a_{ij} of matrix **A** by its cofactor C_{ij} and then transposing. For example, if **A** is given by (II.2), then

$$adj\mathbf{A} = \begin{bmatrix} \begin{vmatrix} 6 & 4 \\ 4 & 8 \end{vmatrix} & -\begin{vmatrix} -2 & 4 \\ 1 & 8 \end{vmatrix} & \begin{vmatrix} -2 & 6 \\ 1 & 4 \end{vmatrix} \\ -\begin{vmatrix} -2 & 1 \\ 4 & 8 \end{vmatrix} & \begin{vmatrix} 3 & 1 \\ 1 & 8 \end{vmatrix} & -\begin{vmatrix} 3 & -2 \\ 1 & 4 \end{vmatrix} \\ \begin{vmatrix} -2 & 1 \\ 6 & 4 \end{vmatrix} & -\begin{vmatrix} 3 & 1 \\ -2 & 4 \end{vmatrix} & \begin{vmatrix} 3 & -2 \\ -2 & 6 \end{vmatrix} \end{bmatrix}^T$$

$$= \begin{bmatrix} 32 & 20 & -14 \\ 20 & 23 & -14 \\ -14 & -14 & 14 \end{bmatrix}^T = \begin{bmatrix} 32 & 20 & -14 \\ 20 & 23 & -14 \\ -14 & -14 & 14 \end{bmatrix} \quad \text{(II.4)}$$

Elements of Matrix Analysis

Singular and nonsingular matrices. A square matrix is called singular if its associated determinant is zero and nonsingular if its associated determinant is nonzero.

Rank of a matrix. A matrix **A** is said to have rank r if there exists an $r \times r$ submatrix of **A** which is nonsingular and all other $q \times q$ submatrices (where $q \geqslant r + 1$) are singular.

For example, consider a 4×4 matrix

$$\mathbf{A} = \begin{bmatrix} 1 & 2 & 3 & 4 \\ 0 & 1 & -1 & 0 \\ 1 & 0 & 1 & 2 \\ 1 & 1 & 0 & 2 \end{bmatrix}$$

det $\mathbf{A} = 0$

and

$$\begin{vmatrix} 1 & 2 & 3 \\ 0 & 1 & -1 \\ 1 & 0 & 1 \end{vmatrix} \neq 0$$

Hence the rank of matrix **A** is 3.

Conjugate matrix. The conjugate of a matrix **A** (denoted as **A***) is the matrix whose each element is the complex conjugate of the corresponding element of **A**.

Real matrix. If $\mathbf{A} = \mathbf{A}^*$, then the matrix **A** has all real elements and **A** is said to be a real matrix.

II.2 Elementary Matrix Operations

Equality of matrices. Two matrices **A** and **B** are said to be equal if they have the same number of rows and columns and the elements of the corresponding orientations are equal.

Multiplication of a matrix by a scalar. A matrix is multiplied by a scalar K if all the mn elements are multiplied by K, i.e., if **A** is given by eqn. (II. 1), then

$$K\mathbf{A} = \begin{bmatrix} Ka_{11} & Ka_{12} & \cdots & Ka_{1n} \\ Ka_{21} & Ka_{22} & \cdots & Ka_{2n} \\ \vdots & \vdots & & \vdots \\ Ka_{m1} & Ka_{m2} & \cdots & Ka_{mn} \end{bmatrix}$$

Further

$$(K\mathbf{A})^T = K\mathbf{A}^T$$

Addition and subtraction of matrices. Addition of two matrices **A** and **B** results in a new matrix **C** with its element c_{ij} equal to the sum of the corresponding elements a_{ij} and b_{ij}. We then denote

$$\mathbf{C} = \mathbf{A} + \mathbf{B}$$

The addition of two 3×3 matrices is illustrated below:

$$\mathbf{C} = \begin{bmatrix} a_{11} & a_{12} & a_{13} \\ a_{21} & a_{22} & a_{23} \\ a_{31} & a_{32} & a_{33} \end{bmatrix} + \begin{bmatrix} b_{11} & b_{12} & b_{13} \\ b_{21} & b_{22} & b_{23} \\ b_{31} & b_{32} & b_{33} \end{bmatrix}$$

$$= \begin{bmatrix} a_{11} + b_{11} & a_{12} + b_{12} & a_{13} + b_{13} \\ a_{21} + b_{21} & a_{22} + b_{22} & a_{23} + b_{23} \\ a_{31} + b_{31} & a_{32} + b_{32} & a_{33} + b_{33} \end{bmatrix}$$

It is evident that addition of two matrices is possible only if they have the same number of rows and columns. Further

$$(\mathbf{A} + \mathbf{B})^T = \mathbf{A}^T + \mathbf{B}^T$$

Similar arguments apply for subtraction of matrices.

It should be noted that any square matrix **A** may be written as the sum of a symmetric matrix $\mathbf{A}_s$ and a skew-symmetric matrix $\mathbf{A}_{sk}$, as is shown below.

Let

$$\mathbf{A} = \mathbf{A}_s + \mathbf{A}_{sk} \tag{II.5}$$

Taking transpose of both sides,

$$\mathbf{A}^T = \mathbf{A}_s^T + \mathbf{A}_{sk}^T$$

$$= \mathbf{A}_s - \mathbf{A}_{sk} \tag{II.6}$$

Solving eqns. (II.5) and (II.6) simultaneously, we obtain

$$\mathbf{A}_s = \frac{\mathbf{A} + \mathbf{A}^T}{2}; \quad \mathbf{A}_{sk} = \frac{\mathbf{A} - \mathbf{A}^T}{2}$$

Multiplication of matrices. The multiplication of two matrices is possible only if number of columns of the first matrix is equal to number of rows of the second. Such matrices are called *conformal*. If a matrix **A** is of order $m \times n$ and **B** is of order $n \times q$, then the product **C** is of order $m \times q$. The elements c_{ij} of the product

$$\mathbf{C} = \mathbf{AB}$$

are given by

$$c_{ij} = \sum_{k=1}^{n} a_{ik} b_{kj}$$

In other words the elements c_{ij} are found by multiplying the elements of the *i*th row of **A** with the elements of the *j*th column of **B** and then summing these element products. For example

Elements of Matrix Analysis

$$\begin{bmatrix} a_{11} & a_{12} & a_{13} \\ a_{21} & a_{22} & a_{23} \\ a_{31} & a_{32} & a_{33} \end{bmatrix} \begin{bmatrix} b_{11} & b_{12} & b_{13} \\ b_{21} & b_{22} & b_{23} \\ b_{31} & b_{32} & b_{33} \end{bmatrix} = \begin{bmatrix} c_{11} & c_{12} & c_{13} \\ c_{21} & c_{22} & c_{23} \\ c_{31} & c_{32} & c_{33} \end{bmatrix}$$

where

$$c_{11} = a_{11}b_{11} + a_{12}b_{21} + a_{13}b_{31}$$
$$c_{12} = a_{11}b_{12} + a_{12}b_{22} + a_{13}b_{32}$$
$$\vdots$$

It is important to note that in general matrix multiplication is not *commutative*, i.e.,

$$\mathbf{AB} \neq \mathbf{BA}$$

If $\mathbf{AB} = \mathbf{BA}$, we say matrices $\mathbf{A}$ and $\mathbf{B}$ commute.

Multiplication is *associative*; i.e.,

$$(\mathbf{AB})\mathbf{C} = \mathbf{A}(\mathbf{BC})$$

Multiplication is *distributive* with respect to addition, i.e.,

$$\mathbf{A}(\mathbf{B} + \mathbf{C}) = \mathbf{AB} + \mathbf{AC}$$

Multiplication of any matrix by a unit matrix results in the original matrix; i.e.,

$$\mathbf{AI} = \mathbf{A}$$

The transpose of the product of two matrices is the product of their trasnposes in reverse order, i.e.,

$$(\mathbf{AB})^T = \mathbf{B}^T \mathbf{A}^T$$

The concept of matrix multiplication assists in the solution of simultaneous linear equations. Consider a set of linear algebraic equations

$$\begin{aligned} a_{11}x_1 + a_{12}x_2 + \ldots + a_{1n}x_n &= c_1 \\ a_{12}x_1 + a_{22}x_2 + \ldots + a_{2n}x_n &= c_2 \\ \vdots \quad\quad \vdots \quad\quad\quad\quad \vdots \quad\quad &\;\; \vdots \\ a_{m1}x_1 + a_{m2}x_2 + \ldots + a_{mn}x_n &= c_m \end{aligned} \quad (\text{II.7})$$

or

$$\sum_{j=1}^{n} a_{ij}x_j = c_i; \; i = 1, 2, \ldots, m$$

Using the rules of matrix multiplication defined above, eqns. (II.7) can be written in the compact notation as

$$\mathbf{Ax} = \mathbf{c} \quad (\text{II.8})$$

where

$$\mathbf{A} = \begin{bmatrix} a_{11} & a_{12} & \cdots & a_{1n} \\ a_{21} & a_{22} & \cdots & a_{2n} \\ \vdots & \vdots & & \vdots \\ a_{m1} & a_{m2} & \cdots & a_{mn} \end{bmatrix}$$

$$\mathbf{x} = \begin{bmatrix} x_1 \\ x_2 \\ \vdots \\ x_n \end{bmatrix}; \; \mathbf{c} = \begin{bmatrix} c_1 \\ c_2 \\ \vdots \\ c_m \end{bmatrix}$$

It is evident that vector-matrix eqn. (II.8) is a useful shorthand representation of the set of eqns. (II.7).

Matrix inversion. The inverse of a square matrix $\mathbf{A}$ is written as $\mathbf{A}^{-1}$ and is defined by the relation

$$\mathbf{A}^{-1}\mathbf{A} = \mathbf{A}\mathbf{A}^{-1} = \mathbf{I}$$

$\mathbf{A}^{-1}$ is determined by the relation

$$\mathbf{A}^{-1} = \frac{\text{adj } \mathbf{A}}{\det \mathbf{A}} \tag{II.9}$$

For example, if $\mathbf{A}$ is given by eqn. (II.2), then from eqns. (II.3), (II.4) and (II.9), we get

$$\mathbf{A}^{-1} = \frac{\text{adj } \mathbf{A}}{\det \mathbf{A}} = \frac{1}{42} \begin{bmatrix} 32 & 20 & -14 \\ 20 & 23 & -14 \\ -14 & -14 & 14 \end{bmatrix}$$

The inverse of the product of two nonsingular square matrices of the same order is the product of their inverses in reverse order, i.e.,

$$(\mathbf{AB})^{-1} = \mathbf{B}^{-1}\mathbf{A}^{-1}$$

Also $$(\mathbf{A}^T)^{-1} = (\mathbf{A}^{-1})^T$$

II.3 Definiteness and Quadratic Forms

In this section, we present the concept of sign definiteness of a scalar function V of a vector $\mathbf{x}$.

Norm of a vector. A norm is any function that assigns to every vector $\mathbf{x}$ of real numbers, a real number K satisfying the following four properties:

1. $\|\mathbf{x}\| \geqslant 0$
2. $\|\mathbf{x} + \mathbf{y}\| \leqslant \|\mathbf{x}\| + \|\mathbf{y}\|$
3. $\|\alpha\mathbf{x}\| = \alpha \|\mathbf{x}\|$ for every scalar α
4. $\|\mathbf{x}\| = 0$ if and only if $\mathbf{x} = \mathbf{0}$

Examples of norms are

$$\|\mathbf{x}\| = \sum_{i=1}^{n} |x_i| \; ; \; \mathbf{x} \text{ is } n \times 1 \text{ vector} \tag{II.10}$$

$$\|\mathbf{x}\| = \max_{i} \{|x_i|\} \tag{II.11}$$

$$\|\mathbf{x}\| = (\mathbf{x}^T\mathbf{x})^{1/2} \tag{II.12}$$

$$\|\mathbf{x}\| = (\mathbf{x}^T\mathbf{A}\mathbf{x})^{1/2} \tag{II.13}$$

where $\mathbf{A}$ is positive definite matrix (definition given later in this section).

The norm in eqn. (II.12) represents magnitude or Euclidean length of $\mathbf{x}$. This norm is most commonly used in explicit calculations.

Positive (negative) definite function. A scalar function $V(\mathbf{x})$ is positive (negative) definite if for all $\mathbf{x}$ such that $\|\mathbf{x}\| \leqslant K$,

(i) $V(\mathbf{x}) > 0 \, (< 0); \mathbf{x} \neq \mathbf{0}$
(ii) $V(\mathbf{0}) = 0$

Positive (negative) semidefinite function. A scalar function $V(\mathbf{x})$ is positive (negative) semidefinite if for all $\mathbf{x}$ such that $\|\mathbf{x}\| \leqslant K$,

(i) $V(\mathbf{x}) \geqslant 0 \, (\leqslant 0); \mathbf{x} \neq \mathbf{0}$
(ii) $V(\mathbf{0}) = 0$

In the above definitions if K is made arbitrarily large, the definitions hold in the entire state-space and are said to be global.

Indefinite function. A scalar function $V(\mathbf{x})$ is indefinite if $V(\mathbf{x})$ assumes both positive and negative values within the region $\|\mathbf{x}\| \leqslant K$, no matter how small K is made.

Quadratic form. An expression such as

$$V(x_1, \ldots, x_n) = \sum_{i=1}^{n} \sum_{j=1}^{n} k_{ij} x_i x_j$$

involving terms of second degree in x_i and x_j is known as quadratic form of n variables.

Quadratic form can be compactly expressed in the vector-matrix form as

$$V(\mathbf{x}) = \mathbf{x}^T \mathbf{Q} \mathbf{x} \qquad (\text{II}.14)$$

where $\mathbf{Q}$ is a constant matrix.

In the expanded form

$$V(\mathbf{x}) = [x_1 \; x_2 \; \ldots \; x_n] \begin{bmatrix} q_{11} & q_{12} & \cdots & q_{1n} \\ q_{21} & q_{22} & \cdots & q_{2n} \\ \vdots & \vdots & & \vdots \\ q_{n1} & q_{n2} & \cdots & q_{nn} \end{bmatrix} \begin{bmatrix} x_1 \\ x_2 \\ \vdots \\ x_n \end{bmatrix} \qquad (\text{II}.15)$$

Note that we can always assume $\mathbf{Q}$ to be symmetric. If $\mathbf{Q}$ is not symmetric, it can be proved that

$$\mathbf{x}^T \mathbf{Q} \mathbf{x} = \mathbf{x}^T \mathbf{Q}_s \mathbf{x}$$

This easily follows from eqn. (II.5) and the property $\mathbf{x}^T \mathbf{Q} \mathbf{x} = \mathbf{x}^T \mathbf{Q}^T \mathbf{x}$, which is evident from eqn. (II.15).

Thus only the symmetric portion of $\mathbf{Q}$ is of importance. We shall therefore tacitly assume that $\mathbf{Q}$ is symmetric.

If a scalar function $V(\mathbf{x})$ is of the quadratic form given by eqn. (II.14), the definiteness of $V(\mathbf{x})$ is attributed to matrix $\mathbf{Q}$. Thus we may speak of *definiteness of a matrix.* An important test of definiteness* of a matrix is given by the Sylvester's theorem

―――――――――
*Matrix $\mathbf{Q}$ is positive definite if all the eigenvalues of $\mathbf{Q}$ are positive.

Sylvester's theorem. The necessary and sufficient conditions for a matrix

$$\mathbf{Q} = \begin{bmatrix} q_{11} & q_{12} & q_{13} & \cdots & q_{1n} \\ q_{21} & q_{22} & q_{23} & \cdots & q_{1n} \\ \vdots & \vdots & \vdots & & \vdots \\ q_{n1} & q_{n2} & q_{n3} & \cdots & q_{nn} \end{bmatrix} \tag{II.16}$$

to be positive definite are that all the successive principal minors of $\mathbf{Q}$ be positive, i.e.,

$$q_{11} > 0; \quad \begin{vmatrix} q_{11} & q_{12} \\ q_{21} & q_{22} \end{vmatrix} > 0; \quad \begin{vmatrix} q_{11} & q_{12} & q_{13} \\ q_{21} & q_{22} & q_{23} \\ q_{31} & q_{32} & q_{33} \end{vmatrix} > 0; \ldots; \det[\mathbf{Q}] > 0 \tag{II.17}$$

The matrix $\mathbf{Q}$ is semidefinite if any of the above determinants is zero.

The matrix $\mathbf{Q}$ is negative definite (semidefinite) if the matrix $-\mathbf{Q}$ is positive definite (semidefinite).

If $\mathbf{Q}$ is positive definite, so is $\mathbf{Q}^2$ and $\mathbf{Q}^{-1}$.

It should be noted that the definiteness of a quadratic form scalar function is global.

II.4 Matrix Calculus

1. *Integration*:

   (a) $\mathbf{x} = \mathbf{x}(t);\ \int \mathbf{x}(t)\, dt = \left[\int x_1(t)\, dt \quad \int x_2(t)\, dt \ldots \int x_n(t)\, dt \right]^T$
   ($n \times 1$ vector)

   (b) $\mathbf{A} = \mathbf{A}(t);\ \int \mathbf{A}(t)\, dt = \begin{bmatrix} \int a_{11}(t)\, dt & \int a_{12}(t)\, dt & \ldots & \int a_{1n}(t)\, dt \\ \vdots & \vdots & & \vdots \\ \int a_{m1}(t)\, dt & \int a_{m2}(t)\, dt & \ldots & \int a_{mn}(t)\, dt \end{bmatrix}$
   ($m \times n$ matrix)

2. *Differentiation with respect to a scalar (time)*:

   (a) $\mathbf{x} = \mathbf{x}(t);\ \dfrac{d\mathbf{x}}{dt} = \left[\dfrac{dx_1}{dt} \quad \dfrac{dx_2}{dt} \ldots \dfrac{dx_n}{dt} \right]^T$

   (b) $\mathbf{A} = \mathbf{A}(t);\ \dfrac{d\mathbf{A}}{dt} = \begin{bmatrix} \dfrac{da_{11}}{dt} & \dfrac{da_{12}}{dt} & \cdot & \dfrac{da_{1n}}{dt} \\ \vdots & \vdots & & \vdots \\ \dfrac{da_{m1}}{dt} & \dfrac{da_{m2}}{dt} & \ldots & \dfrac{da_{mn}}{dt} \end{bmatrix}$

3. *Differentiation with respect to a vector*:

   (a) $\mathbf{x} = \mathbf{x}(t);\quad f = f(\mathbf{x});\quad \dfrac{\partial f}{\partial \mathbf{x}} = \left[\dfrac{\partial f}{\partial x_1} \cdot \dfrac{\partial f}{\partial x_2} \ldots \dfrac{\partial f}{\partial x_n} \right]^T \tag{II.18}$
   $n \times 1$ vector   (scalar)

$\dfrac{df}{\partial \mathbf{x}}$ is often called the *gradient* of f with respect to $\mathbf{x}$.

(b) $\mathbf{x} = \mathbf{x}(t)$; $\mathbf{f} = \mathbf{f}(\mathbf{x})$
$(m \times 1$ vector$)$

$$\dfrac{\partial \mathbf{f}}{\partial \mathbf{x}} = \begin{bmatrix} \dfrac{\partial f_1}{\partial x_1} & \dfrac{\partial f_1}{\partial x_2} & \cdots & \dfrac{\partial f_1}{\partial x_n} \\ \vdots & \vdots & & \vdots \\ \dfrac{\partial f_m}{\partial x_1} & \dfrac{\partial f_m}{\partial x_2} & \cdots & \dfrac{\partial f_m}{\partial x_n} \end{bmatrix} \qquad (\text{II}.19)$$

This is sometimes called the *Jacobian matrix*.

4. *Derivative of product and sum:*

(a) $\mathbf{A} = \mathbf{A}(t)$; $\mathbf{B} = \mathbf{B}(t)$
$(m \times n$ matrix$)$ $(n \times q$ matrix$)$
$\mathbf{F} = \mathbf{AB}$
$(m \times q$ matrix$)$

$$\dfrac{d\mathbf{F}}{dt} = \dfrac{d\mathbf{A}}{dt}\mathbf{B} + \mathbf{A}\dfrac{d\mathbf{B}}{dt}$$

(b) $\mathbf{A} = \mathbf{A}(t)$; $\mathbf{B} = \mathbf{B}(t)$; $\mathbf{F} = \mathbf{A} + \mathbf{B}$
$(m \times n$ matrix$)$ $(m \times n$ matrix$)$

$$\dfrac{d\mathbf{F}}{dt} = \dfrac{d\mathbf{A}}{dt} + \dfrac{d\mathbf{B}}{dt}$$

5. *Derivatives of linear and quadratic forms*:

(a) $\dfrac{d\mathbf{x}^T}{d\mathbf{x}} = \mathbf{I}$; $\mathbf{x}$ is $n \times 1$ vector (II.20)

(b) $\dfrac{d(\mathbf{x}^T \mathbf{b})}{d\mathbf{x}} = \mathbf{b}$; $\mathbf{b}$ is $n \times 1$ vector (II.21)

(c) $\dfrac{d(\mathbf{x}^T \mathbf{x})}{d\mathbf{x}} = 2\mathbf{x}$ (II.22)

(d) $\dfrac{d(\mathbf{x}^T \mathbf{A} \mathbf{x})}{d\mathbf{x}} = (\mathbf{A}^T + \mathbf{A})\mathbf{x}$; $\mathbf{A}$ is $n \times n$ matrix (II.23)

(e) $\dfrac{d(\mathbf{x}^T \mathbf{A} \mathbf{x})}{d\mathbf{x}} = 2\mathbf{A}\mathbf{x}$; $\mathbf{A}$ is $n \times n$ symmetric matrix (II.24)

Further Reading

1. Nering, E.D., *Linear Algebra and Matrix Theory*, John Wiley, New York, 1963.
2. Pipes, L.A., *Matrix Methods for Engineering*, Prentice-Hall, Englewood Cliffs, N.J., 1963.
3. Bellman, R., *Introduction to Matrix Analysis*, McGraw-Hill, New York, 1960.

APPENDIX III

Determination of Roots of Algebraic Equations

In this appendix, two methods for determining the roots of algebraic equations are presented. The first is a graphical method based on root locus approach and the other is Lin's method. By root locus method, the inaccuracy in root determination may be around 10%, while Lin's method is far more accurate.

III.1 Root Locus Method

Steps given below are followed for determination of roots by root locus method.

1. Write the original equation such that linear, quadratic, cubic, quartic, etc., terms are individually identified.

For a fourth-order algebraic equation

$$s^4 + a_3 s^3 + a_2 s^2 + a_1 s + a_0 = 0 \qquad (III.1)$$

the rearrangement is as follows:

$$\{[(s + a_3)s + a_2]s + a_1\}s + a_0 = 0 \qquad (III.2)$$

2. Replace the quadratic term in eqn. (III.2) by a product of two factors. Thus

$$[(s + a_3)s + a_2] = (s + \alpha_1)(s + \alpha_2) \qquad (III.3)$$

From eqns. (III.2) and (III.3) we get

$$[(s + \alpha_1)(s + \alpha_2)s + a_1]s + a_0 = 0 \qquad (III.4)$$

3. Consider next the cubic equation

$$(s + \alpha_1)(s + \alpha_2)s + a_1 = 0$$

or

$$1 + \frac{a_1}{s(s + \alpha_1)(s + \alpha_2)} = 1 + G_1(s) = 0$$

Roots of this equation may be found by making a root locus plot of

$$G_1(s) = \frac{a_1}{s(s + \alpha_1)(s + \alpha_2)}$$

Let the roots thus determined be β_1, β_2' and β_3.

Equation (III.4) may now be written as

$$(s + \beta_1)(s + \beta_2)(s + \beta_3)s + a_0 = 0 \qquad (III.5)$$

or
$$1 + \frac{a_0}{s(s+\beta_1)(s+\beta_2)(s+\beta_3)} = 1 + G_2(s) = 0$$

4. Determine the roots of eqn. (III.5) by making the root locus plot of

$$G_2(s) = \frac{a_0}{s(s+\beta_1)(s+\beta_2)(s+\beta_3)} \tag{III.6}$$

The above steps are best illustrated with the help of a numerical example. Let us determine the roots of the cubic equation

$$s^3 + 4s^2 + 6s + 4 = 0 \tag{III.7}$$

Rearranging this equation to yield linear, quadratic and cubic terms, we get

$$[(s+4)s + 6]s + 4 = 0 \tag{III.8}$$

Now

$$[(s+4)s + 6] = (s + 2 + j\sqrt{2})(s + 2 - j\sqrt{2})$$

Therefore eqn. (III.8) may be written as

$$(s + 2 + j\sqrt{2})(s + 2 - j\sqrt{2})s + 4 = 0$$

Rearranging, we get

$$1 + \frac{4}{s(s+2+j\sqrt{2})(s+2-j\sqrt{2})} = 0 \tag{III.9}$$

The root locus plot of eqn. (III.9) is given in Fig. III.1. From this plot, it is found that the roots corresponding to a gain of 4 are

$$s_{1,2} = -1 \pm j1;\ s_3 = -2$$

Therefore eqn. (III.7) can be written in the factored form as

$$(s + 1 + j1)(s + 1 - j1)(s + 2) = 0$$

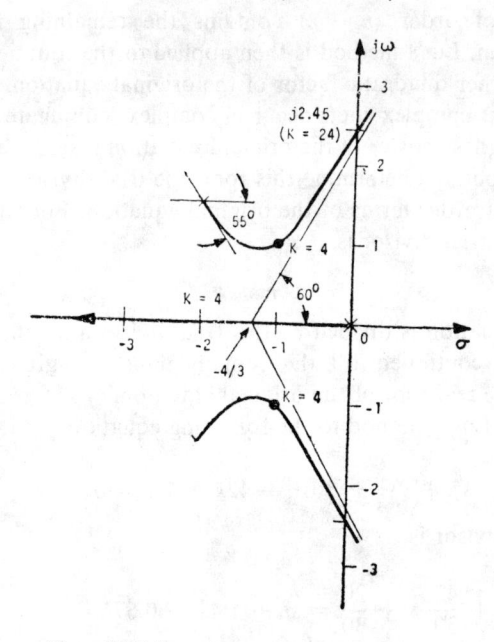

Figure III.1 Root locus plot of eqn. (III.9).

III.2 Lin's Method

Consider the algebraic equation

$$s^n + a_{n-1}s^{n-1} + a_{n-2}s^{n-2} + \ldots + a_2s^2 + a_1s + a_0 = 0 \qquad (\text{III.10})$$

The first trial factor uses the three lowest order terms of the original equation. For eqn. (III.10), this trial factor is

$$s^2 + \frac{a_1}{a_2}s + \frac{a_0}{a_2}$$

The original equation is divided by this first trial factor as follows:

$$
\begin{array}{r}
s^{n-2}+\cdots \\
s^2 + \dfrac{a_1}{a_2}s + \dfrac{a_0}{a_2} \overline{\smash{\big)}\, s^n + a_{n-1}s^{n-1} + a_{n-2}s^{n-2} + \ldots + a_2s^2 + a_1s + a_0}\\
\vdots \qquad \vdots \qquad \vdots \qquad\qquad \vdots \qquad \vdots \qquad \vdots \\
b_2s^2 + b_1s + b_0\\
c_2s^2 + c_1s + c_0\\
\hline
\text{Remainder}
\end{array}
$$

If the remainder is too large, the next trial used is

$$s^2 + \frac{b_1}{b_2}s + \frac{b_0}{b_2}$$

This procedure is continued until the remainder is negligible. The last trial factor is a quadratic factor of the original equation. The quotient polynomial, which is of order $(n-2)$ contains the remaining factors of the original equation. Lin's method is then applied to the quotient polynomial to obtain the other quadratic factor of the original equation.

We know that complex roots occur in complex conjugate pairs. Therefore, if the highest power of the original equation is odd, there must be at least one real root. To determine this root, the trial divisor is chosen from the two lowest order terms of the original equation. For eqn. (III.10), if n is odd, the first trial divisor is

$$s + a_0/a_1$$

The original equation is divided by this trial divisor and the process outlined above is continued till the remainder is negligible. The last trial divisor gives the real root of the original equation.

Let us apply Lin's method to the following equation

$$s^4 + 9s^3 + 30s^2 + 42s + 20 = 0 \qquad (\text{III.11})$$

The first trial divisor is

$$s^2 + \frac{42}{30}s + \frac{20}{30} = s^2 + 1.4s + 0.67$$

Determination of Roots of Algebraic Equations

The division is performed as follows:

$$\begin{array}{r}
s^2+7.6s+18.69\\
s^2+1.4s+0.67\overline{\smash{\big)}\,s^4+9s^3+30s^2+42s+20}\\
\underline{s^4+1.4s^3+0.67s^2}\\
7.6s^3+29.33s^2+42s\\
\underline{7.6s^3+10.64s^2+5.092s}\\
18.69s^2+36.9s+20\\
\underline{18.69s^2+26.16s+12.52}\\
10.74s+7.48
\end{array}$$

Comparing the coefficients of the remainder with those of the first divisor, it is found that the remainder is not negligible.

The second trial divisor is

$$s^2 + \frac{36.9}{18.69}s + \frac{20}{18.69} \approx s^2 + 2s + 1.1$$

Division by the second trial factor is performed as follows:

$$\begin{array}{r}
s^2+7s+14.9\\
s^2+2s+1.1\overline{\smash{\big)}\,s^4+9s^3+30s^2+42s+20}\\
\underline{s^4+2s^3+1.1s^2}\\
7s^3+28.9s^2+42s\\
\underline{7s^3+14s^2+7.7s}\\
14.9s^2+34.3s+20\\
\underline{14.9s^2+29.8s+16.4}\\
4.5s-3.6
\end{array}$$

Third trial factor is

$$s^2 + \frac{34.3}{14.9}s + \frac{20}{14.9} = s^2 + 2.3s + 1.34$$

By continuing the process it can be shown that the fourth trial factor of $(s^2 + 2.5s + 1.5)$ yields a quotient of $(s^2 + 6.5s + 12.2)$ and a remainder of $(1.7s + 1.7)$ which can be considered negligible. Equation (III.11) may therefore be written as

$$(s^2 + 2.5s + 1.5)(s^2 + 6.5s + 12.2) = 0$$

or $(s + 1)(s + 1.5)(s + 3.25 + j\sqrt{1.6})(s^2 + 3.25 - j\sqrt{1.6}) = 0$

More trials may improve accuracy though not necessarily, as the remainder may exhibit oscillatory behaviour.

Further Reading

Towill, D.R., *Transfer Function Techniques for Control Engineers*, Iliffe Books Ltd., London, 1970.

Appendix IV

Answers to Problems
(*Marked answers are approximate)

2.1 (a) $\dfrac{M_1 s^2 + (f_1 + f)s + K_1 + K}{\{M_1 M_2 s^4 + [M_1 f_2 + M_2 f_1 + f(M_1 + M_2)]s^3 + [M_2 K_1 + K(M_1 + M_2) + f_1 f_2 + f(f_1 + f_2)]s^2 + [K_1(f_1 + f_2) + K(f_1 + f_2)]s + K_1 K\}}$

(b) $\dfrac{\dfrac{K}{J_1 J_2}}{s\left[s^3 + \dfrac{f}{J_2} s^2 + \dfrac{K(J_1 + J_2)}{J_1 J_2} s + \dfrac{fK}{J_1 J_2}\right]}$

2.2 $F(t) = M_1 \ddot{y}_1 + f \dot{y}_1 + K_1 y_1 + K_{12}(y_1 - y_2)$
$0 = M_2 \ddot{y}_2 + K_{12}(y_2 - y_1)$

2.3 $F(t) = M_1 \ddot{x}_1 + f_1 \dot{x}_1 + K_1 x_1 + f_{12}(\dot{x}_1 - \dot{x}_2)$
$0 = M_2 \ddot{x}_2 + f_2 \dot{x}_2 + K_2 x_2 + f_{12}(\dot{x}_2 - \dot{x}_1)$

2.4 $\dfrac{k_2}{\{RLCMs^4 + L(M + RCf)s^3 + [RM + Lf + RC(2LK + k_1 k_2)]s^2 + (Rf + 2LK + k_1 k_2)s + 2RK\}}$

2.5 $\dfrac{\theta(s)}{\theta_i(s)} = \dfrac{1}{RCs + 1}$; $\theta(t) = \theta_i\{1 - \exp(-t/RC)\}$

2.6 $\dfrac{C_0(s)}{X(s)} = \dfrac{KC_i/V}{s + Q_0/V}$; $C_0(t) = \dfrac{KC_i x_0}{Q_0}\{1 - \exp(-Q_0 t/V)\}$
where $K = \Delta Q_i / \Delta x$

2.7 $\dfrac{\omega(s)}{E_i(s)} = \dfrac{50}{s + 10.375}$; $\omega(t) = 481.8(1 - \exp(-10.375 t))$

2.8 Closed-loop T.F. $= \dfrac{1}{s(0.1s + 1)(0.2s + 1) + 1}$;

Open-loop T.F. $= \dfrac{1}{s(0.1s + 1)(0.2s + 1)}$

2.9 (a) $\dfrac{G_1(G_2 G_3 + G_4)}{1 + (G_2 G_3 + G_4)(G_1 + H_2) + G_1 H_1 G_2}$

Answers to Problems

(b) $G_4 + \dfrac{G_1G_2G_3}{1 + G_2G_3H_2 + G_2H_1(1 - G_1)}$

2.10 (i) $\dfrac{G_1G_2G_3}{1 + G_2H_3 + G_3H_2 + G_1G_2G_3H_1}$

(ii) $\dfrac{G_3(1 + H_3G_2)}{1 + H_3G_2 + G_3(G_1H_1G_2 + H_2)}$

2.11 $C_1/R_1 = \dfrac{G_1G_2G_3(1 + G_4)}{(1 + G_1G_2)(1 + G_4) - G_1G_4G_5H_1H_2}$

$C_2/R_1 = \dfrac{G_1G_4G_5G_6H_2}{(1 + G_1G_2)(1 + G_4) - H_1H_2G_1G_4G_5}$

2.12 $\dfrac{K_2}{sRC + 1}$

2.13 $\dfrac{(1 + G_4H_2)(G_2 + G_3)G_1}{1 + G_1H_1H_2(G_2 + G_3)}$

2.14 $\dfrac{[G_2G_4G_6(1 + G_5H_2) + G_3G_5G_7(1 + G_4H_1) + G_3G_8G_6 + G_2G_1G_7 - G_3G_8H_1G_1G_7 - G_2G_1H_2G_8G_6]}{1 + G_4H_1 + G_5H_2 - H_1G_1H_2G_8 + G_4H_1G_5H_2}$

2.15 $C_1 = \{[G_1(1 - G_2H_4) + G_3H_4G_4]R_1 + [G_4(1 - G_3H_2) + G_2H_2G_1]R_2\}/\Delta$

$C_2 = \{[G_3(1 - G_4H_1) + G_1H_1G_2]R_1 + [G_2(1 - G_1H_3) + G_4H_3G_3]R_2\}/\Delta$

where $\Delta = 1 - (G_1H_3 + G_2H_4 + G_3H_2 + G_4H_1 + G_1G_2H_1H_2 + G_3G_4H_3H_4) + G_1H_3G_2H_4 + G_3H_2G_4H_1)$

C_1 is independent of R_2 if $H_2 = \dfrac{-G_4}{G_1G_2 - G_3G_4}$

C_2 is independent of R_1 if $H_1 = \dfrac{-G_3}{G_1G_2 - G_3G_4}$

2.16 $\dfrac{X(s)}{U(s)} = \dfrac{\beta_3(s^2 + a_1s + a_2) + \beta_2 s + \beta_1}{s^2 + a_1s + a_2}$

3.1 (b) 0.1, −15 volts, 53 volts (c) 25 volts, −30 volts

3.2 $|S_G^T| = 0.029, |S_H^T| = 1.02$

3.3 (a) $\dfrac{5}{s^2(s + 6) + 10}$ (b) 0.5

3.4 0.04 sec, $K = 0.1$

3.5 (a) $\omega_0(t) = 9.62(1 - e^{-130t})$

(b) $\omega_0(t) = 9.62(1 - e^{-5t})$

(c) For all ω,

$|S_{K_A}^{\omega_0 \text{(with } fb)}| < |S_{K_A}^{\omega_0 \text{(without } fb)}|$

$|S_{\omega_t}^{\omega_0 \text{(with } fb)}| < |S_{\omega_t}^{\omega_0 \text{(without } fb)}|$

3.6 $\dfrac{K}{sRC + 1 + K}$; for all ω, $|S_R^T| \leq |S_R^G|$; time constant reduced by a factor $(1 + K)$ with feedback.

Appendix IV

3.7 (i) $\dfrac{KK_cC_i/V}{s + (Q_0 + KK_cC_i)/V}$

(ii) Steady-state error reduced by a factor $(Q_0 + KK_cC_i)/Q_0$ with feedback.

(iii) Steady-state error reduced by a factor $(Q_0 + KK_cC_i^0)/Q_0$ with feedback.

3.8 (a) $\dfrac{-K_RK_f}{K_2 + K_fK_R}$

3.9 (a) 2×10^{-4} (b) -0.01 rad

3.10 (i) $\dfrac{K_1R}{sRC + 1 + RK_2K_x}$, $\dfrac{K_1}{sRC + 1 + RK_2K_x}$

(ii) Steady-state error is reduced by a factor $(1 + RK_2K_x)$ with feedback.

3.11 (a) $\dfrac{K_GK_B(s\tau_2 + 1)}{s[(s\tau_1 + 1)(s\tau_2 + 1) - K_BK_C(1 - K_T)] + K_GK_B(s\tau_2 + 1)}$

(b) 0

(c) $\dfrac{K_B(s\tau_2 + 1)}{(s\tau_1 + 1)(s\tau_2 + 1) - K_BK_C(1 - K_T)}$

3.12 (a) $S_K^i = 1$, $S_K^T = \dfrac{1}{1 + KK_t}$

(b) For open-loop case, $\Delta\theta(t) = K_2(1 - e^{-t/\tau'})$

For closed-loop case, $\Delta\theta(t)$

$= \dfrac{K_2}{1 + K_1K_t}\left[1 - \exp(-(1 + K_1K_t)\,t/\tau')\right]$

where $K_1 = 2RK_s^2 e_0/(Q_0\rho c + Ah)$, $K_2 = Ah/(Q_0\rho c + Ah)$

$\tau' = Mc/Q_0\rho c + Ah)$

3.13 Open-loop case: $20(1 - e^{-t})$, $50(1 - e^{-0.4t})$

Closed-loop case: $\dfrac{20}{21}(1 - e^{-21t})$, $\dfrac{20}{20.4}(1 - e^{-20.4t})$

4.1 $\dfrac{2.73}{s(0.0157s + 1)}$

4.2 $K_A = 286.4$, $v_C = 50$ volts

4.3 $\dfrac{\theta_C(s)}{\theta_R(s)} = \dfrac{1500}{s(0.01s + 1)(0.02s + 1)(0.25s + 1) + 1500}$

$e_{ss} = 0.667 \times 10^{-3}$ rad

4.4 $\dfrac{X(s)}{Y(s)} = \left(\dfrac{sT_1 + 1}{sT_2 + 1}\right)\dfrac{a}{1 + b}$

where $T_1 = \dfrac{A}{K}\left(\dfrac{a + b}{a}\right)$; $T_2 = \dfrac{A}{K(1 + b)}$;

$K = \dfrac{\text{rate of oil-flow to piston}}{\text{valve opening}}$

Answers to Problems

4.5 $\dfrac{Y_c(s)}{X_r(s)} = \dfrac{\dfrac{bK_rK_1A}{(a+b)K_2}}{Ms^2 + \left(f + \dfrac{A^2}{K_2}\right)s + K + \dfrac{aAK_1}{(a+b)K_2}}$

4.6 $\dfrac{2}{s^2 + 0.02s + 2}$, $\dfrac{1}{0.01s + 1}$

4.8 $\dfrac{\dfrac{bK_p}{a+b}(sT+1)}{sT + 1 + \dfrac{aAK_p}{(a+b)K}}$

5.1 (a) $\theta_0(t) = 1 - 1.355e^{-12.5t} \sin(13.7t + 0.828)$, 13.7 rad/sec, 0.229 sec, 5.6%
  (b) 0.44°
  (c) 1°

5.2 (a) $K_A = 2.1$, $\zeta = 0.22$, $t_s = 1.2$ sec
  (b) $K_D = 0.077$, no effect on steady-state error, $t_s = 0.52$ sec

5.3 (b) $\dfrac{6K_A}{s^2 + 100KK_{AS} + 6K_A}$
  (c) 2.67 amps/volt, 0.024

5.4 (a) 0.0916, 2.87°, (b) 0.458
  (c) Integral-error control reduces the steady-state error to zero.

5.5 20

5.6 (a) 20 rad/sec, (b) 0.025 kg-m², (c) 0.4 newton-m/rad/sec

5.7 (a) $\dfrac{600}{(s+10)(s+60)}$,
  (b) 24.5 rad/sec, 1.43

5.8 100, 0.8 sec, 16.2%, 0.364 sec

5.9 $K_e = 0.116$
  With feedback: $t_s = 2.53$ sec, $M_p = 16.2\%$, $e_{ss} = 0.2$ rad
  Without feedback: $t_s = 4$ sec, $M_p = 35.2\%$, $e_{ss} = 0.2$ rad

5.10 (a) 0.316, 3.16 rad/sec, 0.2 rad
  (b) 1.8, 0.38 rad
  (c) ($K_A = 36$, $K_0 = 5.2$) yield the desired result

5.11 0, 1, ∞, 0.567, 1.323 rad/sec

5.12 $K_p = \infty$, $K_v = 10$, $K_a = 0$, $e_{ss} = \infty$

5.13 $e(t) = 0.1(a_1 + a_2 t)$

5.14 0.0042

5.15 $IAE = 2/\omega_n$, $ITAE = 3/\omega_n^2$

5.16 (i) $\sqrt{K}$, (ii) $K \to \infty$

5.17 $ISE = \zeta + \dfrac{1}{4\zeta}$, $ITSE = \zeta^2 + \dfrac{1}{8\zeta^2}$
  Optimal ζ to minimize $ISE = 0.5$, Min. $ISE = 1$
  Optimal ζ to minimize $ITSE = 0.595$, Min. $ITSE = 0.71$

5.18 $K_A = 0.77$, $K = 0.15$, $\alpha = 4.4$

Appendix IV

6.1 (a) $s_1 = s_2 = -3$, (b) $s_1 = -0.683$, $s_2 = -7.317$, $s_3 = -3$
(c) $s_1 = -3$, $s_2 = 0.5 + j1.66$, $s_3 = 0.5 - j1.66$

6.2
$$a_1 > 0$$
$$a_1 a_2 - a_0 a_3 > 0$$
$$(a_1 a_2 - a_0 a_3)a_3 - a_1^2 a_4 > 0$$
$$a_4 > 0$$

6.3 (a) Stable (b) Two roots in right half s-plane
(c) Two roots in right half s-plane
(d) Four roots on the imaginary axis

6.4 (a) $\infty > k > 0.528$ (b) $36 > k > 0$
(c) Always unstable

6.5 $K = 666.25$; $\omega = 4.06$ rad/sec

6.6 $K = 2$; $a = 0.75$

6.7 (a) $70 > K > 0$ (b) Yes

6.8 Largest time constant is equal to 1 sec

6.9 Maximum value of $K = 7.5$

6.10 $\infty > K > 0.53$

7.1 (i) $-4/3$; 3; $60°, 180°, 300°$
(ii) $-33.7°$
(iii) No breakaway point
(iv) 52, 3.6 rad/sec

7.2* $3.1 < K < 48$, 48, 2.83 rad/sec, 8.34
$$\frac{8.34}{(s + 0.67 + j1.16)(s + 0.67 - j1.16)(s + 4.56)}$$

7.3 No, $1 < K < \infty$, 1, 1 rad/sec, 3, $-1 \pm j\sqrt{2}$

7.4 $\alpha = 1$, $\dfrac{1}{(s + 0.5 + j0.866)(s + 0.5 - j0.866)}$

7.5 (a) With $\alpha = 0$, $e_{ss} = 1$, $\zeta = 0.5$, $t_s = 8$ sec
(b) With $\alpha = 0.2$, $e_{ss} = 1.2$, $\zeta = 0.6$, $t_s = 6.67$ sec
(c) 1

7.6* (c) 4, 0.2, 1.76 rad/sec

7.7 (a) 64, 3.46 rad/sec
(b) 1, 16.2%, 1.815 sec, 4 sec

7.8 $35.7 > K > 23.3$

7.9* $\pm j1.5$, $-1.67 \pm j1.3$

7.10 4 cm

7.11* (b) $0 \leqslant a < 4$
(c) $\dfrac{1}{2aK_A}$
(d) $(a = 0, \zeta = 0.316)$, $(a = 1, \zeta = 0.238)$, $(a = 2, \zeta = 0.15)$, $(a = 4, \zeta = 0)$

7.12* (a) 2 sec, 0.4
(b) One solution is $a = 0.124$, $b = 0.05$

7.13 $S_\alpha^{-r} = 0.125 \angle -143.1°$, $-r = -0.6 + j0.8$

Answers to Problems

7.14 For dominant root $-r = -0.375 + j1.96$,
  (i) $S^-_{K+} = 1.21 \angle 96.4°$
  (ii) $S^{-r}_{p+} = 0.96 \angle -133°$
  (iii) $S^{-r}_{z+} = 0.6 \angle 22.5°$

8.1 (a) Does not cross (b) 0.71 rad/sec, 0.67 (c) Does not cross (d) Crosses the real axis twice at frequencies 16 rad/sec and 375 rad/sec with corresponding amplitudes 8.8×10^{-4} and 5.9×10^{-6} respectively.

8.2 0.683 rad/sec, $-159°$

8.3 (a) 4.1 rad/sec (b) 0.75 rad/sec

8.4 (a) 0.056 (b) 28.5

8.5 (a) $\dfrac{0.625(1 + 0.4s)(1 + 0.1s)}{s(1 + 0.04s)}$ (b) $\dfrac{250}{s(1 + 0.4s)(1 + 0.025s)}$
  (c) $\dfrac{79.8s^2}{(1 + 2s)(1 + s)(1 + 0.2s)}$

8.6* $\dfrac{5(1 + 0.232s)^2}{s(1 + 2s)}$, Type 1

8.7 (a) 475, 26.2 (b) 0.305 sec 25.1 rad/sec

8.8* (a) 2.6, 6.66 rad/sec (b) 0.195, 6.93 rad/sec
  (c) 10.5 rad/sec

8.9* 1.11, 13.25 rad/sec, 24.6 rad/sec

9.1 (a) Two roots in right half s-plane
  (b) Stable
  (c) Two roots on $j\omega$-axis

9.2 (a) Stable (b) Unstable (c) Stable

9.3 $\infty > K > 6$

9.4 0.385, unstable for all values of K

9.5 0.75

9.6 0.168 sec

9.7 $11.36 \times 10^2 < K < 16.95 \times 10^4$

9.8 $K = \dfrac{1}{G_m}\left(\dfrac{1}{T_1} + \dfrac{1}{T_2}\right)$

9.9 $\zeta = 0.01\ \phi_{pm}$;
$$K = \dfrac{1}{D^2}\left[2 \times 10^{-2}\phi_s^2\tau - D \pm 2 \times 10^{-2}\sqrt{(10^{-4}\phi^4\tau^2 - \phi_s^2 D\tau)}\right]$$

9.10 $\omega_c = \sqrt{(\omega_1\omega_2)}$; 90°

9.11* Gain to be changed by a factor of 1/3.5, 42.5°, 0.387, 0.394

9.12* (a) 3.86 db, 10°
  (b) By a factor of 0.156
  (c) By a factor of 0.147

9.13* 1.2, by a factor of 0.77, 1.07 rad/sec

9.14* (a) 0.5 (b) 0.446 (c) 0.63, 0.5 rad/sec 1 rad/sec, (d) 1.35

9.15* 2.35, 2.8 rad/sec, 1.15, 2 rad/sec

9.16* $\left|S_G^T(j\omega)\right|_{peak} = 7$ at $\omega = 7$ rad/sec,

$\left.S_H^T(j\omega)\right|_{peak} = 1$ at $\omega = 0$,

$\omega_b = 1.6$ rad/sec

10.1* Uncompensated system: $K = 7.0$, $K_v = 1.4$, $t_s = 11.6$ sec
Compensated system: $K = 6.0$, $K_v = 12.0$, $t_s = 12.7$ sec

10.2* Uncompensated system: PM = 2.6°, GM = 1.6 db
Compensated system: PM = 37.6°, GM = 18 db

10.3* One solution is

$$G_c(s) = \frac{1.61s + 1}{0.214s + 1},$$ Lead-network attenuation cancelled by amplification $A = 7.55$

10.4* One solution is $G_c(s) = \dfrac{10s + 1}{80s + 1}$,

$C = 10\mu F$, $R_2 = 1M\Omega$, $R_1 = 7M\Omega$

10.5* (a) 56 (b) 3.77 sec, 2.08

(c) One solution is $G_c(s) = \dfrac{(s + 3)(s + 0.9)}{(s + 13.6)(s + 0.15)}$

10.6* One solution is $G_c(s) = \dfrac{(s + 1.5)^2}{(s + 3.5)^2}$

10.7* One solution is $G_c(s) = \dfrac{0.38s + 1}{0.084s + 1}$, Lead-network attenuation cancelled by amplification $A = 4.5$

10.8* One solution is $G_c(s) = \dfrac{6.67s + 1}{66.67s + 1}$

ω_c (uncompensated) = 2.6 rad/sec
ω_c (compensated) = 0.64 rad/sec

10.9* One solution is

(i) $G_c(s) = \dfrac{(s + 1.3)^3}{(s + 4.3)^3}$; $K_v = 0.14$

(ii) $G_c(s) = \dfrac{(s + 2.2)^3}{(s + 14)^3}$

10.10* One solution is $G_c(s) = \dfrac{0.67s + 1}{2s + 1}$,

$\omega_b = 11$ rad/sec, $t_s = 1$ sec

10.11* Uncompensated system: PM = 18°, $\omega_b = 4.83$ rad/sec
Compensated system: PM = 28°, $\omega_b = 7.42$ rad/sec

10.12* One solution is
$K_t = 0.038$, $\omega_r = 3$ rad/sec

10.13* GM = 10 db, PM = 38°

10.14* (a) $\dfrac{3(1 + 1.25s)}{s(1 + 5.95s)(1 + 0.213s)}$

(b) 5

10.15 $G_{ac}(j\omega) = \dfrac{(j\omega)^2 R_1 R_2 C^2 + j2\omega (R_1 C) + 1}{(j\omega)^2 R_1 R_2 C^2 + j\omega C (2R_1 + R_2) + 1}$

$G_{dc}(j\omega) = \alpha \left(\dfrac{1 + j\omega T}{1 + j\omega \alpha T} \right),$

$\alpha = \dfrac{2R_1}{2R_1 + R_2}, \quad T = \dfrac{1}{R_1 C \omega_c^2},$

$\omega_c^2 = \dfrac{1}{R_1 R_2 C^2}$

11.1 (i) $c(k+1) + (1 - 2e^{-T})c(k) = (1 - e^{-T})r(k)$

(ii) $2c(k+2) - (4 - T^2)c(k+1) + (2 + T^2)c(k)$
$= T^2 r(k+1) + T^2 r(k)$

11.2 (i) $\dfrac{z}{z-1} X(z)$

(ii) $X(e^a z)$

11.3 (i) $\dfrac{z(z+1)}{(z-1)^3}$

(ii) $\dfrac{z}{(z-a)^2}$

(iii) $\dfrac{z(z+a)}{(z-a)^3}$

(iv) $e^{az^{-1}}$

(v) $\dfrac{z \sinh \beta}{z^2 - 2z \cosh \beta + 1}$

(vi) $\dfrac{z(z - \cosh \beta)}{z^2 - 2z \cosh \beta + 1}$

(vii) $\dfrac{z}{z + a}$

11.4 (i) $\dfrac{T^2 z(z+1)}{(z-1)^3}$

(ii) $\dfrac{z}{z - e^{-aT}}$

(iii) $\dfrac{Tze^{-aT}}{(z - e^{-aT})^2}$

(iv) $\dfrac{ze^{aT} \sin \omega T}{e^{2aT} z^2 - 2ze^{aT} \cos \omega T + 1}$

(v) $\dfrac{e^{aT} z [e^{aT} z - \cos \omega T]}{e^{2aT} z^2 - 2ze^{aT} \cos \omega T + 1}$

11.5 (i) $\dfrac{aTe^{-aT} z}{[z - e^{-aT}]^2}$

(ii) $\dfrac{z(z - \cos \omega T)}{z^2 - 2z \cos \omega T + 1}$

(iii) $\dfrac{ze^{bT}\sin aT}{e^{2bT}z^2 - 2ze^{bT}\cos aT + 1}$

(iv) $\dfrac{z\sinh aT}{z^2 - 2z\cosh aT + 1}$

(v) $\dfrac{e^{bT}z[e^{bT}z - \cos aT]}{e^{2bT}z^2 - 2ze^{bT}\cos aT + 1}$

11.6 (i) $(-a)^k$

(ii) $(-a)^{k-1}u(k-1)$

(iii) $(a)^{k-1}u(k-1)$

(iv) $3\delta(k) - 6u(k-1) + 17(2)^{k-1}u(k-1)$

(v) $3\delta(k) + 2(-1)^{k-1}u(k-1) - 9(-2)^{k-1}u(k-1)$

(vi) $k\left(\dfrac{1}{2}\right)^k$

(vii) $-\dfrac{2}{\sqrt{7}}(-\sqrt{2})^k\sin\theta k + \dfrac{0.8}{\sqrt{7}}(-\sqrt{2})^{k-1}\sin\theta(k-1)u(k-1)$

$\theta = \tan^{-1}\sqrt{7}$

(viii) $\dfrac{1}{a}ka^k$

(ix) $-3u(k-1) + 3(2)^{k-1}u(k-1) - (k-1)(2)^{k-1}u(k-1)$

11.7* $i_n = 4\cosh\beta n - 4.48\,\dfrac{\sinh\beta n}{\sinh\beta}$; $\beta = 0.9624$

11.8 $\dfrac{C(z)}{R(z)} = H(z) = \dfrac{z-1}{z^2 + 3z + 4}$

$h(k) = -\dfrac{2}{\sqrt{7}}(-2)^k\sin\theta k + \dfrac{2}{\sqrt{7}}(-2)^{k-1}\sin\theta(k-1)u(k-1);$

$\theta = \tan^{-1}\dfrac{\sqrt{7}}{3}$

11.9 $c(k) = \dfrac{1}{6} - \dfrac{3}{2}(-1)^k + \dfrac{7}{3}(-2)^k$

11.11 $\left(\dfrac{z}{z - 0.5}\right)^2$

11.12 Without ZOH: $c(k) = \dfrac{1}{(1 - e^{-aT})}[1 - \exp\{-aT(k+1)\}]$

With ZOH: $c(k) = \dfrac{1}{a}[1 - \exp(-aTk)]$

11.13 $\dfrac{G_0G_2(z)RG_1(z)}{1 + G_0G_1G_2H(z)}$

11.15 $\dfrac{G_1(z)G_2(z)}{1 + G_1(z)HG_2(z)}$

11.16 $\dfrac{G_0G(z)}{1 + H(z)G_0G(z)}$

11.17 $c(k) = 0.5[1 - (-0.264)^k]$

11.18 $c(k) = K[1 - \exp\{-aT(k-1)\}]u(k-1)$

11.19 $c(k) = K[1 - \exp(-aT\Delta)\exp\{-aTk\}]u(k-1)$

11.20 $c(k + 0.5) = [0.631 - 0.356(-0.264)^{k-1}]u(k-1)$
11.21 $0.5 < K < 3.78$
11.22 (i) No (ii) Yes (iii) No
11.23* $K < 1.15$
11.24* $G_c(r) = \dfrac{2.8r + 1}{0.617r + 1}$

$D(z) = \dfrac{3.8z - 1.8}{1.617z + 0.383}$

11.25 $G_c(s) = \dfrac{0.235s - 0.36}{s - 3.6}$

Not realizable

12.1 $\mathbf{A} = \begin{bmatrix} -1 & 1 \\ 1 & -2 \end{bmatrix}; \ \mathbf{B} = \begin{bmatrix} 0 \\ 1 \end{bmatrix}$

12.2 (i) $\begin{bmatrix} \dot{x}_1 \\ \dot{x}_2 \\ \dot{x}_3 \end{bmatrix} = \begin{bmatrix} 0 & 1 & 0 \\ 0 & -f/J & K_T/J \\ 0 & -K_b/L & -R/L \end{bmatrix} \begin{bmatrix} x_1 \\ x_2 \\ x_3 \end{bmatrix} + \begin{bmatrix} 0 \\ 0 \\ 1/L \end{bmatrix} u$

(ii) $\begin{bmatrix} \dot{x}_1 \\ \dot{x}_2 \\ \dot{x}_3 \end{bmatrix} = \begin{bmatrix} 0 & 1 & 0 \\ 0 & 0 & 1 \\ 0 & -\dfrac{1}{JL}(K_b K_T + Rf) & -\dfrac{1}{JL}(JR + fL) \end{bmatrix} \begin{bmatrix} x_1 \\ x_2 \\ x_3 \end{bmatrix}$

$+ \begin{bmatrix} 0 \\ 0 \\ \dfrac{K_T}{JL} \end{bmatrix} u$

12.3 For state variables
$\mathbf{x}^T = [y_1 \ y_2 \ \dot{y}_1 \ \dot{y}_2]$
the state equation is

$\begin{bmatrix} \dot{x}_1 \\ \dot{x}_2 \\ \dot{x}_3 \\ \dot{x}_4 \end{bmatrix} = \begin{bmatrix} 0 & 0 & 1 & 0 \\ 0 & 0 & 0 & 1 \\ -K_1/M_1 & K_1/M_1 & -f_1/M_1 & f_1/M_1 \\ K_1/M_2 & -\dfrac{K_1+K_2}{M_2} & f_1/M_2 & -\dfrac{(f_1+f_2)}{M_2} \end{bmatrix} \begin{bmatrix} x_1 \\ x_2 \\ x_3 \\ x_4 \end{bmatrix}$

$+ \begin{bmatrix} 0 \\ 0 \\ 1/M_1 \\ 0 \end{bmatrix} u$

12.4 For state variables
$\mathbf{x}^T = [y \quad \dot{y} \quad \ddot{y}]$, the stable model is

$$\begin{bmatrix} \dot{x}_1 \\ \dot{x}_2 \\ \dot{x}_3 \end{bmatrix} = \begin{bmatrix} 0 & 1 & 0 \\ 0 & 0 & 1 \\ -6 & -11 & -6 \end{bmatrix} \begin{bmatrix} x_1 \\ x_2 \\ x_3 \end{bmatrix} + \begin{bmatrix} 0 \\ 0 \\ 1 \end{bmatrix} u$$

$$y = x_1$$

12.5 $\dot{\mathbf{x}} = \mathbf{A}\mathbf{x} + \mathbf{B}u; \quad c = \mathbf{C}\mathbf{x}$

(i) Phase-variable form

$$\mathbf{A} = \begin{bmatrix} 0 & 1 & 0 \\ 0 & 0 & 1 \\ 0 & -3 & -4 \end{bmatrix}; \quad \mathbf{B} = \begin{bmatrix} 0 \\ 0 \\ 1 \end{bmatrix}; \quad \mathbf{C}^T = \begin{bmatrix} 40 \\ 10 \\ 0 \end{bmatrix}$$

(ii) Normal form

$$\mathbf{A} = \begin{bmatrix} 0 & 0 & 0 \\ 0 & -1 & 0 \\ 0 & 0 & -3 \end{bmatrix}; \quad \mathbf{B} = \begin{bmatrix} 40/3 \\ -15 \\ 5/3 \end{bmatrix}; \quad \mathbf{C}^T = \begin{bmatrix} 1 \\ 1 \\ 1 \end{bmatrix}$$

(iii) Normal form

$$\mathbf{A} = \begin{bmatrix} 0 & 0 & 0 \\ 0 & -1 & 0 \\ 0 & 0 & -3 \end{bmatrix}; \quad \mathbf{B} = \begin{bmatrix} 1 \\ 1 \\ 1 \end{bmatrix}; \quad \mathbf{C}^T = \begin{bmatrix} 40/3 \\ -15 \\ 5/3 \end{bmatrix}$$

12.6 One solution is

$$\begin{bmatrix} \dot{x}_1 \\ \dot{x}_2 \\ \dot{x}_3 \end{bmatrix} = \begin{bmatrix} 0 & 1 & 0 \\ 0 & 0 & 1 \\ -1 & -3 & -2 \end{bmatrix} \begin{bmatrix} x_1 \\ x_2 \\ x_3 \end{bmatrix} + \begin{bmatrix} 0 \\ 0 \\ 1 \end{bmatrix} u$$

$$y = [3 \quad 3 \quad 1] \begin{bmatrix} x_1 \\ x_2 \\ x_3 \end{bmatrix}$$

12.7 $\begin{bmatrix} e^{\sigma t} \cos \omega t & e^{\sigma t} \sin \omega t \\ -e^{\sigma t} \sin \omega t & e^{\sigma t} \cos \omega t \end{bmatrix}$

12.8 $\begin{bmatrix} 0 & 1 \\ -2 & -3 \end{bmatrix}; \begin{bmatrix} 2e^{-t} - e^{-2t} & e^{-t} - e^{-2t} \\ -2e^{-t} + 2e^{-2t} & -e^{-t} + 2e^{-2t} \end{bmatrix}$

12.9 (a) $\mathbf{x}(t) = \begin{bmatrix} e^t \\ te^t \end{bmatrix}$ (b) $\mathbf{x}(t) = \begin{bmatrix} e^t \\ e^t(t+1) - 1 \end{bmatrix}$

12.10
$$\begin{bmatrix} \dot{z}_1 \\ \dot{z}_2 \\ \dot{z}_3 \end{bmatrix} = \begin{bmatrix} -1 & 0 & 0 \\ 0 & -2 & 0 \\ 0 & 0 & -3 \end{bmatrix} \begin{bmatrix} z_1 \\ z_2 \\ z_3 \end{bmatrix} + \begin{bmatrix} 1 \\ -2 \\ 1 \end{bmatrix} u$$

$$y = \begin{bmatrix} 1 & 1 & 1 \end{bmatrix} \begin{bmatrix} z_1 \\ z_2 \\ z_3 \end{bmatrix}$$

$$\mathbf{x}(t) = \begin{bmatrix} \tfrac{1}{3} - e^{-2t} + \tfrac{2}{3}e^{-3t} \\ 2(e^{-2t} - e^{-3t}) \\ 2(-2e^{-2t} + 3e^{-3t}) \end{bmatrix}, \quad y = x_1(t)$$

12.11 (a)
$$\begin{bmatrix} \dot{x}_1 \\ \dot{x}_2 \\ \dot{x}_3 \end{bmatrix} = \begin{bmatrix} 0 & 1 & 0 \\ 0 & 0 & 1 \\ -1 & 0 & -3 \end{bmatrix} \begin{bmatrix} x_1 \\ x_2 \\ x_3 \end{bmatrix} + \begin{bmatrix} 0 \\ 0 \\ 1 \end{bmatrix} u$$

$$c = x_1$$

(b) $\lambda_{1,2} = 0.052 \pm j0.565$, $\lambda_3 = -3.104$

$$\Lambda = \begin{bmatrix} \lambda_1 & 0 & 0 \\ 0 & \lambda_2 & 0 \\ 0 & 0 & \lambda_3 \end{bmatrix}$$

(c) Unstable (d) $\dfrac{1}{s^3 + 3s^2 + 1}$

12.12 $\mathbf{x}^T(0) = [5k \quad -15k \quad 15k]$

where k is a non-zero multiplier

12.13 (i) $\mathbf{P} = \begin{bmatrix} 1 & 1 & 1 \\ \lambda_1 & \lambda_2 & \lambda_3 \\ \lambda_1^2 & \lambda_2^2 & \lambda_3^2 \end{bmatrix}$

(ii) $\mathbf{P} = \begin{bmatrix} 1 & 1 & 1 \\ \lambda_1 + a_1 & \lambda_2 + a_1 & \lambda_3 + a_1 \\ \lambda_1^2 + a_1 \lambda_1 + a_2 & \lambda_2^2 + a_1 \lambda_2 + a_2 & \lambda_3^2 + a_1 \lambda_3 + a_2 \end{bmatrix}$

12.16
$$\begin{bmatrix} \dot{x}_1 \\ \dot{x}_2 \\ \dot{x}_3 \end{bmatrix} = \begin{bmatrix} 0 & 0 & 1 \\ -2 & -3 & 0 \\ 0 & 2 & -3 \end{bmatrix} \begin{bmatrix} x_1 \\ x_2 \\ x_3 \end{bmatrix} + \begin{bmatrix} 0 \\ 2 \\ 0 \end{bmatrix} u$$

$$y = x_1$$

System is completely controllable and observable

12.17 The system is not completely observable

12.18 One solution is
$$K = 44, k_2 = 0.13, k_3 = 0.03$$

12.19 $K_A = 24$, $\mathbf{k}^T = [1 \quad 3/8 \quad 1/8]$, system is stable

12.20 $K_A = 36$, $k_2 = 1/3$, third pole at $s = -9$

12.22 (i) $\mathbf{A} = \begin{bmatrix} 0 & 1 & 0 \\ 0 & 0 & 1 \\ 2 & -5 & 4 \end{bmatrix}$; $\mathbf{B} = \begin{bmatrix} 0 \\ 0 \\ 1 \end{bmatrix}$

$\mathbf{C} = [1 \quad -7 \quad 4]$

(ii) $\mathbf{A} = \begin{bmatrix} 1 & 1 & 0 \\ 0 & 1 & 0 \\ 0 & 0 & 2 \end{bmatrix}$; $\mathbf{B} = \begin{bmatrix} 0 \\ 1 \\ 1 \end{bmatrix}$

$\mathbf{C} = [2 \quad 1 \quad 3]$

12.23 (i) $\mathbf{A} = \begin{bmatrix} -2 & 0 \\ 0 & -3 \end{bmatrix}$; $\mathbf{B} = \begin{bmatrix} 1 \\ 1 \end{bmatrix}$; $\mathbf{C} = [1 \quad -1]$

(ii) $\begin{bmatrix} (-2)^k & 0 \\ 0 & (-3)^k \end{bmatrix}$

(iii) $\frac{1}{4}(-3)^k - \frac{1}{3}(-2)^k + \frac{1}{12}$

12.24 $\mathbf{x}(k+1) = \begin{bmatrix} 0.6 & 0.232 \\ -0.465 & -0.097 \end{bmatrix} \mathbf{x}(k) + \begin{bmatrix} 0.2 \\ 0.233 \end{bmatrix} u(k)$

12.26 $T = n\pi$ (n = positive integer)

13.1 $\alpha = 2$
13.2 $K_1 = 2$, $K_2 = 20$
13.3 $K = 4.63$
13.4 (i) $E(z) = \dfrac{z^3 - (a+1)z^2 + az}{z^3 + z^2(K - a - 2) + z(1 - K) - a}$

Use Table 13.2 to obtain J

(ii) $K > 0$, $-1 + \dfrac{K}{2} < a < 1$

13.5 $T^*(s) = \dfrac{200}{s^2 + 20s + 200}$

13.6 $D^*(z) = \dfrac{1.32(z - 0.368)}{(z + 3.33)}$

13.7 $u^*(t) = -0.236[x_1(t) + x_2(t)]$
13.8 (i) $K = 2$
(ii) $J_{\min} = 1.5$

(iii) $S_k^{opt} = 0.1$
13.10 (i) $K = \infty$, $J_{min} = 0.5$
   (ii) $K = 1$, $J_{min} = 1$
13.11 (i) $K_1 = K_2 = \infty$
   (ii) $K_1 = 1$, $K_2 = \sqrt{3}$

13.12 $u^*(k) = -\begin{bmatrix} 0.395 & 0.687 \end{bmatrix} \begin{bmatrix} x_1(k) \\ x_2(k) \end{bmatrix}$

13.13 $u^*(0) = -0.177x(0)$
$u^*(1) = -0.177x(1)$
$u^*(2) = -0.177x(2)$
$u^*(3) = -0.164x(3)$
$u^*(k) = -0.177x(k)$ when $N \to \infty$

13.14 $u^*(0) = -0.8$, $u^*(1) = 0$, $u^*(2) = 0$

13.15 $x_1(k+1) = x_1(k) + 0.02x_2(k)$
$x_2(k+1) = -x_1(k) + 0.96x_2(k) + 0.02u(k)$

$J = 10x_1^2(50) + 0.01\sum_{k=0}^{49}[x_1^2(k) + 2x_2^2(k) + u^2(k)]$

13.16 $u^*(t) = \left[\dfrac{\exp\{-\sqrt{2}(t-1)\} - \exp\{\sqrt{2}(t-1)\}}{(\sqrt{2}+1)\exp\{-\sqrt{2}(t-1)\} + (\sqrt{2}-1)\exp\{\sqrt{2}(t-1)\}}\right]x(t)$

13.17 $\dot{p}_{11} = 2p_{12}^2 - 2$; $p_{11}(5) = 1$
$\dot{p}_{12} = -p_{11} + 2p_{12}p_{22} - 1$; $p_{12}(5) = 0$
$\dot{p}_{22} = -2p_{12} + 2p_{22}^2 - 4$; $p_{22}(5) = 2$

13.18 $u^*(t) = -\begin{bmatrix} 1 & \sqrt{2} \end{bmatrix} \begin{bmatrix} x_1(t) \\ x_2(t) \end{bmatrix}$

13.19 $K = 1$

14.1 Stable focus
14.3* Steady-state error $= -0.2$ rad
14.4* Limit cycle of amplitude 0.2 and time period 3.56 sec exists
14.8 $\ddot{e} + (1 - e^2)\dot{e} + e = 0$, stable focus, unstable limit cycle
14.9* $t_s = 2.6$ sec, steady-state error $= 0.25$

14.10 $K_N(X) = \begin{cases} k_1 & ; X < S \\ k_2 + \dfrac{2}{\pi}(k_1 - k_2)\left\{\sin^{-1}\dfrac{S}{X} + \dfrac{S}{X}\sqrt{\left[1 - \left(\dfrac{S}{X}\right)^2\right]}\right\} & ; X > S \end{cases}$

14.11 $K_N(X) = \begin{cases} 0 & ; X < D \\ \dfrac{2K}{\pi}\left(\dfrac{\pi}{2} - \alpha + \dfrac{\sin 2\alpha}{2}\right) & ; X > D \end{cases}$

where $\alpha = \sin^{-1} D/X$

14.12 Stable, $K' = 0.114$
14.13* Unstable limit cycle, $\omega = 0.25$ rad/sec, $b/X = 1.9$

14.14 $x_1^2 + x_2^2 \leqslant 1$

14.16 The origin is asymptotically stable in-the-large.

14.17 $K > 0$

14.18 $-1 < K < 8$

14.19 The origin is asymptotically stable in-the-large.

14.20 The origin is asymptotically stable in-the large.

14.23 Nonlinear function is restricted to first and third quadrants.

Index

Absolute integrable functions, 664
Absolutely control asymptotic system, 665
Absolutely output asymptotic system, 665
Absolutely stable system, 185, 664
A.C. servomotor, 84-88
A.C. tachometer, 88-89
Acceleration error, 149-150
Acceleration error constant, 149
Accelerometer, 17-18, 31-32
Actuating signal, 39, 61, 82
Actuator,
  hydraulic, 110-114
  pneumatic, 119
Adaptive control systems, 160
Adjoint matrix, 688
Algebra of matrices, 689-692
All-pass systems, 265
Amplidyne, 89-92
Amplifiers, 89, 98
Analog controller, 385
Analogous systems, 22-23, 29-30
Analog-to-digital converter, 386
Analytic function, 275
Analytical method, construction of phase trajectory, 609-611
Angle of arrival, 218
Angle criterion, 204
Angle of departure, 218
Applications,
  industrial, 8-9
  military, 7
  non-engineering, 11-12
Argument principle, 274-277
Armature control, 34-37
Asymptote,
  Bode plot, 255-256
  root locus, 210-212
Asymptotic stability, 604-606
  in-the-large (*see* global)
  in-the-small (*see* local)

Attenuation, 329
Automatic control, history of, 5-9
Automatic control system, 2
Automatic tank-level control system, 3
Automobile driving system, 1-2
Autonomous system, 595
Autopilot system, 10-11
Autosyn, 92
Auxiliary polynomial, 193

Back emf, 34-35
Backlash nonlinearity, 590-591
  describing function, 632-633
Bandwidth, 247, 298
Bang-Bang control systems, 611
Barbashin's method, 671
Batch chemical reactor, 8-9
Bellman's dynamic programming, 555-559, 561-565
Bellows, pneumatic, 116-117
Bilear transformation, 430-432
Biological control systems, 12
Block, 31
Block diagram, 38
  of closed-loop system, 39-40
  reduction, 41-45
  terminology, 39
Bode plots, 252-264
  construction procedure, 261-264
Bounded input bounded output stability, 181-183
Branch,
  complex-root, 213
  real-root, 210
  signal flow graph, 45
Break frequency, 255
Breakaway directions, 217
Breakaway point, 213-217
  complex, 217
  real, 215-216

Index

Bridged T network, 383
Brown's construction, 305
Bush form of matrices, 462

Calculus of matrices, 694-695
Cancellation compensation, 339
Canonical form, Jordan, 470-471
Canonical state model, 466-471
Capacitance,
  fluidic, 27
  hydraulic, 28
  pneumatic, 29
  thermal, 24
Carrier control systems, 87, 96
Carrier frequency, 87
Cascade compensation,
  in frequency domain, 351-366
  in time domain, 335-351
Cascaded elements, transfer function of, 32-33
Cauchy's theorem, 274
Cayley-Hamilton theorem, 490
Centre point, 601-602
Centroid, 210-212
Characteristic,
  equation, 139, 475
  polynomial, 139
  roots, 139
Chattering, 591, 612-613
Chemical reactor, 8-9
Circles,
  constant magnitude, 298-300
  constant phase, 300-301
Closed-loop control system, 2, 61, 82
  frequency response, 297-310
  transfer function, 39-40
Cofactor, 688
Column vector, 686
Command input, 61, 82
Companion form of matrices, 462
Compensating networks,
  lag, 331-333
  lag-lead, 333-334
  lead, 328-331
Compensation, 98, 153, 320
  cancellation, 339
  derivative error, 154-155
  derivative output, 155-156
  integral error, 156-157
  lag, 345-348, 358-362
  lag-lead, 348-351, 362-366
  lead, 335-345, 352-358
  of a.c. systems, 98, 374-380
  of d.c. systems, 98, 335-374

of sampled-data systems, 435-443
Compensation cascade, 321-322
  in frequency domain, 351-366
  in time domain, 335-351
Compensation feedback, 321-322
  in frequency domain, 369-374
  in time domain, 366-369
Compensator, 320
  lag, 331-333
  lag-lead, 333-334
  lead, 328-331
Complex variable, 65, 274
Compliance, fluid, 25
Compressibility of oil, 107
Computation of time,
  by average speeds, 619-620
  by graphical integration, 620
Computer control systems, 9-10, 385-386
Conditionally stable systems, 185
Conformal matrices, 690
Conjugate matrix, 689
Constant-M circles, 298-300
Constant-N circles, 300-301
Constant-tension reeling system, 8
Constraint, 172-173, 537
Contour, Nyquist, 278
  indented, 281
Control asymptotic,
  absolute, 665
  of degree α, 665
Control elements, 2, 82
Controllable phase variable form, 497-499
Control law, optimal, 553
Control policy, optimal, 558, 559-560
Control sequence, optimal, 554
Control signal, 82
Control systems, 1
  adaptive, 160
  automatic, 2
  autonomous, 595
  bang-bang, 611
  closed-loop, 2, 61, 82
  digital, 9, 386
  discrete-data, 10
  multivariable, 10
  open-loop, 2, 3, 61
  sampled-data, 9, 386
Control vector, 450
Controllability, 493, 522
  output, 493
  state, 493-496
Controlled variable, 39, 82
Controller,
  analog, 385

Index

digital, 385-386
Corner frequency, 255
Corner plots (see Bode plots)
Correlation of frequency and time responses, 244-250
Coulomb friction, 15, 589
Criterion,
   Jury, 425-429
   Hurwitz, 188-189, 654
   Liapunov, 649-650
   Nyquist, 277-286
   Popov, 663-667
   Routh, 190-191
Critically damped system, 140
Cut-off frequency, 87, 247
Cut-off rate, 298

Damped natural frequency, 139
Damping factor, 139
   correlation with phase margin, 290-291
Damping ratio (see damping factor)
Dashpot, 17
D.C. servomotor, 34, 84
   armature-controlled, 34-37
   field-controlled, 37-38
D.C. tachometer, 49
Dead-band see dead-zone)
Dead time, 231 (see also transportation lag)
Dead-zone nonlinearity, 591
   describing function, 629
Decade of frequency, 254
Decibel, 254
Definiteness of a matrix, 693
Degenerative feedback, 69
Delay time, 142
Delta method, 617-618
Demodulation, 99
Derivative error compensation, 154-155
Derivative output compensation, 155-156
Describing function, 626
   for backlash, 632-633
   for dead-zone, 629
   for relay (ideal), 631
   for relay with dead-zone, 631
   for relay with dead-zone and hysteresis, 630-631
   for relay with hysteresis, 632
   for saturation, 627-628
Describing function method, 625-626
   stability analysis, 634-642
Design specifications,
   higher-order systems, 157-159
   second-order systems, 152-153, 298
Determinant, 687

Diagonal matrix, 686
Diagonalization, 472-481
Diagonalizing matrix, 472
Differential equations of physical systems, 15-30
Digital controller, 385-386
Digital control systems, 9, 386
Digital-to-analog converter, 386
Directional gyro, 121
Direct programming method, 466
Discrete-data control systems, 10
Discrete unit impulse, 402
Discrete unit step, 397
Disturbance signals, 25, 31, 66
Dither, 624
Dominant poles, 158
Dual-input describing function, 634
Duality property, 501
Dynamic error coefficient, 179
Dynamic programming, 555-559, 561-565

Economic inflation problem, model of, 11
Eigenvalues, 475
Eigenvectors, 475
   generalized, 478
Electrical systems, 21-23, 32-38, 84-99
Encircled, 275
Enclosed by contour, 275
Epidemic disease, model of, 12
Equilibrium points, 455-595
Equality of matrices, 689
Equivalent friction, 21
Equivalent moment of inertia, 21
Error, steady-state, 67, 143, 147
Error constants,
   acceleration, 149
   position, 148
   velocity, 149
Error detector, 2, 82
   sensitivity, 96
Error signal, 2
Euclidean norm, 692
Experimental determination of transfer function, 266-270
Exponential matrix function, 482

Factoring of polynomials, (see root determination of algebraic equations)
Feedback,
   control of disturbance signal, 66-69
   control over system dynamics, 65-66
   degenerative, 69
   negative, 39, 69
   positive, 11, 69
   rate, 96

reduction of parameter variations, 62-64
regenerative, 69
state variable, 504-506
unity, 148
Feedback action, 3
Feedback compensation, 321-322
 in frequency domain, 369-374
 in time domain, 366-369
Feedback control systems, 61
Feedback elements, 2, 82
 transfer function of, 39
Feedback signal, 39, 61, 82
Field control, 37-38
Final value theorem, 67, 400, 678
First-order hold, 390
First-order system, time response of, 135-137
Flapper valve, 117-118
Flow,
 compressibility, 107
 laminar, 26
 leakage, 107
 turbulent, 26
Fluid,
 capacitance, 27
 compliance, 25
 resistance, 26
Fluid systems, 25-30
Fly-ball governor, 5
Focus points, 599-601
 stable, 601
 unstable, 601
Force-current analogy, 22-23
Force-voltage analogy, 22-23
Forward path, 46
Forward path gain, 46
Forward path transfer function, 39
Fourier integral (see Fourier transforms),
Fourier transforms,
 definition of, 675
 for periodic functions, 675-677
 inverse, 675
 properties of, 676
Free-body diagram, 17, 19
Free gyro, 121
Frequency domain specifications, 298
Frequency response, 243, 297-310
 correlation with time response, 244-250
Friction,
 coulomb, 15, 589
 equivalent, 21
 viscous, 15, 589

Gain adjustment, 304
 for desired M_r, 310

direct polar plot, 305-307
inverse polar plot, 307-308
Nichols chart, 308
Gain cross-over frequency, 289
Gain margin, 288
Gear ratio, 138
Gear train, 19-21
Generalized eigenvectors, 478
Gilbert's method, 493-495
Global stability, 606
Gyroscopes,
 directional, 121
 free, 121
 integrating, 126
 rate, 125-126
 vertical, 121-123

Hard spring, 584
Harmonics, 586, 626
Hessian matrix, 531
Higher-order systems,
 correlation between time and frequency response, 249-250
 design specifications, 157-159
History of automatic control, 5-9
Hold circuit,
 first-order, 390
 zero-order, 390, 410-411
Homogeneous linear system, 481
Hunting, 4
Hurwitz stability criterion, 188-189, 654
Hydraulic,
 actuator, 110-114
 amplifier, 113
 capacitance, 28
 motor, 104-108
 power steering mechanism, 114-116
 pump, 104-108
 resistance, 26-27
 systems, 104-116
 transmission, 104-105
 valves, 108-110
Hybrid system, 386
Hysteresis nonlinearity, 590-592

IAE criterion, 162
Identity matrix, 687
Impulse response, 65
Impulse sampling, 411
Impulse signal, 65, 134
Incidental nonlinearity, 588
Indefinite function, 693
Indented Nyquist contour, 281
Index of performance, 504, 555
Inertial Navigation, 121-126

Index

Infinite-time regulator problem, 566-569
Initial value theorem, 400, 678
Innerwise positive matrix, 427
Input,
   command, 61, 82
   impulse, 65, 134
   matrix, 452
   parabolic, 134
   ramp, 133
   random, 132
   reference, 2, 39
   sinusoidal, 241
   step, 133
Input node, 45
Instability, 66, 69, 651
Integrable functions,
   absolute, 664
   square, 664
Integral of absolute magnitude of error, 162
Integral error compensation, 156-157
Integral square control, 537
Integral square error, 160
Integral of time absolute error, 162
Integral of time square error, 162
Integrating gyro, 126
Intentional nonlinearity, 588
Inverse Fourier transform, 675
Inverse Laplace transform, 677
Inverse matrix, 692
Inverse polar plots, 252, 302-304
   gain adjustment, 307-308
Inverse z-transform, 396
ISE criterion, 160
Isocline, 613
Isocline method, 613-614
ITAE criterion, 162
ITSE criterion, 162

Jacobian matrix, 456
Jordan block, 471
Jordan canonical form, 470-471
Jordan matrix, 471
Jump resonance, 585, 642-646
Jury's stability test, 425-429

Kalman's test, 496, 501
Kirchhoff's laws, 21
Krasovskii's method, 657

$\mathcal{L}_1$ and $\mathcal{L}_2$ classes of functions, 664
Lag compensation, 345-348, 358-362
Lag compensator, 331-333
Lag-lead compensation, 348-351, 362-366
Lag-lead compensator, 333-334
Lag-lead network, 333-334

Lag network, 331-333
Laminar flow, 26
Laplace expansion formula, 688
Laplace transform,
   definition of, 677
   inverse, 677
   pairs, 679
   properties of, 678
   relation with z-transform, 423
Lead compensation, 335-345, 352-358
Lead compensator, 328-331
Lead network, 328-331
Leakage flow, 107
Liapunov function, 649
Liapunov function construction by,
   Barbashin's method, 671
   Krasovskii's method, 657
   variable gradient method, 660-661
Liapunov's direct method (*see* Liapunov's stability criterion),
Liapunov's stability criterion, 649
   for linear systems, 653-654
   instability theorem, 651-652
   stability theorems, 649-650
Limit cycles,
   stable, 606
   unstable, 607
Limitedly stable systems, 185
Linear approximation of nonlinear systems, 83-84, 455-457
Linear continuous-time model, 452
Linear continuous-time systems with sampled inputs, 518-522
Linear differential equations, 17, 19
Linear discrete-time model, 512
Linearly independent vectors, 473
Linear output regulator problem, 550, 554, 571-572
Linear state regulator problem, 551-554, 556-571
Linear time-invariant model, 14, 451
Linear time-varying model, 14, 451
Linearization of state equations, 455-457
Link mechanism, 113
Lin's method, 698-699
Liquid-level systems, 27-29
   capacitance, 28
   resistance, 26-27
Locally stable systems, 604
Log-magnitude *vs* phase plots, 270-271
Logarithmic plot, 254
Loop, 46
Loop gain, 46
Loop transfer function, 39

Index

M-circles, 298-300
Magnitude criterion, 204
Manually controlled systems, 2
Mapping, 275
Marginal stability (see limited stability)
Mason's gain formula, 47-48
Mathematical model,
  linear continuous-time, 452
  linear discrete-time, 512
  linear time-invariant, 14, 451
  linear time-varying, 14, 451
Mathematical sampling, 395
Matrix,
  addition, 690
  adjoint, 688
  algebra, 689-692
  Bush form, 462
  calculus, 694-695
  conformal, 690
  conjugate, 689
  determinant, 687
  diagonal, 686
  diagonalizing, 472
  equality, 689
  exponential function, 482
  Hessian, 531
  inversion, 692
  Jacobian, 456
  Jordan, 471
  modal, 472
  multiplication, 689, 690
  negative definite, 694
  negative semi-definite, 694
  non-singular, 689
  null, 687
  positive definite, 694
  positive innerwise, 427
  positive semi-definite, 694
  rank of, 689
  real, 689
  rectangular, 686
  Schwarz, 654, 655
  singular, 689
  skew-symmetric, 687
  square, 686
  subtraction, 690
  symmetric, 687
  transpose, 687
  unit (identity), 687
  Vander Monde, 477, 480-481
Matrix Riccati equation, 564
  reduced, 568
Maximization of performance index, 504
Maximum overshoot (see peak overshoot)
Mechanical accelerometer, 17-18, 31-32
Mechanical rotational systems, 18-21
Mechanical translational systems, 15-18
Memory type nonlinearity, 591
Memoryless nonlinearity, 629
Minimization of performance index,
  504-555
Minimum-phase transfer function, 264
Minimum principle, 555
Minor, 688
  principal, 688
Minor feedback loop, 43
Missile launching and guidance system, 7-8
Modal matrix, 472
Modified z-transform, 422
Modulation, 95, 98-99
  pulse amplitude, 386
  pulse width, 386
Moment of inertia, 19
  equivalent, 21
Motor,
  a.c. servo, 84-88
  d.c. servo, 34-38, 84
  hydraulic, 104-108
  stepper, 99-104
Motor torque constant, 35
Multiple-input-multiple-output system, 10, 40, 452
Multiple loop system, 230, 277
Multivariable control systems, 10
Multivariable nonlinearity, 593

N-circles, 300-301
Natural frequency,
  damped, 139
  undamped, 139
Necessary conditions of stability, 186-188
Negative definite matrix, 694
Negative feedback, 39, 69
Negative semi-definite matrix, 694
Network,
  lag, 331-333
  lag-lead, 333-334
  lead, 328-331
  RC notch type, 379-380
Network compensation,
  of a.c. systems, 98, 374-380
  of d.c. systems, 98, 335-374
Newton's law, 17
Nichols chart, 305
Nodal points, 597-599
  stable, 598
  unstable, 599
Node, 45
Noise, 68
No load test, 88

Index

Non-feedback system, 61
Non-homogeneous linear system, 481
Nonlinear systems,
  behaviour, 583-586
  linear approximation of, 83-84, 455-457
  stability analysis of, 604-608, 649-650, 663-667
Nonlinearities,
  backlash, 590-591
  dead-zone, 591
  friction, 589-590
  hysteresis, 590-592
  incidental, 588
  intentional, 588
  memory type, 591
  memoryless, 629
  multivariable, 593
  relay, 591-592
  saturation, 588-589
Nonminimum-phase transfer function, 265
Nonsingular matrix, 689
Nontouching loops, 46
Nonunity feedback systems, 151
Normal-form state model (see Canonical state model)
Norm of a vector, 692
  euclidean, 692
Null matrix, 687
Nyquist contour, 278
  indented, 281
Nyquist plot, 252, 279
Nyquist stability criterion, 277
  applied to inverse polar plots, 284-286
  for relative stability study, 286-294

Observable phase variable form, 501
Observability, 493, 500-501, 522
Observer systems, 509-511
Octave, 254
Open-loop control system, 2, 3, 61
Optimal control,
  output regulator problems, 550, 554, 571-572
  servomechanism problem, 539-550, 555, 572-573
  state regulator problem, 551-554, 556-571
Optimal control law, 553
Optimal control system, 561, 566
Optimal policy, 558, 559-560
Optimal sequence, 554
Optimization of parameters,
  regulators, 573-577
  servomechanisms, 531-538
Optimum switching curve, 611

Order of the system, 30, 454
Oscillations, 4, 140
  self-sustained, 584, 606-608
Output asymptotic,
  absolute, 665
  of degree α, 665
Output controllability, 493
Output equation, 452
Output matrix, 452
Output node, 45
Output regulator system, 550, 554, 571-572
Output stable, 664
  utput stable of degree, α, 664
Overdamped system, 140
Overshoot, 142

Parabolic input, 134
Parallel compensation (see feedback compensation)
Parameter optimization,
  regulators, 573-577
  servomechanisms, 531-538
Parseval's theorem, 678
Partial fraction expansions, 681-685
Path, 45
Peak overshoot, 142
Peak time, 142
Performance index,
  maximization, 504
  minimization, 504-555
Performance specifications, 141-143, 298
Phase cross-over frequency, 288
Phase margin, 289
  correlation with damping ratio, 290-291
Phase-plane, 594
Phase-plane method, 593-595
  system analysis, 620-625
Phase-portrait, 594
Phase-trajectory, 594
  construction by,
    analytical method, 609-611
    delta method, 617-618
    isocline method, 613-614
Phase variable form,
  controllable, 497-499
  observable, 501
Phase variables, 460, 593
Physical model, 13
Physical systems, 13
Physical variables, 457
Piecewise linear systems, 587
Pitch, 122
Plant, 319
Pneumatic,

actuator, 119
bellows, 116-117
flapper valve, 117-118
position control system, 120
relay, 118-119
Pneumatic systems, 29, 116-120
Polar plots, 250-252
  gain adjustment, 305-307
Pole, 65
Pole placement by state feedback, 504-506, 522-523
Pole-zero form of transfer function, 149
Pole-zero cancellation in transfer functions, 503
Pontryagin's minimum principle, 555
Polynomial,
  auxiliary, 193
  characteristic, 139
Popov's stability criterion, 663
  geometrical interpretation, 666-667
  stability theorem, 665-666
Positional gyro (see directional gyro)
Position control systems, 4-5, 96-97, 114-116, 120, 137-139, 506-509
Position error, 148
Position error constant, 148
Positive definite matrix, 694
Positive feedback, 11, 69
Positive semidefinite matrix, 694
Potentiometer, 3, 137
Power steering mechanism, 114-116
Precission, 121
Principal minor, 688
Principle of argument, 274-277
Principle of invariant imbedding, 556
Principle of optimality, 556
Principle of superposition, 14, 41
Proportional plus derivative controller, 154-156
Proportional plus integral controller, 156-157
Pulse amplitude modulation, 386
Pulse transfer function, 405

Quadratic form of functions, 693
Quadratic performance index, 537, 552-555

Ramp input, 133
Random input, 132
Rank of a matrix, 689
Rate gyro, 96
RC notch type network, 379-380
Real matrix, 689
Rectangular matrix, 686

Reduced matrix Riccati equation, 568
Reduction of parameter variations, 62-64
Reference input, 2, 39
Regenerative feedback, 69
Regulator systems,
  optimal control, 551-571
  output, 550, 554, 571-572
  parameter optimization, 573-577
  state, 551-554, 556-571
Relative stability, 186
  analysis, 196-197, 286-297
Relay nonlinearity, 591-592
  describing function, 630-632
Relay, pneumatic, 118-119
Residue, 140, 681
Resistance,
  fluidic, 26
  hydraulic, 26-27
  pneumatic, 29
  thermal, 24
Resolvant matrix, 485
Reynolds number, 26
Riccati equation, 564
  reduced, 568
Resonant frequency, 298
Resonant peak, 298
Restrained gyro (see integrating gyro)
Rise time, 142
Roll, 122
Root contours, 226-231
Root determination of algebraic equations,
  by Lin's method, 698-699
  by root locus method, 696-697
Root locus,
  angle of arrival, 218
  angle criterion, 204
  angle of departure, 218
  asymptotes, 210-212
  breakaway directions, 217
  breakaway point, 213-217
  cancellation of poles and zeroes, 226
  centroid, 210-212
  complex-root branch, 213
  concept, 202-205
  construction rules, table, 221
  determination of gain, 224
  determination of roots for specified gain, 224
  intersection with imaginary axis, 219-220
  magnitude criterion, 204
  real-root branch, 208-209
  typical plots, 225
Root sensitivity, 233-239
Roots of characteristic equation, 139

Index

Rotating amplifier, 89
Rotational systems, 18-21
Routh array, 190
Routh stability criterion, 190-191
Row vector, 686

Saddle points, 599
Sampled-data control systems, 9, 386, 518-519
Sampler, 387, 410-411
Sampling frequency, 388
Sampling process, 387-390
Sampling rate (*see* sampling frequency)
Sampling theorem, 390
Saturation nonlinearity, 588-589
  describing function, 627-628
Scalar function,
  positive definiteness of, 693
  positive semi-definiteness of, 693
Schwarz matrix, 654, 655
Second-order system,
  design specifications of, 152-153, 298
  time response specifications of, 141-143
Self-sustained oscillations, 584, 606-608
Selsyn, 92
Sensitivity,
  analysis in frequency domain, 310-313
  of error detector, 96
  of optimum system, 535
  of transfer function, 62-63, 236-237
  of roots, 233-239
Series compensation (*see* cascade compensation)
Servo, 4
Servomechanisms, 4
  optimal control, 539-550, 550, 572-573
  parameter optimization, 531-538
Servomotor,
  a.c., 84-88
  d.c., armature-controlled, 34-37
  d.c., field-controlled, 37-38
  hydraulic, 104-108
Servo system, 4
Settling time, 142
Shanon's sampling theorem, 390
Ship stabilization, 78
Signal,
  impulse, 65, 134
  parabolic, 134
  ramp, 133
  reconstruction, 390
  step, 133
Signal flow graph, 45
  application to control systems, 49-51

  construction, 46-47
  gain formula, 47-48
Signal to noise ratio, 68
Single-input-single-output systems, 10, 455
Single variable systems, 10
Singular matrix, 689
Singular points, 595
  centre, 601-602
  focus, 599-601
  nodal, 597-599
  saddle, 599
Sink, 45
Sinusoidal describing function, 626
Sinusoidal input, 241
Sinusoidal transfer function, 31
Skew-symmetric matrix, 687
Soft spring, 584
Solution of state equations by,
  canonical transformation, 486-490, 516
  Laplace transform method, 485-486
  using Cayley-Hamilton theorem, 490-492, 517
  z-transform method, 516
Source, 45
Specifications,
  frequency domain, 298
  time domain, 141-143
Spectral analysis of sampling process, 387-390
Spectral factorization, 541
Speed control system, 6, 49-51
Spirule, 204
Spring,
  hard, 584
  nonlinear, 584
  soft, 584
  torsional, 19
Spring-mass-damper system, 17, 584, 647
Spring-mass-dashpot system, 17, 584, 647
Square integrable functions, 664
Square matrix, 686
Stability,
  criterion Jury, 425-429
  criterion Hurwitz, 188-189, 654
  criterion Liapunov, 649-650
  criterion Nyquist, 277-286
  criterion Popov, 663-667
  criterion Routh, 190-191
  necessary conditions for, 186-188
  relative, 186
Stability definitions,
  absolute, 185, 664
  absolutely control asymptotic, 665
  absolutely output asymptotic, 665

asymptotic, 604-606
bounded input bounded output, 181-183
conditional, 185
control asymptotic of degree α, 665
global, 606
limited, 185
local, 604
output asymptotic of degree α, 665
output stable, 664
output stable of degree α, 664
Stabilizability, 506
Stable focus, 601
Stable limit cycle, 606
Stable node, 598
Stable system, 185
Stall torque test, 88
Standard test signals, 133-134
State, 450
State controllability, 493-496
State equations, 451
linearization, 455-457
State feedback, 504-506
State model, 4
canonical, 456-471
derivation from difference equations, 512
derivation from differential equations, 460
derivation from transfer function, 461-471, 513-515
derivation using signal flow graph, 462-466, 513-515
non-unique, 456-457
State observer, 509-511, 559
State point, 451
State regulator, 551-554, 556-571
State space, 451
State trajectory, 451
State transition matrix, 482, 516
computation, 485-492, 516-517
properties, 483, 516
State variable feedback, 504-506
State variables,
canonical, 466
phase, 460, 593
physical, 457
non-uniqueness, 456-457
State vector, 450
Steady-state error, 67, 143, 147
Step signal, 133
Stepper motor, 99-104
Stiction, 15, 590
Stiffness, 17
Sub-harmonics, 586
Summing point, 39

Sum square control, 537
Sum square error, 535
Superposition principle, 14, 41
Suppressed carrier modulation, 95
Sustained-oscillations, 584, 606-608
Switching curve, 611
Sylvester's theorem, 694
Symmetric matrix, 687
Synchro, 92
Synchro control transformer, 94
Synchro error detector, 94
Synchro transmitter, 93
System matrix, 452
Systems,
electrical, 21-23, 32-38, 84-99
fluid, 25-30
liquid-level, 27-29
mechanical, 15-21
pneumatic, 29, 116-120
thermal, 23-25

Tachometer,
a.c., 88-89
d.c., 49
Tachometer feedback, 96-97
Take off point, 39
Taylor's series, 83, 455
Temperature control system, 70-73
Tension control system, 8
Thermal capacitance, 24
Thermal resistance, 24
Thermal systems, 23-25
Time, computation from phase-trajectory, 619-620
Time constant, 65, 136
Time constant form of transfer function, 149
Time response,
correlation with frequency response, 244-250
of first-order systems, 135-137
of second-order systems, 137-141
specifications, 141-143
Torque constant, 35
Torque-speed curve, 85
Torsional spring, 19
Tracking system (*see* servomechanism)
Traffic control system, 4
Trajectory,
methods for constructing, 609-618
Transducer, 92
Transfer function,
cascaded elements, 32-33
closed-loop, 39-40

definition, 30
derivation of, 32, 471, 515
experimental determination of, 266-270
feedback path, 39
forward path, 39
loop, 39
minimum-phase, 264
for multiple-input-multiple-output systems, 40
nonminimum-phase, 265
pole-zero form, 149
sensitivity, 62-63, 236-237
sinusoidal, 31
time-constant form, 149
uniqueness, 471
Transient response specifications, 141-143
Translational systems, 15-18
Transmission matrix, 453
Transmittance, 45
Transportation lag, 231-233, 266-267, 294-296
Transpose of matrix, 687
Turbulent flow, 26
Two-phase servometer, 84-88
Type-0 system, 149
Type-1 system, 150
Type-2 system, 150

Undamped natural frequency, 139
Underdamped system, 140
Undershoot, 144
Uniform periodic sampling, 387
Uniqueness of transfer function, 471
Unit-impulse function, 134
Unit-ramp function, 133
Unit matrix, 687
Unit-parabolic function, 134
Unit-step function, 133
Unity feedback system, 148
Unstable focus, 601
Unstable limit cycles, 607

Unstable node, 599
Unstable system, 185

Valve,
hydraulic, 108-110
pneumatic, 117-118
Vander Monde matrix, 477, 480-481
Vander Pol equation, 606-607
Variable gradient method, 660-661
Variable reluctance motor, 99
Variables,
canonical, 466
phase, 460, 593
physical, 457
Vector equation, 451
Vector norm, 692
euclidean, 692
Velocity error, 149-150
Velocity error constant, 149
Vertical gyro, 121-123
Viscous friction, 15, 589
equivalent, 21
Vortex point (*see* centre point)

Warren-Ross method, 341-345
Watt's governor, 5-6
Weighting factor, 174, 552
Weighting function, 135
Weighting sequence, 402

Yaw, 122

Zero, 151
Zero-order hold, 390, 410-411
z-transform,
definition, 393
inverse, 396
modified, 422
pairs, 402, 413
properties, 401
relation with s-transform, 423

Printed and bound in Singapore by Kin Keong Printing Co. Pte. Ltd.